曹育群

THEORETICAL SOLID STATE PHYSICS

Volume 2: Non-equilibrium and Disorder

WILLIAM JONES

Lecturer in Physics,
The University of Sheffield

NORMAN H. MARCH

Coulson Professor of Theoretical Chemistry,
University of Oxford
(formerly Professor of
Theoretical Solid State Physics,
Imperial College of Science and Technology,
University of London)

Dover Publications, Inc., New York

Published in Canada by General Publishing Company, Ltd., 30 Lesmill Road, Don Mills, Toronto, Ontario.
Published in the United Kingdom by Constable and Company, Ltd., 10 Orange Street, London WC2H 7EG.

This Dover edition, first published in 1985, is an unabridged and unaltered republication of the work first published by Wiley-Interscience, a division of John Wiley & Sons Ltd., London, in 1973.

Manufactured in the United States of America
Dover Publications, Inc., 31 East 2nd Street, Mineola, N.Y. 11501

Library of Congress Cataloging in Publication Data

Jones, William, 1940–
Theoretical solid state physics.

Reprint. Originally published: London ; New York : Wiley-Interscience, 1973.
Bibliography: p.
Contents: v. 1. Perfect lattices in equilibrium—v. 2. Non-equilibrium and disorder.
1. Solid state physics—Collected works. I. March, Norman H. (Norman Henry), 1927– . II. Title.
[QC176.A1J572 1985] 530.4′1 85-10165
ISBN 0-486-65015-4 (vol. 1)
ISBN 0-486-65016-2 (vol. 2)

Preface

This book has been written with professional theoretical physicists in mind and with a view to the training of postgraduate students in theoretical physics. To this end a fairly wide selection of problems has been included in both Volumes.

The two Volumes of the work are concerned with very different aspects of the subject. Volume I represents an attempt to deal with the fundamental theory of the equilibrium properties of perfect crystalline solids. On the other hand, Volume II deals in detail with non-equilibrium properties and also includes a discussion of defects and disordered systems. This latter field is in a state of very rapid development at the time of writing but we felt it essential to include such an account in a work of the present scale, in order to preserve some balance in our treatment.

The writing began in 1966 and we quickly recognized the Herculean task on which we had embarked. It seemed to us, nonetheless, that what was needed was a fairly detailed treatment of the theory, not by-passing difficult many-body theory, but rather laying down foundations where it seemed clear that further structures could be erected.

We also felt that, whenever limitations of space permitted, we should not leave the theory at these foundations but should carry the discussion as far as current papers on the subject. We are conscious still, in spite of the vastness of this work, of the many areas we have either omitted, or to which we have obviously not done justice.

The theory of solids has, in places, now reached such a high level of sophistication that no theory group that exists at present can claim first-hand knowledge of the whole subject. We can only add that many friends and colleagues who are more expert than us in particular areas have given us the benefit of their criticism and advice. We need hardly say that errors and misunderstandings that remain, as they surely will in an undertaking of this scale, are entirely our own responsibility.

Specifically, in our own laboratory especial thanks are due to Drs J. C. Stoddart and J. W. Tucker for their very generous help. Other friends who have made extremely helpful comments are Drs R. D. Lowde, G. J. Morgan, C. Norris, T. M. Rice, P. Schofield, G. L. Sewell, L. J. Sham, and Professors

G. Rickayzen and W. H. Young. Many other folk have helped us too, and we express here our gratitude to them.

Finally, Miss P. A. Traviss (now Mrs A. Hooper) gave us enormous help with the preparation of the manuscript and our grateful thanks are due to her for her expert typing and for her tolerance and cheerfulness throughout the whole exercise.

Contents

CHAPTER 6

Transport

6.1 Introduction

We have already discussed the basic Boltzmann equation of transport theory in Chapter 2. However, though this is very valuable in transport calculations, its range of validity needs careful discussion. To do this involves one in a full quantum-mechanical discussion of transport phenomena, which will constitute a major part of this chapter.

Nevertheless, before we reach this stage, we must introduce the phenomenological equations of transport theory and discuss what a Boltzmann equation approach can tell us about the macroscopic quantities, such as electrical and thermal conductivity, thermoelectric power, etc., appearing in these equations (see also especially Wilson (1953), Ziman (1960) and Jones (1956)).

6.2 Phenomenology

6.2.1 Effect of electric field and thermal gradient

An electric field $\mathscr{E}$ established in the solid results in an electric current density $\mathbf{j}$ linearly related to the field (provided, of course, the field is not too high) through Ohm's law:

$$\mathbf{j} = \sigma\mathscr{E}, \tag{6.2.1}$$

where σ is the conductivity.‡ Similarly, a thermal gradient ∇T causes a current of thermal energy:

$$\mathbf{U} = -\kappa\nabla T, \tag{6.2.2}$$

where κ is the thermal conductivity, in a specimen on open circuit.

More generally, in the presence of both fields together, we can expect to have relations of the form

$$\begin{aligned}\mathbf{j} &= \sigma\mathscr{E} + \alpha\nabla T,\\ \mathbf{U} &= \beta\mathscr{E} + \gamma\nabla T,\end{aligned} \tag{6.2.3}$$

‡The conductivity is written here as if it were a scalar quantity. The conductivity tensor is dealt with in full generality in section 6.3.

where α, β and γ are phenomenological constants but $\gamma \neq -\kappa$ in general, because the existence of a thermal gradient will itself induce an electric field (see below).

It should be emphasized from the outset that in equation (6.2.3) **U**, the energy current, is not, in general, a heat current, although the two are the same in the usual arrangement for thermal conductivity. This is customarily taken with the specimen on open circuit and in a steady state. It is therefore clear that energy permanently transferred through a cross-section of the specimen must eventually be lost as heat. In that case heat current and energy current are identical. In fact, an examination of any physical situation will always reveal what heat is given out or absorbed. Nevertheless, it is useful to write down an explicit relation between **U** and the heat current **Q**. In writing down a formal expression for changes in the total internal energy E for a system in which the number N of particles is variable, one introduces the chemical potential ζ:

$$\delta E = T\delta S + \sum_i X_i\,\delta x_i + \zeta\delta N, \tag{6.2.4}$$

where S is the entropy, the second term represents external influences (and for which $-P\delta V$ is the commonest form met) and the third shows the variation of energy with particle number. Therefore, if external influences can be discounted, it is evident that an energy current **U** together with a particle current‡ $\mathbf{j}/e$ is associated with an entropy current $(\mathbf{U}+\mathbf{j}\zeta/e)/T$ or, in view of the relation $\delta S = \delta Q/T$, a heat current density

$$\mathbf{Q} = \mathbf{U}+\frac{\mathbf{j}\zeta}{e}. \tag{6.2.5}$$

An example in which the term $\mathbf{j}\zeta/e$ can be said to play a vital role is the Peltier effect, which we will describe later.

It is useful to modify equation (6.2.1) to show explicitly the effect of a variation in the chemical potential through the system. First let us divide a closed system into two parts with respective thermodynamic quantities denoted by subscripts 1 and 2. If sub-system 1 receives δN particles from sub-system 2, equation (6.2.4) shows its associated entropy change to be

$$\delta S_1 = -\frac{\zeta_1}{T_1}\delta N, \tag{6.2.6}$$

while sub-system 2 suffers an entropy change

$$\delta S_2 = \frac{\zeta_2}{T_2}\delta N. \tag{6.2.7}$$

‡ The charge on the current carrier is written as $-e$.

If the two parts are to be in equilibrium with one another, the total entropy change must be zero, and thus

$$\frac{\zeta_1}{T_1} = \frac{\zeta_2}{T_2}, \tag{6.2.8}$$

an equilibrium condition which can readily be generalized to an arbitrary number of systems.

In the case when $T_1 = T_2$, a condition for equilibrium is that the chemical potential be constant throughout the system. Hence any spatial variation in ζ will give rise to particle currents tending to restore equilibrium. We therefore write

$$\mathbf{j} = \sigma\mathscr{E} + \varepsilon\nabla\zeta, \tag{6.2.9}$$

ε being a further phenomenological constant. A similar equation also holds for **U**.

6.2.2 Relation between conductivity and diffusion

Consider a closed system in which is established a static electric potential $\phi(\mathbf{r})$. In equilibrium, $\mathbf{j} = 0$ and so

$$\sigma\mathscr{E} + \varepsilon\nabla\zeta = 0. \tag{6.2.10}$$

The effect of an electric field switched on adiabatically is to decrease the energy per particle by $e\phi$ and it is evident that equation (6.2.4) must be modified to read

$$\delta E = T\delta S + \sum_i X_i\,\delta x_i + (\zeta - e\phi)\,\delta N. \tag{6.2.11}$$

The chemical potential is effectively changed from ζ to $\zeta - e\phi$ and a condition for equilibrium is now

$$\zeta(\mathbf{r}) - e\phi(\mathbf{r}) = \text{constant}, \tag{6.2.12}$$

that is,

$$e\mathscr{E} + \nabla\zeta(\mathbf{r}) = 0. \tag{6.2.13}$$

Using (6.2.10) we conclude that

$$\varepsilon = \sigma/e, \tag{6.2.14}$$

and equation (6.2.9) may be rewritten as

$$\mathbf{j} = \sigma\left(\mathscr{E} + \frac{1}{e}\nabla\zeta\right). \tag{6.2.15}$$

From this, Einstein's relation between diffusion constant and mobility, or in this case the conductivity, may be obtained. We simply write down the phenomenological equation (for $\nabla T = 0$)

$$\mathbf{j} = \sigma\mathscr{E} - D\nabla n(\mathbf{r}), \tag{6.2.16}$$

where n is the particle density and the second term on the right-hand side is the current induced by a concentration gradient, with D the diffusion constant. Since

$$(\nabla\zeta)_T = \left(\frac{\partial\zeta}{\partial n}\right)_T \nabla n(\mathbf{r}), \tag{6.2.17}$$

comparison of equations (6.2.15) and (6.2.16) immediately yields

$$D = -\frac{\sigma}{e}\left(\frac{\partial\zeta}{\partial n}\right)_T, \tag{6.2.18}$$

which is a form of Einstein's result.

6.2.3 Statement of Onsager relations

Including in equation (6.2.9) the effect of a thermal gradient, we may write by comparison with equation (6.2.3)

$$\mathbf{j} = \sigma\left(\mathscr{E} + \frac{\nabla\zeta}{e}\right) + \alpha'\nabla T \tag{6.2.19}$$

and similarly for the energy current $\mathbf{U}$

$$\mathbf{U} = \beta\left(\mathscr{E} + \frac{\nabla\zeta}{e}\right) + \gamma'\nabla T, \tag{6.2.20}$$

where α' and γ' differ in general from α and γ in equation (6.2.3).

These equations can be rearranged to take the form

$$\mathbf{j} = L_{11}\left[\mathscr{E} + \frac{T}{e}\nabla\left(\frac{\zeta}{T}\right)\right] + L_{12}T\nabla\left(\frac{1}{T}\right) \tag{6.2.21}$$

and

$$\mathbf{U} = L_{21}\left[\mathscr{E} + \frac{T}{e}\nabla\left(\frac{\zeta}{T}\right)\right] + L_{22}T\nabla\left(\frac{1}{T}\right). \tag{6.2.22}$$

In this form, the Onsager relations hold, which in this special case reduce to the single result

$$L_{12} = L_{21}. \tag{6.2.23}$$

We shall prove the Onsager relations later in this chapter. It is important to note here that if we view equations (6.2.21) and (6.2.22) as tensor equations, equation (6.2.23) has an obvious generalization for the tensor components [see equation (6.12.27) below].

6.2.4 Thermal conductivity in terms of generalized transport coefficients

Let us, for later reference, write down an expression for the thermal conductivity κ, in terms of L_{ij}. This is defined with the specimen on open circuit and this circumstance is the cause of the essential difference, already remarked upon, between this case and that in which an electric field is applied alone. Here cross-terms must play a part: with the specimen on open circuit $\mathbf{j} = 0$ and equation (6.2.21) becomes

$$\mathscr{E}+\frac{T}{e}\nabla\left(\frac{\zeta}{T}\right) = \frac{L_{12}}{L_{11}}\frac{\nabla T}{T}, \tag{6.2.24}$$

so that

$$\mathbf{Q} = \mathbf{U} = \frac{L_{21}L_{12}}{L_{11}}\frac{\nabla T}{T} - L_{22}\frac{\nabla T}{T}. \tag{6.2.25}$$

Hence, by definition [see equation 6.2.2],

$$\kappa = \frac{1}{T}\left(L_{22} - \frac{L_{21}L_{12}}{L_{11}}\right). \tag{6.2.26}$$

6.2.5 Thermoelectric effects

The thermoelectric effects seem to be three apparently distinct phenomena but are in fact intimately related, as shown below.

(a) *Thomson effect.* This refers to the reversible heat (as distinct from Joule heat) generated or absorbed by a metal in which an electric current and temperature gradient exist. If current I flows from a point at temperature T to a point at temperature $T+\delta T$, the rate of generation of the reversible heat is given by

$$\frac{dQ}{dt} = -\mu I \delta T, \tag{6.2.27}$$

where μ is the Thomson coefficient. μ is found to be temperature dependent and so δT must be taken as small.

(b) *Peltier effect.* If a current I is flowing across the junction of two metals kept at constant temperature, heat is reversibly generated or absorbed at the junction, the rate being given by

$$\frac{dQ}{dt} = \Pi_{12} I. \tag{6.2.28}$$

Π_{12} is the Peltier coefficient depending on both metals and I is considered to flow from metal 1 to metal 2. The reversibility implies

$$\Pi_{12} = -\Pi_{21}. \tag{6.2.29}$$

(c) *Seebeck effect.* For the explanation of this we refer to Figure 6.1, disregarding the capacitor for the present. When $T \neq T'$ (the T's being temperatures) a potential difference $V_A - V_D$ appears across AD and is independent of T_0.

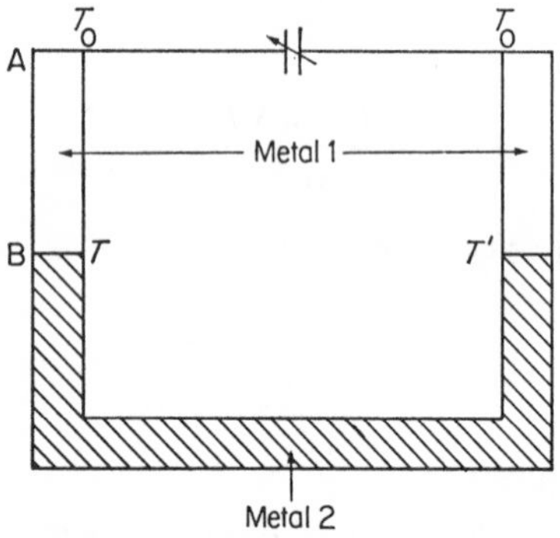

FIGURE 6.1. Arrangement of metals 1 and 2 used to discuss thermoelectric effects.

This is called the *thermoelectric force* Θ_{12}. When T' is kept constant and T is varied it is found that $d\Theta_{12}/dT$ is a function of T only and is defined as the thermoelectric power

$$S_{12} = \frac{d\Theta_{12}}{dT}. \tag{6.2.30}$$

The coefficients S_{12}, Π_{12} and Thomson coefficients μ_1 and μ_2 of the two metals obey simple relationships, first obtained by Lord Kelvin using thermodynamic arguments and thereby bearing his name. We shall reproduce similar arguments here; their rigour may be doubted (since the application is to transport phenomena) but they are illuminating and the results will be verified by using the Onsager relation of equation (6.2.23).

We imagine that by varying the capacitor indicated in Figure 6.1 we can send a charge q round the circuit from A to D, q being small enough to allow us to consider the changes involved as reversible. By the first law of thermodynamics the work $-\Theta_{12}q$ we must then do on the capacitor is equal to the heat given out and so from equations (6.2.27)–(6.2.30),

$$\Theta_{12}+\int_T^{T'}(\mu_1-\mu_2)\,dT''+\Pi_{12}|_T-\Pi_{12}|_{T'}=0. \qquad (6.2.31)$$

The second law of thermodynamics provides a further relation. The entropy change is zero since the process is reversible:

$$\int_T^{T'}\frac{(\mu_1-\mu_2)}{T''}\,dT''+\frac{\Pi_{12}}{T}\bigg|_T-\frac{\Pi_{12}}{T'}\bigg|_{T'}=0. \qquad (6.2.32)$$

Differentiation of these two relationships with respect to temperature T gives

$$-\frac{d\Theta_{12}}{dT}=\mu_1-\mu_2-\frac{d\Pi_{12}}{dT} \qquad (6.2.33)$$

and

$$\frac{\mu_1-\mu_2}{T}=\frac{1}{T}\frac{d\Pi_{12}}{dT}-\frac{\Pi_{12}}{T^2}. \qquad (6.2.34)$$

Hence

$$S_{12}=\frac{d\Theta_{12}}{dT}=\frac{\Pi_{12}}{T} \qquad (6.2.35)$$

and

$$\mu_1-\mu_2=T\frac{d^2\Theta_{12}}{dT^2}=T\frac{dS_{12}}{dT}. \qquad (6.2.36)$$

We shall now verify that the Onsager result (6.2.23) leads to the same relationships and obtain an expression in terms of the L_{ij} for the thermoelectric power, upon which the other thermoelectric quantities depend. We first take the Seebeck effect. When **j** is zero we obtain, from equation (6.2.21) a field due to a temperature gradient of

$$\mathscr{E}=-\frac{T}{e}\nabla\left(\frac{\zeta}{T}\right)+\frac{L_{12}}{L_{11}}\frac{\nabla T}{T}. \qquad (6.2.37)$$

The voltage Θ_{12} across AD is then obtained by taking a line integral from A to D:

$$\Theta_{12}=-\int_D^A\mathscr{E}.d\mathbf{l}=-\frac{1}{e}\int_A^D\nabla\zeta.d\mathbf{l}+\int_A^D\left(\frac{L_{12}}{L_{11}}+\frac{\zeta}{e}\right)\frac{\nabla T}{T}.d\mathbf{l}. \qquad (6.2.38)$$

The first term vanishes since A and D are both at temperature T_0, so that ζ is the same at both points. From the second term we have

$$\Theta_{12} = \int_{A}^{B}\left(\frac{L_{12}^{1}}{L_{11}^{1}}+\frac{\zeta^{1}}{e}\right)\frac{\nabla T}{T}.d\mathbf{l}+\int_{B}^{C}\left(\frac{L_{12}^{2}}{L_{11}^{2}}+\frac{\zeta^{2}}{e}\right)\frac{\nabla T}{T}.d\mathbf{l}+\int_{C}^{D}\left(\frac{L_{12}^{1}}{L_{11}^{1}}+\frac{\zeta^{1}}{e}\right)\frac{\nabla T}{T}.d\mathbf{l}$$

$$= \int_{T'}^{T}\left(\frac{L_{12}^{1}}{L_{11}^{1}}+\frac{\zeta^{1}}{e}\right)\frac{dT}{T}+\int_{T}^{T'}\left(\frac{L_{12}^{2}}{L_{11}^{2}}+\frac{\zeta^{2}}{e}\right)\frac{dT}{T}, \tag{6.2.39}$$

where the superscripts label the metal to which the quantities refer. The form of this expression suggests that we define absolute thermoelectric forces Θ_1 and Θ_2, by writing

$$\Theta_{12} = \Theta_1 - \Theta_2, \tag{6.2.40}$$

where

$$\Theta = \int_{T'}^{T}\left(\frac{L_{12}}{L_{11}}+\frac{\zeta}{e}\right)\frac{1}{T}\,dT. \tag{6.2.41}$$

We differentiate this with respect to T to obtain the *absolute thermoelectric power*

$$S = \frac{1}{T}\left(\frac{L_{12}}{L_{11}}+\frac{\zeta}{e}\right), \tag{6.2.42}$$

in terms of which

$$S_{12} = S_1 - S_2. \tag{6.2.43}$$

The Peltier effect occurs since the energy current associated with a given electrical current differs from metal to metal. A proper discussion of the effect requires a little care, and it will help if we temporarily introduce the heat current defined in equation (6.2.5). If we set $\nabla\zeta$ and $\nabla T = 0$ in equations (6.2.21) and (6.2.22), we obtain

$$\mathbf{Q} = \Pi\mathbf{j}, \tag{6.2.44}$$

where

$$\Pi = \left(\frac{L_{21}}{L_{11}}+\frac{\zeta}{e}\right). \tag{6.2.45}$$

Hence we must have liberated at the junction of metals 1 and 2 a heat per unit time of

$$\frac{dQ}{dt} = Q_1 - Q_2 = (\Pi_1 - \Pi_2)j \tag{6.2.46}$$

and so

$$\Pi_{12} = \Pi_1 - \Pi_2, \tag{6.2.47}$$

where Π is the absolute Peltier coefficient, which equation (6.2.45) shows to be just

$$\Pi = TS \tag{6.2.48}$$

if we use the Onsager relation $L_{12} = L_{21}$ in agreement with the thermodynamically derived equation (6.2.35).

We now see how the Peltier effect arises without relying on a definition of "heat current". With no temperature gradient, equations (6.2.21) and (6.2.22) give

$$\mathbf{U} = \frac{L_{21}}{L_{11}}\mathbf{j} \tag{6.2.49}$$

and so, as the current $\mathbf{j}$ flows from metal 1 to metal 2, there is an energy density change per unit time of

$$\mathbf{U}_1 - \mathbf{U}_2 = \left(\frac{L_{21}^1}{L_{11}^1} - \frac{L_{21}^2}{L_{11}^2}\right)\mathbf{j}, \tag{6.2.50}$$

which must be lost as heat. We have also to note a second effect, the change in chemical potential. Writing

$$j = e\frac{dn}{dt}, \tag{6.2.51}$$

where dn/dt is the particle flux, we see that there must also be a change of energy per unit volume of

$$\frac{dn}{dt}(\zeta_1 - \zeta_2) = (\zeta_1 - \zeta_2)j/e. \tag{6.2.52}$$

The effect is exactly as if the electric potential jumped by an amount $(\zeta_2 - \zeta_1)/e$ as the junction was crossed. In combination with equation (6.2.50) we now see that the rate of production of heat per unit time at the junction is

$$\frac{dQ}{dt} = \left[\left(\frac{L_{12}^1}{L_{11}^1} + \frac{\zeta^1}{e}\right) - \left(\frac{L_{12}^2}{L_{11}^2} + \frac{\zeta^2}{e}\right)\right]j \tag{6.2.53}$$

which is the same as equations (6.2.45) and (6.2.46).

We have now obtained the relation between S_{12} and Π_{12} independently of our earlier thermodynamic argument, and we finally turn to the Thomson effect. This, it must first be noted, is a second-order effect in contrast to the other phenomena we have studied. It concerns heat emitted when energy is drawn from the electric field. We shall not consider such effects elsewhere; when we apply an electric field we shall consider the system as otherwise isolated. Fortunately, however, the energy per unit time drawn from the field is easily written down, and is (for an isotropic material), $j\mathscr{E}$ per unit volume, a very well-known result. We are accustomed to term this the Joule heat, which it is for most practical purposes, but we now see that it must include the Thomson heat when there is a thermal gradient. If we consider all our quantities to vary in the x-direction only, we obtain from equation (6.2.21)

$$j = L_{11}\mathscr{E} + L_{11}\frac{T}{e}\frac{d}{dx}\left(\frac{\zeta}{T}\right) - \frac{L_{12}}{T}\frac{dT}{dx} \tag{6.2.54}$$

or, since $L_{11} = \sigma$ and noting equation (6.2.42),

$$\begin{aligned} \mathscr{E} &= \frac{j}{\sigma} - \frac{T}{e}\frac{d}{dx}\left(\frac{\zeta}{T}\right) - \frac{L_{12}}{T}\frac{dT}{dx} \\ &= \frac{j}{\sigma} - T\frac{dS}{dx}. \end{aligned} \tag{6.2.55}$$

Hence, since in a steady state all the electrical energy must disappear as heat, we have, per unit volume

$$\frac{dQ}{dt} = j\mathscr{E} = \frac{j^2}{\sigma} - jT\frac{dS}{dx}. \tag{6.2.56}$$

The first term on the right-hand side is the Joule heat, and is obviously irreversible, whereas the second term depends in sign on the relative directions of $\mathbf{j}$ and ∇T. This second term must therefore represent the Thomson heat, so that from equation (6.2.27)

$$\mu j\frac{dT}{dx} = jT\frac{dS}{dx} \tag{6.2.57}$$

or

$$\mu = T\frac{dS}{dT}, \tag{6.2.58}$$

in agreement with equation (6.2.36). This equation also gives

$$S = \int_0^T \frac{\mu}{T}\,dT, \tag{6.2.59}$$

there being no constant to be added to the integral, since we can take $S = 0$ at $T = 0$.

We can now see that since we must have $\mu = 0$ in a superconductor we must also have $S = 0$ in a superconductor. Hence S can be measured at very low temperatures by connecting the material under test between superconductors. It is then of interest to note that from equation (6.2.21) we obtain

$$\mathscr{E}+\frac{\nabla\zeta}{e}=S\nabla T. \tag{6.2.60}$$

We saw ζ/e acting as an electric potential in the Peltier effect, and it does so here, for the left-hand side is, by definition of the thermoelectric power, the effective field, when we connect our specimen between superconductors. Although $\mathscr{E}$ is the true field, the measuring apparatus reacts rather to the distribution of electrons and so $\mathscr{E}+(\nabla\zeta/e)$ is obtained.

6.3 Electrical conductivity tensor

The conductivity σ is ordinarily defined by the equation

$$\mathbf{j}=\sigma\mathscr{E}. \tag{6.3.1}$$

Here we have written the conductivity as a scalar, so that $\mathscr{E}$ and the resultant $\mathbf{j}$ are in the same direction. We now wish to show, by considering the dynamics of the electrons in as simple a model as possible, why σ must often be replaced by a tensor, though it *is* in fact scalar for cubic crystals.

Let us first recall the most elementary of the classical considerations involved in Drude's theory of conductivity (1900). Let τ be the mean-free time between collisions of an electron with whatever causes a finite resistivity—phonons, lattice imperfections, etc. During this time the electron will be accelerated by the field to reach, on collision, a velocity $\mathbf{v}=e\mathscr{E}\tau/m$ additional to what it originally had, i.e. a mean additional velocity of $\mathbf{v}=-e\mathscr{E}\tau/2m$. Let us neglect the factor of $\frac{1}{2}$, since we are not being very precise in our statistical analysis, and say that there is an average velocity of $-e\mathscr{E}\tau/m$ imposed on the random motions of the electrons. Then, since

$$\mathbf{j}=ne\mathbf{v}, \tag{6.3.2}$$

where n is the number of electrons per unit volume, we have, from equation (6.3.1),

$$\sigma=\frac{ne^2\tau}{m}. \tag{6.3.3}$$

The same equation could also have been derived by setting up a force equation for $\mathbf{v}$, with a drag term linear in $\mathbf{v}$. The dependence of this drag term on τ is obtainable by saying that, if we wish to write down the deceleration due to collisions, we need only note that the velocity $\mathbf{v}$ is destroyed in a time τ, so that the deceleration is just $\mathbf{v}/\tau$.

6.3.1 Rate of change of crystal momentum

We now turn to a wave-mechanical formulation, retaining the notion of a mean free path and hence of mean free time τ. If the periodic crystal potential were zero, we could immediately write down a force equation for the momentum by the argument outlined above:

$$\hbar\frac{d\mathbf{k}}{dt} = -e\mathscr{E} + \frac{\hbar(\mathbf{k}-\mathbf{k}_0)}{\tau} \tag{6.3.4}$$

($\mathbf{k}_0$ being the momentum of the electron before the field was switched on); the only essential modification to Drude's argument would then be the introduction of Fermi–Dirac statistics to overcome shortcomings we need not mention here. Equation (6.3.4) is in fact a valid equation when the periodic potential is non-zero, as we shall indicate. If H_0 is the Hamiltonian in the absence of the field, the full Hamiltonian is just

$$H = H_0 + e\mathscr{E}.\mathbf{r}. \tag{6.3.5}$$

From the time-dependent Schrödinger equation it follows that the wave function $\psi_{\mathbf{k}}(\mathbf{r})$ at time $t = 0$ evolves as

$$\psi_{\mathbf{k}}(\mathbf{r}, t) = \exp(-iHt/\hbar)\,\psi_{\mathbf{k}}(\mathbf{r}). \tag{6.3.6}$$

Suppose now we switch on the field at $t = 0$, so that $\psi_{\mathbf{k}}(\mathbf{r})$ is just the wave function of the electron in zero electric field, and thus obeys Bloch's theorem:

$$\psi_{\mathbf{k}}(\mathbf{r}+\mathbf{R}) = \psi_{\mathbf{k}}(\mathbf{r})\exp(i\mathbf{k}.\mathbf{R}). \tag{6.3.7}$$

It is now very easy to show that $\psi_{\mathbf{k}}(\mathbf{r}, t)$ obeys Bloch's theorem as well. We have

$$\begin{aligned}\psi_{\mathbf{k}}(\mathbf{r}+\mathbf{R}, t) &= \exp\left\{-i[H_0(\mathbf{r}+\mathbf{R})+e\mathscr{E}.(\mathbf{r}+\mathbf{R})]\frac{t}{\hbar}\right\}\psi_{\mathbf{k}}(\mathbf{r}+\mathbf{R}) \\ &= \exp\left\{-i[H_0(\mathbf{r})+e\mathscr{E}.\mathbf{r}]\frac{t}{\hbar}\right\}\psi_{\mathbf{k}}(\mathbf{r})\exp(i\mathbf{k}.\mathbf{R})\exp(-ie\mathscr{E}.\mathbf{R}t/\hbar),\end{aligned} \tag{6.3.8}$$

i.e.

$$\psi_{\mathbf{k}}(\mathbf{r}+\mathbf{R}, t) = \psi_{\mathbf{k}}(\mathbf{r}, t)\exp\left[i\left(\mathbf{k}-\frac{e\mathscr{E}t}{\hbar}\right).\mathbf{R}\right]. \tag{6.3.9}$$

Thus after time t the crystal momentum $\mathbf{k}$ has become $\mathbf{k}-(e\mathscr{E}t/\hbar)$.

Let us now think of $\psi_{\mathbf{k}}(\mathbf{r}, t)$ as expanded in terms of the eigenfunctions of the unperturbed system, and suppose the band-gaps are those of a typical metal, of the order of electron-volts. It will then be very difficult for the electric field, usually much less than a volt per centimetre, to cause interband transitions, for we have seen that the electric field is only effective over the order of the mean free path (at most some hundreds of angstroms) before a collision occurs. Since $\psi_{\mathbf{k}}(\mathbf{r}, t)$ obeys Bloch's theorem, this must mean

$$\psi_{\mathbf{k}}(\mathbf{r}, t) = \psi_{\mathbf{k}-(e\mathscr{E}t/\hbar)}(\mathbf{r}), \tag{6.3.10}$$

where $\psi_{\mathbf{k}}$ and $\psi_{\mathbf{k}-(e\mathscr{E}t/\hbar)}$ pertain to the same band. It should be noted that every $\mathbf{k}$-vector will be shifted by the same amount, so that the entire distribution in $\mathbf{k}$-space moves rigidly through $-e\mathscr{E}t/\hbar$. This is illustrated in Figure 6.2(a), for simplicity the Fermi surface being taken as a sphere. Figure 6.2(b) illustrates, for square Brillouin zones, why this drift does not

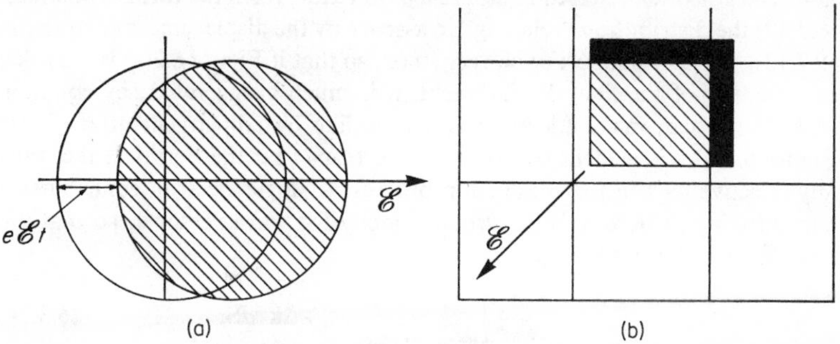

FIGURE 6.2. Displacement of Fermi surface on application of electric field. (*a*) Half-filled band. (b) Completely full band.

produce a net conduction if the band is completely filled. The black regions are entirely equivalent to the newly unoccupied regions of the central BZ because of the periodicity in the reciprocal lattice, and so in this case the drift in reality causes no change whatsoever. This state of affairs is alternatively described very simply as a situation in which the perturbation cannot scatter electrons from one state to another for every available state is already filled.

Relaxation time approximation. The rate of change of $\mathbf{k}$ is, from equation (6.3.10),

$$\hbar \frac{d\mathbf{k}}{dt} = -e\mathscr{E}, \tag{6.3.11}$$

but this neglects scattering from the lattice. Although we cannot justify it completely, we will again, in this preliminary discussion, take the scattering to give rise to a term of the form $(\mathbf{k}-\mathbf{k}_0)/\tau$, so that equation (6.3.4) now holds. It is a trivial matter to verify that the solution of equation (6.3.4) is

$$\hbar\mathbf{k} = \hbar\mathbf{k}_0 - \tau e\mathscr{E}(1-e^{-t/\tau}), \tag{6.3.12}$$

so that the whole distribution in **k**-space is displaced by the electric field by a finite amount [obtained by letting $t\to\infty$ in equation (6.3.12)],

$$\Delta\mathbf{k} = -\frac{\tau e\mathscr{E}}{\hbar}. \tag{6.3.13}$$

This is evidently a realistic picture, in contrast to that given by equation (6.3.9), and the identification of the factor τ in equation (6.3.12) with a mean free time between collisions is quite plausible. For future reference we may note that τ is often referred to as a relaxation time, from the form of equation (6.3.12) the distribution "relaxing" in **k**-space by the displacement of equation (6.3.13). We now take $\Delta\mathbf{k}$ to be very small, so that if Figure 6.2(a) is regarded as a picture of the final displacement, it is much exaggerated; by equation (6.3.13), we can make $\Delta\mathbf{k}$ as small as we like by adjustment of $\mathscr{E}$. From Figure 6.2(a) it is evident that only the electrons near the Fermi surface take any effective part in the conduction process. A single electron contributes a current $e\langle\mathbf{v}_\mathbf{k}\rangle$, where $\mathbf{v}_\mathbf{k}$ is the group velocity for wave vector **k**, so that the total current density is

$$\mathbf{j} = -\frac{e}{4\pi^3}\int_{\text{Fermi surface}} \langle\mathbf{v}_\mathbf{k}\rangle\,\Delta\mathbf{k}.d\mathbf{S}, \tag{6.3.14}$$

where we have taken the density of states in **k**-space to be $1/8\pi^3$ per unit volume. Now we saw in Chapter 1 that

$$\langle\mathbf{v}_\mathbf{k}\rangle = \frac{\hbar}{im}\int \psi_\mathbf{k}^*\,\nabla\psi_\mathbf{k}\,d\mathbf{r} = \frac{1}{\hbar}\frac{\partial E}{\partial\mathbf{k}} \tag{6.3.15}$$

and so, taking equation (6.3.13) for $\Delta\mathbf{k}$, equation (6.3.14) becomes (assuming τ independent of **k**, the generalization when τ is **k**-dependent being obvious)

$$\mathbf{j} = \frac{e^2\tau}{4\pi^3\hbar^2}\int_{\text{Fermi surface}} \frac{\partial E}{\partial\mathbf{k}}\,\mathscr{E}.d\mathbf{S}. \tag{6.3.16}$$

We see that we must write

$$j_\alpha = \sum_\beta \sigma_{\alpha\beta}\mathscr{E}_\beta \tag{6.3.17}$$

as the most general form of Ohm's law, where $\sigma_{\alpha\beta}$ is a conductivity tensor with components, according to equation (6.3.16), given by

$$\sigma_{\alpha\beta} = \frac{e^2\tau}{4\pi^3\hbar^2}\int \frac{\partial E}{\partial k_\alpha}\,dS_\beta. \tag{6.3.18}$$

$\sigma_{\alpha\beta}$ is only, in general, diagonal if the Fermi surface is spherical. In the case of cubic crystals, σ is in fact scalar no matter what shape this surface takes. We shall prove this below, but at the moment we check that it is true for the approximation of equation (6.3.18). Taking the field $\mathscr{E}$ along a cube edge, say the x-axis, the integral in equation (6.3.16) becomes

$$\int_{\text{Fermi surface}} \frac{\partial E}{\partial \mathbf{k}} \mathscr{E} . d\mathbf{S} = \mathscr{E}\left[\hat{\mathbf{x}} \int \frac{\partial E}{\partial k_x} dS_x + \hat{\mathbf{y}} \int \frac{\partial E}{\partial k_y} dS_x + \hat{\mathbf{z}} \int \frac{\partial E}{\partial k_z} dS_x \right]. \tag{6.3.19}$$

The second and third terms evidently vanish since changing the sign of x changes the sign of dS_x without changing the signs or magnitudes of $\partial E/\partial k_y$ or $\partial E/\partial k_z$. We also note that

$$\int \frac{\partial E}{\partial k_x} dS_x = \frac{1}{3} \int \frac{\partial E}{\partial \mathbf{k}} . d\mathbf{S} \tag{6.3.20}$$

so that

$$\sigma = \frac{e^2 \tau}{4\pi^3 \hbar^2} \frac{1}{3} \int_{\text{Fermi surface}} \frac{\partial E}{\partial \mathbf{k}} . d\mathbf{S}. \tag{6.3.21}$$

We regain the classical form of the expression if we can write

$$E = \frac{\hbar^2 k^2}{2m^*}, \tag{6.3.22}$$

where m^* is an effective mass, for then,

$$\frac{1}{3} \int \frac{\partial E}{\partial \mathbf{k}} . d\mathbf{S} = \hbar^2 \frac{4\pi k_f^3}{3m^*} = \frac{4\pi^3 \hbar^2 n}{\Omega m^*} \tag{6.3.23}$$

and

$$\sigma = \frac{ne^2 \tau}{m^*}. \tag{6.3.24}$$

Of course, we can usually at best define a reciprocal mass tensor through

$$\frac{1}{\hbar} \frac{\partial E}{\partial k_\alpha} = \sum_\beta \left(\frac{\hbar}{m^*_{\alpha\beta}}\right) k_\beta. \tag{6.3.25}$$

6.3.2 Consequences of symmetry

We conclude with some comments on the implications of crystal symmetry on $\sigma_{\alpha\beta}$, independent of any approximation to it. For a triclinic crystal, with no rotational symmetry properties, all we can say is that the matrix $\sigma_{\alpha\beta}$

must be Hermitian.‡ However, the presence of symmetry axes enables us to reduce the number of components. Let the symmetry axis be the x-axis, and first let us look at Figure 6.3, in which, as an example, **x** is taken to be a two- or fourfold rotation axis or screw axis and $\mathscr{E}$ is taken parallel to y [Figure 6.3(a)]. Now if we reverse the field, this reverses j_x. Rotation about **x** through π is equivalent to this reversal [Figure 6.3(b)], but this rotation leaves the

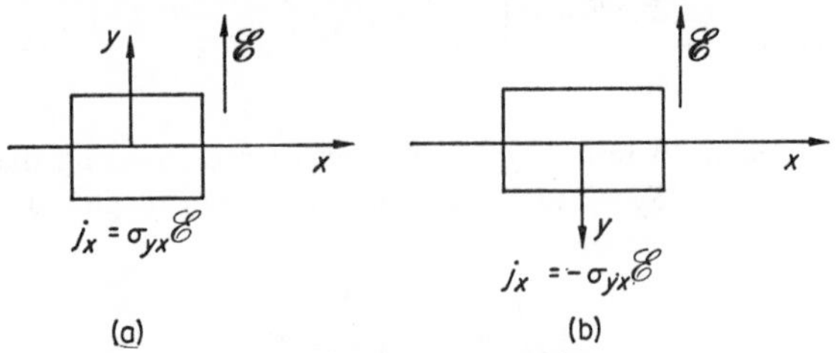

FIGURE 6.3. Illustrating effect of twofold rotation axis or screw axis, with electric field perpendicular to axis, on components of electrical conductivity tensor. In (b) specimen is rotated about x-axis through π.

physical situation completely unchanged, so that j_x is *not* reversed. Hence $j_x = 0$ and this is true for a symmetry axis of any kind, implying

$$\sigma_{xy} = \sigma_{yx} = \sigma_{xz} = \sigma_{zx} = 0, \tag{6.3.26}$$

where x is a symmetry axis.

As our second example, let **x** be a fourfold rotation axis or screw axis. It is then evident that Figures 6.4(a) and (b), in which **x** is taken perpendicular

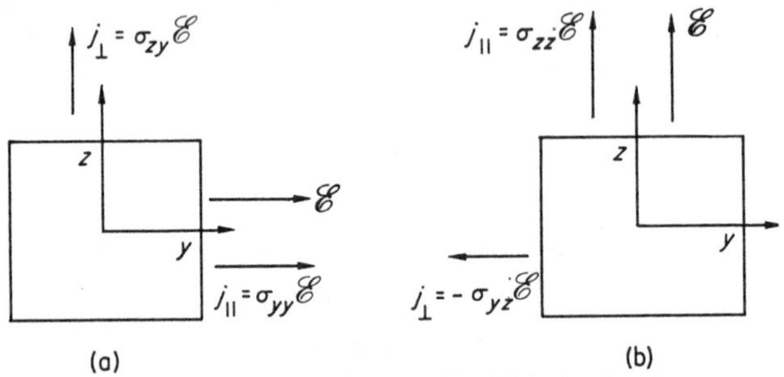

FIGURE 6.4. Illustrating effect of fourfold rotation or screw axis on electrical conductivity tensor, with electric field perpendicular to axis. In (b), field $\mathscr{E}$, perpendicular to x-direction, is rotated through $\pi/2$ relative to (a).

‡ This Hermitian property, which holds if no magnetic field is applied, is proved in section 6.11.

to the paper, represent identical physical situations and neither $j_\perp$ nor $j_\parallel$ changes (except in direction) as $\mathscr{E}$, perpendicular to **x**, is rotated through $\pi/2$. It follows that $j_\perp = 0$ and

$$\sigma_{zy} = \sigma_{yz} = 0, \quad \sigma_{yy} = \sigma_{zz} \tag{6.3.27}$$

if **x** is a fourfold rotation axis or screw axis, so that there are only two independent components of the conductivity tensor. Since cubic crystals have three fourfold rotation axes, it is evident that $\sigma_{\alpha\beta}$ collapses into a scalar for such crystals.

6.4 Solution of Boltzmann equation

We derived the Boltzmann transport equation in Chapter 2, pointing out that the equation holds only if the driving force **F** varies slowly over a distance of the order of the mean free path. This slowly varying assumption is, in fact, also necessary if we are to ascribe physical meaning to the distribution function f for electrons. We have previously seen (Volume 1, Chapter 3) that

$$\mathbf{F} = -e\left(\mathscr{E} + \frac{\mathbf{v} \times \mathscr{H}}{c}\right) = \hbar\dot{\mathbf{k}}, \tag{6.4.1}$$

where **v** is the group velocity defined in equation (6.3.15). If equations (6.4.1) and (6.3.15) are used in the Boltzmann equation (2.9.20), we must replace **p** by $\hbar\mathbf{k}$ throughout, and then $f(\mathbf{kr}t)$ can be interpreted as the local Fermi distribution for the region round **r**. We may note at this point that we can include electron–electron interactions by assuming we are really dealing with the transport of quasi-particles. In Chapter 2 we saw that these are well defined at the Fermi surface, and in the last section (cf. Figure 6.2) we saw that it is only particles at the Fermi surface that are important, at least for fields which are not too high.

Since we can certainly take $\mathscr{E}$ as infinitesimal in calculating the electrical conductivity, it is customary to linearize the Boltzmann equation, i.e. take only terms which are first order in $\mathscr{E}$, when we can replace f by f_0 in the term $(\mathbf{F}/\hbar).(\partial f/\partial\mathbf{k})$.

6.4.1 Collision term

The calculation of the transition probability matrix element for electrons scattered from state $|\mathbf{k}\rangle$ to $|\mathbf{k}'\rangle$ will be discussed below, and we shall now see how it enters the collision term of the Boltzmann equation.

We shall denote by $P(\mathbf{k}'\mathbf{k})$ the intrinsic scattering probability per unit time, which is calculated without regard to the prohibition, by the Pauli principle, of scattering into a state already occupied. The actual probability is thus if f and Π are defined for a specimen of unit volume,

$$\Pi(\mathbf{k}'\mathbf{k}) = f(\mathbf{k})\,[1-f(\mathbf{k}')]\,P(\mathbf{k}'\mathbf{k}). \tag{6.4.2}$$

For doubly filled levels and spins unchanged by scattering, the number of particles scattered into the volume element $d\mathbf{k}'$ from $d\mathbf{k}$ in time δt is then

$$\left(\frac{1}{8\pi^3}\right)\Pi(\mathbf{k}'\mathbf{k})\,d\mathbf{k}\,d\mathbf{k}'\,\delta t.$$

There is a similar expression for scattering from $d\mathbf{k}'$ to $d\mathbf{k}$ and so we obtain

$$\left(\frac{\partial f}{\partial t}\right)_{\text{coll}} = \frac{1}{8\pi^3}\int [\Pi(\mathbf{k}'\mathbf{k})-\Pi(\mathbf{k}\mathbf{k}')]\,d\mathbf{k}'. \tag{6.4.3}$$

If the electron scattering is entirely elastic, the linearized Boltzmann equation can be solved exactly provided $P(\mathbf{k}'\mathbf{k})$ depends only on $\mathbf{k}'-\mathbf{k}$. The collision term is, for this case, already linear, because elastic scattering implies

$$P(\mathbf{k}'\mathbf{k}) = P(\mathbf{k}\mathbf{k}'), \quad f_0(\mathbf{k}') = f_0(\mathbf{k}), \tag{6.4.4}$$

where f_0 is the equilibrium distribution. Writing

$$f = f_0 + f_1, \tag{6.4.5}$$

we find

$$\left[\frac{\partial f}{\partial t}\right]_{\text{coll}} = \frac{1}{8\pi^3}\int P(\mathbf{k}'\mathbf{k})\,[f_1(\mathbf{k}')-f_1(\mathbf{k})]\,d\mathbf{k}'. \tag{6.4.6}$$

In this approximation, the exact solution we mentioned has in fact already been encountered in section 6.3.1, as we shall see later on. We should note however, that equation (6.4.6) is not at all a good approximation at intermediate temperatures.

This is evident from the "weak" scattering form of $\Pi(\mathbf{k}'\mathbf{k})$ for electrons scattered by lattice vibrations, which we shall briefly discuss at this point. Baym (1964) has drawn attention to the similarity between the inelastic scattering of a slow neutron from a metal, which we discussed extensively in Chapter 5, and the inelastic scattering of a conduction electron from the lattice vibrations in a metal. In either case the coupling is to the ion density rather than to individual phonons. In addition to the usual squared matrix element describing the scattering, the scattering probability $P(\mathbf{k}'\mathbf{k})$ is also proportional to the number $S_1(\boldsymbol{\kappa}\omega)$ of available states for the density fluctuations. $S_1(\boldsymbol{\kappa}\omega)$ is just the van Hove correlation function $S(\boldsymbol{\kappa}\omega)$ with the Bragg peaks subtracted out, these latter describing simply the scattering from the perfect lattice.

Thus we find

$$\Pi(\mathbf{k}'\mathbf{k}) = \frac{2\pi}{\hbar} f(\mathbf{k})\,[1-f(\mathbf{k}')]\,NS_1(\boldsymbol{\kappa}\omega)\,|\langle\mathbf{k}'|\,V(\boldsymbol{\kappa})\,|\mathbf{k}\rangle|^2\,\delta[\varepsilon(\mathbf{k})-\varepsilon(\mathbf{k}')-\omega], \tag{6.4.7}$$

where $\boldsymbol{\kappa} = \mathbf{k}'-\mathbf{k}$.

Since $S_1(\boldsymbol{\kappa}\omega)$ obeys the detailed balancing condition (see Appendix A4.2)

$$S_1(\boldsymbol{\kappa}\omega) = e^{\beta\omega}\,S_1(-\boldsymbol{\kappa}, -\omega), \quad (\beta = 1/k_B T) \tag{6.4.8}$$

it can be seen by comparing equation (6.4.8) with equation (6.4.7) that

$$P(\mathbf{k}\mathbf{k}') = e^{-\beta\omega}\,P(\mathbf{k}'\mathbf{k}) \quad [\omega = \varepsilon(\mathbf{k})-\varepsilon(\mathbf{k}')]. \tag{6.4.9}$$

This is actually quite general, being just the result of applying the principle of detailed balancing to $P(\mathbf{k}'\mathbf{k})$. From this result, one sees that $P(\mathbf{k}'\mathbf{k}) \simeq P(\mathbf{k}\mathbf{k}')$ only at elevated temperatures.

From equation (6.4.9) we can prove that $[\partial f_0/\partial t]_{\text{coll}} = 0$ if we write

$$\Pi_0(\mathbf{k}'\mathbf{k}) = f_0(\mathbf{k})\,[1-f_0(\mathbf{k}')]\,P(\mathbf{k}'\mathbf{k}) \tag{6.4.10}$$

for the Π's in equation (6.4.3), f_0 being the equilibrium distribution. We know this *must* be true in the absence of external fields, but it will be helpful later if we show explicitly that

$$\Pi_0(\mathbf{k}'\mathbf{k}) = \Pi_0(\mathbf{k}\mathbf{k}'), \tag{6.4.11}$$

using equation (6.4.9). Since

$$f_0(\mathbf{k}) = \frac{1}{\exp[\beta(\varepsilon_{\mathbf{k}}-\zeta)]+1}, \tag{6.4.12}$$

ζ being the Fermi energy, we have

$$\begin{aligned}\Pi_0(\mathbf{k}'\mathbf{k}) &= \frac{1}{\exp[\beta(\varepsilon_{\mathbf{k}}-\zeta)]+1}\left\{1-\frac{1}{\exp[\beta(\varepsilon_{\mathbf{k}'}-\zeta)]+1}\right\}P(\mathbf{k}'\mathbf{k})\\ &= \frac{\exp[\beta(\varepsilon_{\mathbf{k}'}-\zeta)]}{\{\exp[\beta(\varepsilon_{\mathbf{k}}-\zeta)]+1\}\{\exp[\beta(\varepsilon_{\mathbf{k}'}-\zeta)]+1\}}P(\mathbf{k}'\mathbf{k})\\ &= \exp[\beta(\varepsilon_{\mathbf{k}'}-\zeta)]\,f_0(\mathbf{k}')\,f_0(\mathbf{k})\,P(\mathbf{k}'\mathbf{k}).\end{aligned} \tag{6.4.13}$$

Similarly, for $\Pi_0(\mathbf{k}\mathbf{k}')$ we have

$$\begin{aligned}\Pi_0(\mathbf{k}\mathbf{k}') &= \exp[\beta(\varepsilon_{\mathbf{k}}-\zeta)]\,f_0(\mathbf{k})\,f_0(\mathbf{k}')\,P(\mathbf{k}\mathbf{k}')\\ &= \exp[\beta(\varepsilon_{\mathbf{k}}-\zeta)]\,f_0(\mathbf{k})\,f_0(\mathbf{k}')\exp(-\beta\omega)\,P(\mathbf{k}'\mathbf{k})\\ &= \exp[\beta(\varepsilon_{\mathbf{k}}-\omega-\zeta)]\,f_0(\mathbf{k})\,f_0(\mathbf{k}')\,P(\mathbf{k}'\mathbf{k}).\end{aligned} \tag{6.4.14}$$

Thus we may write from equations (6.4.9) and (6.4.14)

$$\Pi_0(\mathbf{kk}') = \exp[\beta(\varepsilon_{\mathbf{k}'} - \zeta)] f_0(\mathbf{k}) f_0(\mathbf{k}') P(\mathbf{k}'\mathbf{k}) = \Pi_0(\mathbf{k}'\mathbf{k}). \quad (6.4.15)$$

We are now in a position to see how the general form of the collision term given in equation (6.4.3) may be linearized. We first anticipate the solution of the linearized equation when the specialized form in equation (6.4.6) is taken. It is

$$f = f_0 + e\mathscr{E}.\mathbf{v}\tau\frac{\partial f_0}{\partial E}, \quad (6.4.16)$$

which is easily interpreted. It is just an expansion, to order $\mathscr{E}$, in the average energy taken by an electron from the field when a mean-free time (and so mean-free path $l = v\tau$) can be defined.

More generally, we may write f_1 in equation (6.4.5) as

$$f_1 = -\phi\frac{\partial f_0}{\partial E} = +\beta f_0(\mathbf{k})[1 - f_0(\mathbf{k})]\phi. \quad (6.4.17)$$

Let us calculate $\Pi(\mathbf{k}'\mathbf{k})$ to first order in f_1. We have, assuming phonon equilibrium is attained in a time much shorter than the collision time,

$$\Pi(\mathbf{k}'\mathbf{k}) = \Pi_0(\mathbf{k}'\mathbf{k}) + f_1(\mathbf{k})[1 - f_0(\mathbf{k}')]P(\mathbf{k}'\mathbf{k}) - f_0(\mathbf{k}) f_1(\mathbf{k}') P(\mathbf{k}'\mathbf{k}). \quad (6.4.18)$$

Inserting this into equation (6.4.3) we then obtain

$$\left[\frac{\partial f}{\partial t}\right]_{\text{coll}} = \frac{1}{4\pi^3}\int \{f_1(\mathbf{k})[1 - f_0(\mathbf{k}')] - f_0(\mathbf{k}') f_1(\mathbf{k}')\} P(\mathbf{k}'\mathbf{k})$$
$$- \{f_1(\mathbf{k}')[1 - f_0(\mathbf{k})] - f_0(\mathbf{k}')\,\delta_1(\mathbf{k})] P(\mathbf{kk}')\, d\mathbf{k}', \quad (6.4.19)$$

since the terms in Π_0 vanish. Examining the various terms in this equation, taking cognizance of equation (6.4.9) and manipulating terms in f as in equations (6.4.13) and (6.4.14), we see that

$$\begin{aligned} & f_1(\mathbf{k})[1 - f_0(\mathbf{k}')] P(\mathbf{k}'\mathbf{k}) + f_1(\mathbf{k}) f_0(\mathbf{k}') P(\mathbf{kk}') \\ & = f_1(\mathbf{k}) f_0(\mathbf{k}')\{\exp[\beta(\varepsilon_{\mathbf{k}'} - \zeta)]\exp(\beta\omega) + 1\} P(\mathbf{kk}') \\ & = f_1(\mathbf{k})\frac{f_0(\mathbf{k}')}{f_0(\mathbf{k})} P(\mathbf{kk}') \\ & = \beta\phi(\mathbf{k}) f_0(\mathbf{k}')[1 - f_0(\mathbf{k})] P(\mathbf{kk}') \\ & = \beta\phi(\mathbf{k})\,\Pi_0(\mathbf{kk}'). \end{aligned} \quad (6.4.20)$$

Similarly,

$$f_0(\mathbf{k}) f_1(\mathbf{k}') P(\mathbf{k}'\mathbf{k}) + f_1(\mathbf{k}')[1 - f_0(\mathbf{k})] P(\mathbf{kk}') = \beta\phi(\mathbf{k}')\,\Pi_0(\mathbf{k}'\mathbf{k}). \quad (6.4.21)$$

Then by equation (6.4.11), we finally obtain for equation (6.4.19)

$$\left[\frac{\partial f}{\partial t}\right]_{\text{coll}} = \frac{\beta}{4\pi^3}\int [\phi(\mathbf{k})-\phi(\mathbf{k}')]\,\Pi_0(\mathbf{k}'\mathbf{k})\,d\mathbf{k}'. \tag{6.4.22}$$

6.4.2 Conductivity in relaxation time approximation

For elastic collisions, the rate of change of the distribution f with time is given by equation (6.4.6). If we assume that $[\partial f/\partial t]_{\text{coll}}$ can be expressed in terms of a relaxation time τ by

$$-\left[\frac{\partial f}{\partial t}\right]_{\text{coll}} = \frac{f-f_0}{\tau} \equiv \frac{f_1}{\tau}, \tag{6.4.23}$$

then it follows from equations (6.4.6) and (6.4.10) that

$$\tau(\mathbf{k})^{-1} = \frac{1}{4\pi^3}\int P(\mathbf{k}'\mathbf{k})\left[1-\frac{f_1(\mathbf{k}')}{f_1(\mathbf{k})}\right]d\mathbf{k}'. \tag{6.4.24}$$

This is a meaningful result provided it is independent of the applied fields, which in turn means that $f_1(\mathbf{k}')/f_1(\mathbf{k})$ is independent of the fields. This is evidently true in the small field limit. Furthermore, substituting in equation (6.4.16) to be consistent with τ independent of $\mathbf{k}$ we must have

$$P(\mathbf{k}'\mathbf{k}) = P(|\mathbf{k}'-\mathbf{k}|). \tag{6.4.25}$$

It is evident that $P(\mathbf{k}'\mathbf{k}) = P(\mathbf{k}\mathbf{k}')$ is a necessary condition for equation (6.4.25) to be true, which is the same as the requirement of elastic scattering.

(a) *Conductivity in static uniform field.* We now show that the times τ of section 6.3 and the present section are identical. From equation (6.3.14) the electric current is

$$\mathbf{j} = -\frac{e}{4\pi^3}\int \mathbf{v} f\,d\mathbf{k} = -\frac{e^2}{4\pi^3}\int \tau\mathbf{v}(\mathbf{v}\,.\,\mathscr{E})\frac{\partial f_0}{\partial E}\,d\mathbf{k} \tag{6.4.26}$$

and so

$$\sigma_{\alpha\beta} = -\frac{e^2}{4\pi^3}\int \tau v_\alpha v_\beta \frac{\partial f_0}{\partial E}\,d\mathbf{k}. \tag{6.4.27}$$

This can be converted to integrations over shells of **k**-space between constant-energy surfaces, when (cf. the calculation of the density of states in Chapter 1) we obtain

$$\sigma_{\alpha\beta} = -\frac{e^2}{4\pi^3}\int \frac{df_0}{dE}\,dE\int \frac{\tau v_\alpha v_\beta}{|\partial E/\partial \mathbf{k}|}\,dS. \tag{6.4.28}$$

Taking $T = 0$, so that $\partial f_0/\partial E$ is a δ-function at the Fermi energy, we have

$$\sigma_{\alpha\beta} = \frac{e^2}{4\pi^3}\int \frac{\tau v_\alpha v_\beta\, dS}{|\partial E/\partial \mathbf{k}|}. \tag{6.4.29}$$

When we note that

$$\frac{v_\beta\, dS}{|\partial E/\partial \mathbf{k}|} = \frac{v_\beta\, dS}{\hbar v} = \frac{dS_\beta}{\hbar}, \tag{6.4.30}$$

equations (6.4.29) and (6.3.18) are evidently identical.

(b) *Chambers' solution.* When a relaxation time can be defined, the solution of the linearized Boltzmann equation is simple if we have a static electric field. We write $\partial f_1/\partial t = 0$ for the steady state and the answer for f_1 is immediate. We now wish to see what solutions are obtained in more complicated situations, and to do so we formulate the transport problem kinetically in the way originally described by Chambers (1952).

We first interpret the relaxation time in the following way; we suppose that at time $t = 0$ a certain number of electrons n_0 make collisions, and we further suppose that the number not having suffered subsequent collisions during a time t is

$$n(t) = n_0\, \mathrm{e}^{-t/\tau}. \tag{6.4.31}$$

τ is readily shown to coincide with the collision time τ_c. In an interval dt we have $-[dn(t)/dt]\, dt$ electrons making collisions so that

$$\tau_c = -\frac{1}{n_0}\int_0^\infty t\frac{d}{dt}n(t)\, dt = \frac{1}{\tau}\int_0^\infty \mathrm{e}^{-t/\tau}\, t\, dt = \tau. \tag{6.4.32}$$

We have not previously equated τ_c with the relaxation time in the Boltzmann equation, but the kinetic discussion we now give will enable us to do so.

Electrons passing through $\mathbf{r}_0$ with velocity $\mathbf{v}_0$ and energy E at time t_0 follow a certain trajectory before reaching $\mathbf{r}_0$ and the distribution function f is obtained by simply integrating the number scattered into the trajectory over all previous times t, weighted by the probability of reaching $\mathbf{r}_0$; this probability we obtain from equation (6.4.31) and therefore

$$f(\mathbf{r}_0\, \mathbf{v}_0\, t_0) = \int_{-\infty}^{t_0} \frac{dt}{\tau} f_0(E - \Delta E)\, \mathrm{e}^{-(t_0 - t)/\tau}. \tag{6.4.33}$$

$\Delta E(t)$ is the energy a particle acquires from the electric field $\mathscr{E}(\mathbf{r}, t)$ in the duration $t_0 - t$, i.e.

$$\Delta E(t) = -e\int_t^{t_0} \mathscr{E}(\mathbf{r}, s) . \mathbf{v}(s)\, ds. \tag{6.4.34}$$

Hence, writing

$$f_0(E-\Delta E) = f_0(E) - \Delta E \frac{\partial f_0}{\partial E} + \dots \tag{6.4.35}$$

and integrating equation (6.4.33) by parts, we obtain

$$f_1 = f - f_0 = \frac{df_0}{dE} \int_{-\infty}^{t} dt \frac{d\Delta E}{dt} e^{-(t-t_0)/\tau} \tag{6.4.36}$$

or

$$f_1(\mathbf{r}_0 \mathbf{v}_0 t_0) = -e \frac{df_0}{dE} \int_{-\infty}^{t_0} dt\, \mathscr{E}(\mathbf{r}, t) . \mathbf{v}(t) e^{-(t_0-t)/\tau}. \tag{6.4.37}$$

Differentiation with respect to time yields the Boltzmann equation immediately, and moreover shows τ to be just the relaxation time appearing in the term $[\partial f/\partial t]_{\mathrm{coll}} = -f_1/\tau$. We might also note that a change of variable yields

$$f_1 = -e \frac{df_0}{dE} \int_0^{\infty} \mathscr{E}(\mathbf{r}, t_0 - s) . \mathbf{v}(t_0 - s) e^{-s/\tau} dt \tag{6.4.38}$$

and if $\mathbf{v}$ is independent of time and we write $\mathscr{E} = \mathscr{E}_0 e^{i\omega t}$ we find $f_1 \propto e^{i\omega t_0}$.

Finally, if τ is dependent on the time through the velocity, as will happen if a magnetic field is present, the factor $e^{-(t_0-t)/\tau}$ must be replaced by $\exp[-\int_t^{t_0} ds/\tau(s)]$ in equation (6.4.24). The current in this general case is

$$\mathbf{j} = \frac{e^2}{4\pi^3 V} \int \mathbf{v} \frac{df_0}{dE} d\mathbf{k} \int_{-\infty}^{t_0} dt\, \mathscr{E}(\mathbf{r}t) . \mathbf{v}(t) \exp\left[-\int_t^{t_0} ds/\tau(s) \right]. \tag{6.4.39}$$

A situation in which we shall use the type of solution provided by Chambers' method is that where a magnetic field is applied. We can then either incorporate it in the Boltzmann equation by including it in the Lorentz force $\mathbf{F} = e[\mathscr{E} + (\mathbf{v} \times \mathscr{H})/c]$ (which will actually be the method we shall find convenient in discussing the Hall effect) or, alternatively, the steady-state time dependence of the distribution function can be expressed as

$$\frac{\partial f_1}{\partial t} = \frac{\partial \phi}{\partial t} \frac{\partial f}{\partial \phi}, \tag{6.4.40}$$

where ϕ represents the angle $\mathbf{k}$ makes with some chosen direction in a plane perpendicular to $\mathscr{H}$. We then omit the term $(\mathbf{v} \times \mathscr{H})/c$ from $\mathbf{F}$, recognizing that the magnetic field cannot of itself change the energy of an electron. If

$$\frac{\partial \phi}{\partial t} = \omega_c \tag{6.4.41}$$

is constant (ω_c being the cyclotron or gyro-frequency) we can solve the resulting equation by using an integrating factor. The solution is

$$f_1 = \frac{e}{\omega_c}\frac{df_0}{dE}\int_{-\infty}^{\phi_0} \mathscr{E}\,.\,\mathbf{v}(E,\mathbf{k})\exp\left(\frac{\phi-\phi_0}{\omega_c\tau}\right)d\phi, \tag{6.4.42}$$

which we can see is a special case of Chambers' solution with a change of variable from t to the ϕ so that we actually choose a coordinate of the trajectory over which to integrate. Equation (6.4.42) will be used to discuss magneto-resistance in high fields.

6.4.3 Variational solution

The general linearized Boltzmann equation is an integrodifferential equation which we cannot expect to be able to solve directly. We shall now show, however, that there exists a powerful variational principle for this equation, from which we might expect to obtain good results by inserting a trial function from the extremely simple result of the relaxation-time approximation.

First, let us write equation (6.4.3) very formally as

$$\left[\frac{\partial f}{\partial t}\right]_{\text{coll}} = K\phi(k), \tag{6.4.43}$$

where ϕ is defined in equation (6.4.17) and K is a linear (though non-local) operator. It is also *self-adjoint*: for any two real functions ψ and χ, we can write, since $\Pi_0(\mathbf{k}'\mathbf{k}) = \Pi_0(\mathbf{k}\mathbf{k}')$,

$$\int \psi K\chi\,d\mathbf{k} = \frac{1}{8\pi^3}\int \psi(\mathbf{k})\,\Pi_0(\mathbf{k}'\mathbf{k})\,[\chi(\mathbf{k})-\chi(\mathbf{k}')]\,d\mathbf{k}\,d\mathbf{k}'$$

$$= \frac{1}{16\pi^3}\int [\psi(\mathbf{k})-\psi(\mathbf{k}')]\,\Pi_0(\mathbf{k}'\mathbf{k})\,[\chi(\mathbf{k})-\chi(\mathbf{k}')]\,d\mathbf{k}\,d\mathbf{k}'. \tag{6.4.44}$$

Thus it follows that

$$\int \psi K\chi\,d\mathbf{k} = \int \chi K\psi\,d\mathbf{k}. \tag{6.4.45}$$

K is also a *positive definite* operator: for any real function χ,

$$\int \chi K\chi\,d\mathbf{k} \geqslant 0, \tag{6.4.46}$$

which follows from equation (6.4.44) and the fact that $\Pi_0(\mathbf{k}'\mathbf{k})$, being a probability, is necessarily positive definite.

We are now in a position to demonstrate the existence of the variational principle. Let us write the left-hand side of the linearized Boltzmann equation as $X(\mathbf{k})$, so that the Boltzmann equation takes the form

$$X(\mathbf{k}) = K\phi(\mathbf{k}). \tag{6.4.47}$$

For uniform fields, X will not contain ϕ, as we see from its form for an electric field:

$$X(\mathbf{k}) = -e\mathscr{E}.\mathbf{v}_{\mathbf{k}}\frac{\partial f_0(\mathbf{k})}{\partial E}. \tag{6.4.48}$$

From equation (6.4.47),

$$\int \phi X \, d\mathbf{k} = \int \phi K \phi \, d\mathbf{k}. \tag{6.4.49}$$

An equation of this same form can always be written down for a real function ψ, since if

$$\int \psi X \, d\mathbf{k} \neq \int \psi K \psi \, d\mathbf{k}, \tag{6.4.50}$$

multiplication of ψ by a suitable constant will evidently restore the equality

$$\int \psi X \, d\mathbf{k} = \int \psi K \psi \, d\mathbf{k}. \tag{6.4.51}$$

We shall now show that ϕ is the function which, when substituted for ψ in equation (6.4.45), makes each side a maximum.

Let us assume that we have a function ψ which satisfies equation (6.4.51). We first note that equation (6.4.47) implies

$$\begin{aligned}\int [\phi-\psi]\, K\psi \, d\mathbf{k} &= \int \psi K\phi \, d\mathbf{k} - \int \psi K \psi \, d\mathbf{k}\\ &= \int \psi X \, d\mathbf{k} - \int \psi K \psi \, d\mathbf{k} = 0,\end{aligned} \tag{6.4.52}$$

where we have employed equation (6.4.51).

Consider now

$$\begin{aligned}\int \phi X \, d\mathbf{k} - \int \psi X \, d\mathbf{k} &= \int \phi X \, d\mathbf{k} - \int \psi X \, d\mathbf{k} - \int [\phi-\psi]\, K\psi \, d\mathbf{k}\\ &= \int [\phi-\psi]\, K\phi \, d\mathbf{k} - \int [\phi-\psi]\, K\psi \, d\mathbf{k}\\ &= \int [\phi-\psi]\, K(\phi-\psi) \, d\mathbf{k},\end{aligned} \tag{6.4.53}$$

where we have used equation (6.4.52). By equation (6.4.46) the right-hand side of equation (6.4.53) is positive definite and so

$$\int \phi X \, d\mathbf{k} - \int \psi X \, d\mathbf{k} \geqslant 0, \tag{6.4.54}$$

which was to be proved. $\int \phi X \, d\mathbf{k}$ is simply related to the rate of entropy production, and so the result (6.4.54) is not surprising, corresponding to the maximization of this quantity. To see this relation, we note that the rate of entropy production $\dot{S}$ associated with Joulean heating may be expressed as

$$\dot{S} = \frac{\mathscr{E} \cdot \mathbf{j}}{T}. \tag{6.4.55}$$

Using equation (6.4.26) for $\mathbf{j}$ this becomes

$$-4\pi^3 \dot{S} = \frac{1}{T} \mathscr{E} \cdot e \int \mathbf{v}_{\mathbf{k}} f_1(\mathbf{k}) \, d\mathbf{k}. \tag{6.4.56}$$

But from equation (6.4.17), we can eliminate f_1 in favour of ϕ. Finally, using equation (6.4.48), we obtain

$$4\pi^3 \dot{S} = \frac{1}{T} \int \phi X \, d\mathbf{k}, \tag{6.4.57}$$

which is the desired result.

We shall now show that an upper bound to the resistivity can be obtained from this variational principle. The conductivity measured in the direction of the field is

$$\sigma_{11} = \frac{\mathbf{j} \cdot \mathscr{E}}{\mathscr{E}^2} = -\frac{e}{4\pi^3} \frac{\mathscr{E}}{\mathscr{E}^2} \cdot \int \mathbf{v} f \, d\mathbf{k}$$

$$= \frac{e}{4\pi^3 \mathscr{E}^2} \int \mathscr{E} \cdot \mathbf{v} \frac{\partial f_0}{\partial E} \phi \, d\mathbf{k}, \tag{6.4.58}$$

which is just [see equation (6.4.48) above]

$$\sigma_{11} = \frac{e}{4\pi^3 \mathscr{E}^2} \int \phi X \, d\mathbf{k}. \tag{6.4.59}$$

Since σ is independent of $\mathscr{E}$, we now use ϕ and X below as though they correspond to a unit electric field. Using equation (6.4.49), equation (6.4.59) can now be written

$$\sigma_{11} = \frac{e}{4\pi^3} \frac{|\int \phi X \, d\mathbf{k}|^2}{\int \phi K \phi \, d\mathbf{k}}, \tag{6.4.60}$$

which we write in this form because, with ϕ^2 appearing in the top and bottom of the expression, any factor required to make some estimate of ϕ satisfy equation (6.4.47) is irrelevant. We therefore see that, taking equation (6.4.44) for X, the variational principle for the electrical conductivity is

$$\sigma_{11} = \frac{j^2}{P}, \tag{6.4.61}$$

where j is proportional to the current:

$$j = \int \hat{\mathscr{E}} \cdot \mathbf{v}\phi \frac{\partial f_0}{\partial E} d\mathbf{k} \tag{6.4.62}$$

and by equation (6.4.39)

$$P = \frac{1}{2k_B T} \int [\phi(\mathbf{k}) - \phi(\mathbf{k}')]^2 \,\Pi_0(\mathbf{k}'\mathbf{k})\, d\mathbf{k}'\, d\mathbf{k}. \tag{6.4.63}$$

Taking in these expressions any estimate for ϕ whatsoever, equation (6.4.58) gives a lower bound to the conductivity measured in the direction of the field.

We shall now use the variational principle to calculate the relaxation time for isotropic scattering by phonons.

(a) *Isotropic scattering by lattice vibrations.* We take the weak-scattering approximation of equation (6.4.7), make the simplifying assumptions that $\langle \mathbf{k}' | V(\boldsymbol{\kappa}) | \mathbf{k} \rangle$ is a function only of $\boldsymbol{\kappa} = \mathbf{k}' - \mathbf{k}$, and the Fermi surface is a sphere. Such approximations should be good for a metal such as sodium (cf. Chapter 4, section 4.15). Putting

$$\langle \mathbf{k}' | V(\boldsymbol{\kappa}) | \mathbf{k} \rangle = \tilde{V}(\boldsymbol{\kappa}), \tag{6.4.64}$$

we have

$$P = \frac{\pi}{k_B T} \int\int d\mathbf{k}\, d\mathbf{k}' [\phi(\mathbf{k}) - \phi(\mathbf{k}')]^2 |\tilde{V}(\boldsymbol{\kappa})|^2 f(\mathbf{k})\, [1 - f(\mathbf{k}')]\, S_1(\boldsymbol{\kappa}, \omega). \tag{6.4.65}$$

We integrate over shells between surfaces of constant energy, as we did in evaluating equation (6.3.32), to obtain

$$P = \frac{\pi}{k_B T} \int dE \int dE' \int dS \int dS' \left| \frac{\partial E}{\partial \mathbf{k}} \right|^{-1} \left| \frac{\partial E'}{\partial \mathbf{k}'} \right|^{-1} f(\mathbf{k})\, [1 - f(\mathbf{k}')]\, |\tilde{V}(\boldsymbol{\kappa})|^2\, S_1(\boldsymbol{\kappa}\omega). \tag{6.4.66}$$

Since for significant transitions $\hbar\omega \ll \zeta$, the Fermi energy, we can localize the integrations at the Fermi surface. Writing $\partial E/\partial \mathbf{k}$ as $\hbar \mathbf{v}_{\mathbf{k}}$, we then have

$$P = \frac{\pi}{k_B T \hbar^2} \int dE \int dE' \int_{\text{Fermi surface}} \frac{dS}{v_{\mathbf{k}}} \int \frac{dS'}{v_{\mathbf{k}'}} f(\mathbf{k})\, [1 - f(\mathbf{k}')]\, |\tilde{V}(\boldsymbol{\kappa})|^2\, S_1(\boldsymbol{\kappa}, \omega). \tag{6.4.67}$$

We now change the variable of integration E' to $\omega = E'-E$, whereupon we find, writing in the explicit values of $f(\mathbf{k})$ and $f(\mathbf{k}')$ in terms of $E = E_{\mathbf{k}} - \zeta$,

$$P = \beta\pi \int dE \int d\omega \int_{\text{Fermi surface}} \frac{dS}{v_{\mathbf{k}}} \int \frac{dS'}{v_{\mathbf{k}'}} \frac{e^{\beta E} e^{\beta\omega}[\phi(\mathbf{k})-\phi(\mathbf{k}')]^2}{(e^{\beta E}+1)(e^{\beta E}e^{\beta\omega}+1)} |\tilde{V}(\boldsymbol{\kappa})|^2 S_1(\boldsymbol{\kappa},\omega). \tag{6.4.68}$$

We can now perform the integration over E, when we find, with $\hbar = 1$,

$$P = \frac{1}{32\pi^5} \int_{\text{Fermi surface}} \frac{dS}{v_{\mathbf{k}}} \int \frac{dS'}{v_{\mathbf{k}'}} [\phi(\mathbf{k})-\phi(\mathbf{k}')]^2 |\tilde{V}(\boldsymbol{\kappa})|^2 \mathscr{S}_1(\boldsymbol{\kappa}), \tag{6.4.69}$$

where we have introduced the quantity

$$\mathscr{S}_1(\boldsymbol{\kappa}) = \int d\omega \frac{S_1(\boldsymbol{\kappa}\omega)\,\beta\omega}{1-e^{-\beta\omega}}. \tag{6.4.70}$$

In the present approximation $v_{\mathbf{k}} = v_{\mathbf{k}'} = v_{\mathrm{f}}$, and

$$\int_{\text{Fermi surface}} dS \int_{\text{Fermi surface}} dS' [\phi(\mathbf{k})-\phi(\mathbf{k}')]^2 |\tilde{V}(\boldsymbol{\kappa})|^2 \mathscr{S}_1(\boldsymbol{\kappa})$$
$$= \int d\mathbf{k} \int d\mathbf{k}'\, \delta(k-k_{\mathrm{f}})\, \delta(k'-k_{\mathrm{f}})$$
$$\times [\phi(\mathbf{k})-\phi(\mathbf{k}')]^2 |\tilde{V}(\boldsymbol{\kappa})|^2 \mathscr{S}_1(\boldsymbol{\kappa}). \tag{6.4.71}$$

We now convert the integration over $\mathbf{k}'$ to one over $\boldsymbol{\kappa}$:

$$P = \frac{1}{32\pi^5} \int d\boldsymbol{\kappa} \int d\mathbf{k}\, \delta(k-k_{\mathrm{f}})\, \delta(|\mathbf{k}+\boldsymbol{\kappa}|-k_{\mathrm{f}})$$
$$\times [\phi(\mathbf{k})-\phi(\mathbf{k}+\boldsymbol{\kappa})]^2 |\tilde{V}(\boldsymbol{\kappa})|^2 \mathscr{S}_1(\boldsymbol{\kappa}). \tag{6.4.72}$$

Taking $\boldsymbol{\kappa}$ as polar axis for the $\mathbf{k}$-variables, and writing $\mu = \cos\theta = \boldsymbol{\kappa}.\mathbf{k}/\kappa k$, the function $\delta(|\mathbf{k}+\boldsymbol{\kappa}|-k_{\mathrm{f}})$ implies that

$$\mu = \frac{\kappa}{2k_{\mathrm{f}}}, \quad k = k_{\mathrm{f}} \tag{6.4.73}$$

(see Figure 6.5) and

$$P = \frac{1}{32\pi^5 v_{\mathrm{f}}^2} \int_{\kappa<2k_{\mathrm{f}}} d\boldsymbol{\kappa} |\tilde{V}(\boldsymbol{\kappa})|^2 \mathscr{S}_1(\boldsymbol{\kappa}) \int d\phi\, [\phi(\mathbf{k})-\phi(\mathbf{k}+\boldsymbol{\kappa})]^2 \Big|_{\substack{\mathbf{k}.\boldsymbol{\kappa}=2k_{\mathrm{f}}^2 \\ k=k_{\mathrm{f}}}}. \tag{6.4.74}$$

We now insert the simplest approximation to ϕ, consistent with cubic symmetry, viz. (omitting a factor not affecting the results)

$$\phi(\mathbf{k}) = \mathbf{k}.\hat{\mathscr{E}}, \tag{6.4.75}$$

which is the result of the relaxation time approximation. Since the latter approximation becomes exact for isotropic elastic scattering and from the point of view of the electrons the scattering is practically elastic in the present

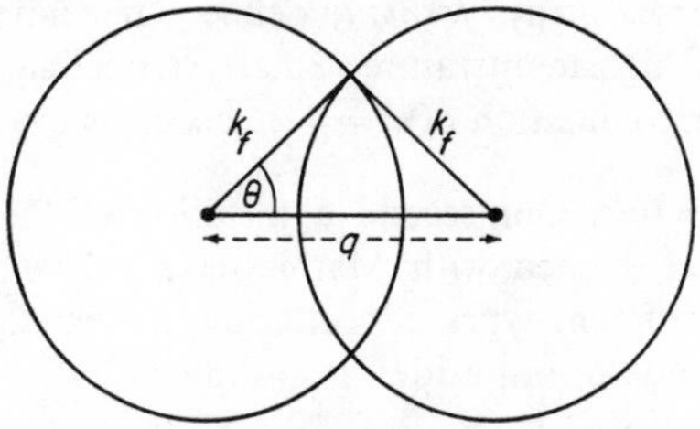

FIGURE 6.5. Two intersecting Fermi spheres, needed in evaluation of integral in equation (6.4.71). We set $q = \kappa$.

case also, equation (6.4.75) should be a good approximation. Then

$$P = \frac{1}{16\pi^4 v_f^2} \int_{\kappa<2k_f} d\boldsymbol{\kappa}\, \boldsymbol{\kappa}.\hat{\mathscr{E}}\, |\tilde{V}(\kappa)|^2 \mathscr{S}_1(\kappa). \tag{6.4.76}$$

This becomes, for cubic crystals (see section 6.2),

$$P = \frac{1}{48\pi^4 v_f^2} \int d\boldsymbol{\kappa}\, \kappa |\tilde{V}(\kappa)|^2 \mathscr{S}_1(\kappa). \tag{6.4.77}$$

In the same approximation we can readily verify from equation (6.4.61) [cf. equations (6.3.21) and (6.3.23)] that

$$j = \frac{k_f^3 e}{3\pi^2 \hbar} \tag{6.4.78}$$

and so we obtain for the resistivity

$$\rho = \frac{1}{\sigma} = \frac{3}{16e^2 v_f^2 k_f^4} \int_{\kappa<2k_f} d\boldsymbol{\kappa}\, \kappa |\tilde{V}(\kappa)|^2 \mathscr{S}_1(\kappa). \tag{6.4.79}$$

At high temperatures ($T \gg \theta_D$), $\beta\omega(1-e^{-\beta\omega})^{-1} \to 1$, and

$$\mathscr{S}_1(\kappa) \to \int d\omega\, S_1(\kappa, \omega) = S_1(\kappa), \tag{6.4.80}$$

the static structure factor. [This approximation inserted in equation (6.4.76) should be good for the molten metal.] In this limit one can define a relaxation time, which from the expression $\sigma = ne^2\tau/m$ is readily given by

$$\tau^{-1} = \frac{m}{12\pi^3}\int_0^{2k_f} k^3\,dk\,|\tilde{V}(\kappa)|^2\,S_1(\kappa). \tag{6.4.81}$$

One can, of course, always *formally* define a relaxation time in this way, but in the limit of high temperatures τ has physical significance in that the relaxation time approximation $\phi(\mathbf{k}) = \mathbf{k}.\hat{\mathscr{E}}$ becomes exact.

(b) *Mathiessen's rule.* Our second application of the variation principle is a simple one, concerned with Mathiessen's rule which states that the resistance due to different types of scattering, for example lattice scattering and the scattering due to impurities, is additive:

$$\rho = \rho_1 + \rho_2. \tag{6.4.82}$$

This is true if relaxation time approximations are additive, for we have as the collision terms due to the types 1 and 2 of scatterers

$$\left[\frac{\partial f}{\partial t}\right]^1_{\text{coll}} = \frac{f_0 - f_1}{\tau_1}, \quad \left[\frac{\partial f}{\partial t}\right]^2_{\text{coll}} = \frac{f_0 - f_1}{\tau_2}, \tag{6.4.83}$$

which, when added together, define a relaxation time through

$$\frac{1}{\tau} = \frac{1}{\tau_1} + \frac{1}{\tau_2}. \tag{6.4.84}$$

The conductivity is proportional to the relaxation time, and so the resistivity is inversely proportional to the relaxation time. Hence equation (6.4.82) follows from equation (6.4.84).

More generally, however, the variational theorem tells us that Mathiessen's rule is not invariably true, even if we accept that we can write

$$\Pi_0(\mathbf{k}'\mathbf{k}) = \Pi_0^1(\mathbf{k}'\mathbf{k}) + \Pi_0^2(\mathbf{k}'\mathbf{k}). \tag{6.4.85}$$

This is consistent with the weak scattering approximation we have previously used for the lattice scattering. We then see that

$$K = K_1 + K_2 \tag{6.4.86}$$

and, from equation (6.4.60),

$$\rho = \frac{4\pi^3}{e}\frac{\int \phi K \phi\,d\mathbf{k}}{\left|\int \phi X\,d\mathbf{k}\right|^2}$$

$$= \frac{4\pi^3}{e}\frac{\int \phi K_1 \phi\,d\mathbf{k}}{\left|\int \phi X\,d\mathbf{k}\right|^2} + \frac{4\pi^3}{e}\frac{\int \phi K_2 \phi\,d\mathbf{k}}{\left|\int \phi X\,d\mathbf{k}\right|^2}. \tag{6.4.87}$$

Now by the variational principle,

$$\rho_1 \leqslant \frac{4\pi^3}{e} \frac{\int \phi K_1 \phi \, d\mathbf{k}}{\left| \int \phi X \, d\mathbf{k} \right|^2}, \quad \rho_2 \leqslant \frac{4\pi^3}{e} \frac{\int \phi K_2 \phi \, d\mathbf{k}}{\left| \int \phi X \, d\mathbf{k} \right|^2} \tag{6.4.88}$$

and Mathiessen's rule must in general be replaced by

$$\rho \geqslant \rho_1 + \rho_2. \tag{6.4.89}$$

6.5 Electron–phonon interaction

Before going on to describe calculations of the electrical conductivity of sodium, based on equation (6.4.79), it will be convenient to discuss in somewhat general terms the nature of the electron–phonon interaction. Whereas, in the earlier discussion, we expressed the scattering of electrons in terms of the ionic density fluctuations, it is often useful to treat the interaction of electrons with individual phonons.

The adiabatic approximation of Chapter 3 leads us to the picture of two non-interacting fields, the phonon field and the system of interacting electrons in a perfect lattice. In order to transcend the adiabatic approximation therefore, we must allow the two systems to interact. If the adiabatic approximation is good, this interaction may be treated by perturbation theory, as we shall now discuss.

6.5.1 Bloch formulation

As in Chapter 2, we denote equilibrium positions by $\mathbf{l}$ and by $\mathbf{u}_\mathbf{l}$ departures of the ions from these. We write the total potential acting on the electrons as

$$V(\mathbf{x}, \mathbf{l}+\mathbf{u}_\mathbf{l}) = V(\mathbf{x}, \mathbf{l}) + \sum_\mathbf{l} \mathbf{u}_\mathbf{l} \cdot \mathbf{V}_\mathbf{l}'(\mathbf{x}) + \dots, \tag{6.5.1}$$

where $\mathbf{x}$ is used to denote all coordinates $(\mathbf{r}_1, \mathbf{r}_2, \dots)$ of the electrons and $\mathbf{l}$ denotes all lattice vectors.

As we saw in Chapter 1, the many-body wave functions which are solutions of

$$-\frac{\hbar^2}{2m} \sum_i \frac{\partial^2}{\partial \mathbf{r}_i^2} \Psi_\kappa(\mathbf{x}) + V(\mathbf{x}, \mathbf{l}) \Psi_\kappa(\mathbf{x}) = E(\kappa) \Psi_\kappa(\mathbf{x}) \tag{6.5.2}$$

may be chosen so that they obey Bloch's theorem in the sense

$$\Psi_\kappa(\mathbf{r}_1 + \mathbf{l}, \mathbf{r}_2 + \mathbf{l}, \dots) = \Psi_\kappa(\mathbf{r}_1, \mathbf{r}_2, \dots) e^{i\kappa \cdot \mathbf{l}}. \tag{6.5.3}$$

The displacement-perturbation of equation (6.5.1) perturbs such a function to

$$\Psi_{\kappa}(\mathbf{x}, \mathbf{u}_l) = \Psi_{\kappa}(\mathbf{x}) + \sum_{\kappa'} \frac{\langle \Psi_{\kappa'} | \sum_{l} \mathbf{u}_l \cdot \mathbf{V}'_l | \Psi_{\kappa} \rangle}{E(\kappa') - E(\kappa)} \Psi_{\kappa'}(\mathbf{x}), \tag{6.5.4}$$

and it is also evident that the matrix element for the probability of transition from state $\boldsymbol{\kappa}$ to $\boldsymbol{\kappa}'$ is

$$M_{\kappa'\kappa} = \sum_{l} \langle \Psi_{\kappa'} | \mathbf{u}_l \cdot \mathbf{V}'_l | \Psi_{\kappa} \rangle. \tag{6.5.5}$$

It may also be noted that this is the appropriate matrix element for transition from $\Psi_{\kappa}(\mathbf{x}, \mathbf{u}_l)$ to $\Psi_{\kappa'}(\mathbf{x}, \mathbf{u}_l)$ as is discussed by Sham and Ziman (1963).

For transitions involving one phonon the energy change shown in the denominator of equation (6.5.4) must obey the selection rule

$$E(\kappa') - E(\kappa) = \pm \hbar\omega(\mathbf{q}, \sigma) \tag{6.5.6}$$

corresponding to the destruction or creation of a phonon, the phonon energy being taken up by, or supplied by, the electrons.

In accordance with Chapter 3 we write

$$\mathbf{u}_l = \sum_{\mathbf{q}\sigma} Q_{\mathbf{q}\sigma} \boldsymbol{\varepsilon}_{\mathbf{q}\sigma} e^{i\mathbf{q}.\mathbf{l}}. \tag{6.5.7}$$

Defining now

$$W_{\mathbf{q}\sigma}(\mathbf{x}) = \sum_{\mathbf{l}} e^{i\mathbf{q}.\mathbf{l}} \boldsymbol{\varepsilon}_{\mathbf{q}\sigma} \cdot \mathbf{V}'_l, \tag{6.5.8}$$

we see that equation (6.5.5) may be written

$$M_{\kappa'\kappa} = \sum_{\mathbf{q}\sigma} Q_{\mathbf{q}\sigma} \langle \Psi_{\kappa'} | W_{\mathbf{q}\sigma} | \Psi_{\kappa} \rangle. \tag{6.5.9}$$

It is easy to ascertain that

$$W_{\mathbf{q}\sigma}(\mathbf{r}_1 + \mathbf{l}, \mathbf{r}_2 + \mathbf{l}, \ldots) = e^{i\mathbf{q}.\mathbf{l}} W_{\mathbf{q}\sigma}(\mathbf{r}_1, \mathbf{r}_2, \ldots). \tag{6.5.10}$$

Now the matrix element $M_{\kappa'\kappa}$ cannot be changed by a shift of origin of the argument of the integrand in equation (6.5.9). Supposing we *do* change the origin by a lattice vector $\mathbf{l}$, conditions (6.5.3) and (6.5.10) yield

$$\langle \Psi_{\kappa'} | W_{\mathbf{q}\sigma} | \Psi_{\kappa} \rangle = \langle \Psi_{\kappa'} | W_{\mathbf{q}\sigma} | \Psi_{\kappa} \rangle e^{i(\mathbf{q}+\kappa-\kappa').\mathbf{l}} \tag{6.5.11}$$

and so we obtain the crystal momentum selection rule

$$\mathbf{q} + \mathbf{K} = \boldsymbol{\kappa}' - \boldsymbol{\kappa}, \tag{6.5.12}$$

where $\mathbf{K}$ is any reciprocal lattice-vector. Just as in Chapter 3 when dealing with phonon–phonon interactions, we call the transition a normal process when $\mathbf{K} = 0$, and an umklapp process when $\mathbf{K} \neq 0$.

6.5.2 *Single-particle approximation*

We now suppose $\Psi_\kappa(\mathbf{x})$ is a determinant of single-particle Bloch functions $\psi_\mathbf{k}(\mathbf{r})$ which satisfy wave equations with a one-body potential $V(\mathbf{r})$, so that the total potential is

$$V(\mathbf{r}_1, \mathbf{r}_2, \ldots) = \sum_i V(\mathbf{r}_i) \tag{6.5.13}$$

and

$$V'_\mathbf{l}(\mathbf{x}) = \sum_i \mathbf{V}'_\mathbf{l}(\mathbf{r}_i) = \sum_i \frac{\partial V(\mathbf{r}_i)}{\partial \mathbf{l}}. \tag{6.5.14}$$

Now all N of the electrons (i) must give the same contribution to the matrix element and so from equations (6.5.8) and (6.5.9)

$$M_{\kappa'\kappa} = N \sum_{\mathbf{q}\sigma} Q_{\mathbf{q}\sigma} \sum_\mathbf{l} e^{i\mathbf{q}.\mathbf{l}} \langle \Psi_{\kappa'} | \boldsymbol{\varepsilon}_{\mathbf{q}\sigma} . \mathbf{V}'_\mathbf{l}(\mathbf{r}_1) | \Psi_\kappa \rangle. \tag{6.5.15}$$

The determinants $\Psi_{\kappa'}$ and Ψ_κ can only differ in one factor if a matrix element is to be non-zero, the differing factors being taken as $\psi_{\mathbf{k}'}$ and $\psi_\mathbf{k}$. Equation (6.5.15) is then

$$M_{\mathbf{k}'\mathbf{k}} = N \sum_{\mathbf{q}\sigma} Q_{\mathbf{q}\sigma} \sum_\mathbf{l} e^{i\mathbf{q}.\mathbf{l}} \langle \psi_{\mathbf{k}'} | \boldsymbol{\varepsilon}_{\mathbf{q}\sigma} . \mathbf{V}'_\mathbf{l}(\mathbf{r}) | \psi_\mathbf{k} \rangle \tag{6.5.16}$$

with selection rules for the single-particle energies

$$E(\mathbf{k}') - E(\mathbf{k}) = \hbar\omega(\mathbf{q}, \sigma) \tag{6.5.17}$$

and for the wave-vectors

$$\mathbf{k}' - \mathbf{k} = \mathbf{q} + \mathbf{K}. \tag{6.5.18}$$

It is convenient to note at this point that a phonon energy $\hbar\omega(\mathbf{q}, \sigma)$ is very small compared to electron energies. Thus the transitions envisaged here can be thought of as between electrons on the Fermi surface.

Suppose we write the perfect-crystal potential in terms of local potentials:

$$V(\mathbf{r}) = \sum_l V_L(\mathbf{r} - \mathbf{l}) \tag{6.5.19}$$

and now take the "dressed-ion" model as sufficient to construct these local potentials, so that

$$\mathbf{V}'_\mathbf{l}(\mathbf{r}) = \frac{\partial V(\mathbf{r})}{\partial \mathbf{l}} = -\nabla V_L(\mathbf{r}_i - \mathbf{l}). \tag{6.5.20}$$

Equation (6.5.16) becomes

$$M_{\mathbf{k}'\mathbf{k}} = -N \sum_{\mathbf{q}\sigma} Q_{\mathbf{q}\sigma} \sum e^{i\mathbf{k}.\mathbf{l}} \langle \psi_{\mathbf{k}'} | \boldsymbol{\varepsilon}_{\mathbf{q}\sigma} . \nabla V_L(\mathbf{r} - \mathbf{l}) | \psi_\mathbf{k} \rangle. \tag{6.5.21}$$

Changing the origin in the integrand by **l**, we get

$$M_{\mathbf{k}'\mathbf{k}} = -N \sum_{\mathbf{q}\sigma} Q_{\mathbf{q}\sigma} \sum_{L} e^{i(\mathbf{q}+\mathbf{k}-\mathbf{k}').\mathbf{l}} \langle \psi_{\mathbf{k}'} | \boldsymbol{\varepsilon}_{\mathbf{k}\sigma} . \nabla V_L(\mathbf{r}) | \psi_{\mathbf{k}} \rangle, \qquad (6.5.22)$$

which by the selection rule (6.5.18) becomes just

$$M^{\sigma}_{\mathbf{k}'\mathbf{k}} = -N^2 Q_{\mathbf{k}\sigma} \boldsymbol{\varepsilon}_{\mathbf{k}\sigma} \langle \psi_{\mathbf{k}'} | \nabla V_L(\mathbf{r}) | \psi_{\mathbf{k}} \rangle. \qquad (6.5.23)$$

If we further suppose the $\psi_{\mathbf{k}}$'s to be well approximated by plane-waves, we have just

$$M^{\sigma}_{\mathbf{k}'\mathbf{k}} = -N^2 Q^2_{\mathbf{q}\sigma} \boldsymbol{\varepsilon}_{\mathbf{q}\sigma} \cdot (\mathbf{k}' - \mathbf{k}) \tilde{V}_L(\mathbf{k}' - \mathbf{k}), \qquad (6.5.24)$$

$\tilde{V}_L$ being the Fourier transform of $V_L(\mathbf{r})$. The connection of this treatment, due to Bloch, with the discussion of the previous section [in particular equation (6.4.7)], is now established by taking the thermal average of the square of this matrix element. Equation (6.4.7) is then regained if, in that equation, $S_1(\mathbf{k}\omega)$ is replaced by its one-phonon approximation. However, it should be stressed that equation (6.5.23) is the correct starting point, the simplification of equation (6.5.24) essentially replacing the scattering of electrons off the localized potential V_L by its Born approximation equivalent.

For normal processes equation (6.5.24) becomes

$$M^{\sigma}_{\mathbf{k}'\mathbf{k}} = -N^2 Q_{\mathbf{q}\sigma} \boldsymbol{\varepsilon}_{\mathbf{q}\sigma} \cdot \mathbf{q} \tilde{V}_L(\mathbf{q}), \qquad (6.5.25)$$

and so in this approximation only longitudinal phonons interact with the electrons *via* normal processes, $\boldsymbol{\varepsilon}_{\mathbf{q}\sigma}$ being perpendicular to **q** for transverse waves, of course. Further, umklapp scattering will be weak in a monovalent metal, at least for reasonably low temperatures and if the FS does not touch the BZ boundary. This is because q must be rather large for an umklapp process to occur, so involving a phonon of rather high energy—which is unlikely to be excited (note that the temperature average of the transition probability $|M_{\mathbf{k}'\mathbf{k}}|^2$ enters through $|Q_{\mathbf{k}\sigma}|^2$).

It is easy to see the meaning of this in physical terms. Equation (6.5.24) presumes the uniform electron-gas model of a metal. Just as this gas, having negligible shear strength, hardly affects the "bare-ion" lattice frequencies when we renormalize them to take account of it (see Chapter 3), so it is unresistant to a transverse wave travelling though the ionic background: it can follow such an ionic motion in (almost) perfect adiabatic fashion.

Two comments need to be made on the validity of equation (6.5.24). First, it is based on free electrons, but we should remember that the pseudo-potential method is available to refine our approach. However, it *cannot* be valid where the Fermi surface touches the BZ boundary (if at all), for we should then use degenerate perturbation theory in the nearly free electron method.

6.6 Resistivities of metals and alloys

6.6.1 Resistivity of sodium

As an example of the application of the variational solution of Boltzmann's equation described earlier (section 6.4.3) we shall now review the calculations made by Greene and Kohn (1965) and Darby and March (1964) on sodium. The calculations of the former authors are based on the result of equation (6.4.79), which, we remind the reader, is an approximate variational solution, with the additional assumptions of the weak-scattering approximation and a spherical Fermi surface. We expect these approximations to be good for the metal in question. Darby and March used the same variational solution, except that they resorted to a numerical procedure at equation (6.4.65), before reduction to equation (6.4.79). We see from equation (6.4.79) that we must calculate $S_1(\boldsymbol{\kappa}, \omega)$, the dynamical structure factor [to obtain $\mathscr{S}_1(\kappa)$], $\tilde{V}(\boldsymbol{\kappa})$, the scattering potential due to an ion, and the Fermi velocity v_{f}.

(a) *Effective mass.* We dispose of the last-mentioned quantity first, writing $v_{\mathrm{f}} = k/m^*$, m^* being an effective mass. Now in discussing an interacting electron gas in a metal we may include in m^* not only the effect of correlations but also the effect of lattice vibrations on the dynamics of the electrons. This electron–phonon interaction has been estimated as changing the effective mass by as much as 33% (Gaumer and Heer, 1960) but it seems clear that in discussing the electron–phonon interaction to lowest order, m^* should be taken as the effective mass before such interaction is included.‡ Accordingly Greene and Kohn took 1.24 as the value of m^*, a figure obtained from cyclotron resonance experiments (see Chapter 7).

We now go on to discuss the calculations of $S_1(\boldsymbol{\kappa}, \omega)$ and $\tilde{V}(\boldsymbol{\kappa})$ in more detail.

(b) *Dynamical structure facture.* $S(\boldsymbol{\kappa}, \omega)$ can in principle be obtained from neutron diffraction experiments, but in the absence of such detailed information one can estimate $S(\boldsymbol{\kappa}, \omega)$, and so $S_1(\boldsymbol{\kappa}, \omega)$, from the phonon spectra revealed by slow neutron-scattering experiments, using the "one-phonon" approximation, which yields equation (5.3.50) for $S_1(\boldsymbol{\kappa}, \omega)$. Again, data are limited, the spectra only being known for high symmetry directions in the Brillouin zone and some isolated points, the measurements being those of Woods and co-workers (1962).

Resort must be made to some method of interpolating these results over the entire BZ. Green and Kohn utilized a model with inter-ionic forces out

‡ An explicit proof of this has been given by Nakajima and Watabe (1963); see also Migdal (1958).

to fifth nearest neighbours, a model to which Woods and colleagues fitted their data. Darby and March, on the other hand, used no model, but exploited the symmetry of the $\omega^2(\mathbf{k})$ relations in the BZ to expand in cubic harmonics, as many terms being used as the number of symmetry directions allows in such a procedure. The relative merits of these two methods must depend on the accuracy of the model chosen in the former procedure; judging by the final results obtained the model actually chosen rendered the methods roughly equivalent.

We should also mention that Woods and co-workers give results for one temperature only, viz. 90 K, and that Darby and March corrected for other temperatures using elastic-constant data, which gives the low-**k** end of the phonon spectrum (see Chapter 3). Such corrections will be discussed later.

(c) *Scattering amplitude*. To calculate $|\tilde{V}(\boldsymbol{\kappa})|^2$ Darby and March proceeded from first principles along lines discussed in problem 3.6 of this book, but the calculations are lengthy and we shall not give details here. Greene and Kohn, on the other hand, adopted a parametric approach, although they did perform certain preliminary calculations to determine the approximate magnitudes of the phase shifts η_l in terms of which they expanded their scattering amplitude $\tilde{V}(\boldsymbol{\kappa})$. On expanding

$$\tilde{V}(\boldsymbol{\kappa}) = \sum_l f_l(\boldsymbol{\kappa})\,\eta_l, \tag{6.6.1}$$

equation (6.4.79) may evidently be written

$$\rho = \sum_{ll'} \rho_{ll'}\,\eta_l\,\eta_{l'}, \tag{6.6.2}$$

where the temperature dependence lies in the factor

$$\rho_{ll'} = \frac{3}{16e^2 k_f^4 v_f^2}\int_{k<2k_f} f_l(\mathbf{k}) f_{l'}(\mathbf{k})\,\mathscr{S}_1(\mathbf{k}). \tag{6.6.3}$$

Greene and Kohn's preliminary calculations showed that four phase shifts were sufficient to determine $\tilde{V}(\boldsymbol{\kappa})$ and also that results were insensitive to η_3, which was therefore fixed at an approximate value $\eta_3 = 0.015$. η_0 and η_1 were then regarded as free, η_2 being fixed by the Friedel sum rule

$$\frac{\pi}{2}\sum_l (2l+1)\,\eta_l = Z, \tag{6.6.4}$$

where Z is the valency. This result is proved in Chapter 10.

Equation (6.6.2) now describes a conic, in practice an ellipse. The ellipses ought all to intersect to provide a unique set of phase shifts. Some difficulties arise in this procedure, and the problem of a unique set of η_l's is not completely solved to date. The "best" set of phase shifts was $\eta_0 = 0.524$, $\eta_1 = 0.258$, $\eta_2 = 0.036$.

(d) *Anharmonicity.* This difficulty we have just described is probably due to deviations from purely harmonic behaviour, and not in the variational trial function used or the phase-shift treatment of the ionic potentials. This is strongly indicated by the fair agreement with experiment obtained by Darby and March who worked within a quasi-harmonic approximation (see Chapter 3), correcting Wood's results to apply to temperatures other than 90 K by use of the known temperature dependence of the elastic constants. We have already seen in Chapter 3 that these reflect the low k end of the spectrum of $\omega^2(\mathbf{k})$. Darby and March were able to draw phonon dispersion relations, reproducing the exact elastic constants, which run smoothly into the measured curves of Woods and colleagues.

(e) *Umklapp processes.* In Darby and March's work, umklapp and normal processes were treated separately whereas in Greene and Kohn's calculation using equation (6.4.79) no such separation had to be made. The separate relative contributions of umklapp and normal processes are of obvious interest, however (and were in fact examined by Greene and Kohn), since only longitudinal phonons interact with electrons in normal processes as we have just shown. Both sets of authors found that at room temperature umklapp processes contribute approximately 70% of the resistivity, increasing to near 80% at 40 K.

(f) *Comparison with experiment.* Our final conclusions are that the quasi-harmonic approximation yields resistivities of solid sodium to within 20%, and much of the present error probably lies not in the treatment of the electronic scattering with individual ions but errors in our estimates of $S_1(\boldsymbol{\kappa}, \omega)$ using the quasi-harmonic approximation. However, we must mention that Greene and Kohn have also calculated the resistivity of the molten metal using equation (6.4.80). The fact that quantitative agreement with experiment is lacking in this case may mean that the electron–ion scattering has still not been quite adequately treated, and some doubt on the relative importance of $S_1(\boldsymbol{\kappa}, \omega)$ and the matrix element must remain. We should stress that whereas, in the liquid metal, it is the form of $\tilde{V}(\boldsymbol{\kappa})$ around $2k_f$ that is crucial, this region is not dominant in the crystal.

6.6.2 High-temperature resistivity and Fermi liquid parameters

Young and Sham (1969) have shown that there is a simple relation between the high-temperature electrical resistivity in simple metals and the electron–phonon contribution to two of the Landau parameters (see also Grimvall, 1969).

As we have discussed in Chapter 2, the many-body effects of low-lying excited states are incorporated in Landau's Fermi liquid theory via the interaction function $f(\mathbf{k}\sigma, \mathbf{k}'\sigma')$ between two quasi-particles with momenta and spins $(\mathbf{k}\sigma)$ and $(\mathbf{k}'\sigma')$ respectively.

The way to get at some of the parameters in the expansion of this interaction in Legendre polynomials is discussed in Chapter 7, via spin waves and magnetoplasma waves. However, the measurements unfortunately do not give the first two coefficients of the spin-symmetric part of the interaction function.

In simple metals, two types of contributions to the function f can be isolated:

(i) from direct Coulomb interaction between electrons,

(ii) from exchange of a phonon between two electrons.

This separation is best effected, not in terms of f, which is, in the microscopic theory, the limit of the electron–hole scattering function Γ as the momentum and energy transfers q and ω tend to zero such that $q/\omega \to 0$, but in terms of the function g (cf. Chapter 2, equation 2.10.83) which is the limit when $q/\omega \to \infty$.

The interaction function g is the sum of a term containing just the Coulomb interaction between electrons and a term involving the exchange of a phonon between the electron–hole pair. The reason for this is very briefly as follows.

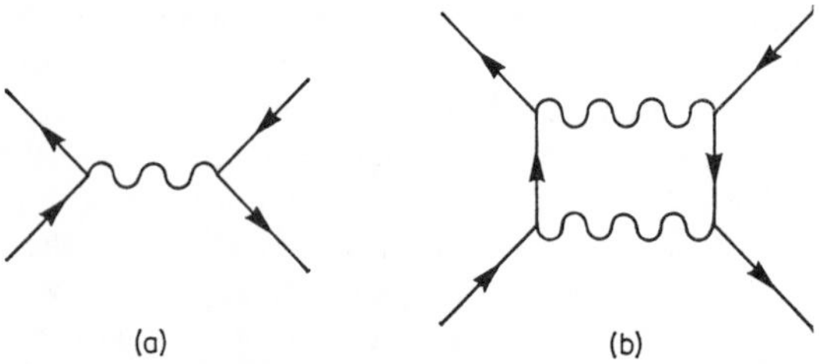

FIGURE 6.6. Electron–hole scattering by exchange of phonons. (a) Scattering to lowest order. Graph is same as that for scattering via Coulomb interaction except that wavy line denotes exchange of phonons. (b) Example of graph which contributes in general case to Γ but not in the limit discussed in text.

Just as the scattering by Coulomb interactions can be represented by graphs, the scattering of electrons by exchange of phonons can be similarly described, as discussed further in Chapter 8. Scattering to lowest order is then as shown in Figure 6.6(a), which is the same as the graph for scattering via Coulomb interaction except that we replace the dotted Coulomb interaction line by a wavy line representing exchange of a phonon. Graphs involving scattering through phonon exchange can be handled in a similar way to the graphs we discussed extensively in Chapter 2. Rice (1968) has shown that if we neglect terms of $O(m/M)^{\frac{1}{2}}$, then graphs such as shown in Figure 6.6(b) do not contribute to the scattering function Γ in the appropriate limit discussed above and only the contribution of graph (a) remains. A corresponding conclusion does not hold for the scattering function f, and that is why g is more convenient here.

Furthermore, the linearized transport equation

$$e\mathscr{E}.\mathbf{v}_{\mathbf{k}} = P(\mathbf{k}), \tag{6.6.5}$$

with $\mathbf{v}_{\mathbf{k}}$ the velocity of a quasi-particle at the Fermi surface, cannot be written in terms of g^{ep}, the phonon contribution to g. In fact, the scattering rate $P(\mathbf{k})$ of the quasi-particle distribution by phonons has a form entirely similar to that for the independent particle model (see section 6.4), except that many-body renormalization is now included. The result may be written as

$$P(\mathbf{k}) = -\frac{4\pi}{\Omega\beta}\sum_{q} g^{ep}(\mathbf{k},\mathbf{k}+\mathbf{q})\,\delta(E_{\mathbf{k}+\mathbf{q}}-E_{\mathbf{k}})\,(\tilde{\Phi}_{\mathbf{k}}-\tilde{\Phi}_{\mathbf{k}+\mathbf{q}}). \tag{6.6.6}$$

In the energy conservation factor $\omega_{\mathbf{q}}$ has been neglected, which is permissible since the ratio v_s/v_f is small. At high temperatures, the electron–phonon scattering is then seen to be approximately elastic. $\tilde{\Phi}_{\mathbf{k}}$ represents the linear deviation of the quasi-particle distribution from local equilibrium [cf. equation 6.4.17].

Using the variational formulation outlined earlier the electrical resistivity ρ at high temperatures is found to be given by

$$\frac{\hbar\beta}{2n\pi}\frac{ne^2\rho}{m} = \frac{m^*}{m}(g_0^{ep}-g_1^{ep}) \tag{6.6.7}$$

where g_0^{ep} and g_1^{ep} are moments of the interaction functions defined by

$$g_l^{ep} = \frac{m^* k_f}{\pi^2}\int\frac{d\Omega_k}{4\pi}\int\frac{d\Omega_{k'}}{4\pi}\,g^{ep}(\mathbf{k}\mathbf{k}')\,P_l(\hat{\mathbf{k}}.\hat{\mathbf{k}}'), \tag{6.6.8}$$

where the angular integrals are over the spherical Fermi surface with radius k_f.

This equation enables estimates of $g_0^{ep}-g_1^{ep}$ to be made from experimentally measured quantities for simple metals (see Table 7.2 below).

6.7 Conductivity and Hall effect for general dispersion relationship

The discussion below will deal with a sufficiently general solution of the Boltzmann equation to enable transport properties to be found for any $E(\mathbf{k})$ relation.

6.7.1 Solution of the Boltzmann equation

If we take the magnetic field $\mathscr{H}$ along the z-axis, and write for the perturbed distribution function

$$f = f_0 - \phi \frac{\partial f_0}{\partial E}, \tag{6.7.1}$$

then the Boltzmann equation takes the form

$$\frac{e\mathscr{H}}{\hbar^2 c}\left(\frac{\partial E}{\partial k_2}\frac{\partial \phi}{\partial k_1} - \frac{\partial E}{\partial k_1}\frac{\partial \phi}{\partial k_2}\right) - \frac{\phi}{\tau} = -\frac{1}{\hbar}\mathbf{P}\,.\,\nabla_{\mathbf{k}} E. \tag{6.7.2}$$

Here $\mathbf{P}$ is given by

$$\mathbf{P} = -e\mathscr{E} + T\nabla_{\mathbf{r}}[(E-\zeta)/T]. \tag{6.7.3}$$

This linear equation for ϕ may now be solved, and it can be verified that the solution is

$$\phi = \frac{\hbar c}{e\mathscr{H}} \exp\left(\frac{\hbar^2 c}{e\mathscr{H}} \int \frac{dk_1}{\tau(\partial E/\partial k_2)}\right) \int \left[\frac{\tau \mathbf{P}\,.\,\nabla_{\mathbf{k}} E}{-\tau(\partial E/\partial k_2)} \exp\left(-\frac{\hbar^2 c}{e\mathscr{H}} \int \frac{dk_1}{\tau(\partial E/\partial k_2)}\right)\right] dk_1, \tag{6.7.4}$$

where the integrations are taken over the curve defined by $E(\mathbf{k}) = \text{constant}$, $k_3 = \text{constant}$, or equivalently over the curve whose gradient is

$$\frac{\partial k_2}{\partial k_1} = -\frac{\partial E}{\partial k_1}\Big/\frac{\partial E}{\partial k_2}. \tag{6.7.5}$$

This solution can in fact be obtained directly using the method of Chambers given in section 6.4.2(b).

The conductivity and Hall effect can be calculated from the perturbed distribution function, which in turn is given by ϕ.

If we now expand in ascending powers of $\mathscr{H}$, then for an isothermal metal we find explicitly, with $\mathbf{\Omega} = \operatorname{grad} E \times \operatorname{grad}_{\mathbf{k}}$,

$$\phi = -\frac{e}{\hbar}\left(\left\{\tau\mathscr{E}\,.\,\frac{\partial E}{\partial \mathbf{k}} - \frac{e}{\hbar^2 c}\tau\mathscr{H}\,.\,\mathbf{\Omega}\left(\tau\mathscr{E}\,.\,\frac{\partial E}{\partial \mathbf{k}}\right)\right.\right.$$

$$\left.\left. - \frac{e}{\hbar^4 c}\tau\mathscr{H}\,.\,\mathbf{\Omega}\left[\tau\mathscr{H}\,.\,\Omega\left(\tau\mathscr{E}\,.\,\frac{\partial E}{\partial \mathbf{k}}\right)\right] + \dots\right\}\right) \tag{6.7.6}$$

and the current $\mathbf{j}$ is given by

$$\mathbf{j} = \frac{e}{4\pi^3 \hbar}\int \frac{\partial E}{\partial \mathbf{k}}\phi\frac{\partial f_0}{\partial E} d\mathbf{k}, \tag{6.7.7}$$

which we obtained earlier.

Neglecting terms of order $\mathscr{H}^3$ and higher, it follows that in an isotropic metal the current is of the form

$$\mathbf{j} = \sigma_0 \mathscr{E} + \lambda \mathscr{E} \times \mathscr{H} + \mu \mathscr{E}\mathscr{H}^2 + \nu \mathscr{H}(\mathscr{E}\,.\,\mathscr{H}). \tag{6.7.8}$$

For cubic metals there are additional terms given by

$$j_x = \xi \mathscr{E}_x \mathscr{H}_x^2, \quad j_y = \xi \mathscr{E}_y \mathscr{H}_y^2, \quad j_z = \xi \mathscr{E}_z \mathscr{H}_z^2. \tag{6.7.9}$$

To get more explicit results, we shall consider such cubic metals, with the axes of coordinates chosen along the crystal axes. If the current is $(j_x, 0, 0)$ and the magnetic field is $(\mathscr{H}_x, 0, \mathscr{H}_z)$, then the electric field is $(\mathscr{E}_x, \mathscr{E}_y, 0)$, where

$$\left.\begin{aligned} j_x &= [\sigma_0 + \mu \mathscr{H}^2 + (\nu + \xi)\mathscr{H}_x^2]\mathscr{E}_x + \lambda \mathscr{E}_y \mathscr{H}_z, \\ 0 &= -\lambda \mathscr{E}_x \mathscr{H}_z + (\sigma_0 + \mu \mathscr{E}^2)\mathscr{H}_y. \end{aligned}\right\} \tag{6.7.10}$$

Hence the Hall coefficient R and the conductivity σ are conveniently written (cf. Wilson, 1953)

$$R = \frac{\mathscr{E}_y}{\mathscr{H} j_x} = \frac{\lambda \sin\theta}{\sigma_0^2} \tag{6.7.11}$$

and

$$\sigma = \frac{j_x}{\mathscr{E}_x} = \sigma_0 \left\{1 + \left[\left(\frac{\mu}{\sigma_0} + \frac{\lambda^2}{\sigma_0^2}\right)\sin^2\theta + \frac{\mu + \nu + \xi}{\sigma_0}\cos^2\theta\right]\mathscr{H}^2\right\}, \tag{6.7.12}$$

where

$$\mathscr{H} = (\cos\theta, 0, \sin\theta)\mathscr{H}. \tag{6.7.13}$$

The constants appearing in the conductivity are easily calculated by substituting for ϕ in the formula (6.7.7) for the current density. We then obtain

$$\sigma_0 = -\frac{e^2}{4\pi^3\hbar^2}\int \tau \left(\frac{\partial E}{\partial k_1}\right)^2 \frac{\partial f_0}{\partial E} d\mathbf{k}, \tag{6.7.14}$$

$$\lambda = \frac{e^4}{4\pi^3\hbar^4 c}\int \tau \frac{\partial E}{\partial k_1}\Omega_3\left(\tau \frac{\partial E}{\partial k_2}\right)\frac{\partial f_0}{\partial E} d\mathbf{k}, \tag{6.7.15}$$

$$\mu = -\frac{e^4}{4\pi^3\hbar^6 c^2}\int \tau \frac{\partial E}{\partial k_1}\Omega_3\left[\tau\Omega_3\left(\tau\frac{\partial E}{\partial k_1}\right)\right]\frac{\partial f_0}{\partial E} d\mathbf{k}, \tag{6.7.16}$$

$$\nu = -\frac{e^4}{4\pi^3\hbar^6 c^2}\int \tau \frac{\partial E}{\partial k_1}\left\{\Omega_1\left[\tau\Omega_2\left(\tau\frac{\partial E}{\partial k_2}\right)\right] + \Omega_2\left[\tau\Omega_1\left(\tau\frac{\partial E}{\partial k_1}\right)\right]\right\}\frac{\partial f_0}{\partial E} d\mathbf{k} \tag{6.7.17}$$

and

$$\mu + \nu + \xi = -\frac{e^4}{4\pi^3\hbar^6 c^2}\int \tau \frac{\partial E}{\partial k_1}\Omega_1\left[\tau\Omega_1\left(\tau\frac{\partial E}{\partial k_1}\right)\right]\frac{\partial f_0}{\partial E} d\mathbf{k}. \tag{6.7.18}$$

It is possible to transform these expressions for μ and $\mu+\nu+\varepsilon$, on integrating by parts, into

$$\mu = \frac{e^4}{4\pi^3 \hbar^6 c^2}\int \tau\left[\Omega_3\left(\tau\frac{\partial E}{\partial k_1}\right)\right]^2 \frac{\partial f_0}{\partial E}\, d\mathbf{k} \tag{6.7.19}$$

and

$$\mu+\nu+\xi = \frac{e^4}{4\pi^3 \hbar^6 c^2}\int \tau\left[\Omega_1\left(\tau\frac{\partial E}{\partial k_1}\right)\right]^2 \frac{\partial f_0}{\partial E}\, d\mathbf{k}. \tag{6.7.20}$$

These show that both these quantities are negative, since $\partial f_0/\partial E$ is negative. As Wilson has emphasized, it follows additionally, by Schwartz's inequality, that $\lambda^2 \leqslant -\mu\sigma_0$, so that σ is less than σ_0 and the resistance is always increased by the presence of a magnetic field.

6.7.2 Special cases

(a) *Spherical energy surfaces and spherical relaxation time.* Without any further assumptions than

$$E(\mathbf{k}) = E(|\mathbf{k}|) \tag{6.7.21}$$

and

$$\tau(\mathbf{k}) = \tau(|\mathbf{k}|) \tag{6.7.22}$$

we can evaluate the Hall coefficient R, and we find that, under these conditions, we regain the free-electron value

$$R_0 = -\frac{1}{nec}. \tag{6.7.23}$$

Thus, we should not necessarily infer parabolic bands in materials where this holds.

(b) *Energy $E(\mathbf{k})$ and $\tau(\mathbf{k})$ expressed as low-order expansions in cubic harmonics.* Davis (1939) pointed out that useful information might be extracted from the general formulae given above in cases where the angularity of $E(\mathbf{k})$ and $\tau(\mathbf{k})$ is not large, and can be corrected by adding low-order cubic harmonic terms K_l. Thus we write for the energy surfaces

$$k = k_0[1+C_1 K_4(\theta,\phi)+\ldots] \tag{6.7.24}$$

and for the relaxation time τ

$$\tau = \tau_0[1+B_1 K_4(\theta,\phi)+B_2 K_6(\theta,\phi)+\ldots] \tag{6.7.25}$$

in materials with cubic symmetry. Then we can obtain explicit expressions for the Hall constant R and the conductivity σ. In particular

$$\sigma = \sigma_0[1+B_1^2+\ldots] \tag{6.7.26}$$

and for spherical energy surfaces and R written as $1/nex$, we find

$$x = 1-\tfrac{4}{21}B_1^2-\tfrac{8}{13}C_1. \tag{6.7.27}$$

These formula appear to be relevant to the study of the variation of the conductivity and the Hall effect of the alkalis under pressure (see Deutsch, Paul and Brooks, 1961).

Whereas, with spherical energy surfaces, x from equation 6.7.27 is less than the free electron value; as a consequence of anisotropic relaxation times, it is found that $R < R_0$ for high pressures for Cs, though not for any other of the alkalis over the range of pressures studied. It seems as if, to understand these results, we must allow appreciable deviations of the Fermi surface from spherical form in Cs. It is not necessary to do so for the other alkalis, though there are indeed some small departures according to Shoenberg and Stiles.

Finally, Raimes and Cooper (1959) have extended the above discussion to deal with alloys, with interesting results.

6.8 Magnetoresistance

6.8.1 Spherical and ellipsoidal Fermi surfaces

In discussing the changes in resistance occasioned by the application of a magnetic field to a sample, we shall again assume that a relaxation time exists. This simplifies our considerations a great deal while at the same time retaining the essential physics of the problem.

The relaxation time will further be assumed to be constant over the Fermi surface, which at first we shall take to be spherical. In this case, the frequency defined in equation (6.4.41) is also constant over the Fermi surface and we can use Chambers' solution of the Boltzmann equation in the simple form of equation (6.4.42). Choosing $\phi_0 = 0$ and $\mathscr{H}$ to lie in the z-direction we have

$$v_x = v_\perp \cos\phi, \quad v_y = v_\perp \sin\phi, \tag{6.8.1}$$

$v_\perp$ being the component of $v_{\rm f}$ perpendicular to z. Then, since

$$\int_{-\infty}^{0} e^{i\phi}\, e^{\phi/\omega_c\tau}\, d\phi = \frac{\omega_c \tau}{1+i\omega_c\tau}, \tag{6.8.2}$$

the magnetoconductivity tensor is

$$\left.\begin{aligned}
\sigma_{xx} &= \sigma_{yy} = \frac{\sigma_0}{1+\omega_c^2\tau^2},\\
\sigma_{xy} &= -\sigma_{yx} = \frac{-\omega_c\tau\sigma_0}{1+\omega_c^2\tau^2},\\
\sigma_{zz} &= \sigma_0,\\
\sigma_{xz} &= \sigma_{zx} = \sigma_{yz} = \sigma_{zy} = 0,
\end{aligned}\right\} \tag{6.8.3}$$

σ_0 being the conductivity in zero field. The equations for $\sigma_{\alpha\beta}$, where α or β is equal to z (or both are equal to z), follow since the electron velocity in the z-direction is unaffected by the magnetic field, and they hold, provided only that ω_c is constant over the Fermi surface, for arbitrary shapes. As we shall see later, the *cyclotron frequency* has this property for ellipsoidal Fermi surfaces.

We should note that, in practice, it is not the conductivity tensor that is measured, but rather the resistivity, with typical components

$$\left.\begin{aligned} \rho_{xx} &= \frac{\sigma_{yy}\sigma_{zz}-\sigma_{yz}\sigma_{zy}}{\det|\sigma|}, \\ \rho_{xy} &= \frac{\sigma_{xz}\sigma_{zx}-\sigma_{xx}\sigma_{zz}}{\det|\sigma|}, \end{aligned}\right\} \tag{6.8.4}$$

and the magnetoresistive effect is usually defined for numerical purposes by taking the ratio $\Delta\rho/\rho$. Both transverse and longitudinal magnetoresistive effects are absent if the Fermi surface is ellipsoidal, the influences of magnetic and Hall fields exactly cancelling to produce zero net perturbation of the electron trajectories. A non-zero result is obtained, however, if there are two types of carrier. The conductivity of each of the bands contributes additively and if we suppose, for purposes of illustration, that both Fermi surfaces are spherical, the transverse magnetoresistance is just

$$\frac{\Delta\rho}{\rho_0} = \frac{\rho_{xx}-\rho_0}{\rho_0} = \frac{(\omega_c^1\tau_1+\omega_c^2\tau_2)^2\sigma_1\sigma_2}{(\sigma_1+\sigma_2)^2+\omega_c^1\tau_1\omega_c^2\tau_2[(n_1-n_2)^2/n_1n_2]}, \tag{6.8.5}$$

where we have taken one band to contain n_1 electrons and the other to contain n_2 holes. Since $\omega_c \propto \mathscr{H}$, the high-field limit is determined by the term of the denominator involving n_1-n_2. If this is non-zero, the magnetoresistance saturates as $\mathscr{H}$ becomes very large, but if $n_1 = n_2$ the dependence on $\mathscr{H}$ of $\Delta\rho/\rho$ is quadratic in all fields. The absence of saturation is, in fact, observed in bismuth and other metals or semi-conductors having equal numbers of electrons and holes.

There are other reasons why the magnetoresistance may not saturate, as discussed in section 6.8.3 below.

6.8.2 *Expansion of conductivity in powers of magnetic field*

Closed analytic forms of the magnetoconductivity tensor may be obtained when the Fermi surface is of any ellipsoidal shape, the above results for spherical surfaces forming a special case of these. However, we wish now to briefly consider situations for which that approximation is inadequate.

A formal expression for the current density may be obtained by expansion in powers of the magnetic field, which, of course, will be directly useful at low fields. Thus, to second order, we have

$$j_i = \sum_j \left[\sigma^0_{ij} + \sum_k \left(\sigma_{ijk} + \sum_i \sigma_{ijkl}\mathscr{H}_i\right)\mathscr{H}_k\right]\mathscr{E}_j. \tag{6.8.6}$$

The coefficients must be chosen to satisfy the reciprocal relations (see section 6.11)

$$\sigma_{ij}(\mathscr{H}) = \sigma_{ji}(-\mathscr{H}) \tag{6.8.7}$$

and, in conjunction with these, the symmetry properties of the crystal can be used to simplify equation (6.8.6). Let us suppose, for example, that the crystal is cubic. The most general form allowed by crystal symmetry for diagonal elements of the conductivity can be verified to be (with i, j and k labelling the crystal axes)

$$\sigma_{ij} = \sigma_0 + \alpha\mathscr{H}_i^2 + \beta(\mathscr{H}_j^2 + \mathscr{H}_k^2) \tag{6.8.8}$$

plus a term linear in $\mathscr{H}_z$, which, however, equation (6.8.7) prohibits. Similarly, the off-diagonal elements are of the form

$$\sigma_{ij} = \gamma\mathscr{H}_k + \eta\mathscr{H}_i\eta\mathscr{H}_j. \tag{6.8.9}$$

Putting $\alpha = b+c+d$, $\beta = b$, $\gamma = a$ and $\eta = d$ equations (6.8.8) and (6.8.9) may be combined to give the expression

$$\mathbf{j} = \sigma_0\mathscr{E} + a(\mathscr{E}\times\mathscr{H}) + b\mathscr{H}^2\mathscr{E} + c(\mathscr{H}.\mathscr{E})\mathscr{H} + d\mathbf{T}.\mathscr{E}, \tag{6.8.10}$$

where $\mathbf{T}$ is a diagonal tensor with non-zero components $\mathscr{H}_x^2, \mathscr{H}_y^2, \mathscr{H}_z^2$.

With the magnetoresistance measured for $\mathbf{j}$ parallel to $\mathscr{E}$ there is evidently no contribution from the term $a(\mathscr{E}\times\mathscr{H})$. However, from equation (6.7.11) the Hall coefficient can be seen to be

$$R_0 = -\frac{a}{\sigma_0^2}. \tag{6.8.11}$$

With $\mathbf{j}$ and $\mathscr{H}$ along cubic symmetry axes, we also find for the transverse magnetoresistance

$$\left(\frac{\Delta\rho}{\rho_0}\right)_{\text{trans}} = -\frac{b}{\sigma_0}\mathscr{H}^2, \tag{6.8.12}$$

while

$$\left(\frac{\Delta\rho}{\rho_0}\right)_{\text{long}} = -\frac{(b+c+d)}{\sigma_0}\mathscr{H}^2. \tag{6.8.13}$$

To obtain the fourth relation necessary to determine a, b, c and d, we can measure the magnetoresistance with $\mathbf{j}$ off the cube axes. If $\mathbf{j}$ is along the (110) direction, for example, we have

$$\left(\frac{\Delta\rho}{\rho_0}\right)_{\text{trans}} = -\frac{(b+d/2)}{\sigma_0}\mathscr{H}^2 \tag{6.8.14}$$

and

$$\left(\frac{\Delta\rho}{\rho_0}\right)_{\text{long}} = -\frac{(b+c+d/2)}{\sigma_0}\mathscr{H}^2. \tag{6.8.15}$$

To obtain the coefficients of the expansion in $\mathscr{H}$ in terms of the $E(\mathbf{k})$ dispersion relation, we take Chambers' solution for the distribution function

$$f_1 = e\frac{df_0}{dE}\int_{-\infty}^{0} \mathscr{E}.\mathbf{v}(t)\,\mathrm{e}^{-t/\tau}\,dt \tag{6.8.16}$$

and expand

$$\mathbf{v}(t) = \mathbf{v}(0)+t\dot{\mathbf{v}}(0)+\frac{t^2}{2}\ddot{\mathbf{v}}(0)+\ldots \tag{6.8.17}$$

to obtain

$$-f_1 = e\frac{df_0}{dE}\mathscr{E}.[\mathbf{v}(0)\,\tau+\dot{\mathbf{v}}(0)\,\tau^2+\ddot{\mathbf{v}}(0)\,\tau^3+\ldots] \tag{6.8.18}$$

and

$$\sigma_{ij} = \frac{e^2}{4\pi^3}\int d\mathbf{k}\frac{df_0}{dE}v_i[v_i(0)\,\tau+\dot{v}_j(0)\,\tau^2+\ddot{v}_j(0)\,\tau^3+\ldots]. \tag{6.8.19}$$

Since

$$\dot{v}_\mu = \dot{\mathbf{k}}.\frac{\partial}{\partial\mathbf{k}}\frac{\partial E}{\partial k_\mu} = \left(\frac{\mathbf{v}\times\mathscr{H}}{c}\right).\frac{\partial}{\partial\mathbf{k}}\frac{\partial E}{\partial k_\mu}, \tag{6.8.20}$$

it is evident that

$$\frac{d^n v_j}{dt^n}\tau^{n+1} \propto \mathscr{H}^n\,\tau^{n+1} \tag{6.8.21}$$

and, for example, to order $\mathscr{H}$, at low temperatures

$$\begin{aligned}\sigma_{xy} &= \sigma^0_{xy}+\frac{e^2}{4\pi^3 c}\int d\mathbf{k}\frac{\partial f_0}{\partial E}v_x(\mathbf{v}\times\mathscr{H}).\frac{\partial}{\partial\mathbf{k}}\frac{\partial E}{\partial k_y}\\ &= \sigma^0_{xy}+\frac{e^2}{4\pi^3 c}\int d\mathbf{k}\frac{\partial f_0}{\partial E}\frac{\partial E}{\partial k_x}\left(\frac{\partial E}{\partial k_y}\frac{\partial^2 E}{\partial k_x\,\partial k_y}-\frac{\partial E}{\partial k_x}\frac{\partial^2 E}{\partial k_y^2}\right)\mathscr{H}.\end{aligned} \tag{6.8.22}$$

It is of interest to note that, since $\sigma_0 \propto \tau$, equation (6.8.21) shows that

$$\frac{d^n v_j}{dt^n}\tau^{n+1} \propto \mathscr{H}^n \sigma_0^{n+1} \tag{6.8.23}$$

and hence, from equation (6.8.2),

$$\frac{\sigma_{ij}(\mathscr{H})}{\sigma_0} = F(\mathscr{H}\sigma_0), \tag{6.8.24}$$

which is known as Kohler's rule. The rule is true outside the possible limits of an expansion in powers of $\mathscr{H}$ as the following argument shows.

Let us assume first that there is no spread in times between electronic collisions but that they occur at exactly equal intervals τ_c. If then an electron suffers a collision at P and orbits under the influence of the magnetic field $\mathscr{H}$ round the Fermi surface to a point Q in time τ_c it must suffer a second collision at point Q. Now let the magnetic field change to $\alpha\mathscr{H}$. An electron moving over the same trajectory from P to Q on the Fermi surface must do so in time τ_c/α. If therefore it suffered collision at the point P and the collision time is changed from τ_c to τ_c/α, the electron would again suffer a second collision at point Q. These considerations make it clear that any **k**-space picture of electrons orbiting and suffering collisions in magnetic field $\mathscr{H}$ and with collision time τ_c is equally valid for a magnetic field $\alpha\mathscr{H}$ and collision time τ_c/α, except that, in the second case, the time scale is shortened by the factor α. In particular, if, on any element of a trajectory, an electron obtains from an applied electric field an energy δE in the former case, it obtains an energy $\delta E/\alpha$ over the same element in the latter. We can now see, on using Chambers' kinetic method [cf. equations (6.4.39) and (6.4.40)] that increase of magnetic field and decrease of collision time by the factor α decreases $\sigma(\mathscr{H})$ in the same proportion. But since $\sigma_0 \propto \tau$, $\sigma(\mathscr{H})/\sigma_0$ is unchanged, as implied by equations (6.4.39) and (6.8.24).

Kohler's rule is often quoted as

$$\frac{\Delta\rho}{\rho} = f\left(\frac{\mathscr{H}}{\rho_0}\right), \tag{6.8.25}$$

an equivalent form to equation (6.8.24). We should note that equation (6.8.5) does not obey this rule, which is true only for metals with a single conduction band.

Equation (6.8.25) shows the advantage of observing magnetoresistance at low temperatures and with samples of low residual resistivity.

6.8.3 High-field magnetoresistance

Equation (6.8.22) shows that the information contained on the Fermi surface from low-field magnetoresistance experiments is likely to be difficult to extract. This is not the case in the high field limit as we shall now show.

When the magnetic field is very strong, an electron orbits the Fermi surface many times without collision. To see just what the orbital frequency is, we start from the equation

$$\hbar \dot{k} = v_{\perp} \mathscr{H}, \tag{6.8.26}$$

where $v_{\perp}$ is the component of electron velocity perpendicular to the field. Hence the time taken to traverse an orbit is

$$\tau = \oint \frac{dl}{\dot{k}} = \frac{c\hbar}{e\mathscr{H}} \oint \frac{dl}{v_{\perp}}, \tag{6.8.27}$$

where dl is an element of the orbit.

Referring to Figure 6.7, we see that, since

$$\mathbf{v} = \frac{1}{\hbar} \frac{\partial E}{\partial \mathbf{k}}, \tag{6.8.28}$$

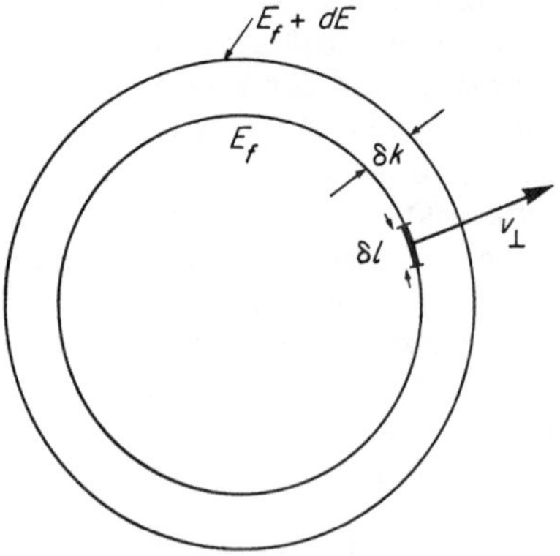

FIGURE 6.7. Cross-sections of volumes by constant energy surfaces E_f and $E_f + dE$.

we can rewrite equation (6.8.27) as

$$\tau = \frac{c\hbar^2}{e\mathscr{H}\,\delta E} \oint dl\, \delta k = \frac{c\hbar^2}{e\mathscr{H}\,\delta E} \oint dA, \tag{6.8.29}$$

where $\oint dA$ is the area between orbits on energy surfaces E_f and $E_f + \delta E$. The orbital or cyclotron frequency is thus

$$\omega_c = \frac{2\pi}{\tau} = \frac{e\mathscr{H}}{c\hbar^2}\frac{dA}{dE}, \tag{6.8.30}$$

where A is the area enclosed by the electron orbit. It can be shown that ω_c is independent of the location of the orbit on the Fermi surface if this is ellipsoidal, but we shall not assume that here.

In high fields, we have $\omega_c \tau \gg 1$ and it is therefore permissible to take Chambers' solution in the form of equation (6.4.42) giving a conductivity

$$\sigma_{ij} = \frac{e^2}{4\pi^3 \hbar^2}\int d\mathbf{k}\frac{\partial f_0}{\partial E} v_i(\mathbf{k}) \int_{-\infty}^{\phi_0} \exp\left[(\phi - \phi_0)/\omega_c \tau\right], \tag{6.8.31}$$

where, in general, ω_c is a function of k.

Since v_j is periodic round an orbit, we can evaluate the integral in steps of 2π, and since $\omega_0 \tau \gg 1$ we can expand the exponential to obtain

$$\frac{1}{\omega_c}\int_0^{2\pi} v_j(\phi - \phi_0) \exp(-\phi/\omega_c \tau)\, d\phi = \frac{1}{\omega_c}\int_0^{2\pi} d\phi\, v_j(\phi)\left(1 - \frac{\phi}{\omega_c \tau} + \frac{\phi^2}{\omega_c^2 \tau^2} + \ldots\right). \tag{6.8.32}$$

This integral is to be multiplied by

$$\sum_{n=0}^{\infty} \exp(-2\pi n/\omega_c \tau) = \frac{1}{1 - \exp(-2\pi/\omega_c \tau)} \approx \frac{\omega_c \tau}{2\pi}. \tag{6.8.33}$$

But the cyclotron frequency $\omega_c \propto \mathscr{H}$ and thus we have an expansion of σ_{ij} in inverse powers of $\mathscr{H}$. The leading term is

$$\sigma_{ij} = \frac{e^2}{4\pi^3 \hbar^2}\int d\mathbf{k}\frac{df_0}{dE} v_i(\mathbf{k}) \frac{\tau}{2\pi}\int_0^{2\pi} v_j(\phi)\, d\phi. \tag{6.8.34}$$

We can see from this result that the velocity v_j of the zero-field conductivity is replaced by its average over an orbit due to the influence of the magnetic field. When $i = j \equiv z$ the term will be non-zero if $\mathscr{H}$ is in the z-direction and σ_{zz} saturates in a magnetic field. However, anistropy of v_z on the Fermi surface means in general that $\sigma_{zz}(\mathscr{H}) < \sigma_{zz}(0)$. On the other hand, the average of v_y and v_z is zero. Furthermore, inspection of equation (6.8.31), or direct use of the reciprocity relations equation (6.8.7) show that σ_{xx} has no term $\propto \mathscr{H}^{-1}$. In fact, as $\mathscr{H} \to \infty$ the leading terms in the conductivity tensor are given by

$$\sigma_{ij}(\mathscr{H}) = \begin{pmatrix} a/\mathscr{H}^2 & b/\mathscr{H} & c/\mathscr{H} \\ -b/\mathscr{H} & d/\mathscr{H}^2 & e/\mathscr{H} \\ -c/\mathscr{H} & -e/\mathscr{H} & f \end{pmatrix}. \tag{6.8.35}$$

The result of inverting equation (6.8.35) to find the resistivity is

$$\rho_{xx} = \frac{df+e^2}{fb^2}, \quad \rho_{xy} = \frac{\mathscr{H}}{b}, \quad \rho_{xz} = \frac{-e}{fb}, \quad \rho_{zz} = \frac{1}{f}, \tag{6.8.36}$$

showing that both transverse and longitudinal magnetoresistance saturate.

So far, however, we have left out of our considerations the possibility that the electron may not pursue closed trajectories. Let us now assume, in fact, that electrons can traverse an open Fermi surface without the trajectory forming a closed orbit. Under the influence of a magnetic field parallel to the z-axis, let us take the general direction of the trajectory to be along the x-axis. Over such an orbit, the mean value of v_x is zero, but the mean value of v_y is non-zero. Equation (6.8.35) then takes the modified form

$$\sigma_{ij}(\mathscr{H}) = \begin{pmatrix} a/\mathscr{H}^2 & b/\mathscr{H} & c/\mathscr{H} \\ -b/\mathscr{H} & g & h \\ -c/\mathscr{H} & h & f \end{pmatrix} \tag{6.8.37}$$

and we now find that

$$\rho_{xx} = \mathscr{H}^2(gf-h^2)(gfa-h^2a+fb^2-2hbc+gc^2)^{-1}, \tag{6.8.38}$$

while $\rho_{xy} \propto \mathscr{H}$ and ρ_{yy} and ρ_{zz} are constants.

On the other hand, if the open trajectory is parallel to neither x nor y, it can contribute to both σ_{xx} and σ_{yy}, neither of which will now go to zero as $\mathscr{H} \to \infty$. In that case, saturation will occur for both x- and y-directions.

We thus conclude that, with the exception of the compensated two-type carriers in 6.8.1, the longitudinal magnetoresistance will *always* saturate and the transverse magnetoresistance will saturate except when measured in the direction of an open trajectory in **k**-space. It therefore becomes possible to investigate the topology and connectivity of Fermi surfaces which allow open trajectories, by studying the marked anisotropy of the magnetoresistance of metals with such Fermi surfaces (see especially Pippard, 1965).

6.9 Thermal transport by electrons

Having discussed the electrical conductivity at some length, we want now to begin the discussion of thermal conduction by considering the contribution the electrons make. We shall, of course, later complete the problem by dealing with the lattice conduction. We again begin with a treatment based on the Boltzmann equation in the relaxation time approximation, and specialized to cubic crystals.

6.9.1 *Boltzmann equation formulation*

When the coupling is weak, we use the equations

$$\left.\begin{aligned} -4\pi^3\mathbf{j} &= e\int \mathbf{v_k} f(\mathbf{k})\,d\mathbf{k}, \\ 4\pi^3\mathbf{U} &= \int \mathbf{v_k} E_\mathbf{k} f(\mathbf{k})\,d\mathbf{k}. \end{aligned}\right\} \tag{6.9.1}$$

To obtain the simultaneous effect of electric potential and thermal gradient, we solve separately to first order. The linearized equation we take when a thermal gradient exists is

$$v_\mathbf{k}.\nabla f_0(\mathbf{k}) = Kf(\mathbf{k}), \tag{6.9.2}$$

where K is a collision operator. As usual

$$f_0(E) = \frac{1}{\exp[\beta(E-\zeta)]+1} \tag{6.9.3}$$

and thus we have

$$\nabla f_0 = \nabla\zeta\frac{\partial f_0}{\partial\zeta}+\nabla T\frac{\partial f_0}{\partial T} = -\left[\nabla\zeta - \frac{\nabla\beta}{\beta}(E-\zeta)\right]\frac{\partial f_0}{\partial E}$$

$$= -\left[\nabla\zeta+\frac{\nabla T}{T}(E-\zeta)\right]\frac{\partial f_0}{\partial E} = \left[T\nabla\left(\frac{\zeta}{T}\right)+\frac{\nabla T}{T}E\right]\frac{\partial f_0}{\partial E}. \tag{6.9.4}$$

Hence from equations (6.9.2) and (6.9.4) we find

$$\frac{\partial f}{\partial} - v_\mathbf{k}.\left[T\nabla\left(\frac{\zeta}{T}\right)+\frac{\nabla T}{T}E_\mathbf{k}\right]\frac{\partial f_0}{\partial E} = Kf. \tag{6.9.5}$$

If a relaxation time exists, so that $f_1 = (f-f_0)/\tau$, we obtain for the steady state

$$f_1(\mathbf{k}) = \frac{\nabla T}{T}\tau v_\mathbf{k} E\frac{\partial f_0}{\partial E}+\frac{\nabla\zeta}{T}\tau v_\mathbf{k}\frac{\partial f_0}{\partial E} \tag{6.9.6}$$

and we can write the coefficients of equation (6.2.22) as

$$L_{21} = L_{12} = \frac{e}{4\pi^3}\int \tau v_\mathbf{k} v_\mathbf{k} v_\mathbf{k} E_\mathbf{k}\frac{\partial f_0}{\partial E}d\mathbf{k} \tag{6.9.7}$$

and

$$4\pi^3 L_{22} = -\int \tau v_\mathbf{k} v_\mathbf{k} E_\mathbf{k}^2\frac{\partial f_0}{\partial E}d\mathbf{k}. \tag{6.9.8}$$

Hence, using the formulae (6.2.26), the thermal conduction can be obtained.

We now turn immediately to a further use of these formulae to obtain the thermoelectric power S defined earlier in this chapter.

6.9.2 Absolute thermopower in terms of conductivity

We now generalize the expression for the conductivity σ in the relaxation time approximation to

$$\sigma(E) = \frac{e^2}{12\pi^3}\int_E \tau v\, ds \tag{6.9.9}$$

which is simply the conductivity when $E = E_f = \zeta$ with the approximation $\partial f_0/\partial E = -\delta(E-E_f)$ taken. To obtain the thermoelectric power S in terms of $\sigma(E)$ we must, in fact, expand $\partial f_0/\partial E$ to order T^2; we have then from equations (6.4.27) and (6.9.3)

$$e^{-2}L_{11} = -\frac{1}{e^2}\int \sigma(E)\frac{\partial f_0}{\partial E}\,dE = \frac{1}{e^2}\sigma(\zeta)+\frac{\pi^2}{6}\left(\frac{k_B T}{e}\right)^2\frac{d^2\sigma(\zeta)}{d\zeta^2} \tag{6.9.10}$$

and

$$e^{-1}L_{12} = -\frac{1}{e^2}\int \sigma(E)\,E\frac{\partial f_0}{\partial E}\,dE = \frac{1}{e^2}\sigma(\zeta)\,\zeta-\frac{\pi^2}{6}\left(\frac{k_B T}{e}\right)^2\left[2\zeta\frac{d\sigma}{d\zeta}+\zeta\frac{d^2\sigma(\zeta)}{d\zeta^2}\right] \tag{6.9.11}$$

and so, to the same order,

$$S = \frac{1}{eT}\left(\frac{eL_{12}+\zeta L_{11}}{L_{11}}\right) = -\frac{\pi^2 k_B^2 T}{3e}\left(\frac{1}{\sigma}\frac{d\sigma(\zeta)}{d\zeta}\right). \tag{6.9.12}$$

If we assume the mean free path $l = v\tau$ is constant over the Fermi surface, we see from equation (6.9.9) that

$$\frac{1}{\sigma(\zeta)}\frac{d\sigma(\zeta)}{d\zeta} = \frac{1}{l}\frac{dl}{d\zeta}+\frac{1}{A}\frac{dA}{\partial\zeta}, \tag{6.9.13}$$

where A is the area of the Fermi surface. This equation is also true if $v\tau$ is not a constant over S_f, when l is to be interpreted as the mean-free path averaged over S_f. In general, we can expect l and A to increase with E; in fact using the Debye model $l \propto E^2$ (see, for example, the discussion given by Jones, 1955) and if the energy surfaces are spherical $A \propto E$, so that we obtain

$$S = -\frac{\pi^2 k_B^2 T}{3e}\frac{1}{\sigma}\frac{d\sigma}{d\zeta} = -\frac{\pi^2 k_B^2 T}{e\zeta}. \tag{6.9.14}$$

We can hardly ever expect this to be quantitatively correct and indeed the noble metals have positive thermopowers in contrast to equation (6.9.14). These metals, as we have seen, have Fermi surfaces contacting the zone boundary so that $dA/d\zeta < 0$. If this negative term outweighs the first in equation (6.9.13), S will be positive. On the other hand, the alkali metals, the heavier ones excluding Li having very spherical Fermi surfaces as we saw in section 4.15 of Chapter 4, have negative thermoelectric powers. The lightest alkali, lithium, is a notable exception and since it seems to be established that its Fermi surface does not contact the BZ boundaries, it must be supposed that l decreases with E. In the following section (a), one sees that the dependence $l \propto E^2$ seems to be assuming s-like behaviour of the wave function and it is this supposition which presumably destroys the validity of the above simple theory in this case. It is also possible that $(1/l)(dl/d\zeta)$ is negative in other metals.

Results for pseudoatom picture of simple metals. To illustrate the above remarks, we shall consider briefly here the work of Young and colleagues (1967) who take, for simple metals, a model in which free electrons are assumed scattered by pseudoatoms. It will be relevant also to mention their work for liquid metals in passing, because their calculations for the solids utilize the Debye model and it is probable that the liquid metal calculations, which use experimental structure factors $S_1(k)$, are more relevant to the solid state at high temperatures than those using the Debye model, particularly for lithium, the electronic properties of which do not change markedly on melting.

Young and co-workers essentially start from equation (6.4.79) where $\mathscr{S}_1$ is replaced by the static structure factor, which we have already argued is permissible at high temperatures. With this substitution we have

$$\frac{1}{\sigma(k)} = \frac{3\pi m^2}{4e^2 k^6}\int_0^{2k_f} |V(\boldsymbol{\kappa})|^2 S_1(\kappa)\,\kappa^3\,d\kappa \quad \left(\frac{k^2}{2m} = E\right), \tag{6.9.15}$$

and it is then readily shown that for a free-electron-like metal

$$S = -\frac{\pi^2 k_B^2 T}{3e}\left(\frac{2m}{k\sigma}\frac{d\sigma(k)}{dk}\right)_{k=k_f} = -\left(\frac{\pi^2 k_B^2 T}{3}\Big/ e\zeta\right) x, \tag{6.9.16}$$

where

$$x = 2 - y - z, \tag{6.9.17}$$

$$y = \int_0^2 |V(k_f\,\xi)|^2 \left[k_f \frac{\partial}{\partial k_f} S_1(k_f\,\xi)\right] \xi^3\,d\xi \Big/ \int_0^2 |V(k_f\,\xi)|^2 S_1(k_f\,\xi)\,\xi^2\,d\xi \tag{6.9.18}$$

and

$$z = \int_0^2 \left\{ k_f \frac{\partial}{\partial k_f} |V(k_f\,\xi)|^2 \right\} S_1(k_f\,\xi)\,\xi^3\,d\xi \Big/ \int_0^2 |V(k_f\,\xi)|^2\, S_1(k_f\,\xi)\,\xi^3\,d\xi. \tag{6.9.19}$$

We have assumed S to be spherically symmetrical here (in the Debye model it is constant out to $2k_f$). The results of their calculations for monovalent metals are shown in Table 6.1, the values tabulated being for x defined in equation (6.9.16).

TABLE 6.1. *Thermoelectric power data for monovalent metals*

		Li	Na	K	Rb	Cs	Cu	Ag	Au
Experimental									
	Solid	−7.2	2.7	3.8	2.5	0.2			
	Liquid	−8.8	2.9	3.5	2.7	−1.3	−3.5	−1.9	−0.6
Theory									
	Solid	−0.7	2.4	3.2	3.3	0.6			
							−2.4	−1.6	−2.2
	Liquid	−4.9		2.2	1.3	−5.1			

The principal contributions to y and z are found to come from values of the integrand near $\xi = 2$. Thus we may write

$$y \approx \left[\frac{\partial S_1(k_f\,\xi)/\partial \xi}{S_1(k_f\,\xi)} \right]_{\xi=2}, \quad z \approx \left[\frac{k_f\,\partial\,|V(k_f\,\xi)|^2/\partial k_f}{|V(k_f\,\xi)|^2} \right]_{\xi=2}. \tag{6.9.20}$$

The value of y is the same in sign for the monovalent metals, $2-y$ being roughly -1 to 5 (see Figure 6.8 for S_1 for liquid metals) while it is z that is responsible for the adjustment of sign of x for sodium, potassium and rubidium. Young and co-workers analysed their potentials into phase shifts, as discussed in Chapter 1, and found the positive value of z for lithium to be largely determined by the p-wave phase shift. The energy dependence of the potential through this phase shift is the important point here.

The change of sign in the thermoelectric power of Cs on melting can be explained as an increase in y as the metal melts, in fact, in the Debye approximation for the solid, y is zero, $S_1(\boldsymbol{\kappa})$ being a constant. We repeat, however, that this is in all probability an approximation poorer than taking the liquid structure factor. The fact that taking the Debye approximation produces the correct sign of the thermoelectric power does not imply, of course, that it is quantitative; Cs has a quite small thermoelectric power and the change in $S_1(\boldsymbol{\kappa})$ on melting does not need to be very drastic to change the sign of S.

To summarize, the investigation we have described provides information on the role of the variation of the mean free path with energy at the Fermi level and indicates its origin. In lithium it is of primary importance since the Fermi surface seems fairly spherical and the negative sign of $dl/d\zeta$ arises from the p-like nature of the wave function. The investigation suggests also,

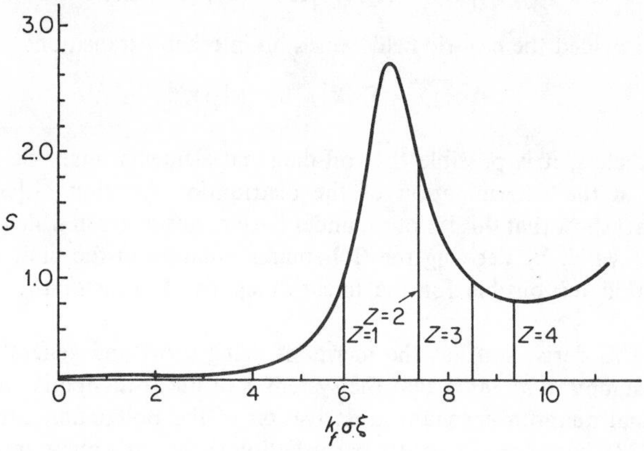

FIGURE 6.8. Structure factors $S(k)$ for liquid metals. Ashcroft and Lekner (1966) hard sphere structure factors are shown, with $2k_f$ marked for valencies $z = 1, 2, 3, 4$, for packing fraction $\eta = 0.45$ (melting point). The hard core diameter σ, the valency z and η are related by $k_f \sigma = (18\pi z\eta)^{1/3}$. ξ along abscissa is $2 \sin \frac{1}{2}\theta$, where θ is angle of scattering.

though the model is, of course, much less well founded in this case, that $dl/d\zeta < 1$ in the noble metals. The markedly aspherical nature of the Fermi surfaces in these metals no doubt also plays a central role in these cases as we have previously indicated.

6.10 Full quantum-mechanical discussion of Boltzmann equation

In using the Boltzmann equation approach, we have dealt entirely with probabilities of states being occupied and probabilities of transitions. In this process, the phases of the probability amplitudes (which appear in the off-diagonal elements of the density matrix) have been ignored. Now it is true that the off-diagonal elements do not appear in the formal expressions for the conductivity, provided interband transitions do not take place. Taking the trace with respect to the Bloch functions of a perfect lattice, we have for the current, for example,

$$\langle \mathbf{j} \rangle = \operatorname{tr}(\rho \mathbf{j}) = \sum_{\mathbf{k}i} \langle \mathbf{k}i | \rho \mathbf{j} | \mathbf{k}i \rangle = \sum_{\substack{\mathbf{k}'\mathbf{k} \\ i'i}} \langle \mathbf{k}i | \rho | \mathbf{k}'i' \rangle \langle \mathbf{k}'i' | \mathbf{j} | \mathbf{k}i \rangle, \quad (6.10.1)$$

where ρ is the density matrix operator, $\mathbf{k}$ and $\mathbf{k}'$ are the wave vectors labelling Bloch functions and i, i' are band indices. It is easy to show that

$$\langle \mathbf{k}'i | \mathbf{j} | \mathbf{k}i \rangle = e\langle \mathbf{k}'i | \mathbf{v} | \mathbf{k}i \rangle = \delta_{\mathbf{k}'\mathbf{k}} \langle \mathbf{k}i | \mathbf{j} | \mathbf{k}i \rangle \quad (6.10.2)$$

and so, provided the electric field causes no interband transitions,

$$\langle \mathbf{j} \rangle = \sum_{\mathbf{k}} \langle \mathbf{k} | \rho | \mathbf{k} \rangle \langle \mathbf{k} | \mathbf{j} | \mathbf{k} \rangle. \quad (6.10.3)$$

Nevertheless, it is possible that off-diagonal elements might be explicitly required in the determination of the distribution function $\langle \mathbf{k} | \rho | \mathbf{k} \rangle$. We shall in fact show that this is not so, under the circumstances in which equation (6.10.3) is valid, by deriving the Boltzmann equation in the limit when the perturbation responsible for the linear-dissipative nature of the system is weak.

While this derivation has the merits of being short and general, it relies on an assumption *ab initio* that the system is of linear-dissipative nature. In the original quantum-mechanical derivation of the Boltzmann equation by Pauli (1928), based on first-order perturbation theory, the phase information was lost in the final equations because of an assumption that the phases of the quantum-mechanical amplitudes, with respect to the eigenfunctions of the unperturbed system, are randomly distributed at all times. This supposition has its counterpart in the hypothesis of molecular chaos in classical kinetic theory.

Again, the derivation is not without difficulty in that no criterion is provided as to the distinctive feature of a perturbation or scattering mechanism that will yield an irreversible approach to equilibrium despite the time-reversal symmetry of the basic equations of quantum mechanics. To throw light on this question, we shall examine the origin of the Boltzmann equation when the conductivity is limited by impurity scattering. In this special case we shall be able to show explicitly what formal property of the perturbation is responsible for dissipative effects.

6.10.1 Density matrix treatment

Because it is a rather general approach, we shall first use the density matrix to show how the Boltzmann equation leads to exact results in the limit of weak scattering. We shall, however, assume, rather than prove, that we are dealing with a linear-dissipative system.

We begin by writing the equation of motion for the density matrix operator as

$$\frac{\partial \rho}{\partial t} = \frac{i}{\hbar}[\rho, H] + \frac{\partial \rho}{\partial t}\bigg|_{\text{ext}} - \frac{\rho - \rho_0}{\tau_s}, \tag{6.10.4}$$

where H is the Hamiltonian of the unperturbed system (density matrix ρ_0) and $\partial\rho/\partial t|_{\text{ext}}$ is the rate of change of ρ due to the external perturbation alone. We shall, at first, leave the nature of this perturbation unspecified, for the same equations will be used later, when we deal with thermal conduction by phonons. We have also added a term representing a weak interaction with the surroundings. Such an interaction is physically necessary in order that the Joule heat be dissipated. In fact, we shall not deal with this effect, working only to first order: but nevertheless, as we shall see, the term $(\rho - \rho_0)/\tau_s$ will be mathematically useful.

We now set $\rho = \rho_0 + \rho_1$ and use the fact that $[\rho_0, H] = 0$. Equation (6.10.4) can then be written as

$$\frac{\partial \rho_1}{\partial t} = \frac{i}{\hbar}[\rho_1, H] + \frac{\partial \rho}{\partial t}\bigg|_{\text{ext}} - \frac{\rho_1}{\tau_s}. \tag{6.10.5}$$

We next separate H into two parts,

$$H = H_0 + V, \tag{6.10.6}$$

V representing the scattering mechanism responsible for finite conductivity, and then we take matrix elements of equation (6.10.5) with respect to eigenfunctions of H_0. Then we find

$$\frac{\partial \rho_1^{ij}}{\partial t} = \frac{i}{\hbar}\sum_l (\rho_1^{il}\langle l|H_0|j\rangle - \langle i|H_0|l\rangle \rho_1^{lj}) + \frac{i}{\hbar}\sum_l (\rho_1^{il}V_{lj} - V_{il}\rho_1^{lj}) + \frac{\partial \rho^{ij}}{\partial t}\bigg|_{\text{ext}} - \frac{\rho_1^{ij}}{\tau_s} \tag{6.10.7}$$

or, if the eigenvalues of H_0 are denoted by ε_i,

$$\frac{\partial \rho_1^{ij}}{\partial t} = \frac{i}{\hbar}\rho_1^{ij}(\varepsilon_j - \varepsilon_i) + \frac{i}{\hbar}\sum_l (\rho_1^{il}V_{lj} - V_{il}\rho_1^{lj}) + \frac{\partial \rho^{ij}}{\partial t}\bigg|_{\text{ext}} - \frac{\rho_1^{ij}}{\tau_s}. \tag{6.10.8}$$

We now assume that V is indeed such that the system reaches equilibrium in the presence of the external field.

(To obtain a non-zero current and equality of chemical potential throughout the system, one can imagine periodic boundary conditions to have been imposed, but we shall not need to use them explicitly.) In equilibrium

$\partial \rho_1^{ij}/\partial t = 0$ and so

$$\frac{i}{\hbar}\left(\varepsilon_i - \varepsilon_j - \frac{i\hbar}{\tau_s}\right)\rho_1^{ij} = \frac{i}{\hbar}\sum_l (\rho_1^{il} V_{lj} - V_{il}\rho_1^{lj}) + \left.\frac{\partial \rho^{ij}}{\partial t}\right|_{\text{ext}} \tag{6.10.9}$$

or

$$\left(\varepsilon_i - \varepsilon_j - \frac{i\hbar}{\tau_s}\right)\rho_1^{ij} = \sum_{l\neq i,j}{}' (\rho_1^{il} V_{lj} - V_{il}\rho_1^{lj}) + V_{ij}(\rho_1^{ii} - \rho_1^{jj})$$

$$+ (V_{jj} - V_{ii})\,\rho_1^{ij} + \frac{\hbar}{i}\left.\frac{\partial \rho^{ij}}{\partial t}\right|_{\text{ext}}. \tag{6.10.10}$$

Because the diagonal elements of V may be absorbed into the ε_i and ε_j, we shall drop the term $(V_{jj} - V_{ii})\,\rho_1^{ij}$.

When $i = j$ we obtain, as $\tau_s \to \infty$

$$0 = \frac{\hbar}{i}\left.\frac{\partial \rho^{ij}}{\partial t}\right|_{\text{ext}} + \sum_{l\neq i,j}{}' (\rho_1^{il} V_{lj} - V_{il}\rho_1^{lj}). \tag{6.10.11}$$

Now in the specific instances we treat later

$$\frac{\hbar}{i}\left.\frac{\partial \rho_1^{ij}}{\partial t}\right|_{\text{ext}} = C_0^{ij} + \lambda C_1^{ij} + \lambda^2 C_2^{ij} + \dots, \tag{6.10.12}$$

where λ is the strength of the interaction V. Hence to lowest order in λ

$$0 = C_0^{ij} + \sum_{l\neq i,j}{}' (\rho_1^{il} V_{lj} - V_{il}\rho_1^{lj}), \tag{6.10.13}$$

showing that $\rho_1^{ij} \propto \lambda^{-1}$ to lowest order. This implies that to lowest order in λ, equation (6.10.10) is

$$\left(\varepsilon_i - \varepsilon_j - \frac{i\hbar}{\tau_s}\right)\rho_1^{ij} = V_{ij}(\rho_1^{ii} - \rho_1^{jj}), \tag{6.10.14}$$

with $\rho_1^{ii} \propto \lambda^{-2}$, all other terms in equation (6.10.10) being of higher order than λ^{-1}. Inserting this into equation (6.10.13) we find

$$C_0^{ii} = \sum_j |V_{ij}|^2 [\rho_1^{ii} - \rho_1^{jj}] \left[\frac{1}{\varepsilon_i - \varepsilon_j + (i\hbar/\tau_s)} - \frac{1}{\varepsilon_i - \varepsilon_j - (i\hbar/\tau_s)}\right]. \tag{6.10.15}$$

Now provided $\hbar/\tau_s \gg \delta\varepsilon$, the spacing between energy levels, the summation may be replaced by an integral in the usual way and we can use the result

$$\lim_{s\to 0+} \frac{1}{x - is} = P\left(\frac{1}{x}\right) + i\pi\delta(x) \tag{6.10.16}$$

to write equation (6.10.15) as

$$-iC_0^{ii} = 2\pi \sum_j |V_{ij}|^2 [\rho_1^{ii} - \rho_1^{jj}]\, \delta(\varepsilon_i - \varepsilon_j), \tag{6.10.17}$$

provided $\hbar/\tau_s$ is much less than the energy interval over which $|V_{ij}|^2 [\rho_1^{ii} - \rho_1^{jj}]$ varies appreciably. This will not in general conflict with the condition $\hbar/\tau_s \gg \delta\varepsilon$, but there is one point deserving attention. This is the apparent inconsistency whereby we dropped the term $(-i\hbar/\tau_s)\,\rho_1^{ii}$ to obtain equation (6.10.11) but retained a similar term in equation (6.10.13). To see that ρ_1^{ii}/τ_s is in fact negligible, we note that if τ_c is the time over which the system relaxes after removal of the external field, we will have

$$\left.\frac{\partial \rho^{ii}}{\partial t}\right|_{\text{ext}} \sim \rho_1^{ii}/\tau_c, \tag{6.10.18}$$

provided $\tau_s/\tau_c \gg 1$. Thus $\rho_{ii}/\tau_s \ll (\partial\rho^{ii}/\partial t)|_{\text{ext}}$, provided $\tau_s \gg \tau_c$.

To proceed further, we shall deal specifically with the electrical conductivity, and show that equation (6.10.17) is equivalent to the Boltzmann equation provided external fields and scattering processes induce no interband transitions. We have

$$\frac{\partial \rho}{\partial t} = \frac{i}{\hbar}[\rho, H + \phi], \tag{6.10.19}$$

where

$$\phi = e\mathscr{E}.\mathbf{r}, \tag{6.10.20}$$

$\mathscr{E}$ being the electric field, and then evidently

$$\left.\frac{\partial \rho}{\partial t}\right|_{\text{ext}} = \frac{ie}{\hbar}\mathscr{E}[\rho_0, \mathbf{r}] \tag{6.10.21}$$

to first order. It will be useful to put this into another form. We first write

$$\rho_0(H) = \sum_{n=0}^{\infty} \mu_n H^n. \tag{6.10.22}$$

Next we take the commutator

$$i[H^n, \mathbf{r}] = i \sum_{j=0}^{n-1} H^j [H, \mathbf{r}]\, H^{n-1-j}, \tag{6.10.23}$$

which, since

$$\frac{\mathbf{p}}{m} = \mathbf{v} = \frac{i}{\hbar}[\mathbf{H}, \mathbf{r}] \tag{6.10.24}$$

and also

$$\frac{\mathbf{p}}{m} = \frac{\partial H}{\partial \mathbf{p}} \tag{6.10.25}$$

(provided all potentials are velocity independent), may be rewritten as

$$\frac{i}{\hbar}[H^n, \mathbf{r}] = \sum_{j=0}^{n-1} H^j \frac{\partial H}{\partial \mathbf{p}} H^{n-1-j} = \frac{\partial H^n}{\partial \mathbf{p}}. \tag{6.10.26}$$

It is now easily seen that

$$\left.\frac{\partial \rho_0}{\partial t}\right|_{\text{ext}} = \frac{ie\mathscr{E}}{\hbar} \cdot \frac{\partial \rho_0}{\partial \mathbf{p}} \tag{6.10.27}$$

and

$$C_0^{i'i} = \frac{ie}{\hbar}\mathscr{E} \cdot \langle \mathbf{k}' | \frac{\partial \rho_0}{\partial \mathbf{p}} | \mathbf{k} \rangle, \tag{6.10.28}$$

where we now assume

$$H_0 | \mathbf{k} \rangle = \varepsilon_{\mathbf{k}} | \mathbf{k} \rangle \tag{6.10.29}$$

and

$$\rho_0(H_0) = \sum_{n=0}^{\infty} \mu_n H_0^n, \tag{6.10.30}$$

since we do not require higher-order terms involving λ. From these last two equations

$$\begin{aligned} \langle \mathbf{k}' | \frac{\partial \rho_0}{\partial \mathbf{p}} | \mathbf{k} \rangle &= \sum_{n=0}^{\infty} \mu_n \langle \mathbf{k}' | \sum_{j=0}^{n-1} H_0^j \frac{\partial H_0}{\partial \mathbf{p}} H_0^{n-j-1} | \mathbf{k} \rangle \\ &= \sum_{n=0}^{\infty} \mu_n \varepsilon_{\mathbf{k}'}^j \varepsilon_{\mathbf{k}}^{n-1-j} \langle \mathbf{k}' | \frac{\partial H_0}{\partial \mathbf{p}} | \mathbf{k} \rangle \\ &= \sum_{n=0}^{\infty} \mu_n \varepsilon_{\mathbf{k}'}^j \varepsilon_{\mathbf{k}}^{n-1-j} \langle \mathbf{k}' | \frac{\mathbf{p}}{m} | \mathbf{k} \rangle. \end{aligned} \tag{6.10.31}$$

By Feynman's theorem the diagonal element is just $(\partial/\partial \mathbf{k}) \langle \mathbf{k} | H_0 | \mathbf{k} \rangle$ and so

$$\begin{aligned} \langle \mathbf{k}' | \frac{\partial \rho_0}{\partial \mathbf{p}} | \mathbf{k} \rangle &= \delta_{\mathbf{k}'\mathbf{k}} \sum_{n=0}^{\infty} \mu_n \varepsilon_{\mathbf{k}}^j \varepsilon_{\mathbf{k}}^{n-1-j} \frac{\partial \varepsilon_{\mathbf{k}}}{\partial \mathbf{k}} \\ &= \delta_{\mathbf{k}'\mathbf{k}} \frac{\partial}{\partial \mathbf{k}} \langle \mathbf{k} | \rho_0 | \mathbf{k} \rangle, \end{aligned} \tag{6.10.32}$$

where off-diagonal elements have been set to zero because we allow no interband transitions.

It is now immediately evident that equation (6.10.17) is in fact equivalent to the Boltzmann equation for the impurity scattering problem (for weak individual scatterers) and holds to lowest order in λ.‡

In Appendix 6.1, it is shown how the Boltzmann equation for electrons scattered by phonons may be derived from the above theory. In that argument it is shown how an equation for the electrons alone can be obtained [cf. equation (6.4.7)], provided the phonons are assumed at all times to be in thermal equilibrium.

The truth of this assumption depends on how quickly the phonon distribution can relax, after an electron–phonon collision, by means of phonon–phonon collisions, collisions with impurities, and so on. We assume throughout most of this chapter that the relaxation is so rapid as to make no direct contribution to transport effects. This is usually true, but not at low temperatures, where the electric field can effect an appreciable displacement of the phonon distribution in momentum space; we shall see later how this leads to anomalous behaviour of thermoelectric powers at low temperatures. For the present, however, we assume this "phonon-drag" effect is negligible.

6.10.2 Impurity scattering and existence of the Boltzmann equation

Up to now the existence of the Boltzmann equation is not completely proved, in that we have given no reason why a system should be dissipative. We shall now discuss the property of the perturbing or scattering potential which gives rise to the dissipative effects, taking the particular example of Bloch waves scattered by randomly distributed impurities.

The crucial term in the Boltzmann equation is the collision term, which in the present instance takes the form

$$\left[\frac{\partial f}{\partial t}\right]_{\text{coll}} = \sum_{\mathbf{k}} P(\mathbf{k}', \mathbf{k})\,[f(\mathbf{k}) - f(\mathbf{k}')], \tag{6.10.33}$$

since we take $\omega = 0$ in equation (6.4.7), the scatterers being static and the scattering elastic.

Let us begin by reminding ourselves of the calculation of transition probabilities such as $P(\mathbf{k}', \mathbf{k})$. We shall confine ourselves at first to low-order perturbation theory. The perturbation is here time independent and if we write the exact wave function as

$$\psi_{\mathbf{k}'}(\mathbf{r}, t) = \sum_{\mathbf{k}} c_{\mathbf{k}}(t)\,\phi_{\mathbf{k}}(\mathbf{r})\,\mathrm{e}^{-iE_{\mathbf{k}'}t/\hbar} \tag{6.10.34}$$

‡ Higher orders in λ are considered by Kohn and Luttinger (1957), who show explicitly how the Joule heat appears at the next order. They also consider the effect of interband transitions.

where the $\phi_{\mathbf{k}}$'s and $E_{\mathbf{k}}$'s are eigenfunctions and eigenvalues of the unperturbed Hamiltonian H, we find to first order

$$c_{\mathbf{k}}(t) = \langle \phi_{\mathbf{k}'}^{*} V \phi_{\mathbf{k}} \rangle \left(\frac{e^{i(E_{\mathbf{k}}-E_{\mathbf{k}'})t/\hbar} - 1}{E_{\mathbf{k}} - E_{\mathbf{k}'}} \right) \quad (\mathbf{k} \neq \mathbf{k}') \tag{6.10.35}$$

if we assume that at $t = 0$

$$c_{\mathbf{k}}(0) = \delta_{\mathbf{k}'\mathbf{k}}. \tag{6.10.36}$$

Hence if the electron was in state $\phi_{\mathbf{k}} \equiv |\mathbf{k}\rangle$ at zero time, the probability that it is in state $|\mathbf{k}'\rangle$ at time t is

$$\begin{aligned} |\langle \mathbf{k}'|V|\mathbf{k}\rangle|^2 \left(\frac{e^{i(E_{\mathbf{k}}-E_{\mathbf{k}'})t/\hbar} - 1}{E_{\mathbf{k}'} - E_{\mathbf{k}}} \right)^2 &= \frac{2\pi t}{\hbar} |\langle \mathbf{k}'|V|\mathbf{k}\rangle|^2 \frac{\sin^2(E_{\mathbf{k}'} - E_{\mathbf{k}})\, t/\hbar}{\pi(E_{\mathbf{k}'} - E_{\mathbf{k}})\, t/\hbar} \\ &= \frac{2\pi t}{\hbar} |\langle \mathbf{k}'|V|\mathbf{k}\rangle|^2 D(E_{\mathbf{k}} - E_{\mathbf{k}'}, t), \end{aligned} \tag{6.10.37}$$

where our definition of D is obvious. This quantity therefore behaves very much like $\delta(E_{\mathbf{k}} - E_{\mathbf{k}'})$, independent of time if t is large enough, in practice when $t > \hbar/\delta E$, where δE is the smallest energy change significant to the problem. We can then write the number of transitions per unit time as

$$P(\mathbf{k}', \mathbf{k}) = \frac{2\pi}{\hbar} |\langle \mathbf{k}'|V|\mathbf{k}\rangle|^2 \, \delta(E_{\mathbf{k}} - E_{\mathbf{k}'}). \tag{6.10.38}$$

To obtain this result, however, we assumed equation (6.10.36), that is, we took the electron to be initially in an unperturbed eigenstate. To be able to discuss development in time in this fashion, we must be able to write, at all times, the density matrix as [cf. equation (2.7.8) of Volume I]

$$\rho(\mathbf{r}', \mathbf{r}) = \sum_{\mathbf{k}} f(\mathbf{k}, t)\, \phi_{\mathbf{k}}^{*}(\mathbf{r}')\, \phi_{\mathbf{k}}(\mathbf{r}), \tag{6.10.39}$$

diagonal in the unperturbed eigenfunctions. We can then say there are $f(\mathbf{k}, t)$ electrons for which $c_{\mathbf{k}'} = \delta_{\mathbf{k}'\mathbf{k}}$ at time t and it follows that

$$\frac{df(\mathbf{k}, t)}{dt} = \sum_{\mathbf{k}} [P(\mathbf{k}', \mathbf{k})\, f(\mathbf{k}') - P(\mathbf{k}, \mathbf{k}')\, f(\mathbf{k})]. \tag{6.10.40}$$

In fact, since the $\psi_{\mathbf{k}}$ are the solutions of the time-dependent Schrödinger equation, Hamiltonian $H + V$, we have by definition

$$\rho(\mathbf{r}', \mathbf{r}) = \sum_{\mathbf{k}''} f(\mathbf{k}'', 0)\, \psi_{\mathbf{k}''}^{*}(\mathbf{r}', t)\, \psi_{\mathbf{k}}(\mathbf{r}, t). \tag{6.10.41}$$

In order to reconcile equations (6.10.39) and (6.10.41), Pauli was led to his "repeated random phases" assumption. We shall show that in the present case of a dilute concentration of impurities distributed at random this assumption is not necessary.

The number n of impurities will be taken to be large ($n \to \infty$ as $N \to \infty$) but $n/N \ll 1$, where N is the number of atoms in the crystal. We write the scattering potential as

$$V(\mathbf{r}) = \sum_j v(\mathbf{r} - \mathbf{R}_j), \tag{6.10.42}$$

where the $\mathbf{R}_j$ constitute a random sample of n lattice vectors. We can see from equation (6.10.35) that a typical cross-term in the expansion $\rho(\mathbf{r}', \mathbf{r})$ as given by equation (6.10.41) in terms of the $\phi_{\mathbf{k}}$, is

$$\sum_{\mathbf{k}''}{}' f(\mathbf{k}'', 0) \langle \mathbf{k}'' | V | \mathbf{k} \rangle \langle \mathbf{k}' | V | \mathbf{k}'' \rangle$$

$$\times \left(\frac{e^{i(E_{\mathbf{k}''} - E_{\mathbf{k}})t/\hbar} - 1}{E_{\mathbf{k}''} - E_{\mathbf{k}}} \right) \left(\frac{e^{i(E_{\mathbf{k}'} - E_{\mathbf{k}''})t/\hbar} - 1}{E_{\mathbf{k}'} - E_{\mathbf{k}''}} \right) \phi^*_{\mathbf{k}'}(\mathbf{r}') \phi_{\mathbf{k}}(\mathbf{r}), \tag{6.10.43}$$

where we do not allow $\mathbf{k}''$ to equal $\mathbf{k}$ or $\mathbf{k}'$.

On using Bloch's theorem, we can put

$$\langle \mathbf{k}'' | V | \mathbf{k} \rangle \langle \mathbf{k}' | V | \mathbf{k} \rangle = \sum_{ij} e^{i(\mathbf{k} - \mathbf{k}'') . \mathbf{R}_i} e^{i(\mathbf{k}'' - \mathbf{k}') . \mathbf{R}_j} \langle \mathbf{k}'' | v | \mathbf{k} \rangle \langle \mathbf{k}' | v | \mathbf{k}'' \rangle. \tag{6.10.44}$$

Let us now examine what the perfect randomness of impurity positions implies. We first propose to take an ensemble average over all possible distributions of impurities (cf. section 10.13.2 in Chapter 10 where a similar procedure is carried through). We shall assume a complete lack of correlation between the positions of different impurities; this allows two or more impurities to be on the same site but since $n/N \ll 1$ we can neglect the resultant error. The ensemble average is now accomplished by letting each $\mathbf{R}_l$ take every possible lattice vector, summing over the result and dividing by N. We shall replace this sum by an integral, an approximation which becomes exact as $N \to \infty$. Thus, taking an ensemble average over the right-hand side of equation (6.10.44) we find, $\mathscr{V}$ being the volume of the system,

$$\frac{1}{\mathscr{V}} \sum_j \int d\mathbf{R}_j \, e^{i(\mathbf{k} - \mathbf{k}') . \mathbf{R}_j} + \frac{1}{\mathscr{V}^2} \sum_{i \neq j}{}' \int d\mathbf{R}_i \, e^{i(\mathbf{k} - \mathbf{k}'') . \mathbf{R}_i} \int d\mathbf{R}_j \, e^{i(\mathbf{k}'' - \mathbf{k}') . \mathbf{R}_j}$$

$$= n \delta_{\mathbf{k},\mathbf{k}'} \quad (\mathbf{k}'' \neq 0). \tag{6.10.45}$$

Denoting the ensemble average by a bar, this shows that

$$\overline{|\langle \mathbf{k} | V | \mathbf{k}'' \rangle|^2} = n |\langle \mathbf{k} | v | \mathbf{k}'' \rangle|^2, \tag{6.10.46}$$

the direct proportionality to the impurity concentration being just what we might expect. However, despite the fact that we have shown the ensemble average of off-diagonal elements to be zero, any such cross-term might very possibly be of order n as well. We test this by taking a root-mean square. If we set

$$M_{\mathbf{k}',\mathbf{k}} = \sum_{ij} e^{i(\mathbf{k}-\mathbf{k}'').\mathbf{R}_j} e^{i(\mathbf{k}''-\mathbf{k}').\mathbf{R}_j} \quad (\mathbf{k}'' \neq \mathbf{k}, \mathbf{k}'), \tag{6.10.47}$$

we have

$$\overline{|M|^2} = \overline{\sum_{ij} e^{i(\mathbf{k}-\mathbf{k}'').(\mathbf{R}_i-\mathbf{R}_j)} \sum_{mn} e^{i(\mathbf{k}'-\mathbf{k}'').(\mathbf{R}_m-\mathbf{R}_n)}}. \tag{6.10.48}$$

By inspection, we find the principal terms to be

$$\overline{|M|^2} = \overline{\sum_{\substack{ij\\mn}} 1} + \overline{\sum_{\substack{i=n\\j=m}} e^{i(\mathbf{k}-\mathbf{k}').(\mathbf{R}_i-\mathbf{R}_j)} e^{i(\mathbf{k}'-\mathbf{k}).(\mathbf{R}_i-\mathbf{R})}}, \tag{6.10.49}$$

and hence we have

$$\overline{|M_{\mathbf{k}'\mathbf{k}}|^2} = n^2(1+\delta_{\mathbf{k}'\mathbf{k}}). \tag{6.10.50}$$

This expansion shows that, for a substantial fraction of configurations, cross-terms are of order n. However, we have not yet taken into account the summation over $\mathbf{k}''$ in equation (6.10.43). Van Hove (1958) has identified the necessary property of a perturbation giving rise to irreversible approach to equilibrium in dissipative systems as

$$\langle \mathbf{k}'|VAV|\mathbf{k}\rangle = X(\mathbf{k})\,\delta(\mathbf{k}'-\mathbf{k}) + Y(\mathbf{k}',\mathbf{k}) \quad (N \to \infty), \tag{6.10.51}$$

where Y has no delta-function singularity and A is an operator diagonal in the unperturbed representation. To show how this property of V arises in the present case, and is of consequence in the approach we have adopted, we first note that equation (6.10.43) is in fact a sum of the form

$$\sum_{\mathbf{k}''} a(\mathbf{k}'')\langle \mathbf{k}'|V|\mathbf{k}''\rangle\langle \mathbf{k}''|V|\mathbf{k}\rangle, \tag{6.10.52}$$

the form of which is of course equivalent to the left-hand side of equation (6.10.51). We shall show that (6.10.43) can be taken to be zero if $\mathbf{k}' \neq \mathbf{k}$.

Let us take a cell c, in $\mathbf{k}$-space, containing n allowed values of $\mathbf{k}''$. Since $n/N \ll 1$, these values of $\mathbf{k}''$ lie very close to one another, so that $a(\mathbf{k}'')$ is sensibly constant over c. In fact, only sums such as $\sum_i e^{i(\mathbf{k}-\mathbf{k}'').\mathbf{R}_i}$ change markedly over the cell and we accordingly consider

$$M_c = \frac{1}{n}\sum_{\mathbf{k}''}^{(c)} \sum_{ij} e^{i(\mathbf{k}-\mathbf{k}'').\mathbf{R}_i} e^{i(\mathbf{k}''-\mathbf{k}').\mathbf{R}_j}, \tag{6.10.53}$$

the average of M over the cell c. It is readily shown that if $\mathbf{k} \neq \mathbf{k}'$

$$\overline{|M_c|^2} = \frac{1}{n^2}\left[n^2 \sum_{\mathbf{k}_1'',\mathbf{k}''}^{(c)} \delta(\mathbf{k}'',\mathbf{k}_1'') + n^2 \sum_{\mathbf{k}_1'',\mathbf{k}''}^{(c)} \delta(\mathbf{k}'-\mathbf{k}_1'', \mathbf{k}-\mathbf{k}'')\right]. \tag{6.10.54}$$

The second term on the right-hand side will almost always be zero, but in any case we can see that

$$\sqrt{|\overline{M_c}|^2} \sim n^{\frac{1}{2}} \quad (\mathbf{k}' \neq \mathbf{k}), \tag{6.10.55}$$

whereas diagonal terms contribute as before. Hence, as $n \to \infty$, the root-mean squares of off-diagonal elements become negligible, showing diagonal elements are insufficient for a vanishingly small fraction of configurations only.

We have thus justified the use of equation (6.10.39) at all times, with the concomitant conclusion that equation (6.10.33) correctly describes the decay of the system to equilibrium. We should also note that, while equation (6.10.50) shows that $\overline{|M|^2} = 2(\bar{M})^2$ if $\mathbf{k} = \mathbf{k}'$, we can show, by a similar method to that by which we treated off-diagonal elements, that equation (6.10.46) can in fact be used for the vast majority of impurity configurations.

Finally, it must be remembered that equation (6.10.38) can only be used when the perturbing potential v (which is the difference between the impurity potential and the potential of the atom the impurity replaces) is weak. This is not usually the case. Nevertheless, the conclusions above still hold provided multiple scattering can be ignored (see Chapter 10, section 10.4.2), which is so if the scatterers are sufficiently dilute. Our arguments hinge on the presence of the phase-factors in equation (6.10.44), and so we must show that these same phase-factors appear when the exact single-centre transition probability amplitudes replace the first-order expressions we have used up till now.

Neglect of multiple scattering means that the coefficient $c_{\mathbf{k}}(t)$ in equation (6.10.34) may be written as the sum

$$c_{\mathbf{k}}(t) = \sum_i c_{\mathbf{k}}(\mathbf{R}_i, t), \tag{6.10.56}$$

where $c_{\mathbf{k}}(\mathbf{R}_i, t)$ would be the coefficient if only the scattering centre at $\mathbf{R}_i$ were present. It is easy to see that

$$c(\mathbf{R}_i, t) = c(0, t)\, e^{i\theta(\mathbf{R}_i)}, \tag{6.10.57}$$

where θ is independent of t; it is now evident that we must show that

$$\theta(\mathbf{R}_i) = (\mathbf{k}' - \mathbf{k}) . \mathbf{R}_i. \tag{6.10.58}$$

Since

$$(H + V)\,\psi_{\mathbf{k}'} = i\hbar \frac{\partial \psi_{\mathbf{k}'}}{\partial t}, \tag{6.10.59}$$

we find, on inserting equation (6.10.34) into this equation,

$$i\hbar \dot{c}_{\mathbf{k}}(\mathbf{R}_i, t) = \langle \phi_{\mathbf{k}}^*\, v(\mathbf{r} - \mathbf{R}_i)\, \psi_{\mathbf{k}'} \rangle\, e^{iE_{\mathbf{k}'}t/\hbar} \tag{6.10.60}$$

or, since $\psi_{\mathbf{k}'} = \phi_{\mathbf{k}'}$ at $t = 0$,

$$i\hbar\dot{c}_{\mathbf{k}}(\mathbf{R}_i, 0) = \langle\phi_{\mathbf{k}}^* v(\mathbf{r}-\mathbf{R}_i)\,\phi_{\mathbf{k}'}\rangle. \tag{6.10.61}$$

On using Bloch's theorem, this becomes

$$\begin{aligned} i\hbar\dot{c}_{\mathbf{k}}(\mathbf{R}_i, 0) &= \langle\phi_{\mathbf{k}}^* v(\mathbf{r})\,\phi_{\mathbf{k}'}\rangle\, e^{i(\mathbf{k}'-\mathbf{k})\cdot\mathbf{R}_i} \\ &= i\hbar\dot{c}_{\mathbf{k}}(0,0)\, e^{i(\mathbf{k}'-\mathbf{k})\cdot\mathbf{R}_i}, \end{aligned} \tag{6.10.62}$$

and comparison of this equation with equation (6.10.57) shows that

$$e^{i\theta(\mathbf{R}_i)} = e^{i(\mathbf{k}'-\mathbf{k})\cdot\mathbf{R}_i}. \tag{6.10.63}$$

Hence we can justify a Boltzmann equation in the collision term of which the exact single-centre transition probabilities are used, along the same lines as we have set out in detail using first-order transition probabilities.

6.10.3 Limit of validity of Boltzmann's equation

To obtain a criterion for the validity of Boltzmann's equation we first note that, to obtain equation (6.10.38) from equation (6.10.37), we took t to be large enough for us to be able to assume energy conservation as expressed by the δ-function in equation (6.10.38). In the present context this implies $t \gg \hbar/\zeta$, where ζ is the Fermi energy. On the other hand, since our derivation excluded significant multiple scattering, we cannot take $t > \tau_c$, where τ_c is the collision time, that is, the mean free time between collisions. We are thus led to the criterion

$$\tau_c \gg \frac{\hbar}{\zeta}. \tag{6.10.64}$$

An objection which might well be levelled at the above considerations is that we have overlooked the way in which the transition probability appears in the final expression (6.10.33). This contains the distribution function $f(\mathbf{k})$, which varies rapidly over a region of extent $k_B T$ when $\varepsilon_{\mathbf{k}} \simeq \zeta$ and it can accordingly be argued that the uncertainty in the energy should be considerably less than $k_B T$. This implies $\tau_c \gg \hbar/k_B T$, a condition we shall, however, show to be far too restrictive.

To accomplish this, we appeal to the form of the general expression for the conductivity when the scattering is elastic. We deal in particular with scattering by static impurities, but the purely elastic scattering by phonons, which effectively occurs at temperatures well above the Debye temperature, is also covered.

In section 6.11 we shall show that, quite generally within an independent particle framework, the conductivity is given by an expression of the form

$$\sigma = \int Q(E) \frac{\partial f_0}{\partial E} dE, \tag{6.10.65}$$

where Q is some function independent of temperature (or, in phonon scattering, depends on temperature in a way irrelevant to the present argument). We now invoke an old argument of Landau, who proposed the above expression on intuitive grounds (Peierls, 1955). From section 6.4 we note that the Boltzmann equation also yields an expression for the conductivity of the form (6.10.65), and therefore provides an explicit form for $Q(E)$ which is certainly valid provided $\hbar/\tau_0 \ll k_B T$. However, $Q(E)$ is independent of temperature and so no approximation to it can be limited in validity by a condition involving the temperature. It follows that instead of $k_B T$, the other energy of significance in the problem, namely ζ, must be involved. Equation (6.10.64) is the correct limiting condition, which is almost always satisfied in practice.

6.11 General expressions for the conductivity

It is quite possible to write down a general expression for the conductivity without resort to a Boltzmann equation or iterative approach, and therefore without loss of phase information in off-diagonal elements of the density matrix. Since for the purpose of calculating the conductivity we need only the first-order change ρ_1, from ρ_0, occasioned by a perturbation V on H, we solve the linearized equation, with $\hbar = 1$,

$$\frac{\partial \rho_1}{\partial t} = i[\rho_1, H] + i[\rho_0, V(t)] - \frac{\rho_1}{\tau}. \tag{6.11.1}$$

The last term on the right-hand side represents a weak interaction with the surroundings and gives rise to the convergence factor in the integral solution already used in Chapter 3, viz.

$$\rho_1(t) = i\,e^{-t/\tau} \int_{-\infty}^{t} e^{s/\tau} e^{iH(s-t)} [\rho_0, V(s)] e^{-iH(s-t)} ds. \tag{6.11.2}$$

On changing the variable of integration to $t-s$ we obtain

$$\rho_1(t) = i \int_0^{\infty} e^{-s/\tau} e^{-iHs} [\rho_0, V(t-s)] e^{iHs} ds, \tag{6.11.3}$$

and introducing a uniform field with explicit time dependence through

$$V(t) = e\mathscr{E}.\mathbf{r}\, e^{i\omega t} \tag{6.11.4}$$

we have

$$\rho_1(t) = e\mathscr{E}.\mathrm{e}^{-i\omega t} i \int_0^\infty \mathrm{e}^{-s/\tau}\,\mathrm{e}^{i\omega s}\,\mathrm{e}^{-iHs}\,[\rho_0, \mathbf{r}]\,\mathrm{e}^{iHs}\,ds. \tag{6.11.5}$$

We allow τ to go to infinity in what follows, on the understanding that we can always reintroduce the factor $\mathrm{e}^{-s/\tau}$ to induce convergence if necessary.

Using the theorem

$$\frac{1}{i}[A, \rho_0] = \rho_0 \int_0^\beta \mathrm{e}^{\eta H}\,\dot{A}\,\mathrm{e}^{-\eta H}\,d\eta, \tag{6.11.6}$$

equation (6.11.5) may be written as

$$-\rho_1(t) = \mathrm{e}^{i\omega t}\,\mathscr{E}.\int_0^\infty \mathrm{e}^{-i\omega s}\,\mathrm{e}^{-iHs}\,\rho_0 \int_0^\beta \mathrm{e}^{\eta H}\,\mathbf{j}\,\mathrm{e}^{-\eta H}\,d\eta\,dt, \tag{6.11.7}$$

where we have introduced the current operator $\mathbf{j} = -e\mathbf{v}$.

To obtain a steady current we imagine periodic boundary conditions are applied and take unit volume, so that the current density is

$$\langle \mathbf{j} \rangle = \mathrm{tr}\,\{\rho \mathbf{j}\} \tag{6.11.8}$$

whereupon use of equation (6.11.7) results in Ohm's law

$$j_\alpha = \sum_\beta \sigma_{\alpha\beta}\,\mathscr{E}_\beta \tag{6.11.9}$$

with the conductivity tensor given by

$$\sigma_{\alpha\beta} = \int_0^\infty \mathrm{e}^{-i\omega t} \int_0^\beta \langle j_\beta(0)\,j_\alpha(t+i\eta)\rangle\,d\eta\,dt, \tag{6.11.10}$$

where we have used the cyclic property of the trace. This is the correlation-function formula emphasized by Kubo. Other forms are

$$\sigma_{\alpha\beta}(\omega) = ei \int_0^\infty \mathrm{e}^{-i\omega t}\,\mathrm{tr}\,\{[\rho_0, x_\beta]\,j_\alpha(t)\}\,dt \tag{6.11.11}$$

and

$$\sigma_{\alpha\beta}(\omega) = e \int_0^\infty \mathrm{e}^{-i\omega t}\,\mathrm{tr}\left\{\frac{\partial \rho_0}{\partial p_\beta}\,j_\alpha(t)\right\} dt. \tag{6.11.12}$$

Equation (6.11.11) follows directly from equation (6.11.5), and equation (6.11.12) then follows on using the expression

$$\frac{\partial \rho_0}{\partial \mathbf{p}} = i[\rho_0, \mathbf{r}]. \tag{6.11.13}$$

It should be emphasized that the above is a many-body theory: $j_\alpha = ev_\alpha$ may be regarded as a shorthand form of $e\sum_i v_\alpha^i$, and $\mathbf{r} \equiv \sum_i \mathbf{r}_i$ etc., where the summations are over all particles.

We now investigate the form of the complex conjugate of the conductivity. From equation (6.11.11),

$$\sigma^*_{\alpha\beta}(\omega) = -ei \int_0^\infty e^{i\omega t} \operatorname{tr}\{[\rho_0, x_\beta] j_\alpha(t)\}^* \, dt. \tag{6.11.14}$$

But taking the trace with respect to a complete set of orthonormal functions $|s\rangle$ we find

$$\begin{aligned}\operatorname{tr}\{[\rho_0, x_\beta] j_\alpha(t)\} &= \sum_s \langle s|[\rho_0, x_\beta] j_\alpha(t)|s\rangle = -\sum_s \langle s|j_\alpha(t)[\rho_0, x_\beta]|s\rangle^* \\ &= -\operatorname{tr}\{[\rho_0, x_\beta] j_\alpha(t)\}^*.\end{aligned} \tag{6.11.15}$$

It follows immediately that

$$\sigma^*_{\alpha\beta}(\omega) = \sigma_{\alpha\beta}(-\omega). \tag{6.11.16}$$

To show the conductivity tensor is symmetric we first note the invariance in time of equilibrium correlation functions:

$$\langle A(t_1) B(t_2)\rangle = \langle A(t_1+\tau) B(t_2+\tau)\rangle \tag{6.11.17}$$

for any two Heisenberg operators A and B; this is easily shown by use of the cyclic property of the trace.

We also note that if the $|s\rangle$ are eigenfunctions of the Hamiltonian H,

$$\langle A(0) B(t)\rangle = \operatorname{tr}\{\rho_0 A(0) B(t)\} = \frac{1}{Z}\sum_{ss'} e^{-\beta\varepsilon_s} \langle s|A|s'\rangle \langle s'|B(t)|s\rangle \tag{6.11.18}$$

and also, if B is Hermitian,

$$\langle s'|B(t)|s\rangle = \langle s|B(t^*)|s'\rangle^* = \langle s^*|B^*(t^*)|s^{*\prime}\rangle. \tag{6.11.19}$$

But $|s^*\rangle$ is also an eigenfunction, eigenvalue ε_s, of H, provided we reverse any magnetic field present at the same time as we take the complex conjugate. Hence, provided A and B are either both even or both odd (j is odd), on taking complex-conjugates and reversing magnetic fields we obtain

$$\langle A(0) B(t)\rangle = \sum_{ss'} e^{-\beta\varepsilon_s} \langle s'^*|A(0)|s^*\rangle \langle s^*|B(-t)|s^{*\prime}\rangle, \tag{6.11.20}$$

so that

$$\langle A(0) B(t)\rangle_{\mathscr{H}} = \langle B(-t) A(0)\rangle_{-\mathscr{H}} = \langle B(0) A(t)\rangle_{-\mathscr{H}}, \tag{6.11.21}$$

where we have used equation (6.11.17). Application of this result to equation (6.11.10) immediately yields

$$\sigma_{\alpha\beta}(\omega, \mathscr{H}) = \sigma_{\beta\alpha}(\omega, -\mathscr{H}). \tag{6.11.22}$$

This may be regarded as an example of Onsager's reciprocal relations.

6.11.1 Independent particle formulation

If the Hamiltonian H can be reduced to a sum of N one-electron Hamiltonians H_1, we can obtain an expression for the conductivity in terms of the one-electron density matrix. The solution for f_1 exactly parallels that for ρ and so we merely replace ρ_0 in equation (6.11.11) by

$$f_0 = \frac{1}{e^{\beta(H_1-\zeta)}+1}, \tag{6.11.23}$$

ζ being chosen so that

$$\mathrm{tr}\{f_0\} = N. \tag{6.11.24}$$

An obvious case within the independent electron model where the above applies occurs when the scattering is due to static impurities or imperfections. A similar substitution is possible for purely elastic scattering by lattice vibrations. If only elastic scattering occurs, the electron energy is a constant of the motion

$$[H_1, H] = 0, \tag{6.11.25}$$

so that on writing

$$H = \sum_i H_i(\mathbf{r}_i) + H_{\mathrm{ph}}, \tag{6.11.26}$$

H_{ph} being the part of the Hamiltonian which contains phonon coordinates only, we can put

$$\rho_0 = \frac{e^{-\beta H}}{Z} = \frac{1}{Z}\exp(-\beta H_{\mathrm{ph}})\exp\left[-\beta\sum_i H_1(\mathbf{r}_i)\right] = \rho_{\mathrm{ph}}\,\rho_{\mathrm{el}}. \tag{6.11.27}$$

Equation (6.11.11) may then be written in the form

$$\sigma_{\alpha\beta} = Ne^2 i\int_0^\infty e^{-i\omega t}\,\mathrm{tr}_{\mathrm{ph}}(\rho_{\mathrm{ph}}\,\mathrm{tr}_{\mathrm{el}}\{[f_0, x_\beta]j_\alpha(t)\})\,dt, \tag{6.11.28}$$

where $\mathrm{tr}_{\mathrm{ph}}$ indicates that the trace is taken over phonon states only and $\mathrm{tr}_{\mathrm{el}}$ is similarly defined. To show this we note that, in full,

$$\sigma_{\alpha\beta} = Ne^2 i\int_0^\infty e^{-i\omega t}\,\mathrm{tr}_{\mathrm{ph}}(\rho_{\mathrm{ph}}\,\mathrm{tr}_{\mathrm{el}}\{[\rho_{\mathrm{el}}, x_\beta]j_\alpha(t)\})\,dt, \tag{6.11.29}$$

where, because of the form of H [equation (6.11.26)],

$$\rho_{el} = e^{-\beta\Sigma H(\mathbf{r}_i)}/Z_{el}. \tag{6.11.30}$$

But

$$\int_0^\infty e^{-i\omega t}\,\mathrm{tr}_{el}\{[\rho_{el}, x_\beta] j_\alpha(t)\}\,dt = \int_0^\infty e^{-i\omega t}\,\mathrm{tr}_{el}\{[f_0, x_\beta] j_\alpha(t)\}\,dt, \tag{6.11.31}$$

where the operators on the right are one-particle operators. This equation must be true since the expression on the left is exactly the same in form as the result one would obtain, for the conductivity of electrons encountering static scatterers, using the many-particle density matrix. This reduces to the one-particle solution if the particles are independent.

For certain purposes, it will be convenient to cast the solution for a static field into a rather different form. To do so we write [cf. equation (6.11.12)]

$$\sigma_{\alpha\beta} = \frac{e}{2}\int_{-\infty}^{\infty} dt\,\mathrm{tr}\left\{j_\alpha(t)\frac{\partial f_0}{\partial p_\beta}\right\} dt \tag{6.11.32}$$

and insert into this an expansion for f_0 of the form

$$f_0 = \sum_{n=0}^{\infty} \mu_n H^n. \tag{6.11.33}$$

Then using the equation

$$\frac{\mathbf{p}}{m} = \mathbf{v} = \frac{\partial H}{\partial \mathbf{p}} \tag{6.11.34}$$

and taking the trace of equation (6.11.32) with respect to eigenfunctions $|s\rangle$ of H,

$$\begin{aligned}
\sigma_{\alpha\beta} &= -\frac{1}{2}\int dt \sum_{ss'} \langle s|e^{iHt} j_\alpha e^{-iHt}|s'\rangle \sum_{n=0}^{\infty} \mu_n \sum_{j=0}^{n-1} \langle s'|H^j j_\beta H^{n-1-j}|s\rangle \\
&= -\frac{2\pi}{2}\sum_{ss'} \langle s|j_\alpha|s'\rangle\langle s'|j_\beta|s\rangle \sum_{n=0}^{\infty} \mu_n \sum_{j=0}^{n-1} \varepsilon_{s'}^j\,\varepsilon_s^{n-1-j}\,\delta(\varepsilon_s - \varepsilon_{s'}) \\
&= -\frac{1}{2}\sum_{ss'} \int dt\,\langle s|j_\alpha(t)|s'\rangle\langle s'|j_\beta|s\rangle \sum_{n=0}^{\infty} \mu_n n \varepsilon_s^{n-1},
\end{aligned} \tag{6.11.35}$$

that is, using equation (6.11.33),

$$\sigma_{\alpha\beta} = -\tfrac{1}{2}\,\mathrm{tr}\left\{\frac{\partial f_0}{\partial H}\int_{-\infty}^{\infty} j_\alpha(t) j_\beta(0)\,dt\right\}, \tag{6.11.36}$$

the desired result.

6.11.2 Weak scattering limit

Inserting the result $\partial\rho/\partial\mathbf{p} = i[\rho, \mathbf{r}]$ into equation (6.11.5) and taking matrix elements, we find

$$\langle\mathbf{k}_1|\rho_1|\mathbf{k}\rangle = e\mathscr{E}(t)\int_0^\infty dt\,\mathrm{e}^{-i\omega t}\int\langle\mathbf{k}_1|\mathrm{e}^{-iHt}|\mathbf{k}'\rangle\langle\mathbf{k}'|\frac{\partial\rho_0}{\partial\mathbf{p}}|\mathbf{k}''\rangle \times\langle\mathbf{k}''|\mathrm{e}^{iHt}|\mathbf{k}\rangle\,d\mathbf{k}'\,d\mathbf{k}''. \tag{6.11.37}$$

In the limit as the scattering goes to zero, $\rho_0 = \rho_{\mathrm{el}}\,\rho_{\mathrm{ph}}$, where ρ_{ph} is the density matrix of the scattering system. We have already discussed $\langle\mathbf{k}'|\partial\rho_{\mathrm{el}}/\partial\mathbf{p}|\mathbf{k}\rangle$ and have seen that off-diagonal elements can be taken as zero. In the weak scattering limit, therefore, off-diagonal elements contribute negligibly in the right-hand side of equation (6.11.37) and we can write

$$\langle\mathbf{k}'|\rho_1|\mathbf{k}\rangle = e\mathscr{E}(t)\int_0^\infty \mathrm{e}^{-i\omega t}\int\langle\mathbf{k}'|\mathrm{e}^{-iHt}|\mathbf{k}''\rangle\langle\mathbf{k}''|\frac{\partial\rho_0}{\partial\mathbf{p}}|\mathbf{k}''\rangle \times\langle\mathbf{k}''|\mathrm{e}^{iHt}|\mathbf{k}\rangle\,d\mathbf{k}''. \tag{6.11.38}$$

By a change of variable [cf. equations (6.11.2) and (6.11.3)] we can rewrite this as

$$\langle\mathbf{k}'|\rho_1|\mathbf{k}\rangle = \frac{ie}{m}\int_{-\infty}^{t} ds\int\langle\mathbf{k}'|\mathrm{e}^{-iH(s-t)}|\mathbf{k}''\rangle\langle\mathbf{k}''|\mathrm{e}^{iH(s-t)}|\mathbf{k}'\rangle \times\mathscr{E}(s).\langle\mathbf{k}''|\frac{\partial\rho_0}{\partial\mathbf{p}}|\mathbf{k}''\rangle\,d\mathbf{k}'' \tag{6.11.39}$$

and on differentiation with respect to time we obtain

$$\frac{\partial}{\partial t}\langle\mathbf{k}'|\rho_1|\mathbf{k}\rangle + e\mathscr{E}.\langle\mathbf{k}'|\frac{\partial\rho_0}{\partial\mathbf{k}}|\mathbf{k}\rangle\,\delta_{\mathbf{k}'\mathbf{k}} = i\langle\mathbf{k}'|[H,\rho_1]|\mathbf{k}\rangle. \tag{6.11.40}$$

Inspection shows this to be equivalent to equation (6.10.17) together with equation (6.10.28) from which we have seen that the Boltzmann equation follows.

It is worthwhile mentioning how Lax (1958) makes contact with the Boltzmann equation. The diagonal part of equation (6.11.39) is

$$f_1(\mathbf{k}) = \langle\mathbf{k}|\rho_1|\mathbf{k}\rangle = \frac{ie\mathscr{E}}{m}.\int_0^\infty \mathrm{e}^{-i\omega t}\int d\mathbf{k}'\,|\langle\mathbf{k}'|\mathrm{e}^{-iHt}|\mathbf{k}\rangle|^2\frac{\partial f(\mathbf{k}')}{\partial\mathbf{k}'}, \tag{6.11.41}$$

where we have used equation (6.10.32). On the other hand, the Boltzmann equation

$$\frac{\partial f_1}{\partial t} - \frac{e\mathscr{E}}{m}.\frac{\partial f_0}{\partial\mathbf{k}} = Kf_1 \tag{6.11.42}$$

(where K is a collision operator) may be formally solved by introducing the Green function W satisfying

$$\frac{\partial W}{\partial t} - KW = \delta(\mathbf{k}' - \mathbf{k})\, \delta(t' - t). \tag{6.11.43}$$

The solution is

$$f_1(\mathbf{k}) = \frac{e\mathscr{E}(t)}{m} \int \frac{\partial f_0(\mathbf{k}')}{\partial \mathbf{k}'} d\mathbf{k}' \int_0^\infty W(\mathbf{k}', \mathbf{k}, t)\, e^{-i\omega t}\, dt, \tag{6.11.44}$$

identical to equation (6.11.41) if

$$W(\mathbf{k}', \mathbf{k}, t) = |\langle \mathbf{k} | e^{-iHt} | \mathbf{k}' \rangle|^2. \tag{6.11.45}$$

The physical definitions of the quantities on the left and right are in fact the same, being the probability of finding the system in state $|\mathbf{k}\rangle$ at time t if it was earlier in state $|\mathbf{k}'\rangle$ at time $t = 0$. That the probability does indeed satisfy equation (6.11.43) is shown directly in the work of Van Hove (1955, 1957) on the so-called master equation governing irreversible relaxation to equilibrium. The central assumption of van Hove's work has already been discussed in section (6.10.2).

We shall also record here for later reference the general form for the transport coefficients obtained using the Green function $W(\mathbf{k}', \mathbf{k}, t)$ defined in equation (6.11.43). We have

$$\begin{aligned} f_1(\mathbf{k}) = e\mathscr{E} \cdot \int_0^\infty dt\, W(\mathbf{k}', \mathbf{k}, t) \frac{\partial f_0(\mathbf{k}')}{\partial \mathbf{k}} d\mathbf{k}'\, dt \\ + \frac{\nabla T}{T} \int_0^\infty dt\, W(\mathbf{k}', \mathbf{k}, t) \frac{\partial f_0(\mathbf{k}')}{\partial E} E_{\mathbf{k}'} v_{\mathbf{k}'}\, d\mathbf{k}' \\ + T\nabla\left(\frac{\zeta}{T}\right) \int_0^\infty dt\, W(\mathbf{k}', \mathbf{k}, t) \frac{\partial f_0(\mathbf{k}')}{\partial E} v_{\mathbf{k}'}\, d\mathbf{k}' \end{aligned} \tag{6.11.46}$$

and

$$\left.\begin{aligned} L_{11} &= -\frac{e^2}{4\pi^3} \int d\mathbf{k}\, d\mathbf{k}'\, v_{\mathbf{k}} \int_0^\infty dt\, W(\mathbf{k}', \mathbf{k}, t) \frac{\partial f_0(\mathbf{k}')}{\partial E} v_{\mathbf{k}'}, \\ L_{12} = L_{21} &= -\frac{e}{4\pi^3} \int d\mathbf{k}\, d\mathbf{k}'\, E_{\mathbf{k}} v_{\mathbf{k}} \int_0^\infty dt\, W(\mathbf{k}', \mathbf{k}, t) \frac{\partial f_0(\mathbf{k}')}{\partial E} v_{\mathbf{k}'}, \\ L_{22} &= -\frac{1}{4\pi^3} \int d\mathbf{k}\, d\mathbf{k}'\, E_{\mathbf{k}} v_{\mathbf{k}} \int_0^\infty dt\, W(\mathbf{k}', \mathbf{k}, t) \frac{\partial f_0(\mathbf{k}')}{\partial E} E_{\mathbf{k}'} v_{\mathbf{k}'}. \end{aligned}\right\} \tag{6.11.47}$$

We also note that if the scattering is purely elastic we can put $E_{\mathbf{k}'} = E_{\mathbf{k}}$ in these formulae since $W(\mathbf{k}', \mathbf{k}, t) = 0$ if $\mathbf{k}'$ and $\mathbf{k}$ refer to different energies.

6.11.3 Nyquist's theorem

In Chapter 4, the form of the fluctuation–dissipation theorem applied to magnetic spin-systems was discussed, and Nyquist's theorem, to be dealt with below, constitutes another application. In its original form (Nyquist, 1928) it provides an expression for the Johnson noise, which is the noise generated by the spontaneous fluctuations in a resistance R in equilibrium with its surroundings. The expression is

$$\overline{V^2} = 4k_B TR\Delta f, \tag{6.11.48}$$

$\overline{V^2}$ being the mean-square voltage across a resistor at temperature T, the voltage fluctuations being measured within a band of frequency width Δf.

Below we shall prove an extended Nyquist's theorem for quantum-mechanical systems and show that it reduces to equation (6.11.48) in the classical limit. The formalism will be written in such a way that the application of the fluctuation–dissipation theorem to other cases is also evident.

Let us first write down an expression giving the noise-power per unit frequency range. We take the square of the Fourier component $I(f)$ of the current to define

$$G(f) = \lim_{\tau\to\infty}\frac{2}{\tau}\left|\int_{-\tau/2}^{\tau/2} I(t)\,e^{-2\pi ift}\,dt\right|^2. \tag{6.11.49}$$

The actual average power dissipated through fluctuations, per unit frequency range, is $G(f)/R$, G being known as the power spectrum (cf. section 4.6.2 of Volume 1). The limit $\tau\to\infty$ is taken to ensure that we do indeed obtain the average power and the factor 2 incorporated to conform with the convention that the total power dissipated is obtained from

$$\lim_{\tau\to\infty}\frac{1}{\tau}\int_0^{\tau} I^2(t)\,dt = \int_0^{\infty} G(f)\,df, \tag{6.11.50}$$

so that negative frequencies do not enter into consideration.

In a quantum-mechanical treatment, an ensemble average must be taken, in addition to the time average of equation (6.11.49), since $I(t)$ must be regarded as an operator (in a classical theory $G(f)$ is, of course, unchanged by taking such an average). For a reason that will soon become clear below, let us omit the factor of 2 from equation (6.11.49) and define

$$\begin{aligned}\mathscr{G}(\omega) &= \lim_{\tau\to\infty}\frac{1}{\tau}\int_{-\tau/2}^{\tau/2} dt_2\int_{-\tau/2}^{\tau/2} dt_1\langle I(t_1)\,I(t_2)\rangle\, e^{-i\omega(t_1-t_2)}\\ &= \lim_{\tau\to\infty}\frac{1}{\tau}\int_{-\tau/2}^{\tau/2} dt_2\int_{-\tau/2}^{\tau/2} dt_1\langle I(t_1+t_2)\,I(t_2)\rangle\, e^{-i\omega t_1}.\end{aligned} \tag{6.11.51}$$

On changing the variable of integration we have left the limits of integration untouched under the assumption that the correlation function in the integrand ensures convergence.

Since the correlation function is invariant with respect to time displacement, equation (6.11.51) is just

$$\mathscr{G}(\omega) = \int_{-\infty}^{\infty} \langle I(t)\, I(0)\rangle\, \mathrm{e}^{-i\omega t}\, dt. \tag{6.11.52}$$

This equation embodies what is known as the Wiener–Khintchine theorem, relating the power dissipation to a current–current correlation function. More generally

$$\mathscr{G}_{\alpha\beta}(\omega) = \int_{-\infty}^{\infty} \langle I_\alpha(t)\, I_\beta(0)\rangle\, \mathrm{e}^{-i\omega t}\, dt \tag{6.11.53}$$

for a three-dimensional specimen. It will be seen that off-diagonal elements measure correlations between noise in the α- and β-directions. It is also evident that, because $I_\alpha(t_1)$ and $I_\beta(t_2)$ will not in general commute, the physical quantities involved will be given by a symmetric combination of their product, namely

$$G_{\alpha\beta}(f) = \mathscr{G}_{\alpha\beta}(\omega) + \mathscr{G}_{\beta\alpha}(-\omega) \quad (\omega = 2\pi f). \tag{6.11.54}$$

To obtain Nyquist's theorem, we turn to our earlier expressions for the conductivity. We first remark that equation (6.11.20) implies, for A and B Hermitian,

$$\langle AB(t)\rangle = \langle B(t^*)\, A\rangle^*, \tag{6.11.55}$$

so that equation (6.11.10) can be written as

$$\begin{aligned} \sigma_{\beta\alpha}(\omega) &= \int_0^\infty dt\, \mathrm{e}^{-i\omega t} \int_0^\beta \langle j_\beta(t - i\hbar\eta)\, j_\alpha(0)\rangle^*\, d\eta \\ &= \int_{-\infty}^0 dt\, \mathrm{e}^{i\omega t} \int_0^\beta \langle j_\beta(0)\, j_\alpha(t + i\eta\hbar)\rangle^*\, d\eta, \end{aligned} \tag{6.11.56}$$

where $\hbar$ has been inserted explicitly because we shall want to take a classical limit later on. From this equation we obtain

$$\sigma_{\alpha\beta}(\omega) + \sigma^*_{\beta\alpha}(\omega) = \int_{-\infty}^{\infty} dt\, \mathrm{e}^{-i\omega t} \int_0^\beta \langle j_\beta(0)\, j_\alpha(t + i\hbar\eta)\rangle\, d\eta. \tag{6.11.57}$$

If we suppose the specimen is in the form of a rectangular parallelepiped, dimensions l_1, l_2 and l_3, we can define the current operator as

$$I_\alpha = \frac{\mathscr{V}}{l_\alpha} j_\alpha \tag{6.11.58}$$

and the admittance as

$$Y_{\alpha\beta}(\omega) = \frac{\mathscr{V}}{l_\alpha l_\beta}\sigma_{\alpha\beta}(\omega). \tag{6.11.59}$$

Then, we can rewrite equation (6.11.57) as (setting $\mathscr{V} = 1$ again)

$$Y_{\alpha\beta}(\omega) + Y^*_{\beta\alpha}(\omega) = \int_{-\infty}^{\infty} e^{-i\omega t}\, dt \int_0^\beta \langle I_\beta(0)\, I_\alpha(t+i\hbar\eta)\rangle\, d\eta. \tag{6.11.60}$$

Now by changing the order of integration and the variable t we find

$$\int_{-\infty}^{\infty} e^{-i\omega t}\, dt \int_0^\beta \langle I_\beta(0)\, I_\alpha(t+i\hbar\eta)\rangle\, d\eta$$

$$= \int_0^\beta e^{-\omega\hbar\eta}\, d\eta \int_{-\infty+i\hbar\eta}^{\infty+i\hbar\eta} e^{-i\omega t}\langle I_\beta(0)\, I_\alpha(t)\rangle\, dt. \tag{6.11.61}$$

Provided $\langle I_\beta(0)\, I_\alpha(t)\rangle$ has no singularities between the line $(-\infty+i\hbar\eta, \infty+i\hbar\eta)$ and the real axis we therefore have

$$\int_{-\infty}^{\infty} e^{-i\omega t}\, dt \int_0^\beta \langle I_\beta(0)\, I_\alpha(t+i\hbar\eta)\rangle\, d\eta$$

$$= \frac{1-e^{-\beta\hbar\omega}}{\hbar\omega}\int_{-\infty}^{\infty} e^{-i\omega t}\langle I_\beta(0)\, I_\alpha(t)\rangle\, dt. \tag{6.11.62}$$

Equations (6.11.59), (6.11.60) and (6.11.62) express the fluctuation–dissipation theorem. It should be noted that it does not depend on the precise physical significance of I_α and I_β, and is equivalent to equation (4.6.38); equations (6.11.61) and (6.11.62) in fact constitute a concise proof of that equation.

We now have, combining equations (6.11.53), (6.11.60) and (6.11.62),

$$\frac{\hbar\omega}{1-e^{-\beta\hbar\omega}}[Y_{\alpha\beta}(\omega) + Y^*_{\beta\alpha}(\omega)] = \mathscr{G}_{\beta\alpha}(-\omega). \tag{6.11.63}$$

We note that $\mathscr{G}$ obeys the detailed balance condition

$$\mathscr{G}_{\alpha\beta}(\omega) = \mathscr{G}_{\beta\alpha}(-\omega)\, e^{\beta\hbar\omega} \tag{6.11.64}$$

and therefore, from equation (6.11.54),

$$G_{\alpha\beta}(f) = 2k_B T[Y_{\alpha\beta}(\omega) + Y^*_{\beta\alpha}(\omega)]\left(\frac{\hbar\beta\omega}{2}\coth\frac{\hbar\beta\omega}{2}\right); \quad \omega = 2\pi f. \tag{6.11.65}$$

The factor $(\hbar\beta\omega/2)\coth(\hbar\beta\omega/2)$ is clearly quantum-mechanical: as $\hbar \to 0$ it goes to unity, whereupon we obtain equation (6.11.48) on writing

$$R^2 G_{\alpha\alpha}(f) = \overline{V^2}.$$

An interesting circumstance arises when a magnetic field is applied. While it is evident that $G_{\alpha\alpha}(f)$ is real, an obvious physical requirement on the power spectrum, off-diagonal elements are not necessarily so. In fact, since

$$\sigma_{\alpha\beta}(\omega, \mathscr{H}) = \sigma_{\beta\alpha}(\omega, -\mathscr{H}) \tag{6.11.66}$$

we have

$$G_{\alpha\beta}(f) = 2k_B T[Y_{\alpha\beta}(\omega, \mathscr{H}) + Y^*_{\alpha\beta}(\omega, -\mathscr{H})]\left(\frac{\hbar\beta\omega}{2}\coth\frac{\hbar\beta\omega}{2}\right) \tag{6.11.67}$$

and off-diagonal elements are real if $\mathscr{H} = 0$ but in general there is an out-of-phase contribution in the correlations of the noise. This can readily be understood in simple terms by referring to Figure 6.9.

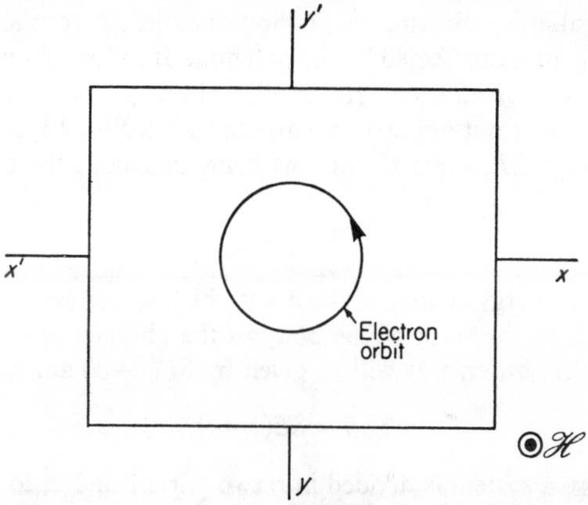

FIGURE 6.9. Illustrating how orbiting electrons in a magnetic field can give out-of-phase contributions to noise. An electron oscillating in y-direction in zero field performs classical circular orbit in the magnetic field shown, thus producing noise across the x–x' terminals out of phase with that across the y–y' terminals.

6.12 General expressions for thermal transport coefficients

Since the temperature is a statistical property of the system, one cannot represent thermal gradients by perturbations on the Hamiltonian. Accordingly, general formulae for the transport coefficients were originally obtained by an analysis of the decay of spontaneous fluctuations, with assumptions made about the local equilibrium matrices during fluctuations and the way these fluctuations decayed. While this approach is very interesting, we shall not discuss it here, for it has been pointed out by Luttinger (1964) how a much shorter derivation can be given. A gravitational potential $c^2\psi(\mathbf{r})$ has the effect of adding to the Hamiltonian a term

$$V_g = \int E(\mathbf{r})\,\psi(\mathbf{r})\,d\mathbf{r} \tag{6.12.1}$$

because an energy density $E(\mathbf{r})$ has an equivalent mass-density $E(\mathbf{r})/c^2$. Such a perturbation will cause thermal currents to flow and, just as one can relate the conductivity to the diffusion constant, so one can relate the response to ∇T to the usual thermal transport coefficients and thus, by solving for the response, find what these coefficients are.

Imagine a system to which is applied a uniform gravitational potential. Since the potential is uniform, the gravitational field is zero and the internal state of the system is unaffected by the potential. It follows from the statistical definitions of entropy and temperature that these quantities are the same as if the potential were not applied. In particular, if a closed system receives an amount of energy δE, other conditions being constant, the entropy change will be given by

$$T\delta S = \delta E_0, \tag{6.12.2}$$

where δE_0 is the energy change of the body which would accompany the same internal change in the state of the body in the absence of the gravitational potential. To first order, this will be given by $\delta E(1-\psi)$ and so

$$T\delta S = \delta E(1-\psi). \tag{6.12.3}$$

Next suppose a system is divided into two parts 1 and 2, to the first part a uniform potential ψ_1 being applied. If an amount of energy δE passes from sub-system 1 to sub-system 2, we find the entropy changes to be

$$\delta S_1 = -\frac{\delta E(1-\psi_1)}{T_1}, \quad \delta S_2 = \delta E\frac{(1-\psi_2)}{T_2}. \tag{6.12.4}$$

If the two parts are in thermodynamic equilibrium, the total entropy change must be zero and we see that rather than the two temperatures being equal we will have

$$\frac{1-\psi_1}{T_1} = \frac{1-\psi_2}{T_2}. \tag{6.12.5}$$

This relation can be extended to apply to any number of systems in thermodynamic equilibrium. In particular, by imagining a solid to be divided into small microscopic sub-systems to which we apply the equilibrium conditions, we find

$$\frac{1-\psi(\mathbf{r})}{T(\mathbf{r})} = \text{const.} \tag{6.12.6}$$

Equilibrium conditions on the chemical potential have already been discussed and from equations (6.2.8) and (6.2.11) we find

$$\frac{\zeta(\mathbf{r})-e\phi(\mathbf{r})}{T(\mathbf{r})} = \text{const}, \tag{6.12.7}$$

where $e\phi(\mathbf{r})$ is the electrical potential energy of a particle.

To first order in small potentials, therefore, equations (6.12.6) and (6.12.7) yield‡

$$\nabla\left(\frac{1}{T}\right)-\frac{\nabla\psi}{T} = 0, \quad \frac{e\mathscr{E}}{T}+\nabla\left(\frac{\zeta}{T}\right) = 0. \tag{6.12.8}$$

Next we examine the significance of these conditions on the phenomenological transport equations [cf. equations (6.2.21) and (6.2.22)]

$$\left.\begin{aligned} \mathbf{j} &= L_{11}\left[\mathscr{E}+\frac{T}{e}\nabla\left(\frac{\zeta}{T}\right)\right]+L_{12}T\nabla\left(\frac{1}{T}\right)-\tilde{L}_{12}\nabla\psi, \\ \mathbf{U} &= L_{21}\left[\mathscr{E}+\frac{T}{e}\nabla\left(\frac{\zeta}{T}\right)\right]+L_{22}T\nabla\left(\frac{1}{T}\right)-\tilde{L}_{22}\nabla\psi, \end{aligned}\right\} \tag{6.12.9}$$

where we have added terms to represent the gravitational driving forces. At the equilibrium to which equation (6.12.8) refers, **j** and **U** will be zero, and we immediately see that

$$L_{12} = \tilde{L}_{12}, \quad L_{22} = \tilde{L}_{22}. \tag{6.12.10}$$

We now solve for ρ_1, when the perturbation V_g is applied. From equations (6.11.3) and (6.11.6) we have

$$\rho_1 = \int_0^\infty e^{-t/\tau} e^{-iHt} \rho_0 \int_0^\beta e^{\eta H} \dot{V}_g e^{-\eta H} d\eta \, e^{iHt} dt. \tag{6.12.11}$$

We shall take periodic boundary conditions to ensure steady currents in the constant gravitational field

$$\psi(\mathbf{r}) = -\mathbf{g}.\mathbf{r}. \tag{6.12.12}$$

We also write the energy density as

$$E(\mathbf{r}) = \frac{1}{2}\sum_j [E_j\,\delta(\mathbf{r}-\mathbf{r}_j)+\delta(\mathbf{r}-\mathbf{r}_j)\,E_j], \tag{6.12.13}$$

‡ For a more rigorous approach than that adopted here, see Luttinger (1964).

where the total Hamiltonian (in the absence of the perturbation)

$$H = \sum_j E_j \tag{6.12.14}$$

has been symmetrically divided into contributions E_j from each particle j. For example, if

$$H = \sum_i \frac{p_i^2}{2m} + \sum_i V(\mathbf{r}_i) + \sum_{ij}' v(\mathbf{r}_{ij}) \tag{6.12.15}$$

we take

$$E_j = \frac{p_i^2}{2m} + V(\mathbf{r}_j) + \frac{1}{2}\sum_i' v(\mathbf{r}_{ij}). \tag{6.12.16}$$

Inserting equations (6.12.12) and (6.12.13) into equation (6.12.1) we find

$$V_{\mathbf{g}} = -\tfrac{1}{2}\mathbf{g}.\sum_j (E_j \mathbf{r}_j + \mathbf{r}_j E_j). \tag{6.12.17}$$

The right-hand side is very closely related to the energy current. To see this we take the continuity equation

$$\dot{E}(\mathbf{r}) + \nabla\mathbf{U}(\mathbf{r}) = 0 \tag{6.12.18}$$

or, in terms of Fourier components,

$$\dot{E}_{\mathbf{k}} = i\mathbf{k}.\mathbf{U}_{\mathbf{k}}. \tag{6.12.19}$$

Now from equation (6.12.13)

$$E_{\mathbf{k}} = \frac{1}{2}\sum_j (E_j e^{i\mathbf{k}.\mathbf{r}_j} + e^{i\mathbf{k}.\mathbf{r}_j} E_j) \tag{6.12.20}$$

and

$$\dot{E}_{\mathbf{k}} = \frac{i\mathbf{k}}{2}.\sum_j \Bigg[\frac{E_j}{2}(\mathbf{v}_j e^{i\mathbf{k}.\mathbf{r}_j} + e^{i\mathbf{k}.\mathbf{r}_j}\mathbf{v}_j) + (\mathbf{v}_j e^{i\mathbf{k}.\mathbf{r}_j} + e^{i\mathbf{k}.\mathbf{r}_j}\mathbf{v}_j)\frac{E_j}{2}$$

$$+\tfrac{1}{2}(\dot{E}_j e^{i\mathbf{k}.\mathbf{r}_j} + e^{i\mathbf{k}.\mathbf{r}_j}\dot{E}_j)\Bigg]. \tag{6.12.21}$$

On noting from equation (6.12.14) that $\sum_j \dot{E}_j = 0$ we find, by taking the limit $\mathbf{k}\to 0$, the volume average $\mathbf{U}_0/\mathscr{V}$ of the energy current is given by

$$\mathbf{U}_0 = \frac{1}{2}\sum_j [(E_j \mathbf{v}_j + \mathbf{v}_j E_j) + (\dot{E}_j \mathbf{r}_j + \mathbf{r}_j \dot{E}_j)] \tag{6.12.22}$$

and hence from equations (6.12.11) and (6.12.17)

$$\rho_1 = \mathbf{g}.\int_0^\infty e^{-t/\tau}\rho_0 \int_0^\beta \mathbf{U}_0(-t-i\eta)\,d\eta\,dt. \tag{6.12.23}$$

It is now evident that

$$L_{12}^{\alpha\beta} = \frac{1}{\mathscr{V}} \int_0^\infty e^{-t/\tau} \int_0^\beta \langle U_0^\alpha j^\beta(t+i\eta) \rangle \, d\eta \, dt, \tag{6.12.24}$$

$$L_{22}^{\alpha\beta} = \frac{1}{\mathscr{V}} \int_0^\infty e^{-t/\tau} \int_0^\beta \langle U_0^\alpha U_0^\beta(t+i\eta) \rangle \, d\eta \, dt. \tag{6.12.25}$$

Since we can solve for the linear response to $\mathscr{E}$ independently of that to $\nabla\psi$, we also note from equation (6.11.8) that

$$L_{12}^{\alpha\beta} = \frac{1}{\mathscr{V}} \int_0^\infty e^{-t/\tau} \int_0^\beta \langle j_\beta U_0^\alpha(t+i\eta) \rangle \, d\eta \, dt. \tag{6.12.26}$$

6.12.1 Symmetry of transport coefficients

The Onsager reciprocal relations

$$L_{12}^{\alpha\beta} = L_{21}^{\beta\alpha}, \quad L_{22}^{\alpha\beta} = L_{22}^{\beta\alpha} \tag{6.12.27}$$

can readily be obtained from equations (6.12.24), (6.12.25) and (6.12.26) in just the same way as we proved equation (6.11.16) for the electrical conductivity.

Finally, we note the single-particle forms for the transport coefficients which are readily proved from the above in a similar fashion to equation (6.11.36). These are

$$L_{ij} = L_{ji} = -\tfrac{1}{2} \operatorname{tr} \left[\frac{\partial f_0}{\partial H} \int_{-\infty}^\infty L_i(0) \, L_j(t) \, dt \right], \tag{6.12.28}$$

with, if $\mathbf{U}$ is the current density operator $\mathbf{U}_0/\mathscr{V}$,

$$L_1 = \mathbf{j}, \quad L_2 = \mathbf{U}. \tag{6.12.29}$$

6.12.2 Wiedemann–Franz law

The law of Wiedemann and Franz (1853) states that the ratio of thermal to electrical conductivities is a constant times the temperature, the constant being independent of the metal and known as the "Lorenz number" L. The law is fairly accurately obeyed by most metals at room temperature, and we shall now show that the Wiedemann–Franz law follows from the present theory if

(i) the scattering is purely elastic and

(ii) the electrons are independent and constitute a degenerate assembly.

The theory predicts

$$\frac{\kappa}{\sigma T} = L = \frac{\pi^2}{3} \left(\frac{k_B}{e} \right)^2, \tag{6.12.30}$$

which is, again, close to the observed value.

It is convenient to derive the law separately for two ranges of coupling:

(i) The intermediate and strong coupling régime defined such that $\hbar/\tau_c \geqslant \zeta_0(\hbar/t_0\,\zeta_0)^{\frac{1}{2}}$ where τ_c is the electron collision time and t_0 is a characteristic time over which the scattering configuration changes appreciably. This latter time will be $\sim\omega^{-1}$ where ω is a typical vibrational frequency of the lattice or impurities. Hence $t_0 \gtrsim 10^{-13}$ sec and $(\hbar/t_0\,\zeta_0)^{\frac{1}{2}} < \frac{1}{10}$. We then have

$$\frac{\hbar}{\tau_c} \geqslant \zeta_0\left(\frac{\hbar}{t_0\,\zeta_0}\right)^{\frac{1}{2}} = \left(\frac{\zeta_0\,t_0}{\kappa}\right)^{\frac{1}{2}}\frac{\hbar}{t_0} \gg \frac{\hbar}{t_0}.$$

In other words, $\tau_c \ll t_0$ and the implication is that the scattering centres do not move appreciably during an interval of the order of the collision time and we can calculate the transport coefficients assuming the scatterers remain fixed.

(ii) The weak coupling régime is defined by

$$\frac{\hbar}{\tau_c} < \zeta_0\left(\frac{\hbar}{t_0\,\zeta_0}\right)^{\frac{1}{2}} \ll \zeta_0, \tag{6.12.31}$$

implying the Boltzmann equation approach to be valid. Elastic scattering by phonons almost always falls into this category.

The most general derivation of the law based on the Boltzmann equation was given by Kohler (1941), while the extension to all scattering strengths was made by Chester and Thellung (1961). We shall follow the latter authors in discussing both régimes, particularly in the use of a generating function.

(a) *Weak coupling*. We define a generating function $\mathscr{L}(s)$ by the equation

$$4\pi^3\mathscr{L}(s) = -\int \frac{\partial f_0(\mathbf{k})}{\partial E} e^{-sE(\mathbf{k})}\,Q(\mathbf{k})\,d\mathbf{k}, \tag{6.12.32}$$

where, in terms of the Green function defined by equation (6.11.43),

$$Q(\mathbf{k}) = \int v_{\mathbf{k}} v_{\mathbf{k}'} \int_0^\infty W(\mathbf{k}',\mathbf{k},t)\,d\mathbf{k}'. \tag{6.12.33}$$

Then equation (6.11.47) shows that (remembering $E_{\mathbf{k}} = E_{\mathbf{k}'}$ since the scattering is elastic)

$$\left.\begin{aligned} L_{11} &= e^2\mathscr{L}(0), \\ L_{12} &= L_{21} = \frac{-e\partial\mathscr{L}(0)}{\partial s}, \\ L_{22} &= \frac{\partial^2\mathscr{L}(0)}{\partial s^2}. \end{aligned}\right\} \tag{6.12.34}$$

From equation (6.2.26),

$$\frac{\kappa}{\sigma T} = \frac{L_{11}L_{22}-L_{12}L_{21}}{(TL_{11})^2}, \tag{6.12.35}$$

since $\sigma = L_{11}$, and so in terms of the generating function

$$L = \frac{\kappa}{\sigma T} = \frac{1}{e^2 T^2}\frac{\partial^2 \ln \mathscr{L}(s)}{\partial s^2}\bigg|_{s=0}. \tag{6.12.36}$$

Now equation (6.12.32) may be written as

$$\mathscr{L}(s) = -\int_0^\infty \frac{\partial f_0(k)}{\partial E} e^{-sE} G(E)\, dE, \tag{6.12.37}$$

where

$$G(E) = n(E)\,\bar{Q}(E). \tag{6.12.38}$$

Here $n(E)$ is the density of states in energy and $\bar{Q}$ is the average of $Q(\mathbf{k})$ over the quantum numbers $\mathbf{k}$ at constant energy. Evidently, if $\mathbf{k}$ merely represents the usual $\mathbf{k}$-vector, $\bar{Q}(E)$ is just the integral over a constant energy surface in $\mathbf{k}$-space.

Now, as is well known and easily verifiable, for the degenerate situation $(k_B T \ll \zeta_0)$

$$\int_0^\infty \phi(E)\frac{\partial f_0}{\partial E}dE = -\phi(\zeta_0) - \frac{\pi^2}{6}(k_B T)^2 \left(\frac{\partial^2 \phi}{\partial E^2}\right)_{\zeta_0} \tag{6.12.39}$$

to good accuracy if $\phi(E)$ is slowly varying relative to $\partial f_0/\partial E$. Thus equation (6.12.37) becomes

$$\mathscr{L}(s) = G(\zeta_0)\, e^{-s\zeta_0} + \frac{\pi^2}{6}(k_B T)^2 e^{-s\zeta_0}\left[\frac{\partial^2 G(\zeta_0)}{\partial \zeta_0^2} - \frac{2s\,\partial G(\zeta_0)}{\partial \zeta_0} + s^2 G(\zeta_0)\right]. \tag{6.12.40}$$

To the same order of accuracy we may use $\ln(1+x) = x + \ldots$ to find

$$\ln \mathscr{L}(s) = -s\zeta_0 + \ln G(\zeta_0) + \frac{\pi^2}{6}(k_B T)^2 G(\zeta_0)^{-1}\left[\frac{\partial^2 G(\zeta_0)}{\partial \zeta_0^2} - 2s\frac{\partial G(\zeta_0)}{\partial \zeta_0} + s^2 G(\zeta_0)\right] \tag{6.12.41}$$

and equation (6.12.35) becomes

$$L = \tfrac{1}{3}\pi^2\left(\frac{k_B}{e}\right)^2, \tag{6.12.42}$$

which was to be demonstrated.

(b) *Strong coupling*. In the strong coupling case we must discard a Boltzmann equation approach, but on the other hand we may use the generalized single-particle formula of equation (6.11.36). Taking the trace with respect to a complete set of eigenfunctions of H, we have

$$L_{ij} = -\pi \sum_{\mathbf{k}\mathbf{k}'} \langle \mathbf{k}' | \frac{\partial f_0}{\partial H} | \mathbf{k} \rangle \langle \mathbf{k} | L_i | \mathbf{k}' \rangle \langle \mathbf{k}' | L_j | \mathbf{k} \rangle \, \delta(\varepsilon_{\mathbf{k}'} - \varepsilon_{\mathbf{k}}). \quad (6.12.43)$$

The generating function is now taken to be

$$\mathscr{L}(s) = -\sum_{\mathbf{k}} \frac{\partial f_0}{\partial E} e^{-sE} Q(\mathbf{k}), \quad (6.12.44)$$

where

$$Q(\mathbf{k}) = \sum_{\mathbf{k}'} |\langle \mathbf{k} | L_1 | \mathbf{k}' \rangle|^2 \, \delta(\varepsilon_{\mathbf{k}} - \varepsilon_{\mathbf{k}'}), \quad (6.12.45)$$

and it is easily verified that equation (6.12.34) and (6.12.36) again hold. We assume that the sum over $\mathbf{k}$ is replaceable by an integral and the proof of the Wiedemann–Franz law proceeds exactly as before.

(c) *Boltzmann statistics*. Unlike the conduction electrons in a metal, those in a semi-conductor often obey Boltzmann statistics. It may then be readily shown, under conditions that turn up in experimental situations, that $\kappa/\sigma T$ is roughly independent of temperature. However, the Wiedemann–Franz law is not valid in the sense we have defined it previously, for $\kappa/\sigma T$ varies from one semi-conductor to another. It is easy to show that this is so by the methods used above.

We now approximate the Fermi function by $e^{-\beta(E-\zeta)}$, and the Lorenz number is again given by equation (6.12.36), but the generating function now has the form:

$$\mathscr{L}(s) = \beta \int_0^\infty e^{-\beta(E-\zeta)} e^{-sE} G(E) \, dE$$

$$= \beta e^{\beta\zeta} \phi(\beta + s). \quad (6.12.46)$$

It is easy to see that, in terms of L as in equation (6.12.36)

$$\frac{\partial^2 \ln \phi(\beta)}{\partial \beta^2} = \left(\frac{Le^2}{k_B^2}\right) \beta^{-2} = \gamma \beta^{-2}, \quad (6.12.47)$$

and if γ is a pure number the most general solution is

$$\phi(\beta) = A\beta^{-\gamma} e^{\alpha\beta}, \quad (6.12.48)$$

A and a being arbitrary constants. Now from equation (6.12.37), the quantity $G(E)$ is, for $E+a>0$, the inverse Laplace transform

$$G(E) = \frac{1}{2\pi i}\int_{\sigma-i\infty}^{\sigma+i\infty} \phi(\beta)\, e^{\beta E}\, d\beta$$

$$= \frac{A}{2\pi i}\int_{\sigma-i\infty}^{\sigma+i\infty} e^{\beta(E+a)}\, \beta^{-\gamma}\, dE = \frac{A(E+a)^{\gamma-1}}{(\gamma-1)!}. \tag{6.12.49}$$

Thus taking the energy spectrum to start at zero, $a = 0$ and

$$G(E) \propto E^{\gamma-1}, \tag{6.12.50}$$

where $\gamma = Le^2/k_B^2$ is independent of the semi-conductor. It is readily verified by using the Boltzmann equation that this is not so, and γ is not a pure number.

6.13 Lattice effects

6.13.1 Heat conduction by lattice

So far we have concentrated almost exclusively on electronic conduction, assuming the lattice to be in thermal equilibrium, independent of the electronic states. There are, however, circumstances where this assumption is not permissible. Moreover, although in metals the lattice conductivity is usually swamped by the electronic contribution, the lattice must carry the whole of the heat conducted through an electrical insulator; the lattice conductivity is also, in these materials, generally much greater than in metals since in a metal electron–phonon collisions inhibit heat conduction by the lattice.

(a) *Kinetic theory.* Without examining the mechanism involved, let us enquire what can be said about lattice conductivity from elementary considerations. Then we would view the phonons as a gas, with a thermal conductivity given by

$$\kappa = \tfrac{1}{3}Cvl, \tag{6.13.1}$$

where C is the specific heat and v and l respectively the mean velocity and mean free path. We might expect to obtain the right order of magnitude for v by setting it equal to the velocity of sound v_s in the lattice when we obtain

$$\kappa = \tfrac{1}{3}Cv_s l. \tag{6.13.2}$$

At low temperatures, as we saw in Chapter 3, the lattice specific heat varies as T^3 and l becomes independent of phonon–phonon collisions, depending instead on scattering by impurities or boundaries, so that, from equation (6.13.2) we expect $\kappa \propto T^3$, which is observed.

In the high-temperature régime, C becomes independent of T and if we suppose the mean free path of the phonons varies as T^{-1}, as does that of the electrons, we get $\kappa \propto T^{-1}$, which is roughly correct. Actually, if the conductivity is limited by phonon–phonon collisions, the use of a mean free path is dubious, as we shall see below, but there are situations where the concept has some validity, as, for example, when the thermal conductivity is partly determined by collisions of phonons with magnons in a magnetic system. In any case, it is better to write $v\tau$ instead of l, τ being a relaxation time, and with this substitution understood we shall now justify equation (6.13.2) in a semi-classical framework.

(b) *Boltzmann equation.* We apply considerations similar to those used in setting up the Boltzmann equation in Chapter 2, section 2.9, following the presentation of Peierls (1955). This now means that we are implicitly dealing with wave packets of phonons which travel with group velocity $d\omega/d\mathbf{k}$, with frequency spread, say, of $\Delta\omega$. We shall actually consider the occupation numbers for pure phonon states and so the first limitation on our considerations is that the occupation numbers must not change appreciably with energy in the range $\hbar\Delta\omega$. In fact, such an appreciable change takes place over an energy range $k_B T$ so that our considerations will be valid provided

$$k_B T > \hbar\Delta\omega = \hbar\frac{\partial\omega}{\partial\mathbf{k}}.\Delta\mathbf{k}. \tag{6.13.3}$$

On the other hand, the wave packet will extend over a distance Δl given by the uncertainty relation

$$\Delta l \Delta k \sim 1 \tag{6.13.4}$$

and so we must have

$$\Delta l > \frac{\hbar v_s}{k_B T}, \tag{6.13.5}$$

where we have written the velocity of sound as an estimate of the group velocity. The second restriction we must impose is that the temperature does not vary appreciably over the extent Δl of the wave packet, in fact equation (6.13.5) and this condition can both be satisfied by a considerable margin in all reasonable cases.

We now suppose the phonon occupation number $n(\mathbf{k})$ to be a slowly varying function of position in the presence of the thermal gradient ∇T. The rate of change occasioned by the motion of the phonons of group velocity $\mathbf{v}_\mathbf{k}$ is therefore

$$\frac{\partial n}{\partial t} = -\frac{\partial n}{\partial \mathbf{r}}.\mathbf{v}_\mathbf{k} = -\frac{\partial n}{\partial \mathbf{r}}.\frac{\partial\omega}{\partial\mathbf{k}}. \tag{6.13.6}$$

We now write

$$n = n_0 + n_1, \tag{6.13.7}$$

where n_1 is proportional to ∇T and n_0 is the distribution in the absence of a thermal gradient:

$$n_0 = \frac{1}{e^{\beta\omega}-1} \quad \left(\beta = \frac{\hbar}{k_B T}\right). \tag{6.13.8}$$

Then, to first order in ∇T

$$\frac{\partial n}{\partial \mathbf{r}} = \nabla T \frac{\partial n_0}{\partial T} = \nabla T \frac{e^{\beta\omega}}{(e^{\beta\omega}-1)^2} \frac{\hbar\omega}{k_B T^2} = \nabla T n_0(n_0+1) \frac{\hbar\omega}{k_B T^2} \tag{6.13.9}$$

we now have

$$\frac{\partial n}{\partial t} = -\frac{\partial n_0}{\partial T} v_{\mathbf{k}} . \nabla T. \tag{6.13.10}$$

This is cancelled by the rate of change due to collisions. If a relaxation time can be defined, we may write

$$n_1 = \tau \frac{\partial n}{\partial t} = -\nabla T . \mathbf{v}_{\mathbf{k}} \frac{\partial n}{\partial t} \tau, \tag{6.13.11}$$

so that the thermal current becomes

$$\mathbf{Q} = \sum_{\mathbf{k}} n_1 \hbar\omega_{\mathbf{k}} \mathbf{v}_{\mathbf{k}} = -\nabla T . \sum_{\mathbf{k}} \mathbf{v}_{\mathbf{k}} \frac{\partial n}{\partial t} \hbar\omega_{\mathbf{k}} v_{\mathbf{k}} \tau. \tag{6.13.12}$$

Since

$$\mathbf{Q} = -\kappa \nabla T, \tag{6.13.13}$$

equation (6.13.12) shows that, for an isotropic medium such as a cubic crystal,

$$\kappa = \frac{1}{3} \sum_{\mathbf{k}} v_{\mathbf{k}}^2 \tau \frac{\partial n}{\partial T} \hbar\omega_{\mathbf{k}} = \frac{1}{3} \sum_{\mathbf{k}} \left| \frac{\partial \omega}{\partial \mathbf{k}} \right| \frac{\partial n}{\partial T} \hbar\omega_{\mathbf{k}} l, \tag{6.13.14}$$

where $l = v_{\mathbf{k}} \tau$. In the Debye approximation $v_{\mathbf{k}} \equiv |\partial\omega/\partial\mathbf{k}| = v_s$ and if we put

$$C = \sum_{\mathbf{k}} \frac{\partial n(\mathbf{k})}{\partial T} \hbar\omega_{\mathbf{k}} \tag{6.13.15}$$

we obtain equation (6.13.2).

(c) *Phonon–phonon collisions.* We do not know, *a priori*, that a perfect lattice forms a linear dissipative system; the observed finite conductivity might be a boundary effect, say. If not, it must be due to phonon–phonon collisions. It is easy to see that normal phonon–phonon processes provide no thermal resistance. Equation (6.13.13) shows that κ would be infinite if a heat current $\mathbf{Q}$ were possible in the absence of a thermal gradient. In that case, for example, a heat current established in a ring which is subsequently thermally isolated would circulate for ever. To obtain a finite conductivity, we require the energy in the current to be eventually dissipated through the system with, in particular, the total crystal momentum

$$\mathbf{P} = \sum n(\mathbf{k})\,\mathbf{k} \tag{6.13.16}$$

returning to its equilibrium value of zero.‡ By definition, normal processes conserve $\mathbf{P}$ and so umklapp processes must be involved [cf. equations (6.13.21) and (6.13.22) below].

To illustrate in detail the role of phonon–phonon collisions, we shall use the Boltzmann equation which we can write as

$$\left.\frac{\partial n_{\mathbf{k}}}{\partial t}\right|_{\text{coll}} = \nabla T n_0(n_0+1)\frac{\hbar\omega}{k_B T^2}\frac{\partial\omega}{\partial\mathbf{k}}. \tag{6.13.17}$$

To calculate the rate of change due to collisions, we suppose that the anharmonic perturbation V will induce transitions from state i to state f according to the golden rule

$$\frac{dP}{dt} = \frac{2\pi}{\hbar}|\langle i|V|f\rangle|^2\,\delta(\varepsilon_i-\varepsilon_f). \tag{6.13.18}$$

It then follows that

$$\left.\frac{dn_{\mathbf{k}}}{dt}\right|_{\text{coll}} = \frac{2\pi}{\hbar}\sum_{if} p_i|\langle i|V|f\rangle|^2(n_{\mathbf{k}}^f-n_{\mathbf{k}}^i)\,\delta(\varepsilon_i-\varepsilon_f), \tag{6.13.19}$$

where p_i is the probability of the state $|i\rangle$ being occupied and $n_{\mathbf{k}}^i$ and $n_{\mathbf{k}}^f$ are the numbers of phonons, of wave vector $\mathbf{k}$, in states i and f respectively.

We shall work to lowest order in the perturbation, which from section 3 of Chapter 3 we recall involves three-phonon processes: equation (3.1.10) gives

$$V = \sum_{\substack{\mathbf{k}_1\mathbf{k}_2\mathbf{k}_3\\ \sigma_1\sigma_2\sigma_3}} V\begin{pmatrix}\sigma_1\,\sigma_2\,\sigma_3\\ \mathbf{k}_1\,\mathbf{k}_2\,\mathbf{k}_3\end{pmatrix}(a_{\mathbf{k}_1\sigma_1}+a^\dagger_{-\mathbf{k}_1\sigma_1})(a_{\mathbf{k}_2\sigma_2}+a^\dagger_{-\mathbf{k}_2\sigma_2})(a_{\mathbf{k}_3\sigma_3}+a^\dagger_{-\mathbf{k}_3\sigma_3}). \tag{6.13.20}$$

We first note that $|n_{\mathbf{k}}^f-n_{\mathbf{k}}^i| = 1$ if non-zero, and secondly, since the energy change in the processes we consider must be zero, processes in which all three phonons are created or destroyed can be disregarded. We are left with four possibilities; the phonon $\mathbf{k}\sigma$ merges with a second phonon to produce a third,

‡ Note that at long wavelengths $\omega = v_s k$, implying $Q \propto P$ at low temperatures.

a phonon splits into a phonon $\mathbf{k}\sigma$ and a third phonon, two phonons produce the phonon $\mathbf{k}\sigma$ and the phonon $\mathbf{k}\sigma$ splits into two other phonons. The former two processes carry double weight, as Figure 6.10 shows.

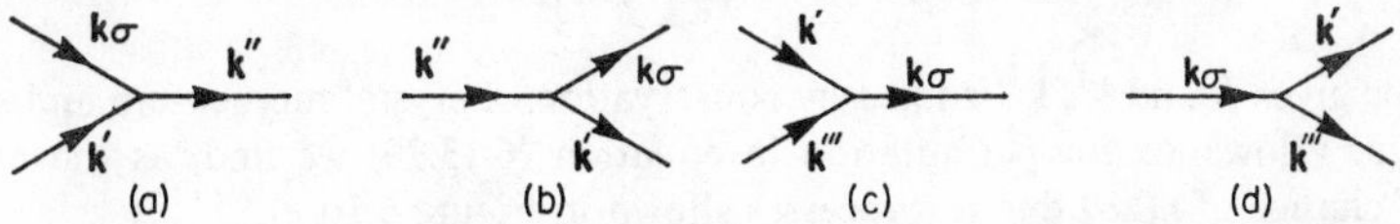

FIGURE 6.10. Three-phonon processes contributing to collision term in the Boltzmann equation. (a) Two phonons merge to produce third. (b) Phonon splits into two. (c) Two phonons combine to produce phonon $\mathbf{k}\sigma$. (d) Phonon $\mathbf{k}\sigma$ splits into two. (a) and (b) carry double weight because label $\mathbf{k}\sigma$ can be assigned in two ways.

We should also note that, in accordance with the rules of section 3.9, for processes shown in (a) and (b)

$$\mathbf{k}+\mathbf{k}'+\mathbf{k}'' = \mathbf{K}, \tag{6.13.21}$$

whereas for those shown in (c) and (d)

$$\mathbf{k}'+\mathbf{k}'''+\mathbf{k} = \mathbf{K}. \tag{6.13.22}$$

We select, to exemplify the evaluation of the contributions to the collision term, the processes of Figure 6.10(a). We have

$$\begin{aligned}
2\pi \sum_{if} |\langle f|a^{\dagger}_{\mathbf{k}'\sigma'}a_{\mathbf{k}\sigma}a_{\mathbf{k}''\sigma''}|i\rangle|^2 p_i \left|V\begin{pmatrix}\sigma\,\sigma'\,\sigma''\\ \mathbf{k}\,\mathbf{k}'\,\mathbf{k}''\end{pmatrix}\right|^2 \delta(\omega_{\mathbf{k}\sigma'}-\omega_{\mathbf{k}\sigma'}-\omega_{\mathbf{k}''\sigma''})\\
= 2\pi \sum_{if} \langle i|a^{\dagger}_{\mathbf{k}''\sigma''}a^{\dagger}_{\mathbf{k}\sigma}a_{\mathbf{k}'\sigma'}|f\rangle\langle f|a^{\dagger}_{\mathbf{k}'\sigma'}a_{\mathbf{k}\sigma}a_{\mathbf{k}''\sigma''}|i\rangle\\
\times p_i \left|V\begin{pmatrix}\sigma\,\sigma'\,\sigma''\\ \mathbf{k}\,\mathbf{k}'\,\mathbf{k}''\end{pmatrix}\right|^2 \delta(\omega-\omega'-\omega'')\\
= 2\pi \sum_{if} \langle i|(a^{\dagger}_{\mathbf{k}''\sigma''}a_{\mathbf{k}''\sigma''})(a^{\dagger}_{\mathbf{k}\sigma}a_{\mathbf{k}\sigma})(a_{\mathbf{k}'\sigma'}a^{\dagger}_{\mathbf{k}\sigma})|i\rangle\\
\times p_i \left|V\begin{pmatrix}\sigma\,\sigma'\,\sigma''\\ \mathbf{k}\,\mathbf{k}'\,\mathbf{k}''\end{pmatrix}\right|^2 \delta(\omega-\omega'-\omega'')\\
= 2\pi \sum_{if} \langle i|n_{\mathbf{k}''\sigma''}n_{\mathbf{k}\sigma}(n_{\mathbf{k}'\sigma'}+1)|i\rangle p_i \left|V\begin{pmatrix}\sigma\,\sigma'\,\sigma''\\ \mathbf{k}\,\mathbf{k}'\,\mathbf{k}''\end{pmatrix}\right|^2 \delta(\omega-\omega'-\omega''),
\end{aligned} \tag{6.13.23}$$

where we have introduced the number operators

$$n_{\mathbf{k}\sigma} = a^{\dagger}_{\mathbf{k}\sigma}a_{\mathbf{k}\sigma} = a_{\mathbf{k}\sigma}a^{\dagger}_{\mathbf{k}\sigma}-1. \tag{6.13.24}$$

In equation (6.13.23) we have the thermal average of products of occupation number operators, which, if none of the three vectors $\mathbf{k}$, $\mathbf{k}'$ and $\mathbf{k}''$ are equal to either of the others, can be replaced by products of the thermal averages to give

$$2\pi n_{\mathbf{k}'\sigma'} n_{\mathbf{k}\sigma}[n_{\mathbf{k}'\sigma'}+1]\left|V\begin{pmatrix}\sigma\,\sigma'\,\sigma''\\ \mathbf{k}\,\mathbf{k}'\,\mathbf{k}''\end{pmatrix}\right|^2 \delta(\omega-\omega'-\omega''). \tag{6.13.25}$$

For given $\mathbf{k}$ and $\mathbf{k}'$, $\mathbf{k}''$ is fixed by conservation of crystal momentum and so, after allowance for permutation in equation (6.13.20) we find, as the rate of change of $n(\mathbf{k}\sigma)$ due to processes shown in Figure 6.10(a),

$$-\frac{3}{\pi^2}\sum_{\sigma'\sigma''}\int d\mathbf{k}'\left|V\begin{pmatrix}\sigma\,\sigma'\,\sigma''\\ \mathbf{k}\,\mathbf{k}'\,\mathbf{k}''\end{pmatrix}\right|^2 n''n(n'+1)\,\delta(\omega-\omega'-\omega''), \tag{6.13.26}$$

where we have replaced the sum over $\mathbf{k}'$ by $(1/8\pi^3)\int d\mathbf{k}'$ and written $n(\mathbf{k}'\sigma') = n'$, etc. It should be noted that $n = n(\mathbf{k}\sigma)$ diminishes in consequence of the processes we have considered.

The contributions of other processes are similarly calculable and we obtain

$$\begin{aligned}\left.\frac{dn\,(\mathbf{k}\sigma)}{dt}\right|_{\text{coll}} = \frac{3}{\pi^2}\int d\mathbf{k}'\Bigg\{\sum_{\sigma'\sigma''}&\left|V\begin{pmatrix}\sigma\,\sigma'\,\sigma''\\ \mathbf{k}\,\mathbf{k}'\,\mathbf{k}''\end{pmatrix}\right|^2[(n+1)(n'+1)\,n''-nn'(n''+1)]\\ &\times\delta(\omega+\omega'-\omega'')-\frac{1}{2}\sum_{\sigma'\sigma'''}\left|V\begin{pmatrix}\sigma\,\sigma'\,\sigma'''\\ \mathbf{k}\,\mathbf{k}'\,\mathbf{k}'''\end{pmatrix}\right|^2\\ &\times[(n+1)\,n'n'''-n(n'+1)\,(n'''+1)]\,\delta(\omega-\omega'-\omega''')\Bigg\},\end{aligned} \tag{6.13.27}$$

where $\mathbf{k}''$ and $\mathbf{k}'''$ are given in terms of $\mathbf{k}$ and $\mathbf{k}'$ by equations (6.13.21) and (6.13.22), $\mathbf{K}$ taking all possible values.

Since in equilibrium, $dn/dt|_{\text{coll}}$ must vanish, we shall follow Peierls (1955) in writing

$$n_1 = n_0(n_0+1)\,g(\mathbf{k}\sigma) \tag{6.13.28}$$

when to first order in g the Boltzmann equation (6.13.17) becomes

$$\begin{aligned}&n_0(n_0+1)\frac{\hbar\omega}{k_B T^2}\frac{\partial\omega}{\partial\mathbf{k}}.\nabla T\\ &\quad=\frac{3}{\pi^2}\int d\mathbf{k}'\Bigg\{\sum_{\sigma'\sigma''}\left|V\begin{pmatrix}\sigma\,\sigma'\,\sigma''\\ \mathbf{k}\,\mathbf{k}'\,\mathbf{k}''\end{pmatrix}\right|^2(n_0+1)\,(n_0'+1)\,n_0''(g''-g-g')\,\delta(\omega-\omega'-\omega'')\\ &\qquad-\frac{1}{2}\sum_{\sigma'\sigma'''}\left|V\begin{pmatrix}\sigma\,\sigma'\,\sigma'''\\ \mathbf{k}\,\mathbf{k}'\,\mathbf{k}'''\end{pmatrix}\right|^2(n_0+1)\,n_0'n_0'''(g'+g'''-g)\,\delta(\omega-\omega'-\omega''')\Bigg\}.\end{aligned} \tag{6.13.29}$$

(i) *High temperatures.* At high temperatures, (n_0+1), etc. can be replaced by n_0, etc. throughout equation (6.13.29) and since $n_0 \propto T$ in this régime we see that we must have $n_1 \propto 1/T$. Then $\mathbf{Q} = \sum n_1 \hbar\omega_{\mathbf{k}}\, \partial\omega/\partial\mathbf{k}$ is also proportional to T^{-1} and hence $\kappa \propto T^{-1}$, a result we got at very simply earlier. On the other hand, a similar argument, using instead of the three-phonon processes the four-phonon processes of the next term in the perturbation theory will give $\kappa \propto T^{-2}$. Actually, three-phonon processes will presumably dominate and experiment does indicate the rough validity of $\kappa \propto 1/T$, but we can see that the decrease with temperature must be more rapid than this, lying between T^{-1} and T^{-2}.

(ii) *Low temperatures.* As the temperature is lowered, the frequency of umklapp processes decreases much more rapidly than that of normal processes, for most phonons become long wavelength ones. Any umklapp processes in equation (6.13.21), for example, must involve a phonon with k (or k' or k'')$>|\mathbf{K}|/4$ and so we can write $n_0 \simeq e^{-\beta\hbar\omega}$. In the Debye model, $n_0 \simeq \exp(-\gamma\Theta_D/T)$; $\gamma \simeq \frac{2}{3}$. The frequency of umklapp processes therefore also decreases roughly as $\exp(-\gamma\Theta_D/T)$ with temperature. At low temperatures, we thus expect the distribution to be pretty well in equilibrium as far as normal processes are concerned. In the absence of driving forces (zero left-hand side) and umklapp processes, equation (6.13.29) has the solution, if we deal specifically with the x-direction,

$$g = Ck_x \quad (C \text{ a constant}), \tag{6.13.30}$$

because of equations (6.13.21) and (6.13.22), and so, at low temperatures, this should be a good approximation for lattices of cubic symmetry with ∇T in the x-direction. The constant C may be obtained by including umklapp processes in the right-hand side of equation (6.13.29), the contribution of normal processes vanishing by virtue of the solution for g adopted. C will then depend (weakly, it is hoped) on the value of $\mathbf{k}$ chosen, and Peierls suggests that one obtains C by multiplying through equation (6.13.29) by k_x and summing over all $\mathbf{k}$. C is then found by considering the rate of change of P; this method is doubly satisfactory in that the contribution of normal processes now vanishes not only for our choice of g but for the exact solution as well.

We see clearly from the discussion above that while normal processes do not, of themselves, give a finite conductivity, they are nevertheless important in maintaining equilibrium against fluctuations for which P is conserved. Other instances are the following. An effect which becomes important at low temperatures, when umklapp processes are rare, is the scattering of crystal vibrations by isotopes and other crystal imperfections. Neglecting normal phonon–phonon processes we find $\kappa \to \infty$ because the imperfections are inefficient as scatterers of long waves. This difficulty disappears when the normal processes are taken into account because these processes limit the amount of energy, otherwise unbounded, that the long waves can carry. Our other example is the size effect already mentioned, where the appropriate mean free path for insertion into equation (6.13.14) depends on the dimension of the crystal. This also becomes noticeable at low temperatures. Since normal processes are much more frequent than umklapp processes, the former should again have appreciable effect and it is indeed found that the character of the size effect is considerably modified by their presence (Klemens, 1951).

6.13.2 Phonon drag

At low temperatures phonon–phonon umklapp processes become too infrequent for us to suppose the phonon system relaxes back to equilibrium between each electron–phonon collision, and the phonon distribution can be said to be dragged along by the electronic current. At very low temperatures indeed we can presumably neglect phonon–phonon umklapp processes altogether and in such circumstances the phonon distribution can be expected to have a drift velocity equal to the electronic drift velocity $\langle\mathbf{v}\rangle$, because the electrons and phonons will then have the same relation to one another as in equilibrium in the absence of a field, collisions being as likely to accelerate as to decelerate the electrons. If C_L is the lattice specific heat, it is easy to see that there will be a lattice heat current of the order of

$$\mathbf{Q}_L \sim C_L T\langle\mathbf{v}\rangle. \tag{6.13.31}$$

Since $\mathbf{Q} = \Pi\mathbf{j}$ and $\mathbf{j} = ne\langle\mathbf{v}\rangle$ we obtain a Peltier coefficient due to the lattice:

$$\Pi_L \sim \frac{C_L T}{ne} \tag{6.13.32}$$

and a contribution to the thermoelectric power of

$$S_L = \frac{\Pi_L}{T} \sim \frac{C_L}{ne}. \tag{6.13.33}$$

This effect of phonon drag is known as the Gurevich effect. From equation (6.13.33) we expect, in normal metals, a large and negative thermoelectric power (the sign being fixed by the sign of e) and varying as T^3 instead of linearly with T as predicted by equation (6.9.14). The thermoelectric power does show anomalous behaviour at low temperatures—in fact often behaving far more remarkably than we have suggested, as Figure 6.11 shows. The positive thermoelectric powers of rubidium and caesium can be understood as originating in electron–phonon umklapp processes, for which one has

$$\mathbf{k}-\mathbf{k}'-\mathbf{q}=\mathbf{K},$$

$\mathbf{q}$ being the phonon wave-vector.

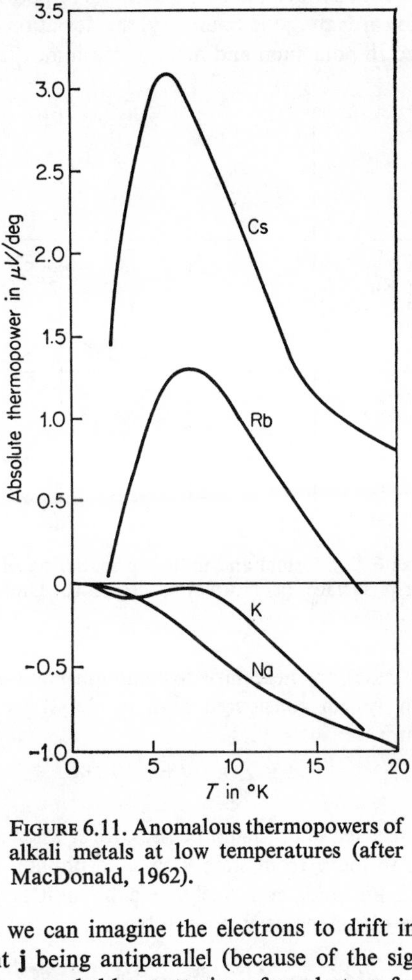

FIGURE 6.11. Anomalous thermopowers of alkali metals at low temperatures (after MacDonald, 1962).

In Figure 6.12 we can imagine the electrons to drift in the direction of $\mathbf{k}'-\mathbf{k}$, the current $\mathbf{j}$ being antiparallel (because of the sign of e) to $\mathbf{k}-\mathbf{k}'$. The current will be retarded by scattering of an electron from $\mathbf{k}$ to $\mathbf{k}'$. In (a), a normal process, $\mathbf{q}$ is antiparallel to $\mathbf{j}$, and so phonon drag results in a thermoelectric power of the usual sign. In (b), however, we see an umklapp process resulting in the emission of a phonon, carrying energy in the *same* direction as $\mathbf{j}$. The effect will clearly be sensitive to the shape of the Fermi surface, the nearer the Fermi surface approaches the zone boundary, the more likely an umklapp process. Thus we conclude from Figure 6.11 that while sodium has a spherical Fermi surface the Fermi surfaces of potassium, rubidium and caesium bulge towards the zone boundary, the deviation from a sphere being least pronounced in potassium and most in caesium.

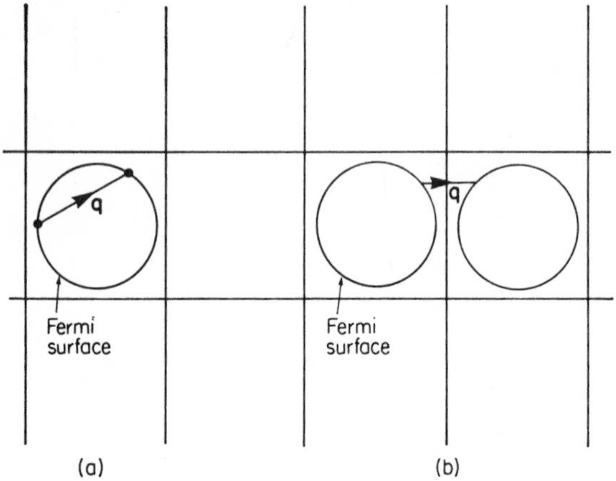

FIGURE 6.12. Normal and umklapp scattering of electrons at Fermi surface. (a) Normal processes. (b) Umklapp processes.

Umklapp processes are necessary to maintain equilibrium of the entire electron–phonon system considered as a whole. If we consider the total crystal momentum

$$\mathbf{P} = \sum_{\mathbf{q}} n(\mathbf{q})\,\mathbf{q} + \sum_{\mathbf{k}} f(\mathbf{k})\,\mathbf{k}, \tag{6.13.34}$$

$\mathbf{P}$ is evidently conserved in normal processes, for if when an electron interacts with the lattice, to move from state $\mathbf{k}$ to $\mathbf{k}'$, a phonon of wave-vector $\mathbf{q}$ is absorbed, $\mathbf{k}' = \mathbf{k}+\mathbf{q}$, and, conversely, if a phonon is created, $\mathbf{k}' = \mathbf{k}-\mathbf{q}$. Further, the rate of change of $\mathbf{P}$ is given by

$$\frac{\partial \mathbf{P}}{\partial t} = \sum_{\mathbf{k}} \mathbf{k}\frac{\partial f}{\partial t} = \frac{e\mathscr{E}}{\hbar}\cdot\sum_{\mathbf{k}} \frac{\partial f_0}{\partial \mathbf{k}}\mathbf{k} = -\frac{ne\mathscr{E}}{\hbar} \tag{6.13.35}$$

and no steady state is obtained.

The umklapp processes we would immediately expect to explain equilibrium are the electron–phonon processes we have spoken of already, but we are then in difficulties with metals such as sodium. Actually there are other non-conserving crystal-momentum processes, viz. electron–electron umklapp processes. Treating the Coulombic interactions as a small perturbation, the transition matrix element for electrons scattered as $\mathbf{k}_1 \to \mathbf{k}'_1$ and $\mathbf{k}_2 \to \mathbf{k}'_2$ is

$$\int \psi^*_{\mathbf{k}_1}(\mathbf{r}_1)\, \psi^*_{\mathbf{k}_2}(\mathbf{r}_2) \frac{e^2}{r_{12}} \psi'_{\mathbf{k}_1}(\mathbf{r}_1)\, \psi'_{\mathbf{k}_2}(\mathbf{r}_2)\, d\mathbf{r}_1\, d\mathbf{r}_2. \qquad (6.13.36)$$

As we discussed in Chapter 2, in a treatment of electrons by an independent particle method the Coulombic interaction e^2/r_{12} should presumably be replaced by a screened interaction, say something like $(e^2/r_{12}) \exp(-\lambda r_{12})$ and, in fact, this is what we require to be able to treat the electronic interaction as a small perturbation. However, the exact form of the interaction we would use is not material here: whatever function $V(r_{12})$ we substitute instead of e^2/r_{12} in the expression (6.13.36), we can see, on using Bloch's theorem, that it vanishes unless

$$\mathbf{k}'_1 + \mathbf{k}'_2 - \mathbf{k}_1 - \mathbf{k}_2 = \mathbf{K}.$$

In passing we see that in a free electron gas, where only normal processes are possible, the collisions have no effect on the conductivity, which is the same as if the electrons were non-interacting, but when a lattice of ions is introduced this is not true.

To see how the frequency of collisions depends on temperature, we suppose the $\mathbf{k}$'s label quasi-particle states with a Fermi distribution significantly different from a step function over a region, near the Fermi energy $\zeta, k_B T$ in extent. Evidently since energy must be conserved and the Pauli principle observed, both colliding electrons must lie within this region, which contains $N(\zeta) k_B T$ states, N being the electronic density of states. The number of combinations of electrons we can take to collide is therefore proportional to T^2, and this is what we expect for the temperature dependence of the frequency dependence of collision. On the other hand, Bloch, on the basis of the Debye theory, has estimated the electrical conductivity of a system of non-interacting electrons in the presence of phonons to vary as T^5 (see, for example, Wilson, 1953). This is in good agreement with experiment, and is an indication that the electron–electron umklapp processes are not the dominant processes here.

We can see, then, that, while the manifestations of phonon drag can be understood, there remains the problem of what processes are principally responsible for equilibrium. We have considered electron–phonon umklapp processes and electron–electron umklapp processes, but mention should also be made of phonon–phonon umklapp processes, for it is maintained by Klemens (1951) that even at low temperatures the inclusion of anharmonicity, if properly treated, is sufficient to maintain a steady state.

6.13.3 Equivalence of the Boltzmann equation and correlation function theory

To conclude the chapter, we shall briefly comment here on the relation between the approach based on the linearized Boltzmann equation

$$\nabla T n_0(n_0+1)\frac{\hbar\omega}{k_B T^2}\frac{\partial\omega}{\partial\mathbf{k}} = \frac{2\pi}{\hbar}\sum_{if} p_i |\langle i|V|f\rangle|^2 (n_{\mathbf{k}}^f - n_{\mathbf{k}}^i)\,\delta(\varepsilon_i - \varepsilon_f) \tag{6.13.37}$$

and the correlation function theory. In particular, we shall now show that the Boltzmann equation approach leads to the same value of the thermal conductivity as the approach due to Mori, Kubo and others when anharmonicity can be treated as a weak perturbation.

We follow the approach used for electrical conductivity in section 6.10.1, where we use the equation for ρ_1, the first-order deviation of the density matrix operator from its equilibrium value ρ_0. However, instead of the potential gradient applied there, we apply a temperature gradient in the present case, which cannot be treated as a perturbation on the original Hamiltonian. Nevertheless, a similar treatment is possible. We formally write

$$\mathbf{Q} = -\kappa\nabla T = \mathrm{tr}\{\rho_1 \mathbf{U}\} \tag{6.13.38}$$

and we compare this with the exact expression

$$\kappa = \frac{1}{T}\int_0^\infty dt\, \mathrm{e}^{-t/\tau}\int_0^\beta \langle U(0)\, U(t+i\eta)\rangle\, d\eta \tag{6.13.39}$$

obtained in equation (6.12.25).

These two expressions are equivalent if we take

$$\rho_1 = -\frac{1}{T}\int_0^\infty dt\, \mathrm{e}^{-t/\tau}\int_0^\beta \rho_0\, \mathbf{U}(t+i\eta)\,.\,\nabla T\, d\eta. \tag{6.13.40}$$

We should emphasize that our considerations depend only on this formal definition of ρ_1, which we see leads to the linearized Boltzmann equation.

Equation (6.13.40) for ρ_1 may be written in the alternative form

$$\rho_1 = \mathrm{e}^{-t/\tau}\int_{-\infty}^{t} ds\, \mathrm{e}^{s/\tau}\, \mathrm{e}^{iH(s-t)}\, \rho_0 \int_0^\beta \mathrm{e}^{-\eta H}\, \mathbf{U}\,.\,\frac{\nabla T}{T}\, \mathrm{e}^{\eta H}\, \mathrm{e}^{-iH(s-t)}\, d\eta \tag{6.13.41}$$

and from this form it is easy to obtain

$$0 = \frac{\partial \rho_1}{\partial t} = -\frac{\rho_1}{\tau} + \rho_0 \int_0^\beta e^{-\eta H}\,\mathbf{U}.\frac{\nabla T}{T} e^{\eta H}\,d\eta + \frac{i}{\hbar}[H, \rho_1]. \tag{6.13.42}$$

If we put

$$\left.\frac{\partial \rho}{\partial t}\right|_{\text{ext}} = \rho_0 \int_0^\beta e^{-\eta H}\,\mathbf{U}.\frac{\nabla T}{T} e^{\eta H}\,d\eta \tag{6.13.42a}$$

then equation (6.13.42) is identical with equation (6.10.5), leading to equation (6.10.13). In the present case

$$H = H_0 + V, \tag{6.13.43}$$

where V is the anharmonic perturbation of strength λ and we can also write

$$\mathbf{U} = \mathbf{U}_0 + \lambda \mathbf{U}_1, \tag{6.13.44}$$

where $\mathbf{U}_0$ is the purely harmonic part of the energy operator

$$\mathbf{U}_0 = \sum_{\mathbf{k}\sigma} a^\dagger_{\mathbf{k}\sigma} a_{\mathbf{k}\sigma} \hbar\omega \frac{\partial \omega}{\partial \mathbf{k}}. \tag{6.13.45}$$

A formal proof of this is left to the reader. Then by the definition of C_0 in equation (6.10.12)

$$C_0^{ii} = -i\hbar \langle i| \int_0^\beta e^{-\eta H_0} \mathbf{U}_0.\frac{\nabla T}{T} e^{\eta H_0} \rho_0\, d\eta\, |i\rangle = \frac{i\hbar}{k_B T^2} \mathbf{U}_0 . \nabla T \rho_0, \tag{6.13.46}$$

where we have used the fact that $\mathbf{U}_0$ and H_0 commute.

Equation (6.10.13) can now be written

$$0 = \frac{\nabla T}{k_B T^2} \langle i| \rho_0 \mathbf{U}_0 |i\rangle + \frac{2\pi}{\hbar} \sum_{j'} |V_{ij}|^2 (\langle j| \rho_1 |j\rangle - \langle i| \rho_1 |i\rangle). \tag{6.13.47}$$

We multiply through by $n_{\mathbf{k}}^i$, the number of phonons of wave-vector $\mathbf{k}$ belonging to state $|i\rangle$ and sum over all i:

$$0 = \frac{\nabla T}{k_B T^2} \sum_i \langle i| \rho_0 \mathbf{U}_0 |i\rangle n_{\mathbf{k}}^i + \frac{2\pi}{\hbar} \sum_{ij} |V_{ij}|^2 (\langle j| \rho_1 |j\rangle - \langle i| \rho_1 |i\rangle\} n_{\mathbf{k}}^i. \tag{6.13.48}$$

The second term on the right-hand side is simply

$$\frac{2\pi}{\hbar} \sum_{ij} |V_{ij}|^2 (\langle j| \rho_1 |j\rangle - \langle i| \rho_1 |i\rangle) n_{\mathbf{k}}^i$$

$$= \frac{2\pi}{\hbar} \sum_{ij} |V_{ij}|^2 \langle i| \rho_1 |i\rangle (n_{\mathbf{k}}^i - n_{\mathbf{k}}^j). \tag{6.13.49}$$

Since

$$p_i = \langle i|\rho|i\rangle = \langle i|\rho_0|i\rangle + \langle i|\rho_1|i\rangle \tag{6.13.50}$$

and replacement of ρ_1 by ρ_0 in equation (6.13.49) is readily shown to give zero, we can see that equation (6.13.47) is equivalent to equation (6.13.37) provided

$$\frac{\nabla T}{k_B T^2}\sum_i \langle i|\rho_0 \mathbf{U}_0|i\rangle n_{\mathbf{k}}^i = \nabla T . n_0(n_0+1)\frac{\hbar\omega}{k_B T^2}\frac{\partial \omega}{\partial \mathbf{k}}, \tag{6.13.51}$$

which we now go on to prove.

Inserting equation (6.13.45) into the left-hand side of equation (6.13.51) we find

$$\begin{aligned}\frac{\nabla T}{k_B T^2}\sum_i \langle i|\rho_0 \mathbf{U}_0|i\rangle n_{\mathbf{k}}^i &= \frac{\nabla T}{k_B T^2}\sum_i \langle i|n_{\mathbf{k}}\rho_0 \mathbf{U}_0|i\rangle \\ &= \frac{\nabla T}{k_B T^2}\sum_{\mathbf{k}'}\frac{\partial \omega_{\mathbf{k}'}}{\partial \mathbf{k}'}\hbar\omega_{\mathbf{k}'}\sum_i \langle i|n_{\mathbf{k}}n_{\mathbf{k}'}\rho_0|i\rangle \\ &= \frac{\nabla T}{k_B T^2}\sum_{\mathbf{k}'}\frac{\partial \omega_{\mathbf{k}'}}{\partial \mathbf{k}'}\hbar\omega_{\mathbf{k}'}\langle n_{\mathbf{k}}n_{\mathbf{k}'}\rangle.\end{aligned} \tag{6.13.52}$$

Now whereas

$$\langle n_{\mathbf{k}}n_{\mathbf{k}'}\rangle = \langle n_{\mathbf{k}}\rangle\langle n_{\mathbf{k}'}\rangle \quad \text{if} \quad \mathbf{k}' \neq \mathbf{k}, \tag{6.13.53}$$

we have, using the cyclic property of the trace

$$\begin{aligned}\langle n^2\rangle &= \frac{1}{Z}\text{tr}\{e^{-\beta H_0} a^\dagger a a^\dagger a\} \\ &= \frac{1}{Z}\text{tr}\{a a^\dagger a\, e^{-\beta H_0} a^\dagger\} \\ &= \frac{1}{Z}\text{tr}\{e^{-\beta H_0} a a^\dagger a\, e^{-\beta H_0} a^\dagger e^{\beta H_0}\} \\ &= \frac{1}{Z}\text{tr}\{e^{-\beta H_0} a a^\dagger a a^\dagger\} e^{-\beta\hbar\omega}.\end{aligned} \tag{6.13.54}$$

Since also $aa^\dagger = a^\dagger a + 1$, we have further that

$$\begin{aligned}\langle n^2\rangle &= \frac{e^{-\beta\hbar\omega}}{Z}[\text{tr}\{e^{-\beta H_0} a^\dagger a a^\dagger a\} + 2\,\text{tr}\{e^{-\beta H_0} a^\dagger a\} + \text{tr}\{e^{-\beta H_0}\}] \\ &= e^{-\beta\hbar\omega}(\langle n^2\rangle + 2\langle n\rangle + 1).\end{aligned} \tag{6.13.55}$$

That is,

$$\langle n^2\rangle = \frac{e^{-\beta\hbar\omega}}{1-e^{-\beta\hbar\omega}}(2\langle n\rangle+1). \tag{6.13.56}$$

Using equation (6.13.53) and the facts that $\partial\omega/\partial\mathbf{k}$ is odd in $\mathbf{k}$ and

$$\langle n_{\mathbf{k}'}\rangle = \langle n_{-\mathbf{k}'}\rangle, \tag{6.13.57}$$

terms for which $\mathbf{k}'\neq\mathbf{k}$ or $-\mathbf{k}$ vanish in equation (6.13.52), which can therefore now be written as

$$\begin{aligned}\frac{\nabla T}{k_B T^2}\sum_i \langle i|\,\rho_0\,\mathbf{U}_0\,|i\rangle\, n_{\mathbf{k}}^i &= \frac{\nabla T}{k_B T^2}\hbar\omega_{\mathbf{k}}\frac{\partial\omega_{\mathbf{k}}}{\partial\mathbf{k}}[\langle n_{\mathbf{k}}\rangle(2\langle n_{\mathbf{k}}\rangle+1-\langle n_{-\mathbf{k}}\rangle)]\\ &= \frac{\nabla T}{k_B T^2}\frac{\partial\omega}{\partial\mathbf{k}}\hbar\omega_{\mathbf{k}}\langle n_{\mathbf{k}}\rangle(\langle n_{\mathbf{k}}\rangle+1).\end{aligned} \tag{6.13.58}$$

Hence equation (6.13.47) is in fact just the linearized Boltzmann equation for phonons and the result we obtain for $\langle i|\,\rho_1\,|i\rangle$, when inserted into equation (6.13.38) in the form

$$\mathbf{Q} = -\kappa\nabla T = \sum_{i,j} n_{\mathbf{k}}^i\,\hbar\omega_{\mathbf{k}}\frac{\partial\omega_{\mathbf{k}}}{\partial\mathbf{k}}\langle i|\,\rho_1\,|i\rangle, \tag{6.13.59}$$

gives a result equivalent to equation (6.13.37) to lowest order in the anharmonic perturbation.

CHAPTER 7

Optical Properties

7.1 Introduction

The study of optical properties of a solid can yield important information concerning band-structure, electron–electron interactions, etc. This is especially so of metals and semi-conductors, to which much of our theory will be oriented in this chapter.

We begin with Maxwell's equations

$$\operatorname{curl}\mathscr{H} = \frac{4\pi\mathbf{j}}{c}+\frac{1}{c}\frac{\partial\mathscr{D}}{\partial t}, \quad \operatorname{div}\mathscr{D} = 4\pi\rho,$$

$$\operatorname{curl}\mathscr{E} = -\frac{1}{c}\frac{\partial\mathbf{B}}{\partial t}, \tag{7.1.1}$$

where

$$\mathscr{D} = \varepsilon\mathscr{E}, \quad \mathbf{B} = \mu\mathscr{H}, \tag{7.1.2}$$

ε being the dielectric constant and μ the permeability. We assume $\mu = 1$.

The question now arises as to what we take for $\mathbf{j}$.‡ One is tempted to reply immediately that it is the current due to the "conduction" electrons, but in fact the point needs rather careful discussion. Thus, in equation (7.1.1) $\mathbf{j}$ is the current produced by charges not considered as part of the medium, and there are always other currents present if $\partial\mathscr{D}/\partial t$ is not trivially equal to $\partial\mathscr{E}/\partial t$. We note that

$$\mathscr{D} = \mathscr{E}+4\pi\mathscr{P}. \tag{7.1.3}$$

The polarization $\mathscr{P}$ results from a charge density ρ' given by

$$\operatorname{div}\mathscr{P} = \rho' \tag{7.1.4}$$

associated with which there exists a current $\mathbf{j}'$ given by

$$-\operatorname{div}\mathbf{j}' = \frac{\partial\rho'}{\partial t}. \tag{7.1.5}$$

‡ For a discussion of this question when $\mu\neq 1$ see Appendix 4.3 of Volume 1.

Thus from equations (7.1.4) and (7.1.5) we can write

$$-\mathbf{j}' = \frac{\partial \mathscr{P}}{\partial t}. \tag{7.1.6}$$

This current is additional to the one we associated with the conduction electrons, and may be termed the "bound current".

Conversely, we can absorb $\mathbf{j}$ into $\partial\mathscr{D}/\partial t$; we see that, if we regard the entire solid as the medium through which an electromagnetic wave passes, we might for consistency take $\mathbf{j} = 0$ in equation (7.1.1), obtaining

$$\operatorname{curl} \mathscr{H} = \frac{\varepsilon}{c}\frac{\partial \mathscr{E}}{\partial t},$$

$$\operatorname{curl} \mathscr{E} = -\frac{1}{c}\frac{\partial \mathscr{H}}{\partial t}. \tag{7.1.7}$$

We shall first analyse the problem in this way. We consider a wave with time variation $e^{i\omega t}$ travelling in the z-direction, with $\mathscr{E}$ and $\mathscr{H}$ parallel to x- and y-directions respectively.
Then

$$-\frac{\partial \mathscr{H}_y}{\partial z} = \frac{i\omega\varepsilon}{c}\mathscr{E}_x,$$

$$\frac{\partial \mathscr{E}_x}{\partial z} = -\frac{i\omega}{c}\mathscr{H}_y, \tag{7.1.8}$$

giving the wave equation

$$\frac{\partial^2 \mathscr{E}_x}{\partial z^2} = -\frac{\omega^2 \varepsilon}{c^2}\mathscr{E}_x, \tag{7.1.9}$$

with solution

$$\mathscr{E} = \mathscr{E}_0 \exp i\omega\left(t - \frac{\mu z}{c}\right) \quad (\mu = \sqrt{\varepsilon}), \tag{7.1.10}$$

which may be written

$$\mathscr{E} = \mathscr{E}_0 \exp i\omega\left(t - \frac{nz}{c}\right) \exp\left(-\frac{\omega k z}{c}\right), \tag{7.1.11}$$

where n and k are real, the former being the refractive index and the latter the extinction coefficient governing the damping of the wave in the medium. We see on comparison of equations (7.1.10) and (7.1.11) that

$$\sqrt{\varepsilon} = \mu = n + ik. \tag{7.1.12}$$

That is, if ε is split into its real and imaginary parts:

$$\varepsilon = \varepsilon_1 + i\varepsilon_2, \tag{7.1.13}$$

then

$$\varepsilon_1 = n^2 - k^2,$$
$$\varepsilon_2 = 2nk. \tag{7.1.14}$$

We might also note that the reflection coefficient for normal incidence is shown in elementary text books to be given by

$$R = \left|\frac{\mu-1}{\mu+1}\right|^2 = \frac{(n-1)^2+k^2}{(n+1)^2+k^2}. \tag{7.1.15}$$

The analysis we have just given links the optical constants n and k with the transverse dielectric constant very clearly but sheds no light on how they are to be linked with the detailed theory of the electronic states of the solid. For this purpose it is easier to treat the medium as free-space containing electrons and thus obtain ε as its free-space value plus terms due to all the electrons. We therefore initially put $\varepsilon = 1$ (unless the polarizability of the lattice must be taken into account), writing

$$\text{curl}\,\mathscr{H} = \frac{4\pi\mathbf{j}}{c} + \frac{\partial\mathscr{E}}{\partial t}. \tag{7.1.16}$$

$\mathbf{j}$ thus contains the current given by equation (7.1.5). If we again take the time dependence of the electric field as $e^{i\omega t}$ and write

$$\mathbf{j} = \sigma(\omega)\,\mathscr{E}, \tag{7.1.17}$$

we have

$$\text{curl}\,\mathscr{H} = \left[\frac{4\pi}{c}\sigma(\omega) + \frac{i\omega}{c}\right]\mathscr{E} \tag{7.1.18}$$

and so by comparison with equation (7.1.7) we see that

$$\varepsilon(\omega) = 1 + \frac{4\pi i}{\omega}\sigma(\omega). \tag{7.1.19}$$

The importance of this formula lies in the fact that, provided the spatial variation of $\mathscr{E}$ does not enter the problem, σ is exactly the conductivity given by the general formulae of the last chapter, since we discussed there the *total* current, making no division of the electrons into "bound" and "conduction" electrons.

We have written σ and ε as if they were scalars in the foregoing, but they may equally well be regarded as tensors, in which case equation (7.1.19) is replaced by

$$\varepsilon_{\alpha\beta}(\omega) = \delta_{\alpha\beta} + \frac{4\pi i}{\omega}\sigma_{\alpha\beta}(\omega) \tag{7.1.20}$$

defining the transverse dielectric tensor.

7.2 Drude–Zener theory

We now apply the above analysis to quasi-free electrons treated using the relaxation-time approximation. This theory is incomplete in that interband transitions are ignored (most obviously, as we shall see, it cannot explain the colour of a metal), but otherwise gives a broadly correct picture of the optical properties of metals.

The linearized Boltzmann equation is, in the relaxation time approximation,

$$\frac{\partial f_1}{\partial t} - e\mathscr{E}.\mathbf{v}\frac{\partial f_0}{\partial E} + \mathbf{v}.\frac{\partial f_1}{\partial \mathbf{r}} = -\frac{f_1}{\tau}. \tag{7.2.1}$$

The wavelength of $\mathscr{E}$ is usually large enough for us to be able to neglect its spatial variation over a macroscopic region of the solid; we further assume f_1 has the same time dependence $e^{i\omega t}$ as $\mathscr{E}$, when we have

$$i\omega f_1 - e\mathscr{E}.\mathbf{v}\frac{\partial f_0}{\partial E} = -\frac{f_1}{\tau} \tag{7.2.2}$$

and so the expression for the d.c. conductivity σ_0 is modified simply by replacing τ^{-1} by $\tau^{-1} + i\omega$. Since $\sigma_0 \propto \tau$, we therefore have

$$\sigma(\omega) = \frac{\sigma_0}{1 + i\omega\tau}, \tag{7.2.3}$$

i.e., if σ_1 and σ_2 are real and imaginary parts repectively,

$$\sigma_1 = \frac{\sigma_0}{1+\omega^2\tau^2}, \quad \sigma_2 = -\frac{\omega\tau\sigma_0}{1+\omega^2\tau^2}. \tag{7.2.4}$$

We can now derive the optical constants from equations (7.1.14) by taking $\varepsilon(\omega)$ to be given by equation (7.1.19). Actually, since equation (7.2.3) refers only to the conduction electrons, and does not include the conductivity for the "bound current", the medium is strictly not that of free space and we ought to take

$$\varepsilon(\omega) = \varepsilon^0(\omega) + \frac{4\pi i}{\omega}\sigma(\omega). \tag{7.2.5}$$

We can in some cases ignore the polarizability of rest of the crystal and take $\varepsilon^0 = 1$, as is the case, for example, in the alkali metals and we will do so at present. Then we find

$$n^2-k^2 = 1+\frac{4\pi}{\omega}\sigma_2,$$

$$nk = \frac{2\pi}{\omega}\sigma_1. \tag{7.2.6}$$

We now take the expression

$$\sigma_0 = \frac{Ne^2\tau}{\mathscr{V}m^*}, \tag{7.2.7}$$

where m^* is an effective mass, which we term the "optical mass". It is not necessarily the effective mass we would use if we were to calculate the d.c. conductivity, for the phonon dressing we expect the electron to have at low frequencies should disappear in the optical region, the lattice being unable to follow the electronic motions. Equations (7.2.4) and (7.2.6) now give

$$n^2-k^2 = 1-\frac{\omega_p^2\tau^2}{1+\omega^2\tau^2}, \tag{7.2.8}$$

$$2nk = \frac{\omega_p^2\tau}{\omega(1+\omega^2\tau^2)},$$

where

$$\omega_p = \sqrt{\frac{4\pi Ne^2}{\mathscr{V}m^*}} = \sqrt{\frac{4\pi\sigma_0}{\tau}} \tag{7.2.9}$$

is the plasma frequency for long wavelengths, as we saw in Chapter 2. For copper $\tau = 2.4\times 10^{14}$ sec while $\omega_p = 1.6\times 10^{16}$ sec^{-1}. We first look at the low-frequency range where $\omega\tau \ll 1$ (the non-relaxation region). This is the radio-frequency region. We can see that

$$(n+k)(n-k) \approx 1-\omega_p^2\tau^2, \quad 2nk \approx \frac{\omega_p^2\tau}{\omega}. \tag{7.2.10}$$

Thus n and k increase indefinitely as $\omega \to 0$ and since $1-\omega_p^2\tau^2$ is independent of ω, the two optical constants become equal:

$$n \simeq k \approx \sqrt{\frac{\omega_p^2\tau}{2\omega}} = \sqrt{\frac{2\pi\sigma_0}{\omega}}. \tag{7.2.11}$$

The skin-depth δ is of interest in this region. This is the distance at which the incident wave is attenuated to 1/e times its original value. From equation (7.1.11) this is related to k by

$$\delta = \frac{c}{\omega k} \tag{7.2.12}$$

or, using equation (7.2.11),

$$\delta = \sqrt{\frac{c^2}{2\pi\sigma_0\,\omega}}, \tag{7.2.13}$$

a standard result of electromagnetic theory.

The region of greater interest to us here is the relaxation region ($\omega\tau \gg 1$) which for copper includes the plasma frequency. From equation (7.2.8) we then obtain

$$\varepsilon_1 = n^2-k^2 \approx 1-\left(\frac{\omega_p}{\omega}\right)^2,$$

$$\varepsilon_2 = 2nk \approx \frac{\omega_p^2}{\omega^3\tau}. \tag{7.2.14}$$

At the plasma edge, $\omega = \omega_p$, we have

$$n^2-k^2 = 0, \quad 2nk = \frac{1}{\omega\tau} \ll 1, \tag{7.2.15}$$

that is, both n and k are sensibly zero and the reflection coefficient $R = 1$. For $\omega \leqslant \omega_p$, we still have $2nk \simeq 0$, but $n^2-k^2 < 0$ and so k must increase more quickly than n as the frequency is lowered; R remains very close to unity. On the other hand, for $\omega > \omega_p$, $n^2-k^2 > 0$ while $k = 0$. Thus for frequencies higher than the plasma frequency the metal is almost completely transparent. This occurs in the far ultraviolet and, as Figure 7.1 shows, the effect is marked though the reflection coefficient does not decrease as abruptly as k.

We ought to note that onset of transparency at the plasma frequency is not an accident. n and k become zero (quite generally, independent of the model) when

$$\varepsilon(\omega) = 0 \tag{7.2.16}$$

and as we saw in Chapter 2 the plasma oscillates spontaneously when

$$\varepsilon_{\text{long}}(\omega, \mathbf{q}) = 0. \tag{7.2.17}$$

Strictly, the dielectric constant in equation (7.2.16) is the transverse dielectric constant, but at long wavelengths

$$\varepsilon_{\text{long}}(\omega, \mathbf{q}) = \varepsilon_{\text{trans}}(\omega, \mathbf{q}), \tag{7.2.18}$$

since in a small region of an isotropic substance there is then no difference between longitudinal and transverse fields as far as the conductivity, through which ε has been defined, is concerned. On the other hand, a transverse field cannot excite a longitudinal plasma oscillation, at least in a uniform electron gas, and so there will be no additional absorption due to the

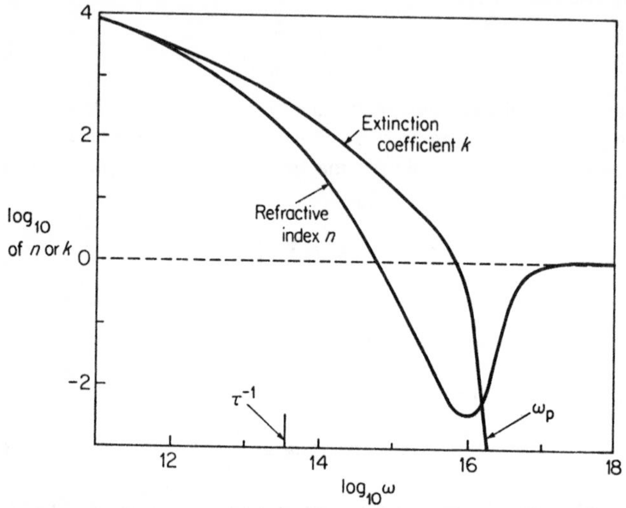

FIGURE 7.1. Optical constants for Cu according to quasi-free electron theory. Absorption coefficient is zero beyond plasma frequency ω_p (after Donovan, 1967).

mechanism at the plasma edge. We shall discuss this point again in section 7.9 where it will be shown that the presence of the lattice actually does allow a weak plasma excitation.

One point, of particular importance in semi-metals and significant in transition and noble metals, should finally be noted. If, in treating the conduction electrons, equation (7.2.5) has to be used since $\varepsilon^0(\omega) \neq 1$, we obtain instead of equation (7.2.14)

$$\varepsilon_1(\omega) = \varepsilon_1^0(\omega) - \left(\frac{\omega_p}{\omega}\right)^2. \tag{7.2.19}$$

Since in general $\varepsilon_1^0(\omega) > 1$, the plasma frequency given by $\varepsilon(\omega) = 0$ will be depressed from its quasi-free electron value $\sqrt{(4\pi Ne^2/\mathscr{V} m^*)}$. In addition the resonance, as described by the energy-loss function $\text{Im}\,\varepsilon^{-1}$, discussed in

Chapter 5, section 5.6.2, will be broadened. A clear example of these effects will be shown to occur in graphite (section 7.5), but before looking at the optical properties of this substance we shall derive some formal relationships of considerable use in analysis of spectra.

7.3 Kramers–Krönig analysis

Having set the scene by the elementary discussion of optical properties in the last section, we wish, before proceeding further in our description of the interpretation of optical phenomena, to establish formal relations between real and imaginary parts of certain physical quantities; these are termed "Kramers–Krönig relations".

We shall first describe the analysis for the conductivity, because it appears for this quantity in its simplest form, and because this particular application of the Kramers–Krönig relations is of considerable importance. Although in Chapter 6 we gave complete formal expressions for the complex conductivity, it proves most practicable to calculate only its real part in terms of the unperturbed electronic states of the system. From our discussion of Nyquist's theorem we can see, by combining equations (6.11.57) and (6.11.62), that

$$\mathrm{Re}\,[\sigma_{\alpha\beta}(\omega)] = \frac{2}{\hbar\omega\coth(\beta\hbar\omega/2)}\int_0^\infty dt\cos\omega t\,\langle j_\alpha(0)j_\beta(t)\rangle, \qquad (7.3.1)$$

from which follows the alternative formula

$$\mathrm{Re}\,[\sigma_{\alpha\beta}(\omega)] = \frac{\pi}{\omega}\sum_{mn}\langle m|j_\alpha|n\rangle\langle n|j_\beta|m\rangle\,[\rho_m-\rho_n]\,\delta(E_m-E_n-\hbar\omega) \qquad (7.3.2)$$

($\rho_m = e^{-\beta E_m}/\sum_m e^{-\beta E_m}$), relating the conductivity, and so the optical absorption, to transitions between stationary states of the system. We now show how the imaginary part of $\sigma(\omega)$, which also plays an important role in optical phenomena, is calculable directly in terms of equation (7.3.2).

We begin our discussion by noting from equation (7.1.17) that the Fourier components of electric field and current are related by

$$j(\omega) = \sigma(\omega)\mathscr{E}(\omega). \qquad (7.3.3)$$

(Throughout this section we shall write σ as if it were scalar; our final results, however, hold for the individual components of the conductivity tensor.) In this equation we allow all three functions to be complex—in particular

$$\sigma = \sigma_1 + i\sigma_2, \qquad (7.3.4)$$

where σ_1 and σ_2 are real. On the other hand, this complex nature is purely formal and the actual current

$$j(t) = \frac{1}{2\pi}\int_{-\infty}^{\infty} \sigma(\omega)\, e^{i\omega t}\mathscr{E}(\omega)\, d\omega \tag{7.3.5}$$

must be real on the application of a real electric field; thus in the convolution of this expression

$$j(t) = \int_{-\infty}^{\infty} \sigma(t-t')\mathscr{E}(t')\, dt' \tag{7.3.6}$$

the quantity

$$\sigma(t) = \frac{1}{2\pi}\int_{-\infty}^{\infty} \sigma(\omega)\, e^{-i\omega t}\, dt \tag{7.3.7}$$

must be real. This implies that

$$\sigma(-\omega) = \sigma^*(\omega), \tag{7.3.8}$$

i.e.

$$\sigma_1(-\omega) = \sigma_1(\omega), \quad \sigma_2(-\omega) = -\sigma_2(\omega). \tag{7.3.9}$$

We come now to the crucial point in obtaining the Kramers–Krönig relations: by the "causality principle"—that the effect cannot precede the cause—$j(t)$ cannot depend on $\mathscr{E}(t')$ if $t<t'$. This implies

$$\sigma(t) = 0 \quad \text{if } t<0 \tag{7.3.10}$$

and so we can write the inverse Fourier transform of $\sigma(t)$ as

$$\sigma(\omega) = \int_0^{\infty} \sigma(t)\, e^{i\omega t}\, dt. \tag{7.3.11}$$

We may note from equation (7.3.6) that if $\mathscr{E}(t')$ is a delta-function impulse, $j(t) = \sigma(t)$, i.e. $\sigma(t)$ is the current produced by a "unit impulse" of electric field at $t = 0$. This establishes, on physical grounds, that $\sigma(t)$ is "well behaved". We shall also use the result that

$$\lim_{\omega\to\infty} \sigma(\omega) = 0. \tag{7.3.12}$$

The good behaviour of $\sigma(t)$ has the consequence that, admitting complex values of ω, $\sigma(\omega)$ has no zeros or singularities in the upper half of the complex-plane,‡ and since equation (7.3.12) holds, $\sigma(\omega)$ vanishes over the

‡ For a discussion of the analytic properties in the upper half-plane see Landau and Lifshitz (1958), § 122.

semi-circle at infinity in this plane. Consider therefore the contour C shown in Figure 7.2, and the integral

$$\int_C \frac{\sigma(\xi)\,d\xi}{\xi-\omega} = 0. \tag{7.3.13}$$

The integral is zero since the contour contains no poles and, since the integrand vanishes over the semi-circle at infinity, can also be written

$$⨍_{-\infty}^{\infty} \frac{\sigma(\xi)}{\xi-\omega}\,d\xi - i\pi\sigma(\omega) = 0, \tag{7.3.14}$$

the second term on the left-hand side being the contribution to the integral over the infinitesimal semi-circle about the simple pole at $\xi=\omega$, and the

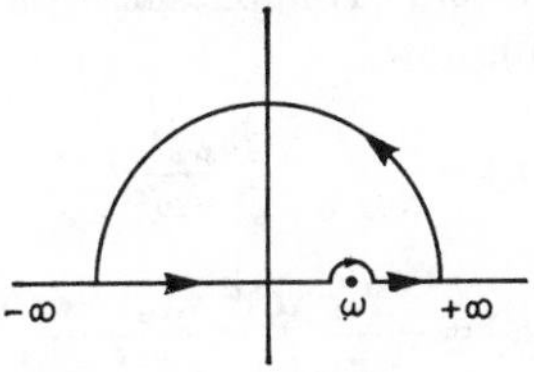

FIGURE 7.2. Form of contour in integral (7.3.13).

bar on the integral sign showing that the principal value is taken. Taking real and imaginary parts of equation (7.3.14) we immediately find

$$\sigma_1(\omega) = \frac{1}{\pi} ⨍_{-\infty}^{\infty} \frac{\sigma_2(\xi)}{\xi-\omega}\,d\xi,$$

$$\sigma_2(\omega) = -\frac{1}{\pi} ⨍_{-\infty}^{\infty} \frac{\sigma_1(\xi)}{\xi-\omega}\,d\xi. \tag{7.3.15}$$

These are the Kramers–Krönig relations, often rewritten as

$$\sigma_1(\omega) = \frac{2}{\pi} ⨍_0^{\infty} \frac{\xi\sigma_2(\xi)}{\xi^2-\omega^2}\,d\xi,$$

$$\sigma_2(\omega) = -\frac{2\omega}{\pi} ⨍_0^{\infty} \frac{\sigma_1(\xi)}{\xi^2-\omega^2}\,d\xi, \tag{7.3.16}$$

which forms are easily obtained from equation (7.3.15) by using the fact that $\sigma_1(\omega)$ is even and $\sigma_2(\omega)$ is odd.

The above analysis is applicable, with possible minor modifications, to any linear relation to the type (7.3.3); for example, the analysis yields relations between real and imaginary parts of the dielectric constant defined by $\mathscr{D} = \varepsilon\mathscr{E}$. While these are readily obtained by using the relation between ε and σ, we shall show how the modifications to equation (7.3.16) come about quite generally. First, $\varepsilon(t)$ has a δ-function singularity at $t = 0$:

$$\varepsilon(t) = \frac{1}{2\pi}\int_{-\infty}^{\infty} \varepsilon'(\omega)\, e^{i\omega t}\, dt + \varepsilon_1(\infty)\, \delta(t), \tag{7.3.17}$$

because a δ-function impulse in $\mathscr{E}(t)$ induces a δ-function impulse in $\mathscr{D}(t)$. As we have indicated, this means that $\varepsilon_1(\infty) \neq 0$ (although $\varepsilon_2(\infty) = 0$ still),

$$\varepsilon(\omega) = \varepsilon'(\omega) + \varepsilon_1(\infty) \quad [\varepsilon'(\infty) = 0] \tag{7.3.18}$$

and the analysis applied to $\sigma(\omega)$ is applicable to $\varepsilon'(\omega)$, so that we obtain, instead of equation (7.3.16),

$$\left.\begin{aligned} \varepsilon_1(\omega) - \varepsilon_1(\infty) &= \frac{2}{\pi} \fint_0^{\infty} \frac{\xi\varepsilon_2(\xi)}{\xi^2 - \omega^2}\, d\xi, \\ \varepsilon_2(\omega) &= -\frac{2\omega}{\pi} \fint_0^{\infty} \frac{\varepsilon_1(\xi) - \varepsilon_1(\infty)}{\xi^2 - \omega^2}\, d\xi, \end{aligned}\right\} \tag{7.3.19}$$

There are corresponding modifications to the forms shown in equation (7.3.15). It should be noted that the modifications will include, in this particular case, a term A/ω, in addition to the integral, in the expression for ε_2, since from the expression for ε in terms of σ it will be seen that ε has a pole at the origin if the solid has finite d.c. conductivity.

We emphasize again that the above analysis is not specialized to the dielectric constant: it holds, for example, for the magnetic susceptibility χ defined through $\mathscr{M} = \chi\mathscr{H}$, and the reciprocal of the dielectric constant defined through $\mathscr{E} = \varepsilon^{-1}\mathscr{D}$. We wish to turn now, however, to a situation to which the above application of Kramers–Krönig analysis is less immediate. From equation (7.1.15) we can write the reflection coefficient as

$$R(\omega) = |r(\omega)|^2, \tag{7.3.20}$$

where

$$r(\omega) = \frac{n + ik - 1}{n + ik + 1}. \tag{7.3.21}$$

This is the amplitude reflection coefficient [actually for the magnetic component of the light wave; that for the electric field differs from equation (7.3.21) by a factor -1] and so the Kramers–Krönig analysis based on the causality principle is again applicable. The analysis is exactly like that for the conductivity since the reflection coefficient cannot go above unity and, according to the analysis of the last section,

$$r(\infty) = 0. \tag{7.3.22}$$

We shall see in the following section that the asymptotic forms for ε_1 and ε_2 predicted in the Drude–Zener model are the correct limiting forms and so equation (7.3.22) is correct in general.

The relations thus obtained are not, however, very useful. On the other hand, let us write

$$r(\omega) = |r(\omega)|\, e^{i\theta(\omega)}, \tag{7.3.23}$$

so that

$$n = (1-|r|^2)/(1-|r|^2-2|r|\cos\theta), \tag{7.3.24}$$

$$k = 2|r|\sin\theta/(1-|r|^2-2|r|\cos\theta). \tag{7.3.25}$$

If we put

$$\ln r = \ln|r| + i\theta \tag{7.3.26}$$

and apply the Kramers–Krönig relations (7.3.16) directly to real and imaginary parts of $\ln r$, we obtain

$$\theta(\omega) = -\frac{\omega}{\pi}\int_0^\infty \frac{\ln R(\xi)}{\xi^2-\omega^2}\,d\xi. \tag{7.3.27}$$

Using this equation we can obtain both independent constants n and k solely from reflectivity data. There are, however, difficulties in justifying the above formula for $\theta(\omega)$, for $\ln r$ is not the ratio between two physical quantities, one of which is the cause and the other the effect. In particular, since $r(\infty) = 0$, $\ln r(\infty) = -\infty$ and so the integral $\int_C [\ln r(\xi)/\xi-\omega]\,d\xi$ over the contour of Figure 7.2 will not converge. However, this problem is overcome by evaluating (we avoid the pole at $\xi = \nu$ just as we avoid that at ω)

$$\int_C \ln r(\xi)\left(\frac{1}{\xi-\omega}-\frac{1}{\xi-\nu}\right)d\xi = \int_C \ln r(\xi)\frac{\omega-\nu}{(\xi-\omega)(\xi-\nu)}\,d\xi. \tag{7.3.28}$$

This integral will, in general, converge and vanish over the semi-circle at infinity, so that we obtain, under the assumption that there are no singularities inside the contour,

$$\int_{-\infty}^{\infty} \ln r(\xi)\left(\frac{1}{\xi-\omega}-\frac{1}{\xi-\nu}\right)d\xi - i\pi(\ln r(\omega)-\ln r(\nu)) = 0. \tag{7.3.29}$$

Taking real parts we therefore have

$$\theta(\omega)-\theta(\nu) = -\frac{1}{2\pi}\int_{-\infty}^{\infty} \ln R(\xi)\left(\frac{1}{\xi-\omega}-\frac{1}{\xi-\nu}\right)d\xi. \tag{7.3.30}$$

We shall choose $\nu = 0$, when we obtain, on using the fact that $R(\omega)$ is even,

$$\theta(\omega)-\theta(0) = -\frac{\omega}{\pi}\int_0^\infty \frac{\ln R(\xi)}{\xi^2-\omega^2}\,d\xi. \tag{7.3.31}$$

There is a fundamental reason for obtaining the difference between θ at two points rather than θ itself. Equation (7.3.23) shows that we can add $2\pi n$ (n any integer) to θ without changing the value of $r(\omega)$. In other words, $\theta(\omega)$ is a many-valued function, and we must settle on the branch most convenient to us if we are to obtain $\theta(\omega)$ alone. Since $\operatorname{Im} r(\omega)$ is odd, and passes through zero at $\omega = 0$, we choose $\theta(0) = 0$, whereupon we obtain equation (7.3.27) from equation (7.3.31). Since θ will be continuous we can then fix its range of variation on the positive half of the real axis as

$$0 \leqslant \theta(\omega) \leqslant \pi \quad (\omega > 0). \tag{7.3.32}$$

For a system in equilibrium with its surroundings the extinction coefficient k will always be positive and then equation (7.3.24) immediately gives equation (7.3.30). Further, as $\omega \to 0$ equation (7.2.14) shows that n approaches unity from below‡ so that in equation (7.3.24) $\cos\theta$ must be negative. We therefore have

$$\theta(0) = 0, \quad \theta(\infty) = \pi. \tag{7.3.33}$$

Using the asymptotic relation for $R(\omega)$, we can indeed show $\theta(\infty) = \pi$ from equation (7.3.31).

We finally remark on the assumptions we have made concerning the analyticity of $\ln r(\omega)$. It will be seen that we have assumed $\ln r(\omega)$ to have neither poles nor branch points in the upper half-plane. As was stated for $\sigma(\omega)$, $r(\omega)$ must be analytic and have no zeros in the upper half-plane. This means that $(d/d\omega)\ln r(\omega) = [dr(\omega)/d\omega]/r(\omega)$ is defined everywhere, so that $\ln r(\omega)$ can have no singularities. Further, there cannot be any branch points, for $\ln r(\omega)$ changes by $2\pi i n$ on circling a branch point, whereas

$$\int_C \frac{d}{d\omega}\ln r(\omega)\,d\omega = 0 \tag{7.3.34}$$

over any contour in the upper half-plane.

We should mention that we can avoid taking the principal part in equation (7.3.31) by writing

$$\theta(\omega) = -\frac{\omega}{\pi}\int_0^\infty \frac{\ln R(\xi)-\ln R(\omega)}{\xi^2-\omega^2}\,d\xi = \frac{1}{2\pi}\int_0^\infty \ln\left[\frac{R(\xi)}{R(\omega)}\right]\left(\frac{1}{\xi+\omega}-\frac{1}{\xi-\omega}\right)d\xi. \tag{7.3.35}$$

‡ The asymptotic relations for ε will be shown to be rigorous in the following section.

In applications this expression is often transformed by parts into

$$\theta(\omega) = -\frac{1}{2\pi}\int_0^\infty \frac{d}{d\xi}\ln R(\xi)\ln\left|\frac{\omega+\xi}{\omega-\xi}\right| d\xi. \tag{7.3.36}$$

Following this general discussion, there are some further basic properties of the dielectric constant that must be established: the so-called "sum rules".

7.4 Sum rules

There are two important sum rules for the dielectric constant which are often used in the analysis of absorption spectra. These may be established in a rigorous fashion from the general expressions for dielectric constant, or, equivalently, complex conductivity. We shall here, however, employ an intuitive argument which considers the response of the electrons to a δ-function impulse $\mathscr{E}(t) = \delta(t)$ of the electric field. This momentary infinite field will swamp completely any other interactions whatsoever and the initial response of the electrons will be as if they were perfectly free.

First, therefore, let us consider the conductivity of a free electron gas. We can get this by letting $\tau\to\infty$ in our equations of section 7.1 but it is interesting to utilize the general formula of equation (6.5.1) which we put in the form

$$\sigma_{\alpha\beta}(\omega) = \frac{e^2}{im\mathscr{V}}\int_0^\infty dt\, e^{-i\omega t}\operatorname{tr}\{[r_\alpha, p_\beta(t)]\,\rho_0\}. \tag{7.4.1}$$

It should be noted that $\mathbf{r} = \sum_i \mathbf{r}_i$ and $\mathbf{p} = \sum_i \mathbf{p}_i$. Independent of the Coulombic interactions, $\mathbf{p}$ and H commute if the external potential is zero and so $\mathbf{p}(t) = \mathbf{p}(0)$. From this it follows that the conductivity of a free-electron gas including interactions is

$$\sigma_{\alpha\beta}(\omega) = \delta_{\alpha\beta}\frac{Ne^2}{m}\int_0^\infty e^{-i\omega t}\,dt. \tag{7.4.2}$$

It is immediately obvious from the general relation (7.3.8) that

$$\begin{aligned}\sigma_{\alpha\beta}(t) &= \frac{Ne^2}{\mathscr{V}m}\delta_{\alpha\beta}, \quad t>0,\\ &= 0, \quad t<0.\end{aligned} \tag{7.4.3}$$

We should therefore interpret the integral of equation (7.4.2) such that

$$\sigma_{\alpha\beta}(\omega) = \frac{Ne^2}{\mathscr{V}m}\left[\pi\delta(\omega) - i\frac{1}{\omega}\right] \tag{7.4.4}$$

and we can note, in passing, that, even if interactions are included, the optical mass is just the bare mass for a translationally invariant electron system (whereas when interactions *and* a non-uniform external potential exist together, both affect the optical mass, as we shall see later on).

Turning now to non-translationally invariant systems, since the immediate response of the electrons to $\mathscr{E}(t) = \delta(t)$ is exactly the same as if they were free, we have that equation (7.4.3) is true at $t = +0$ for any electron system, and so equation (7.3.8) gives, on using equation (7.3.8),

$$\delta_{\alpha\beta}\frac{Ne^2}{\mathscr{V}m} = \frac{1}{\pi}\int_{-\infty}^{\infty}\sigma_{\alpha\beta}(\omega)\,d\omega = \frac{2}{\pi}\int_0^{\infty}\mathrm{Re}\,[\sigma_{\alpha\beta}(\omega)]\,d\omega. \tag{7.4.5}$$

This is the sum rule for the conductivity tensor. We can express it in terms of the imaginary part by letting $\omega \to \infty$ in the Kramers–Krönig relations (7.3.16), when we find

$$\lim_{\omega\to\infty}\mathrm{Im}\,[\omega\sigma_{\alpha\beta}(\omega)] = \delta_{\alpha\beta}\frac{Ne^2}{\mathscr{V}m}. \tag{7.4.6}$$

Since $\varepsilon = 1+(4\pi/\omega)\,i\sigma(\omega)$, equation (7.4.5) also yields a sum rule for the imaginary part of the dielectric constant:

$$\int_0^{\infty}\omega\varepsilon_2(\omega)\,d\omega = \frac{2N\pi^2 e^2}{\mathscr{V}m}. \tag{7.4.7}$$

There is a second sum rule, which reads

$$\int_0^{\infty}\omega\,\mathrm{Im}\left(\frac{1}{\varepsilon}\right)d\omega = -\frac{2N\pi^2 e^2}{\mathscr{V}m} \tag{7.4.8}$$

which we shall now show can be obtained very easily from equation (7.4.7). Using the Kramers–Krönig relations (7.3.19) we can see that since $\varepsilon_1(\infty) = 1$, equation (7.4.7) is equivalent to

$$\lim_{\omega\to\infty}\omega^2[\varepsilon_1(\omega)-1] = -\frac{4N\pi e^2}{\mathscr{V}m}. \tag{7.4.9}$$

That is, at very high frequencies

$$\varepsilon_1(\omega) = 1-\frac{4N\pi e^2}{\mathscr{V}m}\bigg/\omega^2 \quad (\omega\to\infty), \tag{7.4.10}$$

which is just the expression for free electrons [see equation (7.4.4)]. In fact we could, by reversal of the above arguments, derive the sum rules (7.2.14) and (7.4.7) as consequences of the fact that the electrons respond to very high frequencies as if they were perfectly free.

From equation (7.4.10) we can readily obtain the large ω behaviour of

$$\varepsilon^{-1} = \frac{\varepsilon_1 - i\varepsilon_2}{\varepsilon_1^2 + \varepsilon_2^2}. \tag{7.4.11}$$

We find

$$\left.\begin{aligned} \operatorname{Im}\left[\frac{1}{\varepsilon(\infty)}\right] &= 0, \\ \operatorname{Re}\left[\frac{1}{\varepsilon(\infty)}\right] &= 1 \end{aligned}\right\} \tag{7.4.12}$$

and

$$\lim_{\omega\to\infty} \omega^2\left[\operatorname{Re}\left(\frac{1}{\varepsilon}\right) - 1\right] = -\frac{4N\pi e^2}{\mathscr{V} m}. \tag{7.4.13}$$

To obtain equation (7.4.8) it only remains to remark that since $\mathscr{E} = \varepsilon^{-1}\mathscr{D}$ the Kramers–Krönig analysis is applicable to ε^{-1}; hence just as equation (7.4.7) may be obtained from equation (7.4.9), so equation (7.4.8) can be got from equation (7.4.13).

The sum rule for ε_2 is often used to estimate the effective number of electrons contributing to processes up to a certain frequency, in which it is more useful than the sum rule for $\operatorname{Im} \varepsilon^{-1}$. We can see this by taking, say, the conduction electrons of a semi-metal. Below a certain threshold frequency at which the photons begin to kick up electrons from deeper levels only the conduction electrons will contribute to the absorption, and so to ε_2. This is untrue of $\operatorname{Im} \varepsilon^{-1}$, however, since this quantity involves $\varepsilon_1 - 1$ to which the inner electrons contribute at all frequencies. We shall illustrate this particular use of the sum rules by specific reference to the optical properties of graphite below.

7.4.1. Separation of conduction electron and interband effects

Another use of the sum rule concerns the relative contribution of conduction and interband effects. In the former effect the response of the electrons is confined to only the conduction band, an assumption made in section 7.2; in the latter, as already remarked, phonons cause transitions from one band to another. Separating the contributions of the two processes by writing

$$\varepsilon = \varepsilon^{\text{inter}} + \varepsilon^{\text{c}} \tag{7.4.14}$$

we can estimate the contribution of ε^{c} from equation (7.2.14), when we find

$$\lim_{\omega\to\infty} \omega^2(\varepsilon_1^{\text{c}} - 1) = -\frac{4\pi N_{\text{c}} e^2}{\mathscr{V} m^*} \tag{7.4.15}$$

and so, by the use of the Kramers–Krönig relations which is by now familiar

$$\int_0^\infty \omega\varepsilon_2^{c}(\omega)\,d\omega = \frac{2N_c\pi^2e^2}{\mathscr{V}m^*}, \tag{7.4.16}$$

where N_c is the number of conduction electrons and m^* the optical mass. It then follows immediately that

$$\int_0^\infty \omega\varepsilon_2^{\text{inter}}(\omega)\,d\omega + \frac{2N_c\pi^2e^2}{\mathscr{V}m^*} = \frac{2N\pi^2e^2}{\mathscr{V}m}. \tag{7.4.17}$$

Hence separating completely the contribution of the conduction electrons from the rest of ε_2 we can write $N = N_c$ on the right-hand side of equation (7.4.17), $\varepsilon_2^{\text{inter}}$ referring now only to conduction electrons. We see that the closer m^* is to m, the less the influence of transitions from the conduction band to other bands. Silver has an effective (optical) mass very close to unity so the influence of interband transitions should be almost non-existent. On the other hand, m^* is effectively infinite if a band is full so that all contributions to optical behaviour from such a band are from interband transitions.

One point worth remarking on is that even when an effective mass (different from m) is defined, the plasma frequency ω_p may actually be $\sqrt{(4\pi Ne^2/\mathscr{V}m)}$ rather than $\sqrt{(4\pi Ne^2/\mathscr{V}m^*)}$ for the plasma oscillation may occur at a frequency at which the free-electron asymptotic behaviour (7.4.10) has set in.

7.4.2 Optical behaviour of graphite

We single out graphite for discussion here since we can use it to illustrate several points made previously; we show in this particular case

(a) a clear separation, confirmed by the sum rule, of contributions of different bands to the optical properties;

(b) the correlation of optical properties with plasma resonances as revealed by fast-electron energy-loss experiments;

(c) the plasma resonance of electrons of a band depressed in frequency, from the free-electron value, by the contribution of other electrons to the real part of the dielectric constant, and, in contrast, a second resonance where the asymptotic behaviour of equation (7.4.10) can be assumed.

Graphite consists of sheets of carbon atoms, each sheet forming a two-dimensional hexagonal lattice; the sheets are rather far apart and are bound together by means of van der Waals forces. We obtain a fair idea of its band structure by considering a single sheet in the x–y plane by the tight binding method, where it is found that the bands of relevance may be placed in a natural correspondence with molecular orbitals denoted by σ and π, the former and latter being even and odd in z respectively. There are three occupied σ bands and an occupied π band (this being highest in energy). Each holds two electrons, so there are four electrons per atom, there being two atoms per unit cell. We refer the reader to the literature for further details, especially the papers by Coulson and Taylor (1958) and Corbato (1959). It is unnecessary for us to go into further detail here, except to say that overlapping bands in the three-dimensional case make the π-electrons conducting, for the

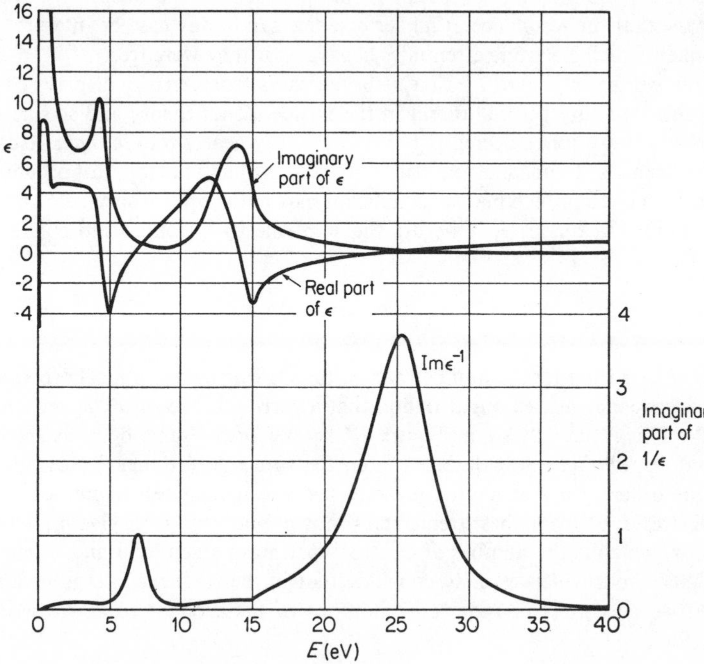

FIGURE 7.3. Optical constants of graphite. Results obtained by Kramers–Krönig analysis of reflectivity data (after Philipp, 1956).

analysis we shall describe relies no more on the above description than to take it as a framework within which to relate the experimental results. These results were obtained by Philipp (1965) and are shown in Figure 7.3. This figure was obtained by Philipp using a Kramers–Krönig analysis of reflectivity data [see equation (7.3.27)] and the further analysis of these data we describe below is also due to this author. There is one point that ought to be made before we describe this analysis, which is that whereas in section 7.2 we only identified longitudinal and transverse dielectric constants at long wavelength for isotropic substances, graphite is markedly anisotropic. Philipp has compared reflectance data for graphite and glassy carbon, and found them to be remarkably similar in structure. Further, the energy-loss function $-\text{Im}(1/\varepsilon)$ shown in Figure 7.3 agrees well with the results of fast-electron energy-loss experiments by Leder and Juddeth (1960) who find a weak loss at 7.2 eV and a much stronger and rather broad loss at 24.7 eV (these numbers varying somewhat from specimen to specimen). We shall later see that the resonance at higher energies can in any case be interpreted as one in which electrons essentially behave as if they were free.

Now we see in Figure 7.3 that as ω increases from zero ε_2 displays considerable structure before (absent in the Drude–Zener theory and so due to interband transitions), decreasing to become very nearly zero, at $\hbar\omega \simeq 10$ eV.

This leads us to imagine that below 10 eV ε_2 is due to the topmost occupied π-electron levels, which play no significant part in the optical spectrum above this energy. In this case, if we use the sum rule by writing, from equation (7.4.7),

$$\int_0^{\omega_c} \omega\varepsilon_2(\omega)\, d\omega = 2n_{\text{eff}}\,\pi^2 e^2/m, \tag{7.4.18}$$

where n_{eff} is the effective number of electrons taking part in optical processes up to frequency ω_c, we ought to find that n_{eff} rises to one as $\hbar\omega_c$ approaches 10 eV. This is actually so, as Figure 7.4 shows. In the same figure is shown the corresponding result obtained from the sum rule for $\text{Im}\,\varepsilon^{-1}$. This gives a quite different n_{eff} at low frequencies because it includes the influence of ε_1. However, at the highest frequencies shown, both sum rules give n_{eff} close to four, which is the number of electrons per atom given by σ and π bands together, from which we may conclude that in the frequency range investigated we need only consider contributions of these bands to ε. We write, therefore,

$$\varepsilon = \varepsilon^{\pi} + (\varepsilon^{\sigma} - 1), \tag{7.4.19}$$

taking the total medium as a superposition of separate media composed of σ and π electrons. To obtain quantitative results for ε^{π} and ε^{σ}, Philipp took $\varepsilon^{\sigma} = 1$ for $\hbar\omega < 9$ eV and let $\varepsilon_2^{\pi} \to 0$ in a similar fashion to the original ε_2 at energies above ~ 25 eV. The Kramers–Krönig relations were then used to obtain $\varepsilon_1^{\sigma}, \varepsilon_1^{\pi}$ being then obtained from equation (7.5.2). The energy loss function $\text{Im}(1/\varepsilon^{\pi})$ is plotted in Figure 7.5, and a plasma resonance is predicted at $\hbar\omega_p \simeq 11$ eV. Since we have seen that the sum rule for ε_2^{π} has almost saturated at this energy, giving one electron per atom, we ought to find

$$\varepsilon \simeq 1 - \left(\frac{\omega_p}{\omega}\right)^2 \tag{7.4.20}$$

in accordance with the asymptotic form of equation (7.3.9), with a free electron plasma frequency $\omega_p = (4\pi Ne^2/m)^{\frac{1}{2}}$. This yields $\hbar\omega = 12.6$ eV, close to the value predicted from Figure 7.5.

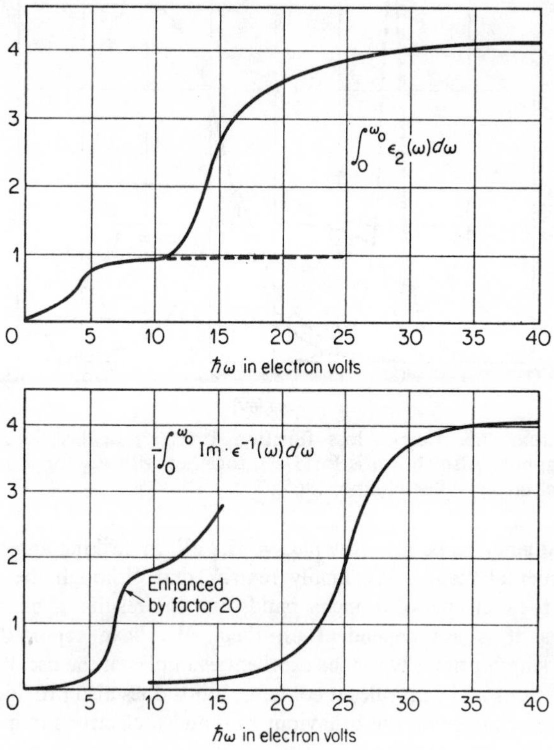

FIGURE 7.4. Effective number of electrons n_{eff} as function of frequency for graphite (after Philipp, 1965).

The influence of ε^σ shifts and broadens the peak in $\text{Im}(1/\varepsilon^\pi)$ to give the resonance as shown in Figure 7.3 at $\hbar\omega \simeq 7$ eV. We can say that the π-electrons are screened by a frequency-dependent dielectric constant arising from the behaviour of the σ-electrons. On the other hand, Figure 7.4 shows that the sum rule for ε_2 given by σ and π electrons *combined* has almost saturated at $n_{\text{eff}} = 4$, at $\hbar\omega \simeq 26$ eV, the position of the second resonance shown in Figure 7.3. The asymptotic form (7.4.20), with $\omega_p = (4\pi e^2 n/m)^{\frac{1}{2}}$ and n set equal to four per atom, should therefore yield the second resonance without modification, to a good approximation. In fact, we find the free-electron

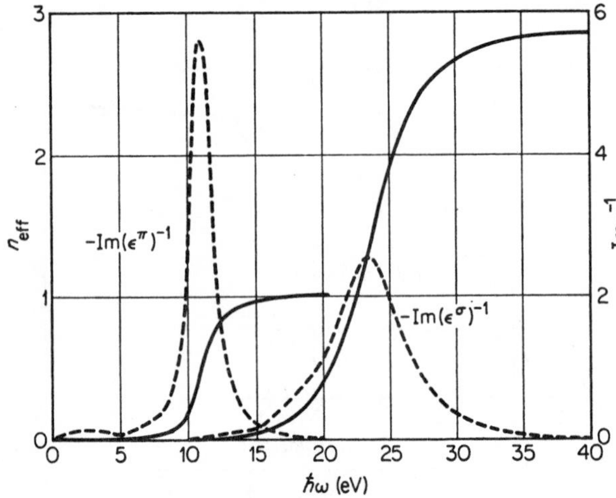

FIGURE 7.5. Energy loss function $\text{Im}(1/\varepsilon^{\pi})$ against ω for graphite. Also shown is $\text{Im}(1/\varepsilon^{\sigma})$, together with n_{eff} for π and σ electrons (after Philipp, 1965).

plasma frequency to be given by $\hbar\omega_p = 25.2$ eV, in striking agreement with the experimental results. We finally remark that although the analysis is guided by the concepts of σ and π bands and the results of band structure calculations, it is not dependent on them. We have reported Philipp's analysis at length primarily for the excellent examples of the use of Kramers–Krönig relations and sum rules it contains, but it does also provide considerable evidence concerning the behaviour of σ and π electrons in graphite.

7.5 Interband transitions

7.5.1 Transition matrix elements

We come now to investigate the "interband" contribution to the absorption. We shall assume, at least at first, that neither phonons nor electron–electron interactions play any part in the processes considered. We accordingly take the independent-electron form of equation (7.3.2), which, as we saw in the last chapter, is obtainable by writing

$$\rho_m \equiv f_0(\varepsilon_m) = \frac{1}{e^{\beta(\varepsilon_m-\zeta)}+1}, \tag{7.5.1}$$

when the interband contribution to σ_1 is obtained by taking the $|m\rangle$ and $|n\rangle$ to be one-electron functions belonging to different bands. Since $\mathbf{j}=e\mathbf{p}$ we must therefore evaluate the matrix element

$$\int \psi_{\mathbf{k}'}^{l*} \frac{\partial}{\partial \mathbf{r}} \psi_{\mathbf{k}}^{m}\, d\mathbf{r} = i\mathbf{k} \int \psi_{\mathbf{k}'}^{l*} \psi_{\mathbf{k}}^{m}\, d\mathbf{r} + \int u_{\mathbf{k}'}^{l*} \frac{\partial}{\partial \mathbf{r}} u_{\mathbf{k}}^{m}\, e^{i(\mathbf{k}-\mathbf{k}')\cdot\mathbf{r}}\, d\mathbf{r}. \tag{7.5.2}$$

Since $u_{\mathbf{k}'}^{l}$ and $u_{\mathbf{k}}^{m}$ are periodic in the crystal it is easily shown that the right-hand side is zero unless $\mathbf{k}'=\mathbf{k}+\mathbf{K}$, where $\mathbf{K}$ is a reciprocal lattice vector, and so

$$\int \psi_{\mathbf{k}'}^{l*} \frac{\partial}{\partial \mathbf{r}} \psi_{\mathbf{k}}^{m}\, d\mathbf{r} = \delta_{\mathbf{k}',\mathbf{k}+\mathbf{K}} N \int_{\Omega} u_{\mathbf{k}}^{l*} \frac{\partial}{\partial \mathbf{r}} u_{\mathbf{k}}^{m}\, d\mathbf{r}, \tag{7.5.3}$$

where N is the number of unit cells of volume Ω in the crystal. We see that crystal momentum is conserved in an interband transition, subject to the validity of the assumptions we made above. We may note in passing that the matrix element here is an approximation to the more exact expression $\int \psi_{\mathbf{k}}^{l*}(\partial/\partial\mathbf{r})\, \psi_{\mathbf{k}}^{m}\, e^{i\mathbf{q}\cdot\mathbf{r}}\, d\mathbf{r}$, where $\mathbf{q}$ is the photon wave number, and so the selection rule is properly

$$\mathbf{k}'=\mathbf{k}+\mathbf{K}+\mathbf{q}. \tag{7.5.4}$$

The wavelength of the light is usually long enough for us to safely ignore $\mathbf{q}$.

Let us take the simplest situation, where the dielectric constant is a scalar. With the substitutions (7.5.1) and (7.5.3) made in (7.3.2) we obtain

$$\varepsilon_2(\omega) = \frac{4\pi}{\omega}\sigma_1 = \frac{\hbar}{\pi^2}\left(\frac{e}{m}\right)^2 \sum_{lm} \int_{\Omega_B} d\mathbf{k} \frac{|\mathbf{p}_{lm}(\mathbf{k})|^2}{\omega^2} f_0(E_{\mathbf{k}}) \times \left[\delta\left(\omega - \frac{E_{lm}}{\hbar}\right) - \delta\left(\omega + \frac{E_{lm}}{\hbar}\right)\right], \tag{7.5.5}$$

where we have converted a sum over $\mathbf{k}$ to an integral in the usual way, and defined

$$E_{lm}(\mathbf{k}) = E_{\mathbf{k}}^{l} - E_{\mathbf{k}}^{m}, \tag{7.5.6}$$

$$\mathbf{p}_{lm}(\mathbf{k}) = \frac{\hbar}{\sqrt{(3)}\, i} \int_{\Omega} u_{\mathbf{k}}^{l*} \frac{\partial}{\partial \mathbf{r}} u_{\mathbf{k}}^{m}\, d\mathbf{r} = \frac{\hbar}{\sqrt{(3)}\, i} \int_{\Omega} \psi_{\mathbf{k}}^{l*} \frac{\partial}{\partial \mathbf{r}} \psi_{\mathbf{k}}^{m}\, d\mathbf{r}. \tag{7.5.7}$$

If ε_2 is not a scalar, we abstract from this formula the relevant component (deleting the factor $1/\sqrt{3}$).

These matrix elements can be written in a variety of ways. Most frequently they are written as "oscillator strengths" (a name originating in the Lorentz theory of insulators). We note that if $l=m$

$$\frac{\hbar^2}{m}\int_\Omega \psi_{\mathbf{k}}^{l*}\frac{\partial}{\partial \mathbf{r}}\psi_{\mathbf{k}}^m\,d\mathbf{r} = \int_\Omega \psi_{\mathbf{k}}^{l*}[\mathbf{r}, H]\psi_{\mathbf{k}}^m\,d\mathbf{r} = (E_{\mathbf{k}}^m - E_{\mathbf{k}}^l)\int_\Omega \psi_{\mathbf{k}}^{l*}\mathbf{r}\psi_{\mathbf{k}}^m\,d\mathbf{r} \tag{7.5.8}$$

and we define the oscillator strength as

$$f_{lm}(\mathbf{k}) = \frac{2m}{3\hbar^2}E_{lm}\left|\int_\Omega \psi_{\mathbf{k}}^{l*}\mathbf{r}\psi_{\mathbf{k}}^m\,d\mathbf{r}\right|^2 = \frac{2E_{lm}}{3m}\left|\frac{\mathbf{p}_{lm}(\mathbf{k})}{E_{lm}}\right|^2. \tag{7.5.9}$$

Of course, the precise form in which the matrix elements are used will depend on the type of solution of the band theoretical problem obtained, and in this connection we shall only quote one further general result. We note that

$$\int_\Omega \psi_{\mathbf{k}}^{l*}\frac{\partial}{\partial \mathbf{k}}\psi_{\mathbf{k}}^m\,d\mathbf{r} = i\int_\Omega \psi_{\mathbf{k}}^{l*}\mathbf{r}\psi_{\mathbf{k}}^m\,d\mathbf{r} + \int_\Omega u_{\mathbf{k}}^{l*}\frac{\partial}{\partial \mathbf{k}}u_{\mathbf{k}}^m\,d\mathbf{r}. \tag{7.5.10}$$

It is easily shown that

$$\int_\Omega \psi_{\mathbf{k}}^{l*}\frac{\partial}{\partial \mathbf{k}}\psi_{\mathbf{k}}^m\,d\mathbf{r} = 0 \tag{7.5.11}$$

and it then follows that

$$\int_\Omega \psi_{\mathbf{k}}^{l*}\mathbf{r}\psi_{\mathbf{k}}^m\,d\mathbf{r} = \frac{1}{i}\int_\Omega u_{\mathbf{k}}^{l*}\frac{\partial}{\partial \mathbf{k}}u_{\mathbf{k}}^m\,d\mathbf{r}. \tag{7.5.12}$$

Not nearly as much is known about the above matrix elements as one could wish, and, of course, accurate knowledge can for the most part be obtained only by detailed calculation except that symmetry considerations show certain matrix elements must be zero. We must content ourselves here with three illustrations of the general character of the matrix elements in different circumstances.

(a) *Transition of inner electron to nearly free electron band.* The most obvious characteristic of the matrix elements when examined in the form on the right-hand side of the equation

$$\int_\Omega \psi_{\mathbf{k}}^{l*}\frac{\partial}{\partial \mathbf{r}}\psi_{\mathbf{k}}^m\,d\mathbf{r} = \int_\Omega u_{\mathbf{k}}^{l*}\frac{\partial}{\partial \mathbf{r}}u_{\mathbf{k}}^m\,d\mathbf{r} \quad (l \neq m) \tag{7.5.13}$$

is that the closer one or both of the wavefunctions are to a plane-wave the smaller the matrix will become. The approximation of a single plane-wave for $\psi_{\mathbf{k}}^m$, which we take as the upper wave function, gives different results in the left- and right-hand sides of equation (7.5.13), since

$$\int_\Omega \psi_{\mathbf{k}}^{l*} \frac{\partial}{\partial \mathbf{r}} e^{i\mathbf{k}.\mathbf{r}}\, d\mathbf{r} = i\mathbf{k} v_l^*(\mathbf{k}), \tag{7.5.14}$$

where $v(\mathbf{k})$ is the momentum eigenfunction (see Chapter 1).

The left- and right-hand sides of equation (7.5.14) do become orthogonal if we orthogonalize $e^{i\mathbf{k}.\mathbf{r}}$ to $\psi_{\mathbf{k}}^l$. The plane-wave should in fact be orthogonal to all states lower than itself and be approximated by

$$\psi_{\mathbf{k}}^m = e^{i\mathbf{k}.\mathbf{r}} - \sum_i b^i(\mathbf{k})\, \psi_{\mathbf{k}}^i(\mathbf{r}), \tag{7.5.15}$$

where the orthogonality condition gives

$$\int_\Omega \psi^{j*} \left[e^{i\mathbf{k}.\mathbf{r}} - \sum_i b^i(\mathbf{k})\, \psi^i(\mathbf{r}) \right] d\mathbf{r} = 0, \tag{7.5.16}$$

i.e.

$$b^j(\mathbf{k}) = \Omega v_j^*(\mathbf{k}). \tag{7.5.17}$$

Equation (7.5.15) gives us

$$\int_\Omega \psi_{\mathbf{k}}^{l*} \frac{\partial}{\partial \mathbf{r}} \psi_{\mathbf{k}}^m\, d\mathbf{r} = ik\Omega v_l^*(\mathbf{k}) - \sum_i b^i(\mathbf{k}) \int_\Omega \psi_{\mathbf{k}}^{l*} \frac{\partial}{\partial \mathbf{r}} \psi_{\mathbf{k}}^i\, d\mathbf{r}. \tag{7.5.18}$$

Hence, since (using atomic units)

$$\int_\Omega \psi_{\mathbf{k}}^{l*} \frac{\partial}{\partial \mathbf{r}} \psi_{\mathbf{k}}^l\, d\mathbf{r} = i \frac{\partial E_l(\mathbf{k})}{\partial \mathbf{k}}, \tag{7.5.19}$$

orthogonalization of $e^{i\mathbf{k}.\mathbf{r}}$ to $\psi_{\mathbf{k}}^l$ means that (7.5.14) is replaced by

$$\int_\Omega \psi_{\mathbf{k}}^{l*} \frac{\partial}{\partial \mathbf{r}} [e^{i\mathbf{k}.\mathbf{r}} - b^i(\mathbf{k})\, \psi_{\mathbf{k}}^l]\, d\mathbf{r} = i \left[\mathbf{k} - \frac{\partial E_l(\mathbf{k})}{\partial \mathbf{k}} \right] \Omega v_l^*(\mathbf{k}). \tag{7.5.20}$$

Only if $\psi_{\mathbf{k}}^{l*}$ pertains to a band arising from deep-lying atomic levels will the band be very narrow and $\partial E_l/\partial \mathbf{k} \ll \mathbf{k}$, when we regain equation (7.5.14). However, equation (7.5.20) shows that other terms have, in general, been left out of account which can be expected to modify equation (7.5.14) considerably.

(b) *Tight-binding transitions.* In terms of Wannier functions $w(\mathbf{r})$

$$\begin{aligned} \int_\Omega \psi_{\mathbf{k}}^{l*} \frac{\partial}{\partial \mathbf{r}} \psi_{\mathbf{k}}^m\, d\mathbf{r} &= \sum_{\mathbf{RS}} \int_\Omega w_i^*(\mathbf{r}-\mathbf{R}) \frac{\partial w_j(\mathbf{r}-\mathbf{S})}{\partial \mathbf{r}}\, d\mathbf{r}\, e^{i\mathbf{k}.(\mathbf{S}-\mathbf{R})} \\ &= \sum_{\mathbf{RS}} \int_\Omega w_i^*(\mathbf{r}-\mathbf{R}) \frac{\partial}{\partial \mathbf{r}} w_j(\mathbf{r}-\mathbf{S}-\mathbf{R})\, e^{i\mathbf{k}.\mathbf{S}} \\ &= \sum_{\mathbf{S}} \int_{\text{all space}} w_i^*(\mathbf{r}) \frac{\partial w_j(\mathbf{r}-\mathbf{S})}{\partial \mathbf{r}}\, e^{i\mathbf{k}.\mathbf{S}}\, d\mathbf{r}. \end{aligned} \tag{7.5.21}$$

The final expression shows that the more localized the Wannier function, the weaker we expect the **k**-dependence to be. In the tight-binding approximation we take the w's to be atomic orbitals ϕ. Then if overlap terms are neglected we have

$$\int_{\Omega} \psi_{\mathbf{k}}^{l*} \frac{\partial}{\partial \mathbf{r}} \psi_{\mathbf{k}}^{m}\, d\mathbf{r} = \int_{\text{all space}} \phi_i^*(\mathbf{r}) \frac{\partial}{\partial \mathbf{r}} \phi_j(\mathbf{r})\, d\mathbf{r}, \tag{7.5.22}$$

independent of **k**. It must be noted that **k**-dependence may be badly given (apart from order of magnitude relative to the right-hand side of the above) with such a simple tight-binding approximation, because the overlap of the true Wannier functions will in general be quite different to that of $\phi_i(\mathbf{r})$ and $\phi_j(\mathbf{r}-\mathbf{R})$, since the Wannier functions have to satisfy.

$$\int w^*(\mathbf{r})\, w(\mathbf{r}-\mathbf{R})\, d\mathbf{r} = \delta(\mathbf{R}, 0). \tag{7.5.23}$$

Equation (7.5.22) is the relevant matrix element in soft X-ray transitions (*q.v.*) between core states and also appears in equation (7.5.5). Transitions between core states are not otherwise of interest to us in optical phenomena, though, since the Pauli principle forbids such transitions. However, there is a rather different situation where tight binding serves as a first approximation. In ionic crystals cohesion obtains as a result of attraction between oppositely charged ions rather than as a result of the presence of an electron gas. In consequence the equilibrium positions of the ions of an alkali halide such as NaCl are much further apart, in units of ionic radii, than those of a metal. The tight-binding approximation is therefore much closer to the truth than a nearly free-electron model. One interesting point in this connection is that bands are associated with either the Na^+ ion or Cl^- ion as Figure 7.6, illustrating the formation of bands as the constituent ions of a solid are brought together, makes immediately clear. The conduction band arises from the Na 3s level and, being empty, can receive electrons from lower bands. Thus an optical transition from valence to conduction band will have the matrix element

$$\sum_{\mathbf{R}} \int \phi_{\text{Na 3s}}(\mathbf{r}) \frac{\partial}{\partial \mathbf{r}} \phi_{\text{Cl 3p}}(\mathbf{r}-\mathbf{R})\, d\mathbf{r}, \tag{7.5.24}$$

where the sum is over the sites of nearest neighbours to a sodium ion. Actually the above expression is only very roughly right, for the Cl^- continuum will contribute to the formation of the conduction band (Slater and Shockley, 1936), but neglecting this we see that the matrix element is independent of **k**. Incidentally, the energy gap between valence and conduction band is so great in sodium chloride that interband transitions do not begin until the frequency of the light is increased into the far ultraviolet. Below the threshold frequency for interband transitions the crystal is quite transparent.

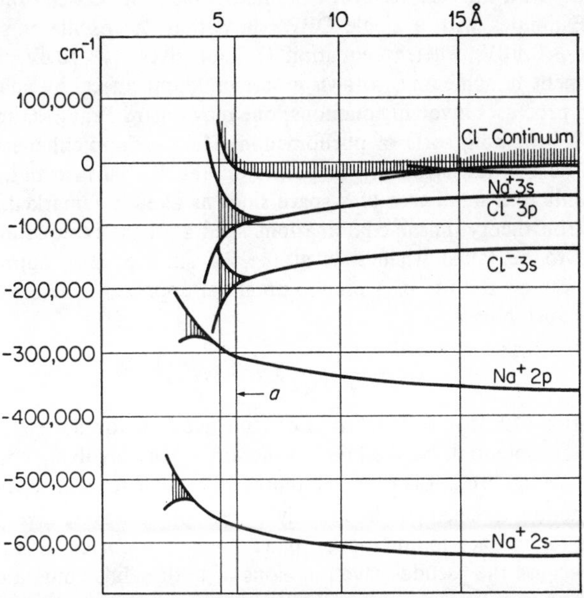

FIGURE 7.6. Energy levels of NaCl as function of interionic spacing: *a* is observed interatomic distance (after Slater and Shockley, 1936).

(c) *Transitions between nearly free-electron bands.* These transitions will in general be quite weak. Approximate expressions in terms of single orthogonalized plane-waves are easily obtained and we shall not discuss them. We wish only to draw attention to the strong enhancement when **k** is near a BZ face. As is well known, a first-order perturbation treatment of the effect of a periodic potential on a free electron leaves the energy unchanged except when **k** is at a point on the BZ such that **k**+**K** (**K** a reciprocal lattice vector) is also. Then degenerate perturbation theory must be used, the zeroth-order wave functions being taken as the linear combinations

$$\psi_{\mathbf{k}}^1 = \frac{1}{\sqrt{(2\Omega)}}(e^{i\mathbf{k}.\mathbf{r}} + e^{i(\mathbf{k}+\mathbf{K}).\mathbf{r}}),$$

$$\psi_{\mathbf{k}}^2 = \frac{1}{\sqrt{(2\Omega)}}(e^{i\mathbf{k}.\mathbf{r}} - e^{i(\mathbf{k}+\mathbf{K}).\mathbf{r}}). \qquad (7.5.25)$$

Then

$$\int_{\Omega_B} \psi_{\mathbf{k}}^{1*} \frac{\partial}{\partial \mathbf{r}} \psi_{\mathbf{k}}^2 \, d\mathbf{r} = -i\mathbf{K} \tag{7.5.26}$$

in contrast with the zero result for the matrix element between single plane-waves. Estimates with a single OPW show that the oscillator strength is typically ~ 0.1 eV, whereas equation (7.5.26) gives $f_{12} \sim 10$ eV. This large enhancement is sometimes known as an umklapp effect, by analogy with umklapp processes involving phonons: one may regard Bragg scattering to be responsible for both sorts of phenomenon. The umklapp enhancement of f will not be restricted entirely to points on the BZ surface but, still, will pertain only to a small area of **k**-space since as already remarked, in nearly free-electron theory, linear combinations such as those of equation (7.5.25) are only to be taken when **k** is at the BZ surface. It is appropriate to remark here on the use of a pseudo-potential approach where we write an OPW expansion as

$$\psi_{\mathbf{k}}^{m} = \frac{1}{N}\left[\phi_{\mathbf{k}} - \sum_i b^i(\mathbf{k})\, \psi_{\mathbf{k}}^i(\mathbf{r})\right] \tag{7.5.27}$$

[cf. equation (7.5.15)]. $\phi_{\mathbf{k}}$ is the part of the wave function determined by use of a pseudo-potential. Now while in general the part involving core orbitals $\psi_{\mathbf{k}}^i$ cannot be ignored, it is found (Phillips, 1966) that for silicon the pseudo-wave function alone, containing only a few plane-waves, gives oscillator strengths for valence to conduction band transitions correct to within 10%. This is because the pseudo-wave functions of both bands contain essentially the same plane-waves, the difference being that the valence band $\phi_{\mathbf{k}}$ is even with respect to exchange of two atoms in the unit cell while the corresponding conduction-band function is odd under the same operation. The situation is similar to that depicted in equation (7.5.25) (though for different reasons, of course) and the oscillator strength is quite large.

7.5.2 Structure of optical spectra

The most obvious thing we can say about the structure of the interband contribution to the optical spectrum of a solid is that it will be zero below a certain threshold frequency, as we can see by reference to Figure 7.7, where we indicate in a one-dimensional example the types of direct interband transition possible. The lowest energy, $\hbar\omega_1$, involved in a transition from an inner band to the conduction band, will obtain for the transition marked A, the transition being to the Fermi level; the lowest energy involved in transitions from the conduction band to the unoccupied band is indicated as $\hbar\omega_2$. Which energy will be the lower cannot be decided *a priori*, and, of course, the situation may be complicated by overlapping bands. But, clearly, a threshold energy occurs. This will often be closely connected with the colour of the metal. For example, interband contributions to the optical behaviour of copper begin with a strong absorption peak at $\hbar\omega \sim 2.5$ eV, and this absorption gives the metal its characteristic colour. The reason why a peak occurs so near to the threshold frequency will be examined later.

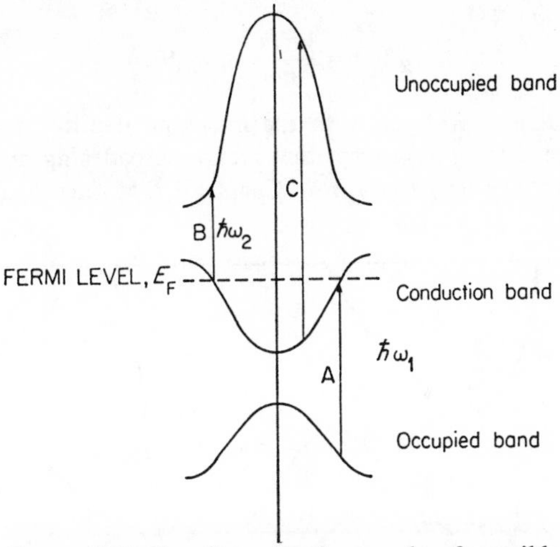

FIGURE 7.7. One-dimensional example of possible types of interband transition. Transition A is from occupied band to Fermi level; transition B from Fermi level to unoccupied band; transition C from conduction state to unoccupied band.

To investigate the structure of interband effects after the onset of absorption we turn to equation (7.5.5). We assume the temperature is effectively zero, so that there are no downward transitions, and write as the contribution to ε_2 by transitions from an occupied band l to an unoccupied band m

$$\varepsilon_2^{lm}(\omega) = \frac{4\pi^2 e^2 \hbar^2}{m\omega\Omega_B} \int_{\Omega_B} d\mathbf{k} f_{lm}(\mathbf{k})\, \delta\left(\omega - \frac{E_{lm}}{\hbar}\right). \tag{7.5.28}$$

We now change variables so that $\int_{\Omega_B} d\mathbf{k}$ is represented as an energy integration and integration over a surface of constant energy:

$$\begin{aligned}\varepsilon_2^{lm}(\omega) &= \frac{4\pi^2 e^2 \hbar^2}{m\omega\Omega_B} \int dE_{lm} \int_S dS f_{lm}(\mathbf{k}) \left|\frac{\partial E_{lm}}{\partial \mathbf{k}}\right|^{-1} \delta\left(\omega - \frac{E_{lm}}{\hbar}\right) \\ &= \frac{4\pi^2 e^2 \hbar^2}{m\omega\Omega_B} \int_{E_{lm}=\hbar\omega} dS f_{lm}(\mathbf{k}) \left|\frac{\partial E_{lm}}{\partial \mathbf{k}}\right|^{-1}. \end{aligned} \tag{7.5.29}$$

Suppose we now take $f_{lm}(\mathbf{k})$ to be independent of the direction of $\mathbf{k}$; by definition of an interband density of states

$$n_{lm}(E_{lm}) = \frac{1}{\Omega_B}\int_{E_{lm}} dS\left|\frac{\partial E_{lm}}{\partial \mathbf{k}}\right|^{-1}, \tag{7.5.30}$$

we can write

$$\varepsilon_2^{lm}(\omega) = \frac{4\pi^2\hbar e^2}{m\omega} f_{lm}\, n_{lm}(\hbar\omega). \tag{7.5.31}$$

We can see that any structure in the interband density of states will be directly reflected in $\varepsilon_2^{lm}(\omega)$ except that lifetime broadening may mask some effects. Further, we may recall from Chapter 1 that since $E_{lm}(\mathbf{k})$ is periodic

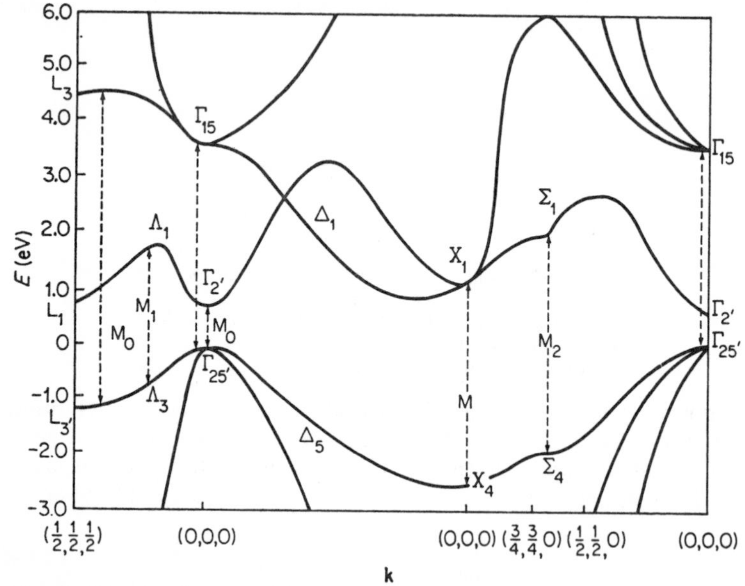

FIGURE 7.8. Pseudo-potential band structure of germanium along principal symmetry lines in BZ.

in $\mathbf{k}$-space, $n_{lm}(\hbar\omega)$ will indeed possess considerable structure. Umklapp effects will aid the structural effect we have just mentioned since the former occurs in a small region of $\mathbf{k}$-space surrounding a point for which $\partial E/\partial\mathbf{k} = 0$. This umklapp enhancement can turn an edge (a van Hove singularity) into a maximum.

Homopolar semi-conductors provide excellent examples of the above, and we shall take germanium in particular. In Figure 7.8 we reproduce pseudopotential energy bands for this material with transitions responsible for the structure marked.

The notation M_0, M_1, M_2, M_3 refers, in the notation of Chapter 1, to minima, the two kinds of saddle point and maxima. We have added superscripts, where necessary, to distinguish different transitions for the same kind of van Hove singularity. Figure 7.9 shows the resulting contributions to

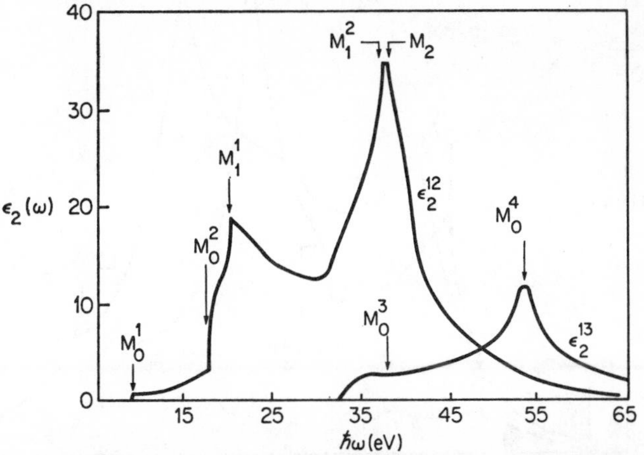

FIGURE 7.9. Interband contributions to ε_2 for germanium (after Phillips, 1965).

ε_2, ε_2^{12} referring to transitions from band 1 to 2 and ε_2^{13} being similarly defined. The structure, due to van Hove singularities, in the interband contribution ε_2^{12} faithfully reflects that in n_2^{12}, and so the reader will see, without needing to refer back to Chapter 1, that a minimum in E_{lm} cannot itself produce a peak. The transitions M_0^4 nevertheless produce a peak in ε_2^{13}, as Figure 7.10 shows, and this illustrates the effect umklapp enhancement (in this case occurring about the point L) can have. The superposition of the two contributions in Figure 7.10 provides most of the structure in the theoretical $\varepsilon_2(\omega)$ for germanium, which is in excellent agreement with experiment, as Figure 7.10 shows.

We want to make a few remarks about copper, with its abrupt increase in ε_2 from zero as shown in Figure 7.11. It is stressed by Ehrenreich (1966) that this sharp peak is not a manifestation of a van Hove singularity, but rather a reflection of the flatness of the d band from which transitions take place, as can be seen by inspection of Figure 7.12. In equation (7.5.29) the integral is restricted not only by the condition $\hbar\omega = E_{lm}$ but also the requirement that the upper energy E_m lies above the Fermi level. Transitions from the flat band come into play very rapidly in accordance with this, resulting in the

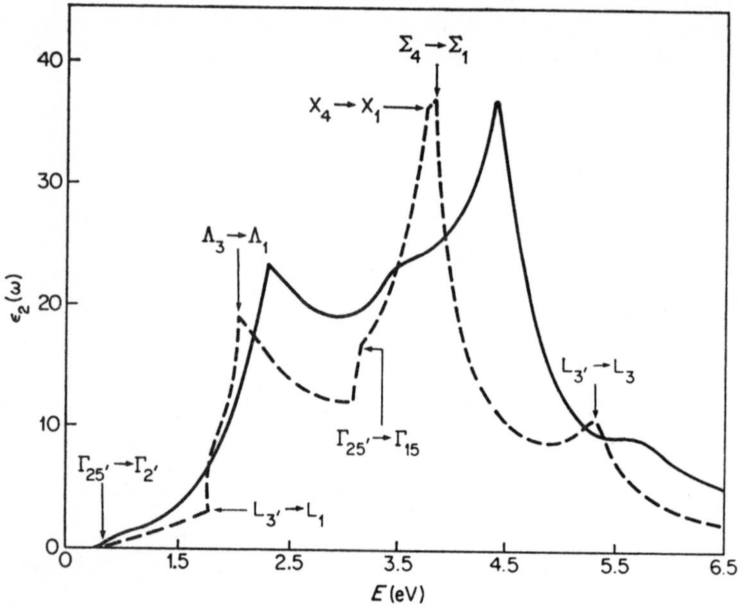

FIGURE 7.10. Comparison of theory and experiment for $\varepsilon_2(\omega)$ for germanium. Dashed line is theoretical curve (after Phillips, 1965).

sharp increase in ε_2. This is illustrated by Figure 7.13 where surfaces of constant $\hbar\omega = E_{lm}$ are indicated, which, because of the flatness of the lower band, are coincident with the constant-energy surfaces of the upper band. Thus the parts of the surface of constant $\hbar\omega$ cross the Fermi surface simultaneously, transitions for every **k**-direction taking place, as the phonon energy is increased, and a sudden rise in ε_2 occurs. A similar explanation may be offered for the initial form of ε_2 for silver.

The second peak has a possible explanation as the reflection of a van Hove singularity in ε_2, the relevant transitions being $L_1 \to L_2$ (see Figure 7.12). We shall return to copper in section 7.6.2.

7.5.3 Indirect transitions

The work of Philipps and others on Ge, which we discussed above, exemplified in detail the structure in the optical spectra arising from inter-band contributions to ε_2 because the semi-conductors show these effects most clearly. The data for metals, in contrast, do not reveal at all clearly the effects we have discussed. Indeed, although one can tentatively connect some of the structure in spectra to van Hove singularities, other aspects can

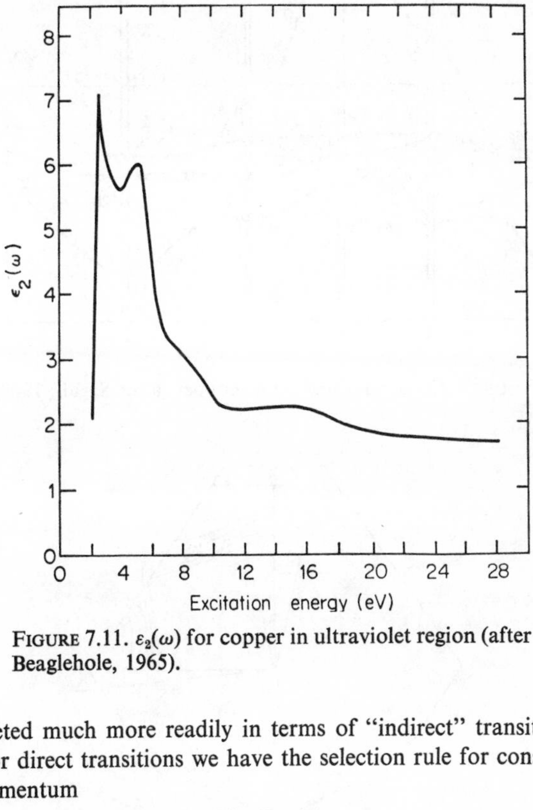

FIGURE 7.11. $\varepsilon_2(\omega)$ for copper in ultraviolet region (after Beaglehole, 1965).

be interpreted much more readily in terms of "indirect" transitions. Thus, whereas for direct transitions we have the selection rule for conservation of crystal momentum

$$\mathbf{k}' = \mathbf{k} + \mathbf{K}, \tag{7.5.32}$$

"indirect" transitions do *not* obey this rule.

One possible mechanism for indirect transitions is provided by the electron–phonon interaction. From Chapter 6, we recall that, following Bloch's

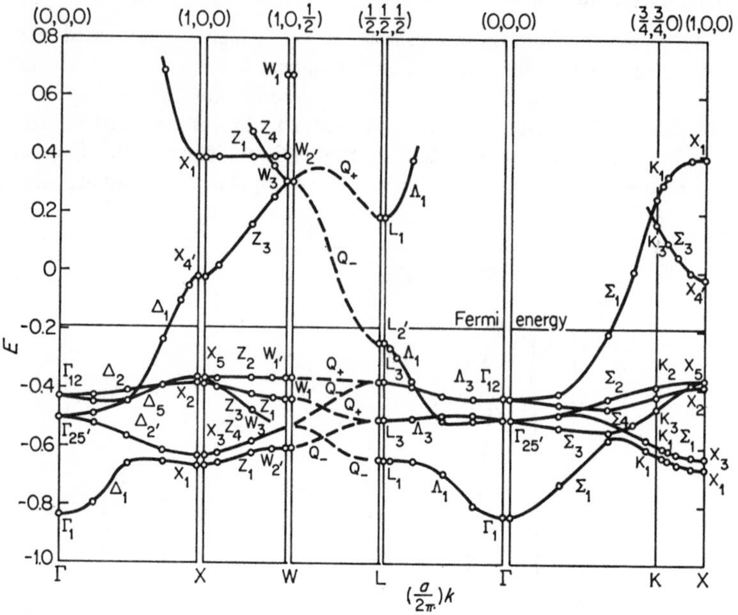

FIGURE 7.12. Band structure of copper (after Segall, 1962).

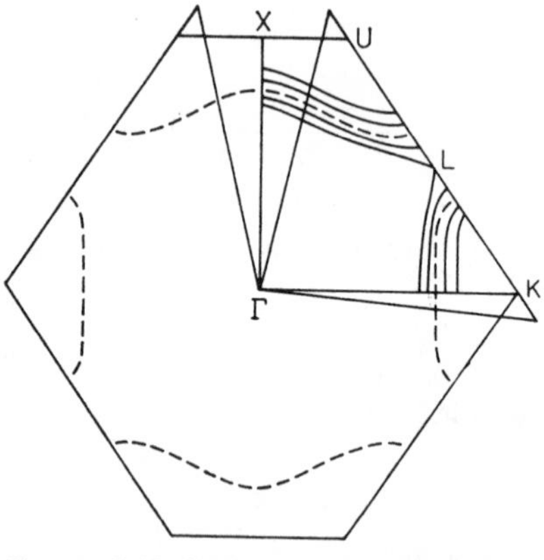

FIGURE 7.13. (110) cross-section of BZ of Cu. Dotted curves: Fermi surface. Full curves: $\hbar\omega = E_{lm}$.

approach, wè can write down a perturbed wave function as

$$\Psi = \psi_{\mathbf{k}} + \sum_{\mathbf{k}'}{}' \frac{M_{\mathbf{k}'\mathbf{k}}}{E_{\mathbf{k}'} - E_{\mathbf{k}}} \psi_{\mathbf{k}'}, \tag{7.5.33}$$

where the $\psi_{\mathbf{k}}$'s are Bloch functions and $M_{\mathbf{k}'\mathbf{k}}$ is the electron–phonon matrix element. It is readily seen that, to lowest order, we obtain

$$\langle \Psi^{l}_{\mathbf{k}'} \nabla \Psi^{m}_{\mathbf{k}} \rangle = \delta_{\mathbf{k}'+\mathbf{K},\mathbf{k}} \langle \psi^{l*}_{\mathbf{k}l} \nabla \psi_{\mathbf{k}} \rangle + \left(\frac{M_{\mathbf{k}'\mathbf{k}}}{E_{\mathbf{k}} - E_{\mathbf{k}'}} \langle \psi^{l*}_{\mathbf{k}'} \nabla \psi_{\mathbf{k}} \rangle + \text{complex conjugate} \right). \tag{7.5.34}$$

The subsequent calculation of ε_2 is straightforward, since ε_2 contains the matrix element in the form $|\langle \Psi^{l*}_{\mathbf{k}'} \nabla \Psi_{\mathbf{k}} \rangle|^2$; it evidently also contains $|M_{\mathbf{k}'\mathbf{k}}|^2/(E_{\mathbf{k}} - E_{\mathbf{k}'})^2$, which can be handled as in Chapter 6. We saw there that a simple treatment, having some validity for the alkali metals at least, gave

$$|M_{\mathbf{k}'\mathbf{k}}|^2 \propto \frac{|V_{\mathbf{q}}|^2}{|\varepsilon(\mathbf{q}, 0)|^2} \quad (\mathbf{q} = \mathbf{k}' - \mathbf{k}), \tag{7.5.35}$$

where $V_{\mathbf{q}}$ is the bare-ion potential. It has been shown by Hopfield (1965), as we shall shortly discuss, that, to this order, the relevant transition rate is not given by equation (7.5.35) but rather by

$$\frac{|V_{\mathbf{q}}|^2}{|\varepsilon(\mathbf{q}, \omega)|^2}, \tag{7.5.36}$$

where ω is the photon frequency. At the plasma frequency, as we saw in Chapter 2, $\varepsilon(\mathbf{q}, \omega) = 0$ and thus, in the ultraviolet, there will be a large lattice-induced absorption, just as is observed.

Role of dynamical screening. Hopfield derived equation (7.5.36) by considering a uniform gas, into which the ions are inserted as perturbing agents, and he calculated the change, to second order in $V_{\mathbf{q}}$, of ε_2 (without division into intraband and interband terms) from its free electron value. In calculating $\varepsilon_2(\omega)$, one then considers the oscillations of the charge density at frequency ω and thus, in calculating the linear response to $V_{\mathbf{q}}$, the relevant dielectric constant is $\varepsilon(\mathbf{q}, \omega)$ rather than $\varepsilon(\mathbf{q}, 0)$.

There are other resonances (peaks in the optical spectra), the origins of which we shall discuss later on. However, we ought, at this point, to remark on the role of electron–electron interactions in inducing indirect transitions. These interactions do indeed produce resonances, as we shall see later. But these are collective effects, which can in no way be wedded to the independent electron description with which we are concerned here. If we consider the excitation of a single particle, a reference to Chapter 8, section 8.3 shows immediately that in no sense can we think of electron–electron interactions inducing indirect transitions. It is, however, possible that electrons excited by the photon field might suffer collisions to give what appear to be indirect transitions in the optical spectrum. Electron–electron collisions are very significant in photoemission, as we shall see.

7.6 Photoemission

7.6.1 Surface and volume effects

Photoemission is the ejection of electrons from a metal by the action of light. Let us first see what information on this effect the free-electron model of a metal yields. As indicated in Figure 7.14 the electrons are supposed to

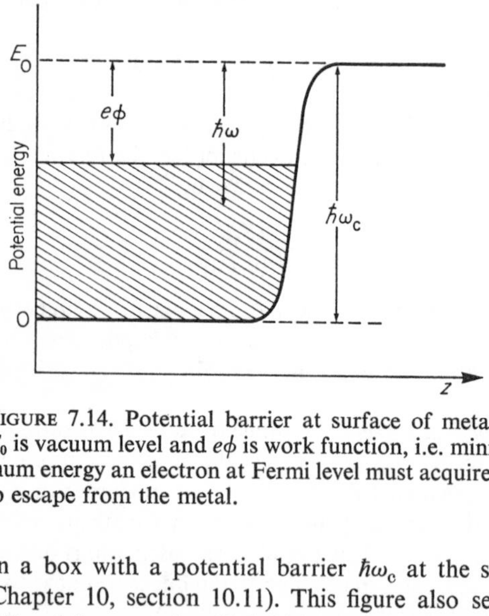

FIGURE 7.14. Potential barrier at surface of metal. E_0 is vacuum level and $e\phi$ is work function, i.e. minimum energy an electron at Fermi level must acquire to escape from the metal.

be contained in a box with a potential barrier $\hbar\omega_c$ at the surface (cf. the discussion in Chapter 10, section 10.11). This figure also serves to define "work-function" and "vacuum level" (which may be defined quite generally without particular reference to the free-electron model). The photocurrent I, proportional to the number of electrons per second escaping from the surface of the metal, is

$$I = \mathrm{e} \int_0^\infty p_z n(p_z)\, D(p_z)\, dp_z, \tag{7.6.1}$$

where we have assumed the surface to lie in the (x–y plane) and $n(p_z)$ is the number of electrons with momentum p_z in the range dp_z. D expresses the probability of escape. When light of frequency ω is shone on the metal, we assume it is possible for an electron to escape provided

$$\frac{p_z^2}{2m}+\hbar\omega \geqslant E_0 \tag{7.6.2}$$

and we shall further assume that if this equality holds

$$D = \text{constant}. \tag{7.6.3}$$

Since the electrons are distributed in energy according to Fermi–Dirac statistics,

$$n(p_z) = \frac{2}{h^3}\int_{-\infty}^{\infty} dp_x\, dp_y \left\{\exp\left[\left(\frac{p^2}{2m}-\zeta\right)\Big/k_{\mathrm{B}}T\right]+1\right\}^{-1} \tag{7.6.4}$$

from which we obtain

$$I = \frac{4\pi me(k_{\mathrm{B}}T)^2 D}{h^3}\int_{-\infty}^{x} \ln(1+e^y)\, dy, \quad x = \frac{\hbar\omega - e\phi}{k_{\mathrm{B}}T} \tag{7.6.5}$$

or

$$I = AT^2 f\left(\frac{\hbar\omega - e\phi}{k_{\mathrm{B}}T}\right), \tag{7.6.6}$$

where

$$f(x) = \int_{-\infty}^{x} \ln(1+e^y)\, dy, \tag{7.6.7}$$

that is,

$$\left.\begin{aligned} f(x) &= e^x - \frac{e^{2x}}{2^2}+\frac{e^{3x}}{3^2}-\dots \quad (x \leqslant 0) \\ &= \frac{\pi^2}{6}+\frac{x^2}{2}-e^{-x}+\frac{e^{-2x}}{2^2}-\frac{e^{-3x}}{3^2}+\dots \quad (x \geqslant 0). \end{aligned}\right\} \tag{7.6.8}$$

Equation (7.6.6), first derived by Fowler (1931), is in excellent agreement, in functional form, with the observed photocurrents of such metals as silver and gold between, say, 300 and 1000 K. It also gives at large T and $\omega = 0$ (no applied field)

$$I = A'T^2 e^{-e\phi/k_{\mathrm{B}}T}, \tag{7.6.9}$$

which is Richardson's equation for thermionic emission. The observed constants A and A' in equations (7.6.6) and (7.6.9) are naturally not equal, because the A of equation (7.6.6) includes such things as probability of excitation of an electron.

Quantum-mechanical description of excitation. We accordingly turn to the quantum-mechanical description of excitation of electrons by photons. We note, first of all, that free electrons cannot be excited for the matrix element $\langle e^{-i\mathbf{k}'.\mathbf{r}} \nabla e^{i\mathbf{k}.\mathbf{r}} \rangle$ is zero. Put another way, a free electron cannot take up the energy of a photon and simultaneously obey the momentum conservation law. For an electron to do this, a third body, capable of giving the electron the necessary momentum, is needed. The potential barrier can play the role of such a third body with the wave function going to zero exponentially outside it; the equation we have to solve for the electron is

$$-\frac{\hbar^2}{2m}\nabla^2\psi+\frac{\hbar}{i}\frac{\partial\psi}{\partial t}-V\psi=-\frac{ie\hbar}{mc}(\mathscr{A}.\nabla\psi) \tag{7.6.10}$$

with, as the simplest approximation possible,

$$\left.\begin{aligned} V &= -E_0 \quad (x<0) \\ &= 0 \quad (x>0). \end{aligned}\right\} \tag{7.6.11}$$

This problem is soluble (see, for example, Sokolov, 1967) but we shall not treat it here because conditions at the surface of a real metal are much more complex than indicated by equation (7.6.11). Moreover, this "surface effect" in photoemission appears experimentally to be much less important than the "volume effect", where the crystal itself is the third body ensuring momentum conservation. One way of investigating this question is to obtain the photoelectric current I from the thin films as a function of film thickness. The results obtained by Thomas (1957) for potassium evaporated into quartz are shown in Figure 7.15. The predominance of volume over surface effect is strongly indicated by linearity of the photocurrent with thickness up to saturation. That saturation is reached is explained by the degradation of energy in electron collisions and, of particular importance in the alkali metals, plasma excitation by the electrons. The forms of the curves could be explained otherwise, as due to the dependence of the surface condition and work function on thickness; the saturation of I at different thicknesses for different photon energies might be a result of the finite penetration depth, of the light into the specimen, varying with wavelength—we remember the amplitude decays exponentially with a decay constant which is frequency dependent. However, it is difficult to see how these explanations of different aspects of the curves of Figure 7.15 fit together, and in any case the absorption of light in the alkalis is known to be negligible to first order even to thicknesses of some hundred ångströms.

Other tests that the phenomenon is a volume effect are possible, for example by correlation of the photoelectron energy spectrum, for given photon energy, with the form expected on the basis of a band theory treatment of the electrons, but we shall simply assume the surface effect is negligible in what follows.

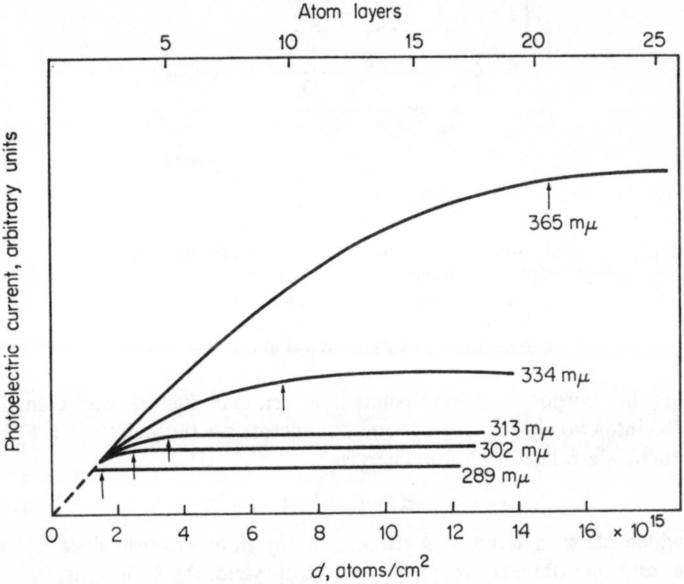

FIGURE 7.15. Photoelectric yield from potassium films. Saturation indicated by arrows (after Thomas, 1957).

It should be noted that although in the full quantum-mechanical treatment of the free electron model photoemission is entirely a surface effect, Fowler's law does not rely on this, for in its derivation a free-electron model was used solely to approximate the distribution of the electrons over energy states and no assumption was made as to the excitation mechanism except such as is entailed in setting D equal to a constant as in equation (7.6.3).

A word must be said about semi-conductors, where even at absolute zero the Fermi energy and topmost occupied level do not coincide. ζ is properly the chemical potential appearing in the Fermi–Dirac distribution $\{\exp[\beta(\varepsilon-\zeta)]+1\}^{-1}$ and so lies in the energy gap between fully occupied valence band and empty conduction band, as shown in Figure 7.16. It can be seen that the threshold photon energy for photoelectrons is no longer $e\phi$ but $e\phi+\delta$.

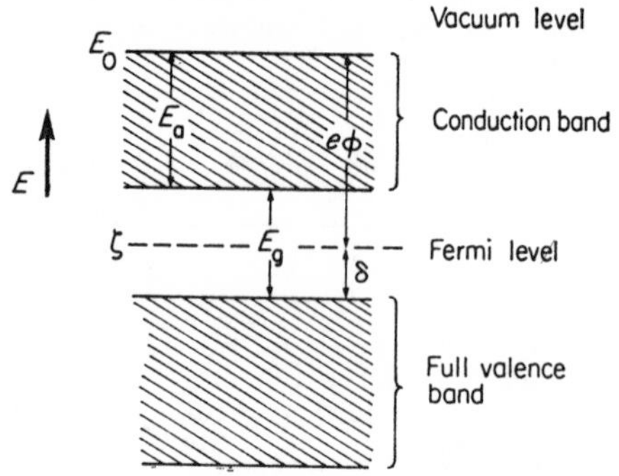

FIGURE 7.16. Situation in semiconductor, showing difference E_a between bottom of conduction band and vacuum level (E_a = "electron affinity"). δ measures difference between top of valence band and Fermi level.

For the purposes of photoemission, semi-conductors are commonly divided into two classes. Type 1 semi-conductors are those for which $E_a < E_g$, and there is a range of photon energies

$$e\phi+\delta = E_a+E_g < h\nu < E_a+2E_g \tag{7.6.12}$$

for which electron–electron scattering of the photoelectron does not take place, and one obtains a high photoelectron yield. As $h\nu$ increases beyond E_a+2E_g and it is possible for the original excited electron to excite an electron–hole pair, the energy distribution changes abruptly to a low value. Type II semi-conductors, on the other hand, have $E_a > E_g$ and characteristically have low yields, since any electron which has enough energy to escape also has enough energy to knock a second electron across the energy gap between valence and conduction bands.

7.6.2 Interpretation of spectra

To interpret the observed energy distribution of photoelectrons for a given incident frequency of light we again discuss the problem within a one-electron framework. Eventually many-body effects must be invoked for a complete explanation and we shall make some brief comments on these later.

We might expect, as a first approximation, that the energy spectrum of the photoelectrons will be related to the density of states within the metal. This is indeed roughly so in aluminium, as Figure 7.17, taken from Wooten and

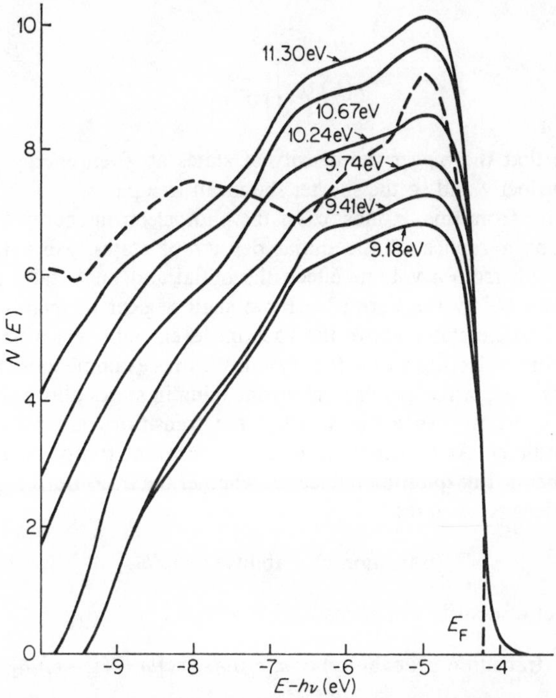

FIGURE 7.17. Photoelectron energy spectra from aluminium, labelled according to incident photon energy. Dotted curve is density of states calculated by Ashcroft (after Wooten *et al.*, 1966).

co-workers (1966) shows. However, the collisions an electron suffers as it makes its way to the surface of the metal will blur structure in the density of states as reflected in the photoelectron spectrum. To see this, let us suppose the mean free paths for collision of the electrons with phonons and other electrons are respectively

$$\left.\begin{aligned} l_{ep} &= v\tau_{ep}, \\ l_{ee} &= v\tau_{ee}, \end{aligned}\right\} \tag{7.6.13}$$

where τ_{ep} and τ_{ee} are independent of the group velocity $\mathbf{v}$, of the electrons, given by

$$\mathbf{v} = \frac{1}{\hbar}\frac{\partial E}{\partial \mathbf{k}}. \qquad (7.6.14)$$

Now since

$$n(E) \propto \int \frac{dS}{|\nabla_{\mathbf{k}} E|}, \qquad (7.6.15)$$

we can see that the bigger the density of states at given energy, the lower the group velocity and so the shorter the mean free paths.

Quite apart from this, in most cases the photoelectron energy spectrum is by no means a replica of the initial density of states. Since the photoelectrons result from a volume effect, the initial and final states are single-particle states below the Fermi level (we shall neglect temperature effects) and single-particle states above the vacuum level, both sets of states being calculated for an electron within the crystal. We thus examine those transitions contributing to ε_2 which result in electrons lying in states above the vacuum level. The question also arises whether the transitions are predominantly direct or indirect. At its simplest, assuming constant transition probability matrix elements, this question reduces to whether the transitions are governed by a joint density of states

$$(\text{transition probability}) \propto n_{if}(\hbar\omega) \qquad (7.6.16)$$

or a product of densities of states,

$$(\text{transition probability}) \propto n_i(\varepsilon_i)\, n_f(\varepsilon_f) \quad [\omega = (\varepsilon_i - \varepsilon_f)/\hbar]. \qquad (7.6.17)$$

These alternatives represent the two extremes where direct transitions are all important or of no importance whatsoever.

(a) *Photoemission and band structure of GaAs.* As should be clear from the discussion given above, a good deal of basic information on band structure can be obtained by detailed photoemission studies. We shall choose the semi-conductor GaAs to illustrate how we can tie up the photoemission with theoretical predictions of band structure, important features connected with critical points of the band structure showing up through direct transitions.

As the photon energy is increased, the optical absorption coefficient α generally increases also and excitation occurs closer and closer to the surface. Specifically, in GaAs, at 1.4 eV near the band edge, α is around 10^3/cm and rises above 10^6/cm at photon energies in the 4–10 eV photon energy range. In consequence, a significant number of electrons are able to escape without appreciable energy loss for the higher values of the photon energy, even in the presence of strong scattering processes.

A typical result (Bell and Spicer, 1970) for the electron energy distribution in GaAs is shown in Figure 7.18, for a high value of photon energy. The

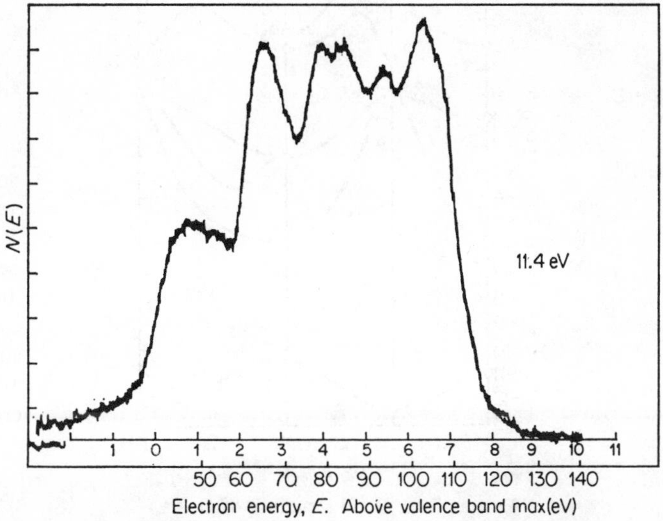

FIGURE 7.18. Typical photoelectron energy distribution from GaAs (after Bell and Spicer, 1970).

details of the structure in this figure, and similar ones for different photon energies, are intimately linked with the electronic band structure.

In Figure 7.19, we have reproduced the band structure of Herman and co-workers (1968). This band structure was obtained basically from OPW calculations of the type described in Chapter 1, though certain empirical adjustments were found to be necessary.

In Table 7.1, we collect together the information which can be deduced from photoemission studies about the energy of certain points in the Brillouin zone, together with the relevant results from the band structure calculation of Herman and colleagues. The states labelled in the table are also shown in Figure 7.19, the calculated values shown in Table 7.1 corresponding to those in the figure.

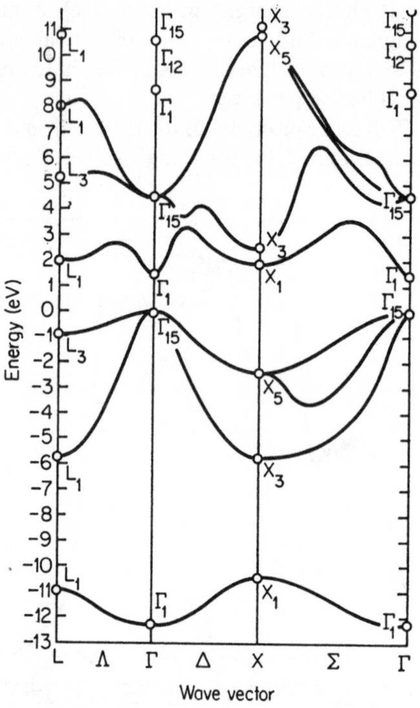

FIGURE 7.19. Band structure of GaAs.

It is possible to show that direct transitions are also responsible for the results from Si, for example (Gobeli and co-workers, 1966), which is to be expected in view of our remarks on homopolar semi-conductors in section 7.5.

(b) *Noble metals.* One of the most important observations of photoemission in metals concerned the d bands in Cu (Berglund and Spicer, 1964, 1965). These authors performed and analysed experiments for a range of photon energies from 0 to 11.7 eV, the analysis being based on equation (7.6.17). More recent work has suggested, however, that direct transitions play a central role in photoemission from Cu (Smith, 1969).

To begin the analysis of the photoemission spectrum, it is important to identify what is characteristic of the electronic band structure, as opposed to effects of collisions. In Figure 7.20, the energy distribution (normalized to total photoelectron yield) as observed by Spicer (1966) is plotted, for a number of incident light frequencies, against

$$E' = E - \hbar\omega + e\phi = E_i + e\phi, \tag{7.6.18}$$

TABLE 7.1. Energy eigenvalues at symmetry points for GaAs

Energy level transition	Experiment‡	Calculation§
Γ_{1c}	1.35 to 1.4 1.45 to 1.5	1.54
Γ_{15c}	4.6 to 4.8 4.8 to 5.0	4.6
Γ_{1c}	9.0 9.1	9.0
Γ_{12c}	10.2 to 10.8 10.3 to 10.9	10.6
Γ_{1v}	−6.0 to −6.7 (−6.5) −5.9 to −6.6	−5.6
L_{3v}	−0.9 to −1.1 −0.9 to −1.1	−0.9
L_{1c}	1.8 to 1.85 1.9 to 1.95	2.00
L_{3c}	5.0 to 5.2 5.1 to 5.3	5.3
X_{5v}	−2.4 to −2.6 −2.3 to −2.5	−2.3
X_{1c}	1.7 to 1.75 1.8 to 1.85	1.90
X_{3c}	2.4 to 2.5	2.5

‡ Table from compilation of Bell and Spicer (1970). Spread of values is due to differences in experimental results of different groups; c and v indicate conduction and valence band states.

§ Calculated values taken from Herman and Spicer (1968).

where E_i is the initial energy of the electron. Since in E' the initial photon energy is subtracted out, we expect the structure which results directly from the band theoretic states to remain stationary. On the other hand, peaks arising from processes in which electrons are inelastically scattered can be expected to move with E', as the photon energy is varied. In Figure 7.20, the first peak in the spectrum does in fact move and disappears at photon energies less than 6 eV. We therefore interpret it as due to scattered electrons, while the other characteristics of the curves reflect the band structure. This interpretation is supported by Figure 7.21, where the spectrum is plotted

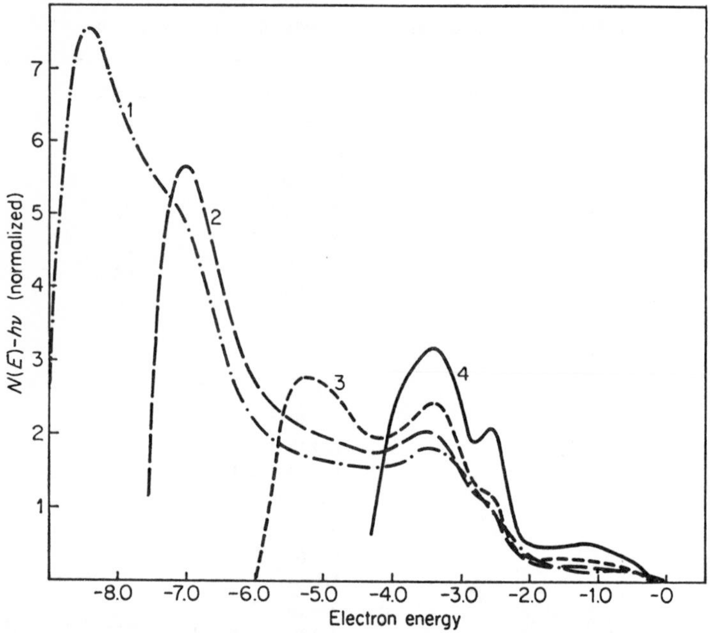

FIGURE 7.20. Photoemission spectrum of caesiated Cu plotted against electron energy referred to initial state. (1) $\hbar\omega$ = 10.4 eV. (2) $\hbar\omega$ = 8.9 eV. (3) $\hbar\omega$ = 7.1 eV. (4) $h\omega$ = 5·6 eV (after Spicer, 1966).

against the actual energy of the photoemitted electrons. In this plot, the first peak does *not* move with photon energy, although it grows and broadens at higher frequencies. This is what might be expected, for an increase in the photon energy increases the number of electrons which undergo inelastic scattering, rather than increasing their final energies. Berglund and Spicer (1964) have treated the effect quantitatively, confirming that the first peak is a result of electron–electron scattering.

After allowance for this scattering, the spectra were analysed by Berglund and Spicer using the hypothesis of non-direct transitions. Obviously, measurements for a single photon frequency do not suffice to yield n_i and n_f separately, but curves at a sufficient number of frequencies will yield these curves uniquely, provided that equation (7.6.17) is correct. In fact, this procedure does appear to apply to the original experiments of Berglund and Spicer on copper, except that the small shaded regions of Figure 7.22 must be attributed to direct transitions in the neighbourhood of $L_{2'} \to L_{12}$ (see Figure 7.12). It

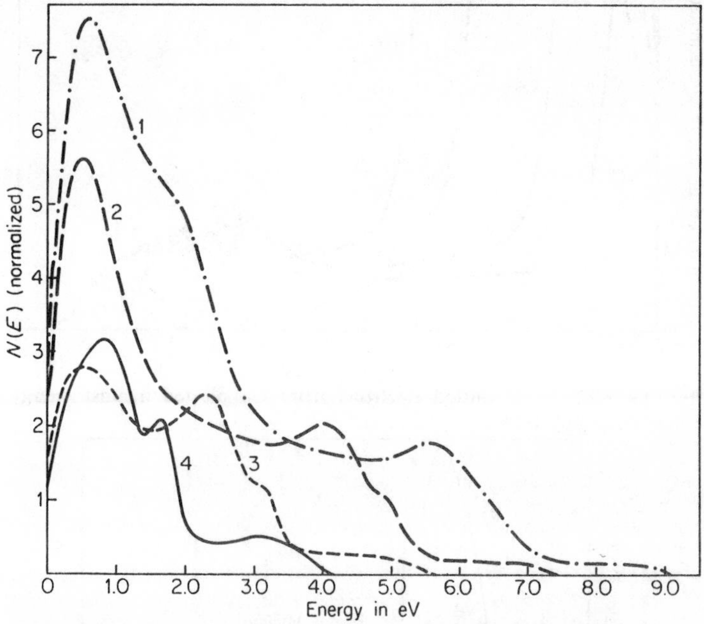

FIGURE 7.21. Photoemitted electrons from caesiated Cu plotted against energy. (1) $\hbar\omega$ = 10.4 eV. (2) $\hbar\omega$ = 8.9 eV. (3) $\hbar\omega$ = 7.1 eV. (4) $\hbar\omega$ = 5.6 eV (after Spicer, 1966).

should be noted that these direct transitions do not dominate the curves making up only a relatively small proportion of the transitions, and are not required to explain any part of the spectrum for $\hbar\omega < 4.4$ eV (which agrees with the energy gap shown in Figure 7.13). This conclusion is supported by the work of Smith (1969), which we shall refer to again below.

The density of states deduced by Berglund and Spicer is shown in Figure 7.23. It should be noted that there is a peak in the density of states 1.8 eV

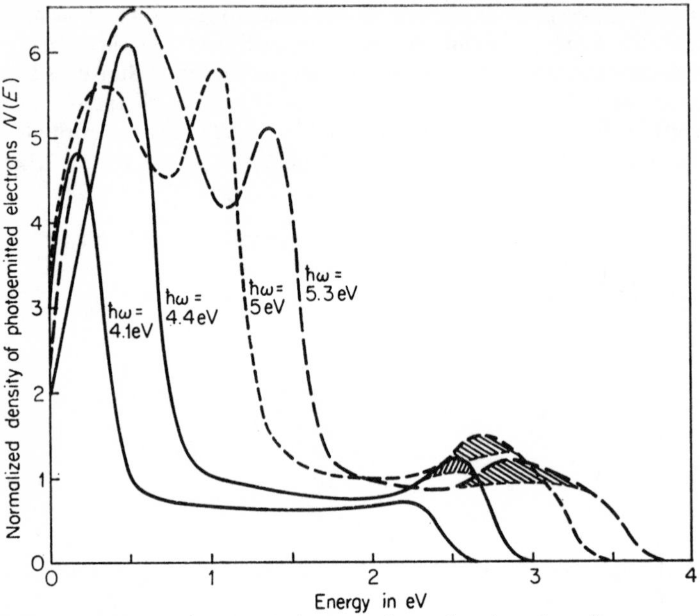

FIGURE 7.22. Photoemitted electrons from Cu plotted against energy (after Berglund, 1966).

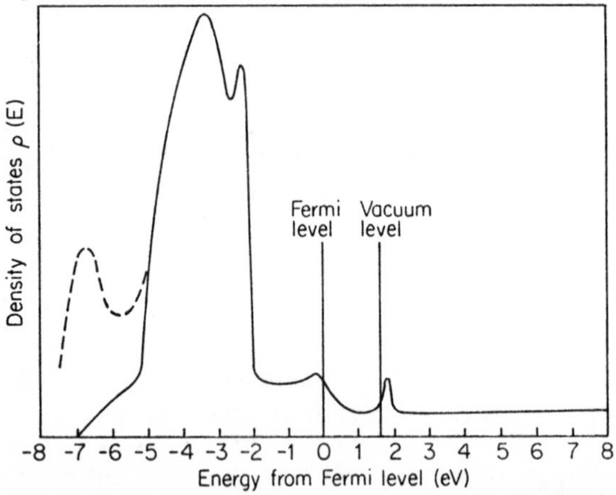

FIGURE 7. 23. Density of states of Cu (after Berglund and Spicer, 1964). Vacuum level was lowered to 1·55 eV above Fermi energy by depositing a monolayer of Cs on the surface of the metal.

above the Fermi level. This may be attributed to a van Hove singularity about X_4' (see Figure 7.13). In general, the photoemission results on Cu are in agreement with band structure studies, but although structure appertaining to transitions near to L_2' and L_{12} and $X_5 \rightarrow X_4'$ are seen, other features one might expect, namely those associated with transitions from around X_1, W_3 or $W_{2,1}$ are absent.

The work of Smith, referred to above, has extended the measurements on Cu, the work function being lowered by caesiating the specimen, following Berglund and Spicer. Higher resolution proved possible because of technical advances since the early work and new results were obtained in the photon energy range between 6.5 and 8.2 eV.

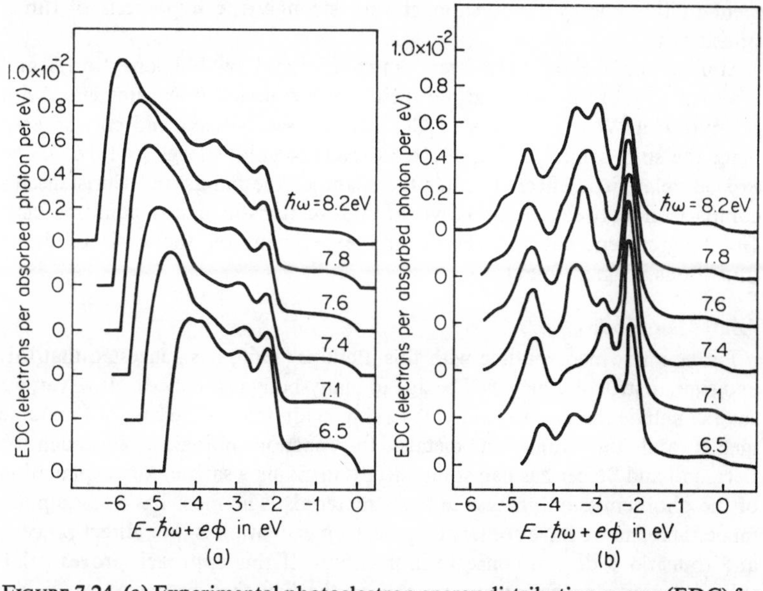

FIGURE 7.24. (a) Experimental photoelectron energy distribution curves (EDC) for caesiated Cu in photon energy range 6.5—8.2 eV. Abscissa shows initial state energy. Zero of energy is taken at Fermi level. (b) Predicted energy distribution curves assuming direct transitions (after Smith, 1969).

The experimental photoelectron energy distribution curves for caesiated Cu in this photon energy range are shown in Figure 7.24(a). The curves are plotted against the energy of the initial state and the zero of energy is at the Fermi level. A peak due to transitions from the uppermost band is seen at

-2.3 eV, together with additional structure at lower energies, associated with the lower bands. This structure was not resolved in the earlier experiments.

It is possible, on the assumption of direct transitions, to use band theory to calculate the predicted energy distribution curves. Smith has used the interpolation scheme of Hodges, Ehrenreich and Lang (1966), fitting the parameters to the APW calculation of Burdick (1963). The results are shown in Figure 7.24(b) and it is interesting to compare these results with the measurements of Figure 1.23(a) in the energy range shown by the vertical lines. In both figures, for $\hbar\omega = 6.5$ eV there are two peaks. On increasing the photon energy, the left-hand peak disappears, while the right-hand peak broadens until, at $\hbar\omega = 7.8$ eV, there is just a broad structure. On going to 8.2 eV this becomes a doublet. These striking similarities between theory and experiment in this energy range support very strongly the hypothesis of direct transitions.

Similar conclusions have been drawn for gold by Nilsson, Norris and Wallden (1970), by comparing their photoemission measurements with relativistic APW calculations of Christensen (see Nilsson and co-workers). Since the spin–orbit splitting of the 5d levels of gold is large (~ 1.6 eV), one expects relativistic effects to alter the shape of the bands and Christensen's comparison between non-relativistic and relativistic bands confirms this. Good agreement between theory and experiment on the basis of direct transitions is again found.

7.6.3 Many-body effects

Perhaps not incompatible with this, Phillips (1966) has suggested that this and similar structure may well be due to many-body resonances. However, we shall conclude the discussion by observing again that in the case of the noble metals and the transition metals, the phenomenological approach of Berglund and Spicer has had some success in giving a satisfactory explanation of the photoemission process in these materials. The most basic assumption made there, as we have seen, is that electron excitation is an indirect process, and transitions do not conserve momentum. If this approach proves valid, then we would have a strong pointer to the fact that processes which are not one-electron in character play a basic role.

Doniach and Šunjić (1970)‡ have therefore built up a many-body theory of photoemission. They treat this as a second-order process where the optical excitation and the surface potential both act as a first-order perturbation to the system. The transition probability is expressed in terms of a correlation function which includes all (Coulomb) scattering inside the solid. If the emitted electron is fast, then its interaction with the rest of the solid can be treated by perturbation theory.

‡ Also Doniach (1970) and private communication from Doniach and Šunjić.

Doniach and Šunjić exclude the possibility of surface resonances and vertex corrections in noble and transition metals at ultraviolet photon energies. They also show that Coulomb scattering of a relatively fast outgoing electron is not likely to produce any self-energy effects leading to the lifetime broadening, but when treated as a first-order perturbation can give recognizable structures in the measured photoelectron spectrum.

But the most basic question is that of direct versus indirect transitions. The analysis of the experimental evidence for indirect processes points to the fact that it is the deep-hole relaxation which is involved in any explanation of a large indirect contribution to the photoelectron spectra from the narrow bands in metals. Doniach and Šunjić propose that the hole relaxation process is a many-pair excitation, but we must refer the reader to the original work for details.

7.7 Soft X-ray emission

We shall give now a fairly brief discussion of a method which has the link with photoemission that both are related to the density of electron states $N(E)$. The above discussion shows that there is a great deal more to the interpretation of photoemission than simply the shape of the density of states, and the same, in fact, is true of soft X-ray emission. Nevertheless, valuable information is being gained continually from this method, and also there has been considerable interest in some many-body effects occurring in the process. We shall only deal here with the one-body theory in a quantitative way, the many-body aspects being touched on at the end.

The basic process of emission follows the ejection of an electron from a core state in the crystal. The entire system then makes a transition from such an excited state to one of lower energy, with an electron falling from a higher energy state into the core vacancy, and energy and angular momentum being conserved by the emission of an X-ray photon.

If we suppose that the wave functions of the initial and final states are Ψ_i and Ψ_f, then we can write down immediately the probability per unit time, per unit solid angle, for the emission of a photon polarized in the x-direction as

$$P = \frac{\nu}{c^3}\left|\int\cdots\int \Psi_i^* \frac{\partial}{\partial x_m}\Psi_f \, d\mathbf{r}_2 \, d\mathbf{r}_3 \ldots d\mathbf{r}_N\right|^2, \tag{7.7.1}$$

where ν is the frequency of the emitted radiation. In writing equation (7.7.1) we assume that the ground state of the N electrons is described by filling up N one-electron states $\phi_1, \ldots, \phi_N$, and taking a single particle product, and that in final and initial states we have to change the wave function only because, at first, there is no electron in the core orbital that the hole occupies, and in the final state we have the mth one-electron state vacant and the core hole filled by the mth electron. It is then a straightforward matter to carry out the integration in equation (7.7.1) to obtain

$$P = \frac{\nu}{c^3} \int \int \phi_1^*(\mathbf{r}) \phi_1(\mathbf{r}') \frac{\partial}{\partial x} \frac{\partial}{\partial x'} \phi_m(\mathbf{r}) \phi_m^*(\mathbf{r}') \, d\mathbf{r} \, d\mathbf{r}', \tag{7.7.2}$$

where ϕ_1 is the wave function of the core state, and ϕ_m is the wave function of the higher energy state which is vacated on emission of the photon. If we now calculate the total intensity emitted for a single transition, averaged over directions and polarizations, then we find

$$I(\nu) = \frac{(4\pi)^2 \nu^2}{3c^3} \int \int \phi_1^*(\mathbf{r}) \phi_1(\mathbf{r}') \left(\frac{\partial}{\partial x} \frac{\partial}{\partial x'} + \frac{\partial}{\partial y} \frac{\partial}{\partial y'} + \frac{\partial}{\partial z} \frac{\partial}{\partial z'} \right) \phi_m(\mathbf{r}) \phi_m^*(\mathbf{r}') \, d\mathbf{r} \, d\mathbf{r}'. \tag{7.7.3}$$

7.7.1 Application to metals

Though the method is rather general, let us specialize to the case of a metal. Then ϕ_1 is a core electron wave function for a low energy state, in a narrow band, while, of course, ϕ_m is a Bloch wave function for the conduction band, which as usual we write as $\psi_{\mathbf{k}}(\mathbf{r})$.

What we will now be measuring is the intensity, $I(E)\,dE$, say, for all transitions in the energy range E to $E+dE$. Then we can write that the intensity $I(E)$ is given by

$$I(E)\,dE = \frac{16\pi^2 \nu^3}{3c^2} \frac{\mathscr{V}}{8\pi^3} \frac{d}{dE} \int_{E(\mathbf{k}) \leqslant E} \left| \int \phi_1^*(\mathbf{r}) \nabla_{\mathbf{r}} \psi_{\mathbf{k}}(\mathbf{r}) \, d\mathbf{r} \right|^2 d\mathbf{k} \, dE. \tag{7.7.4}$$

This argument makes a number of approximations, but in the relatively few cases in which it has been confronted with experiment, it seems to contain many of the gross features of the observed spectrum. One (over) simplification which was made for a long time in this field was to make the assumption that the quantity $\int \phi^*(\mathbf{r}) \nabla_{\mathbf{r}} \psi_{\mathbf{k}}(\mathbf{r}) \, d\mathbf{r}$ in equation (7.7.5) was only slowly varying in energy, and then the intensity would take the form

$$I(E)\,dE = \text{factor independent of energy} \times N(E)\,dE. \tag{7.7.5}$$

This is then the reason for the original belief that measurement of the soft X-ray intensity would yield the conduction band density of states. However, all workers now agree that one is not only concerned with the density of states, but that also we must know the wave functions in order to calculate the matrix element in equation (7.7.4).

It is still useful nevertheless to factor out the density of states. If the square of the matrix element in equation (7.7.4) is constant over constant-energy surfaces in the occupied part of **k**-space, then we can write

$$I(E)\,dE = \frac{16\pi^2\nu^2\mathscr{V}}{3c^3}\left|\int \phi_1^*(\mathbf{r})\,\nabla_\mathbf{r}\,\psi_\mathbf{k}(\mathbf{r})\,d\mathbf{r}\right|^2 N(E)\,dE, \tag{7.7.6}$$

where **k** in the matrix element is given by inverting $E(\mathbf{k})$ and, as usual, the density of states is given in terms of $E(\mathbf{k})$ by equation (1.7.5) of Chapter 1. This can then be regarded, since ν^2 varies very little over the occupied part of the band, as the product of $N(E)$ and a transition probability proportional to $|\int \phi_1^*(\mathbf{r})\,\nabla_\mathbf{r}\,\psi_\mathbf{k}(\mathbf{r})|^2$.

Though a number of calculations of transition probabilities have been carried out, for example, on Li by a number of different workers (Allotey, 1968; Schuey, 1968; Stott and March, 1968), there is still not clear evidence in this metal whether or not the one-electron theory can explain the spectrum, and in this case many-body effects have been invoked. We shall restrict ourselves here to pointing out that, while, as emphasized above, the transition probability must be calculated as a function of energy from the Bloch wave functions, the van Hove singularities in the density of states show up in the soft X-ray emission spectrum, a good example of this being given by the work of Rooke (1968) shown in Figure 7.25. The positions of the van Hove singularities as given by the band structure calculation on aluminium of Segall (1961) are marked by the vertical lines. It can be seen that there is a very nice agreement between the energies at which the experimental discontinuities occur and Segall's predictions.

Stott and March [1966; see also the book on soft X-ray band spectra edited by Fabian (1968); also Fabian, Watson and Marshall, 1971] have pointed out that if the above one-electron theory can be used, then, in a relatively simple metal like Li, it ought to be possible to get information on the wave functions of the conduction band directly from the soft X-ray emission. Actually in their work, they used the momentum eigenfunction $v(\mathbf{k})$ of the band (see Chapter 1, section 1.10) as the basic tool for the analysis when the square of the matrix element of equation (7.7.6) can be expressed as

$$\left|\int \phi^*(\mathbf{r})\,\nabla_\mathbf{r}\,\psi_\mathbf{k}(\mathbf{r})\,d\mathbf{r}\right|^2$$

$$= \sum_{\mathbf{K}_n\mathbf{K}_m} \phi^*(\mathbf{k}+\mathbf{K}_n)\,\phi(\mathbf{k}+\mathbf{K}_m)\,(\mathbf{k}+\mathbf{K}_n).(\mathbf{k}+\mathbf{K}_m)\,v(\mathbf{k}+\mathbf{K}_n)\,v^*(\mathbf{k}+\mathbf{K}_m), \tag{7.7.7}$$

the sum being over reciprocal lattice vectors **K**, and $\phi(\mathbf{k})$ being the wave function of the core state in momentum space.

However, recent work has suggested that the influence of the hole in the core state can have important effects on the soft X-ray spectrum and the final position is unclear. We have therefore to refer the reader to the many-body studies, especially those of Nozières and colleagues (1969) for further details (see problem 7.10 for a brief example from their work; see also Hopfield, 1969).

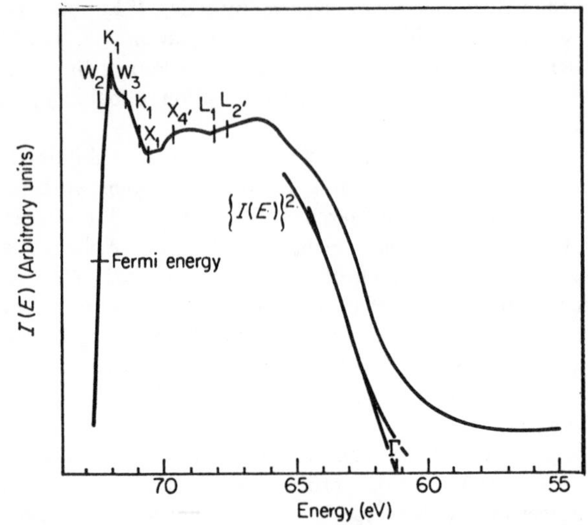

FIGURE 7.25. Aluminium L_{23} emission spectrum with background subtracted as given by Rooke (1968). Vertical lines show predicted positions of Van Hove singularities (Segall, 1961).

7.8 Anomalous skin effect

We are concerned in this section with the microwave region of the electromagnetic spectrum, so that the effects considered are not optical in the sense of relating to visible light, but it is convenient to deal with them here.

It is well known that currents in a conductor are localized near its surface and for this reason the "surface impedance" is sometimes a more suitable quantity than the optical constants we have so far used: unlike the complex refractive index it is physically meaningful when the wave disturbance in the medium does not have the exponential form of equation (7.1.11); more importantly for present purposes, it is directly measurable in the microwave region and is the natural quantity to use in discussing the power absorption by a metal.

The surface impedance Z is defined as (with $\mathbf{z}$ normal into metal)

$$Z = \frac{\text{Electric field at surface}}{\text{Total current per unit area}} = \frac{\mathscr{E}_x(0)}{\int_0^\infty j\,dz}. \tag{7.8.1}$$

Now if we rewrite equation (7.1.9) in terms of the current we have

$$\frac{\partial^2 \mathscr{E}_x}{\partial z^2} + \frac{\omega^2}{c^2}\mathscr{E}_x = \frac{4\pi i\omega}{c^2} j \tag{7.8.2}$$

and neglecting the displacement term (the second term on the left-hand side), which is permissible in the microwave region, we obtain

$$j = \frac{c^2}{4\pi i\omega}\frac{\partial^2 \mathscr{E}_x}{\partial z^2} \tag{7.8.3}$$

and hence

$$\int_0^\infty j\,dz = \frac{c^2}{4\pi i\omega}\int_0^\infty \frac{\partial^2 \mathscr{E}_x}{\partial z^2}\,dz = -\frac{c^2}{4\pi i\omega}\left(\frac{\partial \mathscr{E}_x}{\partial z}\right)_0 = \frac{c}{4\pi}\mathscr{H}_y(0), \tag{7.8.4}$$

where Maxwell's equations have been used to relate $\partial \mathscr{E}_x/\partial z|_0$ to the magnetic field at the surface. We can now see that

$$Z = R + iX = \frac{4\pi}{c}\frac{\mathscr{E}_x(0)}{\mathscr{H}_y(0)} = -\frac{4\pi i\omega}{c^2}\frac{\mathscr{E}_x(0)}{(\partial \mathscr{E}_x/\partial z)_0}. \tag{7.8.5}$$

R, the surface resistance, determines the power absorption of the metal; X is known as the surface reactance.

We can obtain the reflexion coefficient and power absorption quite simply by using equation (7.1.11) to evaluate $\partial \mathscr{E}/\partial z|_0$; we find

$$Z = \frac{4\pi\omega}{ic^2}\frac{\mathscr{E}_x(0)}{(\partial \mathscr{E}_x/\partial z)_0} = \frac{4\pi}{\sqrt{(\varepsilon)}\,c}, \tag{7.8.6}$$

so that the reflexion coefficient is

$$R_r = \left|\frac{\sqrt{(\varepsilon)}-1}{\sqrt{(\varepsilon)}+1}\right|^2 = \left|\frac{Z-Z_0}{Z+Z_0}\right|^2, \tag{7.8.7}$$

where

$$Z_0 = \frac{4\pi}{c}, \tag{7.8.8}$$

which is the intrinsic impedance of free space. Equation (7.8.6) has been found by taking the field to decay exponentially in the metal but, as remarked already, Z is not restricted to this circumstance.

The absorption coefficient is given by

$$A = 1 - R_r = \frac{4Z_0 R}{(Z_0+R)^2+X^2}. \tag{7.8.9}$$

Using the quasi-free electron theory of section 7.2 in the non-relaxation region $\omega\tau \ll 1$, which suffices in the treatment of microwaves, it is very easy to show that equation (7.8.6) becomes

$$Z = (1+i)\frac{2\pi\omega}{c^2}\left(\frac{c^2}{\sigma_0}\right)^{\frac{1}{2}}. \tag{7.8.10}$$

We note that $\sqrt{(c^2/2\pi\sigma_0\omega)} = \delta$ is the classical skin depth, and so

$$R = X = \left(\frac{2\pi\omega}{c^2\sigma_0}\right)^{\frac{1}{2}} = \frac{2\pi\omega}{c^2}\delta. \tag{7.8.11}$$

The absorption coefficient is, by equation (7.8.9),

$$A = \left(\frac{2\omega}{\pi\sigma_0}\right)^{\frac{1}{2}}, \tag{7.8.12}$$

the Hagen–Rubens relation.

7.8.1 Low-temperature behaviour

Equations (7.8.9) to (7.8.12) are the basic equations of the normal skin effect. However, anomalous behaviour is obtained at low temperatures. This was first shown by Pippard (1957) by experiment on an oscillatory circuit in which the conductors were made of the material under investigation.‡ The results are typified by Figure 7.26. The reason for the behaviour shown is not far to seek. In the discussion of the normal skin effect we have effectively used Ohm's law, with the concomitant assumption that the current at a point is completely determined by the electric field there. We should more properly use a non-local relation between **j** and $\mathscr{E}$. We shall show what form this relation takes later but we can first discuss the essential physics involved without the complications of a full mathematical treatment.

The relation between **j** and $\mathscr{E}$ becomes non-local when the mean free path of the electrons is of the same order as a length over which $\mathscr{E}$ changes appreciably. This occurs when electrons can come up into the skin depth and respond to the electric field without suffering scattering while in the skin depth. Our interest below lies in the extreme anomalous region where the mean free

‡ The resonant circuit was a wire loop regarded as a double transmission line closed at one end and resonating when the wavelength was approximately four times its length.

path is many times larger than the skin depth: $l/\delta \gg 1$, and this region can be investigated using Pippard's "ineffectiveness" concept (Pippard, 1960). Electrons are only effective in contributing to the current if they travel almost parallel to the surface, so as to remain long enough in the skin depth to absorb a significant amount of energy from the electric field. One easily sees that the directions of motion of these electrons lie in an angle of the order of

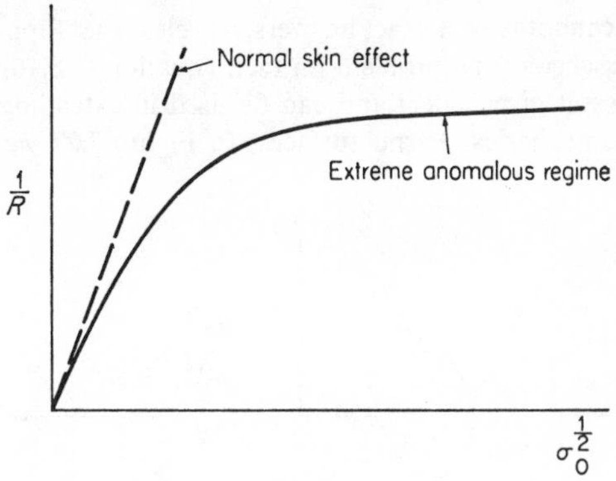

FIGURE 7.26. Anomalous skin effect. R is real part of surface impedance and σ_0 is d.c. conductivity (after Donovan, 1967).

δ/l, and we shall have a proportion $\beta\delta/l$, where $\beta \sim 1$, of the electrons contributing to the conductivity. Assuming the previous expressions are valid for their contribution and the ineffective electrons make no contribution whatever, we find an effective conductivity

$$\sigma_{\text{eff}} = \beta\frac{\delta}{l}\sigma_0. \tag{7.8.13}$$

We then obtain a new skin depth from equation (7.2.13) (with σ_{eff} replacing σ_0), and this is given by

$$\delta = c\sqrt{\frac{l}{2\pi\beta\delta\sigma_0\omega}}, \tag{7.8.14}$$

that is,

$$\delta^3 = \frac{c^2 l}{2\pi\beta\sigma_0\omega}. \tag{7.8.15}$$

From equation (7.8.10) we now obtain, as the asymptotic expression for the surface resistance in the non-relaxation region,

$$R_\infty = \left(\frac{4\pi^2 \omega^2 l}{\beta c^4 \sigma_0}\right)^{\frac{1}{3}}. \tag{7.8.16}$$

We also obtain $X_\infty = R_\infty$, whereas we shall see later that the exact result is

$$X_\infty = \sqrt{(3)}\, R_\infty, \tag{7.8.17}$$

but while one cannot expect exact answers, it is clear that Pippard's argument contains the essence of the problem [in fact, equation (7.8.16) is of the form of the exact result given later] and can be used in extending the results to metals with non-spherical Fermi surfaces. In Figure 7.27 we show a cross-

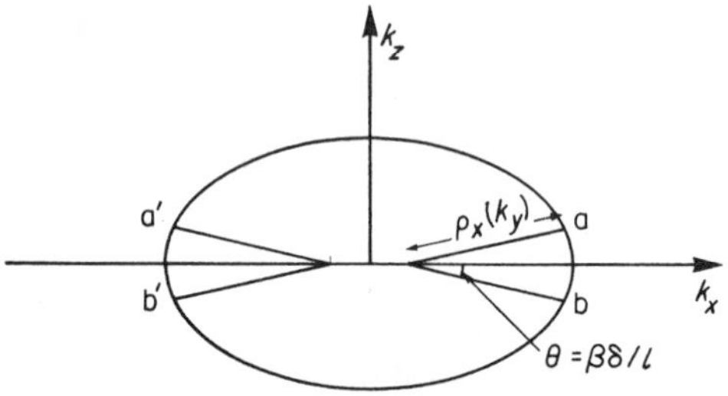

FIGURE 7.27. Cross-section of Fermi surface in x–z plane for some value of y.

section of the Fermi surface in the x–z plane for some value of k_y. Following the method of section 6.3 we have a current density per unit volume of

$$j_x = \frac{e}{4\pi^3 \hbar} \int \tau e \mathscr{E}_x v_{\mathrm{f}}\, dS \tag{7.8.18}$$

[cf. equation (6.4.29)]. The surface over which this integral is taken is localized within the angle $\beta(\delta/l)$ about the x-direction, and if $\rho_x(k_y)$ is the local radius of curvature, in the x–z plane, of the Fermi surface, we can see, on assuming δ/l to be very small, that

$$j_x = \frac{1}{4\pi^3 \hbar} \oint v_{\mathrm{f}} e^2 \mathscr{E}_x \tau |\rho_x(k_y)| \beta \frac{\delta}{l}\, dk_y$$

$$= \frac{e^2 \mathscr{E}_x \beta\delta}{4\pi^3 \hbar} \oint |\rho_x(k_y)|\, dk_y. \tag{7.8.19}$$

The integral over dk_y is taken round the line of the Fermi surface where the surface is cut by the x–y plane.

Inserting the effective conductivity, given by equation (7.8.13), into equation (7.8.19) we find

$$R_\infty = \left(\frac{3^{\frac{3}{2}} \pi^4 \omega^2}{c^4 \hbar e^2 \oint dk_y |\rho_x(k_y)|} \right)^{\frac{1}{3}}, \tag{7.8.20}$$

where we have inserted the value of β which an exact treatment yields, as is discussed below.

7.8.2 Shape of Fermi surface

One sees that the shape of the Fermi surface may be inferred from measurements of the surface resistance of crystals cut at a number of different orientations. Such measurements of surface resistance were first performed by Pippard (1957) with specimens of copper.

The criterion for the above results to be independent of relaxation effects is

$$\omega \ll \frac{v_f}{\delta}. \tag{7.8.21}$$

That is, the period of the electric field must be much greater than the time taken by an electron to traverse the skin depth. That this condition overrides that defining the non-relaxation region, viz., $\omega\tau \ll 1$, is not immediately evident but will become clear shortly.

We now turn to a more formal discussion of the anomalous skin effect, examining the consequences of introducing explicitly a non-local relation between current $\mathbf{j}(\mathbf{r})$ and field $\mathscr{E}(\mathbf{r})$, and including also the effects of reflexion of the electrons at the surface.

We write

$$\mathbf{j}_\mathbf{q}(\mathbf{r}) = \mathbf{j}(\mathbf{q})\, e^{i(\omega t - qz)} \tag{7.8.22}$$

for a single Fourier component of the current, and relate this to the field

$$\mathscr{E}(q\mathbf{r}) = \mathscr{E}(q) e^{i(\omega t - qz)} \tag{7.8.23}$$

by

$$j(q) = \sigma(q) \mathscr{E}(q). \tag{7.8.24}$$

We shall only discuss a situation in which reflexion of electrons at the surface is specular, because the treatment is very lengthy otherwise (see Reuter and Sondheimer, 1948). The problem of completely specular reflexion is soluble by regarding the surface as the interface of two pieces of metal which are mirror images of one another in every detail. The electric field decays from the interface in both $+z$ and $-z$ directions so that the boundary condition on the electric field is

$$\left.\frac{\partial \mathscr{E}}{\partial z}\right|_{+0} = -\left.\frac{\partial \mathscr{E}}{\partial z}\right|_{-0} \tag{7.8.25}$$

which may be introduced into equation (7.8.2) by rewriting this equation as

$$\frac{\partial^2 \mathscr{E}(z)}{\partial z^2} + \frac{\omega^2}{c^2}\mathscr{E}(z) = \frac{4\pi i\omega}{c^2} j(z) + 2\left(\frac{\partial \mathscr{E}}{\partial z}\right)_{+0} \delta(z). \tag{7.8.26}$$

After Fourier transformation this becomes

$$-q^2\mathscr{E}(q) + \left(\frac{\omega}{c}\right)^2 \mathscr{E}(q) = \frac{4\pi i\omega}{c^2} j(q) + \left(\frac{1}{\pi}\right)\left.\frac{\partial \mathscr{E}}{\partial z}\right|_{+0} \tag{7.8.27}$$

or, using equation (7.8.24) and neglecting the displacement term,

$$-\left[\frac{4\pi i}{c^2}\omega\sigma(q) + q^2\right]\mathscr{E}(q) = \left(\frac{1}{\pi}\right)\left.\frac{\partial \mathscr{E}}{\partial z}\right|_{+0} \tag{7.8.28}$$

and hence

$$\mathscr{E}(z) = \frac{1}{\pi}\left(\frac{\partial \mathscr{E}}{\partial z}\right)_{+0} \int_{-\infty}^{\infty} dq \frac{e^{-iqz}}{-q^2 - i\gamma\sigma(q)} \quad \left(\gamma = \frac{4\pi\omega}{c^2}\right). \tag{7.8.29}$$

Thus

$$Z_\infty = \frac{4\pi i\omega}{c^2} \frac{\mathscr{E}(0)}{\partial \mathscr{E}/\partial z|_{+0}} = \frac{2i\gamma}{\pi}\int_0^\infty dq \frac{1}{q^2 + i\gamma\sigma(q)}. \tag{7.8.30}$$

Substituting

$$\gamma\sigma(q) = \frac{s^3}{q}, \quad \xi = \frac{q}{s}, \tag{7.8.31}$$

we find

$$Z_\infty(\omega) = \frac{2\gamma}{\pi s}\int_0^\infty \frac{\xi \, d\xi}{1 - i\xi^3}, \tag{7.8.32}$$

which can be integrated to give

$$Z_\infty(\omega) = \frac{8\pi\omega}{3\sqrt{(3)}\, c^2 s}[1 + \sqrt{(3)}\, i], \tag{7.8.33}$$

which embodies equation (7.8.17). It is found by Reuter and Sondheimer that with completely diffuse reflection of the electrons at the metal surface equation (7.8.33) is modified only by multiplying the right-hand side by 9/8.

We now have to evaluate $s = [\gamma q\sigma(q)]^{-\frac{1}{3}}$ and will do so under the assumption that a relaxation time can be defined. Under the influence of the field $\mathscr{E}(q)e^{iqz}$ we take the f_1 of equation (7.2.1) to propagate according to e^{iqz}. The equation then becomes

$$i\omega f_1(q) - e\mathscr{E}.\mathbf{v}\frac{\partial f_0}{\partial E} + i\mathbf{q}.\mathbf{v}f_1(q) = \frac{f_1}{\tau} \quad (\mathbf{q} = q\hat{\mathbf{z}}). \tag{7.8.34}$$

If we look back to section 7.2 we can see that equation (7.2.3) is now modified to read

$$\sigma(q) = \frac{e^2}{4\pi^3\hbar}\int\frac{\tau v_x\,dS_x}{1+i(\omega-\mathbf{q}.\mathbf{v})\,\tau}, \tag{7.8.35}$$

which is written here for unit volume. Suppose we make $qv_f \gg \omega$. The integrand is then concentrated about the region for which $\mathbf{v}$ is perpendicular to $\mathbf{q}$, i.e. $\mathbf{v}\perp\mathbf{z}$, and is significant only over a very small part of the Fermi surface. The situation may again be represented by Figure 7.27, (except that $\beta\delta/l$ is no longer of any significance). The integrand is only appreciable within angles which we can take as defined by ab and a′b′ in this figure, the angles being very small, and we can write

$$\sigma(q) = \frac{e^2}{4\pi^3\hbar}\oint dk_y\,|\,\rho_x(k_y)|\int_{-\pi/2}^{+\pi/2}\frac{\tau v_\perp\cos\theta\,d\theta}{1+i(\omega-qv_\perp\sin\theta)\,\tau}, \tag{7.8.36}$$

where we have extended the angular integration from that defined by ab to $(-\frac{1}{2}\pi, \frac{1}{2}\pi)$, with negligible loss in accuracy since, as already remarked, the integrand is only appreciable within the former angle. The angular region defined by a′b′ is also included since the integration over k_y is over a line right round the Fermi surface. The definition of $v_\perp$ is evident: it is the magnitude of the component of the velocity in the x–z plane.

Equation (7.8.36) may also be written, on change of variables, as

$$\sigma(q) = \frac{e^2}{4\pi^3\hbar}\oint dk_y\,|\,\rho_x(k_y)|\int_{-1}^{1}\frac{\tau v_\perp\,d\mu}{1+i(\omega-qv_\perp\mu)\,\tau} \tag{7.8.37}$$

or

$$\sigma(q) \approx \frac{e^2}{4\pi^3\hbar q}\oint dk_y\,|\,\rho_x(k_y)|\int_{-\infty}^{\infty}\frac{d\xi}{1+i\xi}. \tag{7.8.38}$$

Here we have extended the range of integration from $(-1, -1)$ to $(-\infty, \infty)$ on changing the variable from μ to ξ, since, again, the first range includes all significantly non-zero values of the integrand. On performing the integral we find

$$\sigma(q) = \frac{e^2}{4\pi^3 \hbar q} \oint dk_y |\rho_x(k_y)| \tag{7.8.39}$$

and hence, from equation (7.8.31),

$$s = \left[\frac{e^2 \omega}{\hbar \pi^2 c^2} \oint dk_y |\rho_x(k_y)|\right]^{\frac{1}{3}}. \tag{7.8.40}$$

From equation (7.8.30) we finally obtain

$$R_\infty(\omega) = \frac{8\pi\omega}{3\sqrt{(3)}\, c^2} \left[\frac{\pi c^2}{e^2 \hbar\omega \oint dk_y |\rho_x(k_y)|}\right]^{\frac{1}{3}} = \frac{8}{9} \left[\frac{\pi^4 \omega^2 3^{\frac{3}{2}}}{e^2 c^4 \hbar \oint dk_y |\rho_x(k_y)|}\right]^{\frac{1}{3}}, \tag{7.8.41}$$

which is equation (7.8.20) apart from the factor 8/9 which, as remarked before, disappears in the solution for diffuse scattering from the surface. The exact solutions for a spherical Fermi surface are readily found from the more general forms we have given here.

We have assumed $qv_{\mathrm{f}} \gg \omega$ for the significant values of q in the problem. Clearly, the wavelengths of significance are those less than the order of δ, that is $q > 1/\delta$ and so $qv_{\mathrm{f}} \gg \omega$ is equivalent to equation (7.8.21). It now also becomes clear why the condition $\omega\tau \ll 1$ is not the operative one. In fact at very low temperatures we find typical values of $v_{\mathrm{f}}/\delta = 10^{13}\ \mathrm{sec}^{-1}$ and $\tau = 10^{-11}$ sec so that the extreme anomalous region includes frequencies much higher than those specified by $\omega\tau \ll 1$.

The way in which we were able to treat specular reflexion strongly indicates that the type of reflexion which actually occurs does not affect the results for the non-relaxation region very much, and Reuter and Sondheimer's calculations confirm this. The situation is very different in the extreme relaxation region ($\omega\tau \gg 1$). Evidently, the conductivity will revert to its local form for ω sufficiently high, but this does not necessarily imply that normal conditions are regained. As an electron comes to the surface it will, on average, gain energy during the last half-cycle of the electric field before the electron strikes the surface. If reflexion is specular the electron loses its energy to the field again, and equation (7.8.11) of the normal skin effect does indeed hold for very high frequencies. On the other hand, if reflexion is diffuse the resulting randomization of the motion leads to a heating of the electron gas and so a net absorption of energy from the field: it has been shown by Holstein (1954) that for free electrons we have, as $\omega \to \infty$,

$$R_\infty = \frac{3\pi v_{\mathrm{f}}}{4c^2}. \tag{7.8.42}$$

It is believed that diffuse reflexion is more likely than specular since the de Broglie wavelength is comparable to the size of irregularities on an atomic scale one expects at the surface. Measurement of R_∞ at very low temperatures should provide evidence bearing on this issue. This process does not, however, constitute the total absorption mechanism, for Holstein has pointed out that the electrons can also absorb energy from the field by emitting a phonon.

A proof of equation (7.8.42), together with a discussion of the effects we have referred to, is given by Pippard (1965).

7.9 Cyclotron resonance

7.9.1 Magnetic field normal to surface

If we can regard the absorption by a solid of electromagnetic energy as a bulk effect the power absorbed is proportional to $\operatorname{Re}(\sigma\mathscr{E}^2)$. In the following section we extend the elementary theory of section 7.2 to cases where a magnetic field is applied. We see from equation (7.10.11) that we should obtain a resonance in the absorption at $\omega = \omega_c$ when a magnetic field is applied normal to the surface of the solid. According to equation (7.10.11) the resonance obtains for the positive circularly polarized wave only, but if there are also holes acting as carriers one obtains a resonance in the negative circularly polarized wave as well. This cyclotron resonance has been extensively used in the investigation of semi-conductors, where the carriers usually occupy small ellipsoidal pockets in **k**-space. By varying the orientation of the face on a single crystal one may then obtain the variation of the effective masses over the ellipsoids.

This is all we shall have to say about cyclotron resonance in semi-conductors except that the condition $\omega_c\tau \gg 1$ is necessary to obtain a sharp resonance, as equation (7.10.11) shows.

In contrast to the bulk absorption by semi-conductors, for which it can be assumed the electric field is uniform throughout the specimen, we must calculate the absorption by metals using equation (7.8.9), because of the small skin depth in these materials. If the calculation of Z as found in equation (7.8.10) is repeated using σ_+ as given by equation (7.10.11) instead of $\sigma = \sigma_0(1+i\omega\tau)^{-1}$ we find

$$R_+(\omega,\mathscr{H}) = \left(\frac{2\pi\omega}{c^2\sigma_0}\right)^{\frac{1}{2}} \{[1+(\omega-\omega_c)^2\tau^2]^{\frac{1}{2}}-(\omega-\omega_c)\tau\}^{\frac{1}{2}} \tag{7.9.1}$$

and for microwave frequencies the absorption is directly proportional to this quantity, it being easy to show that in the limit $\omega \gg \omega_p$ or $\omega \gg \tau^{-1}$

(whichever is the greater) $R, X \ll Z_0$ and so equation (7.8.8) reduces to

$$A = \frac{cR}{\pi}. \tag{7.9.2}$$

We thus see that resonant absorption is not found for metals with a magnetic field applied perpendicular to the surface, and in fact this configuration is hardly worth further discussion with reference to normal metals. However, on differentiating equation (7.9.1) twice it will be discovered that the absorption curve as a function of ω_c has a point of inflexion, at

$$\omega_c = \omega + \tau^{-1}/\sqrt{3}, \tag{7.9.3}$$

which is observable in semi-metals, for which τ is large although σ_0 is small. One can analyse bismuth, for example, assuming different energy surfaces for electrons and holes. The two types of carrier are distinguishable by their different effects on the two kinds of circularly polarized light.

7.9.2 *Azbel'–Kaner resonance*

It was pointed out by Azbel' and Kaner (1956) that resonant absorption of a plane-polarized wave by a metal can obtain if the magnetic field is *parallel* to the surface of the metal and reference to Figure 7.28 makes this

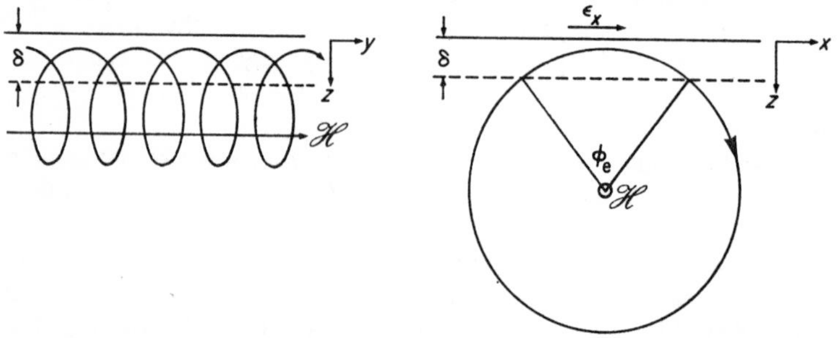

FIGURE 7.28. Trajectory of electron in Azbel'–Kaner resonance. Projection of electron orbit shown will in general have shape of cross-section of Fermi surface, rather than circular form shown, but since ϕ_c is very small, we ignore this complication.

clear. Electrons spiral in the magnetic field, traversing one turn of the helix with a frequency ω_c; if this is also the frequency of the electric field $\mathscr{E}$ the electrons experience at the same phase of their cycle each of the short periods of time they are in the skin depth and so gain energy. It is to be observed that resonance will occur if

$$\omega = n\omega_c, \tag{7.9.4}$$

where n is any positive integer, in contrast to the cyclotron resonance in semi-conductors we discussed previously, where the only possibility is $n = 1$.

To obtain some idea of how the form of Z_∞ is modified by the presence of the magnetic field we argue much as we did in section 7.8, the electrons only being effective, in this case, at the parts of their orbits within the skin depth. We take $\mathscr{E}$ to be static for each brief time the electron experiences the field, noting that on each revolution the field will change in phase by $2\pi\omega/\omega_c$. We also recall from the discussion in the last chapter (particularly that of Chambers' formulation of the resistivity problem) that the number of electrons in a particular orbit will decay in time according to $e^{-t/\tau}$, owing to collisions. This means the number decays by a fraction $e^{-2\pi/\omega_c\tau}$ each revolution and we ought therefore to multiply our expression for the effective conductivity by a sum of the decay factors

$$1+e^{-w}+e^{-2w}+\ldots = \frac{1}{1-e^{-w}}, \tag{7.9.5}$$

where

$$w = \frac{2\pi}{\omega_c\tau}+\frac{i2\pi\omega}{\omega_c}. \tag{7.9.6}$$

It would therefore seem that the surface impedance is of the form

$$Z_\infty(\mathscr{H}) = Z_\infty(0)\,(1-e^{-w})^{\frac{1}{3}}, \tag{7.9.7}$$

which shows oscillations because of the complex nature of w, in the extreme anomalous limit where

$$\phi_e^2 \approx \frac{\beta\delta}{R} \approx \beta\delta\omega_c/v_f \tag{7.9.8}$$

is very small. Typically $R = 500\delta$.

Calculation of conductivity. We can confirm equation (7.9.7) by a less intuitive argument than the above, obtaining $Z(\mathscr{H})$ from the values of $\sigma(q)$, as we did in the treatment of the anomalous skin effect. We begin here with Chambers' expression

$$\mathbf{j}(\mathbf{r}t) = \frac{e^2}{4\pi^3\hbar}\int_{\text{Fermi surface}} \mathbf{v}\left[\int_{-\infty}^{t} \mathbf{v}(t').\mathscr{E}(\mathbf{r}'t')\exp\left(-\frac{t-t'}{\tau}\right)dt'\right]\frac{dS}{v} \tag{7.9.9}$$

into which we insert

$$\mathscr{E}(\mathbf{r}t) = \mathscr{E}(q)\exp[i(\omega t - \mathbf{q}.\mathbf{r})] \quad (\mathbf{q}.\mathbf{r} = qz). \tag{7.9.10}$$

If $qR \gg 1$, i.e. $qv_f \gg \omega_c$, $\mathscr{E}$ will vary so rapidly over an electron's orbit that the integral with respect to t' will be significantly affected only when $\mathbf{v}_\perp$, the component of the velocity perpendicular to $\mathscr{H}$, is nearly parallel to $\mathscr{E}$. This means we can pretend the orbits are circular, and change the variable according to $dt' = d\phi'/\omega_c$ (see Figure 7.29).

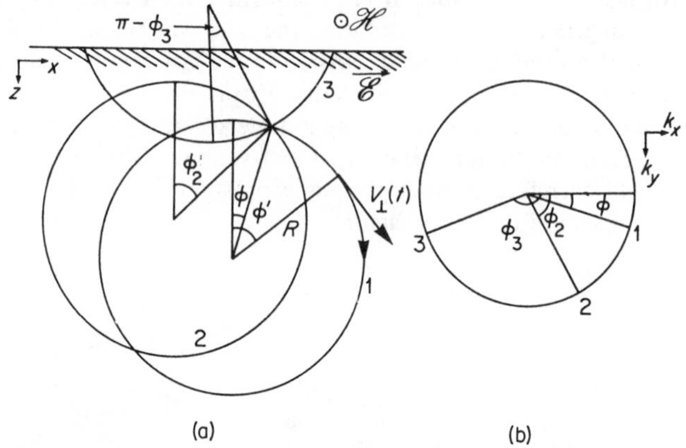

FIGURE 7.29. (a) Shows the projection of the paths of electrons, contributing to $j(\mathbf{r}, t)$, under the influence of the magnetic field. While the orbits round the Fermi surface shown in (b) are the same, the centres of the projected orbits in real space are different. The orbit labelled 3 is incomplete, the electron suffering reflexion at the surface. Note that for this electron, which does not contribute to the cyclotron resonance, $\phi_3 > \pi/2$.

Referring to Figure 7.29, we can see that

$$\mathscr{E}(z't') = \mathscr{E}(zt)\exp\left[\frac{i\omega(\phi'-\phi)}{\omega_c}\right]\exp[-iqR(\cos\phi' - \cos\phi)] \tag{7.9.11}$$

and hence

$$\sigma(q) = \frac{e^2}{4\pi^3\hbar}\int\frac{dS_x}{\omega_c}\int_{-\infty}^{\phi} d\phi' v_\perp \cos\phi' \exp\left[\frac{i\omega(\phi'-\phi)}{\omega_c}\right] \times \exp[-iqR(\cos\phi' - \cos\phi)]\exp[-(\phi-\phi')/\omega_c\tau]. \tag{7.9.12}$$

This is readily obtainable directly as the solution of the transport equation but in reaching it *via* Chambers' formulation we obtain a valuable interpretation of it, in terms of orbiting electrons, which we now exploit.

The integration over ϕ' becomes appreciable only when $\cos\phi' \simeq 1$, which is twice per orbit, once when the electron is near its point of closest approach to the surface, and once when it is near it position most distant from the surface. It is clear that since $R \gg \delta$ the effect of the field near the latter position may be neglected in the evaluation of any quantity bearing on the problem. As far as the evaluation of $\sigma(q)$ is concerned, since $qR \gg 1$ the relative phases of the contributions from near and distant regions vary very rapidly with q and so by neglecting the contribution of the latter region we evaluate $\bar{\sigma}(q)$, the average of $\sigma(q)$ over a small range Δq.

We shall see later that ϕ is taken very close to zero, and so in its very last orbit the electron may very well not traverse the entire region over which we take the integrand of the integration with respect to ϕ' to be significant. However, as ϕ' goes through 2π, the integrand suffers a fractional decay of $\exp(-\phi'/\omega_c\tau) = \exp(-2\pi/\omega_c\tau)$, which is close to unity if $\omega_c\tau \gg 1$, and so we can neglect this last traverse. Since the contribution to the integral when ϕ' is near $2\pi n$ is accompanied by the factor $\exp(-\phi'/\omega_c\tau) = \exp(-2\pi n/\omega_c\tau)$, equation (7.9.12) then becomes

$$\bar{\sigma}(q) = \frac{e^2}{4\pi^3\hbar}\int dS_x \frac{v_\perp}{\omega_c} 2\int_0^{\pi/2} d\phi' \cos\phi' \frac{\exp[-iqR(\cos\phi' - \cos\phi)]}{1-\exp(-w)}, \tag{7.9.13}$$

where $[1-\exp(-w)]$ is the decay factor of equation (7.9.6). The upper limit to the integration over ϕ' has here been set at $\pi/2$ since all that is required is that the significant range of integration is covered. Now since the significant values of ϕ' and ϕ are close to zero we can write

$$2\int_0^{\pi/2} \cos\phi' \exp[-iqR(\cos\phi' - \cos\phi)]\, d\phi' = 2\int_0^{\pi/2} \exp\left[\frac{-iqR(\phi'^2 - \phi^2)}{2}\right] d\phi'. \tag{7.9.14}$$

We can further write

$$2\int_0^{\pi/2} \exp\left[\frac{-iqR\phi'^2}{2}\right] d\phi \simeq 2\left(\frac{\pi}{qR}\right)^{\frac{1}{2}} \int_0^{\infty} \exp\left[\frac{-i\pi\xi^2}{2}\right] d\xi \tag{7.9.15}$$

where, on changing the variable to $\xi = (qR/\pi)^{\frac{1}{2}}\phi'$, we obtain so large an upper limit that we can set it at infinity.

Now the Fresnel integrals have the values

$$\left.\begin{aligned}\int_0^\infty \cos\left(\frac{\pi\xi^2}{2}\right)d\xi &= \tfrac{1}{2},\\ \int_0^\infty \sin\left(\frac{\pi\xi^2}{2}\right)d\xi &= \tfrac{1}{2}\end{aligned}\right\} \tag{7.9.16}$$

and so we have

$$2\int_0^{\pi/2} \exp\left(\frac{-iqR\phi'^2}{2}\right)d\phi' = \left(\frac{\pi}{qR}\right)^{\frac{1}{2}}(1-i) = \left(\frac{2\pi}{qR}\right)^{\frac{1}{2}}\exp\left(-\frac{i\pi}{4}\right) \tag{7.9.17}$$

and

$$\bar{\sigma}(q) = \frac{e^2}{4\pi^3\hbar}\exp\left(-\frac{i\pi}{4}\right)\int_{\text{Fermi surface}} dS_x \frac{v_\perp \exp(iqR\cos\phi)}{\omega_c[1-\exp(-w)]}\left(\frac{2\pi}{qR}\right)^{\frac{1}{2}}. \tag{7.9.18}$$

To evaluate the integral over the Fermi surface we put

$$dS_x = dk_y|\rho_x(k_y)|\cos\phi\, d\phi. \tag{7.9.19}$$

The integrand is significant around $\phi = 0$ and $\phi = \pi$. We ignore one of these regions, for the following reason. Figure 7.29 shows the integration over the Fermi surface to be equivalent to a counting of orbits contributing to the cyclotron resonance, (R, ϕ) being the position z referred to the centre of an electron's orbit. The same figure also shows that when ϕ is nearer π than zero the corresponding orbit cuts the surface of the metal, so that in fact the electron suffers reflexion and does not contribute to the resonance. We therefore omit such orbits and carry out the integral over dk_y over but half a line round the Fermi surface.

Since

$$\int_0^{\pi/2} \exp(iqR\cos\phi)\,d\phi = \exp\left(\frac{i\pi}{4}\right)\left(\frac{2\pi}{qR}\right)^{\frac{1}{2}} \quad (qR \gg 1) \tag{7.9.20}$$

and $R = v_\perp/\omega_c$, we finally obtain

$$\bar{\sigma}(q) = \frac{e^2}{2\pi^2\hbar}\int dk_y \frac{|\rho_x(k_y)|}{1-\exp(-w)}. \tag{7.9.21}$$

Remembering $E(\mathbf{k}) = E(-\mathbf{k})$, this result is the same as equation (7.8.39) apart from the factor $[1-\exp(-w)]^{-1}$.

Our subsequent analysis follows that of section 7.8 to obtain Z in terms of $\sigma(q)$ when reflexion is specular. This procedure may appear invalid for the present problem, because to simulate specular reflexion by replacing the surface by an interface (cf. Figure 7.30) the magnetic field ought to be reversed as the interface is crossed. However, in calculating $\bar{\sigma}(q)$ we have cut out the contributions of electrons suffering reflexion, as already explained. In terms of the model where the surface is replaced by an interface *without* reversal of magnetic field, we have ignored the contributions to $\bar{\sigma}(q)$ of all electrons

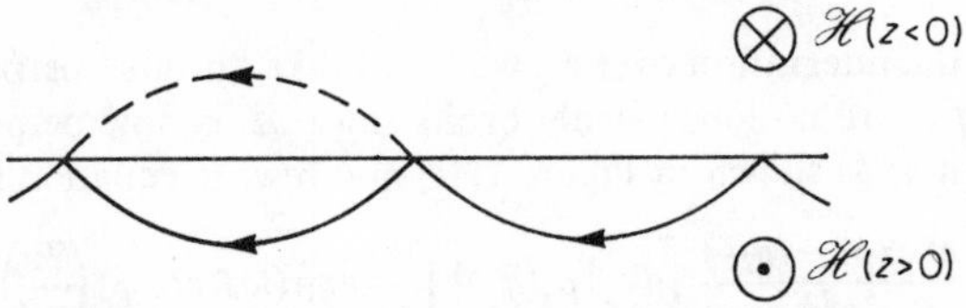

FIGURE 7.30. Simulation of specular reflexion by replacing surface by interface. Note that magnetic field is reversed as we cross interface.

in orbits cutting the interface, on the grounds that they cannot contribute to the resonance anyway. We therefore expect the resulting expression for ellipsoidal Fermi surfaces, *viz.*, equation (7.9.7), to be reasonably accurate in the limit $v_f \gg \omega_c \delta$, $\omega_c \tau \gg 1$. This is indeed confirmed by the much more lengthy and careful analysis of Azbel' and Kaner (1963). The only formal modification is the inclusion of extra small oscillatory terms in the expression for $Z_\infty(\mathscr{H})$.

When the Fermi surface is not ellipsoidal, the cyclotron effective mass m_c, and so $\omega_c = e\mathscr{H}/cm_c$, varies with k_y, and the factor $[1-\exp(-w)]^{-1}$ cannot be taken out of the integral of equation (7.9.21). The contributions to $Z_\infty(\mathscr{H})$ of different parts of the Fermi surface may then be difficult to sort out, but one would expect extremal orbits to contribute most, and closer examination shows that under certain circumstances the contributions of these orbits are in fact dominant. The argument uses a phase-amplitude diagram like the Cornu spiral of wave optics, as is discussed in Appendix 7.1.

We now examine the situation in which the electric and magnetic fields are parallel, in the same anomalous limit as before ($\omega_c \tau \gg 1, v_f \gg \omega_c \delta$). The semi-quantitative "ineffectiveness" argument suggests we obtain the same results as when the electric and magnetic fields are perpendicular, at least when ω_c is invariant over the Fermi surface, and we shall find this to be true.

Taking $\mathscr{H} \| \mathbf{x}$, the electron orbits are in the y–z plane, and equation (7.9.12) is trivially modified to

$$\sigma(q) = \frac{e^2}{4\pi^3 \hbar} \int dS_x v_x \int_{-\infty}^{\phi} d\phi' \exp\left[\frac{i\omega(\phi'-\phi)}{\omega_c}\right] \exp[-iqR(\cos\phi' - \cos\phi)]$$
$$\times \exp\left[-\frac{\phi-\phi'}{\omega_c \tau}\right]. \tag{7.9.22}$$

The integration over ϕ' is exactly as before, and we replace equation (7.9.18) by

$$\overline{\sigma(q)} = \frac{e^2}{4\pi^3\hbar}\int dS_x \frac{v_x}{\omega_c}\exp(iqR\cos\phi)\left(\frac{2\pi}{qR}\right)^{\frac{1}{2}}. \tag{7.9.23}$$

To perform the integration over S_x we again take "orbits" perpendicular to k_y (these paths are no longer truly orbits since $\mathscr{H}$ is now perpendicular to the k_y direction) as shown in Figure 7.31, and rewrite equation (7.9.23) as

$$\overline{\sigma(q)} = \frac{e^2\exp(-i\pi/4)}{4\pi^3\hbar}\int dk_y|\rho_x(k_y)|\int\frac{v_x}{\omega_c}\exp(iqR\cos\phi)\left(\frac{2\pi}{qR}\right)^{\frac{1}{2}}d\theta, \tag{7.9.24}$$

where θ is defined in the figure. Orbits (a) and (b) have a common direction, k_z, and

$$v_z = v_\perp\sin\phi = v'_\perp\sin\theta, \tag{7.9.25}$$

where $v'_\perp$ is the component of $\mathbf{v}$ in the x–z plane. Since ϕ is restricted to very small values, so is θ, and equation (7.9.25) may be rewritten for our purpose as

$$v_\perp\phi = v_x\theta \tag{7.9.26}$$

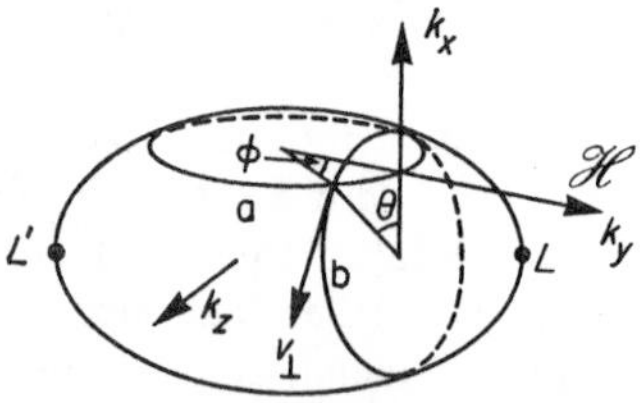

FIGURE 7.31. Showing evaluation of integrals for $\overline{\sigma(q)}$ over Fermi surface. Orbit a is an electron orbit; b is not actually an orbit but a path over which one integrates.

since $v_x \simeq v_\perp$ for the range of values of θ we are interested in. We now have the integral

$$\exp(-i\pi/4)\,2\int_0^{\pi/2} v_x\exp\left(\frac{iqR\phi^2}{2}\right)d\theta \simeq \exp\left(-\frac{i\pi}{4}\right)\int_0^{\pi/2} v_x\exp\left[iqR\left(\frac{v_x}{v_\perp}\right)^2\theta^2/2\right]d\theta \tag{7.9.27}$$

and since $R = v_\perp/\omega_c$ equation (7.9.24) becomes

$$\overline{\sigma(q)} = \frac{e^2}{2\pi^2\hbar}\int dk_y\frac{|\rho_x(k_y)|}{1-\exp(-w)}. \tag{7.9.28}$$

The integral over dk_y is carried out half-way round the Fermi surface for the same reason as before. Note that despite first appearances, we have not neglected orbits, cut by the path (b), for which θ is close to π.

The formula (7.9.28) looks exactly the same as before, but there is actually a very important difference in that, since the paths (b) are in the plane of the field, the oscillations of Z are no longer dominated by the value of ω_c for extremal orbits, but by the value of ω_c for the point L of Figure 7.31 [which is on an "extremal" path of the kind (b)]. The relevant cyclotron effective mass m_c is then no longer the property of a line, for the orbit shrinks to a point at L. We can easily find m_c in terms of the principal radii of curvature ρ_1 and ρ_2 at L. A very small orbit about L is easy to visualize: as indicated in Figure 7.32 it is an ellipse. To calculate $m_c = (1/2\pi)(dS/dE)$ we draw the surface S'

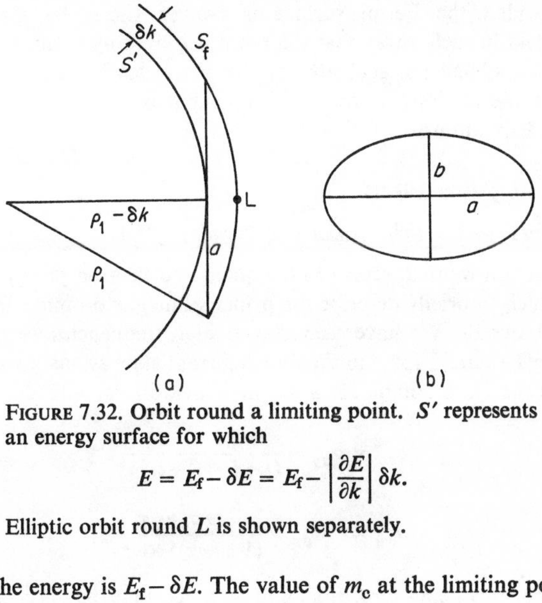

FIGURE 7.32. Orbit round a limiting point. S' represents an energy surface for which

$$E = E_f - \delta E = E_f - \left|\frac{\partial E}{\partial k}\right| \delta k.$$

Elliptic orbit round L is shown separately.

on which the energy is $E_f - \delta E$. The value of m_c at the limiting point is then obtained as

$$m_c = \frac{1}{2\pi}\frac{(\pi ab)}{\delta E}, \tag{7.9.29}$$

where a and b are lengths of semi-major and minor axes of the elliptical orbit in the tangent plane. From Figure 7.32

$$(\rho_1 - \delta k)^2 + a^2 = \rho_1^2, \tag{7.9.30}$$

that is,

$$a \approx \sqrt{(2\rho_1\,\delta k)} = \left(2\rho_1 \frac{\delta E}{|\partial E/\partial \mathbf{k}|}\right)^{\frac{1}{2}}$$

$$= \left(\frac{2\rho_1\,\delta E}{v_f}\right)^{\frac{1}{2}} \tag{7.9.31}$$

and so

$$m_c = \frac{2\pi}{v_f}(\rho_1\,\rho_2)^{\frac{1}{2}}. \tag{7.9.32}$$

Given the shape of the Fermi surface by other means, such as the anomalous skin effect, cyclotron resonance then enables us to determine v_f at any point—provided the Fermi surface is convex there—by arranging the magnetic field in such a way that the point is a limiting point.

We have assumed the geometry of the Fermi surface to be very simple throughout the analysis of this section, but it is easily extended to more complicated situations.

7.10 Magneto-plasma effects

7.10.1 Form of conductivity tensor and Faraday effect

In this section we shall return to the quasi-free electron theory as a framework in which to briefly describe the principal magneto-optic effects in non-magnetic materials. We have seen that at high frequencies we then merely substitute $\tau(1+i\omega\tau)^{-1}$ for τ in the direct-current expressions. From equation (7.2.10) we therefore obtain, for a magnetic field in the z-direction,

$$\sigma_{xx} = \sigma_{yy} = \frac{\sigma_0(1+i\omega\tau)}{(1+i\omega\tau)^2+\omega_c^2\,\tau^2},$$

$$\sigma_{xy} = -\sigma_{yx} = \frac{-\sigma_0\,\omega_c\,\tau}{(1+i\omega\tau)^2+\omega_c^2\,\tau^2}, \tag{7.10.1}$$

other components being zero. It should be noted that the Onsager relations, $\sigma_{ij}(\omega,-\mathscr{H}) = \sigma_{ji}(\omega\mathscr{H})$, are obeyed. In fact, using these relations, we can write quite generally for a substance isotropic in the absence of a magnetic field,

$$\left.\begin{aligned} \varepsilon_{xx} &= \varepsilon_{yy}, \\ \varepsilon_{xy} &= -\varepsilon_{yx}, \\ \varepsilon_{xz} &= \varepsilon_{yz} = \varepsilon_{zx} = \varepsilon_{xy} = 0, \end{aligned}\right\} \tag{7.10.2}$$

since the system must have cylindrical symmetry about $\mathscr{H}$. The tensor equation $\mathscr{D} = \varepsilon\mathscr{E}$ now becomes

$$\left.\begin{aligned} \mathscr{D}_x &= \varepsilon_{xx}\mathscr{E}_x + \varepsilon_{xy}\mathscr{E}_y, \\ \mathscr{D}_y &= \varepsilon_{xx}\mathscr{E}_y - \varepsilon_{xy}\mathscr{E}_x, \\ \mathscr{D}_z &= \varepsilon_{zz}\mathscr{E}_z. \end{aligned}\right\} \tag{7.10.3}$$

We note that

$$\mathscr{D}_x \pm i\mathscr{D}_y = (\varepsilon_{xx} \mp i\varepsilon_{xy})(\mathscr{E}_x \pm i\mathscr{E}_y), \tag{7.10.4}$$

which we shall write as

$$\mathscr{D}_\pm = \varepsilon_\pm \mathscr{E}_\pm. \tag{7.10.5}$$

Not only does this express the results most compactly but if we write

$$\mathscr{E}_x \pm i\mathscr{E}_y = \mathscr{E}_\pm = \exp\left[i\omega\left(t - \mu_\pm \frac{z}{c}\right)\right] \tag{7.10.6}$$

it is evident that $\mathscr{E}_+$ and $\mathscr{E}_-$ respectively denote left and right circularly polarized light, with different propagation constants. Plane-polarized light, therefore, becomes elliptically polarized in transmission through the medium—the Faraday effect. Assuming

$$i\varepsilon_{xy} = g\mathscr{H}, \tag{7.10.7}$$

which is true of the expressions (7.10.1) if we neglect attenuation, we obtain, to first order in $\mathscr{H}$, the angle through which the plane of polarization is rotated as the light traverses unit length,

$$\alpha = \frac{\pi g}{n\lambda}\mathscr{H}, \tag{7.10.8}$$

where n is the refractive index and λ the wavelength of the light. This equation represents a law which is experimentally verified. Magneto-optical properties are fully discussed by Mavroides (1972).

7.10.2 Reflectivity

Writing

$$J_\pm = J_x \pm iJ_y \tag{7.10.9}$$

we can define conductivities for polarized waves by the equation

$$J_\pm = \sigma_\pm(\mathscr{H})\mathscr{E}_\pm \tag{7.10.10}$$

which, according to equation (7.10.1), are

$$\sigma_\pm = \frac{\sigma_0}{1 + i(\omega \mp \omega_c)\tau}. \tag{7.10.11}$$

From equation (7.2.8) the complex refractive indices are given by

$$(n_\pm - ik_\pm)^2 = \varepsilon_0\left\{1 - \frac{i\omega_p^2\tau}{\omega[1+i(\omega\mp\omega_c)\tau]}\right\}, \tag{7.10.12}$$

where, since our interest extends to semi-metals, we have refrained from writing $\varepsilon_0 = 1$, and we have also redefined the classical plasma and cyclotron frequencies to be given by

$$\omega_p^2 = \frac{4\pi ne^2}{m^*\varepsilon_0}, \quad \omega_c = \frac{e\mathscr{H}}{m^* c\varepsilon_0^{\frac{1}{2}}}. \tag{7.10.13}$$

We shall discuss here only the positive polarized wave, for which we obtain from equation (7.10.12), in the limit as $\tau\to\infty$,

$$\left.\begin{aligned} n_+^2 - k_+^2 &\simeq \varepsilon_0\left[1 - \frac{\omega_p^2}{\omega(\omega-\omega_c)}\right], \\ n_+k_+ &\simeq 0. \end{aligned}\right\} \tag{7.10.14}$$

From equation (7.10.14), $n_+^2 - k_+^2 > 0$ at $\omega = 0$, so that $n_+ \neq 0$. Thus

$$R_+ = \frac{(n_+-1)^2}{(n_++1)^2}, \tag{7.10.15}$$

which increases to unity as $\omega\to\omega_c$. At $\omega = \omega_c$, $n_+^2 - k_+^2$ changes sign, giving $n_+ = 0$, and $R_+ = 1$. This continues until

$$\omega(\omega-\omega_c) = \omega_p^2, \tag{7.10.16}$$

when n_+ increases from zero and tends asymptotically to $\sqrt{\varepsilon_0}$. In this increase, R_+ drops to zero as n_+ rises to unity, which is when

$$\omega(\omega-\omega_c) = \omega_p^2\left(\frac{\varepsilon_0}{\varepsilon_0-1}\right). \tag{7.10.17}$$

Since for semi-metals ε_0 is large ($\varepsilon_0 \sim 100$ for bismuth) the values of ω given by (7.10.16) can be very close together. Thereafter R_+ tends to $|\sqrt{(\varepsilon_0)}-1/\sqrt{(\varepsilon_0)}+1|^2$. All this is indicated in Figure 7.33. Essentially the same behaviour is to be observed whether $\omega_c \gtrless \omega_p$.

7.10.3 Helicon modes

Let us suppose that

$$\omega_c\tau \gg 1, \quad \omega_c \gg \omega, \quad \omega_p^2 \gg \omega\omega_c. \tag{7.10.18}$$

Then from equation (7.10.12) we find the extinction coefficients of the circularly polarized waves to be

$$k_+ = \frac{\varepsilon_0^{\frac{1}{2}}\omega_p}{2\omega^{\frac{1}{2}}\omega_c^{\frac{3}{2}}\tau}, \quad k_- = 2\omega_c\tau k_+, \tag{7.10.19}$$

and k_+ will become very small in sufficiently strong fields, since $\omega_c = e\mathscr{H}/m^*c\varepsilon_0^{\frac{1}{2}}$. As we shall see, the frequency region of principal interest is such that $\omega\tau \ll 1$,

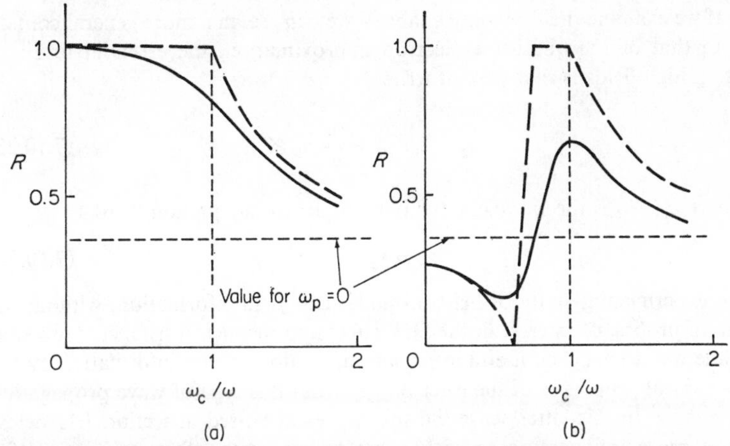

FIGURE 7.33. Reflectivity for a conductor as a function of reduced frequency. Broken curves show the limit as relaxation time τ goes to infinity. Full curves show effect of finite τ. (a) $\omega < \omega_p$. (b) $\omega > \omega_p$ (after Donovan, 1967).

and in this limit the zero-field attenuation is given by equation (7.2.11), so that

$$\frac{k_+}{k_-} = \frac{1}{2^{\frac{1}{2}}(\omega_c\tau)^{\frac{3}{2}}}, \quad \frac{k_-}{k} = \left(\frac{2}{\omega_c\tau}\right)^{\frac{1}{2}}. \tag{7.10.20}$$

Hence the effect of the magnetic field is to reduce the attenuation of both circularly polarized waves, but one very much more than the other, the ratio of extinction coefficients being

$$\frac{k_+}{k} = \frac{1}{2\omega_c\tau} \ll 1. \tag{7.10.21}$$

Under the same conditions, we obtain from equations (7.10.14) and (7.10.12)

$$n_+ = \omega_p\left(\frac{\varepsilon_0}{\omega\omega_c}\right)^{\frac{1}{2}}, \quad n_- = \frac{n_+}{2\omega_c\tau}. \tag{7.10.22}$$

Taking a typical field of $\sim 10^4$ oersteds and $\omega_p \sim 10^{16}$ sec^{-1} we find a phase-velocity of $c/n^+ \sim 10$ cm sec^{-1}.

We thus find, under the conditions indicated in equation (7.10.18), a single circularly polarized wave, of very low phase velocity, with wavelength of the order of 1 cm at frequencies of the order of 10 cycles sec^{-1}. This is the "helicon". Such helicon modes arise from carriers of one sign under the influence of a uniform magnetic field, and have $\mathscr{E}$ nearly perpendicular to $\mathscr{B}$ which is also perpendicular to the propagation vector **k**.

If we examine the foregoing analysis, we can see, in a more general context than that of the quasi-free electron approximation, the essential result of using high fields is that we can write

$$\varepsilon_{\alpha\beta} = \varepsilon_0\, \delta_{\alpha\beta} + \frac{4\pi i}{\omega}\sigma_0(\mathscr{H}), \tag{7.10.23}$$

where $\sigma_0(\mathscr{H})$ is the *static* conductivity. Since from equation (7.10.4)

$$\varepsilon_- = \varepsilon_{xx} + i\varepsilon_{xy}, \tag{7.10.24}$$

the experimental study of helicon modes can yield information, without the use of probes, of interest in the Hall effect and magnetoresistance. However, while we can get such useful information, we do not gain information on the dynamical properties of the electron gas from this type of wave propagation in metals. In this latter sense the spin waves discussed in section 7.11 below have more information to yield. Interactions with other excitations, for example phonons, can be studied for helicon waves.

7.10.4 Alfvén waves

An Alfvén wave is a disturbance, in a charged medium, similar to that of a sound wave, but where the restoring forces come not from the resulting variations of pressure but rather from a magnetic field applied to the system. To see how this comes about let us consider a classical plasma with unit charge on unit mass, and a mass density ρ_0, in a constant magnetic field $\mathscr{H}_0$. If $\mathbf{v}$ is the velocity of a particle, we have a current

$$\mathbf{j} = \rho\mathbf{v} \tag{7.10.25}$$

and so the equation $-\partial\rho/\partial t = \operatorname{div}\mathbf{j}$ immediately gives us

$$\frac{\partial\rho}{\partial t} + \nabla.(\rho\mathbf{v}) = 0 \tag{7.10.26}$$

as an equation of motion of the plasma when it is disturbed from equilibrium.

If we take the equilibrium velocity to be zero, the linearization of equation (7.10.26) produces just

$$\frac{\partial \rho_1}{\partial t}+\rho_0 \nabla . \mathbf{v}=0. \tag{7.10.27}$$

In the absence of electromagnetic effects, we have a second equation

$$\rho_0 \frac{\partial \mathbf{v}}{\partial t}+s^2 \nabla \rho_1=0, \tag{7.10.28}$$

where s is the velocity of sound. In the presence of a magnetic field, however, there is an extra acceleration given by the Lorentz force

$$\rho_0 \frac{e}{c}(\mathbf{v} \times \mathscr{H})=\frac{1}{c}(\mathbf{j} \times \mathscr{H}). \tag{7.10.29}$$

At the moment we shall neglect polarization effects and write

$$\operatorname{curl} \mathscr{H}=4\pi \mathbf{j}, \tag{7.10.30}$$

so that the modification to equation (7.10.29) is

$$\rho_0 \frac{\partial \mathbf{v}_1}{\partial t}+s^2 \nabla \rho_1+\frac{e\mathscr{H}_0}{4\pi} \operatorname{curl} \mathscr{H}_1=0, \tag{7.10.31}$$

where $\mathscr{H}_1$ is the additional field created by the deviation of $\rho=\rho_0+\rho_1$ from ρ_0. Finally, we write equation (7.1.7) as (see, e.g., Gartenhaus, 1964)

$$\frac{\partial \mathscr{H}}{\partial t}=\operatorname{curl}(\mathbf{v} \times \mathscr{H}), \tag{7.10.32}$$

which, again, we linearize, so that it becomes

$$\frac{\partial \mathscr{H}_1}{\partial t}=\operatorname{curl}(\mathbf{v} \times \mathscr{H}_0). \tag{7.10.33}$$

Using this and equation (7.10.27), we can differentiate equation (7.10.31) through with respect to t and rewrite it in the form

$$\frac{\partial^2 \mathbf{v}_1}{\partial t^2}-s^2 \nabla(\nabla . \mathbf{v})+\mathbf{v}_A \times \nabla \times[\nabla \times(\mathbf{v} \times \mathbf{v}_A)]=0, \tag{7.10.34}$$

where

$$\mathbf{v}_A=\frac{\mathscr{H}_0}{\sqrt{(4\pi\rho_0)}}. \tag{7.10.35}$$

We now investigate the plane-wave solutions

$$\mathbf{v}(\mathbf{r}t)=\mathbf{v}_0 \mathrm{e}^{i(\mathbf{k}.\mathbf{r}-\omega t)}, \tag{7.10.36}$$

when equation (7.10.34) becomes

$$-\omega^2 \mathbf{v}_0 + (s^2 + v_A^2)(\mathbf{k}.\mathbf{v}_0)\mathbf{k} + \mathbf{v}_A.\mathbf{k}[(\mathbf{v}_A.\mathbf{k})\mathbf{v}_0 - (\mathbf{v}_A.\mathbf{v}_0)\mathbf{k} - (\mathbf{k}.\mathbf{v}_0)\mathbf{v}_A] = 0. \tag{7.10.37}$$

There is a simple solution (a "magnetosonic" wave) for $\mathbf{k}$ perpendicular to $\mathbf{v}_A$ and so $\mathscr{H}$, but we shall only look at those for $\mathbf{k}$ parallel to $\mathscr{H}$, when (7.10.37) reduces to

$$(k^2 v_A^2 - \omega^2)\mathbf{v} + \left(\frac{s^2}{v_A^2} - 1\right) k^2 (\mathbf{v}_A.\mathbf{v})\mathbf{v}_A = 0. \tag{7.10.38}$$

With $\mathbf{v}$ parallel to $\mathscr{H}$ we obtain just the ordinary longitudinal sound wave, velocity s. However, for the transverse wave we get

$$k^2 v_A^2 - \omega^2 = 0. \tag{7.10.39}$$

This is the Alfvén wave, with phase velocity

$$v_A = \left(\frac{\mathscr{H}_0^2}{4\pi\rho_0}\right)^{\frac{1}{2}}. \tag{7.10.40}$$

One can see that $\mathscr{H}_0^2/8\pi$ plays the role of a "magnetic pressure".

As already remarked, we have neglected the polarization term $(1/c)(\mathbf{v}\times\mathscr{H})$ which strictly ought to be added to equation (7.10.38). However, it is readily shown that if $v_A \ll c$, which is the case of interest, this term does alter the main conclusions.

We can excite an Alfvén wave by an electromagnetic wave; from this point of view, we see that we have a refractive index given by

$$n_A^2 = \frac{c^2}{v_A^2} = \frac{4\pi\rho_0 c^2}{\mathscr{H}_0^2}. \tag{7.10.41}$$

If we compare this with equation (7.10.22) we shall see that the Alfvén wave will not usually show up on excitation by a low-frequency electromagnetic wave, the helicon term in n_+ dominating. However, suppose we take a material in which there are both electrons and holes, with respective number densities and effective masses given by ρ_e, m_e^* and ρ_n, m_n^*. Instead of equation (7.10.22) we shall then obtain, under the same conditions as we investigated the helicon modes,

$$\varepsilon = \varepsilon_0\left[1 + \frac{\omega_{pe}^2}{\omega(\omega_{ce} - \omega)} + \frac{\omega_{ph}^2}{\omega(\omega_{ch} - \omega)}\right], \tag{7.10.42}$$

where we have added additional subscripts upon ω_p and ω_c to denote plasma and gyro-frequencies for electrons and holes. Expanding to second order in $\mathscr{H}^{-1}$ we find

$$\varepsilon = \varepsilon_0 + \frac{4\pi ce}{\mathscr{H}\omega}(\rho_h - \rho_e) + \frac{4\pi c^2}{\mathscr{H}^2}(\rho_e m_e^* + \rho_h m_h^*) + \dots \quad (7.10.43)$$

In a "compensated" material, such as bismuth, $\rho_h = \rho_e$ and the helicon term vanishes, and we find

$$n^2 \simeq \varepsilon = \varepsilon_0 + \frac{4\pi e^2}{\mathscr{H}^2}(\rho_e m_e^* + \rho_h m_h^*) + \dots \quad (7.10.44)$$

ε_0 is usually negligible compared with the second term, and we then see that equation (7.10.44) is directly comparable with equation (7.10.41), and we have an Alfvén velocity given by

$$v_A^2 = \frac{\mathscr{H}^2}{8\pi(\rho_e m_e^* + \rho_h m_h^*)}. \quad (7.10.45)$$

Actually equation (7.10.41) should have a term ε_0 added to its right-hand side, omitted when we neglected the polarization term, and so there is no approximation involved in obtaining equation (7.10.45) from equation (7.10.44).

Alfvén waves have been observed in bismuth in particular, a favourable case because of the compensation already remarked upon. The above treatment should be modified to take account of the anisotropy of the Fermi surface (the effective mass being a tensor quantity). We shall not carry out this extension here as it is quite straightforward (cf. remark on the corresponding extension for helicon waves). The utility of Alfvén wave experiments in the exploration of the Fermi surface is obvious; in particular we may mention their use to show in a direct fashion the shift of the Fermi surface with magnetic field in the de Haas–van Alphen effect (Williams and Smith, 1964).

We should also notice the effect of alloying which upsets the compensation (making $\rho_h \neq \rho_e$) and allows helicon mode propagation as well as Alfvén waves. For a full discussion, reference can be made to Baynham and Boardman (1971) and Platzman and Wolff (1972).

7.11 Conduction electron spin resonance and spin waves

We shall assume the basic experimental facts of electron spin resonance are known to the reader, and our main interest here will be with spin waves. These have the important property, as we shall see, that they can be propagated as waves in metals only because of electron–electron interactions, and we can therefore hope to learn more about these correlation effects. In fact, as we have already said in Chapters 2 and 4, by studying the propagation of these spin waves we can obtain information about the Landau parameters B_0 and B_1.

We want to stress that we are talking here about the spin of the electrons as it is reflected in the weak Pauli-like magnetism that results in simple metals like Na and K, and the discussion is totally different from that of the spin waves discussed in Chapter 4. Also, to avoid another possible confusion, these are not at all related to the spin density waves involved in itinerant antiferromagnetism.

Though, as we have said, we assume knowledge of the basic facts of electron spin resonance, it is of some interest historically that while the phenomenon of conduction electron spin resonance was first observed by Griswold, Kip and Kittel (1952) and the essential features exposed in the theory of Dyson (1955), it was some ten years before the spin wave propagation associated with the conduction electron spin resonance was observed (Schultz and Dunifer, 1966).

7.11.1 Random phase approximation

In section 11 of Chapter 4, we began a discussion of a microscopic theory which will lead us, as we shall see, to obtain the dispersion relation for spin waves. Though this was not the first determination of this spin wave dispersion relation, it will be convenient to take up the discussion begun in Chapter 4 at this point, and later to discuss briefly the original work of Platzman and Wolff (1966) based on Fermi liquid theory.

(a) *Spin excitations for non-magnetic metals.* The excitation spectrum will now be discussed for non-magnetic metals for $\hbar\omega_c \lesssim k_B T \ll E_f$. This is helpful because the calculations can essentially be done at $T=0$, and oscillatory terms occurring for $k_B T \lesssim \hbar\omega_c$ can be neglected.

To explicitly obtain the susceptibility $\chi_{-+}(\mathbf{q}\omega)$ from equations (4.11.101) and (4.11.102), we replace ω by $\omega - i\varepsilon$ and then let $\varepsilon \to 0+$.

Using the identity

$$\frac{1}{\omega_0-\omega+i\varepsilon} = P\frac{1}{\omega_0-\omega} + \pi i\delta(\omega_0-\omega), \tag{7.11.1}$$

equation (4.11.101) can be written as

$$\Gamma = \Gamma' + i\Gamma'', \tag{7.11.2}$$

where Γ and Γ' come respectively from the principal value and delta function terms in equation (7.11.1). Hence

$$\chi_{-+}(\mathbf{q},\omega) = \chi' + i\chi'', \tag{7.11.3}$$

where

$$\chi'' = \Gamma''\left[\left(1-\frac{I}{L^3}\Gamma'\right)^2 + \left(\frac{I}{L^3}\Gamma''\right)^2\right]^{-1}. \tag{7.11.4}$$

Now, as discussed, for example, by Pines and Nozières (1966), the imaginary part of the susceptibility χ'' is directly related to the energy transfer to the system from an oscillating transverse magnetic field. For $\omega > 0$, this energy transfer is physically due to excitations in which an $\uparrow$ spin flips to $\downarrow$ spin (Edwards, 1969).

If the equation

$$1 - \frac{I}{L^3}\Gamma'(\mathbf{q}, \omega) = 0 \tag{7.11.5}$$

has a real root $\omega(\mathbf{q})$ outside the region of single-particle excitations, that is, where $\Gamma''(\mathbf{q}, \omega) = 0$, then χ'' contains a factor $\delta[\omega - \omega(\mathbf{q})]$. This corresponds to a collective mode of infinite lifetime. If, however, equation (7.11.5) has a root $\omega(\mathbf{q})$ such that $\Gamma''(\mathbf{q}, \omega)$ is non-zero but small, then χ'' will exhibit a sharp resonance within the single-particle continuum. We have then a damped collective mode.

(b) *Single-particle excitations.* Single-particle excitations occur for values of ω and $\mathbf{q}$ such that the denominators in equation (4.11.101) vanish. Using equation (4.11.85) we can express the excitation energies in the form

$$\hbar\omega = \frac{\hbar^2}{m^*}(k_z q_z + \tfrac{1}{2}q_z^2) + \mu\hbar\omega_c + g\mu_B \mathscr{H} + \Delta. \tag{7.11.6}$$

Thus, the single-particle or Stoner excitations consist of branches labelled by the integer μ. If we put $g\mu_B \mathscr{H} + \Delta = \Delta'$, then for small q and $\hbar\omega_c \lesssim k_B T \ll E_f$ each branch is bounded approximately by the lines

$$\hbar\omega = \pm\frac{\hbar^2}{m^*}k_f q \cos\delta + \mu\hbar\omega_c + \Delta' \tag{7.11.7}$$

in the ω, q plane, δ being the angle between $\mathbf{q}$ and the direction of the static field, so that

$$\mathbf{q} = q(\sin\delta, 0, \cos\delta) \tag{7.11.8}$$

and k_f is the Fermi wave-vector in zero field. The shaded areas in Figures 7.34(a) and (b) indicate the Stoner continuum for $\delta = 0$ and for a more general value of δ respectively.

In the special case $\delta = \frac{1}{2}\pi$, plotted in Figure 7.35, the single-particle continuum is reduced to a series of straight lines, parallel to the q-axis, at intervals of $\hbar\omega_c$ in energy.

The probability of excitation is determined by $|Q_{n,n+\mu}(q_x)|^2$ and this can be shown [see, for example, Mermin and Canel (1964)] to vary like $q_x^{2|\mu|}$ or as $(q\sin\delta)^{2|\mu|}$ as $q_x \to 0$, so that for $\delta = 0$ only the branch corresponding to

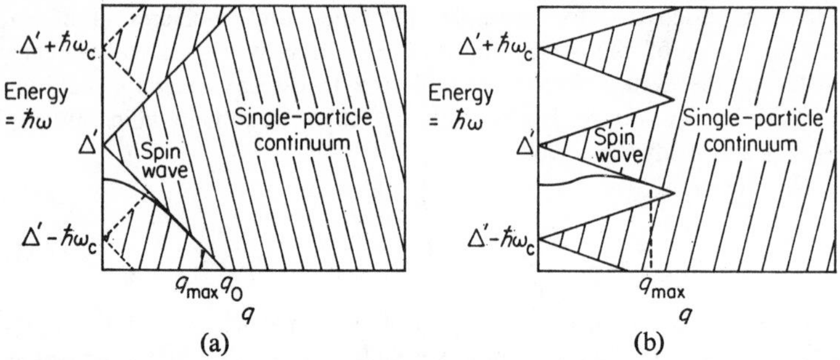

FIGURE 7.34. Illustrates schematically the single-particle excitations, the boundaries shown being determined by equation (7.11.7). (a) Angle δ between $\mathbf{q}$ and direction of static field given by $\delta = 0$. (b) A more general case for $\delta < \frac{1}{2}\pi$ (after Edwards, 1969).

$\mu = 0$ is excited. This branch is the only one excited at $q = 0$ for arbitrary δ, and other branches are only weakly excited for small q. For $k_B T \ll \hbar\omega_c$, additional structure in the Stoner excitation spectrum is resolved.

(c) *Spin waves.* A spin wave with zero wave-vector is generated by acting on the ground state with the operator S^-. The resulting state differs from the ground state only in that the z-component of total spin is reduced by $\hbar$.

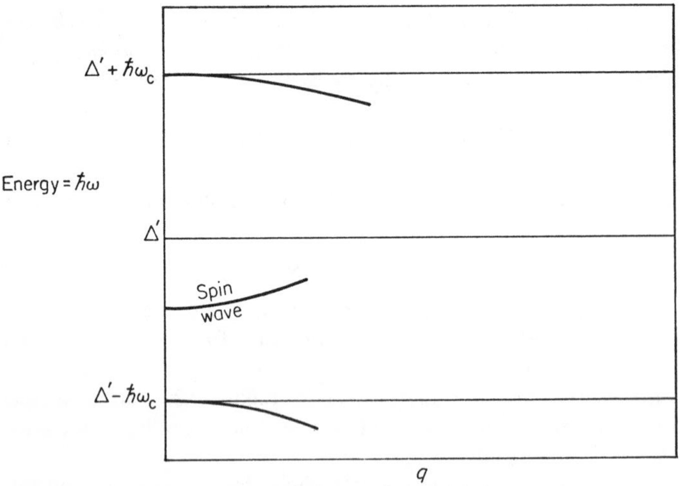

FIGURE 7.35. Same as Figure 7.34 except that $\delta = \frac{1}{2}\pi$. In this case the q-dependent term in equation (7.11.7) vanishes and the single-particle excitations reduce to horizontal lines (after Edwards, 1969).

Hence the excitation energy of this state, which is the principal one excited in electron spin resonance experiments, is just the Zeeman energy $\hbar\omega_s = g\mu_B\mathscr{H}$.

Thus, we expect that the frequency of the long wavelength spin waves will take the form

$$\omega(\mathbf{q}) = \omega_s + \omega'(\mathbf{q}) = \omega_s + D(\delta)q^2 + O(q^4). \tag{7.11.9}$$

We shall therefore seek a solution of equation (7.11.5) of this form, and hence find $D(\delta)$. The behaviour of shorter wavelength spin waves can also be found from the theory.

The spin wave spectrum (7.11.9) in general lies outside the Stoner excitation region for small q. Thus the spin wave is undamped, i.e. $\Gamma'' = 0$ and Γ' in equation (7.11.5) is simply Γ. Furthermore, to $O(q^2)$, the small q result for $|Q_{n,n+\mu}(q_x)|^2$ used above tells us that only terms with $\mu = 0, \pm 1$ need to be included in finding Γ. Explicitly, we can then write

$$1 - \frac{I}{L^3}(\Gamma_0 + \Gamma_{-1} + \Gamma_1) = 0, \tag{7.11.10}$$

where

$$\Gamma_\mu(\mathbf{q}, \omega) = \sum_{k_y k_z n} \frac{|Q_{n,n+\mu}(q_x)|^2 (f_{k_z n\uparrow} - f_{k_z + q_z, n+\mu\downarrow})}{\alpha_n(\mathbf{q}) + \Delta}, \tag{7.11.11}$$

with

$$\alpha_\mu(\mathbf{q}) = \frac{\hbar^2}{m^*}(k_z q_z + \tfrac{1}{2}q_z^2) + \mu\hbar\omega_c - \hbar\omega'(\mathbf{q}). \tag{7.11.12}$$

Equation (7.11.10) can be solved for $\omega'(\mathbf{q})$ to order q^2 by expanding $(\alpha_0 + \Delta)^{-1}$ in equation (7.11.11) for Γ_0 in powers of α_0/Δ. Then one finds

$$\omega'(q) = NI\Delta \frac{\hbar}{2m^*}\left(\frac{q_z^2}{\Delta^2} + \frac{q_x^2}{\Delta^2 - (\hbar\omega_c)^2}\right)$$
$$- \frac{I}{L^3}\frac{\hbar}{m^*}\left[\frac{q_z^2}{\Delta^2}\sum_{k_y k_z n}\frac{\hbar^2 k_z^2}{m^*}(f_{k_z n\uparrow} - f_{k_z n\downarrow})\right.$$
$$\left. + \frac{q_x^2}{\Delta^2 - (\hbar\omega_c)^2}\sum_{k_y k_z n}(n + \tfrac{1}{2})\hbar\omega_c(f_{k_z n\uparrow} - f_{k_z n\downarrow})\right]. \tag{7.11.13}$$

This result is true for both ferromagnetic and non-magnetic metals. In the latter case, the Fermi functions restrict the states that occur in the summations to energies very near to the Fermi energy E_f.

For $k_B T \gtrsim \hbar\omega_c$, the Fermi functions vary but slowly over an energy scale $\hbar\omega_c$ and the summations may be replaced by sums over $\mathbf{k}$ as in zero field. Hence

$$L^{-3}\sum \frac{\hbar^2 k_z^2}{m}(f_\uparrow - f_\downarrow) \simeq \tfrac{1}{3}L^{-3}\sum_{\mathbf{k}} \frac{\hbar^2 k^2}{m}(f_\uparrow - f_\downarrow)$$

$$\approx \tfrac{2}{3}E_f(N_\uparrow - N_\downarrow) \tag{7.11.14}$$

and the second summation in equation (7.11.13) can be shown to lead to the same result. Then, using these two results and also that $NI = \frac{4}{3}E_f \bar{I}$ we obtain

$$\omega'(q) = Dq^2, \tag{7.11.15}$$

where D is given by

$$D(\delta) = \tfrac{1}{3}\hbar^2 v_f^2 \bar{I}\omega_s \left[\frac{\sin^2\delta}{(\hbar\omega_c)^2 - \Delta^2} - \frac{\cos^2\delta}{\Delta^2}\right] \tag{7.11.16}$$

and v_f is the Fermi velocity. This then establishes the dispersion relation of the spin waves at small q, and, as we said earlier, the dispersion relation depends crucially on the angle δ between the direction of the applied field and that of the spin-wave propagation. We shall now show briefly how Fermi liquid theory can deal with the same problem.

7.11.2 Fermi liquid theory of spin wave dispersion relation

We have seen from the RPA treatment above how $\omega(\mathbf{q})$, the dispersion relation for spin waves, has the form (7.11.9). We want now to express D in terms of the Landau parameters, following Platzman and Wolff (1967). To do this, the treatment using the transport equation which we gave in section 2.9.4 can be readily generalized, and since we gave a full microscopic discussion of spin waves above we shall only summarize the main results here.

To get the susceptibility $\chi(\mathbf{k}\omega)$ again, the transport equation for the magnetization M separates into three scalar uncoupled equations for the quantities

$$M^+ = M_x \pm iM_y \tag{7.11.17}$$

and M_z. Since M^+ turns out to be the only quantity with a resonance at the electron precession frequency, it will suffice to deal only with this. The transport equation is then, for the non-interacting case,

$$\frac{\partial M^+}{\partial t} + \left[\mathbf{v}.\nabla + \frac{e}{c}(\mathbf{v}\times\mathscr{H}_0).\frac{\partial}{\partial \mathbf{p}} + i\omega_s\right] M^+$$

$$= \left.\frac{\partial M^+}{\partial t}\right|_{\text{coll}} + \tfrac{1}{2}\gamma_0[\mathbf{v}.\nabla + i\omega_s]\left(-\frac{\partial n_0}{\partial \varepsilon}\right) h^+, \tag{7.11.18}$$

where $\mathscr{H}_0$ is the steady applied field and

$$h^+ = h_x + ih_y, \tag{7.11.19}$$

h being the r.f. field; ω_s is the spin resonance frequency. Treating $\partial M^+/\partial t|_{\text{coll}}$ in a relaxation time approximation and writing

$$M^+ = -\frac{\partial n_0}{\partial E^0(p)} g, \tag{7.11.20}$$

the linearized transport equation of the Landau theory becomes

$$\frac{\partial g}{\partial t} + \left[\mathbf{v}.\nabla + \frac{e}{c}(\mathbf{v}\times\mathscr{H}_0)\frac{\partial}{\partial \mathbf{p}} + i\Omega_0\right](g+\delta\varepsilon_2)$$

$$= -\frac{1}{\tau}\left[g - \int \frac{g\,d\Omega}{4\pi}\right] + \tfrac{1}{2}\gamma_0(\mathbf{v}.\nabla + i\Omega_0)\,h^+, \tag{7.11.21}$$

where Ω_0 is a renormalized spin resonance frequency, given in Landau theory by

$$\Omega_0 = \frac{\omega_s}{1+B_0}, \tag{7.11.22}$$

B_0 being the usual Landau parameter and $\gamma_0 = ge\hbar/2mc$; $\hbar\Omega_0$ is simply the energy required to reverse the spin in the presence of an external field and an isotropic exchange field from the other electrons.

The interaction terms on the right-hand side of equation (7.11.21) are phenomenological in character. The term in $1/\tau$ is an orbital relaxation time characteristic of the d.c. resistance measurement. The collision term is chosen so that the system relaxes to local equilibrium compatible with the local density of magnetization and quasi-particle number.

The quantity $\delta\varepsilon_2$ is the change in the quasi-particle energy due to a change in the distribution function, and is given by

$$\delta\varepsilon_2 = \frac{2}{(2\pi)^3}\int d\mathbf{p}'\,\zeta(\mathbf{p},\mathbf{p}')\,g(p')\,\delta[\mu - E^0(p')], \tag{7.11.23}$$

where $\zeta(\mathbf{p},\mathbf{p}')$ is defined in equation (2.9.41).

For disturbances varying like $e^{i(\mathbf{k}.\mathbf{x}-\omega t)}$, we can now solve the transport equation exactly to order k^2. We then find for the susceptibility

$$\chi(\mathbf{k}\omega) = -(m^*/m)\,\chi_{\text{non-interacting}}\,\Omega_0/\mathbf{P}(k\omega), \tag{7.11.24}$$

where

$$\frac{1}{(2\pi)^3}\int M^+(p,k,\omega)\,d\mathbf{p} \equiv h\chi(k\omega) \tag{7.11.25}$$

and

$$P(\mathbf{k}\omega) = \omega - \omega_s + \tfrac{1}{3}k^2\left(\frac{p_f}{m^*}\right)^2 (1+B_0)(1+B_1)(\bar{\omega}_s - \bar{\Omega}_0)$$

$$\times \left[\frac{\sin^2\delta}{\bar{\omega}_c^{*2} - (\bar{\omega}_s - \bar{\Omega}_0)^2} - \frac{\cos^2\delta}{(\bar{\omega}_s - \bar{\Omega}_0)^2}\right]$$

$$\equiv \omega - \omega_s + iD^* k^2. \tag{7.11.26}$$

Here $\bar{\omega}_c^* = (e\mathscr{H}_0/m^* c)(1+B_1)$, $\bar{\Omega}_0 = \Omega_0(1+B_1)$, $\bar{\omega}_s = \omega_s + i/\tau$, and δ is the angle the wave vector $\mathbf{k}$ makes with the z-axis.

As B_0 and $B_1 \to 0$, the coefficient D^* in equation (7.11.26) becomes a real number and the susceptibility is a purely diffusive kernel with an anisotropic diffusion coefficient. In this limit, the results reduce, for small $\omega\tau$, to those of Dyson (1955) referred to above.

But if $|(B_0 - B_1)\,\omega\tau/(1+B_0)| \gg 1$, D^* approaches a pure imaginary number and $\chi(k,\omega)$ exhibits a branch of singularities along the curve $P(k\omega) = 0$ in ω, k space, which are, of course, the spin waves.

In Chapter 4, we saw that the magnetization $\mathbf{M}$ obeyed the equation

$$\frac{d\mathbf{M}}{dt} = \gamma \mathbf{M} \times \mathscr{H}. \tag{7.11.27}$$

Adding to this a diffusion term (see Chapter 4) and a term to account for spin-lattice relaxation,‡ with characteristic time T_2, we find

$$\frac{\partial \mathbf{M}}{\partial t} = \gamma(\mathscr{H}_0 + \mathscr{H}) \times \mathbf{M} + D\nabla^2\mathbf{M} + \frac{1}{T_2}(\mathbf{M}_0 - \mathbf{M}). \tag{7.11.28}$$

This equation determines the transverse magnetism as

$$\chi_+ \sim \frac{1}{\omega - \omega_s + i/T_2 + iDk^2}. \tag{7.11.29}$$

The above expression for χ_+, valid for an infinite medium, can now be used to effect a solution of the finite slab transmission problem. We can then find the transmitted magnetic field as a function of $\alpha \equiv [1 - (\omega_s/\omega)]\,\omega T_2$ in the neighbourhood of the first spin-wave side-band and results obtained by Platzman and Wolff are shown in Figure 7.36.

The beautiful experiments of Schultz and co-workers, referred to earlier, confirm the above theory and allow the Landau parameters to be determined. We shall shortly bring together the information we now possess on the

‡ For a discussion of the exchange of energy between spins and the lattice, see, for example, Slichter (1965).

Landau parameters, but before doing so we want to comment on the relation between the Fermi liquid theory above and the RPA treatment.

In fact, the result of Platzman and Wolff above is obtained if we put $B_1 = 0$ and $B_0 = -\bar{I}$ in equation (7.11.16) of the RPA theory ($D \equiv -iD^*$).

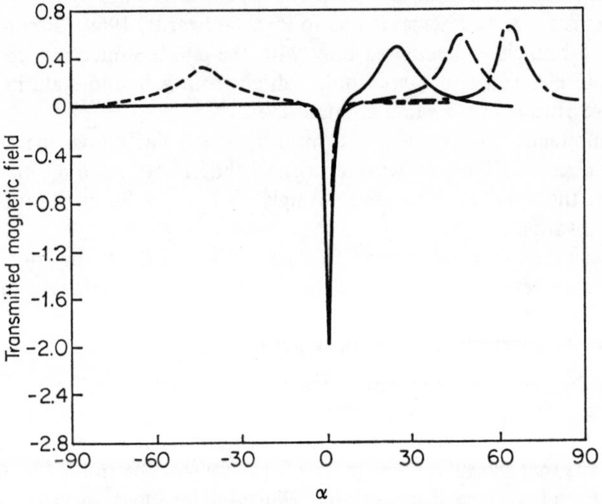

FIGURE 7.36. Calculated values of transmitted magnetic field as function of $\alpha = [1-(\omega_s/\omega)]\,\omega T_2$ in neighbourhood of first spin-wave side band. Landau parameters were taken as $B_0 = -0.2$ and $B_1 = 0.01$. Curves correspond to four different angles which the d.c. magnetic field makes with the normal to the surface of the specimen, namely: - - - - - -, 65°; ———, 75°; – – 80°; – – – 90°. Other parameters were taken to be such that Fermi velocity = 10^8 cm/sec, slab thickness = 0.32 mm, $\omega = 10^{11}$ cycle/sec, $\omega\tau = 60$, $\omega T_2 = 6 \times 10^4$. τ is relaxation time characteristic of d.c. resistance measurements while T_2 is a phenomenological spin relaxation time (after Platzmann and Wolff, 1967).

The absence of p-wave effects is due to the use of constant interactions in **k** space, and I is seen to be the effective interaction relating to s-wave quasi-particle scattering.

I can be estimated for Na from the data of Schultz and Dunifer (1967). They find $\delta_c = 69.7 \pm 0.4°$, where δ_c is defined by

$$D(\delta_c) = 0, \tag{7.11.30}$$

and then it follows that

$$\cos \delta_c = \frac{\bar{I}}{1-\bar{I}} \frac{g}{2} \left(\frac{m^*}{m}\right).$$

If we put $g = 2$ and $m^*/m = 1.24$, then $\bar{I} = 0.219$ for Na. For this metal where $\Delta < \hbar\omega_c$, the sign of D changes as δ passes through δ_c.

This effect can be interpreted as follows (Edwards, 1969). For $\delta = 0$, the spin-wave branch is associated only with the $\mu = 0$ Stoner branch; that is, the $\downarrow$ spin electron and $\uparrow$ spin hole, which form a bound state in the spin wave, are always in the same Landau level.

The interaction with the $\mu = 0$ branch, which lies above the spin-wave branch, depresses the spin-wave energy so that $D < 0$. As δ approaches $\frac{1}{2}\pi$, however, the spin wave couples strongly to $\mu = -1$ Stoner branch and is forced upwards.

We have seen that the Landau parameters B_0 and B_1 can be got from such spin wave observations and we shall now discuss the values thereby obtained.

7.12 Role of interactions and Landau parameters

From the transport equation of Fermi liquid theory for the quasi-particle distribution, it is possible to prove that transport properties that are independent of time are not affected by the interactions between quasi-particles. This means that the galvanomagnetic effects are still determined by the shape of the Fermi surface, for example. Time-independent properties depend generally on the Landau interaction function $f(k, k')$, but this is not true of the surface impedance in the anomalous limit, which we discussed in section 7.8, so that the anomalous skin effect still measures the curvature of the Fermi surface and, as it turns out, the Azbel'–Kaner resonance the effective mass.

Thus, in collecting together information on the Landau parameters and seeing, if possible, how they help to interrelate experiments, let us begin by comparing the mass from specific heat measurements with that referred to above from the Azbel'–Kaner resonance. In the first part of Table 7.2, we therefore show the effective masses found from these two experiments and it is seen that for the three metals shown, Na, K and Rb, the differences between the results are small and it is perhaps significant that they are largest for Rb. For, in section 4.15.3 we saw that from the de Haas–van Alphen experiments, the Fermi surface is more distorted for Rb than for Na and K, and therefore we expect any discrepancies to be in this direction.

Secondly, as we have discussed at length, B_0 and B_1 can be obtained from spin-wave observations, and the results obtained are also recorded in Table 7.2.

Next, it is possible to compare the values of B_0 and m^* with direct measurements of the Pauli spin susceptibility, from electron spin resonance experiments of the type of Schumacher and Slichter. The second part of Table 7.2 shows the value of B_0 obtained from the spin wave spectrum discussed above by Schultz and Dunifer (1967) and combining this with m^* from the Azbel'–Kaner cyclotron resonance experiment we find the ratio χ/χ_0 in Na. This is compared with the results of the power absorbed in the conduction electron spin resonance line as measured by Schumacher and Vehse (1963). The second entry is obtained by measuring the fractional shift in electron spin resonance due to the polarization of the nuclei. The value obtained in this way by Ryter is shown. There is seen to be quite good agreement between the three values. Similar values are also given for K.

TABLE 7.2

	Azbel'–Kaner m^*	Specific heat m^*
Na	1.24 ± 0.02	1.23 ± 0.02
K	1.21 ± 0.02	1.25 ± 0.02
Rb	1.20 ± 0.05	1.26 ± 0.03

Note: Azbel'–Kaner m^* values are from Grimes and Kip (1963) and Grimes (1967). Specific heat m^* values are from Martin (1961) and Lien and Phillips (1964).

Values of B_0 and B_1 from spin-wave measurements

	B_0	B_1
Na	-0.21 ± 0.05	$+0.01 \pm 0.03$
K	-0.2 ± 0.1	$+0.1 \pm 0.2$

Values of χ/χ_0

Na	1.74 ± 0.1
	1.50 ± 0.07
K	1.58 ± 0.1
	1.68 ± 0.2

Electron–phonon contribution to g from Young–Sham method

	$g_0^{ep}-g_1^{ep}$	Ashcroft–Wilkins approach
Na	0.092	0.08
K	0.067	0.06

We want then to say a little more at this point about the way the electron–phonon interaction affects the problem. We saw in the discussion of high-temperature electrical resistivity in Chapter 6 that Young and Sham have related the contribution of the electron–phonon interaction to the Landau parameters, to the measured high-temperature resistivity and we show their results in the fourth part of Table 7.2 (cf. also Grimvall, 1969).

Their theory yields, as we saw in equation (6.4.10), the quantity $g_0^{ep}-g_1^{ep}$ and the results thus obtained are recorded in the table. We gave earlier a very brief description of what microscopic theory has to say about the effects of the electron–phonon interaction.

Arguing along similar lines, Ashcroft and Wilkins (1963) used pseudo-potentials and the observed phonon spectrum to calculate the mass enhancement due to electron–phonon effects. The coefficients g_0^{ep} and g_1^{ep} calculated in this way for Na, using Ashcroft's pseudo-potential for Na and that of Lee and Falicov for K, are shown for comparison with the results of the Young–Sham analysis in Table 7.2. In these calculations, the electron states are taken as single plane-waves and the results are found to depend only rather weakly on the shape of the pseudo-potential used. The agreement again is encouraging.

7.13 Infrared optical properties of insulators

In Chapter 3, we set up a microscopic description of the dynamical matrix in an insulator, and separated the irregular and the regular parts as a function of $\mathbf{q}$. The frequency of the acoustic modes was shown to tend to zero linearly with q as $q \to 0$ (cf. Sherrington, 1971 for a zero-gap semiconductor).

We shall conclude the Chapter by discussing here the optical modes in the long wavelength limit (Sham, 1969).

We shall not include the retardation effect, and hence we shall not find the hybrid mode of optical phonons and photons: the so-called "polariton" (see, for example, Ziman, 1972); q will be assumed small compared with the distance to the zone boundary, but large compared with ω/c.

We can write for the normal modes

$$\sum_{\lambda'}\sum_{i'} D_{\lambda\lambda'}(\mathbf{q}ii')\, \varepsilon_{i'\sigma}^{\lambda'} = \omega_{\mathbf{q}\sigma}^2\, \varepsilon_{i\sigma}^{\lambda}, \tag{7.13.1}$$

where D is the dynamical matrix, while $\boldsymbol{\varepsilon}_{i\sigma}$ and $\omega_{\mathbf{q}\sigma}$ are respectively the polarization vector and frequency of the $(\mathbf{q}\sigma)$ mode.

Now Born and Huang (1954) in their treatment of optical properties start from the macroscopic energy density

$$U = \frac{1}{2}\sum_{kk'} f(k\lambda, k'\lambda')\, u_\lambda(k)\, u_{\lambda'}(k') + \sum_{\mathbf{k}} f(k\lambda, \mu)\, u_\lambda(k)\, \mathscr{E}_\mu - \tfrac{1}{2} f(\lambda\mu)\, \mathscr{E}_\lambda\, \mathscr{E}_\mu, \tag{7.13.2}$$

where $\mathscr{E}$ is the macroscopic electric field given by

$$\mathscr{E}_\lambda = -\sum_{\lambda'} (4\pi q_\lambda q_{\lambda'}/q^2)\mathscr{P}_{\lambda'} \tag{7.13.3}$$

and $\mathscr{P}$ is the macroscopic polarization or dipole moment per unit volume when the crystal is vibrating with wave-vector q. We shall shortly obtain general expressions for the quantities $f(k\lambda, k'\lambda')$ from the microscopic theory, whereas Born and Huang used rigid-ion and polarizable-ion models.

In order to obtain $\mathscr{P}$ from the microscopic theory, we first note that the contribution from the bare nuclei per unit amplitude of the normal mode $\mathbf{q}j$ is

$$\mathscr{P}^I_\lambda = \Omega_0^{-1} \sum_k Z_k e\varepsilon^\lambda_{k\mathbf{q}j} M_k^{-\frac{1}{2}}. \tag{7.13.4}$$

The change to first order of the electron density distribution can be written in Fourier transform in terms of the response function $\mathscr{F}$ as

$$\rho^{(1)}(\mathbf{q}) = -i\Omega_0^{-1} \sum_{\mathbf{kK}} \mathscr{F}(\mathbf{q}, \mathbf{q}+\mathbf{K})\, \mathrm{e}^{-i\mathbf{K}.\mathbf{x}_k} v(\mathbf{q}+\mathbf{K}, \mathbf{k})\,(\mathbf{q}+\mathbf{K}).\boldsymbol{\varepsilon}_{kqi} M_k^{-\frac{1}{2}}. \tag{7.13.5}$$

If we now use the small q forms of the response function derived in Chapter 3, we find for the leading term at small q the result

$$-e\rho^{(1)}(\mathbf{q}) = -i\mathbf{q}.\mathscr{P}^{\mathrm{E}}, \tag{7.13.6}$$

where the electronic contribution to the macroscopic polarization is

$$\mathscr{P}^{\mathrm{E}}_\mu = -e\Omega_0^{-1} \sum_k \{Z_{\mathbf{k}}\,\delta_{\mu\lambda} - [Z_k\,\delta_{\mu\lambda} - Z_{\mu\lambda}(k)]/\varepsilon_\infty(\mathbf{q})\}\, \varepsilon^\lambda_{k\mathbf{q}j} M_k^{-\frac{1}{2}}. \tag{7.13.7}$$

Hence

$$\mathscr{P}_\mu = \frac{e}{\Omega_0\,\varepsilon_\infty(q)} \sum_k [Z_{\mathbf{k}}\,\delta_{\mu\lambda} - Z_{\mu\lambda}(k)]\, \varepsilon^\lambda_{k\mathbf{q}j} M_k^{-\frac{1}{2}}. \tag{7.13.8}$$

For the acoustic modes, $\varepsilon^\lambda_{k\mathbf{q}j} M_k^{-\frac{1}{2}}$ is independent of k and from the sum rule (3.5.10) we reach the conclusion that no macroscopic polarization is induced.

Referring to the specific form of the dynamical matrix, it is possible to rewrite equation (7.13.1) in the form

$$\omega^2_{\mathbf{q}j}\, \varepsilon^\lambda_{k\mathbf{q}j} = \sum_{k'} \bar{D}_{\lambda\lambda'}(\mathbf{q}, kk')\, \varepsilon^{\lambda'}_{k\mathbf{q}j} - e[Z_k\,\delta_{\lambda\mu} - Z^\dagger_{\lambda\mu}(k)]\, M_k^{-\frac{1}{2}} \mathscr{E}_\mu, \tag{7.13.9}$$

where this equation defines $\bar{D}$.

From the energy density of equation (7.13.2) we can get the equations of motion and polarization which we then compare with equations (7.13.8) and

(7.13.9). This leads straightway to the identification

$$f(k\lambda, k'\lambda') = \bar{D}_{\lambda\lambda'}(0, kk')(M_k M_{k'})^{\frac{1}{2}}/\Omega_0, \qquad (7.13.10)$$

$$f(k\lambda, \mu) = -e[Z_k\,\delta_{\lambda\mu} - Z_{\lambda\mu}(k)]/\Omega_0, \qquad (7.13.11)$$

$$f(\lambda\mu) = -e^2\,\mathscr{P}^{(2)}_{\lambda\mu}. \qquad (7.13.12)$$

As a consequence, $\bar{D}$ determines a set of normal modes with frequencies ω_σ and polarization vectors $\boldsymbol{\varepsilon}_{i\sigma}$, three of which have zero frequency, which are the acoustic branches. The macroscopic dielectric tensor, as a function of frequency, is given by

$$\varepsilon_{\lambda\mu}(\omega) = \delta_{\lambda\mu} - 4\pi\left(e^2 P^{(2)}_{\lambda\mu} + \sum_j \frac{M_\lambda(j)\,M^\dagger_\mu(j)}{\omega^2 - \omega_j^2}\right), \qquad (7.13.13)$$

where the electric dipole moment $M_\lambda(j)$ is given by

$$M_\lambda(j) = \Omega_0^{-\frac{1}{2}}\,e\sum_k [Z_k\,\delta_{\lambda\mu} - Z_{\lambda\mu}(k)] \times M_k^{-\frac{1}{2}}\,\varepsilon_{\mu kj}. \qquad (7.13.14)$$

7.13.1 Diatomic crystals with tetrahedral symmetry

We shall conclude this section by summarizing, following Sham (1969), the results for the class of diatomic crystals with tetrahedral symmetry, which includes NaCl, ZnS and diamond among others. Symmetry arguments set out by Born and Huang yield the results (cf. equation 3.12.15)

$$Q_{\lambda\lambda'}(0, kk') = \delta_{\lambda\lambda'} \times 4\pi/3\Omega_0, \qquad (7.13.15)$$

$$S_{\lambda\lambda'}(0, kk') = \delta_{\lambda\lambda'}s, \quad k \neq k', \qquad (7.13.16)$$

$$Z_{\lambda\mu}(k) = \delta_{\lambda\mu}\,z_k, \qquad (7.13.17)$$

$$\mathscr{P}^{(2)}_{\lambda\mu} = \delta_{\lambda\mu}\,p. \qquad (7.13.18)$$

Hence we find that the transverse optical mode frequency is given by

$$\omega_0^2 = M^{-1}[s - (4\pi e^2/3\Omega_0)Z_1 Z_2], \qquad (7.13.19)$$

where M is the reduced mass of the two nuclei in the unit cell. The high-frequency dielectric constant is

$$\varepsilon_\infty = 1 - 4\pi e^2 p \qquad (7.13.20)$$

and the infrared form is given by

$$\varepsilon(\omega) = \varepsilon_\infty + 4\pi e^2(Z_1 - z_1)^2/\Omega_0\,M(\omega_0^2 - \omega^2). \qquad (7.13.21)$$

The appearance of $(Z_1 - z_1)\,e$ in the dielectric constant shows that it can be regarded as the dynamical charge of the ions in the sub-lattice $k = 1$. The sum rule (3.1.10) ensures that the dynamical charge on the other sub-lattice is compensatory. It follows readily then that, for the diamond lattice, the dynamical charge is zero; i.e, it is not infrared active.

It follows from equation (7.13.9) that we have the longitudinal optic mode frequency

$$\omega_l = (\varepsilon_0/\varepsilon_\infty)^{\frac{1}{2}}\,\omega_0. \tag{7.13.22}$$

This affords an alternative, and microscopic, proof of the Lyddane–Sachs–Teller relation which we proved macroscopically in Appendix 3.5.

CHAPTER 8

Superconductivity

8.1 Introduction

We have referred on a number of occasions to normal metals. These are reasonably well described, in gross terms, by the properties of the electron gas which we discussed fully in Chapter 2. While electron interactions are significant, we are still led to a single-particle spectrum which has the basic feature in common with the non-interacting gas that there are states available for particle excitation immediately outside the Fermi surface. This is, in fact, precisely what we mean when we speak of the normal state of a metal.

In that discussion, in Chapter 2, we were dealing with purely repulsive interactions; the Coulomb forces. However, imagine for a moment that we had another type of interaction, and that, under certain circumstances, this could be attractive and lead to some kind of bound state. Then it is clear that the situation would be qualitatively changed, and the level spectrum of the particle excitations would no longer be normal. But how could such an attractive interaction arise? We shall see that it is possible, but that we obviously then have to go outside the framework of Chapter 2, and deal with the electrons moving through a lattice of vibrating ions. It is this vibrating ionic lattice, affecting the electrons in their relative motions, which will prove the key to the solution of the problem.

One other point is worth making, by way of introduction. It was clear to the London brothers, in the 1930's, that what one was seeing in the phenomenon of superconductivity was essentially an illustration of a quantum state on a macroscopic scale. We shall see this explicitly in section 8.13 below.

8.2 Specific heat

It is worth following up the point made above, on the nature of the single-particle excitations in a normal metal, by noting the fact that we are then led, as we have seen, to an electronic specific heat that at low temperatures is proportional to T. This reflects the fact that only those electrons in a narrow

From a physical point of view, we can say that as an electron moves through the lattice it causes polarization, and a second electron is attracted towards such polarization. Thus, the motions of two electrons are coupled attractively. If ω is the zero-point oscillator frequency of an ion, the normal electron gas is perturbed by a lattice with energy $\hbar\omega \sim 10^{-1}$ eV per ion. This, then, must be a rough measure of the distance from the Fermi level over which the interactions can be attractive.

8.5 Binding in momentum space

So far then, we see the need for an attractive interaction and the isotope effect tells us quite clearly that the vibrating lattice is involved. But we want to sharpen up at this stage what we mean by binding. We can do this by reverting to the basic property of a superconductor, namely its vanishingly small electrical resistance below T_c.

Let us return to the conventional description of transport in a normal metal. Figure 8.2(a) shows the appropriate model of a displaced Fermi

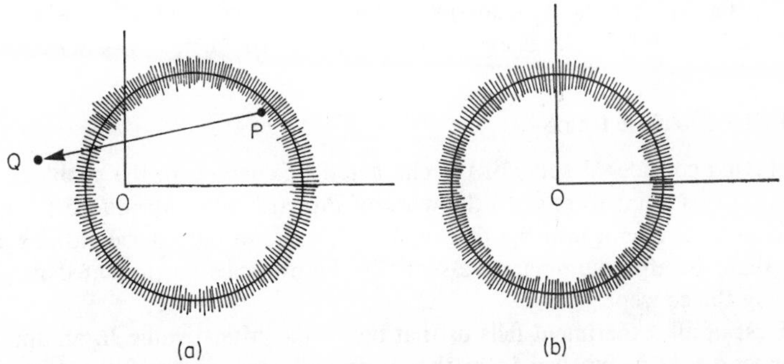

FIGURE 8.2(a). Illustrating current-carrying state in normal metal. Before application of electric field, distribution is centred at 0. Single-particle transitions of type $P \to Q$ are energetically favoured when field is switched off, in contrast to situation in superconductors. (b) Zero current state.

sphere, blurred out a bit by interactions and by temperature effects. As we saw in Chapter 6, when the electric field is switched off, the effect of the scattering from defects or phonons is that there is a transfer of electrons from the leading edge of the Fermi sphere to the other edge. Clearly, after some characteristic time, we return to the undisplaced Fermi sphere of Figure 8.2(b).

But in the superconducting state, a state which is carrying current, such as is shown in Figure 8.2(a), is stable even when there is no field. Single-particle transitions of the type $P \rightarrow Q$ in Figure 8.2(a) cannot then be favoured energetically. The conclusion, bearing in mind our remarks about attractive interactions and binding between electrons, seems clear: the particle at P must be bound in momentum space to the other particles.

Thus with particles bound together in such a way in momentum space, the degree of binding being represented by a finite energy which is required to excite electrons to higher energy states, we are led to the idea of a gap in the excitation spectrum in the superconductor. This gap is dependent on the electronic distribution and must not be confused with the gap in an insulator, which is generated by the (static) periodic potential of the ions.

Given this gap, we can return to the energy of 10^{-8} eV per atom representing the difference between normal and superconducting states. We can now understand this qualitatively as follows. Obviously the energy associated with the transition temperature is of order $k_B T_c$. Now in the normal state just above T_c, the electrons involved in the transition are confined to a narrow energy region of order $k_B T_c$ around the Fermi level. If the density of states is ρ say, then the total energy change in the transition is of order $(k_B T_c)(\rho k_B T_c)$, which typically is again of the order of 10^{-8} eV.

8.6 Characteristic lengths

Having considered some of the characteristic energies in the problem, we want at this point to refer to the work of Pippard, who introduced the idea of coherence length into the theory. This is basic to our considerations and we shall therefore summarize essentially the evidence he presented in proposing the concept.

First of all, experiment tells us that below the critical value H_c an applied magnetic field is expelled from the superconductor, except for a very thin temperature-dependent region at the surface, having a thickness λ_L, say, $\sim 10^{-5}$ cm. This Meissner effect, showing almost perfect diamagnetism, is one of the very interesting problems for the microscopic theory and we shall return to it below. However, Pippard focused attention on three points in introducing the coherence length:

(i) The sharpness of the normal–superconducting transition.

(ii) The fact that the penetration depth is very insensitive to the application of a magnetic field comparable with $\mathscr{H}_c$.

(iii) The existence of a large surface energy at a boundary between normal and superconducting phases in the presence of the critical field.

Pippard's observations were then that point (i) suggested that the phenomenon was one of cooperative behaviour in which local fluctuations (e.g. persistence of local order above the transition temperature) were more or less removed. The second point suggested, from the fact that a field equal to the critical $\mathscr{H}_c$ may produce a change of only a per cent or so, that the disturbing effect of the field is not confined simply to the penetration layer, but is acting over a greater distance $\sim 10^{-4}$ cm. Thirdly, if we write the surface energy in the form $l\mathscr{H}_c^2/8\pi$, the length l is found to be $\sim 10^{-4}$ cm.

Thus we have a further length, a distance over which strong correlations exist in a conductor, the coherence length ξ, of order 10^{-4} cm. We should expect that if, as in the cases discussed above, the penetration depth λ_L is small compared with the coherence length ξ, we should have a very different situation from that when $\lambda_L/\xi \gg 1$. This is indeed so and we shall have to deal differently with the soft or type I superconductors (λ_L/ξ small) and the hard or type II ($\lambda_L/\xi \gg 1$). The intermediate régime when the two lengths are about the same is more difficult and must await a discussion of the Ginzburg–Landau theory. But next we must see exactly how the attractive interaction between electrons, known to be connected intimately with the lattice vibrations, can arise (see also the books by Schrieffer, 1964, and Rickayzen, 1965).

8.7 Fröhlich Hamiltonian

We shall now set up Fröhlich's Hamiltonian for electron–phonon coupling, after which we shall show that the electron–phonon term can lead to an attractive force between the electrons. Basically, the idea is that an electron distorts or polarizes the positive ions in its vicinity, so that it carries with it an associated displacement field and this displacement field $\mathbf{P}(\mathbf{r})$ can be such that a net attraction between electrons is possible. In very simple terms, when one electron leaves a region of the lattice, it leaves behind an excess of positive ion density, which in turn attracts another electron into this region. In this way, it appears that there is an effective coupling of the electrons, via the lattice.

We write the Hamiltonian in the form

$$H = H_{el} + H_f + H_{int}, \tag{8.7.1}$$

where H_{el} is the electron Hamiltonian, H_f that of the field and H_{int} represents the interaction. Because we are primarily concerned with the electron–phonon interaction here, we shall ignore electron–electron interactions and simply write in terms of annihilation operators $a_{\mathbf{k}}$

$$H_{el} = \sum_{\mathbf{k}} \varepsilon_{\mathbf{k}} a_{\mathbf{k}}^{\dagger} a_{\mathbf{k}}. \tag{8.7.2}$$

Correspondingly we ignore phonon–phonon interactions and put

$$H_f = \sum_{\mathbf{k}} \hbar\omega_{\mathbf{k}}(b_{\mathbf{k}}^{\dagger} b_{\mathbf{k}} + \tfrac{1}{2}), \tag{8.7.3}$$

where the $b_{\mathbf{k}}^{\dagger}$ are the phonon creation operators, the phonons of course being the quanta of the field.

The question remains as to what to take for H_{int}. Consider a free-electron model of a metal and let the model be strained uniformly, the dilatation being Δ (the fractional change in volume). The ions of charge Z will increase in density by amount $-N\Delta$ (N being the density of ions) and electrons will follow this change with an increase $-ZN\Delta$ in the electron density. Their Fermi energy suffers a change, to first order, of

$$\delta E_f = -\frac{ZN\Delta}{n(E_f)}, \tag{8.7.4}$$

$n(E)$ being the density of states. Suppose, on the other hand, that the total volume is kept constant, but a long wavelength longitudinal phonon mode propagates, causing a local dilatation $\Delta(\mathbf{r})$. The electrons will again flow to maintain charge neutrality, but there can be no local change of Fermi energy as given by equation (8.7.4) because the Fermi energy (or chemical potential) must be constant throughout the metal. Instead, the distribution of electrons must be such as to produce a potential balancing the energy change in this equation. The potential will be

$$\Delta V(\mathbf{r}) = \frac{ZN}{n(E_f)}\Delta(\mathbf{r}). \tag{8.7.5}$$

In passing, we note that the above argument immediately yields the rule that the local potential V_{L}^{s} produced by electrons screening an ion immersed in the free electron gas is such that, in the long wavelength limit,

$$\tilde{V}_{\text{L}}^{s}(0) = -\tfrac{2}{3}E_f. \tag{8.7.6}$$

However, the present purpose of the above argument is to introduce the idea of calculating the electron–phonon interaction by means of a *deformation potential*. In Appendix 8.1 we discuss this further without recourse to a particular model.

We take it, then, that if we have a long wavelength disturbance in the lattice, producing a local dilatation $\Delta(\mathbf{r})$, the effect on the electrons is to perturb the crystal potential by the deformation potential, which, in the simple local theory, is

$$\Delta V(\mathbf{r}) = C\Delta(\mathbf{r}), \tag{8.7.7}$$

In general C will be an integral operator.

If $\mathbf{P}(\mathbf{r}, t)$ represents a displacement of a lattice, it is evident that

$$\Delta(\mathbf{r}, t) = \operatorname{div} \mathbf{P}(\mathbf{r}, t). \tag{8.7.8}$$

It can be seen that transverse displacements cannot contribute to Δ and $\mathbf{P}(\mathbf{r}, t)$ is taken to represent a longitudinal displacement field, obeying the condition

$$\operatorname{curl} \mathbf{P} = 0. \tag{8.7.9}$$

If the one-electron eigenstates are represented by $\phi_{\mathbf{k}}(\mathbf{r})$, we can define field operators of the form

$$\psi(\mathbf{r}) = \sum_{\mathbf{k}} a_{\mathbf{k}} \phi_{\mathbf{k}}(\mathbf{r}). \tag{8.7.10}$$

Then using equations (8.7.7) and (8.7.8), the interaction Hamiltonian may immediately be written down as

$$H_1 = C \int \psi^\dagger \psi \operatorname{div} \mathbf{P} d\mathbf{r}. \tag{8.7.11}$$

To express the displacement field $\mathbf{P}$ in terms of the phonon operators, we note that, by its very definition, $\mathbf{P}(\mathbf{r})$ must be slowly varying over many lattice spacings, for a lattice is not a continuum and thus $\mathbf{P}(\mathbf{r})$ is strictly defined only at the lattice points $\mathbf{l}$, where it takes the values $\mathbf{u}_{\mathbf{l}}$, the displacements of the ions from equilibrium. Evidently then, H_1 strictly expresses only the interaction with the electrons of long wavelength longitudinal acoustic phonons, in which limit we can take

$$\omega_{\mathbf{q}} = v_s q, \tag{8.7.12}$$

v_s being the velocity of sound, and the polarization vectors as

$$\boldsymbol{\varepsilon}_{\mathbf{q}} = \mathbf{q}/q. \tag{8.7.13}$$

These restrictions being observed, we can obtain $\mathbf{P}(\mathbf{r})$ from $\mathbf{u}(\mathbf{l})$ as given in terms of the $b_{\mathbf{q}}^\dagger, b_{\mathbf{q}}$ by equations (3.3.2) and (3.7.3)‡, by replacing $\mathbf{l}$ by the continuous variable $\mathbf{r}$, to obtain

$$\mathbf{P}(\mathbf{r}) = \sum_{\mathbf{q}} \frac{\mathbf{q}}{q} \left(\frac{1}{2NM\omega_{\mathbf{q}}}\right)^{\frac{1}{2}} (b_{\mathbf{q}} e^{i\mathbf{q}\cdot\mathbf{r}} + b_{\mathbf{q}}^\dagger e^{-i\mathbf{q}\cdot\mathbf{r}}) \tag{8.7.14}$$

and hence

$$\operatorname{div} \mathbf{P} = i \sum \left(\frac{1}{2NM\omega_{\mathbf{q}}}\right)^{\frac{1}{2}} q (b_{\mathbf{q}} e^{i\mathbf{q}\cdot\mathbf{r}} + b_{\mathbf{q}}^\dagger e^{-i\mathbf{q}\cdot\mathbf{r}}), \tag{8.7.15}$$

whereupon we find, with $D_{\mathbf{q}}$ related to C in an obvious way,

$$H_1 = i \sum_{\mathbf{kq}} D_{\mathbf{q}} (b_{\mathbf{q}} a_{\mathbf{k}+\mathbf{q}}^\dagger a_{\mathbf{k}} - b_{\mathbf{q}}^\dagger a_{\mathbf{k}-\mathbf{q}}^\dagger a_{\mathbf{k}}). \tag{8.7.16}$$

‡ In Chapter 3 of Volume 1 the phonon operators were denoted by a's. We use b's in this chapter to avoid confusion with the electron annihilation operators.

Thus the total Hamiltonian may be written

$$H = \sum_{\mathbf{k}} \varepsilon_{\mathbf{k}} a_{\mathbf{k}}^{\dagger} a_{\mathbf{k}} + \sum_{\mathbf{q}} \hbar\omega_{\mathbf{q}}(b_{\mathbf{q}}^{\dagger} b_{\mathbf{q}} + \tfrac{1}{2}) + i \sum_{\mathbf{q}} D_{\mathbf{q}}(b_{\mathbf{q}} \rho_{\mathbf{q}}^{\dagger} - b_{\mathbf{q}}^{\dagger} \rho_{\mathbf{q}}), \tag{8.7.17}$$

where $\rho_{\mathbf{q}}$ represents the electron density operator

$$\rho_{\mathbf{q}} = \sum_{\mathbf{k}} a_{\mathbf{k}-\mathbf{q}}^{\dagger} a_{\mathbf{k}} = \rho_{-\mathbf{q}}^{\dagger}. \tag{8.7.18}$$

8.7.1 Canonical transformation of Hamiltonian

Equation (8.7.16) does not immediately show us that there is an indirect electron–electron interaction between phonons. To make this interaction apparent, we can either treat the problem by diagrammatic techniques or follow Fröhlich in making a canonical transformation. We shall discuss the first method shortly; at the moment we make a canonical transformation which is such that

$$\tilde{Q} = e^{-S} Q e^{S}, \tag{8.7.19}$$

Q being any operator. The unitary transformation is the special case where $S^{\dagger} = -S$.

Then it is readily shown that

$$\tilde{Q} = Q + \sum_{n=1}^{\infty} \frac{[S_n, Q]}{n!}, \tag{8.7.20}$$

where S_n is defined by the recurrence relation

$$S_n = [S, [S_{n-1}, Q]] \quad (S_1 = S). \tag{8.7.21}$$

We shall take a transformation on equation (8.7.17) in which there are no off-diagonal terms of order D. Writing

$$H = H_0 + \lambda H_1, \tag{8.7.22}$$

we obtain from equation (8.7.20)

$$\tilde{H} = e^{-S} H e^{S} = H_0 + \lambda H_1 + [H_0, S] + \lambda[H_1, S] + \tfrac{1}{2}[[(H_0 + \lambda H_1), S], S] + \dots \tag{8.7.23}$$

from which it can be seen that, if

$$\lambda H_1 + [H_0, S] = 0, \tag{8.7.24}$$

S is of order λ and H has no term linear in λ. If the eigenstates of H_0 are represented by $|m\rangle$, etc., it is also obvious from equation (8.7.24) that

$$\langle n|S|m\rangle = \frac{\lambda \langle n|H_1|m\rangle}{E_m - E_n}. \tag{8.7.25}$$

Let us now apply these results to equation (8.7.17) with

$$H_0 = \sum_{\mathbf{k}} \varepsilon_{\mathbf{k}} a_{\mathbf{k}}^\dagger a_{\mathbf{k}} + \sum_{\mathbf{q}} \hbar\omega_{\mathbf{q}}(b_{\mathbf{q}}^\dagger b_{\mathbf{q}} + \tfrac{1}{2}) \tag{8.7.26}$$

and

$$H_1 = i\sum_{\mathbf{q}} D_{\mathbf{q}}(b_{\mathbf{q}}\rho_{\mathbf{q}}^\dagger - b_{\mathbf{q}}^\dagger \rho_{\mathbf{q}}). \tag{8.7.27}$$

If we now calculate matrix elements of $\tilde{H}$ (which will have the same eigenvalues as H) to order D^2, we can readily show that an effective electron–electron interaction results, of the form

$$H_{\text{el-el}} = \sum_{\mathbf{q}}\sum_{\mathbf{k}}\sum_{\mathbf{k}'} \mathscr{V}_{\mathbf{kq}}\, a_{\mathbf{k}'+\mathbf{q}}^\dagger a_{\mathbf{k}'} a_{\mathbf{k}-\mathbf{q}}^\dagger a_{\mathbf{k}} \tag{8.7.28}$$

with

$$\mathscr{V}_{\mathbf{kq}} = \frac{D_{\mathbf{q}}^2\,\omega_{\mathbf{q}}}{(\varepsilon_{\mathbf{k}}-\varepsilon_{\mathbf{k}-\mathbf{q}})^2-\omega_{\mathbf{q}}^2}. \tag{8.7.29}$$

The electron–electron interaction is attractive for excitation energies such that $|\varepsilon_{\mathbf{k}\pm\mathbf{q}}-\varepsilon_{\mathbf{k}}| < \omega_{\mathbf{q}}$. Taking the average value of $\omega_{\mathbf{q}}$ as ~ 0.025 eV ($\hbar = 1$ here), this is to be compared with the average value of $\varepsilon_{\mathbf{k}}-\varepsilon_{\mathbf{k}+\mathbf{q}}$ at the Fermi surface of about 4×10^{-4} eV. Thus the Fröhlich interaction is attractive in the region of interest. The repulsive region is not important. What does matter is that, where the Fröhlich interaction is attractive, it dominates the Coulombic repulsion, in which case we can expect a superconducting state. The Bardeen–Cooper–Schrieffer interaction, as we shall see below, approximates to equation (8.7.28) by disregarding the repulsive part and replacing $\mathscr{V}_{\mathbf{kq}}$ of equation (8.7.29) by a constant.

8.8 Pairing hypothesis

We shall now proceed to incorporate the features of the discussion above into two assumptions:

(i) We suppose that the mechanism responsible for superconductivity is an attractive interaction between pairs of particles, each occupying states having energy within an energy shell of width 2δ at the Fermi surface. In the case when the isotope effect is valid for the metal in question, it is clear that we can replace 2δ by $\hbar\omega$.

(ii) That we can focus attention on pairs of particles with total momentum zero.

We must then see if these assumptions can afford a basis for constructing a theory of the superconducting state (for some justification of assumption (ii), cf. March, Young and Sampanthar, 1967). We therefore go on to show that such pairs with zero total momentum will bind together to produce the desired energy gap, under the assumption of an attractive force simulating the Fröhlich interaction derived above.

8.8.1 Binding energy of Cooper pair

We start with a pair of particles at the Fermi level and without the interaction the total energy of the pair is $2E_f$. To see whether binding can occur we must now study the ground-state of the Schrödinger equation

$$\left[-\frac{\hbar^2}{2m}\nabla_1^2-\frac{\hbar^2}{2m}\nabla_2^2+v(\mathbf{r}_1\,\mathbf{r}_2)\right]\psi = E\psi. \tag{8.8.1}$$

Here v simulates the attractive interaction we have discussed above, and for binding we have to show that the ground-state eigenvalue is reduced below $2E_f$.

To solve this equation, we expand in terms of the set of states before the interaction is switched on, remembering however that we have electrons occupying states within the Fermi sphere. Thus we write the perturbed wave function (Ω being the volume of the system)

$$\psi = \sum_{|\mathbf{k}'|>k_f} c_{\mathbf{k}'}\frac{e^{i\mathbf{k}'.(\mathbf{r}_1-\mathbf{r}_2)}}{\Omega}, \tag{8.8.2}$$

where we have simply formed two-particle unperturbed states with individual momenta $\mathbf{k}$ and $-\mathbf{k}$, the resulting wave function being just $\Omega^{-1}e^{i\mathbf{k}.(\mathbf{r}_1-\mathbf{r}_2)}$.

At this stage, to keep the details at a minimum, we need not deal with the two spin states and we suppose the states in momentum space to be singly occupied. Then the wave function (8.8.2) may be substituted into the wave equation (8.8.1), when we find

$$\sum_{|\mathbf{k}'|>k_f} c_{\mathbf{k}'}[E_{\mathbf{k}'}-E+v(\mathbf{r}_1\,\mathbf{r}_2)]\frac{e^{i\mathbf{k}'.(\mathbf{r}_1-\mathbf{r}_2)}}{\Omega} = 0, \tag{8.8.3}$$

where $E_{\mathbf{k}} = \hbar^2 k^2/m$. If we now multiply both sides of equation (8.8.3) by $e^{i\mathbf{k}.(\mathbf{r}_1-\mathbf{r}_2)}$ and integrate over $\mathbf{r}_1$ and $\mathbf{r}_2$ we find

$$(E_{\mathbf{k}}-E)\,c_{\mathbf{k}} = -\sum_{|\mathbf{k}'|>k_f} c_{\mathbf{k}'}\,v(\mathbf{k},-\mathbf{k},\mathbf{k}',-\mathbf{k}'), \tag{8.8.4}$$

where $v(\mathbf{k},-\mathbf{k},\mathbf{k}',-\mathbf{k}')$ represents the matrix elements of the interaction between the two-particle states built from plane-waves.

We now have to make a convenient choice of these matrix elements, compatible with the earlier physical discussion of the superconducting interaction. We choose the simplest possible (Hermitian) form

$$v(\mathbf{k}, -\mathbf{k}, \mathbf{k}', -\mathbf{k}') = \begin{cases} -v/\Omega & \text{for } E_{\mathbf{k}} > E_{\mathrm{f}}, E_{\mathbf{k}'} < E_{\mathrm{f}} + \delta, \\ 0 & \text{otherwise,} \end{cases} \tag{8.8.5}$$

thus restricting the scattering states to the energy range $(E_{\mathrm{f}}, E_{\mathrm{f}} + \delta)$.

The choice of the negative sign is, of course, to get an attractive interaction, the constancy of the matrix element is really reflecting some average over the energy shell, and the factor Ω^{-1} on the right-hand side is to make v of order unity.

If we substitute these matrix elements in the equation (8.8.4) for $c_{\mathbf{k}}$ we find

$$c_{\mathbf{k}} = \frac{v}{E_{\mathbf{k}} - E} \frac{1}{\Omega} \sum_{E_{\mathrm{f}} < E_{\mathbf{k}'} < E_{\mathrm{f}} + \delta} c_{\mathbf{k}'}. \tag{8.8.6}$$

Though the solution of such an equation for the $c_{\mathbf{k}}$'s is, in general, difficult, we can determine the eigenvalues in this case by a simple enough procedure. Thus we sum each side of equation (8.8.6) over **k** values in the energy range between E_{f} and $E_{\mathrm{f}} + \delta$, and then a common factor $\sum_{E_{\mathrm{f}} < E_{\mathbf{k}} < E_{\mathrm{f}} + \delta} c_{\mathbf{k}}$ appears. The resulting equation for the eigenvalues E takes the form

$$\frac{1}{v} = \frac{1}{\Omega} \sum_{E_{\mathrm{f}} < E_{\mathbf{k}} < E_{\mathrm{f}} + \delta} \frac{1}{E_{\mathbf{k}} - E}. \tag{8.8.7}$$

It is now a standard matter to find the roots of equation (8.8.7). We simply plot both sides as a function of E and find where the resulting curves intersect (cf. Figure 8.3). While the left-hand side is simply a positive constant, it will be seen that as E goes from E_{f} to $E_{\mathrm{f}} + \delta$ the right-hand side varies very rapidly between plus and minus infinity. During each such fluctuation it takes on the value $1/v$ once. Thus, in this energy range δ above the Fermi level we find a continuum of solutions.

But the basic point is that there is a portion of the curve below E_{f} which separates off from the continuum and decreases monotonically to zero as $E \to -\infty$. This branch of the curve takes on the value v^{-1} at some volume-independent energy below the Fermi surface. This is the bound state we have been searching for.

Now that the presence of this bound-state level, with energy E_0, has been established, it is not a difficult matter to calculate the binding energy of the pair; electrons in such bound states are called Cooper pairs. Thus, if we now apply equation (8.8.7) specifically to the ground state, and go from a sum to an integral via

$$\frac{1}{\Omega} \Sigma \to \int d\varepsilon \rho(\varepsilon), \tag{8.8.8}$$

where ρ is the density of states, we find

$$\frac{1}{v} = \int_{E_f}^{E_f+\delta} \frac{\rho(\varepsilon)\,d\varepsilon}{\varepsilon - E_0}. \tag{8.8.9}$$

We can make an approximate evaluation of the integral by noting that in the range of integration we expect $(\varepsilon - E_0)^{-1}$ to vary much more rapidly than the density of states. Since the energy width δ is known to be small, we

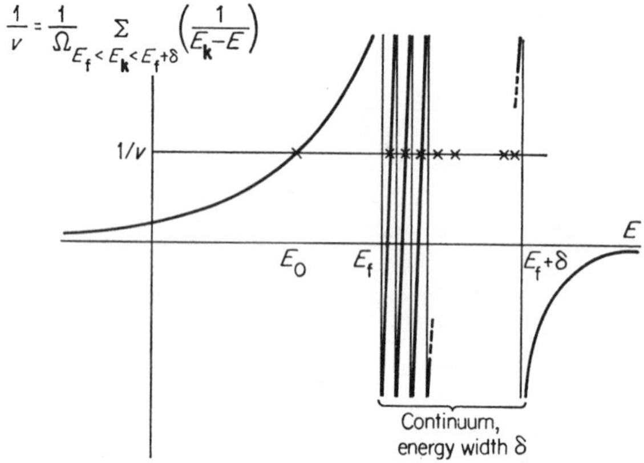

FIGURE 8.3. Showing binding of Cooper pairs. Crosses between Fermi energy E_f and $E_f + \delta$ correspond to solutions of equation (8.8.7) in continuum. Isolated cross shows bound state with volume-independent energy.

approximate $\rho(\varepsilon)$ by its value at the Fermi level. The resulting integral is trivial and we find a binding energy given by

$$E_f - E_0 = \frac{\delta}{e^{1/\rho v} - 1} \sim \delta\, e^{-1/\rho v}, \tag{8.8.10}$$

where we have, in the second step, assumed that $e^{1/\rho v} \gg 1$. If we recall that, in metals, experiment indicates that $E_f - E_0 \sim 10^{-4}$ eV, while $\delta \sim 10^{-1}$ eV, the quantity $e^{1/\rho v}$ is certainly large in this case.

The interesting point about equation (8.8.10) is that, although the binding energy is small, we *cannot* expand it by perturbation theory in powers of v. Thus, in the subsequent generalization of this two-particle theory to the many-body problem we must avoid perturbation theory. In fact, we shall employ the variational method.

8.8.2 *Size of Cooper pair*

The extension of the Cooper pair wave function in space may be estimated as follows. The orbital wave function for a pair may be written,

$$\psi(\mathbf{r}_1\mathbf{r}_2) = \phi_{\mathbf{q}}(\mathbf{r}_1-\mathbf{r}_2)\exp\left(i\mathbf{q}.\frac{\mathbf{r}_1+\mathbf{r}_2}{2}\right), \tag{8.8.11}$$

where $\hbar\mathbf{q}$ is the momentum of the centre of mass. For the singlet spin case $\phi_{\mathbf{q}}$ is symmetric and for $\mathbf{q}=0$ is an eigenfunction of angular momentum. Considering the centre of mass at rest,

$$\psi(\mathbf{r}_1\mathbf{r}_2) = \phi_0(\mathbf{r}) = \sum_{\mathbf{k}} c(\mathbf{k})\, e^{i\mathbf{k}.(\mathbf{r}_1-\mathbf{r}_2)}, \quad \mathbf{r}=\mathbf{r}_1-\mathbf{r}_2, \tag{8.8.12}$$

where the $\mathbf{k}$ sum is over states near the Fermi surface. The mean square radius of a Cooper pair is

$$\langle r^2\rangle = \int|\phi_0(\mathbf{r})|^2 r^2\,d\mathbf{r}\Big/\int|\phi_0(\mathbf{r})|^2\,d\mathbf{r} = \sum_{\mathbf{k}}|\nabla_{\mathbf{k}}c(\mathbf{k})|^2\Big/\sum_{\mathbf{k}}|c(\mathbf{k})|^2. \tag{8.8.13}$$

With the potential (8.8.5), the two-body Schrödinger equation leads to the solution for the momentum eigenfunctions $c(\mathbf{k})$ as

$$c(\mathbf{k}) = \frac{c}{\Delta+2E_{\mathrm{f}}-(\hbar^2k^2/2m)}, \tag{8.8.14}$$

where Δ is the binding energy for the pair relative to the Fermi level and c is a constant independent of $\mathbf{k}$. Substitution into equation (8.8.13) leads to

$$(\langle r^2\rangle)^{\frac{1}{2}} = \frac{2}{\sqrt{3}}\frac{\hbar v_{\mathrm{f}}}{\Delta} \sim 10^{-4}\text{ cm}, \tag{8.8.15}$$

where v_{f} is the velocity at the Fermi surface. This quantity is essentially the coherence length and represents the distance over which the superconducting electrons are correlated, or equivalently the range of local order.

Having seen that the binding energy of a Cooper pair, estimated crudely, has the order of magnitude we need, and that the size of the Cooper pair is related to the coherence length, we can go ahead to construct a many-body theory, based on these ideas.

We want to stress that, to explain superconductivity, it is necessary to take into account that Cooper pairs are all interacting with one another, and one has to account for the interactions between all the electrons at the same time. Thus, while Cooper's idea was crucial to the development of the theory, it is only the starting point.

8.9 Bardeen–Cooper–Schrieffer (BCS) theory

It is clear from the above remarks that we have a truly many-body problem of considerable complexity and it will be necessary to proceed by making reasonable assumptions, which are most conveniently expressed via a simplified Hamiltonian, which we now consider.

8.9.1 Reduced Hamiltonian

The ground state of a set of Fermions, in which both screened Coulomb and superconducting interactions are present, is required. The discussion of the two-body problem given above strongly suggests that in describing the main features of the superconducting phase it is the attractive interactions which play the crucial role. The simplest procedure, then, is to ignore the screened Coulomb interactions completely, the (incomplete) justification being that they contribute equally to both the superconducting and normal states. We shall then consider how they can be incorporated later in section 8.10.

The precise form of the many-body superconducting interaction must of course be specified. Our physical picture is of a gas of independent Fermions except for bound pairs near the Fermi surface. These pairs may scatter against each other, always conserving total momentum and individual spin, but scattering between members of different pairs is precluded as this would mean the breaking of pairing bonds.

We are then led to the following reformulation of the many-body problem. We start from a set of independent Fermions described by the Hamiltonian

$$H_0 = \sum \varepsilon_{\mathbf{k}} a_{\mathbf{k}}^\dagger a_{\mathbf{k}}. \tag{8.9.1}$$

In this section we shall use the notation that K represents a single-particle state of momentum $\mathbf{k}$ and spin σ, and $-K$ a state $-\mathbf{k}$, spin $-\sigma$. The interaction

$$V = \tfrac{1}{2}\sum \mathscr{V}_{KK'} a_K^\dagger a_{-K}^\dagger a_{-K'} a_{K'} \tag{8.9.2}$$

applied at time $t = -\infty$, say, is constructed in accordance with our previous discussion. The summation in equation (8.9.2) is over $\varepsilon_{\mathbf{k}}$ and $\varepsilon_{\mathbf{k}'}$ restricted to lie in the energy shell of thickness 2δ about the Fermi surface. The matrix elements $\mathscr{V}_{KK'}$, as written above, are representing an attractive interaction and are therefore negative. We are thus faced with investigating the wave function which evolves after infinite time and which is a stationary state of the reduced Hamiltonian.

$$H = H_0 + V. \tag{8.9.3}$$

Following the presentation of March, Young and Sampanthar, we can represent the situation by Figure 8.4.

There is a strong resemblance to the Cooper two-particle problem. There, only two particles K and $-K$, were selected at the Fermi surface and their scattering in a shell at the Fermi limit studied. In contrast to Cooper's problem however, many such pairs are now allowed to scatter. This is a

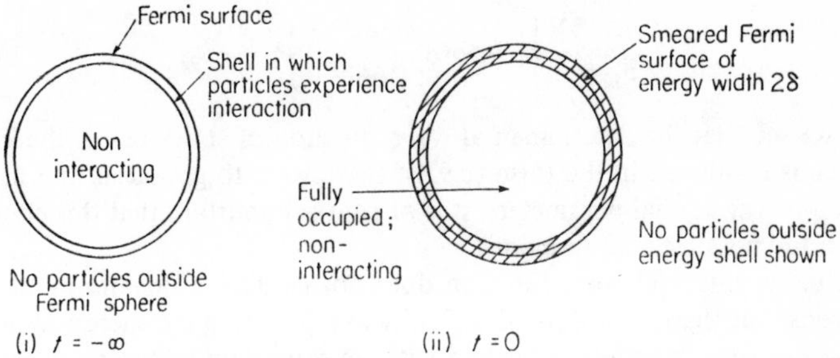

FIGURE 8.4. Evolution of interacting ground state from free-particle state, for model Hamiltonian in which interactions only occur within shell around Fermi surface.

much more complicated situation because, whereas previously we knew that our chosen pair could freely scatter in the energy shell, now such scattering is inhibited by the possibility of the occupation of such states by other particles.

It is now possible to go ahead and solve the problem of finding the ground-state wave function for the Hamiltonian (8.9.3), without making any further approximations, to $O(1/N)$. The wave function thus obtained is that used by Bardeen, Cooper and Schrieffer (1957). However, the philosophy of their original paper was to regard the wave function as a very convenient trial function in a variational calculation. Their trial function had the form

$$|\psi\rangle = \prod_{K>0} (u_K + v_K a_K^\dagger a_{-K}^\dagger)|0\rangle, \tag{8.9.4}$$

where u and v are to be taken as real and such that

$$u_K = u_{-K} = u_{\mathbf{k}}, \quad v_K = -v_{-K} = v_{\mathbf{k}} \tag{8.9.5}$$

and finally, for normalization purposes, we must have

$$u_k^2 + v_k^2 = 1. \tag{8.9.6}$$

The usefulness of such a trial wave function is not by any means clear at this stage, but when its detailed properties are investigated, and their physical consequences calculated, we shall see that it has great merit.

However, in spite of the fact that it is difficult at this stage to justify the choice of wave function (8.9.4), we can see that it certainly correlates in a special manner those pairs of particles with zero total momentum and spin zero. Also, if we made the choice

$$\left.\begin{matrix} u_K = 0 \\ v_K = 1 \end{matrix}\right\} \; k < k_f, \qquad \left.\begin{matrix} u_K = 1 \\ v_K = 0 \end{matrix}\right\} \; k > k_f, \tag{8.9.7}$$

then we see that the determinantal wave function of states below the Fermi surface is contained in the form (8.9.4). If we were to go ahead and use u_K and v_K as variational parameters, it is of course important that this choice is within the possible trial functions.

Actually, this trial wave function does not contain a definite number of electrons, but there is no formal difficulty in calculating the energy. What we would do, since we are not dealing with an eigenstate of the total number operator N_{op}, is to minimize $H - \mu N_{\text{op}}$ with μ chosen such that

$$\langle N_{\text{op}} \rangle = N. \tag{8.9.8}$$

8.9.2 Explicit form of ground-state solution

The expectation value of the quantity $H - \mu N_{\text{op}}$ can be calculated straightforwardly using equation (8.9.4), but we shall use here a less rigorous, but more appealing approach, suggested by Rickayzen (see also Anderson, 1958), to get the answer. Because we have a many-body problem, we expect $a_{-K} a_K$ to be close to its average value, $A_{\mathbf{k}}$ say. Thus we write

$$a_{-K} a_K = A_{\mathbf{k}} + (a_{-K} a_K - A_{\mathbf{k}}) \tag{8.9.9}$$

and, since the term in the brackets is small, we neglect the square of it to obtain from equations (8.7.28) and (8.9.1), with $\varepsilon_{\mathbf{k}}$ measured from μ,

$$H - \mu N_{\text{op}} = \sum_{\mathbf{k}} \varepsilon_{\mathbf{k}} a^{\dagger}_{\mathbf{k}\sigma} a_{\mathbf{k}\sigma} + \sum \mathscr{V}_{\mathbf{kk'}} (a^{\dagger}_{\mathbf{k}\uparrow} a^{\dagger}_{-\mathbf{k}\downarrow} A_{\mathbf{k'}} + A^{*}_{\mathbf{k}} a_{-\mathbf{k'}\downarrow} a_{\mathbf{k'}\uparrow} - A^{*}_{\mathbf{k}} A_{\mathbf{k'}}). \tag{8.9.10}$$

We now go through the calculation, and eventually equate $A_{\mathbf{k}}$ to the average of $a_{-\mathbf{k}\downarrow} a_{\mathbf{k}\uparrow}$ calculated from this solution. If we write

$$\Delta_{\mathbf{k}} = \sum_{\mathbf{k'}} \mathscr{V}_{\mathbf{kk'}} A_{\mathbf{k'}}, \tag{8.9.11}$$

then we can diagonalize by the following canonical transformation

$$a_{\mathbf{k}\uparrow} = u_k \gamma_{k0} + v_k \gamma^*_{k1}, \tag{8.9.12}$$

$$a^*_{-k\downarrow} = -v_k \gamma_{k0} + v_k \gamma^*_{k1}, \tag{8.9.13}$$

the choice of u_k and v_k which makes the Hamiltonian diagonal being readily verified as

$$2\varepsilon_k u_k v_k + \Delta_k (v_k^2 - u_k^2) = 0 \tag{8.9.14}$$

with, of course,

$$u_k^2 + v_k^2 = 1 \tag{8.9.15}$$

for a canonical transformation. If we now write

$$E_{\mathbf{k}} = (\varepsilon_k^2 + \Delta_k^2)^{\frac{1}{2}}, \tag{8.9.16}$$

then we find as the solution

$$u_k^2 = \frac{1}{2}\left(1 + \frac{\varepsilon_k}{E_k}\right), \quad v_k^2 = \frac{1}{2}\left(1 - \frac{\varepsilon_k}{E_k}\right). \tag{8.9.17}$$

It can be shown that

$$H - \mu N_{\text{op}} = \sum_{\mathbf{k}} E_k (\gamma^*_{k0}\gamma_{k0} + \gamma^*_{k1}\gamma_{k1}) + \sum_k (\varepsilon_k - E_k + \Delta_k A_k), \tag{8.9.18}$$

where the first term on the right is as if we have free Fermi particles with energies E_k.

The ground state of this Hamiltonian is given by

$$\gamma_{k0}|\,\rangle = 0 = \gamma_{k1}|\,\rangle \tag{8.9.19}$$

and the solution is equation (8.9.4). Thus the self-consistency condition is

$$\langle\,| a_{-k} a_k |\,\rangle = A_k \tag{8.9.20}$$

and from equation (8.9.11) we find

$$\Delta_{\mathbf{k}} = \sum_{\mathbf{k}'} -\frac{\mathscr{V}_{\mathbf{k}\mathbf{k}'}\Delta_{\mathbf{k}'}}{2E_{\mathbf{k}'}}. \tag{8.9.21}$$

The trivial solution $\Delta_{\mathbf{k}} = 0$ of equation (8.9.21) clearly corresponds to the normal state.

8.9.3 Solution of BCS equation for averaged potential

At this stage we shall choose an interaction simulating that in a superconductor. For calculational convenience, we take such a form to be

$$\mathscr{V}_{KK'} = \begin{cases} -V/\Omega & \text{if } K, K' \text{ have same spin and } |\varepsilon_K|, |\varepsilon_{K'}| < \delta, \\ 0 & \text{otherwise,} \end{cases} \tag{8.9.22}$$

which should be compared with the choice (8.8.5) which we made when treating the Cooper pair. Much of that treatment is relevant here and so we can simply summarize the argument now. It is worth commenting though on the fact that the quantity μ above still is only a Lagrange multiplier. How can we know that $|\varepsilon_K| < \delta$ actually defines a shell around the Fermi surface? The answer is that it is a straightforward calculation to show that for non-interacting particles $\mu = E_f$. For interacting particles in which only small amounts of energy are involved, as is the case here, μ differs little from E_f and so $|\varepsilon_K| < \delta$ defines a shell round the Fermi surface.

With the interaction (8.9.22) inserted in equation (8.9.21) we find

$$\Delta_K = 0 \quad (|\varepsilon_K| > \delta) \tag{8.9.23}$$

and

$$\Delta_K = \frac{V}{2\Omega} \sum_{\substack{|\varepsilon_{K'}|<\delta \\ K',K \text{ same spin}}} \frac{\Delta_{K'}}{E_{K'}} \quad (|\varepsilon_K| < \delta). \tag{8.9.24}$$

But we can solve equation (8.9.23) by noticing that $\Delta_K = \Delta$, independent of K and hence we obtain

$$\frac{2}{V} = \frac{1}{\Omega} \sum_{|\varepsilon_K|<\delta} \frac{1}{\sqrt{(\varepsilon_K^2 + \Delta^2)}}, \tag{8.9.25}$$

where the summation is over singly occupied states. Just as in the discussion of the Cooper pair, we then find

$$\frac{2}{V} = \rho \int_{-\delta}^{\delta} \frac{d\varepsilon}{\sqrt{(\varepsilon^2 + \Delta^2)}} \tag{8.9.26}$$

and completing the integration we find

$$\Delta = \delta \operatorname{cosech} \frac{1}{\rho V} \sim 2\delta\, e^{-1/\rho V}. \tag{8.9.27}$$

(a) *Ground-state properties.* It is now perfectly straightforward to calculate the ground-state energy, and we obtain for the weak coupling case

$$E(\delta) = E(0) - \tfrac{1}{2}\Omega\rho\Delta^2, \tag{8.9.28}$$

which immediately confirms that the solution we have obtained is energetically below the Slater determinant solution by an amount $\frac{1}{2}\Omega\rho\Delta^2$. Here then we see that, since Δ is typically of the order 10^{-4} eV, the total energy is of the order we anticipated earlier.

Secondly, if H_{c0} is the critical magnetic field at $T = 0$, it is clear that we can write

$$\frac{H_{c0}^2}{8\pi} = \frac{E(0)-E(\delta)}{\Omega} = \tfrac{1}{2}\rho\Delta^2 = 2\rho\delta^2 e^{-2/\rho V}. \tag{8.9.29}$$

If M is the ionic mass, and putting

$$\delta = \hbar\omega \propto 1/M^{\frac{1}{2}}, \tag{8.9.30}$$

we find the relation

$$H_{c0} M^{\frac{1}{2}} = \text{const.} \tag{8.9.31}$$

Thirdly, it is worth remarking on the nature of the momentum distribution, which clearly depends on v_K. It is straightforward to show that this is given by

$$\langle\psi| a_k^\dagger a_k |\psi\rangle = v_k^2 = \frac{1}{2}\left(1 - \frac{\varepsilon_k}{E_k}\right)$$

$$= \frac{1}{2}\left(1 - \frac{\varepsilon_k - \mu}{\sqrt{[(\varepsilon_k-\mu)^2 + \Delta_k^2]}}\right) \tag{8.9.32}$$

and this is shown in Figure 8.5. The discontinuity in the Fermi distribution is seen to be removed.

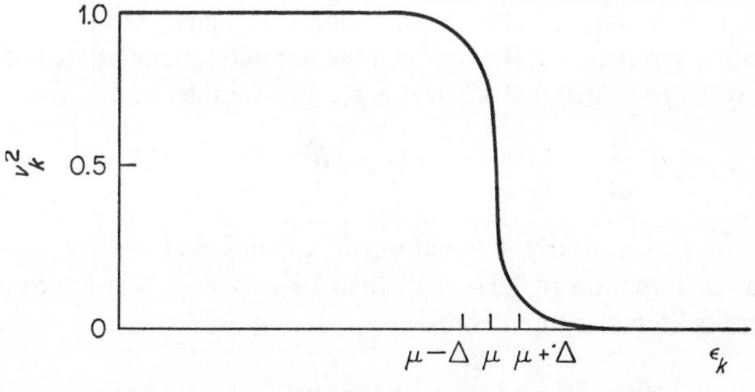

FIGURE 8.5. Momentum distribution in superconducting metal. Note that discontinuity in Fermi distribution is removed.

Finally, in discussing ground-state properties, we can note that, from equations (8.9.18) and (8.9.28), we can write

$$H = \sum_{\mathbf{k}} E_k(\gamma_{k0}^* \gamma_{k0} + \gamma_{k1}^* \gamma_{k1}) - \tfrac{1}{2}\rho\Omega\Delta^2 \tag{8.9.33}$$

with $E_k = (\varepsilon_k^2 + \Delta^2)^{\frac{1}{2}}$. The E_k's evidently are excitation energies and the minimum corresponding to $\varepsilon_k = 0$ is $E_k = \Delta$. Thus a minimum energy Δ is required to create an excitation. The gap in the spectrum is given by 2Δ, since in an absorption or emission process the state of two electrons is altered.

(b) *Elevated temperatures.* We shall not give the argument in any detail, but we can see generally what is happening by noting from the above argument that to form an excited state, we have to break up a pair, and this is why there is the gap. As we increase the temperature, more pairs are lost, the interaction energy of the remainder decreases and u_k and v_k change. The development then proceeds as for $T = 0$ except for self-consistency. This leads to a new gap equation

$$\Delta_K = -\frac{1}{2}\sum_{K'} V_{KK'}\frac{\Delta_{K'}}{E_{K'}}(1-2f_K) \tag{8.9.34}$$

where f_K is the Fermi function

$$f_K = \frac{1}{e^{\beta E_K}+1}, \quad \beta = \frac{1}{k_B T}. \tag{8.9.35}$$

If we use the BCS interaction (8.9.22) as before, we find

$$\frac{2}{V} = \rho\int_{-\delta}^{\delta}\frac{d\varepsilon}{\sqrt{(\varepsilon^2+\Delta^2)}}\tanh\tfrac{1}{2}\beta\sqrt{(\varepsilon^2+\Delta^2)}, \tag{8.9.36}$$

where the summation as before is over singly occupied states.

The critical temperature above which the superconducting state is destroyed is given by putting $\Delta(\beta) = 0$ when $\beta = \beta_c$, and we then find

$$\frac{1}{\rho V} = \int_0^{\delta}\frac{d\varepsilon}{\varepsilon}\tanh\tfrac{1}{2}\beta_c\,\varepsilon = \int_0^{\beta_c\delta}\frac{dx}{x}\tanh\tfrac{1}{2}x. \tag{8.9.37}$$

The last integral is clearly a function of $\beta_c\,\delta$ only and for superconductors this is large. Equation (8.9.37) may then be solved in this limiting case to yield a critical temperature defined by

$$\frac{1}{\beta_c} = 1.14\delta\, e^{-1/\rho V}. \tag{8.9.38}$$

It is then easy to draw an absolute curve for superconductors of $\Delta(\beta)/\Delta(\infty)$ against β_c/β and this is shown in Figure 8.6.

From the form of this curve we expect that the low-temperature properties will be determined by an almost constant energy gap. Near T_c, it is found that the form of Δ behaves as

$$\Delta \sim \left(1-\frac{T}{T_c}\right)^{\frac{1}{2}}. \tag{8.9.39}$$

From equation (8.9.38), if we put $\delta = \hbar\omega$, we obtain the isotope effect in the form

$$\beta_c \omega = \text{const.} \tag{8.9.40}$$

Also, from the energy-gap formula at absolute zero, we find a universal ratio for all BCS type superconductors as

$$\frac{2\Delta(\infty)}{k_B T_c} = 3.52, \tag{8.9.41}$$

while using equation (8.9.29) for the critical field $\mathscr{H}_{c0}$ at zero temperature gives a further absolute ratio

$$\frac{\gamma T_c^2}{\mathscr{H}_{c0}^2} = 0.17, \tag{8.9.42}$$

where $\gamma = \frac{2}{3}\pi^2 \rho k_B^2$. We might remark that departures from the $M^{-\frac{1}{2}}$ law are frequent and this is because the Coulomb interaction comes in. Then V depends on $\hbar\omega$ and T_c is not proportional to $M^{-\frac{1}{2}}$.

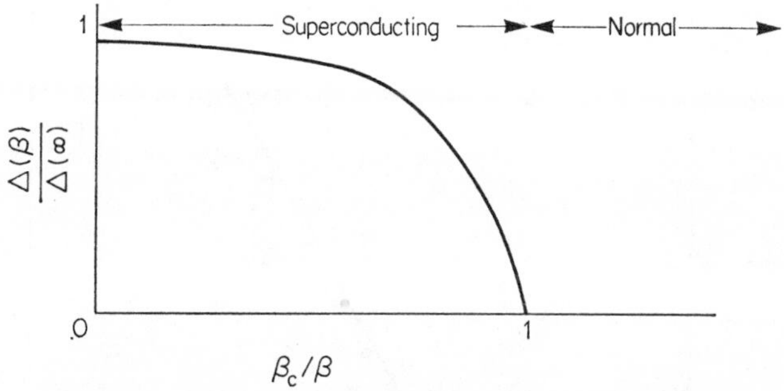

FIGURE 8.6. Gap parameter against reduced temperature.

The theory can also be used to calculate the critical field, when Figure 8.1 is regained, and to obtain the specific heat. The latter, it turns out, can be written in the form

$$c = \frac{k_B \beta^2}{\Omega} \sum_K f_K(1-f_K) \left[E_K^2 + \frac{\beta}{2}\frac{\partial}{\partial\beta}\Delta_K^2 \right]. \tag{8.9.43}$$

The first term in the square brackets gives a contribution continuous for all temperatures. Above the critical temperature this is the only contribution, since Δ is identically zero. On the other hand, the second term is discontinuous as we go through the critical temperature, the fall in the specific heat as the system becomes normal being

$$\frac{k_B \beta_c^3}{2\Omega} \sum_K \left(\frac{\partial}{\partial \beta} \Delta_K^2 \right)_{\beta=\beta_c} \sim 2.43 \gamma T_c. \tag{8.9.44}$$

In this way the general features of the observed specific heat results referred to earlier are seen to follow.

8.9.4 Thermal conductivity of superconductors

The thermal conductivity of superconductors has some interest, and the form of the measurements for Al are shown in Figure 8.7. The points denote experimental results on different samples of Al, and these are plotted

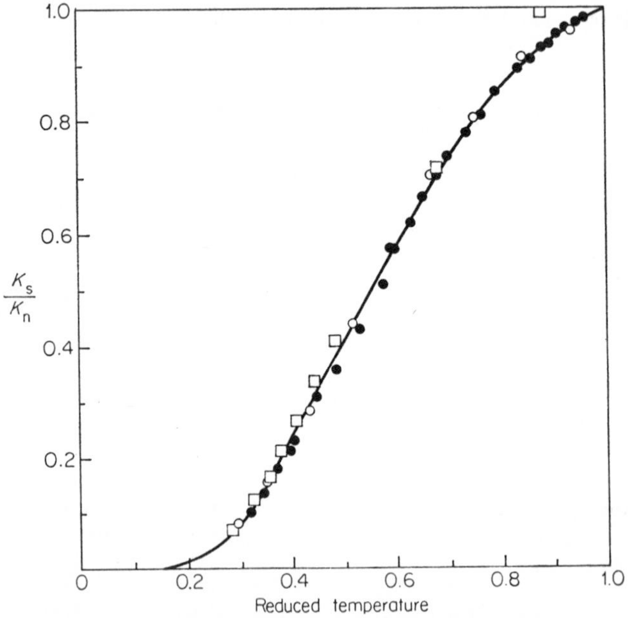

FIGURE 8.7. Ratio of thermal conductivity, in superconducting and normal states, against reduced temperature, for impurity scattering. Points are measurements on different samples of Al (after Satterthwaite, 1962). Solid curve is theory based on equation (8.9.45).

in terms of the ratio of the values in the superconducting and normal states, K_s and K_n respectively. As is usual, a reduced temperature plot is employed.

We shall argue later that, for the results shown in Figure 8.7, scattering by impurities limits the electron mean free paths. We therefore start the discussion by considering this case.

Let us assume that an increase in the temperature of the system does not change the pairing, nor, to first order, does it change the quantities $u_{\mathbf{k}}$ and $v_{\mathbf{k}}$. Then we can focus attention on the change in the distribution function $f(\mathbf{k})$ and we can use the Boltzmann equation to determine this, the details being given in Appendix 8.2.

In this way, the ratio of the thermal conductivity in the superconducting state to that in the normal state is given by

$$\frac{K_s}{K_n} = \frac{\displaystyle\int_0^\infty \varepsilon E(\partial f/\partial E)\, d\varepsilon}{\displaystyle\int_0^\infty \varepsilon^2(\partial f/\partial \varepsilon)\, d\varepsilon} = \frac{\displaystyle\int_\Delta^\infty E^2(\partial f/\partial E)\, dE}{\displaystyle\int_0^\infty E^2(\partial f/\partial E)\, dE}, \tag{8.9.45}$$

where ε is related to E and Δ by equation (8.9.16) and the role of the gap Δ as a cut-off in the integral in the numerator is the crucial point to note. This result was obtained independently by Bardeen, Rickayzen and Tewordt (1959) and Geilikman (1958).

The result of this theory is shown in the solid curve in Figure 8.7. Theory and experiment are in excellent agreement and this confirms that the mean free path of electrons in the presence of impurities is the same in the normal and superconducting states (cf. Appendix 8.2).

However, the case we have discussed, in which the electron mean free path is limited by impurity scattering, is not the only case of physical interest. In Pb and Hg, for example, the lattice scattering is much stronger and the scattering by phonons is the limiting factor on the mean free path. The scattering is obviously no longer elastic and creation and annihilation of excitations by phonons must be included. We shall not go into further detail here (see Tewordt, 1963, for a full discussion).

8.10 Strong coupling superconductors

We have just drawn attention to the fact that there is strong lattice scattering in Pb and Hg. We shall say a little more about that later, but we want here to stress that when the lattice scattering is strong, we need a more powerful theory. To get this, we have to go back to first principles, and look beyond the Fröhlich interaction (8.7.28).

8.10.1 Diagrammatic representation of electron–phonon interaction

The Fröhlich interaction of equation (8.7.28) is formally very similar to the electron–electron interaction via Coulomb forces. Thus, by analogy with the representation of Coulomb scattering in Chapter 2, we can display the mutual scattering of two electrons through the electron–phonon–electron

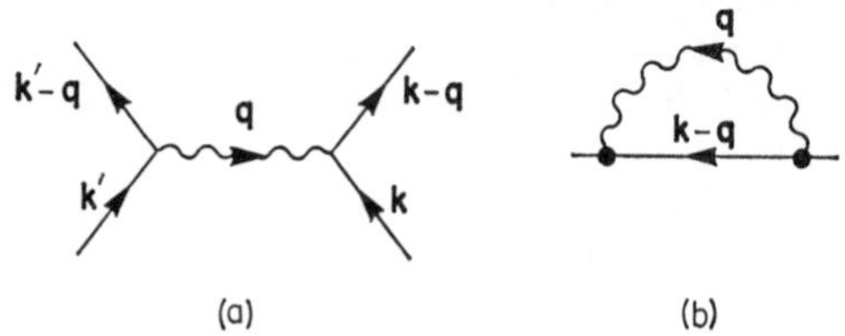

FIGURE 8.8(a) Scattering of two electrons by electron–phonon–electron interaction. (b) Lowest-order correction to self-energy due to phonon interaction.

interaction as in Figure 8.8(a), the wavy line being associated with the factor $\mathscr{V}_{\mathbf{kq}}$ in equation (8.7.29), or with the appropriate factor in the BCS interaction, if we use the simplified form $\mathscr{V}$ = constant. For a normal metal, a graphical analysis can then be carried out as for electron–electron interactions in Chapter 2: for example Figure 8.8(b) shows the lowest-order correction to the electron self-energy $\Sigma(\mathbf{k}\omega)$ due to the phonon field. A difficulty here is that the phase change to the superconducting state renders the perturbation theory developed for the normal metal invalid. However, as we shall discuss later, Nambu (1960) has shown how the formalism may be rewritten in such a way that the diagrams used to deal with the normal state are applicable to the superconducting state also.

The BCS theory turns out to be inadequate for superconductors in which the electron–phonon interaction is strong and a primary reason for this is the instantaneous nature of the BCS interaction. Further, it has been shown by Eliashberg (1960) that the canonically transformed Fröhlich Hamiltonian does not properly represent the retarded nature of the electron–electron interaction by phonons; we have already seen that the Fröhlich interaction of equation (8.7.28) is correct only to lowest order in the canonically transformed Hamiltonian. We may also note that the phonons were treated in the harmonic approximation. We therefore return to the original Hamiltonian of equation (8.7.17) before the canonical transformation was carried out.

The full graphical analysis is now a synthesis of the original diagrammatic methods for electrons and phonons separately; to distinguish Fermion lines from phonon lines we draw the latter as wavy lines. We note from equation (8.7.29) for the Fröhlich interaction that the wavy line of Figure 8.8(a) can be associated with the free-phonon propagator

$$iG_0(\mathbf{q}\omega) = \frac{i}{\omega-\omega_q-i\delta}-\frac{i}{\omega+\omega_q+i\delta}, \qquad (8.10.1)$$

where here we have $\omega = \varepsilon_{\mathbf{k}}-\varepsilon_{\mathbf{k}-\mathbf{q}}$, in agreement with equation (8.7.29). For the electron–phonon interaction, we see that the term $D_{\mathbf{q}}\, b_{\mathbf{q}}\, a^\dagger_{\mathbf{k}+\mathbf{q}}\, a_{\mathbf{k}}$ involves the absorption of a phonon with momentum transfer of $\mathbf{q}$ to the electron, and can be represented graphically as in Figure 8.9.

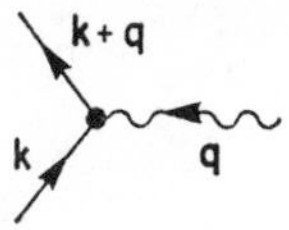

FIGURE 8.9. Absorption of phonon with momentum transfer.

To make the Hamiltonian we are discussing quite explicit, let us adopt the form

$$H_{\text{el-ph}} = \sum_{\mathbf{k}\mathbf{q}\sigma} g_{\mathbf{q}}[b_{\mathbf{q}}+b^\dagger_{-\mathbf{q}}]\, a^\dagger_{\mathbf{k}+\mathbf{q}\sigma}\, a_{\mathbf{k}\sigma}, \qquad (8.10.2)$$

where $g_{\mathbf{q}}$ can be energy or ω-dependent and is proportional to $D_{\mathbf{q}}$ used above.

In addition to the rules for Coulomb interactions formulated in Chapter 2, we have for normal systems:

(i) To associate with each electron–phonon vertex of the kind shown in Figure 8.9 a factor $-ig_{\mathbf{q}}$.

(ii) To associate with each free phonon line (wavy line) the free phonon propagator $iG_0(\mathbf{q}\omega)$ given by equation (8.10.1).

We can now go ahead with the calculation of such properties as the self-energy by summing sub-series of the perturbation expansion. For example, we can dress the phonon lines, in which case in Figure 8.8(b) the wavy line is to be associated with the true phonon propagator including anharmonic effects. The Fermion line of the diagram may also be dressed.

The inclusion of Coulombic interactions causes the electron–phonon interaction to be screened, as can be seen from Figure 8.10. By comparison with Figure 2.43 of Chapter 2 for the dielectric function $\varepsilon(\mathbf{q}\omega)$, it can be seen that the electron–phonon interaction is modified by screening to become $g_{\mathbf{q}}/\varepsilon(\mathbf{q}\omega)$ which can constitute a considerable reduction.

In spite of the strong electron–phonon coupling, which, for example, means that Landau–Fermi liquid theory is not applicable‡, it remains true

‡ This is true when the inverse lifetime $\hbar/\tau$ of the low-lying quasi-particle states is of the order of the excitation energy, which, in turn, is of the order of $\hbar\omega_{\text{D}}$, ω_{D} being the Debye frequency.

that there is a small parameter $(m/M)^{\frac{1}{2}}$ in the theory, as emphasized by Migdal. This also implies that phonon corrections to the electron–phonon vertex, such as shown in Figure 8.11(a) are small (see Scalapino, Schrieffer and Wilkins, 1966). Coulombic corrections such as that shown in Figure 8.11(b)

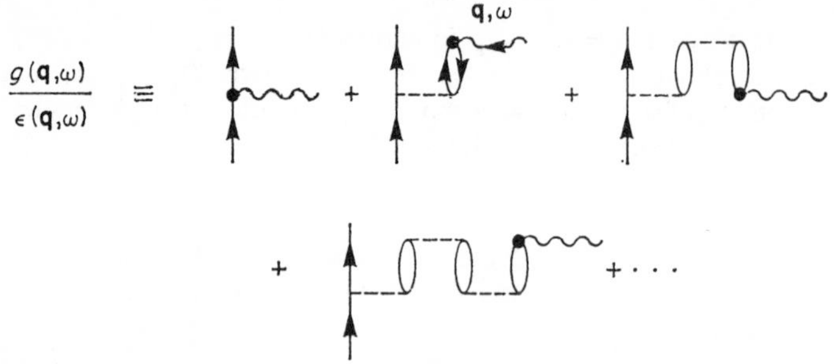

FIGURE 8.10. Electron–phonon interaction, screened by Coulombic repulsion.

are not necessarily small, but are more or less constant factors, as shown by Rice, so that they can be included in $g_{\mathbf{q}}$ as found from experiment (again see Scalapino and co-workers).

In order to apply perturbation methods to superconductors the possibility of the existence of Cooper pairs must be included in the formalism from the

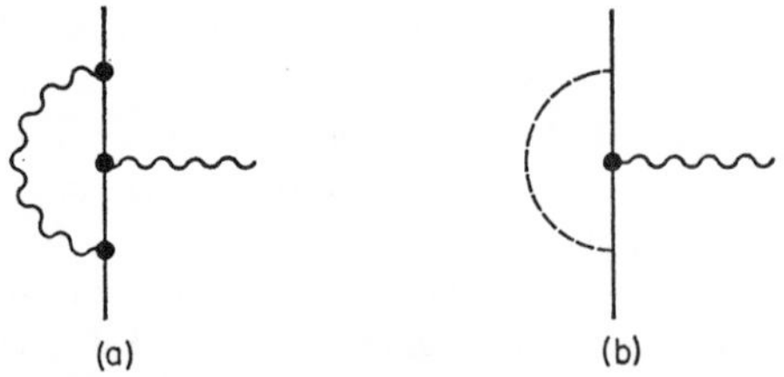

FIGURE 8.11. Corrections to electron–phonon vertex. (a) Lowest-order phonon correction. (b) Lowest-order Coulomb correction.

outset. In the Nambu (1960) method, this is accomplished by taking the "anomalous propagators"

$$\left.\begin{aligned} F(\mathbf{k}t) &= -i\langle\Psi_0|T\{a_{\mathbf{k}\uparrow}(t)\,a_{-\mathbf{k}\uparrow}(0)\}|\Psi_0\rangle, \\ F^{\dagger}(\mathbf{k}t) &= -i\langle\Psi_0|T\{a^{\dagger}_{-\mathbf{k}\downarrow}(t)\,a^{\dagger}_{\mathbf{k}\uparrow}(0)\}|\Psi_0\rangle, \end{aligned}\right\} \tag{8.10.3}$$

which respectively correspond to destruction and creation of a Cooper pair, and combining them with the normal propagators

$$G(\mathbf{k}\uparrow, t) = -i\langle\Psi_0|T\{a_{\mathbf{k}\uparrow}(t)\, a^\dagger_{\mathbf{k}\uparrow}(0)\}|\Psi_0\rangle, \tag{8.10.4}$$

$$G(-\mathbf{k}\downarrow, -t) = -i\langle\Psi_0|T\{a^\dagger_{-\mathbf{k}\downarrow}(t)\, a_{-\mathbf{k}\downarrow}(0)\}|\Psi_0\rangle \tag{8.10.5}$$

to form a matrix propagator **G** with components

$$\left.\begin{aligned} G_{11} &= G(\mathbf{k}\uparrow, t), \quad G_{12} = F(\mathbf{k}t), \\ G_{21} &= F^\dagger(\mathbf{k}, t), \quad G_{22} = G(-\mathbf{k}\downarrow, -t). \end{aligned}\right\} \tag{8.10.6}$$

The free-particle matrix propagator is obtained by substituting the non-interacting ground state for Ψ_0. We then find

$$\mathbf{G}_0(\mathbf{k}\omega) = \begin{pmatrix} \dfrac{1}{\omega-\varepsilon_\mathbf{k}+i\delta} & 0 \\ 0 & \dfrac{1}{\omega+\varepsilon_\mathbf{k}+i\delta} \end{pmatrix}. \tag{8.10.7}$$

The matrix propagator can be compactly written in the form

$$\mathbf{G}(\mathbf{k}t) = -i\langle\Psi_0|T\{\psi(\mathbf{k}t)\,\psi^\dagger(\mathbf{k}0)\}|\Psi_0\rangle, \tag{8.10.8}$$

where the ψ is the spinor

$$\psi = \begin{pmatrix} a_{\mathbf{k}\uparrow} \\ a^\dagger_{-\mathbf{k}\downarrow} \end{pmatrix}, \tag{8.10.9}$$

and $a_\mathbf{k}$ is the annihilation operator for a Bloch state. If one rewrites the Hamiltonian of equation (8.7.17) in terms of ψ's, one obtains the result given in equation (8.10.13) below where, however, Coulombic interactions have also been included. This is formally the same as the original Hamiltonian and so the diagrams used for the normal state may be used here also with G_0 associated with each Fermion line. It should be noted that the presence in equation (8.10.13) of the Pauli spin operator $\sigma_3 = \begin{pmatrix} 1 & 0 \\ 0 & -1 \end{pmatrix}$ may be incorporated in the diagrammatic analysis by associating a factor σ_3 with each scattering vertex. The two additional rules then required to interpret the diagrams are:

(i) To associate the Pauli spin matrix σ_3 with each electron–phonon vertex.

(ii) To associate the matrix propagator (8.10.8) with each Fermion line.

At non-zero temperatures, the imaginary time formalism introduced for phonons in Chapter 3 is to be used. Its application to electrons is illustrated in section 8.10.2 below.

8.10.2 Eliashberg equations

The superconductors Pb and Hg were noticed, at an early stage, to be anomalous in their deviation from the law of corresponding states and from the predictions of the simplest form of the BCS theory given in section 8.9.4 above.

The reasons for this are now rather clear. For these superconductors, the coupling strength is large and it is necessary to use the more realistic retarded electron–electron interaction via the phonons discussed above and to include Coulombic repulsions, within the framework of a strong coupling theory.

The essential step in this programme was due to Eliashberg (1960). In his theory, the energy gap function in coordinate space and time has the form

$$\Sigma(\mathbf{r}-\mathbf{r}', t-t') = F(\mathbf{r}-\mathbf{r}', t-t')\, V(\mathbf{r}-\mathbf{r}', t-t'), \qquad (8.10.10)$$

where V is the retarded potential due to phonons and F is a pair Green function. It turns out that while F has long range, it oscillates with a wavelength of the order of a de Broglie wavelength for an electron at the Fermi momentum k_f, and destructively interferes with V such that the product, giving essentially the gap function, has only a short range in $\mathbf{r}-\mathbf{r}'$. Once again, therefore, the short-range part of the interaction is dominant.

The work we shall describe includes the Coulomb interactions, which was first done by Morel and Anderson (1962). The presentation we adopt below, to obtain the Eliashberg equations, follows quite closely that of Scalapino, Schrieffer and Wilkins (1966). We shall then report briefly the results of accurate computations made by McMillan (1968) on the basis of the Eliashberg equations.

The formalism introduced by Nambu (1960), in which one works with a two-component field operator,

$$\psi_{\mathbf{k}} = \begin{pmatrix} a_{\mathbf{k}\uparrow} \\ a^{\dagger}_{-\mathbf{k}\downarrow} \end{pmatrix} \qquad (8.10.11)$$

where, as usual, $a_{\mathbf{k}\uparrow}$ destroys an electron in a Bloch state $\mathbf{k}$ and $\uparrow$ spin, is convenient for treating strong coupling superconductors.

The bare-phonon field operator

$$\phi_{\mathbf{q}\lambda} = b_{\mathbf{q}\lambda} + b^{\dagger}_{-\mathbf{q}\lambda} \qquad (8.10.12)$$

is a combination of creation and annihilation operators for bare phonons.

The Hamiltonian of the system is then taken to have the form

$$H = \sum_{\mathbf{k}} \varepsilon_{\mathbf{k}} \psi_{\mathbf{k}}^{\dagger} \sigma_3 \psi_{\mathbf{k}} + \sum_{\mathbf{q}\lambda} \Omega_{\mathbf{q}\lambda} b_{\mathbf{q}\lambda}^{\dagger} b_{\mathbf{q}\lambda} + \sum_{\mathbf{k}\mathbf{k}'\lambda} g_{\mathbf{k}\mathbf{k}'\lambda} \phi_{\mathbf{k}-\mathbf{k}'\lambda} \psi_{\mathbf{k}'}^{\dagger} \sigma_3 \psi_{\mathbf{k}}$$

$$+\frac{1}{2} \sum_{\mathbf{k}_1\mathbf{k}_2\mathbf{k}_3\mathbf{k}_4} \langle \mathbf{k}_3 \mathbf{k}_4 | V_c | \mathbf{k}_1 \mathbf{k}_2 \rangle (\psi_{\mathbf{k}_3}^{\dagger} \sigma_3 \psi_{\mathbf{k}_1})(\psi_{\mathbf{k}_4}^{\dagger} \sigma_3 \psi_{\mathbf{k}_2}) + \text{const}, \tag{8.10.13}$$

where $\varepsilon_{\mathbf{k}}$ is the one-electron Bloch energy measured relative to the Fermi energy E_f, while the σ's are, as usual, the Pauli matrices. Ω, g and V_c represent, respectively, the bare phonon frequencies, the strength of the bare electron–phonon coupling and the bare Coulomb repulsion.

Translational invariance of V_c restricts $\mathbf{k}_1+\mathbf{k}_2-\mathbf{k}_3-\mathbf{k}_4$ to be either zero or a reciprocal lattice vector $\mathbf{K}$. The system is taken to be a box of unit volume and, as usual, periodic boundary conditions are imposed. The electrons are described in an extended zone scheme (see Chapter 1, section 1.9.1) and the phonons are described in a reduced zone scheme which is extended periodically throughout $\mathbf{q}$-space to allow Umklapp processes to be handled automatically.

We work with the elevated temperature Green functions, which are defined in the Nambu formalism by

$$\mathbf{G}(\mathbf{k}\tau) = -\langle UT\{\psi_{\mathbf{k}}(\tau)\psi_{\mathbf{k}}^{\dagger}(0)\}\rangle, \tag{8.10.14}$$

$$D_{\lambda}(\mathbf{q}\tau) = -\langle T\{\phi_{\mathbf{q}\lambda}(\tau)\phi_{\mathbf{q}\lambda}^{\dagger}(0)\}\rangle, \tag{8.10.15}$$

where the average is over the grand canonical ensemble

$$\langle Q\rangle \equiv \mathrm{Tr}\,(\mathrm{e}^{-\beta H} Q)/\mathrm{Tr}\,\mathrm{e}^{-\beta H}. \tag{8.10.16}$$

The operators in equations (8.10.14) and (8.10.15) develop with imaginary time $i\tau$ in accordance with

$$\psi_{\mathbf{k}}(\tau) = \mathrm{e}^{H\tau}\psi_{\mathbf{k}}(0)\,\mathrm{e}^{-H\tau}, \tag{8.10.17}$$

$$\phi_{\mathbf{q}\lambda}(\tau) = \mathrm{e}^{H\tau}\phi_{\mathbf{q}\lambda}(0)\,\mathrm{e}^{-H\tau}. \tag{8.10.18}$$

T represents the usual time-ordered product, and the operator U in equation (8.10.14) is given by

$$U = 1 + R^{\dagger} + R, \tag{8.10.19}$$

where $R^{\dagger}$ converts a given state in an N-particle system into the corresponding state in the $N+2$ particle system. For the ground states

$$R^{\dagger}|0,N\rangle = |0,N+2\rangle, \tag{8.10.20}$$

$$R|0,N\rangle = |0,N-2\rangle. \tag{8.10.21}$$

It should be noted that $\mathbf{G}$ is a 2×2 matrix, the diagonal elements G_{11} and G_{22} being the conventional Green functions for $\uparrow$ spin electrons and $\downarrow$ spin holes, while G_{12} and G_{21} describe the pairing properties [related to F in equation (8.10.10) above]. We saw in Chapter 3 that the phonon Green function could be expanded in a Fourier series

$$D_\lambda(\mathbf{q}\tau) = \frac{1}{\beta}\sum_{n=-\infty}^{\infty} \mathrm{e}^{-i\nu_n\tau} D_\lambda(q, i\nu_n) \tag{8.10.22}$$

and similarly we can write

$$\mathbf{G}(\mathbf{k}\tau) = \frac{1}{\beta}\sum_{n=-\infty}^{\infty} \mathrm{e}^{-i\omega_n\tau} G(\mathbf{k}, i\omega_n), \tag{8.10.23}$$

where

$$\nu_n = 2n\pi/\beta, \quad \omega_n = (2n+1)\pi/\beta \tag{8.10.24}$$

with n an integer.

As we have seen earlier, the one-electron Green function for the non-interacting system is given by

$$\mathbf{G}_0(\mathbf{k}, i\omega_n) = [i\omega_n - \varepsilon_{\mathbf{k}}\sigma_3]^{-1}. \tag{8.10.25}$$

Also, apart from the difference that the electronic self-energy $\Sigma(\mathbf{k}, i\omega_n)$ is now a 2×2 matrix, it satisfies the usual Dyson equation (2.10.7) which we can express as

$$[\mathbf{G}(\mathbf{k}, i\omega_n)]^{-1} = [\mathbf{G}_0(\mathbf{k}, i\omega_n)]^{-1} - \Sigma(\mathbf{k}, i\omega_n). \tag{8.10.26}$$

The Nambu formalism is now dealt with straightforwardly by applying the familiar Feynman rules for the perturbation series, in calculating G and D.

The procedure now parallels that for treating electron–phonon interactions in normal metals, introduced by Migdal (see Chapter 6). The electron–phonon interaction is treated to order $(m/M)^{\frac{1}{2}} \sim v_s/v_f \sim \omega_D/E_f$, in the usual notation. Eliashberg worked out the appropriate integral equation to this order in a superconductor at $T=0$, but using the Hamiltonian (8.10.13) this can be generalized to include the Coulombic interactions and to apply at elevated temperatures.

The region of prime interest to us in the self-energy $\Sigma(\mathbf{k}, -i\omega_n)$ is around $k \sim k_f$ and $|\omega_n| \ll E_f$. While, in this range, the Coulomb interaction leads to screening and renormalization effects that are significant, it does not cause significant variations in Σ in a region $\sim\omega_D$ about the Fermi surface. In other words, the Coulomb interaction serves largely to renormalize the bare electron and phonon energy spectra and screen the electron–phonon interaction, as assumed in the Fröhlich model of section 8.7. There then remains a short-range screened interaction which must be treated differently from that used for the strongly retarded phonon interaction between electrons.

We can now express diagrammatically the basic approximation which will be used for the self-energy Σ. In Figure 8.12, the solid lines represent G as given by Dyson's equation (8.10.26) in terms of the self-energy, which is to be determined self-consistently. The dashed line in part (a) of Figure 8.12 is simply the screened electron–electron Coulomb interaction. In part (b) of Figure 8.12, the wavy line represents the phonon Green function or propagator. We also screen the electron–phonon vertex, and put $\bar{g} = g/\varepsilon$.

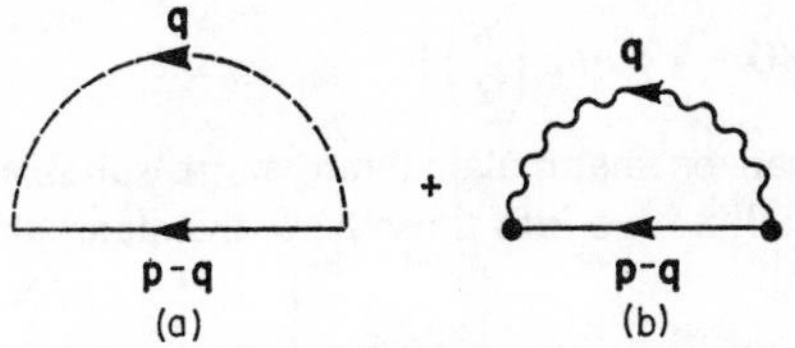

FIGURE 8.12. Electron self-energy to lowest order in (a) Coulomb and (b) electron–phonon interaction.

Thus the basic approximation for $\Sigma(\mathbf{k}, i\omega_n)$ is obtained by writing down the contributions from the two diagrams shown in Figure 8.12. These are

$$\Sigma(\mathbf{k}, i\omega_n) = -\frac{1}{\beta}\sum_{\mathbf{k}'n'} \sigma_3\, G(\mathbf{k}', i\omega_{n'})\, \sigma_3 \times \left[\sum_{\lambda} |\bar{g}_{\mathbf{k}\mathbf{k}'\lambda}|^2\, D_\lambda(\mathbf{k}-\mathbf{k}', i\omega_n - i\omega_{n'}) + V(\mathbf{k}-\mathbf{k}')\right], \tag{8.10.27}$$

where the screened Coulomb interaction V has been taken to depend only on the momentum transfer $\mathbf{k}-\mathbf{k}'$. Introducing the spectral representation of the phonon Green function given by

$$D_\lambda(\mathbf{q}, i\nu_m) = \int_0^\infty d\nu\, A_\lambda(\mathbf{q}\nu)\{[1/(i\nu_m - \nu)] - [1/(i\nu_m + \nu)]\} \tag{8.10.28}$$

the spectral weight function is

$$A_\lambda(\mathbf{q}\nu) = \left(1 - e^{\beta\nu}\sum_{i,j} e^{-\beta E_i} |\langle j|\phi_{\mathbf{q}\lambda}|i\rangle|^2\, \delta(\nu - E_j + E_i) \Big/ \sum_i e^{-\beta E_i}\right), \tag{8.10.29}$$

where

$$H|n\rangle = E_n|n\rangle. \tag{8.10.30}$$

Substituting this into equation (8.10.27) we find, after changing from the summation over n to an integral round the contour C shown in Figure 8.13,

$$\Sigma(\mathbf{k}, i\omega_n) = -\frac{1}{2\pi i}\sum_{k'}\int_C dz\, \sigma_3 G(\mathbf{k}'z)\,\sigma_3$$

$$\times\left[\sum_\lambda \int_0^\infty d\nu\, A_\lambda(\mathbf{k}-\mathbf{k}',\nu)\left(\frac{1}{i\omega_n - z - \nu}\frac{1}{1+e^{-\beta z}} + \frac{1}{i\omega_n - z+\nu}\frac{1}{1+e^{\beta z}}\right)\right.$$

$$\left. -\tfrac{1}{2}V(\mathbf{k}-\mathbf{k}')\tanh\left(\frac{\beta z}{2}\right)\right]. \qquad (8.10.31)$$

This equation can be manipulated into more suitable forms for actual solution, but we shall not go into those here, the manipulations being found

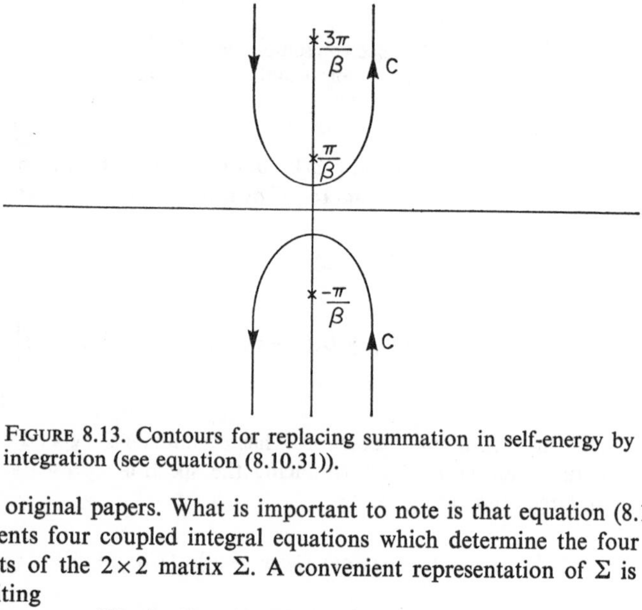

FIGURE 8.13. Contours for replacing summation in self-energy by integration (see equation (8.10.31)).

in the original papers. What is important to note is that equation (8.10.31) represents four coupled integral equations which determine the four components of the 2×2 matrix Σ. A convenient representation of Σ is given by writing

$$\Sigma(\mathbf{k}\omega) \equiv [1 - Z(\mathbf{k}\omega)]\,\omega\mathbf{1} + \phi(\mathbf{k}\omega)\,\sigma_1 + \chi(\mathbf{k}\omega)\,\sigma_3, \qquad (8.10.32)$$

where the choice of phases has been such that the coefficient of the Pauli matrix σ_2 is zero.

Using equation (8.10.32) and Dyson's equation, we find the analytically continued one-electron Green function is given by

$$G(\mathbf{k}\omega) = \left.\frac{\omega Z(\mathbf{k}\omega)\,\mathbf{1} + \bar{\varepsilon}(\mathbf{k}\omega)\,\sigma_3 + \phi(\mathbf{k}\omega)\,\sigma_1}{\omega^2 Z^2(\mathbf{k}\omega) - \bar{\varepsilon}^2(\mathbf{k}\omega) - \phi^2(\mathbf{k}\omega)}\right|_{\mathrm{Im}\,\omega = +\delta}, \qquad (8.10.33)$$

where

$$\bar{\varepsilon}(\mathbf{k}\omega) = \varepsilon_{\mathbf{k}} + \chi(\mathbf{k}\omega). \tag{8.10.34}$$

It is quite clear that the calculation of G is reduced to calculating the three functions Z, ϕ and χ and that, when this is done, the function

$$\Delta(\mathbf{k}\omega) = \phi(\mathbf{k}\omega)/Z(\mathbf{k}\omega) \tag{8.10.35}$$

is taking on the role of the gap parameter of the pairing theory and vanishes in the normal state.

8.10.3 Application of Eliashberg equations

It is worth recording here that, after simplifying the basic equation (8.10.31), Scalapino, Schrieffer and Wilkins (1966) get the following approximate expressions for the energy gap $\Delta(\omega)$ and $Z(\omega)$ at $T = 0$:

$$\Delta(\omega) = \frac{1}{Z(\omega)} \int_0^{\omega_c} d\omega' \,\mathrm{Re}\left\{\frac{\Delta(\omega')}{[\omega'^2 - \Delta^2(\omega')]^{\frac{1}{2}}}\right\} \times [K_+(\omega', \omega) - \rho(0)\, U_c] \tag{8.10.36}$$

and for $Z(\omega)$

$$[1 - Z(\omega)]\,\omega = \int_0^{\infty} d\omega' \,\mathrm{Re}\left\{\frac{\omega'}{[\omega'^2 - \Delta^2(\omega')]^{\frac{1}{2}}}\right\} K_-(\omega', \omega), \tag{8.10.37}$$

where the frequency integral in the expression for $\Delta\omega$ can be cut off, it turns out, at $\omega_0 \sim 10\omega_d$, because of the rapid convergence of the integrand. The kernels $K_\pm(\omega, \omega')$ can be written in the form

$$K_\pm(\omega, \omega') = \sum_\lambda \int_0^\infty d\nu\, \alpha_\lambda^2(\nu)\, F_\lambda(\nu) \left[\frac{1}{\omega' + \omega + \nu + i\delta} \pm \frac{1}{\omega' - \omega + \nu - i\delta}\right], \tag{8.10.38}$$

where $F_\lambda(\nu)$ is the phonon density of states for the λ-mode, given by

$$F_\lambda(\nu) = \int \frac{d\mathbf{q}}{(2\pi)^3} A_\lambda(q\nu), \tag{8.10.39}$$

and $\alpha_\lambda^2(\nu)$ is an effective electron–phonon coupling constant defined by

$$F_\lambda(\nu)\, \alpha_\lambda^2(\nu) = \frac{\rho(0)}{8\pi k_f^2} \int d\Omega_q \int_0^{2k_f} q\, dq\, |\bar{g}_{q\lambda}|^2 A_\lambda(q, \nu). \tag{8.10.40}$$

This represents an average of the electron (λ-mode)–phonon matrix elements over allowed momentum transfers. The quantities which determine the self-energy in a region of order ω_c of the Fermi surface are $\alpha_\lambda^2(\nu)$, $F_\lambda(\nu)$ and

$\rho(0)\,U_c$, U_c being the pseudo-potential representing the Coulomb interaction (cf. Morel and Anderson, 1962). Once these quantities are given, the functions $\Delta(\omega)$ and $Z(\omega)$ can be found by solving these two equations (8.10.36) and (8.10.37). The important results obtained in this way by McMillan (1968) are reported in Appendix 8.3. The difficulties of the BCS theory applied to Pb and Hg are seen there to be removed.

8.11 Meissner effect

We turn now to deal with one of the most difficult problems for the theory of superconductivity, the existence of the Meissner effect, which comes about when a static magnetic field is applied to a superconductor. Classical theory cannot lead to such an effect: for type 1 superconductors, for $T<T_c$, magnetic flux is excluded from the interior, except for a thin layer at the surface. The magnetic susceptibility χ is therefore accurately equal to -1 and the magnetic moment $\mathscr{M}$ per unit volume, produced in the specimen by a field $\mathscr{H}$, is given by $\mathscr{M}=-\mathscr{H}$. However, for type II superconductors this is true only up to a field, say $\mathscr{H}_{c_1}$, which is less than the critical field $\mathscr{H}_{c_2}$ which destroys the superconductivity.

Below, we shall deal with the Meissner effect by considering the application of a general time-dependent electromagnetic field to the superconductor. However, before doing so, it will be useful to summarize the arguments by which the Londons attempted to account for the Meissner effect, and for the existence of persistent currents, by setting up macroscopic equations for the superconductor.

8.11.1 Form of relation between current density and vector potential in superconductors

Historically, the Londons tried to account for the existence of the Meissner effect and for the existence of persistent currents by searching for macroscopic equations to describe the superconductor.

If we assume for a moment that all the conduction electrons accelerate freely in an electric field $\mathscr{E}$, then we must replace Ohm's law by

$$\frac{m}{ne^2}\frac{\partial \mathbf{j}}{\partial t}=\mathscr{E}. \tag{8.11.1}$$

In general, in a superconductor, let us write Λ instead of m/ne^2 (since we do not expect *all* electrons in fact to accelerate freely) and we have

$$\Lambda\frac{\partial \mathbf{j}}{\partial t}=\mathscr{E}. \tag{8.11.2}$$

Taking the curl of this equation, we find

$$\Lambda \operatorname{curl} \frac{\partial \mathbf{j}}{\partial t} = \operatorname{curl} \mathscr{E} = -\frac{1}{c} \frac{\partial \mathscr{B}}{\partial t}, \tag{8.11.3}$$

where we have used the appropriate Maxwell equation in replacing curl $\mathscr{E}$ in favour of the field $\mathscr{B}$.

If we drop the time derivative, we have the further equation

$$\Lambda \operatorname{curl} \mathbf{j} = -\frac{\mathscr{B}}{c}. \tag{8.11.4}$$

It is of interest to ask now what consequences stem from these equations (8.11.2) and (8.11.4). We shall anticipate the range of validity of the London equations by applying them only to physical situations where no space charge builds up. Then we have, from conservation of charge,

$$\operatorname{div} \mathbf{j} = 0, \tag{8.11.5}$$

and, under the same circumstances,

$$\operatorname{div} \mathscr{E} = 0. \tag{8.11.6}$$

Under the conditions we are discussing here we neglect the normal and the displacement currents relative to the supercurrent, $\mathbf{j}_s$ say. Then we can write the field equations in the form

$$\left.\begin{aligned} &\operatorname{div} \mathscr{E} = 0, \quad \operatorname{div} \mathscr{B} = 0, \\ &\operatorname{curl} \mathscr{E} = -\frac{1}{c}\frac{\partial \mathscr{B}}{\partial t}, \quad \operatorname{curl} \mathscr{B} = \frac{4\pi}{c}\mathbf{j}_s \end{aligned}\right\}. \tag{8.11.7}$$

(a) *Existence of Meissner effect.* To understand how the Meissner effect can arise, we shall consider a semi-infinite slab of superconducting material, with a plane surface. Now we switch on a static uniform field $\mathscr{H}$ outside, parallel to the surface in the xy plane.

Then we have to solve the equations

$$\Lambda \operatorname{curl} \mathbf{j}_s = -\frac{\mathscr{B}}{c}, \quad \operatorname{curl} \mathscr{B} = \frac{4\pi}{c}\mathbf{j}_s, \quad \operatorname{div} \mathscr{B} = 0 \tag{8.11.8}$$

within the superconductor. From the second of these equations, we see that the boundary condition that the tangential component of $\mathscr{B}$ is continuous is already implied. The equation div $\mathscr{B} = 0$ is satisfied, since these components are only functions of z, and the other two equations yield

$$\nabla^2 \mathscr{B} = -\operatorname{curl}\operatorname{curl} \mathscr{B} = \frac{4\pi}{\Lambda c^2}\mathscr{B}. \tag{8.11.9}$$

Thus we have

$$\frac{\partial^2 \mathscr{B}_x}{\partial z^2} = \frac{4\pi}{\Lambda c^2}\mathscr{B}_x \tag{8.11.10}$$

with solution

$$\mathscr{B}_x(z) = \mathscr{H}\exp(-z/\lambda), \quad z = 0 \text{ in the surface,} \tag{8.11.11}$$

where

$$\lambda = (\Lambda c^2/4\pi)^{\frac{1}{2}}. \tag{8.11.12}$$

The field is seen then to penetrate a distance λ into the superconductor. If we take as an order of magnitude estimate $m/ne^2 \sim 10^{-31}$ sec^2 as giving Λ, then we find $\lambda \sim 10^{-6}$ cm and the magnetic induction inside macroscopic superconductors is for practical purposes zero. Thus the London macroscopic theory leads to the Meissner effect.

(b) *Infinite conductivity.* Because the supercurrent $\mathbf{j}_s$ is 90° out of phase with $\mathscr{E}$, as follows from (8.11.33) the London macroscopic equations allow no dissipation of energy. If we wish to allow for the possibility of energy dissipation, we must include the normal current $\mathbf{j}_n$ in the theory. However, in a static field, it can be shown that there is no approximation involved in neglecting $\mathbf{j}_n$. The conclusion is that in a static uniform electric field, there is no dissipation of energy, and hence the material will exhibit infinite conductivity in this model.

8.11.2 General conditions for existence of Meissner effect

We have seen that the London equations contain the Meissner effect. However, it turns out that the local relation between the current density $\mathbf{j}$ and the vector potential which they imply is too special and it is therefore of interest to enquire as to the general conditions under which the Meissner effect can exist.

We must consider the effect that a time-dependent transverse field can have on a macroscopic system. We assume that the applied field will arise from some source current density $\mathbf{j}_s$. There will, in addition, be an induced current density given by $\mathbf{j}_i$ where

$$\operatorname{curl}\mathscr{H} = 4\pi(\mathbf{j}_s + \mathbf{j}_i), \tag{8.11.13}$$

provided we neglect the displacement current. If the fields are assumed weak, $\mathbf{j}_i$ will depend linearly on them. Since the fields are assumed to be transverse and are connected by $\operatorname{curl}\mathscr{E} = (-1/c)(\partial\mathscr{B}/\partial t)$, $\mathbf{j}_i$ will be a linear function of just one of the fields, say $\mathscr{B}$.

If we limit ourselves to infinite superconductors which are translationally invariant, we can treat the different Fourier components independently. Since $\mathbf{j}_i$ and $\mathscr{B}$ must be perpendicular,

$$\mathbf{j}_i(\mathbf{q}\omega) \propto \mathbf{q} \times \mathscr{B}(\mathbf{q}\omega). \tag{8.11.14}$$

The proportionality factor in general must depend on $\mathbf{q}$ and ω and in what follows we shall find it convenient to express it in the form

$$\frac{4\pi}{c}\mathbf{j}_i(\mathbf{q}\omega) = -\frac{K(\mathbf{q}\omega)\, i\mathbf{q} \times \mathscr{B}(\mathbf{q}\omega)}{q^2}. \tag{8.11.15}$$

Clearly $K(\mathbf{q}\omega)$ is a linear response function, such as we have introduced earlier in other connections.

We also have from Maxwell's equations

$$\mathbf{q} \times \mathscr{E} = \frac{\omega}{c}\mathscr{B}, \quad i\mathbf{q} \times \mathscr{B} = \frac{4\pi}{c}(\mathbf{j}_s + \mathbf{j}_i) \tag{8.11.16}$$

and the relation between $\mathbf{j}_i$ and $\mathscr{E}$ is

$$\frac{4\pi}{c^2}\mathbf{j}_i(\mathbf{q}\omega) = \frac{iK(\mathbf{q}\omega)}{\omega}\mathscr{E}(\mathbf{q}\omega). \tag{8.11.17}$$

The a.c. conductivity $\sigma(0\omega)$ is clearly

$$\sigma(0\omega) = \frac{ic^2 K(0\omega)}{4\pi\omega} \tag{8.11.18}$$

and for infinite conductivity the right-hand side must tend to infinity as $\omega \to 0$. If the infinite conductivity arises from electrons accelerating freely in an electric field (cf. equation 8.11.2) then

$$-i\omega\mathbf{j}(0\omega) \propto \mathscr{E}(0\omega). \tag{8.11.19}$$

This requires the more stringent condition

$$\lim_{\omega \to 0} K(0\omega) = \text{constant} \neq 0 \tag{8.11.20}$$

for superconductivity.

In an isotropic system K will be solely a function of q^2 and for the current density to be real when the field is real we must have

$$K(|\mathbf{q}|, 0)^* = K(|-\mathbf{q}|, 0) = K(q0), \tag{8.11.21}$$

where $K(q0)$ is real. From Maxwell's equations we may write

$$i(\mathbf{q} \times \mathbf{B})\left[1 + \frac{K(q0)}{q^2}\right] = \frac{4\pi}{c}\mathbf{j}_s. \tag{8.11.22}$$

This shows that the total permeability of the medium is $\{1+[K(q0)/q^2]\}$ and this must be positive. The magnetic induction in **q**-space is given by

$$\mathscr{B}(\mathbf{q}) = \frac{4\pi}{c}\frac{i\mathbf{q}\times\mathbf{j}_s(\mathbf{q})}{q^2+K(q0)}. \tag{8.11.23}$$

Working in coordinate space, we may write

$$\mathscr{B}(\mathbf{r}) = \frac{\Omega 4\pi}{(2\pi)^3 c}\int d\mathbf{q}\frac{i\mathbf{q}\times\mathbf{j}_s(\mathbf{q})}{q^2+K(q0)}\,e^{i\mathbf{q}.\mathbf{r}}. \tag{8.11.24}$$

The behaviour of $\mathscr{B}(\mathbf{r})$ at large **r** is determined by the zeros of $q^2+K(q0)$. If $[q^2+K(q0)]$ has no zeros on the real axis, then $\mathscr{B}(\mathbf{r})$ will decay exponentially. If it has zeros on the real axis, it can decay as an inverse power of r. Since we have seen that $\{1+[K(q0)/q^2]\}$ is positive for real q, $q^2+K(q0)$ can be zero only when q is zero. Hence the condition for a Meissner effect is that $K(q0)$ tend to a non-zero constant (positive) as $q\to 0$. This result is due to Schafroth (1951).

For ω not equal to zero, we can write that the induced current is related to the vector potential through

$$\mathscr{B}(\mathbf{q}\omega) = i\mathbf{q}\times\mathscr{A}(\mathbf{q}\omega) \tag{8.11.25}$$

and

$$\frac{4\pi}{c}\mathbf{j}_i = K(\mathbf{q}\omega)\frac{(\mathbf{q}.\mathscr{A})\,\mathbf{q}-q^2\mathscr{A}}{q^2}. \tag{8.11.26}$$

In the case when the potential is transverse (perpendicular to **q**) we have therefore

$$\frac{4\pi}{c}\mathbf{j}_i(\mathbf{q}\omega) = -K(\mathbf{q}\omega)\,\mathscr{A}(\mathbf{q}\omega). \tag{8.11.27}$$

In the London theory described above, $K(q0)$ is a positive constant. We shall see later that the BCS theory leads to this situation.

Thus, the general condition on the response function K is that a Meissner effect exists if

$$\lim_{q\to 0}\lim_{\omega\to 0} K(q\omega) = \text{non-zero constant}. \tag{8.11.28}$$

We have seen above that it is the limiting process on K, in the reverse order, which is the crucial factor in determining whether the conductivity is infinite.

8.11.3 Microscopic theory of response function $K(\mathbf{q}\omega)$

We shall now extend our discussion to deal with the Meissner effect at a microscopic level, by obtaining an explicit form for the linear response function $K(\mathbf{q}\omega)$ introduced in equation (8.11.15). We need not reproduce a lot of the detail, since we follow the general lines laid down in Chapter 3. In the present case, the perturbation is

$$H_1 = \sum_i \frac{[\mathbf{p}_i - (e/c)\mathscr{A}(\mathbf{r}_i)]^2}{2m} - \sum_i \frac{p_i^2}{2m}, \tag{8.11.29}$$

which we can rewrite in terms of the current operator $\mathbf{j}^0(\mathbf{r}) \equiv e\mathbf{p}/m$ and the mass density $\rho(\mathbf{r})$. We then find

$$H_1 = -\frac{1}{c}\int d\mathbf{r}\, \mathbf{j}^0(\mathbf{r}) . \mathscr{A}(\mathbf{r}) + \frac{e^2}{2m^2 c^2}\int d\mathbf{r}\, \mathscr{A}^2(\mathbf{r})\, \rho(\mathbf{r}) \tag{8.11.30}$$

and working to lowest order in $\mathscr{A}$ we obtain, following the methods of Chapter 3,

$$\langle \mathbf{j}^0(\mathbf{r}t)\rangle_{\mathscr{A}} = \frac{i}{c}\int_{-\infty}^{t} d\mathbf{r}'\, dt' \langle [\mathbf{j}^0(\mathbf{r}t), \mathbf{j}^0(\mathbf{r}'t')]\rangle \mathscr{A}(\mathbf{r}'t'), \tag{8.11.31}$$

where the average on the right is over the unperturbed system. It is important here to note that this gives us only the "paramagnetic" current. The total current is to be found by recalling that in the presence of a magnetic field the velocity $\mathbf{v}_i$ of a particle of canonical momentum $\mathbf{p}_i$ is given by

$$m\mathbf{v}_i = \mathbf{p}_i - \frac{e}{c}\mathscr{A}(\mathbf{r}t). \tag{8.11.32}$$

The total current operator is thus

$$\mathbf{j}(\mathbf{r}) = \mathbf{j}^0(\mathbf{r}) - \frac{e^2}{mc}\mathscr{A}(\mathbf{r}t)\,\rho(\mathbf{r}), \tag{8.11.33}$$

where the second term, the so-called diamagnetic current, is already first-order in $\mathscr{A}$. In the absence of the perturbation, $\rho(\mathbf{r}) = n$, a constant, in a translationally invariant system and deviations from this due to the presence of $\mathscr{A}$ will obviously affect $\langle \mathbf{j}\rangle$ only to higher order. We therefore write

$$\langle \mathbf{j}(\mathbf{r}t)\rangle_{\mathscr{A}} = \langle \mathbf{j}^0(\mathbf{r}t)\rangle_{\mathscr{A}} - \frac{e^2 n}{m}\mathscr{A}(\mathbf{r}t), \tag{8.11.34}$$

where $\langle \mathbf{j}^0(\mathbf{r}t)\rangle_{\mathscr{A}}$ is given by equation (8.11.31).

We shall assume that the vector potential $\mathscr{A}$ has the form

$$\mathscr{A}(\mathbf{r}t) = e^{i(\mathbf{q}.\mathbf{r}-\omega t)}\,\mathscr{A}(\mathbf{q}\omega). \tag{8.11.35}$$

Then defining a susceptibility χ in terms of the current–current correlation function through

$$\chi_{\alpha\beta}(\mathbf{r}-\mathbf{r}', \omega+i\eta) = \int_{-\infty}^{\infty} \frac{d\omega'}{\omega+i\eta-\omega'} \int_{-\infty}^{\infty} dt'\, e^{i\omega'(t-t')} \langle [j^0_\alpha(\mathbf{r}'t), j^0_\beta(\mathbf{r}t')]\rangle, \tag{8.11.36}$$

which depends only on $\mathbf{r}-\mathbf{r}'$ because translational invariance is assumed, we have

$$c\langle j^0_\alpha(\mathbf{q}\omega)\rangle = \sum_\beta \chi_{\alpha\beta}(\mathbf{q}\omega)\mathscr{A}_\beta(\mathbf{q}\omega), \tag{8.11.37}$$

where $\chi_{\alpha\beta}(\mathbf{q}\omega)$ is the Fourier transform of $\chi_{\alpha\beta}(\mathbf{r}\omega)$.

Since $\chi_{\alpha\beta}(\mathbf{q}\omega)$ must transform as a tensor and $\mathbf{q}$ is the only vector it depends on, it can be expressed in the form

$$\chi_{\alpha\beta}(\mathbf{q}\omega) = \frac{q_\alpha q_\beta}{q^2}\chi^{\mathrm{L}}(\mathbf{q}\omega) + \left(\delta_{\alpha\beta} - \frac{q_\alpha q_\beta}{q^2}\right)\chi^{\mathrm{T}}(\mathbf{q}\omega). \tag{8.11.38}$$

The first term, parallel to $\mathbf{q}$ in both indices α and β, is the longitudinal component; the second term, perpendicular to $\mathbf{q}$ in both indices, is the transverse component. Hence, when the vector potential is transverse (i.e. $\mathbf{q}.\mathscr{A} = 0$) we find

$$c\langle \mathbf{j}(\mathbf{q}\omega)\rangle_{\mathscr{A}} = \left[\chi^{\mathrm{T}}(\mathbf{q}\omega) - \frac{e^2 n}{m}\right]\mathscr{A}(\mathbf{q}\omega) \tag{8.11.39}$$

and, comparing this with equation (8.11.22), we see that

$$K(\mathbf{q}\omega) = \frac{4\pi}{c^2}\left[\frac{e^2 n}{m} - \chi^{\mathrm{T}}(\mathbf{q}\omega)\right], \tag{8.11.40}$$

provided $\mathscr{A}(\mathbf{q}\omega)$ is interpreted as the *total* vector potential; if $\mathscr{A}_{\mathrm{ind}}$ is the induced potential and $\mathscr{A}_{\mathrm{ext}}$ the external potential

$$\mathscr{A}(\mathbf{q}\omega) = \mathscr{A}_{\mathrm{ind}}(\mathbf{q}\omega) + \mathscr{A}_{\mathrm{ext}}(\mathbf{q}\omega). \tag{8.11.41}$$

(a) *Induced vector potential.* The application of an external potential $\mathscr{A}_{\mathrm{ext}}$ induces currents which themselves generate magnetic fields. We do not wish here to embark on a basic analysis of the introduction of a vector potential in the quantum-mechanical Hamiltonian, for it is clear from a physical point of view that, neglecting field fluctuations, the quantity $\mathscr{A}$ in equation (8.11.39) is the total potential, because the field determining the behaviour of a small but macroscopic element of a system will consist not only of $\mathscr{A}_{\mathrm{ext}}$ but also of the fields due to the presence of the rest of the system.

We calculate the induced vector potential from the result

$$\left(\frac{1}{c^2}\frac{\partial^2}{\partial t^2}-\nabla^2\right)\mathscr{A}_{\text{ind}}(\mathbf{r}t)=\frac{4\pi}{c}\langle\mathbf{j}(\mathbf{r}t)\rangle \tag{8.11.42}$$

and by using such an averaged vector potential we obtain results accurate to order $(v/c)^2$, v being a typical particle velocity. The above equation reads, in Fourier transform,

$$\left[\frac{(\omega+i\eta)^2}{c^2}-q^2\right]\mathscr{A}_{\text{ind}}=\frac{4\pi}{c}\langle\mathbf{j}\rangle. \tag{8.11.43}$$

Then, using equations (8.11.27) and (8.11.41) we find

$$\mathscr{A}(\mathbf{q}\omega)=\mathscr{A}_{\text{ext}}(\mathbf{q}\omega)\bigg/\left[1-\frac{e^2K(\mathbf{q}\omega)}{(\omega+i\eta)^2-c^2q^2}\right]. \tag{8.11.44}$$

The denominator in equation (8.11.44) can be seen to be like a transverse dielectric function. Actually the form differs from the classical definition of this quantity by the presence of the factor $(\omega+i\eta)^2$ in addition to $-c^2q^2$. The reason for this difference is that the transverse dielectric function relates $\mathscr{B}$ and $\mathscr{A}$ and while

$$\mathscr{B}=\nabla\times\mathscr{A}(\mathbf{r}t) \tag{8.11.45}$$

it is not true, except in the static limit $\omega+i\eta\to 0$, that $\mathscr{H}=\nabla\times\mathscr{A}_{\text{ext}}$.

(b) *Static, long wavelength magnetic field perturbation.* Let us consider now this static case, corresponding to a time-independent vector potential $\mathscr{A}$. We then have, from equations (8.11.40) and (8.11.44),

$$\mathscr{A}(\mathbf{q})=\frac{\mathscr{A}_{\text{ext}}(\mathbf{q})}{1+(4\pi/c^2q^2)\left[(e^2n/m)-\chi^{\text{T}}(\mathbf{q}0)\right]}. \tag{8.11.46}$$

We further simplify the situation by assuming the applied field to be of very long wavelength. We first write, formally,

$$\lim_{q\to 0}\chi^{\text{T}}(q0)=e^2\rho_{\text{n}}/m^2. \tag{8.11.47}$$

(In a two-fluid model of a superfluid ρ_{n} would be the mass density of the normal fluid and χ^{T} the appropriate response function; see Baym, 1969.) From the two equations above, we can see that in the long wavelength limit $q\to 0$

$$\mathscr{A}(\mathbf{q})\sim\frac{q^2}{q^2+\Lambda^{-2}}\mathscr{A}_{\text{ext}}(\mathbf{q}), \tag{8.11.48}$$

where the quantity Λ, which is evidently a length, is given by

$$\frac{1}{\Lambda^2} = \frac{4\pi ne^2}{c^2}\left(\frac{\rho-\rho_n}{\rho}\right), \tag{8.11.49}$$

the mass density mn having been denoted as usual by ρ. If $\rho_n < \rho$, then it follows that Λ is real and it can be seen that in the limit $q \to 0$ the total electromagnetic potential approaches zero as q^2 times the applied potential. Hence slowly varying external magnetic fields will be excluded from the material; this is the Meissner effect.

The length Λ introduced above can be identified with the penetration depth. To see this, we multiply equation (8.11.48) through by $q(q^2+\Lambda^{-2})$ and convert to coordinate space. Then we obtain

$$(\nabla^2 - \Lambda^{-2})\,\mathscr{B}(\mathbf{r}) = \nabla^2 \mathscr{H}(\mathbf{r}) \tag{8.11.50}$$

which is valid when $\mathscr{H}(\mathbf{r})$ is slowly varying in space. For a constant $\mathscr{H}$, the solution for $\mathscr{B}$ obviously falls off exponentially inside the specimen, provided the characteristic length Λ is real, the precise form of the solution depending, of course, on the boundary conditions.

(c) *Relation to London theory.* It is important to note that equation (8.11.39) reduces to the London equation

$$\langle \mathbf{j}(\mathbf{r}t)\rangle = \frac{\rho_n - \rho}{\rho}\frac{ne^2}{mc}\mathscr{A}(\mathbf{r}t) \tag{8.11.51}$$

in the limit of fields that vary slowly in space and time. Equation (8.11.51), as shown by London, leads to a vanishing electrical resistance, since $\mathscr{E} = (-1/c)(\partial\mathscr{A}/\partial t)$ and therefore

$$\frac{\partial}{\partial t}\langle \mathbf{j}(\mathbf{r}t)\rangle = \frac{\rho-\rho_n}{\rho}\frac{ne^2}{m}\mathscr{E}(\mathbf{r}t). \tag{8.11.52}$$

It follows that if $\mathscr{E} = 0, (\partial/\partial t)\langle j\rangle = 0$ and the current is constant in time.

(d) *Connection with normal systems.* In a normal system, $\lim_{q\to 0}\lim_{\omega\to 0} K(\mathbf{q}\omega) = 0$ as already noted. Thus, it follows from equation (8.11.40) that

$$\lim_{q\to 0} \chi^{\mathrm{T}}(q0) = -\frac{e^2 n}{m} \tag{8.11.53}$$

(i.e. $\rho_n = \rho$). But the main interest lies in the next term. In order that equation (8.11.46) yields a finite result as $q \to 0$, this next term must evidently have the form bq^2. It then follows that

$$\mathscr{A}(\mathbf{q}) = \frac{\mathscr{A}_{\text{ext}}(\mathbf{q})}{1+(4\pi be^2/c^2)} \tag{8.11.54}$$

and, with b positive, this just expresses the Landau diamagnetism of normal metals.

(e) *Exponential decay of magnetic field.* From the above, it will be seen that to explain the Meissner effect at a microscopic level we must demonstrate that $\rho_n < \rho$ in a superconductor.

To accomplish this, we use the current–current correlation function

$$G_{\alpha\beta}(\mathbf{rr}'\omega) = \int_{-\infty}^{\infty} dt\, e^{i\omega(t-t')} \langle [j^0_\alpha(\mathbf{r}t), j^0_\beta(\mathbf{r}'t')] \rangle. \tag{8.11.55}$$

For a translationally invariant system, this depends only on $\mathbf{r}-\mathbf{r}'$ and *not* on $\mathbf{r}$ and $\mathbf{r}'$ separately, and we accordingly introduce the transform $G_{\alpha\beta}(\mathbf{q}\omega)$. The connection between G and χ is readily established through equation (8.11.36); in particular

$$\chi_{zz}(\mathbf{q}0) = -\int_0^{\infty} \frac{d\omega}{\omega} G_{zz}(\mathbf{q}\omega). \tag{8.11.56}$$

This is the quantity we shall investigate, for it follows from equation (8.11.38) that

$$\chi^{\mathrm{T}}(\mathbf{q}\omega) = \chi_{zz}(\mathbf{q}\omega) \quad (\mathbf{q} \text{ perpendicular to } z). \tag{8.11.57}$$

By employing detailed balancing (see Chapter 6, equation 6.4.2) we can write

$$\frac{G_{zz}(\mathbf{q}\omega)}{1-e^{-\beta\omega}} = \frac{1}{\mathscr{V}} \int d\mathbf{r}\, d\mathbf{r}'\, e^{-i\mathbf{q}.(\mathbf{r}-\mathbf{r}')} \int_{-\infty}^{\infty} dt\, e^{i\omega(t-t')} \langle j^0_z(\mathbf{r}t) j^0_z(\mathbf{r}'t') \rangle, \tag{8.11.58}$$

since at $T=0$

$$\langle j_\alpha(\mathbf{r}t) j_\alpha(\mathbf{r}'t) \rangle = \sum_n \langle 0 | j^0_\alpha(\mathbf{r}) | n \rangle \langle n | j^0_\alpha(\mathbf{r}') | 0 \rangle e^{i(E_n-E_0)(t-t')}, \tag{8.11.59}$$

and we therefore have

$$\chi_{zz}(\mathbf{q}0) = -\frac{2}{\mathscr{V}} \sum_n \frac{|\langle n | j_{-\mathbf{q}}^0 | 0 \rangle|^2}{E_n - E_0}, \tag{8.11.60}$$

where

$$j^0_{-\mathbf{q}} = \int d\mathbf{r}\, e^{i\mathbf{q}.\mathbf{r}} j^0_z(\mathbf{r}). \tag{8.11.61}$$

We need to understand what happens to the terms, in the sum (8.11.60), as $q \to 0$. To examine these terms, we note first, because of translational invariance, that we can choose the states $|n\rangle$ such that each corresponds to a definite momentum and since the ground state has zero momentum, $\langle n|j_{-\mathbf{q}}|0\rangle \neq 0$ only if $|n\rangle$ has momentum $\hbar\mathbf{q}$, $j_{-\mathbf{q}}$ changing the momentum of any state by $\hbar\mathbf{q}$. Further

$$\lim_{q\to 0} j_{-\mathbf{q}}|0\rangle = 0 \tag{8.11.62}$$

since

$$\lim_{q\to 0} j_{-\mathbf{q}} = j_z. \tag{8.11.63}$$

Hence the only states $|n\rangle$ contributing to the sum in equation (8.11.60) are states of momentum $\mathbf{q}$, the energies of which tend to E_0 as $\mathbf{q} \to 0$. But there are no states of this kind corresponding to the excitation of particles from the Fermi distribution, because of the presence of the energy gap of the BCS theory.‡ Thus at $T = 0$, $\rho_n = 0$ and $\lim_{q\to 0} K(q0)$ is a non-zero constant. At finite temperatures, contributions to χ^T as $q \to 0$ are to be expected, but we can see that the energy gap will still change the long wavelength limit of χ^T crucially and that Λ^2 will be positive. Thus, the energy gap ensures the existence of the Meissner effect.

The above argument has been posed in a way which makes it appear to hinge on translational invariance, and so, as it stands, it is insufficiently general to deal with the introduction of small quantities of impurities, which are known in practice not to destroy the Meissner effect. Ensemble averaging over impurity positions, as discussed elsewhere in the book, may be used to obtain the necessary generalizations (see also Rickayzen, 1965).

8.12 Ginzburg–Landau equations

We must now give some discussion of the basic features of hard (type II) superconductors. We will see that inherent in such a description is the need to allow a spacially varying superfluid density, and the highly successful Ginzburg–Landau theory was designed to meet such a situation.

We shall restrict ourselves, below, to the original phenomenological discussion of Ginzburg and Landau, but we should point out that Gor'kov, using a Green function formulation of the BCS method, was able to provide a microscopic basis for the theory (cf. March, Young and Sampanthar, 1967).

The starting point is to define a function $\Psi(\mathbf{r})$ such that $|\Psi(\mathbf{r})|^2$ is the density of superconducting electrons at position $\mathbf{r}$. Thus Ψ is a kind of

‡ There are also the so-called Anderson (1960) modes which are, however, longitudinal and so make no contribution to χ^T.

effective wave function but in the Gor'kov derivation, for instance, Ψ turned out to be proportional to the local energy gap parameter $\Delta(\mathbf{r})$ (see also sections 8.13 and 8.14). This is a very reasonable result since the ability to create superconducting electrons is directly related to the size of Δ.

In terms of Ψ, the total free energy of the superconducting state in the presence of a magnetic field $\mathscr{H}(\mathbf{r})$ is written plausibly in the form

$$F = \int d\mathbf{r}\, F_H^s(\mathbf{r}) = \int d\mathbf{r} \left[F_0^s(\mathbf{r}) + \frac{1}{2m} \left| -i\hbar\nabla\Psi - \frac{e^*}{c}\mathbf{A}(\mathbf{r})\Psi \right|^2 + \frac{\mathscr{H}^2(\mathbf{r})}{8\pi} \right]. \tag{8.12.1}$$

Here e^* is the charge on one of the "particles" constituting the superfluid; since these are quite evidently Cooper pairs, $e^* = 2e$. The term F_0^s is written

$$F_0^s = \frac{\mathscr{H}_c(T)}{8\pi}\left(1 - 2\left|\frac{\Psi}{\Psi_0}\right|^2 + \left|\frac{\Psi}{\Psi_0}\right|^4\right), \tag{8.12.2}$$

where Ψ_0 denotes the zero field value of Ψ. The right-hand side of equation (8.12.2) may be regarded as the low-order terms in a power series in Ψ and is thus most appropriate near the transition point. The coefficients are chosen to give the expression the properties $F_0^s = \mathscr{H}_c^2(T)/8\pi$ when $\Psi = 0$, and $\partial F_0^s/\partial\Psi = 0 = F_0^s$ when $\Psi = \Psi_0$.

Before proceeding further, it is convenient, following Abrikosov, to use reduced variables. Thus we define the following dimensionless quantities:

$$\frac{\Psi}{\Psi_0} = \psi, \quad \frac{\mathscr{A}}{\sqrt{(2)}\,\mathscr{H}_c(T)} = \mathbf{a}, \quad \frac{\mathscr{H}}{\sqrt{(2)}\,\mathscr{H}_c(T)} = \mathbf{h}, \quad K = \frac{\sqrt{(2)}\,e^*}{\hbar c}\mathscr{H}_c(T)\,\lambda_L^2. \tag{8.12.3}$$

The fundamental length λ_L in equation (8.12.3), given by

$$\lambda_L^2 = \frac{mc^2}{4\pi e^{*2}|\Psi_0|^2}, \tag{8.12.4}$$

is a convenient unit of length to use in rewriting the free energy as

$$\frac{F}{\mathscr{H}_c^2(T)/4\pi} = \int d\mathbf{r}\left[\tfrac{1}{2} - |\psi|^2 + \tfrac{1}{2}|\psi|^4 + \left|\frac{i}{K}\nabla\psi + \mathbf{a}\psi\right|^2 + h^2(\mathbf{r})\right]. \tag{8.12.5}$$

The procedure is now to vary ψ and $\mathbf{a}$ in equation (8.12.5), when the respective Euler equations are found to be

$$\left(\frac{i}{K}\nabla + \mathbf{a}\right)^2\psi = \psi - |\psi^2|\psi \tag{8.12.6}$$

and since $\mathbf{h} = \text{curl}\,\mathbf{a}$,

$$-\text{curl}\,\text{curl}\,\mathbf{a} = |\psi|^2\mathbf{a} + \frac{i}{2K}(\psi^*\nabla\psi - \psi\nabla\psi^*). \tag{8.12.7}$$

Equations (8.12.6) and (8.12.7) are the Ginzburg–Landau equations. We give an alternative derivation from hydrodynamics, due to Fröhlich (1969), in Appendix 8.4. For solutions ψ of these equations, (8.12.5) takes the form

$$\frac{F}{\mathscr{H}_c^2(T)/4\pi} = \int d\mathbf{r}\,[\tfrac{1}{2} - \tfrac{1}{2}|\psi|^4 + h^2(\mathbf{r})]. \tag{8.12.8}$$

The quantities $\mathscr{H}_c(T)$, λ_L and K are the three basic parameters of the theory. $\mathscr{H}_c(T)$, as defined above, is the thermodynamic bulk critical field and is given by

$$\mathscr{H}_c^2(T)/8\pi = -\int_0^\infty \mathscr{M}\,d\mathscr{H}. \tag{8.12.9}$$

For type I superconductors, equation (8.12.9) is really redundant, for $\mathscr{H}_c(T)$ is then the usual critical field, above which $\mathscr{M} = 0$, and below which we have the property $\mathscr{M} = -\mathscr{H}/4\pi$ of a perfect diamagnet. In type II materials, the $\mathscr{M}(\mathscr{H})$ relationship is more complicated and equation (8.12.9) is required in its general form.

8.12.1 Superconductor surface and physical significance of λ_L

To interpret λ_L, we will consider the case of a superconductor occupying the half-space $x>0$ and with a weak applied field $\mathbf{h}_0$ in the region $x<0$. Inside the superconductor, the field and vector potential may be described as shown in Figure 8.14 where $a'(x) = h(x)$. The Ginzburg–Landau equations (8.12.6) and (8.12.7) then reduce to

$$-\frac{1}{K^2}\psi'' + a^2\psi = \psi - |\psi|^2\psi \tag{8.12.10}$$

and if we neglect terms in ψ quadratic in h_0,

$$a'' = a. \tag{8.12.11}$$

Equation (8.12.11) has the elementary solution $a = -h_0\,e^{-x}$, $h = h_0\,e^{-x}$. Thus, $\lambda_L\,(=1)$ is a measure of the field penetration into the superconductor and is, in fact, the usual London penetration depth. This explicit expression for a may now be inserted in equation (8.12.10) to obtain information about ψ. Expanding the latter as a power series in h_0^2 and discarding h_0^4 terms in equation

and the other at $\mathscr{H}_{c2}$ when the penetration is complete. Actually, it has in fact been found (Saint-James and de Gennes, 1963) that because of a very persistent superconducting sheath near the surface, an extremely weak diamagnetic effect can obtain up to some maximum value $\mathscr{H}_{c3}$. We shall not enter into this latter effect here, but will be concerned with exhibiting the

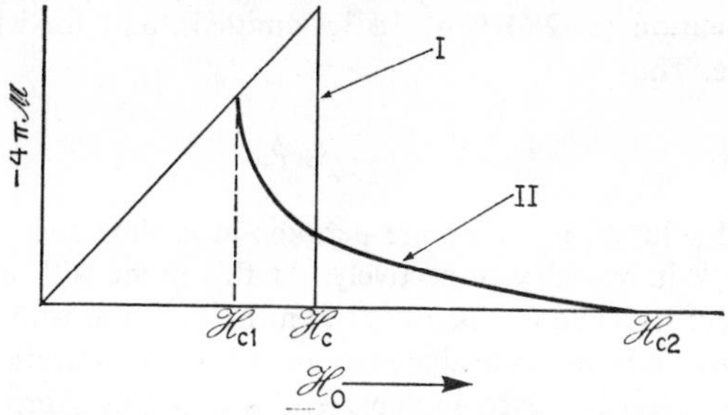

FIGURE 8.16. Observed magnetization versus field curves for typical types I and II superconductors.

essential nature of the region $\mathscr{H}_{c1} < \mathscr{H} < \mathscr{H}_{c2}$. If the two curves in Figure 8.16 enclose the same area, then because of equation (8.12.9), $\mathscr{H}_c$ will be the thermodynamic critical field for the type II case.

It is clear that above $\mathscr{H}_{c1}$, the striations referred to at the end of the previous section have occurred, and we must re-examine the Ginzburg–Landau equations, bearing in mind this new physical effect. Now, for example, the situation depicted in Figure 8.15 is no longer appropriate. It is convenient to examine the two limiting cases when $\mathscr{H}_0$ is first near $\mathscr{H}_{c2}$ and then $\mathscr{H}_{c1}$.

(a) $\mathscr{H}_0 \lesssim \mathscr{H}_{c2}$. The starting point in this regime is to suppose that, as a zeroth approximation, we can write (Figure 8.14) $h = h_0$, where h_0 is a constant. Thus, to within a trivial constant which serves to locate the yz-plane, $a = h_0 x$. We also know that now ψ is a small quantity. Hence, to begin with, we neglect the second Ginzburg–Landau equation, which is homogeneous and quadratic in ψ, and solve the first, omitting the smaller $|\psi|^2\psi$ term.

If we suppose ψ is independent of y, then once more equation (8.12.10) is obtained, where now, however, $a = h_0 x$. Thus, we find a Schrödinger equation appropriate to a simple harmonic oscillator; for bounded solutions, $h_0 = K/(2n+1)$. The largest possible value of h_0 is thus K (when $n = 0$) which implies [from equation (8.12.3) and Figure 8.16] that

$$\mathscr{H}_{c2} = \sqrt{(2)}\, K \mathscr{H}_c. \tag{8.12.13}$$

The corresponding eigenfunction is proportional to $\exp(-\frac{1}{2}K^2 x^2)$.

The above analysis is not yet, however, satisfactory, the ψ we have found being highly localized in the x-direction. This has no physical significance and to resolve this difficulty we note that there is a whole family

$$\psi_n = \exp(ikny)\exp\left[-\tfrac{1}{2}K^2\left(x-\frac{kn}{K^2}\right)^2\right], \tag{8.12.14}$$

satisfying equation (8.12.6) (with $|\psi|^2\psi$ omitted), and for which (8.12.13) is appropriate. Thus

$$\psi = \sum_{n=-\infty}^{\infty} c_n \psi_n \tag{8.12.15}$$

is an allowed solution if the c's are periodic in n, thus giving ψ periods of k/K^2 and $2\pi/k$ in x and y respectively. At this point with $\mathscr{H} = \mathscr{H}_{c2}$, k is quite arbitrary; its value will be fixed when the situation with $\mathscr{H}$ just below $\mathscr{H}_{c2}$ is resolved. (k becomes available as an energy minimization parameter.) This latter problem is solved in considerable detail by Abrikosov, and we shall be content here to summarize, rather briefly, his method.

The next step is to include quadratic terms in ψ by substituting equation (8.12.15) into equation (8.12.3). This leads to revised values of h and a given by

$$h = h_0 - \frac{1}{2K}|\psi|^2 \tag{8.12.16}$$

and

$$a = h_0 x - \frac{1}{2K}\int^x |\psi|^2\, dx. \tag{8.12.17}$$

Then account is taken of third-order terms by returning, with equations (8.12.16) and (8.12.17) to equation (8.12.6) with $|\psi|^2\psi$ now included. This leads to the result (the bar indicating a macroscopic average)

$$\frac{(K-h_0)}{K}\overline{|\psi|^2} + \left(\frac{1}{2K^2}-1\right)\overline{|\psi|^4} = 0. \tag{8.12.18}$$

With these results, we can now compute macroscopic variables of interest. For the magnetic induction, equation (8.12.16) yields

$$b = \bar{h} = h_0 - \frac{(K-h_0)}{(2K^2-1)\beta}, \quad \beta = \frac{\overline{|\psi|^4}}{(\overline{|\psi|^2})^2}, \tag{8.12.19}$$

while the free energy, given by equation (8.12.8), is proportional to

$$f_1 = \tfrac{1}{2} + b^2 - \frac{(K-b)^2}{1+(2K^2-1)\beta}. \tag{8.12.20}$$

To minimize f_1 it is clear that for given b we must make β as small as possible.

So far, the results have been independent of the specific choice of the c's in equation (8.12.15). At this stage, however, because we are now concerned with the minimization of β, as defined in equation (8.12.19), a choice of c_n's (periodic in n) must be made. Strictly speaking, this should be on energy grounds—a complicated problem, but fortunately it seems that the choice when all the c's are equal is best. Under these circumstances

$$\beta = \frac{k}{K\sqrt{(2\pi)}}\left[\sum_n \exp\left(-\frac{k^2}{2K^2}n^2\right)\right]^2 \tag{8.12.21}$$

and the minimum β is 1.18 when $k = K\sqrt{(2\pi)}$. Then, writing $b = h+4\pi m$, for external fields just below $\mathscr{H}_{c2}$, equation (8.12.19) gives

$$-4\pi m = \frac{1}{1.18\,(2K^2-1)}(h_{c2}-h_0). \tag{8.12.22}$$

Both in its linear dependence on $(h_{c2}-h_0)$ (see Figure 8.16) and in its proportionality constant, equation (8.12.22) is found to be in agreement with experiment. The constant $|\psi|^2$ contours in the xy-plane, appropriate to the solution thus calculated, are plotted (suitably normalized) in Figure 8.17. The arrows indicate the current directions. It is clear that we have a configuration of vortices.

(b) *$\mathscr{H}_0$ just greater than $\mathscr{H}_{c1}$.* As $\mathscr{H}_0$ decreases from the situation described above near $\mathscr{H}_{c2}$, the assumption is that the vortices persist, though becoming more and more separated from each other, in qualitative agreement with the magnetization curve shown in Figure 8.16. In the limit $\mathscr{H}_0 = \mathscr{H}_{c1}$, we may consider a completely isolated vortex, at least as a first approximation.

Then, directing $\mathbf{a}$ perpendicular to the radius vector and writing $\psi = fe^{i\theta}$, equations (8.12.6) and (8.12.7) lead to

$$-\frac{1}{K^2 r}\frac{d}{dr}\left(r\frac{df}{dr}\right)+q^2 f = f - f^3, \tag{8.12.23}$$

$$\frac{d}{dr}\left[\frac{1}{r}\frac{d}{dr}(rq)\right] = qf^2, \tag{8.12.24}$$

where $q = |\mathbf{a} - K^{-1}\nabla\theta|$. The field is given by

$$h = -\frac{1}{r}\frac{d}{dr}(rq). \tag{8.12.25}$$

Equations (8.12.23) and (8.12.24) are to be solved subject to the boundary conditions that f (everywhere finite)$\rightarrow 1$ as $r\rightarrow\infty$, while $q\sim|K^{-1}\nabla\theta| = 1/Kr$ as $r\rightarrow 0$, and $\rightarrow 0$ as $r\rightarrow\infty$. This final condition means that at large r, $a\sim 1/Kr$ and thus the magnetic induction per vertex line is

$$b = \int h\,dS = \oint \mathbf{a}.d\mathbf{l} = \frac{1}{Kr}2\pi r = \frac{2\pi}{K}. \tag{8.12.26}$$

The problem is now well-defined, but only when $K\gg 1$ does further analytical progress seem possible. In what follows, therefore, we will consider

FIGURE 8.17. Contours of constant $|\psi|^2$ in the xy-plane, suitably normalized, from Abrikosov solution.

only this case. For $r<1$, we may write q to leading order as $1/Kr$. Inserting this in equation (8.12.23) and solving in series, we find

$$\left.\begin{aligned} f^2 &= 1-\frac{1}{K^2r^2}+\ldots \quad (Kr>1), \\ f &= \text{const.}\,r+\ldots \quad (Kr<1). \end{aligned}\right\} \tag{8.12.27}$$

On returning to equation (8.12.8) and rearranging a little using equations (8.12.25) and (8.12.26), the energy per unit length of filament can be written

$$\varepsilon = \pi \int_0^\infty \left[(1-f^4)\,r - f^2 \frac{d}{dr}(r^2 q^2)\right] dr. \tag{8.12.28}$$

Replacing the range of integration by its most important part, which, in view of the above discussion, is $(1/K, 1)$, integration of equation (8.12.28) gives, in leading order, $(2\pi/K^2)\ln K$. Abrikosov calculated the next order term by numerical methods and found that a more accurate expression is $(2\pi/K^2)$ $(\ln K + 0.081)$. For a transition from the perfectly diamagnetic state to that with vortex structure, $\varepsilon < 2h_{c1}b$. Thus, the transition occurs at the external field value

$$h_{c1} = \frac{1}{2K}(\ln K + 0.081) \quad (K \gg 1), \tag{8.12.29}$$

a result that is in fairly good agreement with experiment. Abrikosov shows how to obtain the interaction energies between vortices and hence the equilibrium filament density. In this way the details of Figure 8.16 immediately above $\mathscr{H}_{c1}$ are found from this theory.

The solution of the Ginzburg–Landau equations we have just discussed, due to Abrikosov (1957), is appropriate to homogeneous type II superconductors in magnetic fields just less than the upper critical field $\mathscr{H}_{c2}$. As we have seen, this solution has the periodicity of a square lattice and was, at first, conjectured to be the solution corresponding to lowest free energy.

Subsequently, Kleiner, Roth and Autler (1964) have shown that there is a solution characterized by magnetic field maxima corresponding to an equilateral triangular lattice and a value of the parameter β equal to 1.16, which has a lower free energy than the square lattice solution with $\beta = 1.18$, discussed above. Experimental tests of this small difference in β are possible, in principle, by measuring the slope of the magnetization curve, which is linear near $\mathscr{H}_{c2}$. Unfortunately, the triangular and square lattices lead to only 2% difference in this slope, and would be hard to distinguish in this way. Furthermore, in real superconductors, periodic arrays of the filaments may well be considerably altered due to crystal imperfections.

We shall content ourselves here with a brief summary of the work of Kleiner and colleagues. The starting point is again Abrikosov's general approximate solution (8.12.15) with the periodicity condition $c_{n+N} = c_n$. The coefficients c_n in equation (8.12.15) as well as the parameter k are to be adjusted to minimize the free energy. Kleiner and co-workers take $N = 2$ as the next simplest case to that treated by Abrikosov. $|\psi|^2$, which as we saw above, is proportional to the density of superconducting electrons, is required to be invariant under a centring translation with respect to the rectangular cell having sides $L_x = 2k/K^2, L_y = 2\pi/k$, and area $4\pi/K^2$ independent of the ratio $L_x/L_y = k^2/\pi K^2$ of the sides. The centring translational symmetry condition imposes on $|\psi|^2$ the translational symmetry of an equilateral triangular lattice when L_x/L_y is suitably adjusted.

This condition, namely that $|\psi|^2$ is unchanged when (x,y) goes into $(x+\frac{1}{2}L_x, y+\frac{1}{2}L_y)$, leads directly to the relation $c_1 = \pm ic_0$ between the coefficients in equation (8.12.15). The geometry of the corresponding solutions $\psi_\pm$ is usefully characterized by the points at which the superconducting electron concentration $|\psi|^2$ is the same. The points $(x+m+\frac{1}{2}pL_x, y+n+\frac{1}{2}pL_y)$, when m, n and p are integers, are translationally equivalent to (x,y). The zeros, which help in clarifying the geometric nature of the solutions, correspond to normal filaments where $\mathscr{H} = \mathscr{H}_0-(1/2K)|\psi|^2$ is a maximum. They are at points translationally equivalent to $(\frac{1}{4}L_x, \frac{1}{4}L_y)$ for ψ_+ and $(\frac{1}{4}L_x, \frac{3}{4}L_y)$ for ψ_-. Additionally, $(\frac{1}{2}L_x, 0)$ and $(0,0)$ are equivalent for $\psi_\pm$; and also $|\psi_+|^2 = |\psi_-|^2$ at $(x,y) = (0,0)$.

The Abrikosov parameter now takes the form

$$\beta = \frac{2}{(2\pi)^{\frac{1}{2}}}\frac{k}{K}\frac{(|c_0|^4+|c_1|^4)f_0^2+4|c_0|^2|c_1|^2 f_0 f_1+2\,\mathrm{Re}\,(c_0^{*2}c_1^2)f_1^2}{(|c_0|^2+|c_1|^2)^2},$$

where the quantities f, related to theta functions, are given by

$$f_n\left(\frac{L_x}{L_y}\right) = \sum_{m=-\infty}^{\infty} \exp\left[-\left(\frac{\pi}{2}\right)\frac{L_x}{L_y}(2m+n)^2\right].$$

The free-energy density decreases monotonically with decreasing values of β for fixed K and $\mathscr{H}_0$ and the minimum of β with respect to c_1/c_0 occurs when the centring translational symmetry is present, being given by

$$\beta\left(\frac{L_x}{L_y}\right) = \left(\frac{L_x}{2L_y}\right)^{\frac{1}{2}}[f_0^2+2f_0f_1-f_1^2] = \beta\left(\frac{L_y}{L_x}\right),$$

which is plotted in Figure 8.18. β is a minimum when $(L_x/L_y) = 3^{\frac{1}{2}}$, and the corresponding contours of $|\psi|^2$ are shown in Figure 8.19. $|\psi|^2$ has hexagonal symmetry and its zeros occur on a triangular lattice with a nearest-neighbour distance $(4\pi/3^{\frac{1}{2}}K^2)^{\frac{1}{2}} = L_y$.

It is of interest to note that the value $\beta = 1.1803$ corresponding to $L_x/L_y = 1$ in Figure 8.18 corresponds to $|\psi|^2$ having square symmetry, with nodes on a square lattice with nearest-neighbour distance $(2\pi/K^2)^{\frac{1}{2}}$. This solution is essentially equivalent to the square lattice solution of Abrikosov.

We shall go on to discuss the quantum-mechanical macroscopic state that a superconductor represents, dealing below with off-diagonal long-range

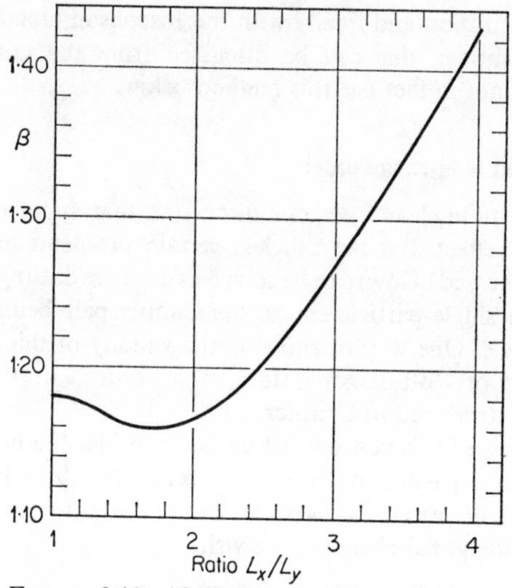

FIGURE 8.18. Abrikosov parameter β versus L_x/L_y.

FIGURE 8.19. Constant $|\psi|^2$ contours for hexagonal symmetry. Zeros lie on triangular lattice. Maximum of $|\psi|^2$ is normalized to unity.

order, flux quantization and finally with the Josephson junction. Though, as Josephson has shown, this can be discussed from the Ginzburg–Landau theory, we shall not in fact use this method below.

8.13 Off-diagonal long-range order

It is by now quite clear from our discussion that superconductivity is a true many-body effect, but nevertheless certain problems (e.g. that of flux quantization discussed below) can be solved as if we are dealing with individual Cooper pairs in single-particle states, the Cooper pair being regarded as a composite particle. One way of showing the validity of this procedure is to use the concept of "off-diagonal long-range order" in the second-order density matrix introduced in Chapter 2.

To fix our ideas of this concept, let us first consider the first-order density matrix $\rho(\mathbf{r}'\mathbf{r})$ of free-particle systems. These certainly exhibit long-range order on the diagonal of $\rho(\mathbf{r}'\mathbf{r})$, for $\rho(\mathbf{rr})$ is simply the constant density ρ_0. To examine the off-diagonal elements, we write

$$\rho(\mathbf{r}'\mathbf{r}) = \sum_{\mathbf{k}} n_{\mathbf{k}} e^{i\mathbf{k}.(\mathbf{r}'-\mathbf{r})} \tag{8.13.1}$$

and since, for Fermions, $0 \leqslant n_{\mathbf{k}} \leqslant 1$, as $\mathbf{r}'-\mathbf{r} \to \infty$, $\rho(\mathbf{r}'\mathbf{r})$ tends to zero; the first-order density matrix for Fermions does not exhibit off-diagonal long-range order. The same conclusion holds true for free bosons at elevated temperatures. However, below the Bose–Einstein condensation temperature, the population of the single-particle ground state $\phi(\mathbf{r})$ is $n_0 = N\alpha$, say, where α is a finite fraction and N is the total number of particles and hence

$$\rho(\mathbf{r}'\mathbf{r}) \to N\alpha\psi^*(\mathbf{r}')\psi(\mathbf{r}), \quad \mathbf{r}-\mathbf{r}' \to \infty. \tag{8.13.2}$$

Now we turn to deal with the second-order density matrix for Fermions. Consider the spin-dependent quantity

$$\rho_2(\mathbf{r}\downarrow\mathbf{r}\uparrow;\mathbf{r}'\uparrow\mathbf{r}'\downarrow) = \int \Psi^*(\mathbf{r}'\uparrow,\mathbf{r}'\downarrow,\mathbf{x}_3\ldots\mathbf{x}_N)\Psi(\mathbf{r}\downarrow,\mathbf{r}\uparrow,\mathbf{x}_3\ldots\mathbf{x}_N)\,d\mathbf{x}_3\ldots d\mathbf{x}_N. \tag{8.13.3}$$

It is readily shown that for a normal system this goes to zero as $\mathbf{r}'-\mathbf{r} \to \infty$. On the other hand, Yang (1962) takes the existence of off-diagonal long-range order in equation (8.13.3) to be characteristic of electrons in the superconducting state. To see how this can arise from Cooper pairing, we expand ρ_2 in two-body orbitals, which we choose to be natural orbitals in the sense of Chapter 2:

$$\rho_2(\mathbf{r}\downarrow\mathbf{r}\uparrow;\mathbf{r}'\uparrow\mathbf{r}'\downarrow) = \sum_p n_p \phi_p^*(\mathbf{r}'\uparrow\mathbf{r}'\downarrow)\phi_p(\mathbf{r}\downarrow\mathbf{r}\uparrow). \tag{8.13.4}$$

Now the Cooper pairs behave like bosons in the sense that n_p is not restricted to lie between 0 and 1, and in fact the occupation numbers n_p for the ground state will be of order N. Thus, as $\mathbf{r}-\mathbf{r}'\to\infty$, ρ_2 takes the product form

$$\rho_2(\mathbf{r}\downarrow\mathbf{r}\uparrow;\mathbf{r}'\uparrow\mathbf{r}'\downarrow)\to\psi(\mathbf{r}')\psi(\mathbf{r}) \tag{8.13.5}$$

and we can define the order parameter $\Delta(\mathbf{r})$ (see section 8.12) such that

$$\Delta(\mathbf{r})\propto\psi(\mathbf{r}). \tag{8.13.6}$$

From the definition (8.13.3) of ρ_2, if the whole system of electrons is given a momentum

$$\Delta(\mathbf{r})\propto e^{i\mathbf{p}_s\cdot\mathbf{r}/\hbar}, \tag{8.13.7}$$

where p_s is the momentum of a pair, one can define then the free energy $F(p_s)$ depending on p_s or on phase, and this will have a minimum at $p_s=0$. We shall now go on to use these considerations in discussing further macroscopic states of superconductors.

8.14 Macroscopic states in superconductors

The order parameter $\Delta(\mathbf{r})$ given in equation (8.13.7) will be written as

$$\Delta(\mathbf{r})\propto e^{i\chi(\mathbf{r})}, \tag{8.14.1}$$

where, from equation (8.13.7),

$$\chi(\mathbf{r})=\frac{\mathbf{p}_s\cdot\mathbf{r}}{\hbar}, \tag{8.14.2}$$

and hence

$$\mathbf{p}_s=\hbar\nabla\chi. \tag{8.14.3}$$

As we have just seen, the free energy F of the superconductor is a function of the momentum of a Cooper pair p_s, and hence may be written

$$F\equiv F(\hbar\nabla\chi), \tag{8.14.4}$$

in the absence of a magnetic field. We now switch on the magnetic field and we have to ensure gauge invariance. If $\mathscr{A}\to\mathscr{A}+\nabla\phi$, then the total wave function Ψ changes in such a way that

$$\Psi(\mathbf{r}_1\mathbf{r}_2\ldots)\to\exp\left[i\sum_i e\phi(\mathbf{r}_i)/c\hbar\right]\Psi. \tag{8.14.5}$$

But from the discussion above of off-diagonal long-range order we see that this implies that the order parameter $\Delta(\mathbf{r})$ is transformed according to

$$\Delta(\mathbf{r})\to\exp\left[\frac{2ie\phi(\mathbf{r})}{c\hbar}\right]\Delta. \tag{8.14.6}$$

Thus, under a gauge transformation, we have

$$\chi \to \chi + \frac{2e\phi}{c\hbar}. \tag{8.14.7}$$

Therefore, physical quantities must depend on

$$\chi - \frac{2e}{c\hbar}\int^{\mathbf{r}} \mathscr{A}(\mathbf{r}').d\mathbf{r}', \tag{8.14.8}$$

or on

$$\nabla\chi - \frac{2e}{c\hbar}\mathscr{A}. \tag{8.14.9}$$

In particular, the free energy F is formally

$$F \equiv F\left(\nabla\chi - \frac{2e}{c\hbar}\mathscr{A}\right). \tag{8.14.10}$$

Now quite generally the current carried by a system in the presence of a magnetic field is given by the functional derivative of F with respect to $\mathscr{A}$:

$$\mathbf{j} = -c\frac{\delta F}{\delta \mathscr{A}} \tag{8.14.11}$$

and from equation (8.14.10) we find

$$\mathbf{j} = \frac{2e}{\hbar}\frac{\partial F}{\partial(\nabla\chi)}. \tag{8.14.12}$$

We want to argue from this as to the value of the current $\mathbf{j}$ in the presence of the magnetic field. To do so, we noted above that in zero field the free energy is a minimum when $p_s = 0$. If n_s is the density of Cooper pairs, it is clear that the free energy will be of the form

$$F = n_s \frac{p_s^2}{2m_{\text{eff}}}, \tag{8.14.13}$$

where m_{eff} is some appropriate mass for the Cooper pair, which we fortunately will not need to specify in getting the main results below. Thus, in terms of χ, we have

$$F = \frac{n_s(\hbar\nabla\chi)^2}{2m_{\text{eff}}}, \tag{8.14.14}$$

and hence from equation (8.14.10) we have in the presence of the magnetic field

$$F = \frac{n_s\{\hbar[\nabla\chi - (2e/c\hbar)\mathscr{A}]\}^2}{2m_{\text{eff}}}. \tag{8.14.15}$$

By taking the functional derivative we find

$$\mathbf{j} = \frac{2en_s\mathbf{P}_s}{m_{\text{eff}}}, \tag{8.14.16}$$

where

$$\mathbf{P}_s = \hbar\nabla\chi - \frac{2e\mathscr{A}}{c}. \tag{8.14.17}$$

8.14.1 Flux quantization

Consider now a cylindrical piece of superconductor, with a hole in the centre.

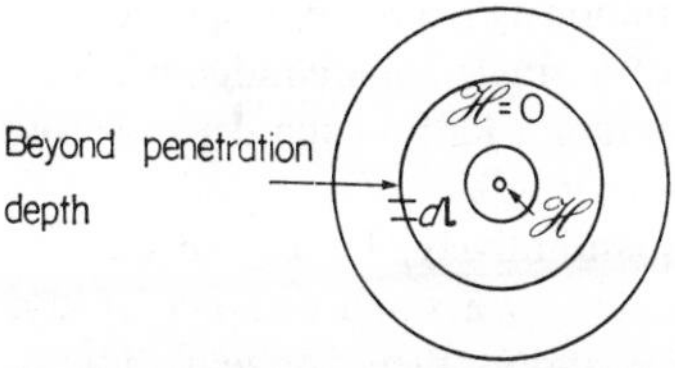

FIGURE 8.20. Hollow piece of superconductor, illustrating configuration used in discussing flux quantization. Field $\mathscr{H}$ exists in hole.

Since χ appears as a phase factor, the change in χ in going round a ring must equal $2\pi\nu$ where ν is an integer. Let us suppose a field $\mathscr{H}$ exists in the hole. There will be penetration of the field to a distance Λ. Within the cylinder beyond the penetration depth, $\mathbf{j} = 0$ and hence from equation (8.14.16) we have $\mathbf{P}_s = 0$. Hence, from equation (8.14.17)

$$\hbar\nabla\chi = \frac{2e\mathscr{A}}{c}. \tag{8.14.18}$$

Thus

$$\oint \mathscr{A} . d\mathbf{l} = \frac{c\hbar}{2e} \quad \times (\text{change in } \chi \text{ in going round loop})$$

$$= \frac{ch}{2e}\nu. \tag{8.14.19}$$

But $\oint \mathscr{A}.d\mathbf{l}$ is simply the flux enclosed, and hence the flux is quantized in units $ch/2e$ ($\equiv 2\times 10^{-7}$ gauss cm^2).

8.14.2 Josephson junction

We want to see next how these same arguments can be used to analyse what can happen at the junction of two such superconductors between which is sandwiched a thin layer of insulating material: a Josephson junction.

If the two superconductors we are considering were completely isolated from each other, then the Ginzburg–Landau equations imply that it would be possible to alter the phases χ in each, independently. This is because, from any chosen solution of these equations, other solutions can be found by changing the phase.

On the other hand, in a single superconductor under given external conditions, all phase differences are fixed. But if we consider a Josephson junction in which the thickness of the insulator is gradually reduced to zero, we can expect to go over continuously from the properties of the two isolated superconductors to those of a single superconductor. Therefore the free energy of the system must contain a term which depends on the relative phases on the two sides of the barrier.

To examine this quantitatively, let us consider two points in a single superconductor, separated by a small distance, d say. If the values of χ at the points 1 and 2 are χ_1 and χ_2, then from equation (8.14.4) we can write the free energy in the form

$$F\left(\hbar\frac{\chi_1-\chi_2}{d}\right). \tag{8.14.20}$$

Hence the current j in the direction 1–2 is given by

$$j = \frac{2ed}{\hbar}\frac{\partial F}{\partial(\chi_1-\chi_2)} = -\frac{\partial N_1}{\partial t}2e, \tag{8.14.21}$$

where N_1 is the number of Cooper pairs in this superconductor.

It might appear from the arguments given above that the theory hinges on the assumption that χ varies slowly in space, whereas one wishes to use it in situations where χ varies rapidly. This point can be covered by showing that χ and N are complementary variables, as set out in Problem 8.3.

Now let us apply these results to the two superconductors separated by the insulating slab. To do so, we assume that we can take for the term in the free energy describing the coupling the form $F(\chi_1-\chi_2)$, where χ_1 and χ_2 are the phases on each side of the insulating slab. But χ is a cyclic variable and a simple form for the coupling term in the free energy might be

$$F(\chi_1-\chi_2) = \text{const}\,[1-\cos(\chi_1-\chi_2)]. \tag{8.14.22}$$

Hence, from equation (8.14.21), the current can be written

$$j = j_m \sin(\chi_1-\chi_2) \tag{8.14.23}$$

and it can be seen that, in this example, there is no interaction between the superconductor, and therefore no current, when $\chi_1 = \chi_2$. In fact, microscopic theory (Josephson, 1962) justifies this result in the limit of weak coupling. For a junction of this kind, it is evident from equation (8.14.23) that tunnelling of a supercurrent occurs up to a maximum value j_m.

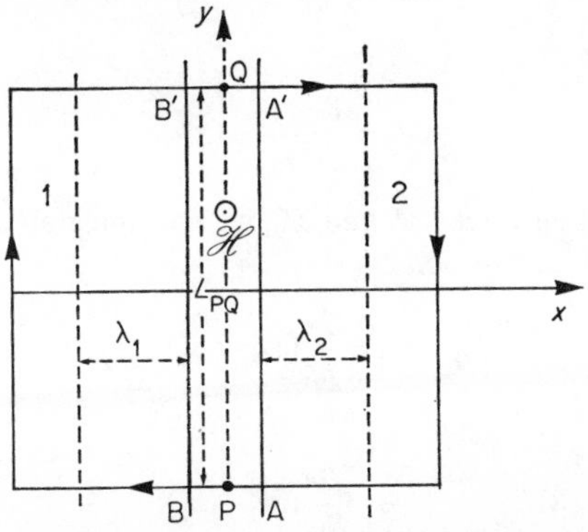

FIGURE 8.21. Contour used in calculating variation of phase over surface insulating slab in Josephson junction.

(a) *Effect of magnetic field on junction.* We must consider the effect of a magnetic field on the junction. Writing as a generalization of equation (8.14.23)

$$j = j_m \sin\phi, \tag{8.14.24}$$

then evidently ϕ is simply $\chi_1-\chi_2$ in zero field. We saw from our earlier argument invoking gauge invariance that physical properties must depend only on $\hbar\nabla\chi-(2e/c)\mathscr{A}$. Thus ϕ has to be replaced by

$$\phi \to \chi_1-\chi_2+\frac{2e}{\hbar c}\int_A^B \mathscr{A}.d\mathbf{l}, \tag{8.14.25}$$

where the line integral is taken between the two superconductors (see Figure 8.21).

Let us choose a gauge such that $\mathscr{A}$ is parallel to the barrier, so that $\mathscr{H} = \operatorname{curl} \mathscr{A}$ becomes

$$\mathscr{H}_z(x) = \frac{\partial}{\partial x}\mathscr{A}_y(x, y). \tag{8.14.26}$$

We remark, first of all, that if we are well away from the junction then $\mathscr{A}_y(x, y)$ has well-defined limiting values. To see this, we note that deep inside superconductor 1 (that is beyond the penetration depth λ_1, shown in Figure 8.21), the Meissner effect is complete, no current flows and so, by equations (8.14.16) and (8.14.17),

$$\frac{2e}{\hbar}\mathscr{A}_y = \frac{\partial \chi_1(y)}{\partial y} \quad (x \ll -\lambda_1). \tag{8.14.27}$$

Similarly we have

$$\frac{2e}{\hbar}\mathscr{A}_y = \frac{\partial \chi_2(y)}{\partial y} \quad (x \gg \lambda_2). \tag{8.14.28}$$

Calling these limiting values $\mathscr{A}_1$ and $\mathscr{A}_2$, we now integrate round the contour $ABB'A'$ of Figure 8.21 to obtain

$$\phi_Q - \phi_P = \frac{2e}{\hbar c}(\mathscr{A}_1 - \mathscr{A}_2) L_{PQ} \tag{8.14.29}$$

or

$$\frac{\partial \phi}{\partial y} = \frac{2e}{\hbar c}(\mathscr{A}_1 - \mathscr{A}_2). \tag{8.14.30}$$

To express this in terms of the magnetic field $\mathscr{H}_0$ in the junction, we note that if we choose $\mathscr{A}_y = 0$ at $x = 0$ then we have

$$\mathscr{A}_1 = \int_0^{-\infty} \mathscr{H}\, dx = -\lambda_1 \mathscr{H}_0, \tag{8.14.31}$$

$$\mathscr{A}_2 = \int_0^{\infty} \mathscr{H}\, dx = \lambda_2 \mathscr{H}_0, \tag{8.14.32}$$

and hence from equation (8.14.30)

$$\frac{\partial \phi}{\partial y} = \frac{2e}{\hbar c}(\lambda_1 + \lambda_2)\mathscr{H}_0. \tag{8.14.33}$$

But, from the Maxwell equation $\operatorname{curl} \mathscr{H} = (4\pi/c)\mathbf{j}$, we have

$$\frac{\partial \mathscr{H}_0}{\partial y} = \frac{4\pi j}{c} = \frac{4\pi}{c} j_m \sin \phi, \tag{8.14.34}$$

where we have used equation (8.14.24). Combining equations (8.14.33) and (8.14.34), we arrive at the result first given by Ferrell and Prange (1963)

$$\frac{\partial^2 \phi}{\partial y^2} = \lambda'^{-2} \sin \phi, \tag{8.14.35}$$

where

$$\lambda'^2 = \frac{\hbar c^2}{8\pi e j_m(\lambda_1+\lambda_2)}. \tag{8.14.36}$$

λ' is typically of the order of 1 mm and has the meaning of a penetration depth, as can be seen by putting $\sin\phi \sim \phi$ in equation (8.14.35).

The detailed behaviour of coupled superconductors depends on the transverse dimensions of the barrier relative to λ'. We shall remark here only on the case of a barrier small compared with λ. Just as with very thin films, the magnetic field inside such a barrier is almost constant. The main interest then is in the field dependence of the critical supercurrent, and we can estimate this as follows.

In the domain of complete penetration, if $|\mathscr{H}| \gg \phi_0/(\lambda_1+\lambda_2)$, where ϕ is the flux quantum, $\mathscr{H}$ is nearly uniform along the junction and from equation (8.14.33) we have

$$\phi_1 \simeq \frac{2\pi y}{L} + \text{const}, \quad L = \frac{\phi_0}{(\lambda_1+\lambda_2)\mathscr{H}}. \tag{8.14.37}$$

We then find

$$i = j_m \sin\phi \simeq j_m \sin\left(2\pi \frac{y-y_0}{L}\right) \tag{8.14.38}$$

and is periodic with period L. If we have a junction of finite length l, then we can write for the total current I

$$I = \int_0^l j_m \sin\phi \, dy \simeq \frac{j_m L}{2\pi} \int d\phi \sin\phi \quad (L \ll \lambda')$$

$$= \frac{j_m L}{2\pi}\left\{\cos\phi(0) - \cos\left[\phi(0) + \frac{2\pi l}{L}\right]\right\}. \tag{8.14.39}$$

The maximum value occurs when $\phi_0 = (\pi/2) - (\pi l/L)$ and then we find

$$I = \left[\frac{I_m \sin(\pi\Phi/\Phi_0)}{(\pi\Phi/\Phi_0)}\right] \sin(\chi_1 - \chi_2), \tag{8.14.40}$$

where Φ_0 is the basic flux quantum and Φ/Φ_0 is the number of flux quanta in the junction. The factor $\sin(\chi_1-\chi_2)$ comes, of course, from the result when $\mathscr{H} = 0$.

The current can be measured as a function of magnetic field, and such oscillatory behaviour is found, confirming the essential features of Josephson tunnelling.

(b) *Junction with voltage across it.* Finally, we want to consider briefly a junction with a voltage across it. Cooper pairs on different sides of the barrier then have energies differing by $\Delta E = 2eV$, V being the potential difference between the two sides of the barrier, and $2e$ appearing as the charge of a Cooper pair.

Tunnelling through the barrier can then take place only as a virtual process, with associated oscillating currents at a frequency $\nu = \Delta E/h$, that is,

$$\nu = \frac{2eV}{h}. \tag{8.14.41}$$

When the interaction between the oscillating currents and the electromagnetic currents is taken into account, one finds that real processes occur, energy being conserved by the emission of a photon (Josephson, 1964). This is illustrated in Figure 8.22. In this case the radiation is coherent, since every photon comes from an identical process.

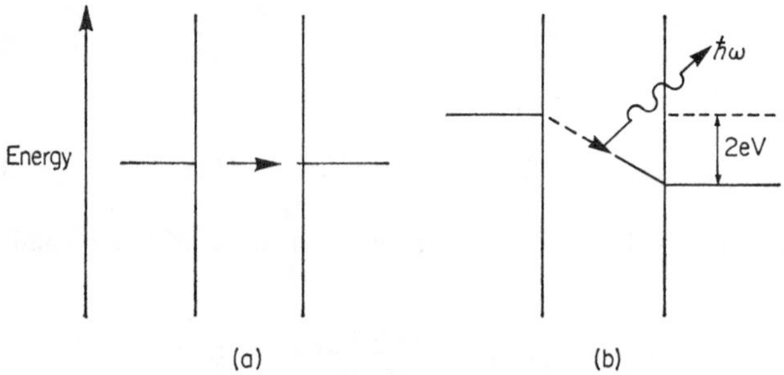

FIGURE 8.22. Illustrates tunnelling of Cooper pairs through an insulating slab (Josephson, 1964). (a) D.C. supercurrents at zero voltage. (b) A.C. supercurrents with applied voltage V, leading to emission of photons. Corresponding to each emitted photon (energy $\hbar\omega$), a Cooper pair tunnels through barrier. Energy conservation then yields $\hbar\omega = 2\,\text{eV}$ since charge of a Cooper pair is $2e$.

Taking into account the fact that the oscillating currents arise from changes in ϕ, we find

$$\dot{\phi} = 2\pi\nu = 2eV/\hbar. \tag{8.14.42}$$

If V is time-independent, then we can write

$$\chi_1-\chi_2=(\chi_1-\chi_2)_0+\frac{2eV}{\hbar}t \tag{8.14.43}$$

and hence, from equation (8.14.24),

$$j=j_m\sin\left[(\chi_1-\chi_2)_0+\frac{2eVt}{\hbar}\right]. \tag{8.14.44}$$

The current oscillates in time and we see that a d.c. voltage induces alternating current.

The effect of r.f. power on the tunnelling supercurrent of Cooper pairs is also of considerable interest. In his original paper, Josephson discussed this effect and predicted the occurrence of regions of zero slope separated by $\hbar\omega/2e$ in the current–voltage characteristics in the presence of the r.f. field. To see how this occurs, let us write instead of $2eV$,

$$2eV+2e\mathscr{V}\sin\omega t, \tag{8.14.45}$$

where ω is the angular frequency of the r.f. power. Then, when $n\omega = 2eV/\hbar$, $n = 1, 2$ etc., we will get a d.c. current as can be seen by averaging over a period $2\pi/\omega$. Again this has been confirmed experimentally (see, for example, Shapiro, 1963).

CHAPTER 9

Polarons and excitons

9.1 Introduction

Conduction in metals and semi-conductors has been discussed at some length in Chapter 6. Here, the mean free path is long relative to the inter-atomic spacing, and the energy bands are well defined in the sense that electron correlation energies do not become comparable with band kinetic energies.

On the other hand, we know from the brief discussion of the metal–insulator transition in Chapter 4 that, when a system with one electron per atom is formally subjected to a lattice expansion, a point is reached where the conductivity changes character. In the band language, the conduction band splits in the middle, due to short-range interactions which forbid two electrons to be found on the same atom. The system becomes insulating, because of the electron interactions.

It turns out that localization can occur by a number of processes other than the one discussed above, namely a reduction in overlap of the electronic wave functions by increasing the lattice spacing. If the electron–lattice interaction becomes very strong, or the system is disordered, or electron wave functions are localized by the application of strong magnetic fields (this latter situation is only of importance at present in favourable materials like n-type InSb) then a new type of conduction with very low mobility, often no more than a few cm^2/volt sec, in contrast to band mobilities which can be of the order of 10^5 or 10^6 cm^2/volt sec, can occur.

This appears to be important in a wide class of materials, including such classic exceptions from the Bloch–Wilson band classification as nickel oxide, or more generally 3d transition metal oxides, amorphous materials and glasses, and some classes of molecular and organic semi-conductors.

We shall focus attention here on the problems centring round the polaron. We discuss how polaron formation can occur and the way in which the adiabatic approximation must be transcended. Small polaron theory is briefly considered, with particular reference to the mobility problem.

The chapter concludes with a discussion of excitons, including some brief remarks about excitonic phases.

9.2 Coupling of electron to ionic lattice

The way an electron moves through an ionic crystal has been of considerable interest over a period of more than 35 years. This is due both to fundamental interest in the many-body problem which it turns out to constitute, and also to its importance in our understanding of the physics of semiconductors and general low-mobility solids referred to above.

As in many aspects of the theory of solids and fluids, early work of outstanding significance was that of Landau (1933) who introduced the notion of the self-trapped electron. The idea is simple enough; namely that an electron, by its Coulomb interaction with the ions in a crystal having appreciable ionicity, produces a polarization. In the potential field resulting from this polarization, the electron could, it was suggested, become bound. Then, as the electron moves in the potential, the ions, being much more massive than the electron, do not take up their equilibrium positions again in a time of the order of the period of the electronic motion.

The next important step came when Fröhlich (1937) gave a quantitative treatment of the way electrons are scattered in ionic crystals. In this work, the interaction of the electrons with the vibrating lattice was treated via the use of the polarization field.

While it is useful, in many cases, to treat the polarization field as a continuum, this becomes a rather poor approximation in the case of strong coupling and this requires a basically different method of treatment.

In developing the theory of the polaron, we shall be guided by the experimental observations which such a theory should explain.

9.2.1 Study of carrier mobility

The properties of an electron (or a hole) in an ionic solid can be investigated experimentally in a diversity of ways. Optical constant measurements, spin resonance of bound electrons or holes, cyclotron resonance of free carriers and some types of infrared studies can all contribute valuable information. Nevertheless, it is still true that the bulk of our knowledge comes from the magnitude and temperature-dependence of carrier mobility, referred to above.

The drift mobility μ is related to a mean free time τ, defined over a large number of collisions by

$$\mu = \frac{e}{m}\tau, \tag{9.2.1}$$

where we are dealing with carriers of charge e.

The orders of magnitude we are talking about are mobilities usually less than 100 $cm^2 volt^{-1} sec^{-1}$ at room temperature (in, for example, alkali or silver halides), and this is due to electrons scattered primarily by longitudinal optical phonons (cf. Chapter 3), through the polarization field associated with this mode, which we discussed above. If we assume a free electron mass in equation (9.2.1), then it immediately follows that the mean free time corresponding to this mobility is 5.7×10^{-14} sec.

9.2.2. Coupling constant of optical mode—electron interaction

As we shall see below, a measure of the interaction between electrons and the optical mode vibrations is given by the coupling constant

$$\alpha = \frac{e}{h}\left(\frac{m}{2h}\right)^{\frac{1}{2}}\left(\frac{1}{\varepsilon_\infty}-\frac{1}{\varepsilon_s}\right)\left(\frac{m^*}{m\omega_l}\right)^{\frac{1}{2}}$$

$$= 1.4\times 10^8\left(\frac{\varepsilon_s-\varepsilon_\infty}{\varepsilon_\infty\,\varepsilon_s}\right)\left(\frac{m^*/m}{\omega_l}\right)^{\frac{1}{2}}, \qquad (9.2.2)$$

where ε_s and ε_∞ represent respectively the static and optical dielectric constants, ω_l is the longitudinal mode frequency $(\varepsilon_s/\varepsilon_\infty)^{\frac{1}{2}}\,\omega_0$ and m^* is the band mass of the electron, which is conveniently written in equation (9.2.2) in units of the free electron mass m. In order to fix our ideas on the orders of magnitudes involved, we show some typical values of α, as well as dielectric constants and frequencies in Table 9.1. The assumption made, unless m^*/m is known, is that $m^*/m = 1$.

Of the cases listed in Table 9.1 (p. 961), it is useful to make a very rough separation into highly polar substances such as the alkali halides, $\alpha > 3$, intermediate coupling cases where $\alpha \sim 1.5$, such as the silver halides, and weakly polar crystals such as the compound semi-conductors. In a material like PbS, α is small because the optical dielectric constant is almost as large as the static constant. There are certain materials such as NiO in which one expects to find highly localized polarons. These, as we remarked above, require different treatment (see section 9.7).

9.3 Polaron theory

We have discussed how an electron polarizes the medium through which it passes in discussing superconductivity. We now want to use a related argument, due to Landau (1933), to discuss how an electron moving in an ionic crystal can be trapped by the polarization cloud which forms round it (see Austin and Mott, 1970).

Denoting the static dielectric constant by ε_s, it is clear that at a sufficiently large distance from the electron, the potential energy in which another electron in the crystal finds itself is $e^2/\varepsilon_s r$. However, if the ions could not move, it would be $e^2/\varepsilon_\infty r$, where ε_∞ is the high-frequency dielectric constant. Thus the potential energy $V_p(r)$ due to the displacement of ions is given by

$$V_p(r) = -\frac{e^2}{r}\left(\frac{1}{\varepsilon_\infty}-\frac{1}{\varepsilon_s}\right). \tag{9.3.1}$$

This is the potential which leads to the "self-trapping" of the electron. If we assume that the potential $V_p(r)$ is valid into a radius r_p and inside that the potential is a constant determined by continuity of V_p at r_p, then the problem is simply to determine r_p, the radius of the well.

9.3.1. Small polaron

There are two limiting approximations. In the first, the effective mass m^* of the electron is assumed to be so high that the kinetic energy due to its localization in the well is negligible. Then r_p must be the order of the inter-ionic distance, generally a little less. As shown in Appendix 9.1, it can be calculated from the phonon spectrum, or roughly estimated if a common frequency is assigned to the optical phonons.

We thus arrive at the concept of the "small polaron", the energy of which can be written as a sum of

(i) The energy required to polarize the medium

$$(\tfrac{1}{2}e^2/r_p)\left(\frac{1}{\varepsilon_\infty}-\frac{1}{\varepsilon_s}\right). \tag{9.3.2}$$

(ii) The lowering of the potential energy of the electron,

$$-\frac{e^2}{r_p}\left(\frac{1}{\varepsilon_\infty}-\frac{1}{\varepsilon_s}\right). \tag{9.3.3}$$

This gives us a rough continuum theory of the polaron energy $-W$ as

$$-W = -\frac{1}{2}\frac{e^2}{r_p}\left(\frac{1}{\varepsilon_\infty}-\frac{1}{\varepsilon_s}\right). \tag{9.3.4}$$

We can develop a more exact method of treating the interaction of the electron with the whole spectrum of optical phonons. If we then make an approximation for the phonon spectrum, we obtain equation (9.3.4), with r_p given by

$$r_p = \tfrac{1}{2}(\pi/6)^{\frac{1}{3}}\, l, \tag{9.3.5}$$

where l^{-3} is the number of centres per unit volume.

9.3.2. Large polaron

This is the other limit when the effective mass m^* is not large. The kinetic energy $\hbar^2\pi^2/2m^*r_p^2$ of an electron confined in a sphere of radius r_p is now an essential feature, and we can write

$$-W=-\frac{1}{2}\frac{e^2}{r_p}\left(\frac{1}{\varepsilon_\infty}-\frac{1}{\varepsilon_s}\right)+\frac{\hbar^2\pi^2}{2m^*r_p^2}. \tag{9.3.6}$$

This is minimized when

$$r_p=2\pi^2\hbar^2\left(\frac{1}{\varepsilon_\infty}-\frac{1}{\varepsilon_s}\right)^{-1}\Big/m^*e^2. \tag{9.3.7}$$

Hence the magnitude of polaron energy is given by

$$W=(\tfrac{1}{4}e^2/r_p)\left(\frac{1}{\varepsilon_\infty}-\frac{1}{\varepsilon_s}\right). \tag{9.3.8}$$

Having sketched in some of the elementary physics, we report first below a more detailed quantitative calculation of the large polaron binding energy.

9.4 Polaron Hamiltonian in continuum model

The Hamiltonian used to describe the electron interacting with the lattice has already been derived in section 8.7. We need only note that, in the present case, we are primarily interested in a single electron with Hamiltonian (neglecting the periodic potential)

$$H_{el}=\frac{p^2}{2m}+c\Delta(\mathbf{r}), \tag{9.4.1}$$

where $\Delta(\mathbf{r})$ is the deformation potential. The free-electron gas model by which we introduced the deformation potential is, of course, inappropriate here in ionic crystals. We note rather that a uniform dilatation Δ will modify the one-electron energies, if isotropy can be assumed, to yield

$$E(\mathbf{k})\simeq E_0(\mathbf{k})+c\Delta, \tag{9.4.2}$$

where $E_0(\mathbf{k})$ is the eigenvalue in the unstrained crystal. For a long wavelength disturbance, equation (9.4.2) is generalized to equation (9.4.1). The justification of the use of a deformation potential is discussed more fully in Appendix 8.1. From equations (8.7.8) and (8.7.15), we write the electron Hamiltonian, in terms of the phonon operators a and $a^\dagger$, as

$$H_{el}(\mathbf{r})=\frac{p^2}{2m}+i\sum_{\mathbf{q}}c_q(a_{\mathbf{q}}e^{i\mathbf{q}\cdot\mathbf{r}}+a_{\mathbf{q}}^\dagger e^{-i\mathbf{q}\cdot\mathbf{r}}). \tag{9.4.3}$$

However, we have not, so far, discussed the physical significance of the constant c in equation (9.4.1) and this we must now do. We shall see below that it is directly related to the coupling constant α we have already introduced in section 9.2.2.

To see how this connection arises, we notice, following Fröhlich (1954) that we can write down the interaction between an excess electron and a dielectric continuum as

$$H_{\text{int}} = -\int \mathscr{D}(\mathbf{r}-\mathbf{r}_{\text{el}})\mathscr{P}(\mathbf{r})\,d\mathbf{r} = e\int \nabla_{\mathbf{r}}(|\mathbf{r}-\mathbf{r}_{\text{el}}|^{-1})\mathscr{P}(\mathbf{r})\,d\mathbf{r}, \tag{9.4.4}$$

corresponding to the classical interaction between an electron with electric displacement $\mathscr{D}(\mathbf{r}-\mathbf{r}_{\text{el}}) = -e\nabla_{\mathbf{r}}(|\mathbf{r}-\mathbf{r}_{\text{el}}|^{-1})$ and the induced longitudinal polarization field $\mathscr{P}(\mathbf{r})$. The interaction vanishes for transverse fields. The total longitudinal polarization of the actual ionic lattice, caused by a slow conduction electron, can be divided into two parts:

$$\mathscr{P}_{\text{total}}(\mathbf{r}) = \mathscr{P}_{\text{ir}}(\mathbf{r}) + \mathscr{P}_{\text{opt}}(\mathbf{r}), \tag{9.4.5}$$

where the first part, describing the displacement polarization of the ionic lattice, is the infrared component, while the second part, taking into account the polarization of each ion, is the optical component. The interaction of the second term $\mathscr{P}_{\text{opt}}(\mathbf{r})$ with the excess electron is not of interest to us, since the core electrons follow adiabatically the motion of the slow electron. Thus the optical polarization is always excited, independent of the electron velocity. This interaction therefore contributes a term periodic in the lattice to the Hamiltonian, and can be accounted for by choosing the mass appropriately, within the effective mass approximation. The term of interest then for our problem is the displacement polarization $\mathscr{P}_{\text{ir}}(\mathbf{r}) = \mathscr{P}(\mathbf{r})$. In order to relate $\mathscr{P}$ and $\mathscr{D}$ in terms of an effective dielectric constant, it is useful (cf. Appel, 1968) to consider:

(i) A test charge at rest, when we can write

$$\mathscr{D} = \varepsilon_{s}\mathscr{E} = \mathscr{E} + 4\pi\mathscr{P}_{\text{total}}. \tag{9.4.6}$$

(ii) A test charge oscillating in a rigid lattice (no displacement) with a frequency higher than the optical frequencies, when

$$\mathscr{D} = \varepsilon_{\infty}\mathscr{E} = \mathscr{E} + 4\pi\mathscr{P}_{\text{opt}}. \tag{9.4.7}$$

Taking the difference between $\mathscr{P}_{\text{tot}}$ and $\mathscr{P}_{\text{opt}}$ from these two equations, we find

$$\mathscr{P}(\mathbf{r}) = (4\pi)^{-1}(\varepsilon_{\infty}^{-1} - \varepsilon_{s}^{-1})\mathscr{D}, \tag{9.4.8}$$

in terms of the static and high-frequency dielectric constants ε_s and ε_{∞}.

Then the interaction term (9.4.4) above can readily be transformed, to relate to the form (9.4.1). In doing so, one must use an equation like (8.7.8) relating Δ to a displacement polarization. In this way, we can find explicitly the value of c in equation (9.4.1) in terms of α. We shall make use of the explicit result for the coupling constant immediately below.

9.5 Ground-state energy and effective mass

From the previous section, the basic Hamiltonian for the polaron problem may be written in the form (see, for example, Pines, 1963)

$$H = \sum_{\mathbf{k}} \hbar\omega a_{\mathbf{k}}^{\dagger} a_{\mathbf{k}} + \frac{p^2}{2m_e} + \sum_{k} (V_{\mathbf{k}} a_{\mathbf{k}} e^{i\mathbf{k}.\mathbf{r}} + V_{\mathbf{k}}^{\dagger} a_{\mathbf{k}}^{\dagger} e^{-i\mathbf{k}.\mathbf{r}}), \tag{9.5.1}$$

where

$$V_{\mathbf{k}} = \frac{i\hbar\omega}{k}\left(\frac{\hbar}{2m_e\omega}\right)^{\frac{1}{4}}\left(\frac{4\pi\alpha}{V}\right)^{\frac{1}{2}}. \tag{9.5.2}$$

The first term describes the phonon field, the second obviously describes the conduction electron, and the third term represents the interaction between the electron and the phonon field of frequency ω that we have been discussing. The dimensionless coupling constant is defined by equation (9.2.2).

We can say at the outset that the simple qualitative consequences of the electron–phonon interaction are threefold. First, it causes a shift in the relative position of the conduction and the valence bands. Secondly, it changes the electron mass m_e and finally it obviously gives rise to scattering of the electron by the phonons and hence determines the mobility of the electron.

9.5.1 Perturbative description of interaction

We consider first the lowest-order perturbation approximation for the effect of the interaction. We shall first consider the system at $T = 0$, in which initially there are no phonons present. In the absence of the interaction, the system can be represented by the wave function

$$\psi_0 = e^{i\mathbf{p}.\mathbf{r}}|0\rangle, \tag{9.5.3}$$

where $|0\rangle$ represents the state with no phonons present. Switching on the interaction now causes transitions between this state and these states of the system in which a single phonon of momentum $\hbar\mathbf{k}$ is present and the electron has momentum $\mathbf{p}-\hbar\mathbf{k}$. The state vector of this system is clearly given by

$$\phi_{\mathbf{p}-\hbar\mathbf{k}} a_{\mathbf{k}}^{\dagger}|0\rangle, \tag{9.5.4}$$

where $\phi_{\mathbf{p}} = e^{i\mathbf{p}.\mathbf{r}}$. The matrix element for the transition is simply $V_k^\dagger$. Thus, to first order, the wave function of the system is

$$\psi = \phi_{\mathbf{p}}|0\rangle - \sum_{\mathbf{k}} V_{\mathbf{k}} \phi_{\mathbf{p}-\hbar\mathbf{k}} a_{\mathbf{k}}^\dagger |0\rangle \bigg/ \left(\hbar\omega + \frac{\hbar^2 k^2}{2m_e} - \frac{\hbar\mathbf{k}.\mathbf{p}}{m_e}\right). \tag{9.5.5}$$

A resonance which corresponds to the possibility of the electron making a transition to a state of momentum $\mathbf{p}$ with emission of a real phonon of momentum $\hbar\mathbf{k}$ (i.e. with conservation of energy and momentum) will occur if $\mathbf{k}.\mathbf{p} = \omega + (hk^2/2m_e)$.

We shall avoid this difficulty by considering only slow electrons for which $p \leqslant (m_e \omega/k) + (\hbar k/2)$, and thus only virtual transitions will be considered.

The lowest-order energy change due to the interaction is

$$\Delta E = -\sum \frac{|\langle n|H_{\text{int}}|0\rangle|^2}{E_n - E_0}$$

$$= -\sum_{\mathbf{k}}^{n} \frac{|V_{\mathbf{k}}|^2}{\hbar\omega - (\hbar^2 k^2/2m_e) - (\hbar\mathbf{k}.\mathbf{p}/m_e)}. \tag{9.5.6}$$

For low electron momentum $\mathbf{p}$, the energy change may be expanded as a power series in $\mathbf{p}$. The result is readily shown to be

$$\Delta E = \sum_k \frac{V_k^2}{\hbar\omega + (\hbar^2 k^2/2m_e)} + O(p^2). \tag{9.5.7}$$

We now substitute for V_k, converting from summations over $\mathbf{k}$ to integrals, when we find

$$\Delta E = -\alpha\hbar\omega - \frac{\alpha}{6}.\frac{p^2}{2m_e}. \tag{9.5.8}$$

The total energy of the final state is therefore

$$\frac{p^2}{2m_e} + \Delta E = -\alpha\hbar\omega + \frac{p^2}{2m_e}\left(1 - \frac{\alpha}{6}\right) + \text{higher order terms} \tag{9.5.9}$$

Thus, the constant shift in the energy referred to above is $-\alpha\hbar\omega$ and the electron mass is changed to

$$m^* = \frac{m_e}{1 - (\alpha/6)}. \tag{9.5.10}$$

For small α, which means small electron–phonon coupling, the changes in the ground-state energy and in the effective mass are both small. On the other hand, as α increases to ~ 6, perturbation theory predicts an enormous change in the effective mass. This is not physically significant, but a consequence of the breakdown of perturbation theory.

9.5.2. Higher approximations

To improve the above perturbation treatment directly proves somewhat troublesome. We shall therefore describe an alternative method proposed independently by a number of authors (Lee and Pines, 1952; Tyablikov, 1952, 1954; Gurari, 1953). The idea is to make an approximation which easily allows for the possibility of any number of virtual phonons in the cloud present around an electron.

Consider the probability amplitude of finding n distinguishable phonons of momenta $\mathbf{k}_1, \mathbf{k}_2, \ldots, \mathbf{k}_n$ in the cloud around the electron. This may be written as $\langle \mathbf{k}_1, \mathbf{k}_2, \ldots, \mathbf{k}_n | \psi \rangle$ where ψ is the total wave function of the system. In the intermediate-coupling theory, this amplitude is assumed to take the separable form

$$\langle \mathbf{k}_1, \mathbf{k}_2, \ldots, \mathbf{k}_n | \psi \rangle = c_n f(\mathbf{k}_1) f(\mathbf{k}_2) \ldots f(\mathbf{k}_n), \tag{9.5.11}$$

where c_n and $f(\mathbf{k})$ are to be found by minimizing the total energy of the system. Thus, in essence, the procedure is a variational one, in which allowance is made for an arbitrary number of phonons, but in which one assumes that all phonons are emitted into states with the same form of wave function $f(\mathbf{k})$. This reminds one of a Hartree-like calculation, in that any correlations between the emission of successive phonons are neglected.

Calculations based on equation (9.5.11) are relatively easy, but the same answers can be obtained by using a canonical transformation (Lee, Low and Pines, 1953) and we shall use this latter technique here.

9.5.3 Method of canonical transformation

We exploit first the fact that the total momentum of the system

$$\mathbf{P}_{\mathrm{op}}^{\mathrm{tot}} = \mathbf{p} + \sum_{\mathbf{k}} a_{\mathbf{k}}^{\dagger} a_{\mathbf{k}} \hbar \mathbf{k} \tag{9.5.12}$$

is a constant of the motion. This means that by suitable choice of representation, the electron coordinates can be eliminated from the Hamiltonian. We therefore consider a canonical transformation such that we write the wave function Φ of the system as

$$\Phi = U_1 \Psi, \tag{9.5.13}$$

where

$$U_1 = \exp\left(-i \sum_{\mathbf{k}} a_{\mathbf{k}}^{\dagger} a_{\mathbf{k}} \mathbf{k} . \mathbf{r}\right). \tag{9.5.14}$$

Φ satisfies the wave equation

$$H\Phi = E\Phi, \tag{9.5.15}$$

while Ψ obeys

$$H_c \Psi = E\Psi,$$

with

$$H_c = U_1^{-1} H U_1. \tag{9.5.16}$$

In order to calculate H_c, we need the transformation of the operators $\mathbf{p}, a_{\mathbf{k}}, a_{\mathbf{k}}^{\dagger}$ and $e^{i\mathbf{k}.\mathbf{r}}$. It is readily shown that

$$\mathbf{p} \rightarrow \mathbf{p}_{\text{new}} \equiv U_1^{-1} \mathbf{p} U_1 = \mathbf{p} + \sum_{\mathbf{k}} a_{\mathbf{k}}^{\dagger} a_{\mathbf{k}} \hbar \mathbf{k} \tag{9.5.17}$$

and

$$a_{\mathbf{k}} \rightarrow (a_{\mathbf{k}})_{\text{new}} = a_{\mathbf{k}} e^{i\mathbf{k}.\mathbf{r}}; \quad a_{\mathbf{k}}^{\dagger} \rightarrow (a_{\mathbf{k}})_{\text{new}}^{\dagger} = a_{\mathbf{k}}^{\dagger} e^{-i\mathbf{k}.\mathbf{r}}, \tag{9.5.18}$$

while $e^{i\mathbf{k}.\mathbf{r}}$ remains unchanged by the transformation. Thus we have

$$H_c = \left[\left(\mathbf{p}_{\text{new}} - \sum_{\mathbf{k}} \hbar \mathbf{k} a_{\mathbf{k}}^{\dagger} a_{\mathbf{k}} \right)^2 \Big/ 2m_e \right] + \sum_{\mathbf{k}} (\hbar\omega a_{\mathbf{k}}^{\dagger} a_{\mathbf{k}} + V_{\mathbf{k}} a_{\mathbf{k}} + V_{\mathbf{k}}^{\dagger} a_{\mathbf{k}}^{\dagger}). \tag{9.5.19}$$

In H_c, as we desired, the electron coordinates have disappeared. Thus the momentum $\mathbf{p}_{\text{new}}$ is a constant of the motion; it is the total momentum $\mathbf{P}$.

Hence we can write the basic Hamiltonian in the form

$$H = \left[\left(\mathbf{P} - \sum_{\mathbf{k}} \hbar \mathbf{k} a_{\mathbf{k}}^{\dagger} a_{\mathbf{k}} \right)^2 \Big/ 2m_e \right] + \sum_{\mathbf{k}} (V_{\mathbf{k}} a_{\mathbf{k}} + V_{\mathbf{k}}^{\dagger} a_{\mathbf{k}}^{\dagger}) + \sum_{\mathbf{k}} \hbar\omega a_{\mathbf{k}}^{\dagger} a_{\mathbf{k}}. \tag{9.5.20}$$

We now consider the trial wave function

$$\Psi = U_2 |0\rangle, \tag{9.5.21}$$

where $|0\rangle$ represents the state with no phonons present, and

$$U_2 = \exp \sum_{\mathbf{k}} [a_{\mathbf{k}}^{\dagger} f(\mathbf{k}) - a_{\mathbf{k}} f^{\dagger}(\mathbf{k})], \tag{9.5.22}$$

with $f(\mathbf{k})$ to be determined. U_2 can be viewed as a unitary transformation which acts simply to displace the phonon coordinates $a_{\mathbf{k}}$ and $a_{\mathbf{k}}^{\dagger}$. Thus one finds

$$a_{\mathbf{k}} \rightarrow U_2^{-1} a_{\mathbf{k}} U_2 = a_{\mathbf{k}} + f(\mathbf{k}), \tag{9.5.23}$$

where we have made use of the result

$$e^{-S} a_{\mathbf{k}} e^{S} = a_{\mathbf{k}} + [a_{\mathbf{k}}, S] + \frac{1}{2!}[[a_{\mathbf{k}}, S], S] \ldots, \tag{9.5.24}$$

here $[a_{\mathbf{k}}, S] = f(\mathbf{k})$ is a c-number so that all terms after the second vanish.

To see the relation between this trial wave function and that considered earlier, let us obtain the probability amplitude $\langle \mathbf{k}_1, \ldots, \mathbf{k}_n | \Psi \rangle$ for finding n phonons, of momenta $\mathbf{k}_1, \ldots, \mathbf{k}_n$, in the phonon field. We can write

$$|\mathbf{k}_1, \ldots, \mathbf{k}_n\rangle = \frac{1}{\sqrt{n!}} a_{\mathbf{k}_1}^{\dagger} \cdots a_{\mathbf{k}_n}^{\dagger} |0\rangle, \tag{9.5.25}$$

so that

$$\langle \mathbf{k}_1, \ldots, \mathbf{k}_n | 0 \rangle = \frac{1}{\sqrt{n!}} \langle 0 | a_{\mathbf{k}_1} \ldots a_{\mathbf{k}_n} U_2 | 0 \rangle. \tag{9.5.26}$$

Now

$$a_{\mathbf{k}} U_2 = U_2 a_{\mathbf{k}} + U_2 f(\mathbf{k}), \tag{9.5.27}$$

and since annihilation operators $a_{\mathbf{k}_1} \ldots a_{\mathbf{k}_n}$ all give zero when operating on the vacuum state $|0\rangle$, we find

$$\langle \mathbf{k}_1, \ldots, \mathbf{k}_n | 0 \rangle = \frac{f(\mathbf{k}_1) \ldots f(\mathbf{k}_n)}{\sqrt{n!}} \langle 0 | U_2 | 0 \rangle. \tag{9.5.28}$$

To obtain $\langle 0 | U_2 | 0 \rangle$, we use the fact that if the commutator $[A, B]$ of two operators A and B is a c-number, then

$$e^{A+B} = e^A e^B e^{-[A,B]/2}. \tag{9.5.29}$$

Hence we can write

$$U_2 = \exp \left[\sum_{\mathbf{k}} a_{\mathbf{k}}^\dagger f(\mathbf{k}) - \sum_{\mathbf{k}} a_{\mathbf{k}} f^\dagger(\mathbf{k}) - \sum_{\mathbf{k}} \frac{|f(\mathbf{k})|^2}{2} \right] \tag{9.5.30}$$

and finally

$$\langle 0 | U_2, 0 \rangle = \exp \left[-\frac{1}{2} \sum_k |f(\mathbf{k})|^2 \right] \langle 0 | 0 \rangle. \tag{9.5.31}$$

Hence

$$\langle \mathbf{k}_1, \ldots, \mathbf{k}_n | 0 \rangle = f(\mathbf{k}_1) \ldots f(\mathbf{k}_n) \left\{ \exp \left[-\frac{1}{2} \sum_{\mathbf{k}} |f(\mathbf{k})|^2 \right] \Big/ \sqrt{n!} \right\}. \tag{9.5.32}$$

We can carry out the variational calculation of the ground-state energy by regarding U_2 as generating a canonical transformation. Thus we can write

$$\Psi = U_2 \chi \tag{9.5.33}$$

and

$$H_{\text{new}} \chi = E \chi, \tag{9.5.34}$$

where

$$H_{\text{new}} = U_2^{-1} H U_2 = H_0 + H_1. \tag{9.5.35}$$

As a variational calculation, we then take $\chi = |0\rangle$ and evaluate the expectation value $E = \langle 0 | H_0 + H_1 | 0 \rangle$.

It is now a straightforward, if lengthy, matter to show that

$$H_0 = \left[\left(\mathbf{P}-\sum_{\mathbf{k}} a_{\mathbf{k}}^{\dagger} a_{\mathbf{k}} \hbar\mathbf{k}\right)^2 \Big/ 2m_e\right] + \sum_{\mathbf{k}} [V_{\mathbf{k}} f(\mathbf{k}) + c.c]$$

$$+\frac{\hbar^2}{2m_e}\left[\sum_{\mathbf{k}} |f(\mathbf{k})|^2 \mathbf{k}\right]^2 + \sum_{\mathbf{k}} |f(\mathbf{k})|^2 \left(\hbar\omega - \frac{\hbar}{m_e}\mathbf{k}.\mathbf{P} + \frac{\hbar^2 k^2}{2m_e}\right)$$

$$+\sum_{\mathbf{k}} a_{\mathbf{k}}^{\dagger} a_{\mathbf{k}} \mathbf{k}.\frac{\hbar^2}{m_e}\sum_{\mathbf{k}'} |f(\mathbf{k}')|^2 \mathbf{k}' + \sum_{\mathbf{k}} \hbar\omega a_{\mathbf{k}}^{\dagger} a_{\mathbf{k}}$$

$$+\sum_{\mathbf{k}} \left\{ a_{\mathbf{k}}^{\dagger} \left[V_{\mathbf{k}}^{\dagger} + f(\mathbf{k}) \left(\hbar\omega - \frac{\hbar\mathbf{k}.\mathbf{P}}{m_e} + \frac{\hbar^2 k^2}{2m_e} + \frac{\hbar^2 \mathbf{k}}{m_e} \sum_{\mathbf{k}'} |f(\mathbf{k}')_{\mathbf{k}'}|^2 \right) \right] + c.c \right\} \tag{9.5.36}$$

and

$$H_1 = \frac{\hbar^2}{2m_e}\sum_{\mathbf{k},\mathbf{k}'} \mathbf{k}.\mathbf{k}'[a_{\mathbf{k}} a_{\mathbf{k}'} f^{\dagger}(\mathbf{k}) f(\mathbf{k}') + c.c + 2a_{\mathbf{k}}^{\dagger} a_{\mathbf{k}'} f(\mathbf{k}) f^{\dagger}(\mathbf{k}')]$$

$$+\frac{\hbar^2}{m_e}\sum_{\mathbf{k},\mathbf{k}'} \mathbf{k}.\mathbf{k}'[a_{\mathbf{k}}^{\dagger} a_{\mathbf{k}} a_{\mathbf{k}'} f^{\dagger}(\mathbf{k}') + a_{\mathbf{k}'}^{\dagger} a_{\mathbf{k}}^{\dagger} a_{\mathbf{k}} f(\mathbf{k}')]. \tag{9.5.37}$$

The ground-state energy $E = \langle 0|\, H_0 + H_1 \,|0\rangle$ is

$$E = \frac{P^2}{2m_e} + \sum_{\mathbf{k}} [V_{\mathbf{k}} f(\mathbf{k}) + c.c] + \frac{\hbar^2}{2m_e}\left[\sum_{\mathbf{k}} |f(\mathbf{k})|^2 \mathbf{k}\right]^2$$

$$+\sum_{\mathbf{k}} |f(\mathbf{k})|^2 \left(\hbar\omega - \frac{\hbar\mathbf{k}.\mathbf{P}}{m_e} + \frac{\hbar^2 k^2}{2m_e}\right). \tag{9.5.38}$$

To find $f(\mathbf{k})$ and $f^{\dagger}(\mathbf{k})$ we must minimize the energy with respect to these quantities. We then find the Euler equation

$$V_{\mathbf{k}} + f^{\dagger}(\mathbf{k}) \left[\hbar\omega - \frac{\hbar\mathbf{k}.\mathbf{P}}{m_e} + \frac{\hbar^2 k^2}{2m_e} + \frac{\hbar^2}{m_e}\left(\sum_{\mathbf{k}} |f(\mathbf{k}')|^2 \mathbf{k}'\right).\mathbf{k} \right] = 0, \tag{9.5.39}$$

together with its complex conjugate. This shows that the linear terms in H_0 vanish. This part of the Hamiltonian can now be solved exactly and we may then treat H as a perturbation.

(a) *Intermediate-coupling solutions.* Since the total momentum **P** defines the only preferred direction in the problem, it is readily seen by symmetry that $\sum_{\mathbf{k}'} f(\mathbf{k}')|^2 \hbar\mathbf{k}'$ can only differ from $\mathbf{P}$ by a scalar factor. Let

$$\sum_{\mathbf{k}'} |f(\mathbf{k}')|^2 \hbar\mathbf{k}' = \eta\mathbf{P}, \tag{9.5.40}$$

and then we have, from the Euler equation,

$$f(\mathbf{k}) = \frac{-V_{\mathbf{k}}^{\dagger}}{h\omega + (h^2 k^2/2m_e) - (h\mathbf{k}\,.\,\mathbf{P}/m_e)(1-\eta)} \tag{9.5.41}$$

and

$$\eta\mathbf{P} = \sum_{\mathbf{k}} \frac{|V_{\mathbf{k}}|^2 h\mathbf{k}}{h\omega - (h\mathbf{k}\,.\,\mathbf{P}/m_e)(1-\eta) + (h^2 k^2/2m_e)}. \tag{9.5.42}$$

We can now substitute the explicit form for $V_{\mathbf{k}}$, and replacing the summation by an integration we find

$$(\eta-1)^2\eta = \frac{\alpha}{2}\left(\frac{2m_e h\omega}{P^2}\right)^{\frac{3}{2}}\left[\sin^{-1}q - \frac{q}{(1-q^2)^{\frac{1}{2}}}\right], \tag{9.5.43}$$

where

$$q = (\eta-1)\left(\frac{P^2}{2m_e h\omega}\right)^{\frac{1}{2}} \tag{9.5.44}$$

and

$$E = \frac{P^2}{2m_e}(1-\eta^2) - \frac{\alpha h\omega \sin^{-1} q}{q}. \tag{9.5.45}$$

For small momentum $\mathbf{P}$, only terms up to order P^2 in energy need be retained and the solutions for η and E may be obtained explicitly.

One finds then

$$\eta = \frac{\alpha/6}{1+\alpha/6} + O(P^2) \tag{9.5.46}$$

and

$$E = -\alpha h\omega + \frac{P^2}{2m_e}\frac{1}{1+\alpha/6}. \tag{9.5.47}$$

The main point is that this intermediate coupling theory gives a polaron mass $m^* = m_e(1+\alpha/6)$, whereas weak coupling, as we saw, gave $m^* = m_e(1-\alpha/6)^{-1}$.

(b) *Range of validity*. As with all approximate theories, we need to try to estimate the range of validity of the variational treatment discussed above.

We can make an estimate as follows. The variational calculation which has been performed is equivalent to a transformation which takes

$$H \to H_0 + H_1. \tag{9.5.48}$$

If we find $f(\mathbf{k})$ from the Euler equation, then H_0 is diagonal and the lowest-state wave function is simply the vacuum state, $|0\rangle$. We can then regard H_1 as small and estimate its size using standard perturbation theory techniques.

The states coupled to $|0\rangle$ are states in which two phonons are present: one finds for the energy shift

$$\Delta E = -\sum_n \left| \frac{(H_1)_{0n}}{E_n - E_0} \right|^2$$

$$= \frac{-h^2}{2m_e} \sum_{\mathbf{k}_1,\mathbf{k}_2} \frac{(\mathbf{k}_1 - \mathbf{k}_2)^2 |f(\mathbf{k}_1)|^2 |f(\mathbf{k}_2)|^2}{2h\omega + (\eta - 1)(h/m_e)(\mathbf{k}_1 + \mathbf{k}_2) . \mathbf{P} + (h^2/2m_e)(\mathbf{k}_1 + \mathbf{k}_2)^2}. \tag{9.5.49}$$

Lee, Low and Pines (1953) have evaluated this expression numerically to second order in P^2, with the result

$$\Delta E = -0.014\alpha^2 h\omega + \frac{0.02\alpha^2}{(1+\alpha/6)^2} \frac{P^2}{2m_e}. \tag{9.5.50}$$

An exact evaluation of the first term by Grosjean (1959) subsequently gave $-0.0159\alpha^2 \hbar\omega$. The corrections to the momentum-independent term are rather smaller than those to the effective mass as calculated in intermediate coupling theory. For $\alpha = 3$, for example, it is found that H_1 alters the momentum-independent term by $\sim 5\%$, while the effective mass is changed by some 6%. Having discussed the energy and effective mass at some length, we shall next consider the polaron mobility.

9.6 Polaron mobility

We shall now describe a calculation of the low-temperature drift mobility of the polaron, using perturbation theory in conjunction with the Kubo formula derived in Chapter 6.

The argument we shall follow is, essentially, due to Langreth and Kadanoff (1964). The motivation for such an approach is that it seems likely that the best intermediate coupling theory of the polaron mobility is likely to have a power-series expansion quite close to the exact form.

The theory then concerns itself with an expansion in the coupling constant α, and in particular in calculating the lowest-order term.

9.6.1 Expansion of self energy

We again use the Fröhlich Hamiltonian given in equation (9.5.1) as the starting point. The electron–phonon interaction is eliminated in favour of a retarded electron–electron interaction (cf. Chapter 8, section 8.10).

$$\left.\begin{aligned} V(1-1') &= V^{>}(1-1') \quad \text{for} \quad it_1 > it_1' \\ &= V^{<}(1-1') \quad \text{for} \quad it_1 < it_1' \end{aligned}\right\}, \tag{9.6.1}$$

with

$$V^{>}(\mathbf{r}t) = V^{<}(\mathbf{r}, -t) = -\frac{i\alpha}{r\sqrt{2}}[(\bar{n}_q+1)\,e^{-it}+\bar{n}_q e^{it}]. \tag{9.6.2}$$

Here $\bar{n}_q$ stands for the equilibrium number of phonons in the state q. The restriction will be imposed that $1/\beta$ is much smaller than the phonon energy.

We shall work with the spectral weight function $A(\mathbf{p}\omega)$ for the one-electron Green function

$$A(\mathbf{p}\omega) = \frac{\Gamma(\mathbf{p}\omega)}{[\omega-(p^2/2)-\operatorname{Re}\Sigma(\mathbf{p}\omega)]^2+[\Gamma(\mathbf{p}\omega)/2]^2} \tag{9.6.3}$$

by expanding the self-energy

$$\Sigma(\mathbf{p}, p_0) = \int \frac{d\omega'\,\Gamma(\mathbf{p}\omega')}{2\pi(p_0-\omega')} \tag{9.6.4}$$

in a power series in α. Since we only assume one electron in the crystal, we can, of course, make simplifications appropriate to the low-density case.

In general, one-electron propagation, at finite temperatures, can be described by the advanced and retarded Green functions. Their respective Fourier transforms we denote by $G^{>}(\mathbf{p}\omega)$ and $G^{<}(\mathbf{p}\omega)$, related to $A(\mathbf{p}\omega)$ by

$$G^{>}(\mathbf{p}\omega) = A(\mathbf{p}\omega)\,[1-f(\omega)], \tag{9.6.5}$$

$$G^{<}(\mathbf{p}\omega) = A(\mathbf{p}\omega) f(\omega), \tag{9.6.6}$$

with

$$f(\omega) = [e^{\beta(\omega-\mu)}+1]^{-1}, \tag{9.6.7}$$

μ as usual being the chemical potential. In the low-density limit $\beta\mu \to -\infty$, $f(\omega) \ll 1$ and we can write

$$G^{>}(\mathbf{p}\omega) \simeq A(\mathbf{p}\omega) \tag{9.6.8}$$

and

$$G^{<}(\mathbf{p}\omega) \simeq A(\mathbf{p}\omega)\,e^{-\beta(\omega-\mu)} \ll G^{>}(\mathbf{p}\omega). \tag{9.6.9}$$

Thus in the calculation of the self-energy Σ we can take $G^{<}(\mathbf{p}\omega)$ to be zero.

The expansion of $\Sigma(1-1')$ can now be made, and the expansion to second order is shown diagrammatically in Figure 9.1. The solid lines represent the

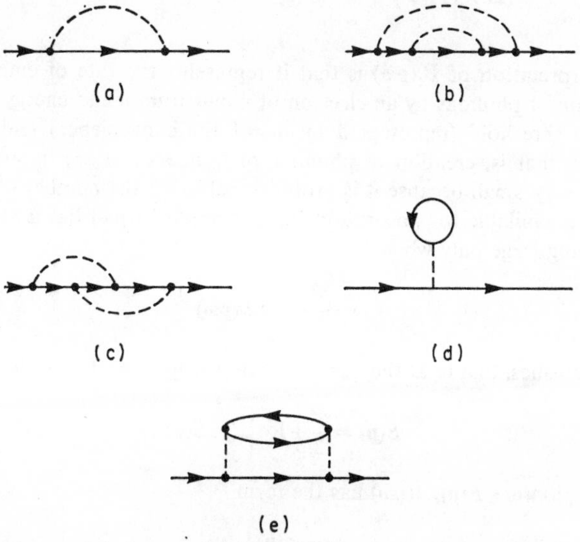

FIGURE 9.1. Lowest order diagrams of self-energy of large polaron. Evaluation of (a), (b) and (c) is discussed in text. (d) and (e) do not contribute to the order of approximation considered (after Langreth and Kadanoff, 1964).

free propagator

$$G_0(11') = -i\int\frac{d\mathbf{p}}{(2\pi)^3}\exp\left[i\mathbf{p}\,.(\mathbf{r}_1-\mathbf{r}_1')-i\frac{p^2}{2}(t_1-t_1')\right] \quad \text{for} \quad it_1>it_1'$$

$$= 0 \quad \text{for} \quad it_1<it_1', \tag{9.6.10}$$

while the dashed lines represent the retarded potential V. The first-order diagram (a) gives a contribution

$$\Sigma_1(1-1') = iV(1-1')\,G_0(1-1') \tag{9.6.11}$$

or, taking the double Fourier transform,

$$\Sigma_1(\mathbf{p},p_0) = \int\frac{d\omega'}{2\pi}\,\frac{\Gamma_1(\mathbf{p}\omega')}{p_0-\omega'}, \tag{9.6.12}$$

with

$$\Gamma_1(\mathbf{p}\omega) = \int \frac{d\mathbf{p}'}{(2\pi)^3} \frac{2^{\frac{3}{2}}\pi\alpha}{(\mathbf{p}-\mathbf{p}')^2} \left[\bar{n}_q\, \delta\left(\omega+1-\frac{p'^2}{2}\right) + (\bar{n}_q+\omega-1)\, \delta\left(1-\frac{p'^2}{2}\right)\right]. \tag{9.6.13}$$

The interpretation of $\Gamma_1(\mathbf{p}\omega)$ is that it represents the rate of emission and absorption of phonons by an electron of momentum $\mathbf{p}$ and energy ω. Below a certain threshold (represented by $\omega = 1$ for convenience) real emission processes, that is, creation of phonons of frequency ω are impossible and $\Gamma(\mathbf{p}\omega)$ is very small because it is proportional to $\bar{n}_q$, the number of phonons which are available for absorption. In this region $A(\mathbf{p}\omega)$ has a very simple form, being large only when

$$\omega - \frac{p^2}{2} - \operatorname{Re}\Sigma(\mathbf{p}\omega)$$

nearly vanishes, that is, at the quasi-particle energy $E(\mathbf{p})$. This is then given by

$$E(\mathbf{p}) = \frac{p^2}{2} + \operatorname{Re}\Sigma(\mathbf{p}, E(\mathbf{p})). \tag{9.6.14}$$

In the region $\omega \sim E(\mathbf{p})$, $A(\mathbf{p}\omega)$ has the form

$$A(\mathbf{p}\omega) = \frac{z(\mathbf{p})\,[\tau(\mathbf{p})]^{-1}}{[\omega - E(\mathbf{p})]^2 + [2\tau(\mathbf{p})]^{-2}}, \tag{9.6.15}$$

where $\tau(\mathbf{p})$ is the quasi-particle lifetime defined by

$$\tau(\mathbf{p}) = z(\mathbf{p})\, \Gamma(\mathbf{p}, E(\mathbf{p})). \tag{9.6.16}$$

The quantity $z(\mathbf{p})$, essentially the wave function renormalization constant, is defined by

$$z(\mathbf{p}) = \left[1 - \frac{\partial}{\partial\omega} \operatorname{Re}\Sigma(\mathbf{p}\omega)\right]^{-1}\Bigg|_{\omega=E(\mathbf{p})}. \tag{9.6.17}$$

But when $\omega < E(0)$, $\omega - (p^2/2) - \operatorname{Re}\Sigma(\mathbf{p}\omega)$ can no longer vanish. Here we can neglect $\Gamma(\mathbf{p}\omega)$ in the denominator of equation (9.6.3) and then we find

$$A(\mathbf{p}\omega) = \frac{\Gamma(\mathbf{p}\omega)}{[\omega - (p^2/2) - \operatorname{Re}\Sigma(\mathbf{p}\omega)]^2} \quad \text{for} \quad \omega < E(0). \tag{9.6.18}$$

The states which are appreciably excited at low temperatures will be our main concern, and these are the low-momentum quasi-particle states ($p^2/2 \sim \beta^{-1} \ll 1$). We can now find the properties of these quasi-particles to first order in α by applying equations (9.6.12) and (9.6.13), when we find for small $\mathbf{p}$ and ω that

$$\operatorname{Re}\Sigma(\mathbf{p}\omega) = -\alpha[1+\tfrac{1}{2}\omega-\tfrac{1}{6}p^2]. \tag{9.6.19}$$

Using equation (9.6.14) we regain the result we derived earlier in the chapter, namely that for $p^2/2 \ll 1$

$$E(\mathbf{p}) = -\alpha+\frac{p^2}{2m^*}+O(p^4), \tag{9.6.20}$$

with

$$m^* = 1+\frac{\alpha}{6}, \tag{9.6.21}$$

while

$$z(0) = 1-\frac{\alpha}{2}. \tag{9.6.22}$$

These three equations will be used below in the calculation of the polaron mobility, which is our main aim here. We shall also require an expression for the density of electrons

$$\begin{aligned} n &= \int \frac{d\mathbf{p}}{(2\pi)^3} \int \frac{d\omega}{2\pi} G^<(\mathbf{p}\omega) \\ &= \int \frac{d\mathbf{p}}{(2\pi)^3} \int \frac{d\omega}{2\pi} e^{-\beta(\omega-\mu)} A(\mathbf{p}\omega), \end{aligned} \tag{9.6.23}$$

which is correct to first order in α. The factor $e^{-\beta(\omega-\mu)}$ highly weights the integrand in equation (9.6.23) for small values of ω and we therefore break up the integral into two parts: for $\omega > E(0)$ we use equation (9.6.15) to express $A(\mathbf{p}\omega)$ because the temperature factor cuts out all contributions from energies appreciably greater than the minimum quasi-particle energy $E(0)$. Then we find

$$\begin{aligned} n = &\int \frac{d\mathbf{p}}{(2\pi)^3} \int_{E(0)}^{\infty} \frac{d\omega}{2\pi} e^{-\beta(\omega-\mu)} z(0) \\ &\times \left\{ \frac{\tau^{-1}(0)}{[\omega-E(\mathbf{p})]^2+[2\tau(0)]^{-2}} \right\} + \int \frac{d\mathbf{p}}{(2\pi)^3} \\ &\times \int_{-\infty}^{E(0)} \frac{d\omega}{2\pi} \frac{e^{-\beta(\omega-\mu)}\,\Gamma(\mathbf{p}\omega)}{[\omega-(p^2/2)-\operatorname{Re}\Sigma(\mathbf{p}\omega)]^2}. \end{aligned} \tag{9.6.24}$$

As $\tau \to \infty$, we can replace the curly bracket in equation (9.6.24) by $2\pi\delta[\omega - E(\mathbf{p})]$. Also, to first order in α, we can replace $\Gamma(\mathbf{p}\omega)$ in the second term by the $\Gamma_1(\mathbf{p}\omega)$ defined by equation (9.6.13) and neglect the $\operatorname{Re}\Sigma(\mathbf{p}\omega)$ in the denominator of the second term. Then we find after some manipulation that

$$n = \int \frac{d\mathbf{p}}{(2\pi)^3} \mathrm{e}^{-\beta[E(\mathbf{p})-\mu]}. \tag{9.6.25}$$

It seems likely, though this has only been proved to first order in α, that it is valid to all orders in α for the low-temperature situation considered here.

It turns out that, while the above calculations to first order in α are quite adequate for our present purposes, $\tau^{-1}(0)$ is in fact needed to $O(\alpha^2)$ and this is a little more troublesome though we shall see, after some discussion, that there are no corrections to this order. We have to get the contributions (b) and (c) of Figure 9.1. Figure 9.1(b) gives

$$\Sigma_{2b}^{\gtrless}(1-1') = -\int d\bar{\mathbf{r}}_1 \int_{t_{1'}}^{t_1} d\bar{t}_1 \int d\bar{\mathbf{r}}_{1'} \int_{t_{1'}}^{t_1} d\bar{t}_{1'}\, V^{>}(\bar{1}-\bar{1}')\, V^{>}(1-1')$$
$$\times G_0^{\gtrless}(1-\bar{1})\, G_0^{\gtrless}(\bar{1}-\bar{1}')\, G_0^{\gtrless}(\bar{1}'-1'), \tag{9.6.26}$$

which has the Fourier transform

$$\Gamma_{2b}(\mathbf{q}\omega) = \int d\mathbf{r} \int_{-\infty}^{\infty} dt\, \mathrm{e}^{-i\mathbf{q}.\mathbf{r}+i\omega t}\, i\Sigma_{2b}^{\gtrless}(\mathbf{r}, t)$$
$$= \operatorname{Re} \int \frac{d\mathbf{p}}{(2\pi)^3} \int \frac{d\mathbf{p}'}{(2\pi)^3} \frac{16\pi^3\alpha^2\bar{n}}{(\mathbf{p}-\mathbf{q})^2(\mathbf{p}-\mathbf{p}')^2}$$
$$\times \left\{ \delta\left(\omega - \frac{p'^2}{2}\right) \left[\frac{1}{[\omega+1-(p^2/2)+i\varepsilon]^2} + \frac{1}{[\omega-1-(p^2/2)+i\varepsilon]^2} \right] \right.$$
$$+ \left[\frac{\partial}{\partial\omega} \delta\left(\omega+1-\frac{p^2}{2}\right) \right] \frac{1}{\omega-(p'^2/2)}$$
$$\left. + \left[\frac{\partial}{\partial\omega} \delta\left(\omega-1-\frac{p^2}{2}\right) \right] \frac{1}{\omega-(p'^2/2)} \right\} + O(\bar{n}^2), \tag{9.6.27}$$

where ε is an infinitesimal. Similarly the contribution of Figure 9.1(c) to the self-energy is given by

$$\Sigma_{2c}^{\gtrless}(1-1') = -\int d\bar{\mathbf{r}}_1 \int_{t_{1'}}^{t_1} d\bar{t}_1 \int d\bar{\mathbf{r}}_{1'} \int_{t_{1'}}^{t_1} d\bar{t}_{1'}\, V^{>}(1-\bar{1}')$$
$$\times V^{>}(\bar{1}-1')\, G_0^{\gtrless}(1-\bar{1})\, G_0^{\gtrless}(\bar{1}-\bar{1}')\, G_0^{\gtrless}(\bar{1}'-1') \tag{9.6.28}$$

and

$$\Gamma_{2c}(\mathbf{q}\omega) = P\int\frac{d\mathbf{p}}{(2\pi)^3}\int\frac{d\mathbf{p}'}{(2\pi)^3}\frac{32\pi^3\alpha^2\bar{n}}{(\mathbf{p}-\mathbf{q})^2(\mathbf{p}'-\mathbf{q})^2}$$
$$\times\left\{\delta\left[\omega-\frac{(\mathbf{p}+\mathbf{p}'-\mathbf{q})^2}{2}\right]\frac{1}{\omega+1-(p^2/2)}\frac{1}{\omega-1-(p'^2/2)}\right.$$
$$+\delta\left(\omega+1-\frac{p^2}{2}\right)\frac{1}{\omega-1-(p'^2/2)}\frac{1}{\omega-[(\mathbf{p}+\mathbf{p}'-\mathbf{q})^2/2]}$$
$$\left.+\delta\left(\omega-1-\frac{p'^2}{2}\right)\frac{1}{\omega+1-(p^2/2)}\frac{1}{\omega-[(\mathbf{p}+\mathbf{p}'-\mathbf{q})^2/2]}\right\}+O(\bar{n}^2),\tag{9.6.29}$$

where P indicates that the principal values of the integrals are to be taken.

To second order in α, $\tau^{-1}(0)$ contains the sum $\Gamma_{2b}(0,0)+\Gamma_{2c}(0,0)$. It should be noted that the first terms in the { } of equations (9.6.27) and (9.6.29) each produce discontinuities at $\omega = 0$, $\mathbf{q} = 0$. Each term contributes for $\omega > 0$ and $\mathbf{q} = 0$, neither contributes for $\omega < 0$. However, as was apparently first shown by Schultz, these discontinuities cancel in the sum $\Gamma_{2b}(\mathbf{q}\omega)+\Gamma_{2c}(\mathbf{q}\omega)$. Therefore we can safely eliminate them by calculating Γ_{2b} and Γ_{2c} for ω just less than zero. We find on carrying out the integrations that

$$\Gamma_{2b}(0,\omega)\big|_{\omega=0^-} = -2\alpha\bar{n}[\tfrac{1}{2}\pi\alpha]\tag{9.6.30}$$

and

$$\Gamma_{2c}(0,\omega)\big|_{\omega=0^-} = 2\alpha\bar{n}[\tfrac{1}{2}\pi\alpha].\tag{9.6.31}$$

The second-order contributions to $\tau^{-1}(0)$ from the diagrams (b) and (c) of Figure 9.1 exactly cancel, so that, to second order, only $\Gamma_1(\mathbf{p}\omega)$ contributes to $\tau^{-1}(0)$. From equation (9.6.16)

$$\tau^{-1}(0) = z(0)\,\Gamma_1(0, E(0))\tag{9.6.32}$$

and near $\omega = 0$, equation (9.6.13) implies that

$$\Gamma_1(0,\omega) = 2\alpha\bar{n}\left[1-\frac{\omega}{2}\right].\tag{9.6.33}$$

Thus

$$\tau^{-1}(0) = 2\alpha\bar{n}+O(\alpha^3).\tag{9.6.34}$$

The inverse quasi-particle lifetime contains no corrections therefore to $O(\alpha^2)$.

Mobility from Kubo formula. To calculate the mobility $\mu_{\text{polaron}} \equiv \mu_{\text{p}}$, we use the Kubo formula (cf. equation 6.11.10)

$$\mu = \frac{e\beta}{6n}\int_{-\infty}^{\infty}dt\int d\mathbf{r}\,\langle\mathbf{j}(\mathbf{r},t)\,.\,\mathbf{j}(\mathbf{0},0)\rangle,\tag{9.6.35}$$

where $\mathbf{j}(\mathbf{r}, t)$ is the momentum current which can be written in terms of the Heisenberg field operators as

$$\mathbf{j}(\mathbf{r}, t) = -\tfrac{1}{2}i[\Psi^{\dagger}(\mathbf{r}t)\nabla\Psi(\mathbf{r}t) - \nabla\Psi^{\dagger}(\mathbf{r}t)\Psi(\mathbf{r}t)]. \tag{9.6.36}$$

The correlation function in equation (9.6.35) can be expressed in terms of the two-particle Green function G_2 as

$$\langle T(\mathbf{j}(\mathbf{r}_1, t_1) . \mathbf{j}(\mathbf{r}_2, t_2))\rangle = \tfrac{1}{4}(\nabla_1 - \nabla_{1'}) . (\nabla_2 - \nabla_{2'})\, G_2(1, 2;\, 1', 2')\big|_{1'=1^+, 2'=2^+}. \tag{9.6.37}$$

Thus, equation (9.6.35) becomes

$$\mu_{\mathrm{p}} = \frac{e\beta}{24n}\int_{-\infty}^{\infty} dt_1 \int d\mathbf{r}_1 (\nabla_1 - \nabla_{1'}) . (\nabla_2 - \nabla_{2'})\, G_2^{\gtrless}(12, 1'2')\big|_{1'=1^+, 2'=2^+=0}. \tag{9.6.38}$$

Here the $G_2^{\gtrless}$ means that we must use the form of G_2 when $it_1 > it_2$.

To evaluate equation (9.6.38), we can expand G_2 in a power series in G and V. This expansion is shown in Figure 9.2. It should be noted that we cannot expand G_2 in terms of G_0 and V because μ_{p} is proportional to α^{-1} and hence the G_0 expansion cannot converge. For most systems, even the expansion

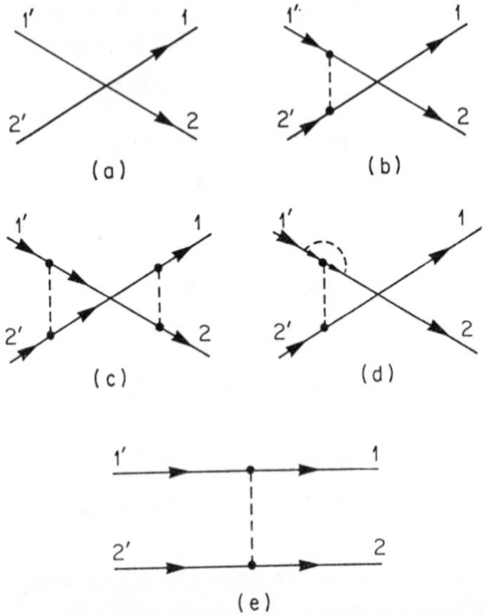

FIGURE 9.2. Diagrams for two-particle Green function. Solid lines represent G's and not G_0's (after Langreth and Kadanoff, 1964).

in V and G will converge very slowly. Fortunately, this convergence is particularly good for the case of low-temperature optical phonon scattering. For small α, Figure 9.2(a) is of order α^{-1}. Figures 9.2(b) and (c) turn out at low temperatures to be of order α^0 and α respectively. Since we wish to calculate μ_p to $O(\alpha^0)$ we must calculate the contributions from Figures 9.2(a) and (b) but we can safely neglect Figures 9.2(c) and (d) and all higher-order diagrams.

Figure 9.2(a) gives a contribution to the mobility of

$$\mu_{pa} = \frac{e\beta}{24n}\int_{-\infty}^{\infty} dt_1 \int d\mathbf{r}_1(\nabla_1-\nabla_{1'}).(\nabla_2-\nabla_2')\,G^>(1-2')\,G^<(2-1')\big|_{1'=1^+,2'=2^+=0}$$

$$= \frac{e\beta}{6n}\int \frac{d\mathbf{p}}{(2\pi)^3}\int \frac{d\omega}{2\pi} p^2[A(\mathbf{p}\omega)]^2\,e^{-\beta(\omega-\mu)}. \tag{9.6.39}$$

The predominant contributions to equation (9.6.39) come for $\mathbf{p}=0$, $\omega \simeq E(\mathbf{p})$. In this region we can write

$$[A(\mathbf{p}\omega)]^2 = \frac{4\pi[z(\mathbf{p})]^2\,[\tau(\mathbf{p})]^{-2}}{\{\omega-E(\mathbf{p})\}^2+\{[2\tau(\mathbf{p})]^{-2}\}^2}, \tag{9.6.40}$$

which becomes, in the limit of large τ

$$[A(\mathbf{p}\omega)]^2 = 4\pi\tau(\mathbf{p})\,[z(\mathbf{p})]^2\,\delta(\omega-E(\mathbf{p})). \tag{9.6.41}$$

Hence equation (9.6.39) may be written as

$$\mu_{pa} = \frac{e\beta}{6n}\int \frac{d\mathbf{p}}{(2\pi)^3} p^2\,2\tau(0)\,[z(0)]^2\,e^{-\beta(E(\mathbf{p})-\mu)}$$

$$= e\tau(0)\,[z(0)]^2\,m^*, \tag{9.6.42}$$

where we have used equation (9.6.25) in going from the first to the second part of equation (9.6.42).

We now go back to equations (9.6.21), (9.6.22) and (9.6.34) to find

$$\mu_{pa} = \left(\frac{e}{2\alpha\bar{n}}\right)[1-\tfrac{5}{6}\alpha]. \tag{9.6.43}$$

The next contribution to the mobility comes from Figure 9.2(b), which gives

$$G_2^b(12;\,1'2') = -i\int_0^{-i\beta} d\bar{1}\int_0^{-i\beta} d\bar{1}'\,G(1-\bar{1})$$

$$\times G(\bar{1}-2')\,V(\bar{1}-\bar{1}')\,G(2-\bar{1}')\,G(\bar{1}'-1'). \tag{9.6.44}$$

For $t_1 = t_{1'}$ and $t_2 = t_{2'} = 0$,

$$G_2^b(1,2,1'2') = -i\int_0^{t_1} d\bar{1} \int_{t_1}^{-i\beta} d\bar{1}'\, G^>(1-\bar{1})$$
$$\times G^>(\bar{1}-2')\, V^<(\bar{1}-\bar{1}')\, G^<(2-\bar{1}')\, G^>(\bar{1}'-1') \qquad (9.6.45)$$

plus terms of order $[G^<]^2$ which can be neglected. Substituting equation (9.6.45) into equation (9.6.38), and after some manipulation, the contribution to the mobility is found to be

$$\mu_{pb} = \frac{e\beta}{3n}\int\frac{d\omega}{2\pi}\int\frac{d\mathbf{p}}{(2\pi)^3}\int\frac{d\mathbf{p}'}{(2\pi)^3}\frac{2^{5/2}\pi\alpha}{(\mathbf{p}-\mathbf{p}')^2}\mathbf{p}.\mathbf{p}'\,\mathrm{e}^{-\beta(\omega-\mu)}$$
$$\times\{\bar{n}A(\mathbf{p}\omega)\,\mathrm{Re}\,G(\mathbf{p}',\omega+1)\,[A(\mathbf{p}\omega)\,\mathrm{Re}\,G(\mathbf{p}',\omega+1)$$
$$+\mathrm{Re}\,G(\mathbf{p}\omega)\,A(\mathbf{p}',\omega+1)]+(\bar{n}+1)\,A(\mathbf{p}\omega)\,\mathrm{Re}\,G(\mathbf{p}',\omega-1)$$
$$\times[A(\mathbf{p}\omega)\,\mathrm{Re}\,G(\mathbf{p}',\omega-1)+\mathrm{Re}\,G(\mathbf{p}\omega)\,A(\mathbf{p}',\omega-1)]\}, \qquad (9.6.46)$$

where

$$\mathrm{Re}\,G(\mathbf{p}\omega) = P\int\frac{d\omega'A(\mathbf{p}\omega')}{2\pi(\omega-\omega')}. \qquad (9.6.47)$$

We neglect the terms in equation (9.6.46) proportional to $\bar{n}$ and notice that the only important contribution to equation (9.6.46) comes from the term involving $[A(\mathbf{p}\omega)]^2$, which we can calculate using equation (9.6.41). Then we find

$$\mu_{pb} = \frac{e\beta}{3n}(2^{5/2}\pi\alpha)\,\tau(0)\int\frac{d\mathbf{p}}{(2\pi)^3}\int\frac{d\mathbf{p}'}{(2\pi)^3}\frac{\mathbf{p}.\mathbf{p}'}{(\mathbf{p}-\mathbf{p}')^2}$$
$$\times\left[\frac{1}{1+(p'^2/2)}\right]^2\mathrm{e}^{-\beta[(p^2/2)-\mu]}+O(\alpha). \qquad (9.6.48)$$

Hence we have finally

$$\mu_{pb} = \tfrac{2}{3}\alpha e\tau(0) = \left(\frac{e}{2\alpha\bar{n}}\right)\tfrac{2}{3}\alpha \qquad (9.6.49)$$

and

$$\mu_p = \mu_{pa}+\mu_{pb} = \mu_0\left(1-\frac{\alpha}{6}\right), \qquad (9.6.50)$$

where μ_0 is the weak coupling mobility ($\mu_0 = e/2\alpha\bar{n}$).

We see that the quasi-particle picture, which would predict

$$\mu = \frac{e\tau(0)}{m^*} = \frac{e}{2\alpha\bar{n}}\left[1-\frac{\alpha}{6}+O(\alpha^2)\right], \tag{9.6.51}$$

is verified. Langreth and Kadanoff speculate that $e\tau(0)/m^*$ is probably correct to all orders in α at low temperatures.

However, the result (9.6.50) is certainly an exact perturbative form for the mobility, appropriate when $\alpha/6 \ll 1$ and at low temperatures $\beta^{-1} \ll 1$. Appel (1968) refers to other calculations of the large polaron mobility, and we only mention in concluding this section that the Low and Pines (1960) formula yields $1-0.5\alpha$ for the ratio μ_p/μ_0 as $\alpha \to 0$. Even such a sophisticated calculation as theirs can hardly be quantitatively reliable in the intermediate coupling range in the light of this comparison.

Having dealt with the perturbative theory of the large polaron mobility at some length, we now turn to discuss the localized polaron.

9.7 Feynman polaron radius and validity of continuum approximation

The approximations involved in a continuum model are clearly no longer satisfactory when the polaron radius becomes of the order of the lattice constant a. A useful measure of the polaron radius for all values of the coupling constant α is given by the so-called Feynman polaron dimension r_f. Feynman treated Fröhlich's model by path integral methods (see, for example, Schultz, 1963) which lead to a picture in which the polaron simulates an electron bound in a harmonic potential to a fictitious particle with mass M, with a characteristic frequency ν. The resulting harmonic oscillator wave function defines a Feynman polaron radius which is

$$r_f = (\tfrac{3}{2}\mu\nu)^{\frac{1}{2}} : \mu^{-1} = M^{-1}+m^{-1}. \tag{9.7.1}$$

Schultz (1959) has calculated r_f by a variational method and a selection of his results is shown in Table 9.1.

TABLE 9.1

Crystal	a (Å)	r_f/a	α	$(\hbar/2m\omega_0)^{\frac{1}{2}}$ (Å)	ϵ_s	ϵ_∞
NaCl	5.63	1.81	5.5	10.9	5.6	2.2
KCl	6.28	1.72	5.6	12.1	4.7	2.1
Cu_2O	2.46	5.93	2.5	8.8	10.5	4.0
AgCl	5.55	8.43	3.6‡	12.5	12.3	4.0

‡ Including m^*/m this value becomes 1.7.

For NaCl, with the assumption $m^* = m$, we get $r_f > a$ and it seems that the continuum approximation leads to small errors in the self-energy and effective mass. On the other hand, if in an alkali halide crystal $m^* \gg m$, the radius $r_f < a$ and the continuum approximation breaks down. This appears to be the case for valence band holes in KCl, for example, and also in other substances such NiO, and if so the large polaron picture is plainly inadequate.

9.7.1 *Molecular crystal model and small polaron mobility*

We shall content ourselves here with describing just one aspect of small polaron theories, viz. that concerned with the "molecular crystal model" introduced by Holstein (1959). The model is highly idealized, but probably contains the essential features of the problem, and enabled Holstein to show very clearly how two separate régimes of polaron motion can arise. These are (see Appel, 1968):

(i) A low-temperature régime in which the small polaron can be described in terms of states of an energy band, and for which the transport theory of band electrons of Chapter 6 can be utilized to calculate the mobility.

(ii) A régime above a transition temperature T_t in which the polaron band picture breaks down, the band width becoming smaller than the energy uncertainty of the polaron band states. The polaron states are then more profitably considered to be localized and polaron motion is essentially a diffusion process. The drift mobility μ is readily calculated in this régime by using Einstein's relation $\mu = eD/k_B T$ where D is the diffusion coefficient for random phonon-activated hopping motion of the polaron.

The molecular crystal model consists of a linear chain of N diatomic molecules, the orientations of which are fixed, as are the centres of gravity $\mathbf{R}_n = n\mathbf{a}$. The lattice vibrations consist of vibrations in the internuclear separations x_n of these molecules, and affect the electrons through the molecular potential $U(\mathbf{r}-\mathbf{R}_n, x_n)$. The molecular crystal model has the same essential content as a linear chain of alternating positive and negative ions, but has the mathematical merit that only one type of potential occurs.

Thus we begin with the Hamiltonian $H = H_{el}+H_{int}+H_{lattice}$, where

$$\left.\begin{aligned} H_{el}+H_{int} &= -\frac{\hbar^2}{2m}\nabla^2 + \sum_{n=1}^{N} U(\mathbf{r}-\mathbf{R}_n, x_n), \\ H_{lattice} &= \sum_{n=1}^{N}\left(-\frac{\hbar^2}{2M}\frac{\partial^2}{\partial x_n^2} + \tfrac{1}{2}M\omega_0^2 x_n^2 + \tfrac{1}{2}M\omega_1^2 x_n x_{n+1}\right), \end{aligned}\right\} \quad (9.7.2)$$

where the presence of the coupling term $x_n x_{n+1}$ gives rise to dispersion in the lattice frequencies. The wave function of the total Hamiltonian is given as a superposition

$$\Psi = \sum_n a_n(x_1 x_2 \ldots x_n)\,\phi_n(\mathbf{r}, x_n), \tag{9.7.3}$$

where the ϕ_n are localized molecular wave functions, satisfying

$$\left[-\frac{\hbar^2}{2m}\nabla^2 + U(\mathbf{r}-\mathbf{R}_n, x_n)\right]\phi_n(\mathbf{r}-\mathbf{R}_n, x_n) = E(x_n)\,\phi_n(\mathbf{r}-\mathbf{R}_n, x_n), \tag{9.7.4}$$

the eigenvalues of which depend on the intermolecular distance in a manner which we take as

$$E(x_n) = -Ax_n + \text{const}, \tag{9.7.5}$$

with the constant set equal to zero for convenience.

One now finds the equation of motion for the amplitudes a_n to be

$$i\hbar\frac{\partial}{\partial t}a_n(x_1, \ldots, x_n) = [H_{\text{lattice}} - Ax_n]\,a_n(x_1, \ldots, x_n) + \sum_{\pm} J(x_n, x_{n\pm1})\,a_{n\pm1}(x_1, \ldots, x_n), \tag{9.7.6}$$

where J is the two-centre overlap integral

$$J(x_m, x_n) = \int \phi^*(\mathbf{r}-\mathbf{R}_m, x_m)\,U(\mathbf{r}-\mathbf{R}_m, x_m)\,\phi(\mathbf{r}-\mathbf{R}_n, x_n)\,d\mathbf{r}. \tag{9.7.7}$$

Numerous approximations are involved in deriving equation (9.7.6). Except to remark that they are all of a tight binding nature, we shall not go into them here; we take equations (9.7.6) and (9.7.7) as defining the molecular crystal model, and go on to explore the consequences of this definition.

We follow Holstein (1959) in using the J of equation (9.7.7) as a perturbation parameter: this is permissible when the electronic band width $2J$ (of the *rigid* lattice) is small compared to the characteristic energy $A^2/2M\omega_0^2$. Because of the nature of the theory thus obtained, as will appear below, this is termed the jump-perturbation approximation.

Let us introduce normal coordinates into equation (9.7.6), to describe the internuclear vibrations in terms of standing waves. We write

$$\xi_q = \left(\frac{2}{N}\right)^{\frac{1}{2}} \sum_{n=1}^{N} x_n \sin\left(q^n + \frac{\pi}{4}\right), \tag{9.7.8}$$

where $q = 2\pi j/N$, j taking all integral values between $\pm N/2$. Equation (9.7.6) is then transformed into

$$i\hbar\frac{\partial a_n}{\partial t} = \sum_q\left[-\frac{\hbar^2}{2M}\frac{\partial^2}{\partial\xi_q^2}+\tfrac{1}{2}M\omega_q^2(\xi_q-\Delta\xi_q^{(n)})^2-\left(\frac{2}{N}\right)^{\frac{1}{2}}A\xi_q\sin\left(q^n+\frac{\pi}{4}\right)\right]a_n(\xi_q).$$
$$-J(a_{n+1}+a_{n+1} \qquad (9.7.9)$$

We are to use J, which is here assumed to be a constant, as a perturbation parameter, and so we first obtain the zeroth-order wave functions by setting $J = 0$. We then have a system of N independent oscillators, with displaced equilibrium positions

$$\Delta\xi_q^{(n)} = \left(\frac{A}{M\omega_q^2}\right)\left(\frac{2}{N}\right)^{\frac{1}{2}}\sin\left(q^n+\frac{\pi}{4}\right) \qquad (9.7.10)$$

and with frequencies given by the dispersion relation

$$\omega_q^2 = \omega_0^2+\omega_1^2\cos q. \qquad (9.7.11)$$

The eigenfunctions are given by

$$a_n(p,\{n_q\}) = \delta_{pn}\chi_p(\{n_q\}) = \delta_{pn}\prod_q\chi_{n_q}\left(\left[\frac{M\omega_q}{\hbar}\right]^{\frac{1}{2}}(\xi_q-\Delta\xi_q^{(p)})\right). \qquad (9.7.12)$$

Here $\{n_q\}$ denotes the entire set of phonon occupation numbers $n_{q_1}, n_{q_2}, \ldots, n_{q_N}$, while the χ_{n_q} are normalized harmonic oscillator wave functions. The corresponding eigenvalues are

$$\mathscr{E}(\{n_q\}) = \sum_q\hbar\omega_q(n_q+\tfrac{1}{2})-E_{\text{b}}, \qquad (9.7.13)$$

with

$$E_{\text{b}} = \sum_q\left[\frac{A^2}{2M\omega_q^2}\right]\left(\frac{2}{N}\right)\sin^2\left(qp+\frac{\pi}{4}\right)\sim\pi^{-1}\int_0^\pi\frac{A^2}{2M\omega_q^2}dq. \qquad (9.7.14)$$

When J is switched on, we write, in terms of the states of equation (9.7.12),

$$a_n = \sum_{p',\{n_q'\}}C(p',\{n_q'\})\,a_n(p',\{n_q'\}) \qquad (9.7.15)$$

and the equation for the C's follows immediately, to lowest order in J, as

$$i\frac{\partial C}{\partial t}(p,\{n_q\}) = \sum_{p',\{n_q'\}}\langle p,\{n_q\}|\,V\,|p',\{n_q'\}\rangle\,C(p',\{n_q'\})\exp\{it[\mathscr{E}(\{n_q\})-\mathscr{E}(\{n_q'\})]\}, \qquad (9.7.16)$$

where

$$\langle p,\{n_q\}|\,V\,|p',\{n_q'\}\rangle = -J\sum_{r=\pm1}\delta_{p',p'\pm r}S(p,\{n_q\};\,p',\{n_q'\}), \qquad (9.7.17)$$

S being the overlap integral between the two products of the corresponding harmonic oscillator wave functions on adjacent sites p and p'. The evaluation of the S's and the solution for C will be found in Appendix (9.3) and here we wish only to summarize the steps and results.

One finds that as the temperature T tends to zero, transitions for which $\{n_q\}$ and $\{n_q'\}$ are different become negligible, and the a_n are given by the product of a plane-wave factor e^{ikn} with the lattice function describing a set of harmonic oscillators:

$$a_n(k, \{n_q\}) = e^{ikn} \prod_q \chi_{n_q}\left(\left[\frac{M\omega_0}{\hbar}\right]^{\frac{1}{2}} [\xi_q - \Delta\xi_q^{(n)}]\right). \tag{9.7.18}$$

The corresponding energies are

$$E(k, \{n_q\}) = \mathscr{E}(\{n_q\}) - 2J\cos k \exp[-S(\{n_q\})], \tag{9.7.19}$$

where

$$S(\{n_q\}) = \pi^{-1}\int_0^\pi (1+2n_q)\gamma_q \, dq. \tag{9.7.20}$$

with

$$\gamma_q = \left(\frac{A^2}{2M\omega_q^2\hbar\omega_q}\right)(1-\cos q). \tag{9.7.21}$$

We evidently have an allowed band of polaron energies, with half-width

$$\Delta E(\{n_q\}) = 2J\exp[-S(\{n_q\})]. \tag{9.7.22}$$

If the lattice vibrations were absent, the band half-width would be $2J$ and a reduction such as shown in equation (9.7.22) is a well-established result of the theory. In equilibrium, $\langle n_q\rangle = (e^{\beta\omega}-1)^{-1}$ where $\beta^{-1} = k_B T$ so that $S(\{n_q\})$ increases with increasing temperature, with a consequent reduction of the band width. Notice that even at absolute zero, when $\{n_q\} \equiv 0$, there remains a reduction (due to the zero-point motion) of the half-width below $2J$. The physical significance of $S(\{n_q\} \equiv 0)$ is that it describes the ratio of the polaron binding energy E_b to the optical phonon energy $\hbar\omega_0$.

So long as band states are well defined we may follow various authors (see, for example, Sewell, 1958) and in calculations of polaron band states simply thermally average over the overlap integral and $S(\{n_q\})$. In fact, transitions between band states k and k' take place because of off-diagonal elements of $\langle p\{n_q\}|V|p'\{n_q'\}\rangle$, i.e. elements for which $\{n_q\} \neq \{n_q'\}$. Standard perturbation theory gives the transition probability per unit time as

$$W(k, \{n_q\} \to k'\{n_q'\}) = \left(\frac{2\pi}{\hbar}\right)|\langle k, \{n_q\}|V|k'\{n_q'\}\rangle|^2 \delta(E(k, \{n_q\}) - E(k', \{n_q'\})), \tag{9.7.23}$$

where the transition matrix element is

$$\langle k,\{n_q\}|\,V\,|k',\{n'_q\}\rangle = \sum_{pp'} e^{-i(k_p-k'_{p'})}\langle p,\{n_q\}|\,V\,|p',\{n'_q\}\rangle. \tag{9.7.24}$$

Obviously, it makes sense to work with polaron band states as long as the inverse of the mean lifetime

$$\tau(k,\{n_q\}) = \left\{\sum_{k',\{n_{q'}\}} W(k,\{n_q\}\to k',\{n'_q\})\right\}^{-1} \tag{9.7.25}$$

is much less than the band half-width:

$$\frac{\hbar}{\tau(k,\{n_q\})} \ll \Delta E(k,\{n_q\}) = 2J\exp\left[-S(\{n_q\})\right]. \tag{9.7.26}$$

As the temperature is increased so that this condition is violated, perturbation theory on these states breaks down. One then expects to be able to use the picture alternative to the above; that of a localized polaron. This is most meaningful if the diagonal transitions giving rise to the band picture become unimportant. Since localized states can be built up from the extended states, one expects their lifetimes to be comparable when non-diagonal transitions become dominant; one is thus led to the condition

$$\frac{\hbar}{\tau(p,\{n_q\})} \gg 2J\exp\left[-S(\{n_q\})\right] \tag{9.7.27}$$

[cf. equation (9.7.26)]. When this is obeyed one can assume the polaron to be localized on site p and ignore diagonal transitions ($\{n_q\}\equiv\{n'_q\}$) in the calculation of its lifetime. We give a more detailed derivation of this condition in Appendix (9.3).

Actually, the transition, from the low-temperature régime of band motion to the high temperature regime of hopping motion, takes place over a narrow temperature range about the transition temperature T_t. This is because the lifetime of a band state decreases rapidly as T_t is approached from below, the inverse lifetime of a transition $k,\{n_q\}\to k',\{n'_q\}$ being proportional to the product of those n_q that are changed because of phonon absorption. The thermal averages of these increase rapidly with temperature.

The respective contributions of diagonal ($\{n_q\}\equiv\{n'_q\}$) and non-diagonal ($\{n_q\}\not\equiv\{n'_q\}$) transitions are considered in detail in Appendix (9.3). One finds that above T_t, so that diagonal transitions may be neglected, matters are simplified considerably if the temperature is sufficiently high that

$$\pi^{-1}\int_0^{\pi} 2\gamma_q \operatorname{cosech}\left(\tfrac{1}{2}\beta\hbar\omega_q\right)dq \gg 1 \tag{9.7.28}$$

and, further, one has $\hbar\omega_q \ll k_B T$ so that the classical limit of the phonon distribution can be used.

After averaging over the phonon distribution, one finds the mean lifetime of a localized polaron (which may hop to sites on either side) is

$$\tau(p,T) = W(p \to p+1, T) + W(p \to p-1, T), \tag{9.7.29}$$

where

$$W(p, p\pm 1, T) = \frac{J^2}{\hbar}\left(\frac{\pi}{4k_B T E_a}\right)^{\frac{1}{2}} e^{-E_a/k_B T}. \tag{9.7.30}$$

Here E_a is the hopping activation energy

$$E_a = \pi^{-1}\int_0^{\pi} \frac{A^2}{4M\omega_q^2}(1-\cos q)\,dq, \tag{9.7.31}$$

which is smaller (by a factor of two usually) than the energy E_b required to thermally dissociate the polaron, this energy being given by the right-hand side of equation (9.7.14).

The transition temperature T_t may be defined by

$$\frac{\hbar}{\tau(p,T)} = \Delta E(T_t) = 2J e^{-S(T_t)}, \tag{9.7.32}$$

where $S(T)$ is the temperature average of $S(\{n_q\})$:

$$S(T) = \sum_q \left(\frac{\gamma_q}{N}\right) \coth\left(\tfrac{1}{2}\beta\hbar\omega_q\right). \tag{9.7.33}$$

Equations (9.7.29), (9.7.30), (9.7.31), (9.7.32) and (9.7.33) together give a transcendental equation for T_t. We ought perhaps to emphasize at this point that the picture we choose, either that of extended states labelled by k or localized states labelled by p, is a matter of convenience, for either set of states can be formed from the other. In fact Holstein shows that to lowest order in J their lifetimes $\tau(k,T)$ and $\tau(p,T)$ are the same, so that the definition (9.7.32) of the transition temperature is consistent with both (9.7.26) and (9.7.27). On the other hand, the usefulness of the extended states is largely limited to the régime in which the band picture is valid.

Holstein has solved for T_t in the case $\gamma = S_0 = 10$, and finds that for $\hbar\omega_0/J$ between 1 and 100, the transition temperature lies between $0.4\theta_0$ and $0.55\theta_0$, where θ_0 is a characteristic temperature given by $\theta_0 = \hbar\omega_0/k_B$.

For polarons associated with holes (rather than electrons, but for which the theory is the same) in the valence band of KCl, Nettel (1962) finds that

S_0 is between 15 and 20 and so T_t is a smaller fraction of θ_0. The high-temperature condition (9.7.28), which, if we neglect dispersion, can be expressed as

$$2S_0 \operatorname{cosech}(\hbar\omega_0/2k_B T) \gg 1 \tag{9.7.34}$$

is satisfied for $T \gtrsim \theta_0/4$ if $S_0 \sim 10$ or larger.

The solution for T_t is shown in Figure 9.3, giving us the transition between band and hopping régimes.

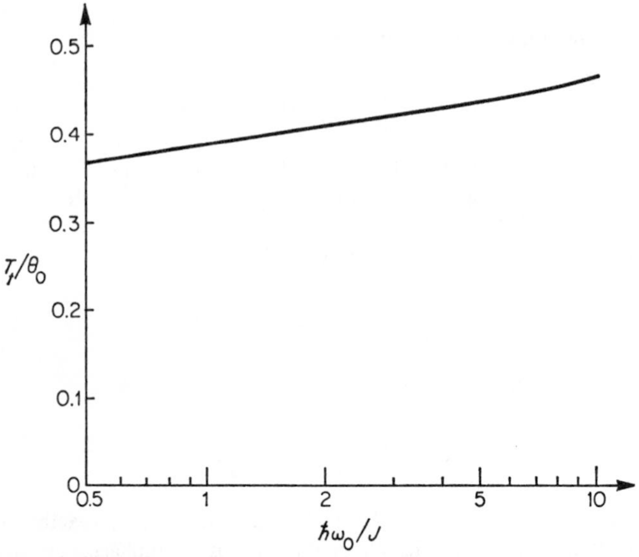

FIGURE 9.3. Reduced transition temperature as function of $\hbar\omega_0/J$. Case shown is $\gamma = S_0 = 10$.

9.7.2 Hopping mobility for small polarons

As we remarked above, when the polaron states become localized the polaron motion is essentially diffusive. For a non-degenerate system of small polarons, mobility and diffusion coefficient are connected by the Einstein relation

$$\mu = eD/k_B T \tag{9.7.35}$$

and $D \sim a^2 W$ where a is $\sim 3 \times 10^{-8}$, corresponding to the approximate cation–cation distance in an alkali halide. Using the result (9.7.30) and equation (9.7.35) we find for the hopping mobility the result

$$\mu_{\text{hop}} = ea^2 W/k_B T$$

$$= \frac{ea^2}{k_B T}\frac{J^2}{\hbar^2 \omega_0}\left[\frac{\pi}{\gamma \operatorname{cosech} \frac{1}{2}(\beta\hbar\omega_0)}\right]^{\frac{1}{2}} \exp[-2\gamma \tan \tfrac{1}{4}(\beta\hbar\omega_0)]. \tag{9.7.36}$$

9.7.3 *Band mobility for small polarons*

At temperatures less than T_t, we can again use the Einstein relation but the diffusion constant is now given by (with now no well defined jump distance)

$$D = \langle v^2(k,T)\,\tau(k,T)\rangle_{a\omega}, \tag{9.7.37}$$

where

$$v(k,T) = \left(\frac{a}{\hbar}\right)\left[\frac{\partial E(k,T)}{\partial k}\right] = \left(\frac{2J}{\hbar}\right)\sin k \exp[-S(T)]. \tag{9.7.38}$$

Then it can be shown that the band mobility is given by

$$\mu_{\text{band}} = \frac{e}{k_B T} 2J^2 a^2 \frac{e^{-2S(T)}}{\hbar}\,\tau(k,T)$$

$$= \frac{e^2 a^2 \omega_0}{k_B T}\left[\frac{\gamma \operatorname{cosech} \frac{1}{2}(\beta\hbar\omega_0)}{\pi}\right]^{\frac{1}{2}} \exp[-2\gamma \operatorname{cosech} \tfrac{1}{2}(\beta\hbar\omega_0)].$$

The range of validity of this result is restricted to $T \gtrsim \theta_0/4$, because of equation (9.7.34). The temperature dependence of the hopping and band mobilities is shown in Figure 9.4, taken from Holstein (1959).

We want to stress finally that the formula for the hopping mobility of small polarons was derived under two restrictions: (i) $2J$ is less than the polaron binding energy and (ii) the energy resonance integral J is so small that the hopping mobility can be found using J to first order in a perturbation theory. Holstein has made an estimate of the upper limit of J below which perturbation theory is valid, but we must refer the reader to the original paper for further details.

9.8 Excitons

The theory of excitons, to which we now turn, was invented by Frenkel (1931, 1936), Peierls (1932) and Wannier (1937). It is appropriate to describe the low-lying excited states of insulators (see the review by Elliot, 1963).

Though we might begin by regarding one particular atom in the lattice as excited, it would not be physically reasonable to characterize an excited electronic state of a perfect crystal in this way, because the choice of the atom which is excited is clearly arbitrary. In the presence of interactions, the resultant degenerate states resonate with each other, and in a proper description the excitations propagate through the crystal. This situation is somewhat analogous to that met in Chapter 4, where the low-lying excited states

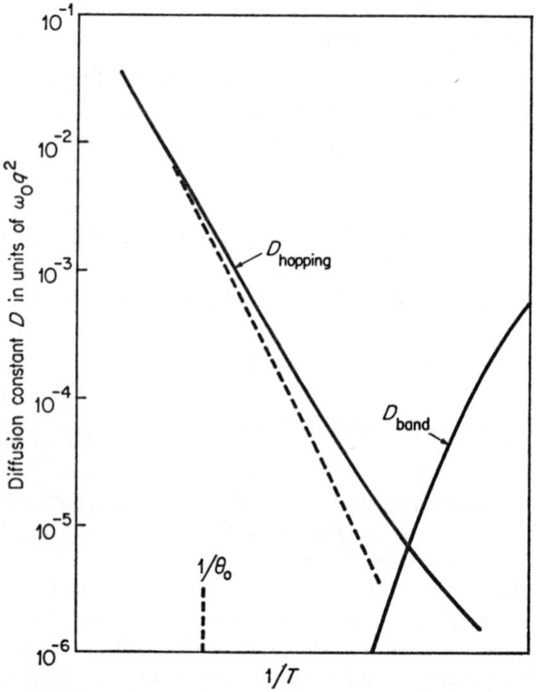

FIGURE 9.4. Illustrating temperature dependence of mobility in hopping and band régime. Diffusion constant is plotted, but is related to mobility by Einstein relation.

of a ferromagnet could be thought of starting with a misaligned spin on one atom, this misalignment propagating through the crystal as a spin wave. Such a wave is, as we saw, characterized by a wave vector **k**. Frenkel described the states in which excitations propagate as excitons; in modern parlance these are the quasi-particles of the running waves (like magnons and phonons).

This was the description used by Frenkel and Peierls. Wannier showed that the excitation could be regarded as a bound electron–hole pair, the hole in the valence band resulting from the excitation of an electron across the gap in the insulator. In this case, the state is described by a total wave vector **k**.

However, Wannier's ideas were formulated with particular reference to weakly bound electron–hole pairs, and therefore the case in which the electron and hole are separated by distances large compared with the lattice constant is nowadays referred to as the Wannier exciton.

Though the idea of Wannier has, in principle, general applicability, the small exciton envisaged by Frenkel is less usefully described mathematically by such an approach. To contrast the two régimes we refer to Figure 9.5. In the large-radius exciton, the problem is obviously simpler since the electron is seeing the hole plus an average lattice potential, whereas when the exciton radius becomes of the order of the lattice spacing, the detail of the lattice potential becomes important, in addition to the effect of the hole.

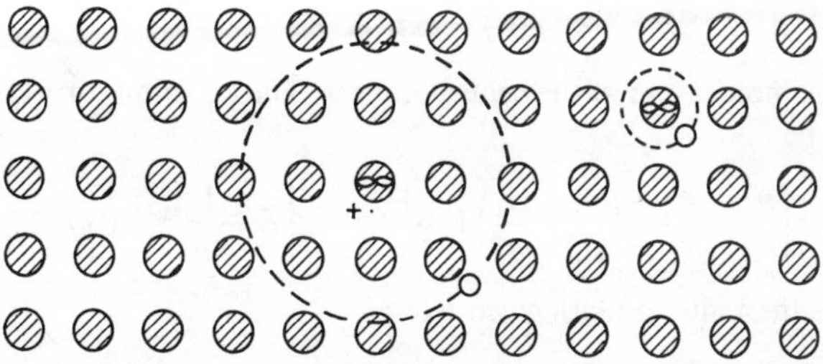

FIGURE 9.5. Shows large exciton, where electron sees hole plus average lattice potential. In contrast, small exciton is also shown and here detail of lattice potential is important.

We shall focus attention first on the Wannier exciton, which we shall now go on to deal with in some detail.

9.8.1 Effective mass equation for Wannier exciton

Though the detailed lattice potential, as we have seen, is not required in this case, we must clearly attribute to the electron and the hole effective masses m_e^* and m_h^* which reflect the character of the conduction and valence bands in the crystal under consideration. A fuller presentation of effective mass theory will be given in the next chapter.

Wannier's idea can then be formulated in a useful approximate way by neglecting the other particles than the excited electron and hole, except through the way they determine the effective masses m_e^* and m_h^* and an effective interaction, which we take as $-e^2/\varepsilon r$, where ε is an appropriate dielectric constant. The Schrödinger equation is therefore written as

$$\left(-\frac{\hbar^2}{2m_e^*}\nabla_e^2-\frac{\hbar^2}{2m_h^*}\nabla_h^2-\frac{e^2}{\varepsilon r_{eh}}\right)\psi = E\psi, \tag{9.8.1}$$

where we have assumed that the conduction and valence bands are non-degenerate and parabolic.

We now separate out the centre-of-mass motion by writing ψ in the form

$$\psi(\mathbf{R},\mathbf{r}) = g(\mathbf{R})f(\mathbf{r}), \tag{9.8.2}$$

where

$$\mathbf{R} = \tfrac{1}{2}(\mathbf{r}_e+\mathbf{r}_h) \quad \text{and} \quad \mathbf{r} = \mathbf{r}_e-\mathbf{r}_h \tag{9.8.3}$$

represent the centre of mass of the pair and the electron–hole separation. $g(\mathbf{R})$ is then found to satisfy a free-particle wave equation, with solution

$$g(\mathbf{R}) = \exp(i\mathbf{k}.\mathbf{R}), \tag{9.8.4}$$

$\mathbf{k}$ being the momentum of the centre of mass. The equation for $f(\mathbf{r})$ is readily shown to be

$$\left[\frac{p^2}{2\mu}-\frac{e^2}{\varepsilon r}-\frac{\hbar}{2}\left(\frac{1}{m_h^*}-\frac{1}{m_e^*}\right)\mathbf{k}.\mathbf{p}\right]f(\mathbf{r}) = \left[E-\frac{1}{8\mu}\hbar^2|\mathbf{k}|^2\right]f(\mathbf{r}), \tag{9.8.5}$$

where μ is the reduced mass given by

$$\frac{1}{\mu} = \frac{1}{m_e^*}+\frac{1}{m_h^*}. \tag{9.8.6}$$

The term $\mathbf{k}.\mathbf{p}$ is just like the one we met in the discussion of section 1.11.7 on electronic band structure. Inverting the argument we used there by writing

$$f(\mathbf{r}) = e^{i\alpha\mathbf{k}.\mathbf{r}}F(\mathbf{r}), \tag{9.8.7}$$

we find that F satisfies the Schrödinger equation

$$\left[\frac{p^2}{2\mu}-\frac{e}{\varepsilon r}\right]F = \left[E-\frac{\hbar^2 k^2}{2(m_e^*+m_h^*)}\right]F, \tag{9.8.8}$$

where we have chosen

$$\alpha = \frac{1}{2}\frac{(m_e^*-m_h^*)}{(m_e^*+m_h^*)} \tag{9.8.9}$$

to eliminate the $\mathbf{k}.\mathbf{p}$ term.

But equation (9.8.8) is just a hydrogenic atom equation, and for each value of **k** we have bound states given by

$$E_n(\mathbf{k}) = \frac{-\mu e^4}{2\hbar^2 \varepsilon^2 n^2} + \frac{\hbar^2 k^2}{2(m_e^* + m_h^*)}. \tag{9.8.10}$$

The corresponding total wave function is

$$\psi_{nlm\mathbf{k}} = e^{i\mathbf{k}\cdot(\mathbf{R}+\alpha\mathbf{r})} F_{nlm}(\mathbf{r}), \tag{9.8.11}$$

and we see that, in essence, we are now working with a centre-of-mass coordinate $\mathbf{R}+\alpha\mathbf{r}$. The $F_{nlm}(\mathbf{r})$ are simply the appropriate hydrogenic wave functions.

For every hydrogenic state (nlm), we clearly have a band of energies by varying **k** in equation (9.8.10). In a lattice, **k** will run over the BZ and we get an exciton band, representing low-lying excited states of the crystal.

This theory can be tested because optical absorption from the ground state can create excitons near $\mathbf{k} = 0$, and the direct absorption spectrum is then expected to be a series of lines below the optical absorption edge of the crystal. Experimental results for Cu_2O are shown in Figure 9.6. In this case, it appears that there are two spherical band edges at $\mathbf{k} = 0$, and the spectrum turns out to be well described by the above theory. The $n = 1$ case is missing

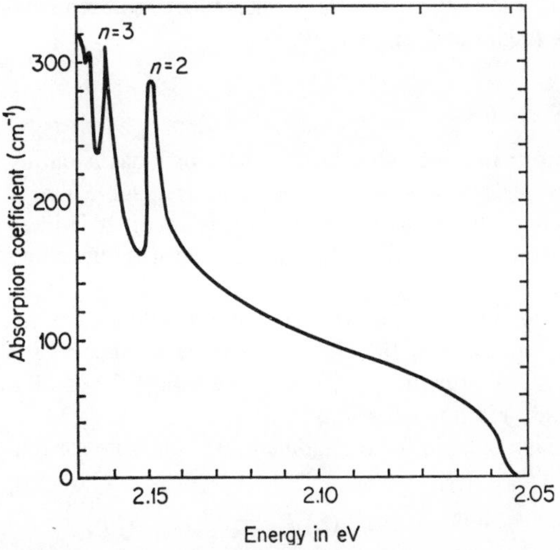

FIGURE 9.6. Cu_2O exciton spectrum. Principal quantum numbers associated with "hydrogenic" spectrum are shown.

(i.e. is very weak) since the band–band transition is forbidden (see, for example, Gross, 1962).

As Knox (1963) has emphasized, the bands need only be parabolic over a region of $\sim 15\%$ of the distance from the centre of the BZ to the boundary in the (100) direction. In general, if $E \propto k^2$ out to a $|\mathbf{k}|$ value $\sim \mu/ma_0 e$, where $a_0 = \hbar^2/me^2$, then even the hydrogenic ground state 1s energy will not be appreciably in error.

The above treatment can be extended to deal with ellipsoidal band edges by using the reciprocal effective mass tensor of sections 1.10.2 and 1.10.3 (see Problem 9.1).

For a more basic approach to the exciton problem than we have been able to give here, we refer the reader to Knox (1963). However, we want to comment finally on the dielectric constant to use in the Schrödinger equation (9.8.1).

The situation depends again on the electron–hole separation. If the electron and hole are close together, their internal kinetic energy is high. The other valence electrons, which can be assigned an effective resonant frequency E_G/h, E_G being the energy gap, cannot follow the internal motion of the pair, and the dielectric constant can be put equal to unity. But as the radius of the exciton becomes large, this situation no longer obtains and eventually even the ions can respond to the motion of the pair: $\varepsilon \to \varepsilon_\infty$. The way in which the transition from $-e^2/r$ to $-e^2/\varepsilon_\infty r$ takes place has been studied by Haken and Schottky (1958) and Englert (1959).

9.8.2 Frenkel exciton

While excitons in which the electron–hole pair has a radius greater than the lattice parameter occur quite frequently, it is of some interest to study the other limit when the radius of the exciton is small. It is likely that such a model is relevant to solid argon, but certain that it is applicable to molecular crystals.

Let us try to build an approximate ground-state wave function of the Hamiltonian representing the electrons in the insulator by starting from ground-state atomic functions $\phi(\mathbf{r})$, since we expect the small radius exciton to occur in tight-binding situations.

Thus, if we ignore spin for the moment, we can write for our ground-state wave function Φ_0

$$\Phi_0 = A\{\phi_{0\mathbf{R}_1}(\mathbf{r}_1)\,\phi_{0\mathbf{R}_2}(\mathbf{r}_2)\ldots\phi_{0\mathbf{R}_N}(\mathbf{r}_N)\}, \tag{9.8.12}$$

where A is the antisymmetrization operator and $\phi_{0\mathbf{R}}(\mathbf{r})$ is the ground-state wave function of the atom at $\mathbf{R}$.

Consider next the promotion of an electron on the atom at $\mathbf{R}_i$ into an excited state $\phi_{n\mathbf{R}_i}(\mathbf{r})$, when we can simply change this one orbital in the wave function (9.8.12) to obtain

$$\Phi_{n\mathbf{R}_i} = A\{\phi_{0\mathbf{R}_1}(\mathbf{r}_1)\,\phi_{0\mathbf{R}_2}(\mathbf{r}_2)\ldots\phi_{n\mathbf{R}_i}(\mathbf{r}_i)\ldots\phi_{0\mathbf{R}_N}(\mathbf{r}_N)\}. \tag{9.8.13}$$

But this wave function puts one atom on a special footing, whereas the translational invariance of the Hamiltonian requires that the true many-body wave functions obey Bloch's theorem, as we showed in Chapter 2. We therefore write a wave function of Bloch form as

$$\Phi_n(\mathbf{k}) = N^{-\frac{1}{2}}\sum_{\mathbf{R}} e^{i\mathbf{k}.\mathbf{R}}\,\Phi_{n\mathbf{R}}, \tag{9.8.14}$$

where N is the number of atoms in the crystal.

Now we calculate the expectation value of the many-body Hamiltonian H_0 with respect to these zero-order wave functions (9.8.14), when we find

$$\begin{aligned} E_n(\mathbf{k}) &= \int \Phi_n^*(\mathbf{k})\,H_0\,\Phi_n(\mathbf{k})\,d\mathbf{r}_1\ldots d\mathbf{r}_N \\ &= E_{0n} + \sum_{\mathbf{R}} e^{i\mathbf{k}.\mathbf{R}}\,V_{\mathbf{R}0}^n. \end{aligned} \tag{9.8.15}$$

This, in fact, is simply the Fourier series expansion of the energy $E_n(\mathbf{k})$. The meaning of $V_{\mathbf{R}0}^n$ is that it represents the coupling between the wave function $\Phi_{n\mathbf{R}}$ on site R and that on site 0, through the Hamiltonian H_0.

The energies in equation (9.8.5) obviously define a band in just the same way as in the one-electron theory of Chapter 1, when we let $\mathbf{k}$ range over N different values. The band width is determined in order of magnitude by the largest of the V's in equation (9.8.15).

At this stage, we wish to get an estimate of the way $V_{\mathbf{R}0}^n$ in equation (9.8.15) depends on the distance $\mathbf{R}$ between the two sites we are considering (always one at the origin). Though it is not essential for this particular result, it will be helpful in showing the structure of the Frenkel exciton theory to effect a generalization of equation (9.8.15) and consider the quantity

$$H_{nn'}(\mathbf{k}) = \int \Phi_{0n}^*(\mathbf{k})\,H_0\,\Phi_{0n'}(\mathbf{k})\,d\mathbf{r}_1\ldots d\mathbf{r}_{2N}. \tag{9.8.16}$$

In the same way that Hartree–Fock energies are obtained, it is straightforward to show that we can write the above result in the form

$$\begin{aligned} H_{nn'}(\mathbf{k}) &= \delta_{nn'}\,E_0 + W_{nn'} \\ &\quad + \sum_{\mathbf{R}\neq 0} e^{i\mathbf{k}.\mathbf{R}}\left[2\int \phi_{l0}^*(\mathbf{r}_1)\,\phi_{0\mathbf{R}}^*(\mathbf{r}_2)\frac{e^2}{r_{12}}\phi_{00}(\mathbf{r}_1)\,\phi_{l'\mathbf{R}}(\mathbf{r}_2)\,d\mathbf{r}_1\,d\mathbf{r}_2 \right. \\ &\quad \left. - \int \phi_{l0}^*(\mathbf{r}_1)\,\phi_{0\mathbf{R}}(\mathbf{r}_2)\frac{e^2}{r_{12}}\phi_{l'\mathbf{R}}(\mathbf{r}_1)\,\phi_{00}(\mathbf{r}_2)\,d\mathbf{r}_1\,d\mathbf{r}_2\right], \end{aligned} \tag{9.8.17}$$

where the term involving the sum over $\mathbf{R}$ has been written in a form appropriate to a singlet state. In the triplet state, the first contribution in the square bracket of this term is in fact zero. We need not say very much about the quantity $W_{nn'}$ as it is of no consequence in the use of the theory we make below, but we can remark that it vanishes in the case $n = n'$ whenever the crystal point symmetry is such that the atomic states ϕ_n and $\phi_{n'}$ do not mix. This would be the case, for example, if n and n' were to refer to two different p functions in a cubic crystal.

It is with the lattice sum that we must now deal. The exchange term in the square bracket of equation (9.8.17) in fact is readily shown to decay exponentially as $\mathbf{R}$ is increased; it essentially has the same range as the atomic orbitals. However, the first term, which as we remarked above appears only in singlet states, requires a much more careful treatment. We can regard it, if somewhat artificially, as the Coulomb interaction between two charge clouds $\phi_n^*(\mathbf{r})\,\phi_0(\mathbf{r})$ and $\phi_0^*(\mathbf{r}-\mathbf{R})\,\phi_l(\mathbf{r}-\mathbf{R})$. As $\mathbf{R}$ increases, these charge clouds overlap less and less with each other, but the first term in the square bracket of equation (9.8.17) does not increase exponentially with $\mathbf{R}$, for it is the interaction between the multipole moments of these two "charge distributions". This type of term is known as an excitation transfer interaction as it looks like a transition matrix element in which initially the atom at the origin is excited, and finally the atom at $\mathbf{R}$ is excited.

In fact the general theory of the Frenkel exciton can now be said to consist of the calculation and diagonalization of the matrix (9.8.17). We shall content ourselves with working out an application of the theory here to the case of the splitting of longitudinal and transverse exciton states near $\mathbf{k} = 0$, which are built from p functions.

Longitudinal–transverse effect in exciton theory. The argument rests on the calculation of the excitation transfer interaction term in equation (9.8.17), and we shall take the specific case in which the ground state is s-like and there exist three possible p-like excited states labelled by n (cf. Knox, 1963). As is well known from the treatment of van der Waals interactions, we now expand the Coulomb interaction (e^2/r_{12}) in a Taylor series n powers of the components of $\mathbf{r}_1$ and $\mathbf{r}_2 - \mathbf{R}$. Because of the orthogonality of ϕ_0 and ϕ_n, only one term of this expansion contributes to equation (9.8.17), namely

$$[(\boldsymbol{\mu}_{n0}\cdot\boldsymbol{\mu}_{0n'})\,R^2 - 3(\boldsymbol{\mu}_{n0}\cdot\mathbf{R})\,(\boldsymbol{\mu}_{0n'}\cdot\mathbf{R})]\,R^{-5}, \tag{9.8.18}$$

where

$$\boldsymbol{\mu}_{n0} = 2^{\frac{1}{2}}\,e\int \phi_n^*(\mathbf{r})\,\mathbf{r}\phi_0(\mathbf{r})\,d\mathbf{r} \tag{9.8.19}$$

and

$$\boldsymbol{\mu}_{0n'} = 2^{\frac{1}{2}} e \int \phi_0^*(\mathbf{r})\, \mathbf{r}\phi_{n'}(\mathbf{r})\, d\mathbf{r}. \tag{9.8.20}$$

We now specialize to cubic crystals, and neglect the short-range part of the lattice sum in equation (9.8.17). We obviously then have to calculate

$$\sum_{\mathbf{R}\neq 0} e^{i\mathbf{k}.\mathbf{R}}\, [(\boldsymbol{\mu}_{n0}\cdot\boldsymbol{\mu}_{0n'})\, R^2 - 3(\boldsymbol{\mu}_{n0}\cdot\mathbf{R})\, (\boldsymbol{\mu}_{0n'}.\mathbf{R})]\, R^{-5}. \tag{9.8.21}$$

This term is zero when **k** is zero, for then it is the interaction between a dipole at the origin and a perfect cubic lattice of dipoles identical to each other. Cohen and Keffer (1955) have dealt in detail with the sum for cubic crystals. It is found that for small **k** such that $0 < ka \ll 1$, where a is the lattice constant, the sum reduces to

$$-\frac{4\pi}{3}\, \rho\mu_{n0}\,\mu_{0n'}(\delta_{nn'} - 3k_n\, k_{n'}\, k^{-2}) + O(k^2\, a^2), \tag{9.8.22}$$

where ρ is the density of lattice points and k_n and $k_{n'}$ are the projections of **k** on the respective dipole moments μ_{0n} and $\mu_{0n'}$. As $k \to 0$, expression (9.8.22) does not approach a unique value, being piecewise discontinuous at $k = 0$ (see Cohen and Keffer, 1955).

We must now choose the most suitable set of p functions with which to work out the theory and we shall follow Knox in the way we do this. The choice will in turn determine the orientations of the μ_{n0}. When a single impurity atom, for example, is placed in a static cubic field and excited to a P level, the three p functions may be oriented in any direction as long as they remain orthogonal. Now equation (9.8.22) will be of simple form if ϕ_n and $\phi_{n'}$ are chosen parallel and/or perpendicular to **k**; that is if the p functions are made longitudinal and transverse to **k**, as shown in Figure 9.7. In fact, except for the terms of order $(ka)^2$, this choice diagonalizes equation (9.8.22). If $n = n'$ and ϕ_n is longitudinal then $k_n = k$ and equation (9.8.22) becomes

$$+\frac{8\pi}{3}|\mu_{n0}|^2\, \rho + O(k^2\, a^2), \tag{9.8.23}$$

which holds then for longitudinal p-like excitons. If $n = n'$ and ϕ_n is transverse to **k** ($\phi_{x'}$ or $\phi_{y'}$) then by definition $k_n = 0$ and equation (9.8.23) takes the form

$$-\frac{4\pi}{3}|\mu_{n0}|^2\, \rho + O(k^2\, a^2), \tag{9.8.24}$$

which is appropriate for transverse p-like excitons. Finally, if $n \neq n'$, then k_n or $k_{n'}$ or both will be zero and equation (9.8.22) is merely of order $(ka)^2$.

It can be seen immediately then from equations (9.8.23) and (9.8.24) that longitudinal and transverse exciton states near $\mathbf{k} = 0$ built up from p functions will be split by an amount $4\pi\rho|\mu_{n0}|^2$ which is proportional to the atomic s–p oscillator strength.

This effect was discovered by Heller and Marcus (1951) and although it is, as we have seen, easily demonstrated using the Frenkel model, it is a

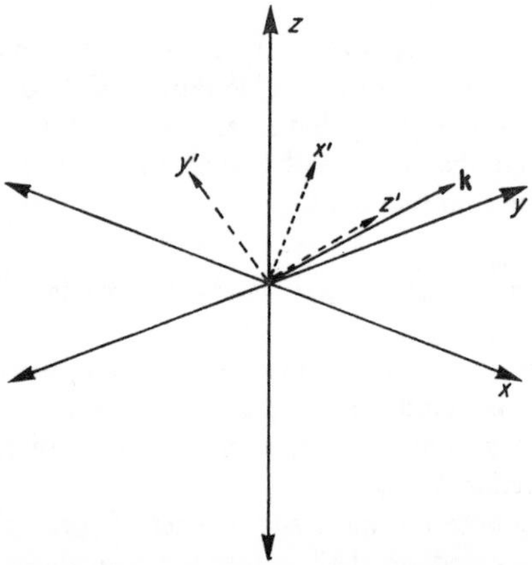

FIGURE 9.7. Coordinate transformation useful in describing transverse and longitudinal states. xyz axes are associated with crystal symmetry axes. k is in an arbitrary direction while z' is parallel to k and x' is is in xy plane.

general phenomenon, and expected to occur in any set of exciton states which can be produced by an allowed electric dipole transition. There is a very close analogy to the Lyddane–Sachs–Teller splitting of the longitudinal and transverse optical branches near the centre of the BZ in the phonon case (compare section 7.13).

We shall go on, from this discussion of large and small radius excitons, to consider briefly the so-called excitonic phases.

9.9 Excitonic phases

It was observed by Mott (1961) that in a semi-metal it could be, under certain conditions, that one could form bound electron–hole pairs, leading then to a non-conducting material. Later on, Knox (1963) pointed out that, in an insulator, the binding energy of an exciton $|E_x|$ might conceivably exceed the energy gap E_g. In this case, the normal insulating state with a fully occupied valence band would become unstable against the formation of excitons.

As a consequence of these proposals, a number of authors have discussed a new phase which can be formed at very low temperatures, and which has been named by Jerome, Rice and Kohn (1967) an "excitonic insulator".

The theory is closely related to the discussion of the exciton in the previous section, and also, as we shall see below, there is a close formal similarity between the excitonic insulating state and the superconducting state. The similarity must not be pressed to detailed physical properties though and, in particular, there is no Meissner effect in an excitonic insulator.

Let us write down a description of the ground state of an excitonic insulator in the case when we have a single valence-band maximum at $\mathbf{k} = 0$ and a single conduction-band minimum at $\mathbf{k} = \mathbf{q}$.

We shall represent the single-particle energies in this case by

$$\varepsilon_a(\mathbf{k}_a) = -\tfrac{1}{2}E_g - (2m_a)^{-1}k_a^2, \tag{9.9.1}$$

$$\varepsilon_a(\mathbf{k}_b) = \tfrac{1}{2}E_g + (2m_b)^{-1}k_b^2, \tag{9.9.2}$$

where $\mathbf{k}_a$ and $\mathbf{k}_b$ refer to the respective band extrema. We have chosen to measure energies from the centre of the gap and E_g may be positive or negative, m_a and m_b being the relevant effective masses.

For the insulating ground state in elementary band theory we can now write down the wave function

$$\Phi = \prod_{\mathbf{k}} a_{\mathbf{k}}^{\dagger} |\,\rangle, \tag{9.9.3}$$

where $|\,\rangle$ is the state with no electrons, $a_{\mathbf{k}}^{\dagger}$, $a_{\mathbf{k}}$ create and destroy electrons in band a with wave vector $\mathbf{k}$ and the values of $\mathbf{k}$ in equation (9.9.3) run over the Brillouin zone.

The possibility of forming bound electron–hole pairs now suggests a new Hartree–Fock ground-state wave function

$$\Psi = \prod_{\mathbf{k}} \alpha_{\mathbf{k}}^{\dagger} |\,\rangle, \tag{9.9.4}$$

where, if the system is indeed unstable against formation of pairs, then we expect that a state with a lower energy than Φ can be obtained by choosing

$$\alpha_{\mathbf{k}} = u_{\mathbf{k}} a_{\mathbf{k}} - v_{\mathbf{k}} b_{\mathbf{k}}, \quad |u_{\mathbf{k}}|^2 + |v_{\mathbf{k}}|^2 = 1, \tag{9.9.5}$$

$b_{\mathbf{k}}^{\dagger}$ and $b_{\mathbf{k}}$ being respectively creation and annihilation operators in band b, with wave vector $\mathbf{q}+\mathbf{k}$.

Writing Ψ' in terms of Φ, we then find

$$\Psi' = \prod_{\mathbf{k}} (u_{\mathbf{k}}^* - v_{\mathbf{k}}^* b_{\mathbf{k}}^{\dagger} a_{\mathbf{k}})\,\Phi \tag{9.9.6}$$

and this shows, in analogy with the Bardeen–Cooper–Schrieffer (BCS) theory discussed in Chapter 8, that in this state a hole in $(a, \mathbf{k})$ and an electron in $(b, \mathbf{k}+\mathbf{q})$ are either both present or both absent.

It now remains to link up the BCS-like description (9.9.6) with equation (9.8.8) for the exciton wave function. To do so, we proceed in standard fashion to minimize the total energy with respect to the quantities $u_{\mathbf{k}}$ and $v_{\mathbf{k}}$. Just as in the BCS theory, we are then led to a gap function $\Delta_{\mathbf{k}}$ which satisfies

$$\Delta_{\mathbf{k}} = \sum_{\mathbf{p}} V(\mathbf{k}-\mathbf{p})\,[\Delta_{\mathbf{p}}/2[\xi_{\mathbf{p}}^2 + |\Delta_{\mathbf{p}}|^2]^{\frac{1}{2}}], \tag{9.9.7}$$

where

$$\xi_{\mathbf{k}} = \tfrac{1}{2}[\varepsilon_b(\mathbf{k}) - \varepsilon_a(\mathbf{k})] \tag{9.9.8}$$

and

$$V(\mathbf{q}) = \frac{4\pi e^2}{\varepsilon(\mathbf{q})q^2} \tag{9.9.9}$$

is the interaction term in the Hamiltonian, $\varepsilon(\mathbf{q})$ being the appropriate dielectric function for the material under discussion.

Abbreviating the quantity $\Delta_{\mathbf{p}}/2[\xi_{\mathbf{p}}^2 + |\Delta_{\mathbf{p}}|^2]^{\frac{1}{2}}$ appearing in equation (9.9.7) by $\phi_{\mathbf{p}}$, the gap equation may be written as

$$\left[\left(E_{\mathrm{g}} + \frac{k^2}{2\mu}\right)^2 + 4|\Delta_{\mathbf{k}}|^2\right]^{\frac{1}{2}} \phi_{\mathbf{k}} = \sum_{\mathbf{p}} V(\mathbf{k}-\mathbf{p})\,\phi_{\mathbf{p}}. \tag{9.9.10}$$

This can now be compared with equation (9.8.8) in Fourier transform for the exciton wave function: namely

$$\left[\frac{k^2}{2\mu} + |E_{\mathrm{x}}|\right]\chi_{\mathbf{k}} = \sum_{\mathbf{p}} V(\mathbf{k}-\mathbf{p})\,\chi_{\mathbf{p}}, \tag{9.9.11}$$

where μ is the reduced mass ($\mu^{-1} = m_a^{-1} + m_b^{-1}$). These two equations together can now be used to verify that $\Delta = 0$ when $E_{\mathrm{g}} \geqslant |E_{\mathrm{x}}|$. However, a non-trivial solution corresponding to $|\Delta| > 0$ exists for all values of E_{g}, either positive or negative, which are $\leqslant |E_{\mathrm{x}}|$.

We ought to elaborate a little on the relationship of the theory outlined above to the theory of superconductivity. In Chapter 8, we saw that a basic classification of a superconductor was that the two-particle density matrix $\langle \mathbf{r}_1' \mathbf{r}_2' | \rho_2 | \mathbf{r}_1 \mathbf{r}_2 \rangle$ remains finite in the limit $|\mathbf{r}_1 - \mathbf{r}_1'| \to \infty$, $\mathbf{r}_1 \sim \mathbf{r}_2$, $\mathbf{r}_1' \sim \mathbf{r}_2'$. This is, of course, the off-diagonal long-range order we stressed in Chapter 8, and is a basic characteristic of the superconducting state.

In the case of the excitonic insulator at $T=0$, Jérome and colleagues (1967) have shown that the two-particle density matrix has no such off-diagonal long-range order. However, a new periodicity in this density matrix, characterized by the wave vector $\mathbf{q}$, does appear and this is a basic feature of the excitonic phase. The d.c. conductivity at $T=0$ can then be shown to be zero and we are dealing with an insulating phase.

9.9.1 Finite temperature resistivity

At finite temperatures, a calculation paralleling that given in Chapter 8 for the superconducting gap leads to the result

$$\Delta_{\mathbf{p}} = \sum_{\mathbf{k}} V(\mathbf{p}-\mathbf{k})(\Delta_{\mathbf{k}}/2E_{\mathbf{k}})\tanh(\tfrac{1}{2}\beta E_{\mathbf{k}}), \tag{9.9.12}$$

where $E_{\mathbf{k}}^2 = \xi_{\mathbf{k}}^2 + |\Delta_{\mathbf{k}}|^2$ and $\beta = (1/k_B T)$ as usual. This should be compared with the BCS result (8.9.34) at elevated temperatures.

To set up the resistivity, we use a formalism which is by now rather familiar. Let us examine the response of the system to a low-frequency electric field $\mathscr{E}(\mathbf{q}\omega)$ which we can represent by a vector potential

$$\mathscr{A}(\mathbf{q}\omega) = -(i/\omega)\,\mathscr{E}(\mathbf{q}\omega). \tag{9.9.13}$$

Then we define the response function K_{ij} by

$$j_i(\mathbf{q}\omega) = \sum_j K_{ij}(\mathbf{q}\omega)\,\mathscr{A}_j(\mathbf{q}\omega) \tag{9.9.14}$$

and the conductivity σ given, for cubic crystals, by

$$\mathbf{j}(\mathbf{q}\omega) = \sigma(\mathbf{q}\omega)\,\mathscr{E}(\mathbf{q}\omega) \tag{9.9.15}$$

is evidently

$$\sigma(\mathbf{q}\omega) = -i\,\mathrm{Re}\left[\frac{1}{\omega}K_{ii}(\mathbf{q}\omega)\right]. \tag{9.9.16}$$

To find the d.c. conductivity in a uniform electric field we take the appropriate limit as q and ω tend to zero.

To calculate the response function K at elevated temperatures, we can proceed via the finite-temperature time-dependent Green's functions defined as usual by

$$\left.\begin{aligned} \mathscr{G}_i(1,1') &= -i\langle T\{\tilde{\psi}_i(1)\,\tilde{\psi}_i^\dagger(1')\}\rangle \\ \mathscr{F}^\dagger(1,1') &= -i\langle T\{\tilde{\psi}_a(1)\,\tilde{\psi}_b^\dagger(1')\}\rangle \end{aligned}\right\} \quad i=a,b, \tag{9.9.17}$$

where $\langle ... \rangle$ is the thermal average of the time-ordered product. It is shown in Appendix 9.2 that, in the case when the masses m_a and m_b are equal, we can write, in Fourier transform,

$$\mathscr{G}_b(\mathbf{p}\varepsilon) = \frac{u_\mathbf{p}^2(1-n_\mathbf{p})}{\varepsilon - E_\mathbf{p} + i\delta} + \frac{v_\mathbf{p}^2 n_\mathbf{p}}{\varepsilon + E_\mathbf{p} + i\delta} + \frac{v_\mathbf{p}^2(1-n_\mathbf{p})}{\varepsilon + E_\mathbf{p} - i\delta} + \frac{u_\mathbf{p}^2 n_\mathbf{p}}{\varepsilon - E_\mathbf{p} - i\delta}, \tag{9.9.18}$$

$$\mathscr{G}_a(\mathbf{p}\varepsilon) = \frac{v_\mathbf{p}^2(1-n_\mathbf{p})}{\varepsilon - E_\mathbf{p} - i\delta} + \frac{u_\mathbf{p}^2 n_\mathbf{p}}{\varepsilon + E_\mathbf{p} - i\delta} + \frac{u_\mathbf{p}^2(1-n_\mathbf{p})}{\varepsilon + E_\mathbf{p} + i\delta} + \frac{v_\mathbf{p}^2 n_\mathbf{p}}{\varepsilon - E_\mathbf{p} + i\delta} \tag{9.9.19}$$

and

$$\mathscr{F}(\mathbf{p}\varepsilon) = \mathscr{F}^\dagger(\mathbf{p}\varepsilon) = u_\mathbf{p} v_\mathbf{p}\left(\frac{1-n_\mathbf{p}}{\varepsilon - E_\mathbf{p} + i\delta} - \frac{n_\mathbf{p}}{\varepsilon + E_\mathbf{p} + i\delta} - \frac{1-n_\mathbf{p}}{\varepsilon + E_\mathbf{p} - i\delta} + \frac{n_\mathbf{p}}{\varepsilon - E_\mathbf{p} - i\delta}\right), \tag{9.9.20}$$

where $n_\mathbf{p} = [\exp(\beta E_\mathbf{p})+1]^{-1}$. The phase of the gap function has been chosen to be real: the total energy is independent of this choice. The finite-temperature response function K is obtained in Appendix 9.2 and the double limit we discussed above, required to get the d.c. conductivity in a uniform electric field, is given by

$$\lim_{\omega\to 0}\left[\lim_{q\to 0} K_{ij}(\mathbf{q}\omega)\right]$$
$$= \frac{e^2}{m^2}\sum_\mathbf{p} p_i p_j \frac{4u_\mathbf{p}^2 v_\mathbf{p}^2}{E_\mathbf{p}}(1-2n_\mathbf{p}) + \frac{e}{m}\sum_\mathbf{p} p_i \gamma_j(\mathbf{p})\frac{u_\mathbf{p} v_\mathbf{p}}{E_p}(u_\mathbf{p}^2 - v_\mathbf{p}^2)(1-2n_\mathbf{p}), \tag{9.9.21}$$

where $\gamma_j(\mathbf{p})$ satisfies the equation

$$\gamma_j(\mathbf{p}) = \sum_\mathbf{k} V(\mathbf{p}-\mathbf{k})\frac{(u_\mathbf{k}^2 - v_\mathbf{k}^2)^2}{2E_\mathbf{k}}(1-2n_\mathbf{k})\gamma_j(\mathbf{k})$$
$$+\frac{e}{m}\sum_\mathbf{k} V(\mathbf{p}-\mathbf{k})\, 2k_j \frac{u_\mathbf{k} v_\mathbf{k}}{E_\mathbf{k}}(u_\mathbf{k}^2 - v_\mathbf{k}^2)(1-2n_\mathbf{k}). \tag{9.9.22}$$

The order of the limits $\mathbf{q}\omega \to 0$ is important away from $T = 0$. The d.c. conductivity is determined by the response to a very low-frequency uniform electric field and thus we must take the limit $\mathbf{q}\to 0$ before letting $\omega \to 0$.

In general the calculation of the conductivity involves the solution of equation (9.9.22). However, in the semi-metal limit the inhomogeneous term in equation (9.9.22) is zero on account of particle–hole symmetry. Therefore

$\gamma(\mathbf{p}) = 0$ is the solution and in this limit it is easy to show that

$$\lim_{\omega\to 0}\lim_{q\to 0} K_{ij}(\mathbf{q}\omega) = \frac{e^2}{m^2}\sum_{\mathbf{p}} p_i p_j \frac{\Delta^2}{E_{\mathbf{p}}^3}(1-2n_{\mathbf{p}}) \tag{9.9.23}$$

$$= \frac{e^2}{3m^2}\sum_{\mathbf{p}} \frac{p^2\Delta^2}{E_{\mathbf{p}}^3}(1-2n_{\mathbf{p}})\,\delta_{ij}. \tag{9.9.24}$$

There is also the diamagnetic contribution to the response function (see Appendix 9.2) given by

$$K_{ij}^{\mathrm{d}}(0,0) = -\frac{e^2}{m}\sum_{\mathbf{p}} [2v_{\mathbf{p}}^2 + (u_{\mathbf{p}}^2 - v_{\mathbf{p}}^2)\,n_{\mathbf{p}}]\,\delta_{ij}. \tag{9.9.25}$$

In the semi-metal limit, the second term on the right-hand side vanishes on account of particle symmetry and combining the above two results we find from equation (9.9.16)

$$\sigma(0,\omega) = -\frac{i}{\omega}\mathrm{Re}\left[\lim_{\omega\to 0}\left\{\lim_{q\to 0}[K_{ii}^{\mathrm{p}}(\mathbf{q}\omega) + K_{ii}^{\mathrm{d}}(\mathbf{q}\omega)]\right\}\right]$$

$$= -\frac{ie^2}{m\omega}\sum_{\mathbf{p}}\left[\frac{p^2\Delta^2}{3mE_{\mathbf{p}}^3}(1-2n_{\mathbf{p}}) - \left(1-\frac{\xi_{\mathbf{p}}}{E_{\mathbf{p}}}\right)\right]. \tag{9.9.26}$$

If we denote the d.c. conductivity by $\sigma_1(0)$ and write this in the usual fashion as

$$\sigma_1(0) = \frac{n_{\mathrm{eff}}\,e^2\,\tau}{m}, \tag{9.9.27}$$

where τ is the relaxation time then, using the Kramers–Kronig relationship, the effective number of carriers n_{eff} may be shown to take the form

$$n_{\mathrm{eff}} = 2n_{\mathrm{c}}\int_0^\infty \frac{dt}{(t^2+1)^{\frac{3}{2}}}\{\exp[\beta\Delta(t^2+1)^{\frac{1}{2}}]+1\}^{-1}, \tag{9.9.28}$$

where n_{c} is the number of carriers which in the semi-metal limit is equal to $k_{\mathrm{f}}^3/3\pi^2$. The temperature dependence of the conductivity can now be written formally as

$$\frac{d\sigma}{dT} = \frac{\tau e^2}{m}\left\{\frac{\partial n_{\mathrm{eff}}}{\partial T} + \frac{\partial n_{\mathrm{eff}}}{\partial \Delta}\frac{\partial \Delta}{\partial T}\right\}. \tag{9.9.29}$$

As $T \to T_c$, the second term on the right-hand side diverges and $d\sigma/dT \to \infty$. On the other hand, as $T \to 0$, $d\sigma/dT \to 0$. The qualitative behaviour thus to be expected is sketched in Figure 9.8. An entirely similar calculation leads

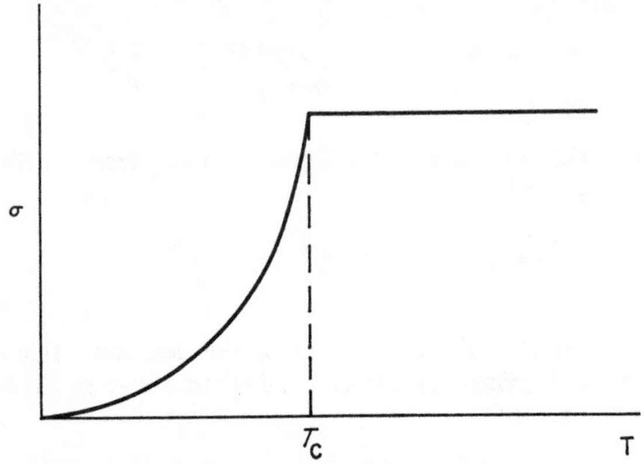

FIGURE 9.8. Schematic form of d.c. conductivity σ as function of temperature in semimetallic limit. T_c is the transition temperature.

to the behaviour of σ at a fixed temperature as the energy gap E_g is changed by varying the external pressure. An extensive account of this general area is given in the article by Halperin and Rice (1968).

CHAPTER 10

Defects and disordered systems

10.1 Introduction

One of the basic questions we must ask about impurities or defects in metals and semi-conductors is: "What potential should we use to represent the effect of the impurity centre on the energy bands?"

Of course, a deeper question can be asked; namely: "Can we define such a potential anyway, in a many-body system such as we have in a solid?" We gave a partial answer to this question in Chapter 1, where we saw that the exact charge density in the ground state of a many-body system could be generated by a suitable local one-body potential. It seems then that the use of potential theory is widely justified in the impurity problem and we shall proceed on that assumption.

We shall deal, at first, with only one aspect of the potential, its range, for charged impurity centres in metals and semi-conductors. The detailed form of the potential depends, of course, on the electronic structure of the cores of the solute and solvent atoms. Here, at the present stage in the development of the theory, some use of pseudo-potentials will be involved. Some detail of pseudo-potential theory [fully discussed in the book by Harrison (1966)] was given in Chapter 1 and it is necessary to use this to make progress in a number of applications [e.g. a vacancy in Si; Callaway and Hughes (1967)].

10.2 Debye shielding by free carriers

10.2.1 Shielding of charged centre in metals

We have, in essence, already dealt with the screening of an excess charge Ze in a degenerate Fermi gas in Chapter 2.

(a) *Semi-classical theory.* If we assume that the defect potential $V(\mathbf{r})$ varies but slowly over a de Broglie wavelength of an electron at the Fermi level, then in linear Hartree theory we have from equation (2.3.13)

$$\nabla^2 V = q^2 V : q^2 = \frac{4k_f}{\pi a_0}, \qquad (10.2.1)$$

where k_f is the Fermi wave number. The appropriate solution for a single charged defect at the origin is clearly, following Mott,

$$V = -\frac{Ze^2}{r}\mathrm{e}^{-qr} \tag{10.2.2}$$

and, in a good metal, the screening radius or Debye length q^{-1} is about 1 Å.

Actually, impurity centres like Zn in Cu are not sufficiently weak to allow the linearization in equation (10.2.1). The non-linear Thomas–Fermi equation has then to be solved, and V in equation (10.2.2) is not then, of course, simply linear in Z. However, the essential features of the problem remain unchanged from the linear solution.

(b) *Wave theory.* More serious than the linearization is the fact that the potential in equation (10.2.2) varies appreciably over a de Broglie wavelength and the semi-classical approximation of the Thomas–Fermi theory is not really valid. In Chapter 2, it was shown that in a linear (Born) approximation, equation (10.2.1) must then be replaced by [see equation (2.3.14)]

$$\nabla^2 V = \frac{2k_f^2}{\pi^2 a_0}\int d\mathbf{r}'\, V(\mathbf{r}')\frac{j_1(2k_f|\mathbf{r}-\mathbf{r}'|)}{|\mathbf{r}-\mathbf{r}'|^2}. \tag{10.2.3}$$

We saw in Chapter 2 that in the case of an extreme short-range potential, the displaced charge round the scattering centre fell off asymptotically like

$$\rho-\rho_0 \sim \frac{\cos 2k_f r}{r^3}, \tag{10.2.4}$$

and this can be shown (cf. Problem 10.8) to be the case for the self-consistent equation (10.2.3). From this equation, the potential $V(r)$ can be written asymptotically as

$$V(r) \sim \frac{\cos 2k_f r}{r^3}. \tag{10.2.5}$$

The potentials obtained from equations (10.2.1) and (10.2.3) are shown in Figure 10.1. The rather rapid screening out of the Coulomb potential is shown in both cases, but the wave theory shows diffraction effects in the tail, and leads to the Friedel oscillations in the charge density and potential. The screening is incomplete and these oscillations, with wavelength π/k_f, reflect the sharp Fermi surface.

If we avoid the linearization in equation (10.2.3), or introduce core structure in the scattering centre, then a phase shift ϕ is introduced into equation (10.2.5), and we have the asymptotic form $\cos(2k_f r+\phi)/r^3$, as shown in detail in section 10.4.4 below.

Experimental evidence for the essential correctness of equation (10.2.5) comes from:

(i) Observation of Knight shifts in dilute alloys, which can be explained using equation (10.2.5) but not equation (10.2.2) (see Kohn and Vosko, 1960).

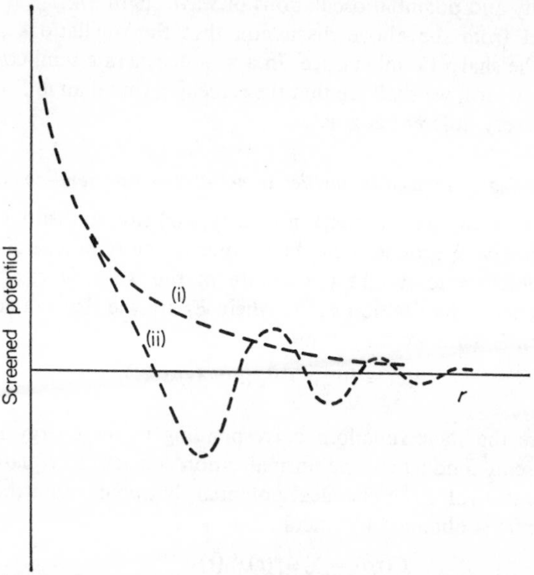

FIGURE 10.1. Screened potentials round point charge in Fermi gas. (i) Semiclassical theory. (ii) Wave theory.

(ii) Less directly, the phonon spectra of some pure metals (e.g. Pb) which cannot be explained with a short-range ion–ion interaction deriving from equation (10.2.2), as discussed further in section 10.12.1, but requires a long-range oscillatory interaction like equation (10.2.5).

We have considered a point-ion model here, to establish the nature and range of the screened potential $V(r)$. If we write the Fourier components of $V(r)$ in terms of the dielectric function $\varepsilon(k)$, namely

$$V(k) = -\frac{4\pi Ze^2}{k^2\,\varepsilon(k)}, \tag{10.2.6}$$

then we can replace the point-ion potential $-4\pi Ze^2/k^2$ by an appropriate pseudo-potential to obtain a more realistic impurity potential, namely

$$V_{\text{pseudo}}(k) = \frac{\Delta V_{\text{bare ion}}}{\varepsilon(k)}. \tag{10.2.7}$$

The presence of $\varepsilon(k)$ as given by the wave theory of equation (2.3.20) again ensures the Friedel oscillations in $V_{\text{pseudo}}(\mathbf{r})$.

These then are the main characteristics determining the range of impurity potentials in metals: a region of localized Mott screening of ~ 1 Å, followed by the density and potential oscillations of wavelength π/k_f.

We expect from the above discussion that the oscillations are present because of the sharp Fermi surface. In a non-degenerate semi-conductor, to which we now turn, we shall see that the screening round an ionized impurity centre has a very different character.

10.2.2 Screening of ionized impurities in non-degenerate semi-conductors

We want now to find the electron density, $n(\mathbf{r})$ say, to distinguish it from the degenerate case, around a charged impurity centre in a semi-conductor. We have therefore to weight the square of the wave functions with the Boltzmann energy distribution $e^{-\beta E_i}$, where E_i are the electron energy levels and $\beta = 1/k_B T$. Thus we may write

$$n(\mathbf{r}) = \text{const} \sum_i \psi_i^*(\mathbf{r})\,\psi_i(\mathbf{r})\,e^{-\beta E_i}, \tag{10.2.8}$$

where ψ_i are the wave functions corresponding to the energy levels E_i in the impure semi-conductor. The normalization constant in equation (10.2.8) will in general involve the chemical potential, but apart from this constant, the density $n(\mathbf{r})$ is obtained by calculating

$$C(\mathbf{r}\beta) = \sum_i \psi_i^*(\mathbf{r})\,\psi_i(\mathbf{r})\,e^{-\beta E_i}, \tag{10.2.9}$$

which is the diagonal element of the Bloch density matrix introduced in Chapter 4, equation (4.15.6). Either we can now use the first-order perturbation theory for C, which is given in Appendix 1.2, or we can use the relation (A1.2.7), between C and the density $\rho(\mathbf{r}E)$ of a degenerate gas with Fermi level E

$$C(\mathbf{r}\beta) = \beta \int_0^\infty \rho(\mathbf{r}E)\,e^{-\beta E}\,dE. \tag{10.2.10}$$

We can thus derive $n(\mathbf{r})$, from the knowledge we already have of $\rho(\mathbf{r}E)$ from Chapter 2. Adopting this later course, it is a straightforward matter to take the Laplace transform of equation (2.3.11), when we find

$$n(\mathbf{r}) - n_0 = -\frac{mn_0}{\pi\hbar^2} \int d\mathbf{r}_1\, V(\mathbf{r}_1)\, \frac{\exp(-2mk_B T|\mathbf{r}_1 - \mathbf{r}|^2/\hbar^2)}{|\mathbf{r}_1 - \mathbf{r}|}. \tag{10.2.11}$$

In this equation, n_0 is the density of free carriers, while m should be interpreted as an effective mass m^*, which will not be written in explicitly for notational convenience. Combining equation (10.2.11) with the Poisson equation

$$\nabla^2 V = -\frac{4\pi e^2}{\varepsilon}[n(\mathbf{r}) - n_0], \tag{10.2.12}$$

where ε is the macroscopic dielectric constant, we find

$$\nabla^2 V = \frac{4me^2 n_0}{\hbar^2 \varepsilon}\int \frac{d\mathbf{r}'\, V(\mathbf{r}')\exp(-2mk_B T|\mathbf{r}'-\mathbf{r}|^2/\hbar^2)}{|\mathbf{r}-\mathbf{r}'|}. \tag{10.2.13}$$

If we assume $V(\mathbf{r}')$ is slowly varying in space, and therefore replace it by $V(\mathbf{r})$ [cf. the argument in Chapter 2, following equation (2.3.11)], then we find from equation (10.2.13) that

$$\nabla^2 V \simeq \frac{4\pi n_0 e^2 V}{\varepsilon k_B T} = q_D^2 V, \tag{10.2.14}$$

where the Debye length q_D^{-1} is evidently determined by equation (10.2.14). As an example, consider one P atom in 10^5 Si atoms at room temperature. Then equation (10.2.14) gives the Debye length $q_D^{-1} \sim 60$ Å, which is, of course, very large compared with the screening length q^{-1} round a charge in a metallic conductor discussed above. The potential in the semi-conductor is seen then to be *long range* and the Coulomb field is only weakly screened.

As shown in Appendix 10.1, equation (10.2.13) can be solved in **k**-space, and is essentially giving the same answer as equation (10.2.14) provided a parameter γ, defined by

$$\gamma = \pi\hbar^2 n_0 e^2/\varepsilon m(k_B T)^2, \tag{10.2.15}$$

is $\ll 1$. Equation (10.2.14) is the basic equation of the so-called Brooks–Herring theory of ionized impurity screening (see also Dingle, 1955; Mansfield, 1956). A brief summary of the main results for impurity scattering in a semi-conductor, based on equation (10.2.14), is given in Appendix 10.1.

10.3 Phase shift analysis for strong scatterers

The analysis of the previous section used the Born approximation and for many physical problems this is not an adequate approximation for quantitative work. We turn now to deal in full with the scattering of plane waves from a spherical defect centre.

In the presence of a spherical scattering centre, we discussed in Chapter 1 the way in which the perturbed wave function may be analysed into partial (distinct l values) spherical waves at large distances from the scatterer, the partial waves being shifted in phase by amounts η_l. It turns out that a variety of physical properties (residual resistance as discussed in Appendix 10.1, vacancy formation energy, diamagnetic susceptibility, etc.) may be expressed in terms of the phase shifts η_l, and these expansions are generally quite rapidly convergent (it is unusual to have to go beyond d-waves, i.e. $l = 2$).

10.3.1 Expressions for phase shifts in terms of scattering potential

From equation (1.11.77) of Chapter 1, we can write the amplitude $f(\theta)$ for scattering through an angle θ as

$$f(\theta) = \frac{1}{k}\sum_{l=0}^{\infty}(2l+1)\,e^{i\eta_l}\sin\eta_l\,P_l(\cos\theta). \tag{10.3.1}$$

We can now readily obtain a formula for the phase shifts in terms of the solutions ϕ_l of the radial wave equation

$$\frac{d^2(r\phi_l)}{dr^2}+2\left[\frac{k^2}{2}-V(r)-\frac{l(l+1)}{2r^2}\right]r\phi_l = 0 \tag{10.3.2}$$

chosen so that $\phi_l(r)\sim(kr)^{-1}\sin(kr+\eta_l-l\pi/2)$. From equations (1.11.66), (1.11.68) and (1.11.75) we obtain

$$f(\theta) = -\sum_{l=0}^{\infty}(2l+1)\,e^{i\eta_l}\int_0^{\infty} j_l(kr)\,V(r)\,\phi_l(r)\,r^2\,dr, \tag{10.3.3}$$

so that by comparison with equation (10.3.1)

$$\sin\eta_l = -k\int_0^{\infty} j_l(kr)\,V(r)\,\phi_l(r)\,r^2\,dr. \tag{10.3.4}$$

Normalization of ϕ_l is sometimes chosen to yield $\tan\eta_l$.

10.3.2 Level shifts and undeformed energy bands

Let us suppose that the unperturbed Schrödinger equation is $H_0\psi_0 = E_0\psi_0$, and in the presence of the defect, $(H_0+V)\psi = E\psi$. Then evidently

$$\langle\psi_0^*\,H_0\psi\rangle+\langle\psi_0^*\,V\psi\rangle = E\langle\psi_0^*\,\psi\rangle, \tag{10.3.5}$$

that is,

$$(E-E_0)\langle\psi_0^*\,\psi\rangle = \langle\psi_0^*\,V\psi\rangle \tag{10.3.6}$$

and if $H_0 = -\nabla^2$

$$E(\mathbf{k})-k^2 = \frac{\int e^{-i\mathbf{k}.\mathbf{r}}\,V(\mathbf{r})\,\psi_{\mathbf{k}}(\mathbf{r})\,d\mathbf{r}}{\int e^{-i\mathbf{k}.\mathbf{r}}\,\psi_{\mathbf{k}}(\mathbf{r})\,d\mathbf{r}}, \tag{10.3.7}$$

where in the presence of the perturbation we have labelled ψ and its energy by $\mathbf{k}$ such that $\psi_{\mathbf{k}} \to e^{i\mathbf{k}.\mathbf{r}}$ as $V(\mathbf{r}) \to 0$. The numerator of equation (10.3.7) has the appearance of a scattering amplitude (cf. equation (1.11.66)), and we might immediately expect to obtain a good approximation to it from equation (10.3.1) with $\theta = 0$ (forward scattering). In fact, if we take the first Born approximation, putting $\psi_{\mathbf{k}} = e^{i\mathbf{k}.\mathbf{r}}$ in both numerator and denominator of the right-hand side of equation (10.3.7) we obtain the *rigid band theorem*:

$$E(\mathbf{k}) - k^2 = \frac{1}{\mathscr{V}} \int V(\mathbf{r})\, d\mathbf{r} = \bar{V}. \tag{10.3.8}$$

The significance of this is that the band moves up (down) without change of shape for repulsive (attractive) potentials.

That care is needed in using equation (10.3.7) is seen by taking $e^{i\mathbf{k}.\mathbf{r}} = e^{ikz}$ and inserting the solution of equation (1.11.68) for $\psi_{\mathbf{k}}$ in the level shift formula of equation (10.3.7). With this substitution, the numerator is certainly exactly the forward scattering amplitude apart from a factor -4π. But since the wave function of equation (1.11.68) appears to have the same energy as $e^{i\mathbf{k}.\mathbf{r}}$ ($\sqrt{E} = k$), the level shift seems to be zero and equation (10.3.7) would give the forward scattering amplitude to be zero!

The resolution of this paradox lies in the boundary conditions. To illustrate matters let us suppose the boundary is a large sphere which has the scatterer at its centre. Then if a spherical wave with wave-number k suffers a phase shift $\eta_l(k)$, the allowed value of k must change in such a way that

$$k'R + \eta_l(k) = kR \tag{10.3.9}$$

and so

$$k'^2 - k^2 \simeq -2k\eta_l/R. \tag{10.3.10}$$

Supposing all phase shifts are equal and small, and making the Born approximation in the sense of replacing $\phi_l(r)$ by $j_l(kr)$ in equation (10.3.4), we can obtain the rigid band theorem (see section 10.3.4 for a proper proof).

The crucial appearance of the boundary conditions here makes us realize that we must proceed with care in approximating in equation (10.3.7). In the first place, the substitution of the forward scattering amplitude on the right-hand side of this equation is tantamount to using (1.11.68) with $c_l = e^{i\eta_l}$, as already remarked, but this equation was obtained with the particular aim of describing the scattering of incoming particles, and not for the problem we describe here where waves are scattered into themselves. For a more precise formulation of the present case the reader may consult Ziman (1965). We can in fact justify the use of the solution of section 1.11.5 by noting that in approximating to the numerator the boundary conditions are not really of great importance, at least for small phase shifts. This is because the range of integration is effectively the range of the potential, which is usually small compared to the volume of the system, and so the value of the integral will not be greatly changed by small changes in $e^{i\mathbf{k}.\mathbf{r}}$ or $\psi_{\mathbf{k}}$ necessitated by boundary effects, and equation (1.11.65) gives us a suitable approximation to $\psi_{\mathbf{k}}$ since it behaves as e^{ikz} at large distances from the scatterer. Approximating to the denominator is a very different matter, and we shall consider it in a moment. We first take steps to avoid the embarrassment of complex estimates of the energy level shift, which can well arise in direct approximations to the right-hand side of equation (10.3.7) since both the numerator and denominator are in general complex.

Since the left-hand side is real, and the equation is exact, one can easily show that equation (10.3.7) can be replaced by

$$E(\mathbf{k})-k^2 = \frac{\mathrm{Re}\left[\int e^{-i\mathbf{k}.\mathbf{r}}\, V(\mathbf{r})\, \psi_{\mathbf{k}}(\mathbf{r})\, d\mathbf{r}\right]}{\mathrm{Re}\left[\int e^{-i\mathbf{k}.\mathbf{r}}\, \psi_{\mathbf{k}}(\mathbf{r})\, d\mathbf{r}\right]} \tag{10.3.11}$$

without any approximation whatever.

Let us now look at the approximation to the denominator. To substitute equation (1.11.65) for $\psi_{\mathbf{k}}$ is both pointless and dangerous, since boundary conditions are all important in orthonormality relations. It is instead much simpler and safer to assume that the perturbation correction to $e^{i\mathbf{k}.\mathbf{r}}$ necessitated by the presence of an impurity in the electron gas is small everywhere except in the vicinity of the impurity. The denominator of equation (10.3.11) then becomes just the volume of the system.

Finally, a finite concentration of scatterers can be easily allowed for on the assumption that their effects do not interfere. There being N scatterers, our approximation to the level shift is

$$E(\mathbf{k})-k^2 = -\frac{N}{\mathscr{V}}\mathrm{Re}\left[\frac{4\pi}{k}\sum_{l=0}^{\infty}(2l+1)\, e^{i\eta_l}\sin\eta_l\right], \tag{10.3.12}$$

where we have now explicitly substituted in the expression for the forward-scattering amplitude in terms of phase shifts, provided these are small, when $\sin\eta_l \simeq \eta_l$.

10.3.3 Friedel sum rule and integrated density of states

The Friedel sum rule relates the phase shifts caused by a charged impurity in an electron gas to its total charge $|e|Z$. Its importance derives from the fact that the impurity is screened by the displaced charge, so that the scattering potential must be calculated self-consistently as discussed in section 10.2.1. It turns out in some cases that adjustment of the potential to give phase shifts satisfying the sum rule is sufficient to take accurate account of the effects of this self-consistency requirement.

We consider then the addition of an excess charge $Z|e|$ with Z electrons to ensure charge neutrality. We take as the unperturbed wave function the plane wave $e^{i\mathbf{k}.\mathbf{r}}$ and the corresponding radial parts of the perturbed wave function to be given by equation (10.3.2). (The small change of k is irrelevant here; it is of order $1/\mathscr{V} \to 0$ and the boundary conditions only enter in the counting of energy levels.) The argument below follows Kittel (1963).

At large distances the wave function must be unchanged except for the phase shifts in the partial waves, and the asymptotic form of $\phi_l(r)$ will be given by equation (1.11.71). We take $c_l = 1$, for since we are only interested in charge densities its complex nature is of no interest.

Consider now a large sphere, radius R, centred on the impurity. We shall find the total screening charge displaced into this sphere by computing

$$\frac{1}{\mathscr{V}}\int_0^{2\pi} d\phi \int_0^{\pi} \sin\theta\, d\theta \int_0^R |\psi(\mathbf{r})|^2 r^2\, dr - \frac{1}{\mathscr{V}}\int_0^{2\pi} d\phi \int_0^{\pi} \sin\theta\, d\theta \int_0^R |e^{i\mathbf{k}.\mathbf{r}}|^2 r^2\, dr$$

$$= \frac{4\pi}{\mathscr{V}}\sum_{l=0}^{\infty}(2l+1)\left[\int_0^R \phi_l^2(r)\, r^2\, dr - \int_0^R j_l^2(kr)\, r^2\, dr\right]. \qquad (10.3.13)$$

Let us denote the solution of equation (10.3.2), with k replaced by k', by $\phi_l'(r)$. By multiplying equation (10.3.2) through by ϕ_l' and the corresponding equation for ϕ_l' by ϕ_l, we find, after subtraction of the two equations

$$r\phi_l' \frac{d^2}{dr^2}(r\phi_l) - r\phi_l \frac{d^2}{dr^2}(r\phi_l') = (k'^2 - k^2)\,\phi_l'(r)\,\phi_l(r)\, r^2. \qquad (10.3.14)$$

Integrating from 0 to R we find

$$\left[r\phi_l' \frac{d}{dr}(r\phi_l) - r\phi_l \frac{d}{dr}(r\phi_l')\right]_0^R = (k'-k)(k'+k)\int_0^R \phi_l'\,\phi_l\, r^2\, dr. \qquad (10.3.15)$$

We now expand $\phi_l'(r)$ in powers of $k'-k$:

$$\phi_l'(r) = \phi_l(r) + (k'-k)\frac{d}{dk}\phi_l(r) + \dots. \qquad (10.3.16)$$

Then, on taking the limit $k' \to k$, equation (10.3.15) yields

$$\left[\frac{d}{dk}(r\phi_l)\frac{d}{dr}(r\phi_l) - r\phi_l \frac{d^2}{dk\, dr}(r\phi_l)\right]_0^R = 2k\int_0^R \phi_l^2(r)\, r^2\, dr. \qquad (10.3.17)$$

The contribution from the lower limit of the left-hand side vanishes, and the contribution of the upper limit is obtained by taking the asymptotic form below equation (10.3.2). We find

$$\frac{1}{k}\left\{R+\frac{d\eta_l}{dk}-\frac{1}{2k}\sin\left[2\left(kR+\eta_l-\frac{l\pi}{2}\right)\right]\right\} = 2k\int_0^R \phi_l^2 r^2\,dr. \qquad (10.3.18)$$

The corresponding result for $j_l(kr)$ is obtained simply by putting η_l and $d\eta_l/dk = 0$, whereupon we obtain equation (10.3.13) in the form

$$\frac{1}{\mathscr{V}}\int_{\text{sphere}}(|\psi|^2-|e^{i\mathbf{k}.\mathbf{r}}|^2)\,d\mathbf{r} = \frac{2\pi}{\mathscr{V}k^2}\sum_{l=0}^{\infty}(2l+1) \times\left[\frac{d\eta_l}{dk}-\frac{1}{k}\sin\eta_l\cos(2kR+\eta_l-l\pi)\right]. \qquad (10.3.19)$$

We might remark that the above shows that, to ensure ϕ_l and $j_l(kr)$ normalize to the same values over the entire volume, c_l must be slightly different from unity if we are to retain the asymptotic form used above but the correction is negligible, as we shall see when we return to this point a little later. Now, taking account of spin, the density of states in **k**-space, will be $\mathscr{V}/4\pi^3$. Hence the total number of particles displaced into the sphere is, from equation (10.3.19)

$$\frac{1}{4\pi^3}\int_0^{2\pi}d\phi\int_0^{\pi}\sin\theta\,d\theta\int_0^{k_f}k^2\,dk\int_{\text{sphere}}\{|\psi|^2-|e^{i\mathbf{k}.\mathbf{r}}|^2\}\,d\mathbf{r}$$
$$=\frac{2}{\pi}\sum_{l=0}^{\infty}(2l+1)\,\eta_l(k_f)-\frac{2}{\pi}\sum_{l=0}^{\infty}(2l+1)\int_0^{k_f}\sin\eta_l\cos(2kR+\eta_l-l\pi)\,dk. \qquad (10.3.20)$$

As R is increased, the magnitude of the second term oscillates, and must die out as R get very large. In fact, we shall find the asymptotic form of the density oscillations to be $A\cos(2k_f r+\phi)/r^3$, as anticipated in section 10.2.1. At very large R, we are left with just the constant term, which must be Z, the number of electrons screening the impurity. We therefore finally obtain

$$Z = \frac{2}{\pi}\sum_{l=0}^{\infty}(2l+1)\,\eta_l(k_f), \qquad (10.3.21)$$

which is the Friedel sum rule (see also Ziman, 1972).

Let us suppose now that we take the boundary of the entire system to be spherical with the impurity at the centre. If we take R to be the radius of the system in equation (10.3.19), we can see how the normalization constant on ϕ_l must be changed from that on $j_l(kr)$ and the difference, as already remarked, will make negligible difference to our analysis. However, what if we apply the entire analysis to the entire system? Whereas before we supposed that the difference $\int_{\text{sphere}}\{|\psi|^2 - |e^{i\mathbf{k}.\mathbf{r}}|^2\}\,d\mathbf{r}$ was compensated outside the sphere to make ψ and $e^{\mathbf{k}.\mathbf{r}}$, normalize in the same fashion, we can no longer do so. However, the result in equation (10.3.21) is undoubtedly applicable, and we can regard the procedure of this section, when applied to the entire system, as constituting a formal device enabling us to avoid explicit consideration of boundary conditions. We also note that Z is the difference in integrated density of states, after and before insertion of the impurity, integrated to the Fermi level. Further, there is no special significance in integration to k_f in equation (10.3.20): if we define

$$Z(k) = \frac{2}{\pi}\sum_{l=0}^{\infty}(2l+1)\,\eta_l(k), \tag{10.3.22}$$

then

$$N_i(k) - N_0(k) = Z(k), \tag{10.3.23}$$

where $N_0(k)$ is the integrated density of states for the free electron gas, and $N_i(k)$ is the corresponding quantity after insertion of the impurity.

We can check our formula for the energy level shift in terms of small phase shifts, from equation (10.3.23). In equation (10.3.23), $k = \sqrt{E}$, of course, so that, if the level shift is small, we have

$$E(\mathbf{k}) = (k - \delta k)^2, \tag{10.3.24}$$

where, to first order in δk,

$$Z(k) = N_0(k+\delta k) - N_0(k) = \frac{k^2}{\pi^2}\,\delta k \mathscr{V}. \tag{10.3.25}$$

Hence

$$E(\mathbf{k}) - k^2 = -2k\,\delta k = -\frac{2\pi^2}{k}\frac{Z(k)}{\mathscr{V}}, \tag{10.3.26}$$

so that from equation (10.3.23), the level shift formula is

$$E(\mathbf{k}) - k^2 = -\frac{4\pi}{\mathscr{V}k}\sum_{l=0}^{\infty}(2l+1)\,\eta_l(k). \tag{10.3.27}$$

Comparison with equation (10.3.12) shows that the two equations agree for small phase shifts, the regime of validity of the two formulae.

10.3.4 Change in sum of eigenvalues due to defect

As we shall discuss briefly below for the case of a lattice vacancy, an important term in the energy of a defect crystal involves the change in the sum of the one-electron eigenvalues due to the presence of the defect potential $V(r)$. If this is spherical, we have in essence obtained this term in the energy change above.

However, let us write down the energy directly for a scattering potential at the centre of a large spherical box of radius R and requiring that, both with and without the defect, the wave functions vanish at $r = R$. This implies that the k-spectrum is discrete and we should distinguish between the allowed values in the perturbed and unperturbed states.

If we denote the unperturbed states by a subscript zero, then the asymptotic forms of the radial wave functions are again given by

$$\phi_l^0(k^0 r) = (k^0 r)^{-1} \sin(k^0 r - \tfrac{1}{2} l\pi) \tag{10.3.28}$$

and

$$\phi_l(kr) = (kr)^{-1} \sin(kr - \tfrac{1}{2} l\pi + \eta_l). \tag{10.3.29}$$

The potential therefore alters the allowed values of k by $\Delta k = k - k^0$.

From the boundary condition at $r = R$, we have

$$k^0 R = n\pi + \tfrac{1}{2} l\pi, \quad n = 1, 2, \ldots, \tag{10.3.30}$$

and

$$kR = n\pi + \tfrac{1}{2} l\pi - \eta_l. \tag{10.3.31}$$

Hence, for a fixed l, we may write

$$\Delta k = -\eta_l / R. \tag{10.3.32}$$

Now a change Δk_l in momentum involves a change in the energy of an electron in state l of

$$\Delta E = k \Delta k = -k\eta_l / R. \tag{10.3.33}$$

(a) *Expression in terms of total displaced charge.* To obtain from this total change in the eigenvalue sum, we must sum over all l states and integrate with respect to k up to the Fermi level. Remembering that the states are distributed with density $2R/\pi$, including spin, per unit energy range of k, we find for the change in the eigenvalue sum

$$\Delta E_{\text{eigenvalue}} = -\int_0^{k_f} \sum_{l=0}^{\infty} (2l+1) \frac{k\eta_l(k)}{R} \frac{2R}{\pi} dk$$

$$= -\int_0^{k_f} Z(k)\, k\, dk, \tag{10.3.34}$$

where $Z(k)$ is defined by equation (10.3.22), and from the Friedel sum argument is the charge displaced below energy $k^2/2$. Equation (10.3.34) is exact.

(b) *Vacancy formation energy in simple metals.* We shall briefly summarize here how equation (10.3.34) can be made the basis of a calculation of vacancy formation energies in simple metals (Fumi, 1955). We first recall that vacancies exist in all crystals at elevated temperatures with an equilibrium density given by (see Flynn, 1972; Howard and Lidiard, 1964)

$$n = AN\exp\left(-\frac{E_v}{k_B T}\right), \tag{10.3.35}$$

where n is the number of vacancies in a crystal of N sites, E_v is the energy of formation of a vacancy and A is a numerical factor of the order of 10. At the melting point of a metal, one site in about 10^4 or 10^5 is usually vacant.

We shall turn immediately to discuss how the energy of formation E_v of vacancies in simple close-packed metals such as Mg, Al and Pb can be estimated, using equation (10.3.34). If we calculate $\Delta E_{\text{eigenvalue}}$ from this equation in Born approximation, then we find (cf. Problem 10.8) the result

$$\Delta E_{\text{eigenvalue}} = \tfrac{2}{3} Z E_f, \tag{10.3.36}$$

where Z is the valency and E_f is the Fermi energy. This result is independent of the potential $V(r)$, provided that the Friedel sum rule (10.3.21) is satisfied.

The simplest theory outlined by Fumi corrects equation (10.3.36) only for the fact that, when we form a vacancy, we increase the volume of the metal and, neglecting relaxation in close-packed metals, we assume the increase to be one atomic volume. Then (cf. equation (1.2.8)), the kinetic energy of the electron gas is reduced by $\tfrac{2}{5} Z E_f$ and we find

$$E_v \doteqdot \tfrac{4}{15} Z E_f. \tag{10.3.37}$$

Table 10.1 shows the measured formation energies E_v for Cu, Mg, Al and Pb and they are seen to be all of the order of 1 eV. But ZE_f for Pb is of the order of 30 eV and equation (10.3.37) is clearly a very bad approximation in this metal.

TABLE 10.1

	Cu	Mg	Al	Pb
E_v in eV	1.2 ± 0.1	0.89	0.75	0.5
E_v/ZE_f	0.17 ± 0.01	0.064	0.021	0.01 (3)

The answer lies partly in calculating the sum of the eigenvalues from the phase shifts $\eta_l(k)$, but, more importantly for the polyvalent metals, in calculating the other terms which must enter into the energy, in particular the self-energy of the displaced charge and the exchange energy.

In keeping with the Born results that the change in the eigenvalue sum is independent of the choice of $V(r)$, provided the Friedel sum rule is satisfied, detailed calculations (Stott, Baranovsky and March, 1970) show that this change is relatively insensitive to the choice of potential for strong scattering such as we have in Pb. Similarly, the exchange energy change was found to be model insensitive. But the self-energy of the displaced charge, in contrast, depends strongly on the model adopted, and requires self-consistent calculations. The results obtained from self-consistent calculations based on point ions are recorded in Table 10.2. Comparing these results with the

TABLE 10.2. Contributions to vacancy formation energy in units of ZE_f

	Cu	Mg	Al	Pb
$\Delta E_{\text{eigenvalue}}$ + kinetic energy change	0.1939	0.1657	0.1560	0.1382
Exchange energy correction	0.1530	0.1298	0.0981	0.0961
Self-energy correction	−0.24	−0.21	−0.22	−0.19
Vacancy formation energy	0.11 ± 0.04	0.09 ± 0.03	0.03 ± 0.02	0.04 ± 0.03

second row in Table 10.1, we see that the main features of the experiments are given by the theory, though it is difficult from the theory to obtain a quantitative result for E_v for Pb, because of the severe cancellation shown on Table 10.2 for this metal.

The way one can transcend the free-electron model in calculating the displaced charge round the vacancy is discussed in section 10.4.5 below. For the moment, we leave the discussion of the repulsive potentials representing the vacancy to consider some qualitative consequences of strongly attractive defect potentials.

10.3.5 Virtual bound states

It is evident from the analysis of section 10.3.1 that the partial wave $j_l(kr)$ is most strongly scattered when $\sin \eta_l$ is at an extremum. This is so when

$$\eta_l = (n + \tfrac{1}{2})\pi, \tag{10.3.38}$$

where n is an integer. It is found in explicit calculations that η_l is rapidly varying near the value of k for which $\eta_l = (n+\frac{1}{2})\pi$, and when viewed as a function of energy, the "resonance" is quite sharp. We have already seen the implication of this for d-bands in pure metals in Chapter 1; here we are interested in impurity effects.

It is possible to gain a little insight without detailed calculation by examining the Friedel sum rule. Suppose $V(r)$ is sufficiently attractive to bind electrons of the Fermi gas. Then, if we wish, we can regard the original ion plus this bound state as the impurity. Thus, if we suppose the bound state to have $l = l'$, we find that the sum rule (10.3.21) becomes

$$Z-2(2l'+1) = \frac{2}{\pi}\sum_{l\neq l'}^{\infty}(2l+1)\,\eta_l. \tag{10.3.39}$$

It is immediately clear that to obtain consistency between equation (10.3.39) and the original sum rule we must put $\eta_{l'} = \pi$. Now, if we reduce the strength of the attractive potential $V(r)$, the bound state will rise in energy until it eventually merges into the continuum. There is no longer then a bound state, but, as one studies $\eta_{l'}$ as a function of k in the band, a strong variation with k must occur in a localized region of energy and the phase shift moves from the bound state value of π through the resonance value of $\pi/2$. Thus, the "resonance" was termed by Friedel a virtual bound state.

We can therefore see that if the virtual level can lie close to the Fermi level, as indeed proves possible, then the properties of an alloy can change markedly with but small changes in the electron concentration. We shall return to a detailed discussion of virtual bound states in connection with the Koster–Slater model of an impurity centre in section 10.4.3, and in connection with localized moments in section 10.7.

10.4 Scattering theory for Bloch waves

Previously, we have considered plane waves scattered by the defect potential. While this is appropriate in some simple metals, where we can assume weak pseudo-potentials, it is often not adequate, and we must deal with Bloch waves scattering off the impurity or defect centre.

The character of the scattering can be usefully discussed in terms of three different types of defect potential:

(i) Weak scatterers.

(ii) Strong scatterers, but with slow spatial variations in the solute potentials.

(iii) Strongly attractive potentials which lead to bound or resonant states.

10.4.1 Weak scatterers

We have already seen in Chapter 3 how to write down the change in electron density in a periodic lattice, due to a change $V(\mathbf{r})$ in the potential, when the perturbation is weak. The answer is given by equation (3.10.54) in terms of the linear response function $F(\mathbf{rr}'E)$ defined in equation (3.10.55). In the present case, the displaced charge below energy E is given by

$$\Delta(\mathbf{r}E) = \int d\mathbf{r}'\, V(\mathbf{r}')\, F(\mathbf{rr}'E), \tag{10.4.1}$$

where $\Delta(\mathbf{r}E) = \rho(\mathbf{r}E) - \rho_{\mathrm{p}}(\mathbf{r}E)$, with ρ the density in the defect crystal and ρ_{p} the density in the periodic lattice. The form of F is given by

$$\frac{\partial F}{\partial E} = 2\,\mathrm{Re}\left[G_{\mathrm{p}}(\mathbf{rr}'E_+)\frac{\partial \rho_{\mathrm{p}}(\mathbf{r}'\mathbf{r}E)}{\partial E}\right], \tag{10.4.2}$$

where G_{p} is the perfect lattice Green function and ρ_{p} is the Dirac density matrix in the periodic crystal. Thus, knowing the perfect lattice solutions and the defect potential, the displaced charge can be calculated by quadrature.

Evidently equation (10.4.1) is the generalization of the plane-wave theory of section (10.2.1). Indeed, if we insert plane-waves into equation (10.4.2) we have immediately

$$G_{\mathrm{p}} \to \frac{\exp[i(2E)^{\frac{1}{2}}|\mathbf{r}-\mathbf{r}'|]}{4\pi|\mathbf{r}-\mathbf{r}'|}, \quad \frac{\partial \rho_{\mathrm{p}}}{\partial E} = \frac{\sin[(2E)^{\frac{1}{2}}|\mathbf{r}-\mathbf{r}'|]}{2\pi^2|\mathbf{r}-\mathbf{r}'|},$$

$$F \to -\frac{k^2}{2\pi^3}\frac{j_1(2k|\mathbf{r}-\mathbf{r}'|)}{|\mathbf{r}-\mathbf{r}'|^2}, \quad E = \frac{k^2}{2}, \tag{10.4.3}$$

regaining the earlier result for the displaced charge.

The integrated density of states can obviously be obtained from the result (10.4.1) and is given in Problem (10.9). It requires for its evaluation knowledge of the local density $\rho_{\mathrm{p}}(\mathbf{r}, E)$ in the perfect crystal, but not the diagonal elements of the Dirac density matrix, which is, of course, a considerable simplification.

Using the linear response function F, required also, as we saw in Chapter 3, in the calculation of phonon dispersion relations, it is, in principle, possible to investigate the effect of anisotropic Fermi surfaces and the related problem of the angular dependence of the displaced charge round a defect or impurity. Such knowledge will clearly be important in a variety of problems in dilute alloys [cf. the work of Drain (1968) on nuclear magnetic resonance studies of dilute alloys].

10.4.2 *Density of states and level shifts for slowly varying potentials*

We have already used the Thomas–Fermi theory extensively, to deal with slowly varying potentials. For free electrons, the local density $\rho_0(\mathbf{r}E)$ is, of course, independent of $\mathbf{r}$ and we have, in the presence of the defect potential,

$$\rho(\mathbf{r}E) = \rho_0(\mathbf{r}, E - V(\mathbf{r})]. \tag{10.4.4}$$

This suggests that we write

$$\rho(\mathbf{r}E) = \rho_0[\mathbf{r}, E - g(\mathbf{r}E)], \tag{10.4.5}$$

where $g(\mathbf{r}E)$ is now a functional of the potential. When the potential V is slowly varying spatially, we expect from comparison of equations (10.4.4) and (10.4.5) that $g(\mathbf{r}E)$ will be a slowly varying function of energy. But now, even for strong potentials, it follows from equations (10.4.4) and (10.4.5) that g is roughly proportional to V and we might expect that $g(\mathbf{r}E)$ can be usefully approximated by first-order theory, even though equations (10.4.4) and (10.4.5) are non-linear in V. It is then straightforward, using the results of section 10.2.1, to obtain the first-order approximation

$$g_1(\mathbf{r}E) = \frac{k}{2\pi}\int d\mathbf{r}_1\, V(\mathbf{r}_1)\frac{j_1(2k|\mathbf{r}-\mathbf{r}_1|)}{|\mathbf{r}-\mathbf{r}_1|^2}; \quad E = \frac{k^2}{2}. \tag{10.4.6}$$

For slowly varying potentials we obtain

$$g_1(\mathbf{r}E) \simeq V(\mathbf{r}), \tag{10.4.7}$$

giving back the Thomas–Fermi limit, and this theory is seen to have the merit that it now includes first-order perturbation theory as well as the semi-classical limit.

The generalization of these arguments to Bloch waves is now clear, if somewhat complicated. We simply write (Stoddart, March and Stott, 1969)

$$\rho(\mathbf{r}E) = \rho_p[\mathbf{r}, E - g(\mathbf{r}E)], \tag{10.4.8}$$

where $\rho_p(\mathbf{r}E)$ is the local density in the pure metal. To first order in the impurity potential, it can then be shown that

$$\begin{aligned} g_1(\mathbf{r}E) = &-\left[\frac{\partial\rho_p(\mathbf{r}E)}{\partial E}\right]^{-1}\int d\mathbf{r}_1\int dE_1\, dE_2 \\ &\times\left[\frac{\delta(E-E_1)+\delta(E-E_2)}{(E_1-E_2)^2} - \frac{2\{\theta(E-E_1)-\theta(E-E_2)\}}{(E_1-E_2)^3}\right] \\ &\times \rho_p(\mathbf{r}\mathbf{r}_1 E_2)\,\rho_p(\mathbf{r}_1\mathbf{r}E_1). \end{aligned} \tag{10.4.9}$$

This theory, though explicit, has not been used in practical calculations beyond a plane-wave framework, because, at present, we lack knowledge of the Dirac density matrix $\rho_p(\mathbf{r}\mathbf{r}_1 E)$ in periodic crystals.

Nevertheless, equation (10.4.8) gives us an explicit method of deforming the pure crystal local density $\rho_p(\mathbf{r}E)$ to account for scattering off solute atoms.

10.4.3 *Virtual and bound states: Koster–Slater model*

Though the approximate theory of the previous section can be applied to defect centres which scatter electrons strongly, for example, lattice vacancies, it is not suitable for treating strong and localized attractive potentials which can lead to resonances or virtual bound states.

The merit of the approach initiated by Koster and Slater (1954) is that it can deal very naturally with bound and resonant states. The price paid for this, however, is to assume a solute potential $V(\mathbf{r})$ which seems very restrictive.

We shall present the Koster–Slater model here in its very simplest form, and later discuss the nature of the potential $V(\mathbf{r})$ underlying the model. Rather than employing the density matrix approach of sections 10.4.1 and 10.4.2 above, we shall deal at first directly with the wave functions in the presence of the defect centre.

(a) *Integral equation for perturbed wave function.* Let us first write down the analogue of the Schrödinger equation for the scattered wave function $\psi_\mathbf{k}(\mathbf{r})$ in terms of the complete set of Bloch functions $\phi_\mathbf{k}^m(\mathbf{r})$ of the unperturbed problem, m being a band index and $\mathbf{k}$ lying in the BZ. When the ϕ's are plane waves, we saw in equation (2.3.7) that the integral equation takes the form

$$\psi_\mathbf{k}(\mathbf{r}) = \mathscr{V}^{-\frac{1}{2}}\, e^{i\mathbf{k}.\mathbf{r}} - \frac{m}{2\pi\hbar^2}\int \frac{e^{ik|\mathbf{r}-\mathbf{r}'|}}{|\mathbf{r}-\mathbf{r}'|}\, V(\mathbf{r}')\,\psi_\mathbf{k}(\mathbf{r}')\,d\mathbf{r}'. \tag{10.4.10}$$

In order to effect the generalization to deal with Bloch waves, let us rewrite the "spherical" wave $e^{ik|\mathbf{r}-\mathbf{r}'|}/|\mathbf{r}-\mathbf{r}'|$ in terms of plane waves. Apart from a multiplying constant, this takes the form

$$\sum_{\mathbf{k}'} \frac{e^{-i\mathbf{k}'.\mathbf{r}}\, e^{i\mathbf{k}'.\mathbf{r}'}}{(\hbar^2 k^2/2m) - (\hbar^2 k'^2/2m) + i\varepsilon}, \tag{10.4.11}$$

the quantity $i\varepsilon$ ensuring that in equation (10.4.10) the second term on the right-hand side represents outgoing scattered waves.

Then for the generalization to Bloch wave scattering,

$$e^{i\mathbf{k}'.\mathbf{r}} \to \phi^m_{\mathbf{k}'}(\mathbf{r}), \quad \frac{\hbar^2 k'^2}{2m} \to E_m(\mathbf{k}')$$

and it is readily verified that the precise form of the integral equation is

$$\psi_{\mathbf{k}}(\mathbf{r}) = \phi_{n\mathbf{k}}(\mathbf{r}) + \sum_{m\mathbf{k}'} \int \frac{\phi^*_{m\mathbf{k}'}(\mathbf{r})\,\phi_{m\mathbf{k}'}(\mathbf{r}')\,V(\mathbf{r}')\,\psi_{\mathbf{k}}(\mathbf{r}')}{E - E_{m\mathbf{k}'} + i\varepsilon}\,d\mathbf{r}'. \tag{10.4.12}$$

To solve this equation exactly, it is at present necessary to take some simple model for the defect potential $V(\mathbf{r})$.

The model of Koster and Slater is somewhat formal but reduces the scattering problem posed by equation (10.4.12) to one band, at least in the simplest form of the model, by making direct assumptions about the impurity potential $V(\mathbf{r})$. In other words, we start from the basic integral equation (10.4.12), and we attempt to solve it exactly, subject to rather drastic assumptions about the potential $V(\mathbf{r})$. These assumptions amount to a potential which is not long range; from the discussion of section 10.2 it appears then more appropriate to *metals* than to semi-conductors. Also the Wannier functions, in terms of which the model is posed, must not be very long range.

To formalize the theory, we assume that the average of $V(\mathbf{r})$ between Wannier functions $w_n(\mathbf{r}-\mathbf{R}_j)$, centred on lattice sites $\mathbf{R}_j$, is given by

$$\int w_n(\mathbf{r}-\mathbf{R}_j)\,V(\mathbf{r})\,w_{n'}(\mathbf{r}-\mathbf{R}_l)\,d\mathbf{r} = V_{nn}\,\delta_{nn'}\,\delta(\mathbf{R}_j-\mathbf{R}_0)\,\delta(\mathbf{R}_l-\mathbf{R}_0), \tag{10.4.13}$$

where the impurity is assumed to be at the site $\mathbf{R}_0$ and the Wannier functions are related to the Bloch functions $\phi_{n\mathbf{k}}(\mathbf{r})$ through

$$\phi_{n\mathbf{k}}(\mathbf{r}) = \sum_{\mathbf{R}_j} e^{i\mathbf{k}.\mathbf{R}_j}\,w_n(\mathbf{r}-\mathbf{R}_j). \tag{10.4.14}$$

Clearly, equation (10.4.13) implies, because of the factor $\delta_{nn'}$, that the impurity potential does not couple different bands, and therefore, by analogy with equation (10.4.14), we can expand the perturbed wave function $\psi_{\mathbf{k}}(\mathbf{r})$ in equation (10.4.12) in the form

$$\psi_{\mathbf{k}}(\mathbf{r}) = \sum_{\mathbf{R}_j} U(\mathbf{R}_j)\,w_n(\mathbf{r}-\mathbf{R}_j). \tag{10.4.15}$$

We now substitute equations (10.4.14) and (10.4.15) into (10.4.12), use the result (10.4.13) defining the potential, and we find for the amplitude $U(\mathbf{R}_0)$ of the Wannier function at the impurity site an expression which must reduce, from comparison of equations (10.4.14) and (10.4.15) when V_{nn} in equation

(10.4.13) is zero, to $e^{i\mathbf{k}.\mathbf{R}_0}$, but which involves basically, from equation (10.4.12), the quantity

$$\sum_{\mathbf{k}'} \frac{1}{E-E_{n\mathbf{k}'}+i\varepsilon} = \int \frac{d\mathbf{k}'}{(2\pi)^3}\left(\frac{1}{E-E_{n\mathbf{k}'}+i\varepsilon}\right). \tag{10.4.16}$$

But we can readily transform the $\mathbf{k}'$ integration in equation (10.4.16) into an integral over energy, with the density of states $n(E)$ in band n introduced. Then equation (10.4.16) becomes

$$\int \frac{n(s)\,ds}{E-s+i\varepsilon} = P(E)-i\pi n(E), \tag{10.4.17}$$

where the function $P(E)$ is given by

$$P(E) = P\int \frac{n(s)\,ds}{E-s}, \tag{10.4.18}$$

P on the right-hand side of equation (10.4.18) denoting the principal value. The function $P(E)$, involving only the density of states in the band under consideration, is the basic quantity in the Koster–Slater theory, as we shall see below. Putting all the pieces together, it is easy to show that

$$U(\mathbf{R}_0) = \frac{e^{i\mathbf{k}.\mathbf{R}_0}}{[1-V_{nn}P(E)]+i\pi V_{nn}n(E)}. \tag{10.4.19}$$

Values of U at other lattice sites than the impurity site $\mathbf{R}_0$ can also be computed, but equation (10.4.19) will suffice for our purposes.

We can see immediately, without going into special cases at this stage, some of the essential features of the solution (10.4.19). Thus, if we find an energy E for which the term $[1-V_{nn}P(E)]$ appearing in the denominator of equation (10.4.19) vanishes, then two situations can arise. If the energy E, say E_0, corresponding to the vanishing of $[1-V_{nn}P(E)]$, lies inside the band, then of course the term $i\pi V_{nn}n(E)$ in the denominator of equation (10.4.19) remains. However, we expect $U(\mathbf{R}_0)$ to become large as E passes through E_0: we shall have a resonant behaviour at E_0. This is the analogue of the virtual bound state of Friedel, which we discussed for plane-wave scattering in terms of rapid variation of the scattering cross-section with energy as we moved through such a state in an energy band.

However, if E_0 lies outside the band, $U(\mathbf{R}_0)$ blows up at E_0, and this is reflecting the fact that the amplitude of the Wannier function on the impurity site is becoming infinite relative to the amplitude of the Bloch wave. This is signalling a real bound state round the impurity.

As a specific case, we shall consider below the form of the conduction band for Cu metal, and the shape of the function $P(E)$ in that case. First, however, let us employ the tight-binding dispersion relation (cf. Problem 1.5), which was taken as an example by Koster and Slater in their original paper.

(b) *Tight-binding band structure.* The band structure

$$E(\mathbf{k}) \propto [\cos k_x a + \cos k_y a + \cos k_z a] \tag{10.4.20}$$

will thus be considered as a model for constructing the central function $P(E)$ of the theory. The form shown schematically in Figure 10.2 is then obtained. Indeed this form is obtained rather generally, because, from the

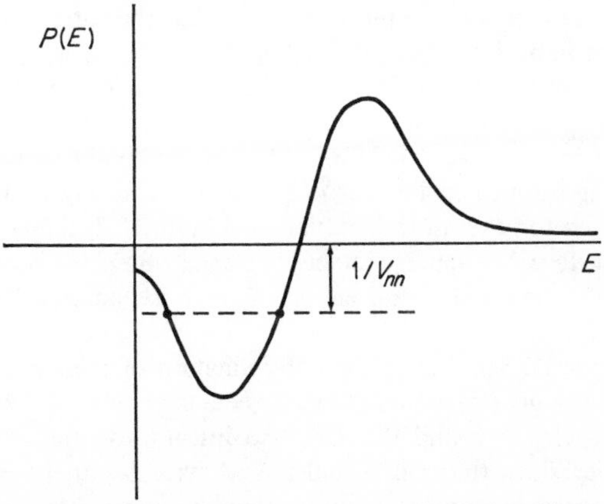

FIGURE 10.2. Schematic form of Koster–Slater function $P(E)$ of equation (10.4.18) for tight-binding dispersion relation (10.4.20).

definition of $P(E)$, it follows that the function approaches $1/E$ for large positive or negative values of E. In the range between, its behaviour is of course determined by the precise form of $n(E)$ for the band in question. If $n(E)$ has a fairly smooth form then $P(E)$ has the shape of a dispersion curve as shown in Figure 10.2 (see also Dawber and Turner (1966) and Figure 10.3 for an example appropriate to the conduction band of metallic Cu).

We have seen that the behaviour of the amplitude $U(\mathbf{R}_0)$ of the Wannier function on the impurity site is such that, when $1 - V_{nn}P(E) = 0$, either this becomes large, or if the energy corresponding to this condition lies *outside* the band then $U(\mathbf{R}_0)$ becomes infinite.

Obviously such behaviour arises from attractive potentials, and the energies at which resonant or bound states occur are determined by the intersections of the line $(1/V_{nn})$ with $P(E)$, as shown in Figure 10.2.

Koster and Slater show that the extreme values of $P(E)$ correspond to energies lying within the band for the dispersion relation (10.4.20), and thus, for appropriate values of V_{nn}, the method leads to a virtual bound state or resonant state.

(c) *Conduction band of Cu.* We shall argue below that the Koster–Slater assumptions made about the impurity potential in the simplest "one-band, one-site" model we are discussing are leading to a local potential $V(\mathbf{r})$ which is short range, with oscillations in the wings [actually these arise from oscillations in the Wannier function $w(\mathbf{r})$]. From the discussion in section 10.2.1, this means that the impurity potential has the correct general features for a charged impurity centre in a metal.

Thus we shall briefly consider here the example of the conduction band of metallic Cu, following the work of Dawber and Turner (1966). Both an approximate form of the density of states in the conduction band, and the corresponding form calculated for $P(E)$ are shown in Figure 10.3. The form of $P(E)$ is seen to have all the features of Figure 10.2, though the fact that the Fermi surface in copper contacts the zone boundary is reflected in the spike in $n(E)$ below the Fermi energy E_f and the kink in $P(E)$ below the Fermi energy.

Dawber and Turner discuss the determination of the strength parameter V_{nn}, and from a generalization of the Friedel sum rule (10.3.21) (i.e. getting the displaced charge round the defect to integrate to the total number of electrons) they show that the virtual bound state lies at the Fermi level E_f. This is too restrictive and they argue that, because of the one-band, one-site approximation, one perhaps ought to count only s-electrons.

Some understanding of the reduction in the electronic specific heat $C_v = \gamma T$ can be gained for increasing Zn concentration in Cu from this model. The coefficient γ is of course related to the density of states at the Fermi level, and the result for this, for small concentrations of impurities, is recorded in Appendix 10.2. The decrease in γ is considerably greater on this theory than the rigid band model predicts (see again Appendix 10.2), and this is in general agreement with experiments on Cu–Zn (see, however, Isaacs and Massalski, 1965).

(d) *Form of Koster–Slater "one-band one-site" potential in metals.* It is not clear at first sight just what spatial variation of the defect potential $V(\mathbf{r})$ is implied by the assumption (10.4.13). In a metal, we have seen that a given density $\rho(\mathbf{r}E_f)$ can be generated by a suitably chosen potential. In this way,

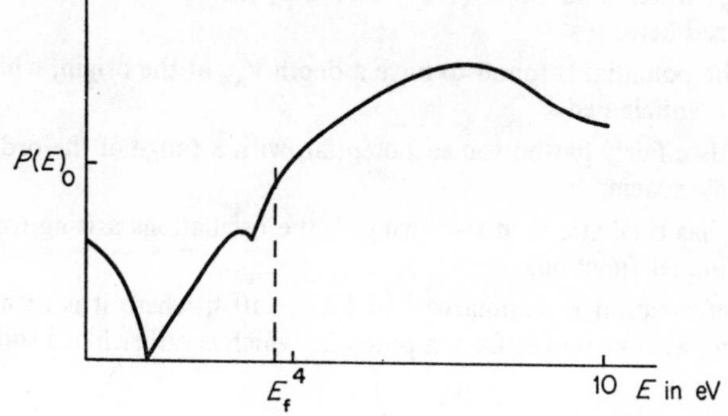

(a)

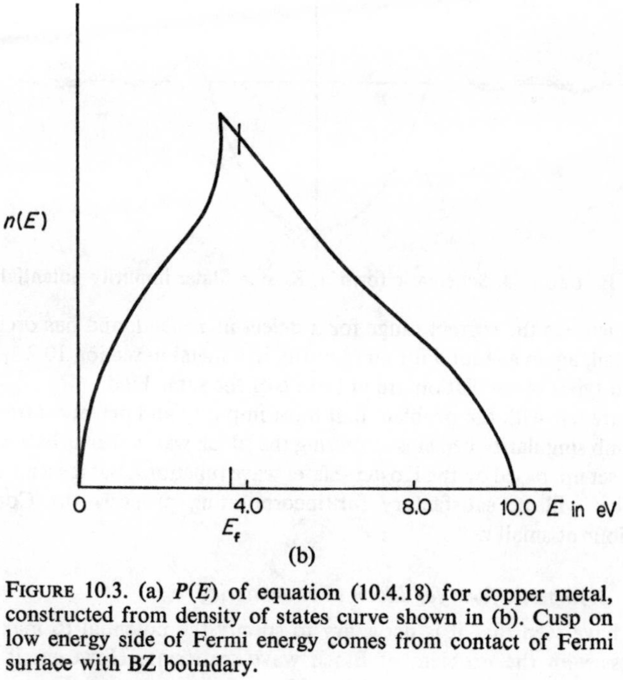

(b)

FIGURE 10.3. (a) $P(E)$ of equation (10.4.18) for copper metal, constructed from density of states curve shown in (b). Cusp on low energy side of Fermi energy comes from contact of Fermi surface with BZ boundary.

a defect potential for a metal can be extracted. The details are given by Stoddart, March and Stott (1969) and only the general features will be summarized here:

(i) The potential is found to have a depth V_{nn} at the origin, which could have been anticipated.

(ii) It is a fairly flat-bottomed potential, with a range of the order of the interatomic spacing.

(iii) It has oscillations in the "wings", the oscillations arising from those in the Wannier function.

This information is summarized in Figure 10.4, where it is seen that we are dealing with scattering from a potential which is rather like a square well,

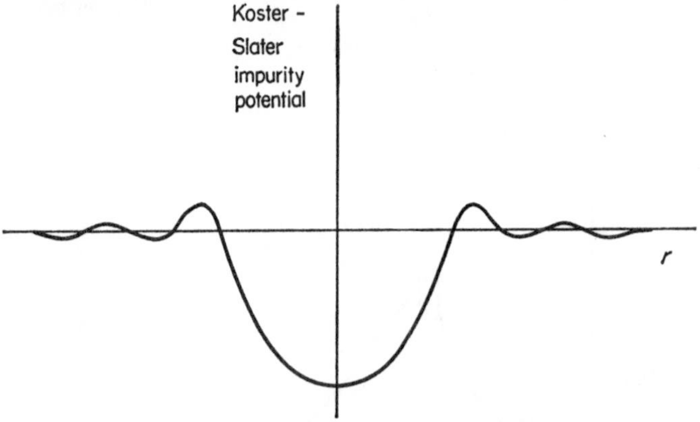

FIGURE 10.4. Schematic form of Koster–Slater impurity potential.

but which has the correct range for a defect in a metal, and has oscillations in the tail, again as found for an impurity in a metal in section 10.2.1, though the two types of oscillation are not quite of the same kind.

We are left with the problem that most impurity and defect centres have a Coulomb singularity which is scattering the Bloch waves. Perturbation theory can be set up, based on the Koster–Slater wave functions, but it seems unlikely that this will be satisfactory for incorporating properly the Coulombic behaviour at small **r**.

10.4.4 *Asymptotic form for Bloch wave scattering*

We have seen that it is necessary to specify the potential to make much progress with the problem of Bloch wave scattering. However, if we ask only about asymptotic properties, and assume a spherical Fermi surface, then some more general progress can be made, as demonstrated by Kohn and Vosko (1960). Their theory is valuable in interpreting nuclear magnetic resonance data for dilute alloys.

Suppose, as above, that the defect potential is $V(r)$ while the periodic potential generating the local density of electrons, $\rho(\mathbf{r}E)$, with energy less than E, is $V_p(\mathbf{r})$. Then we must solve the Schrödinger equation

$$\left[-\frac{\hbar^2}{2m}\nabla^2+V_p(\mathbf{r})+V(\mathbf{r})\right]\psi = E\psi. \tag{10.4.21}$$

For simplicity we shall assume that the conduction band $E(\mathbf{k})$ is spherically symmetrical.

We are now interested in the asymptotic form of the solutions $\psi_\mathbf{k}$ of equation (10.4.21) corresponding to an incident Bloch wave $\phi_\mathbf{k}$ and an outgoing scattered wave. We shall take the solute atom to be situated at the origin. In complete analogy with the theory of plane-wave scattering, one finds‡

$$\psi_\mathbf{k}(\mathbf{r}) = \phi_\mathbf{k}(\mathbf{r})+[f(\mathbf{k},\mathbf{k}')/r]\,\phi_{\mathbf{k}'}(\mathbf{r}), \tag{10.4.22}$$

where $|\mathbf{k}'| = |\mathbf{k}|$ and furthermore

$$\mathbf{k}' = k'(\mathbf{r}/r); \tag{10.4.23}$$

that is, $\mathbf{k}'$ is in the same direction as $\mathbf{r}$ (see Figure 10.5). This property can be seen as follows. An electron arriving at $\mathbf{r}$ after being scattered at the origin

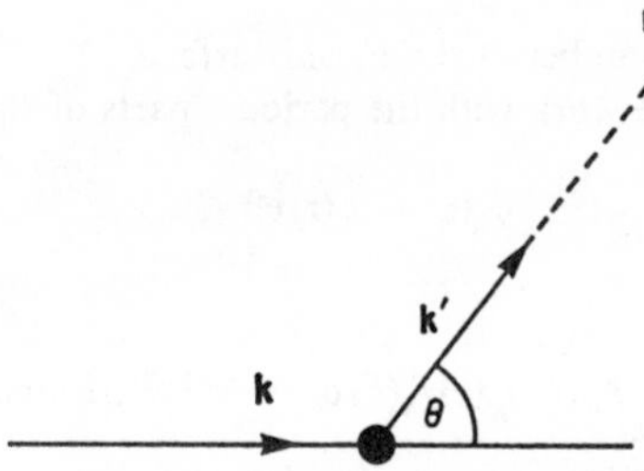

FIGURE 10.5. Illustrating scattering of Bloch waves off a defect centre.

has a velocity vector $\mathbf{v}$ in the direction of $\mathbf{r}$. Because of the assumed spherical symmetry of $E(\mathbf{k})$, its $\mathbf{k}$-vector is parallel to $\mathbf{v}$ and hence also to $\mathbf{r}$.

In the absence of more detailed information we assume that $f(\mathbf{k},\mathbf{k}')$ depends only on the magnitude of $\mathbf{k}$ and the angle θ between $\mathbf{k}$ and $\mathbf{k}'$ or

‡ To calculate the density from equation (10.4.22), we strictly ought to include a higher-order (r^{-2}) term. However, we shall not complicate the argument with this, since the cross-term with $\phi_\mathbf{k}(\mathbf{r})$ cancels with some integrated parts (e.g. equation (10.4.31)).

k and **r** (see Figure 10.5), and shall write it as $f_k(\theta)$. We can then expand $f_k(\theta)$ in spherical harmonics in the usual way (see equation (10.3.1)),

$$f_k(\theta) = \frac{1}{2ik}\sum_{l=0}^{\infty}(2l+1)\,[e^{2i\eta_l(k)}-1]\,P_l(\cos\theta), \tag{10.4.24}$$

the η_l's being, as usual, the phase shifts. For large r, the excess electron density contributed by the wave function $\psi_{\mathbf{k}}$ can be written as

$$\begin{aligned}\delta n_{\mathbf{k}}(\mathbf{r}) &\equiv |\psi_{\mathbf{k}}(\mathbf{r})|^2 - |\phi_{\mathbf{k}}(\mathbf{r})|^2 \\ &= \delta n_{\mathbf{k}}^{(1)} + \delta n_{\mathbf{k}}^{(2)},\end{aligned} \tag{10.4.25}$$

where

$$\delta n_k^{(1)}(\mathbf{r}) = \frac{1}{r}\,[\phi_{\mathbf{k}}^*(\mathbf{r}) f_k(\theta)\,\phi_{\mathbf{k}'}(\mathbf{r}) + \text{complex conjugate}] \tag{10.4.26}$$

and

$$\delta n_k^{(2)}(\mathbf{r}) = \frac{1}{r^2}|f_k(\theta)|^2\,|\phi_{\mathbf{k}'}(\mathbf{r})|^2. \tag{10.4.27}$$

The total excess density at **r** is

$$\delta n(\mathbf{r}) = \frac{2}{(2\pi)^3}\int_{k<k_f} \delta n_{\mathbf{k}}(\mathbf{r})\,d\mathbf{k}, \tag{10.4.28}$$

where k_f is the wave number at the Fermi surface.

To evaluate $\delta n_{\mathbf{k}}^{(1)}$ we work with the periodic parts of the Bloch waves

$$\phi_{\mathbf{k}}(\mathbf{r}) = u_{\mathbf{k}}(\mathbf{r})\,e^{i\mathbf{k}\cdot\mathbf{r}}, \tag{10.4.29}$$

when we obtain

$$\delta n^{(1)}(\mathbf{r}) = \frac{2}{(2\pi)^3}\frac{1}{r}\left[\int_{k<k_f} u_{\mathbf{k}}^*(\mathbf{r})\,u_{\mathbf{k}'}(\mathbf{r}) f_k(\theta)\,e^{-ikr(\cos\theta-1)}\,d\mathbf{k} + \text{complex conjugate}\right]. \tag{10.4.30}$$

For large r only the exponential in equation (10.4.30) is a rapidly varying function of the angle θ between **k** and **r**. This allows us to extract the dominant term by integrating equation (10.4.30) by parts with respect to θ. For, let $F(\theta)$ be a slowly varying function of θ. Then

$$\int_0^{\pi} F(\theta)\,e^{-ikr(\cos\theta-1)}\,d(\cos\theta) = \frac{1}{ikr}F(\pi)\,e^{2ikr} - \frac{1}{ikr}F(0) + O\left(\frac{1}{r^2}\right). \tag{10.4.31}$$

Applying this result to equation (10.4.30) gives

$$\delta n^{(1)}(\mathbf{r}) = \frac{1}{2\pi^2 r^2}\left\{\left[\int_0^{k_f} e^{2ikr}\frac{f_k(\pi)}{i}u^*_{-\mathbf{k}}(\mathbf{r})\,u_{\mathbf{k}}(\mathbf{r})\,k\,dk + \text{complex conjugate}\right]\right.$$
$$\left. - \left[\int_0^{k_f}\frac{f_k(0)}{i}u^*_{\mathbf{k}}(\mathbf{r})\,u_{\mathbf{k}}(\mathbf{r})\,k\,dk + \text{complex conjugate}\right]\right\}. \tag{10.4.32}$$

The integrals are over the magnitude of **k** only, its direction being fixed in the direction of **k**′ or **r**. The second bracket in equation (10.4.32) is precisely cancelled by the contribution of $\delta n^{(2)}_{\mathbf{k}}(\mathbf{r})$, by virtue of the so-called optical theorem

$$\int d\Omega\,|f_k(\theta)|^2 = \frac{4\pi}{ik}\,\mathrm{Im}\,f_k(0), \tag{10.4.33}$$

the integration being over the solid angle $d\Omega$. Further we may use the identity

$$u^*_{-\mathbf{k}}(\mathbf{r}) = u_{\mathbf{k}}(\mathbf{r}) \tag{10.4.34}$$

to obtain

$$\delta n(\mathbf{r}) = \frac{1}{2\pi^2 r^2}\left\{\int_0^{k_f} e^{2ikr}\frac{f_k(\pi)}{i}[u_{\mathbf{k}}(\mathbf{r})]^2\,k\,dk + \text{complex conjugate}\right\}. \tag{10.4.35}$$

Again, for large r, only the exponential is a rapidly varying function of k so that one may integrate by parts as in equation (10.4.31). This gives the final result that

$$\delta n(\mathbf{r}) = -\left(\frac{1}{4\pi^2}\right)\frac{1}{r^3}\{\exp(2ik_f r)\,[u_{\mathbf{k}_f}(\mathbf{r})]^2\,k_f f_{k_f}(\pi) + \text{complex conjugate}\}, \tag{10.4.36}$$

where $\mathbf{k}_f$ points in the direction of **r**. In the special case of plane waves $[u_{\mathbf{k}_f}(\mathbf{r}) \equiv 1]$, equation (10.4.36) reduces to Friedel's result

$$\delta n_{\text{free}}(r) = \frac{1}{2\pi^2 r^3}\sum_l (2l+1)\,[-\sin\eta_l \cos(2k_f r + \eta_l - l\pi)], \tag{10.4.37}$$

which may be expressed as

$$\delta n_{\text{free}}(r) = A\cos(2k_f r + \phi)/r^3, \tag{10.4.38}$$

where

$$A = \frac{1}{2\pi^2}\left[\left\{\sum_l (2l+1)\left[-\sin\eta_l \cos(\eta_l - l\pi)\right]\right\}^2 + \left\{\sum_l (2l+1)\left[-\sin\eta_l \sin(\eta_l - l\pi)\right]\right\}^2\right]^{\frac{1}{2}} \tag{10.4.39}$$

and

$$\phi = \tan^{-1}\frac{\sum_l (2l+1)\sin\eta_l \cos(\eta_l - l\pi)}{\sum_l (2l+1)\sin\eta_l \sin(\eta_l - l\pi)}. \tag{10.4.40}$$

Here the η_l's are the scattering phase shifts at the Fermi surface. This then makes quantitative the result we had anticipated in section 10.2.1.

10.4.5 Kohn–Rostoker–Beeby form of electron density round vacancy in metal

Earlier in section 10.3.4, we discussed how the energy of formation of a vacancy depends on the displaced charge. We want now to describe how the Kohn–Rostoker method we discussed in Chapter 1, section 1.11.3, can be used to calculate the density of electrons in a lattice with a defect.

While it seems likely from the discussion of section 10.3.4 that fully self-consistent calculations of the displaced charge round a vacancy will eventually be necessary, a great simplification occurs if we:

(1) Start from a periodic potential which is a sum of non-overlapping muffin-tin potentials, of the kind we used in the discussion of the KKR method in Chapter 1.

(2) Approximate the vacancy by simply removing a muffin-tin potential from the site at which the vacancy is formed.

Of course, in practice, this is an oversimplified model which ought to be refined later by allowing in careful detail for electronic redistribution in a self-consistent manner, and secondly by allowing relaxation of the surrounding atoms to the vacancy (see section 10.10 below).

Nevertheless, it is a very interesting result that, in the above model, a theory can be developed for the electron distribution round the vacancy in terms of quantities which, at least in principle, are available from KKR calculations on a perfect lattice. We shall summarize the argument, due to Beeby (1967), below, giving some of the details in Appendix 10.3, where we also record the result for the density round an impurity centre.

(a) *t-matrix and Green function.* We shall find it convenient at this point to introduce the t-matrix, since it will not only be used in this application, but also when we come to discuss disordered systems in section 10.13. We have seen that the Green function for the system with potential V is given by

$$G(\mathbf{rr}') = G_0(\mathbf{r}-\mathbf{r}') + \int G_0(\mathbf{r}-\mathbf{r}'')\, V(\mathbf{r}'')\, G(\mathbf{r}''\mathbf{r}')\, d\mathbf{r}'' \tag{10.4.41}$$

where G_0 is the free-electron propagator. If the potential V were genuinely weak, we could replace G by G_0 in the integral term in equation (10.4.41). The total scattering matrix, or T-matrix, allows us to make this replacement, the resulting equation for G remaining exact, by writing

$$G(\mathbf{rr}') = G_0(\mathbf{r}-\mathbf{r}') + \int G_0(\mathbf{r}-\mathbf{r}'')\, T(\mathbf{r}'', \mathbf{r}''')\, G_0(\mathbf{r}'''-\mathbf{r}')\, d\mathbf{r}''\, d\mathbf{r}'''. \tag{10.4.42}$$

Now the muffin-tin potential assumption is that, with v of finite range,

$$V(\mathbf{r}) = \sum_{\mathbf{R}} v_{\mathbf{R}}(|\mathbf{r}-\mathbf{R}|), \tag{10.4.43}$$

where $\mathbf{R}$ denotes the lattice sites and $v_{\mathbf{R}}(\mathbf{r})$ is the type of potential on the site $\mathbf{R}$. The electron density $\rho(\mathbf{r})$ is of course obtained by summing the squares of the wave functions up to the Fermi level E_{f} and is related to the Green function $G(\mathbf{rr}'E)$ by

$$\rho(\mathbf{r}E_{\mathrm{f}}) = -\frac{1}{\pi}\int_{-\infty}^{E_{\mathrm{f}}} \operatorname{Im} G(\mathbf{rr}E)\, dE. \tag{10.4.44}$$

If we generalize the electron density as before and deal with the number of electrons per unit volume at $\mathbf{r}$ with energy less than E, $\rho(\mathbf{r}E)$, then we shall find it convenient to work with the quantity

$$\frac{\partial \rho(\mathbf{r}E)}{\partial E} = \sigma(\mathbf{r}E) = -\frac{1}{\pi} \operatorname{Im} G(\mathbf{rr}E). \tag{10.4.45}$$

It will also be convenient to introduce the scattering matrix for each potential $v(r)$. This can be written in the form

$$\begin{aligned} t_R(\mathbf{rr}') &= v_R(\mathbf{r})\, \delta(\mathbf{r}-\mathbf{r}') + v_R(\mathbf{r})\, G_0(\mathbf{r}-\mathbf{r}')\, v_R(\mathbf{r}') \\ &\quad + \int v_R(\mathbf{r})\, G_0(\mathbf{r}-\mathbf{r}'')\, v_R(\mathbf{r}'')\, G_0(\mathbf{r}''-\mathbf{r}')\, v_R(\mathbf{r}')\, d\mathbf{r}'' + \dots \end{aligned} \tag{10.4.46}$$

which sums up all the scatterings from a single potential. Then, it is straightforward to show that, in terms of the t-matrices, the Green function G for the total potential V can be written in the form

$$G(\mathbf{rr}') = G_0(\mathbf{r}-\mathbf{r}') + \sum_{R} \int G_0(\mathbf{r}-\mathbf{r}'')\, t_R(\mathbf{r}''-\mathbf{R}, \mathbf{r}'''-\mathbf{R})\, G_0(\mathbf{r}'''-\mathbf{r}')\, d\mathbf{r}''\, d\mathbf{r}''' + \dots, \tag{10.4.47}$$

where the $(N+1)$th term in this expansion is

$$\sum_{\mathbf{R}_1 \neq \mathbf{R}_2 \neq \mathbf{R}_3 \neq \ldots \neq \mathbf{R}_N} \int G_0(\mathbf{r}-\mathbf{r}_1)\, t_{\mathbf{R}_1}(\mathbf{r}_1-\mathbf{R}_1, \mathbf{r}_1'-\mathbf{R}_1)\, G_0(\mathbf{r}_1'-\mathbf{r}_2)\, t_{\mathbf{R}_2}(\mathbf{r}_2-\mathbf{R}_2, \mathbf{r}_2'-\mathbf{R}_2) \ldots$$
$$\times G_0(\mathbf{r}_N'-\mathbf{r}')\, d\mathbf{r}_1\, d\mathbf{r}_1' \ldots d\mathbf{r}_N'. \qquad (10.4.48)$$

The restriction on the summation that $\mathbf{R}_i \neq \mathbf{R}_{i+1}$ is because, from equation (10.4.46), t contains all the scattering from a single site and further scattering must therefore be from another site.

There is a physical description which can be given to equation (10.4.48). The electron starts at $\mathbf{r}$ and propagates as a free electron to any point $\mathbf{r}_1$, in the sphere of the potential at $\mathbf{R}_1$. From there it is scattered to any other point $\mathbf{r}_1'$, in the same potential. It then propagates from $\mathbf{r}_1'$ to any point $\mathbf{r}_2$, in another potential at $\mathbf{R}_2$. This process continues for N scatterings, the electron finally reaching $\mathbf{r}'$. In the perfect lattice case, all this turns out to lead to the KKR equation, as we shall see below.

To proceed with the calculation, it is useful next to change variables such that $\mathbf{r}_i - \mathbf{R}_i \to \mathbf{r}_i$, $\mathbf{r}_i' - \mathbf{R}_i \to \mathbf{r}_i'$. This then gets the lattice positions into the propagators and the first part of equation (10.4.48) becomes

$$\sum_{\mathbf{R}_1 \neq \mathbf{R}_2 \neq \mathbf{R}_3 \neq \ldots} \int G_0(\mathbf{r}-\mathbf{r}_1-\mathbf{R}_1)\, t(\mathbf{r}_1, \mathbf{r}_1')\, G_0(\mathbf{r}_1'-\mathbf{r}_2+\mathbf{R}_1-\mathbf{R}_2)\, t(\mathbf{r}_2, \mathbf{r}_2') \ldots d\mathbf{r}_1\, d\mathbf{r}_1'. \qquad (10.4.49)$$

The subscripts to the t-matrices have been left out, for notational convenience.

Next, we exploit the spherical symmetry which we assume for the one-centre potential $v(\mathbf{r})$, and expand the t-matrices in their angular momentum components

$$t(\mathbf{x}, \mathbf{y}) = \sum_L t_l(x, y)\, Y_L(\mathbf{x})\, Y_L(\mathbf{y}), \qquad (10.4.50)$$

where L stands for (lm) and the spherical harmonics $Y_L(\mathbf{x})$ are chosen to be real. At this stage, the angular integrations for all variables except $\mathbf{r}_1$ and $\mathbf{r}_N'$ are performed. The second G_0 factor in equation (10.4.49) becomes, for example,

$$G_0(2) = \int Y_L(\mathbf{r}_1')\, G_0(\mathbf{r}_1'-\mathbf{r}_2+\mathbf{R}_1-\mathbf{R}_2)\, Y_{L'}(\mathbf{r}_2)\, d\Omega_{\mathbf{r}_1'}\, d\Omega_{\mathbf{r}_2}. \qquad (10.4.51)$$

Because of the non-overlapping of the potentials, this can be evaluated by writing $G_0(\mathbf{x})$ in Fourier transform and expanding the plane wave $\exp(i\mathbf{k}.\mathbf{r})$ using Bauer's expansion (1.11.40). Then we find

$$G_0(2) = j_l[\surd(E)\, r_1']\, j_{l'}[\surd(E)\, r_2]\, G^0_{LL'}(\mathbf{R}_1-\mathbf{R}_2), \qquad (10.4.52)$$

:senting a parabolic-like band of width 2Δ with an extra factor a to .e it more ($a>0$) or less ($a<0$) peaked. Figure 10.6(a) shows a plot of) for the three cases $a = \pm 1, 0$ with a choice $\Delta = 1$, while the density σ lotted in Figure 10.6(b). The case $a = 0$ gives a density proportional to

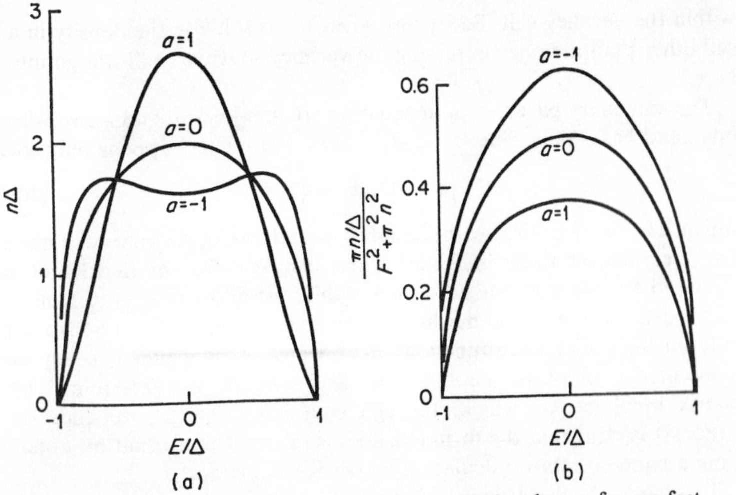

FIGURE 10.6. (a) Model density of states in Beeby theory for perfect crystal. (b) Density in imperfect lattice, from equation (10.4.67).

the density of states, but we see that increasing the peak height reduces the electron density, particularly when n is large.

It is worth noting in conclusion that Harris (1970) has cast Beeby's theory into a form more suitable for numerical computation.

10.4.6 Effective mass theory for shallow impurity levels in semi-conductors

We shall now consider the problem of shallow donor levels in semi-conductors. As we saw in section 10.2.2, the Coulomb potential is often no very effectively screened and we might expect the donor levels to be relate to the energy states in the Coulomb field $-e/\varepsilon r$.

The method we derive below is the so-called effective mass theor shallow states. A somewhat more general presentation of effective mass t is also given in Appendix 10.4. In the course of the following deriv the effective mass equation from the scattering form (10.4.12 Schrödinger equation, it will become quite clear why the theory i to shallow states.

where

$$G^0_{L'}(\mathbf{R}_1-\mathbf{R}_2) = \frac{2}{\pi}\int \frac{e^{i\mathbf{j}\cdot(\mathbf{R}_1-\mathbf{R}_2)}}{E-j^2+i\varepsilon} Y_L(\mathbf{j})\, Y_{L'}(\mathbf{j}) \frac{j_l(jr_1')}{j_l[\sqrt{(E)}\,r_1']} \frac{j_{l'}(jr_2)}{j_{l'}[\sqrt{(E)}\,r_2]}\, i^{l'-l}\, d\mathbf{j} \tag{10.4.53}$$

and is, in fact, independent of r_1' and r_2. The energy E will be taken as positive; the changes are only slight for the case $E<0$.

The radial integrations are now carried out, except for the variables $\mathbf{r}_1$, $\mathbf{r}_1'$, $\mathbf{r}_N$ and $\mathbf{r}_N'$. Each scattering then gives integrals like

$$t_l(E) = \int j_l[\sqrt{(E)}\,r_2]\, j_l[\sqrt{(E)}\,r_2']\, t_l(r_2, r_2')\, r_2^2\, dr_2\, r_2'^2\, dr_2'. \tag{10.4.54}$$

In this way, the term (10.4.48) can be written as

$$\sum_{\mathbf{R}_2\neq\mathbf{R}_3\neq\ldots\neq\mathbf{R}_{N-1}} \sum_{L',L''\ldots L^{N-2}} G^0_{LL'}(\mathbf{R}_1-\mathbf{R}_2)\, t_{l'}(E)\, G^0_{L'L''}(\mathbf{R}_2-\mathbf{R}_3)\, t_{l''}(E)\ldots$$
$$\times\, t_{l^{N-2}}(E)\, G^0_{L^{N-2}L^{N-1}}(\mathbf{R}_{N-1}-\mathbf{R}_N), \tag{10.4.55}$$

where the angular momentum sums are absorbed into the notation by defining matrices $\mathbf{t}$ and $\mathbf{G}^0(\mathbf{R})$ through

$$(\mathbf{t})_{LL'} = t_l(E)\,\delta_{LL'};\quad [\mathbf{G}^0(\mathbf{R})]_{LL'} = G^0_{LL'}(\mathbf{R}). \tag{10.4.56}$$

It is only to be expected from the discussion of the KKR method in Chapter 1 that $\mathbf{t}$ depends only on the single-centre scattering potential v while $\mathbf{G}^0(\mathbf{R})$ depends only on the lattice, both being implicit functions of the energy.

(b) *Density of electrons.* Since we can see from equation (10.4.44) that only the diagonal part $G(\mathbf{r},\mathbf{r})$ enters the determination of the electron density, we can put $\mathbf{r}=\mathbf{r}'$ in equation (10.4.48). Suppose now that $\mathbf{r}$ is inside one of the potentials, since this simplifies the treatment considerably. The centre of this particular sphere may be chosen as the origin of coordinates. One term in the sum is then $\mathbf{R}_1 = 0$, in which case it is necessary that $\mathbf{R}_2\neq 0$. The integrals not performed above are then

$$b_L(\mathbf{r}) = \int G_0(\mathbf{r}-\mathbf{r}_1)\, t_l(r_1, r_1')\, Y_L(\mathbf{r}_1)\, j_l[\sqrt{(E)}\,r_1']\, d\mathbf{r}_1\, r_1'^2\, dr_1'. \tag{10.4.57}$$

Alternatively, when $\mathbf{R}_1\neq 0$ we have

$$\int G_0(\mathbf{r}-\mathbf{r}_1-\mathbf{R}_1)\, t_l(r_1, r_1')\, Y_L(\mathbf{r}_1)\, j_l[\sqrt{(E)}\,r_1']\, d\mathbf{r}_1\, r_1'^2\, dr_1', \tag{10.4.58}$$

which can be shown to take the form

$$\sum_{L'} Y_{L'}(\mathbf{r})\, j_{l'}[\sqrt{(E)}\,r]\, G_{L'L}(\mathbf{R}_1)\, t_l(E) = \sum_{L'} A_{L'}(\mathbf{r})\, G_{L'L}(\mathbf{R}_1)\, t_l(E). \tag{10.4.59}$$

After some detailed manipulation which is recorded in Appendix 10.3, we find that the Green function (10.4.42), on the diagonal, and for the case when all the t's are the same, as in the perfect metal, can be written in the form, with **1** the unit matrix,

$$G(\mathbf{r},\mathbf{r}) = G_0(\mathbf{r}-\mathbf{r})+\int G_0(\mathbf{r}-\mathbf{r}')\,t(\mathbf{r}',\mathbf{r}'')\,G_0(\mathbf{r}''-\mathbf{r})\,d\mathbf{r}'\,d\mathbf{r}''$$
$$+\sum_{\substack{LL'\\L''L'''}}\Omega_B\int_{\Omega_B} d\mathbf{m}[A_L(\mathbf{r})+b_L(\mathbf{r})]\,G_{LL'}(\mathbf{m})\,t_{l'}(E)$$
$$\times[\mathbf{1}-\Omega_B\,\mathbf{G}(\mathbf{m})\,\mathbf{t}]^{-1}_{L'L''}\,G_{L''L'''}(\mathbf{m})\,[A_{L'''}(\mathbf{r})+b_{L'''}(\mathbf{r})]. \tag{10.4.60}$$

Here

$$\mathbf{G}(\mathbf{m}) = \frac{1}{\Omega_B}\sum_{\mathbf{R}_i\neq 0} e^{-i\mathbf{m}.\mathbf{R}_i}\,G(\mathbf{R}_i), \tag{10.4.61}$$

where it should be emphasized that, because in equation (10.4.61), $\mathbf{R}_i\neq\mathbf{R}_{i+1}$, the $\mathbf{R}_i = 0$ term in equation (10.4.61) must be omitted. This implies that

$$\int_{\Omega_B}\mathbf{G}(\mathbf{m})\,d\mathbf{m} = 0. \tag{10.4.62}$$

Some further calculations, again summarized in Appendix 10.3, then show us that the combination $A_L(\mathbf{r})+b_L(\mathbf{r})$ needed in equation (10.4.60) can, in fact, be written in the form

$$[A_L(\mathbf{r})+b_L(\mathbf{r})]/t_l(E) = \frac{1}{v(r)}\,\frac{Y_L(\mathbf{r})\int t_l(r,r_1')\,j_l[\sqrt{(E)}\,r_1']\,r_1'^2\,dr_1'}{\int t_l(r,r_1')\,j_l[\sqrt{(E)}\,r_1']\,j_l[\sqrt{(E)}\,r]\,r^2\,dr\,r_1'^2\,dr_1'} \tag{10.4.63}$$

and the left-hand side has been written in this way because it is in fact simply the radial wave function $R_l(r)$ (within the muffin-tin) which is regular at the origin (the normalization of R_l is discussed in Appendix 10.3).

In this way, we are led to the two basic equations of this treatment for the perfect lattice and the lattice containing a vacancy

$$\sigma_{\text{lattice}}(\mathbf{r}E) = -\frac{1}{\pi}\sum_{LL'} R_L(\mathbf{r})\,R_{L'}(\mathbf{r})\frac{1}{\Omega_B}\,\text{Im}\int_{\Omega_B} d\mathbf{m}[\mathbf{t}^{-1}-\Omega_B\,\mathbf{G}(\mathbf{m})]^{-1}_{LL'} \tag{10.4.64}$$

and

$$\sigma_{\text{vacancy}}(\mathbf{r}E) = -\frac{1}{\pi}\,\text{Im}\left[\sum_{LL'} A_L(\mathbf{r})\left\{\frac{1}{\Omega_B}\int_{\Omega_B}[\mathbf{t}^{-1}-\Omega_B\,\mathbf{G}(\mathbf{m})\right.\right.$$

within the vacancy cell. Beeby also gives the result for the cell other than the one containing the vacancy, but we shall here.

The imaginary part shown in equation (10.4.64) is only non- integrand has a pole, since $\mathbf{t}^{-1}-\Omega_B\,\mathbf{G}(\mathbf{m})$ is real. This happens

$$|\mathbf{t}^{-1}-\Omega_B\,\mathbf{G}(\mathbf{m})| = 0,$$

which is in fact identical with the KKR equation. At the energy i the integral over these imaginary parts behaves like the density $n(E)$ and therefore is non-zero only within the allowed energy b radial distribution of the density is seen from equation (10.4.64) t only on the radial wave functions $R_l(\mathbf{r})$, which are dependent solely one-centre muffin-tin potential, but their weights are determined matrix involving **G**, which, through equation (10.4.61), depends lattice. It is clear that the form (10.4.64) is related to information obta from a band-structure calculation by the KKR method.

Turning to the density in the imperfect lattice, we see again from equa (10.4.65) that the density is only non-zero in the allowed bands of the per crystal, for only then can there be an imaginary part to the integral.

(c) *Model calculation.* We shall conclude the discussion, which has followe closely the method and presentation of Beeby, by considering a simple example he gave, to get some idea of what $\sigma(\mathbf{r}E)$ is like. Thus, suppose we have a potential for which only $t_0(E)$ is non-zero. Then (see Appendix 10.3) the density is given by

$$\sigma(\mathbf{r}E) = \Omega_B[A_0(\mathbf{r})]^2\frac{n(E)}{F^2(E)+\pi^2 n^2(E)}, \tag{10.4.67}$$

where $n(E)$ is the density of states of the host lattice and $F(E)$ is its Hilbert transform.

Next, to see how this behaves, let us choose $n(E)$ to have the form

$$\left.\begin{aligned} n(E) &= \frac{2}{\pi\Delta}\frac{(1-aE^2)}{(1-a\Delta^2/4)}\sqrt{(1-E^2/\Delta^2)} \quad (|E|<\Delta),\\ &= 0 \quad (|E|\geqslant\Delta), \end{aligned}\right\} \tag{10.4.68}$$

rep ma $n(E$ is

To anticipate the results, the theory turns out as though one could use an equivalent Hamiltonian. This, for the nth band with dispersion relation $E_{n\mathbf{k}}$, has the form (which we shall prove, or a quite equivalent result)

$$H_{\text{equiv}} = E_n(-i\nabla) + V(\mathbf{r}), \tag{10.4.69}$$

where $V(\mathbf{r})$ is the impurity potential, if necessary suitably screened. This immediately leads to the wave equation

$$[E_n(-i\nabla) + V(\mathbf{r})]\,F(\mathbf{r}) = EF(\mathbf{r}) \tag{10.4.70}$$

from which to determine the bound-state energy levels. We have written $F(\mathbf{r})$ to connect with the detailed treatment given below.

What we have in mind is a semi-conductor with valence and conduction bands as shown in Figure 10.7. The shallow states we shall consider, in say

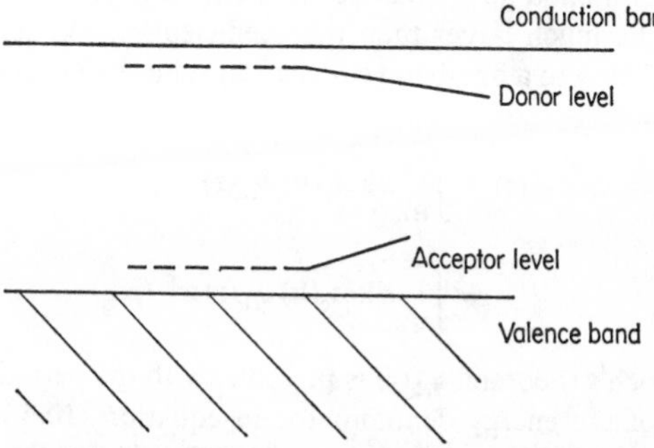

FIGURE 10.7. Shallow donor and acceptor levels in semiconductor.

Ge and Si, have energies measured relative to the band edges of the order of 0.01 to 0.1 eV. We shall see that the corresponding wave functions are quite diffuse, spreading out over many atoms. On the other hand, some impurities in Ge and Si, such as Cu and Au, can bind levels with energies as large as $\frac{1}{2}$ eV. These "deep" impurity states are not well understood and will require a more sophisticated theory than we outline here.

Though the equivalent Hamiltonian method is rather general, we shall derive the theory from the scattering formalism we have set up and used in the Koster–Slater model, for very simple assumptions about the band structure (compare Lukes and Roberts, 1967; also private communication).

(a) *Bound-wave functions for shallow donors.* For bound states, we omit the incident wave in equation (10.4.12) and also, since the state E lies outside the bands, omit the term $i\varepsilon$.

Since the Bloch functions $\phi_{n\mathbf{k}}(\mathbf{r})$ form a complete set, the bound-state wave function $\psi(\mathbf{r})$ may be expanded in the form

$$\psi(\mathbf{r}) = \sum_n \int_{\mathrm{BZ}} d\mathbf{k}\, A_n(\mathbf{k})\, \phi_{n\mathbf{k}}(\mathbf{r}). \tag{10.4.71}$$

Substituting equation (10.4.71) into (10.4.12), we obtain by straightforward calculation the result for the coefficients $A_n(\mathbf{k})$

$$A_n(\mathbf{k}) = \frac{1}{E - E_{n\mathbf{k}}} \int \phi^*_{n\mathbf{k}}(\mathbf{r}')\, V(\mathbf{r}')\, \psi(\mathbf{r}')\, d\mathbf{r}'. \tag{10.4.72}$$

(b) *Conduction band minimum at* $\mathbf{k} = 0$. If the conduction band minimum is at $\mathbf{k} = 0$, then E is nearly equal to $E_c(0)$ for shallow donor levels, where c is the conduction band label. We see from equation (10.4.72) that the term $A_c(\mathbf{k})$ will then be much larger than the coefficients $A_n(\mathbf{k})$ for $n \neq c$ and we can reduce the theory to a one-band calculation. Thus from equation (10.4.71) we have

$$\begin{aligned} \psi(\mathbf{r}) &= \int_{\mathrm{BZ}} d\mathbf{k}\, A_c(\mathbf{k})\, \phi_{c\mathbf{k}}(\mathbf{r}) \\ &= \int_{\mathrm{BZ}} d\mathbf{k}\, A_c(\mathbf{k})\, u_{c\mathbf{k}}(\mathbf{r})\, e^{i\mathbf{k}.\mathbf{r}}, \end{aligned} \tag{10.4.73}$$

where, from Bloch's theorem, $u_{c\mathbf{k}}(\mathbf{r})$ is periodic, with the period of the lattice. Also, because of the energy denominator in equation (10.4.72), $A_c(\mathbf{k})$ falls off in magnitude rapidly as we go away from $\mathbf{k} = 0$ and this suggests that we expand $u_{c\mathbf{k}}(\mathbf{r})$ around $\mathbf{k} = 0$, that is,

$$u_{c\mathbf{k}}(\mathbf{r}) = u_{c0}(\mathbf{r}) + \mathbf{k}.(\nabla_{\mathbf{k}}\, u_{c\mathbf{k}}(\mathbf{r}))_{\mathbf{k}=0} + \dots, \tag{10.4.74}$$

and then neglect the $\mathbf{k}$-dependent term, since $A_c(\mathbf{k})$ is becoming small when that term becomes important in equation (10.4.74). Then we find from equation (10.4.73) that

$$\psi(\mathbf{r}) = u_{c0}(\mathbf{r})\, F(\mathbf{r}), \tag{10.4.75}$$

where

$$F(\mathbf{r}) = \int_{\mathrm{BZ}} d\mathbf{k}\, A_c(\mathbf{k})\, e^{i\mathbf{k}.\mathbf{r}}. \tag{10.4.76}$$

If, in equation (10.4.72), we now make the approximation that we can expand $u_{c0}(\mathbf{r})$ as

$$u_{c0}(\mathbf{r}) = \sum_{\mathbf{K}_n} v_{\mathbf{K}_n}\, e^{i\mathbf{K}_n.\mathbf{r}}, \tag{10.4.77}$$

where

$$G^0_{L'}(\mathbf{R}_1-\mathbf{R}_2) = \frac{2}{\pi}\int \frac{e^{i\mathbf{j}.(\mathbf{R}_1-\mathbf{R}_2)}}{E-j^2+i\varepsilon} Y_L(\mathbf{j})\, Y_{L'}(\mathbf{j}) \frac{j_l(jr_1')}{j_l[\sqrt{(E)}\, r_1']} \frac{j_{l'}(jr_2)}{j_{l'}[\sqrt{(E)}\, r_2]} i^{l'-l}\, d\mathbf{j} \qquad (10.4.53)$$

and is, in fact, independent of r_1' and r_2. The energy E will be taken as positive; the changes are only slight for the case $E<0$.

The radial integrations are now carried out, except for the variables $\mathbf{r}_1$, $\mathbf{r}_1'$, $\mathbf{r}_N$ and $\mathbf{r}_N'$. Each scattering then gives integrals like

$$t_l(E) = \int j_l[\sqrt{(E)}\, r_2]\, j_l[\sqrt{(E)}\, r_2']\, t_l(r_2, r_2')\, r_2^2\, dr_2\, r_2'^2\, dr_2'. \qquad (10.4.54)$$

In this way, the term (10.4.48) can be written as

$$\sum_{\mathbf{R}_2\neq\mathbf{R}_3\neq\ldots\neq\mathbf{R}_{N-1}} \sum_{L',L''\ldots L^{N-2}} G^0_{LL'}(\mathbf{R}_1-\mathbf{R}_2)\, t_{l'}(E)\, G^0_{L'L''}(\mathbf{R}_2-\mathbf{R}_3)\, t_{l''}(E)\ldots$$
$$\times\, t_{l^{N-2}}(E)\, G^0_{L^{N-2}L^{N-1}}(\mathbf{R}_{N-1}-\mathbf{R}_N), \qquad (10.4.55)$$

where the angular momentum sums are absorbed into the notation by defining matrices $\mathbf{t}$ and $\mathbf{G}^0(\mathbf{R})$ through

$$(\mathbf{t})_{LL'} = t_l(E)\, \delta_{LL'}; \quad [G^0(\mathbf{R})]_{LL'} = G^0_{LL'}(\mathbf{R}). \qquad (10.4.56)$$

It is only to be expected from the discussion of the KKR method in Chapter 1 that $\mathbf{t}$ depends only on the single-centre scattering potential v while $\mathbf{G}^0(\mathbf{R})$ depends only on the lattice, both being implicit functions of the energy.

(b) *Density of electrons.* Since we can see from equation (10.4.44) that only the diagonal part $G(\mathbf{r}, \mathbf{r})$ enters the determination of the electron density, we can put $\mathbf{r} = \mathbf{r}'$ in equation (10.4.48). Suppose now that $\mathbf{r}$ is inside one of the potentials, since this simplifies the treatment considerably. The centre of this particular sphere may be chosen as the origin of coordinates. One term in the sum is then $\mathbf{R}_1 = 0$, in which case it is necessary that $\mathbf{R}_2 \neq 0$. The integrals not performed above are then

$$b_L(\mathbf{r}) = \int G_0(\mathbf{r}-\mathbf{r}_1)\, t_l(r_1, r_1')\, Y_L(\mathbf{r}_1)\, j_l[\sqrt{(E)}\, r_1']\, d\mathbf{r}_1\, r_1'^2\, dr_1'. \qquad (10.4.57)$$

Alternatively, when $\mathbf{R}_1 \neq 0$ we have

$$\int G_0(\mathbf{r}-\mathbf{r}_1-\mathbf{R}_1)\, t_l(r_1, r_1')\, Y_L(\mathbf{r}_1)\, j_l[\sqrt{(E)}\, r_1']\, d\mathbf{r}_1\, r_1'^2\, dr_1', \qquad (10.4.58)$$

which can be shown to take the form

$$\sum_{L'} Y_{L'}(\mathbf{r})\, j_{l'}[\sqrt{(E)}\, r]\, G_{L'L}(\mathbf{R}_1)\, t_l(E) = \sum_{L'} A_{L'}(\mathbf{r})\, G_{L'L}(\mathbf{R}_1)\, t_l(E). \qquad (10.4.59)$$

After some detailed manipulation which is recorded in Appendix 10.3, we find that the Green function (10.4.42), on the diagonal, and for the case when all the t's are the same, as in the perfect metal, can be written in the form, with **1** the unit matrix,

$$G(\mathbf{r},\mathbf{r}) = G_0(\mathbf{r}-\mathbf{r}) + \int G_0(\mathbf{r}-\mathbf{r}')\,t(\mathbf{r}',\mathbf{r}'')\,G_0(\mathbf{r}''-\mathbf{r})\,d\mathbf{r}'\,d\mathbf{r}''$$

$$+ \sum_{\substack{LL' \\ L''L'''}} \Omega_B \int_{\Omega_B} d\mathbf{m}[A_L(\mathbf{r})+b_L(\mathbf{r})]\,G_{LL'}(\mathbf{m})\,t_{l'}(E)$$

$$\times [\mathbf{1}-\Omega_B\,\mathbf{G}(\mathbf{m})\,\mathbf{t}]^{-1}_{L'L''}\,G_{L''L'''}(\mathbf{m})\,[A_{L'''}(\mathbf{r})+b_{L'''}(\mathbf{r})]. \qquad (10.4.60)$$

Here

$$\mathbf{G}(\mathbf{m}) = \frac{1}{\Omega_B} \sum_{\mathbf{R}_i \neq 0} e^{-i\mathbf{m}.\mathbf{R}_i}\,G(\mathbf{R}_i), \qquad (10.4.61)$$

where it should be emphasized that, because in equation (10.4.61), $\mathbf{R}_i \neq \mathbf{R}_{i+1}$, the $\mathbf{R}_i = 0$ term in equation (10.4.61) must be omitted. This implies that

$$\int_{\Omega_B} \mathbf{G}(\mathbf{m})\,d\mathbf{m} = 0. \qquad (10.4.62)$$

Some further calculations, again summarized in Appendix 10.3, then show us that the combination $A_L(\mathbf{r})+b_L(\mathbf{r})$ needed in equation (10.4.60) can, in fact, be written in the form

$$[A_L(\mathbf{r})+b_L(\mathbf{r})]/t_l(E) = \frac{1}{v(r)}\,\frac{Y_L(\mathbf{r}) \int t_l(r,r_1')\,j_l[\sqrt{(E)}\,r_1']\,r_1'^2\,dr_1'}{\int t_l(r,r_1')\,j_l[\sqrt{(E)}\,r_1']\,j_l[\sqrt{(E)}\,r]\,r^2\,dr\,r_1'^2\,dr_1'} \qquad (10.4.63)$$

and the left-hand side has been written in this way because it is in fact simply the radial wave function $R_l(r)$ (within the muffin-tin) which is regular at the origin (the normalization of R_l is discussed in Appendix 10.3).

In this way, we are led to the two basic equations of this treatment for the perfect lattice and the lattice containing a vacancy

$$\sigma_{\text{lattice}}(\mathbf{r}E) = -\frac{1}{\pi} \sum_{LL'} R_L(\mathbf{r})\,R_{L'}(\mathbf{r}) \frac{1}{\Omega_B} \operatorname{Im} \int_{\Omega_B} d\mathbf{m}[\mathbf{t}^{-1}-\Omega_B\,\mathbf{G}(\mathbf{m})]^{-1}_{LL'} \qquad (10.4.64)$$

and

$$\sigma_{\text{vacancy}}(\mathbf{r}E) = -\frac{1}{\pi}\text{Im}\left[\sum_{LL'} A_L(\mathbf{r})\left\{\frac{1}{\Omega_B}\int_{\Omega_B}[\mathbf{t}^{-1}-\Omega_B\,\mathbf{G}(\mathbf{m})]^{-1}\,d\mathbf{m}\right\}^{-1}_{LL'} A_{L'}(\mathbf{r})\right], \tag{10.4.65}$$

within the vacancy cell. Beeby also gives the result for the density in a unit cell other than the one containing the vacancy, but we shall not go into that here.

The imaginary part shown in equation (10.4.64) is only non-zero when the integrand has a pole, since $\mathbf{t}^{-1}-\Omega_B\,\mathbf{G}(\mathbf{m})$ is real. This happens only when

$$|\mathbf{t}^{-1}-\Omega_B\,\mathbf{G}(\mathbf{m})| = 0, \tag{10.4.66}$$

which is in fact identical with the KKR equation. At the energy in question, the integral over these imaginary parts behaves like the density of states $n(E)$ and therefore is non-zero only within the allowed energy bands. The radial distribution of the density is seen from equation (10.4.64) to depend only on the radial wave functions $R_l(\mathbf{r})$, which are dependent solely on the one-centre muffin-tin potential, but their weights are determined by the matrix involving $\mathbf{G}$, which, through equation (10.4.61), depends on the lattice. It is clear that the form (10.4.64) is related to information obtainable from a band-structure calculation by the KKR method.

Turning to the density in the imperfect lattice, we see again from equation (10.4.65) that the density is only non-zero in the allowed bands of the perfect crystal, for only then can there be an imaginary part to the integral.

(c) *Model calculation.* We shall conclude the discussion, which has followed closely the method and presentation of Beeby, by considering a simple example he gave, to get some idea of what $\sigma(\mathbf{r}E)$ is like. Thus, suppose we have a potential for which only $t_0(E)$ is non-zero. Then (see Appendix 10.3) the density is given by

$$\sigma(\mathbf{r}E) = \Omega_B[A_0(\mathbf{r})]^2\frac{n(E)}{F^2(E)+\pi^2 n^2(E)}, \tag{10.4.67}$$

where $n(E)$ is the density of states of the host lattice and $F(E)$ is its Hilbert transform.

Next, to see how this behaves, let us choose $n(E)$ to have the form

$$\left.\begin{aligned} n(E) &= \frac{2}{\pi\Delta}\frac{(1-aE^2)}{(1-a\Delta^2/4)}\sqrt{(1-E^2/\Delta^2)} \quad (|E|<\Delta), \\ &= 0 \quad (|E|\geqslant\Delta), \end{aligned}\right\} \tag{10.4.68}$$

representing a parabolic-like band of width 2Δ with an extra factor a to make it more ($a>0$) or less ($a<0$) peaked. Figure 10.6(a) shows a plot of $n(E)$ for the three cases $a=\pm 1, 0$ with a choice $\Delta=1$, while the density σ is plotted in Figure 10.6(b). The case $a=0$ gives a density proportional to

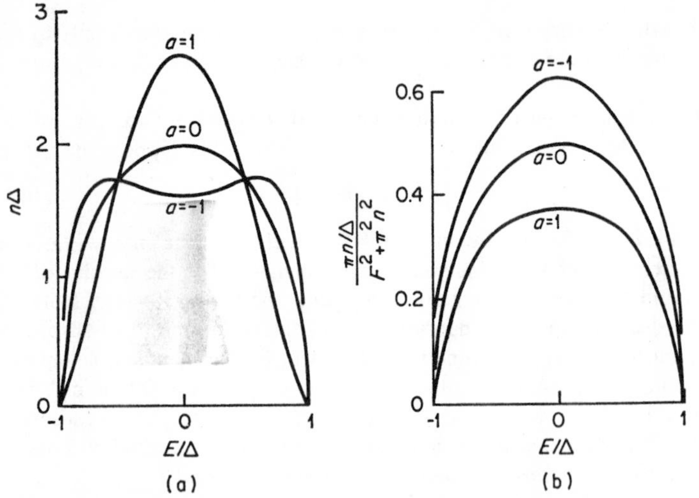

FIGURE 10.6. (a) Model density of states in Beeby theory for perfect crystal. (b) Density in imperfect lattice, from equation (10.4.67).

the density of states, but we see that increasing the peak height reduces the electron density, particularly when n is large.

It is worth noting in conclusion that Harris (1970) has cast Beeby's theory into a form more suitable for numerical computation.

10.4.6 *Effective mass theory for shallow impurity levels in semi-conductors*

We shall now consider the problem of shallow donor levels in semi-conductors. As we saw in section 10.2.2, the Coulomb potential is often not very effectively screened and we might expect the donor levels to be related to the energy states in the Coulomb field $-e/\varepsilon r$.

The method we derive below is the so-called effective mass theory of shallow states. A somewhat more general presentation of effective mass theory is also given in Appendix 10.4. In the course of the following derivation of the effective mass equation from the scattering form (10.4.12) of the Schrödinger equation, it will become quite clear why the theory is restricted to shallow states.

To anticipate the results, the theory turns out as though one could use an equivalent Hamiltonian. This, for the nth band with dispersion relation $E_{n\mathbf{k}}$, has the form (which we shall prove, or a quite equivalent result)

$$H_{\text{equiv}} = E_n(-i\nabla) + V(\mathbf{r}), \tag{10.4.69}$$

where $V(\mathbf{r})$ is the impurity potential, if necessary suitably screened. This immediately leads to the wave equation

$$[E_n(-i\nabla) + V(\mathbf{r})]\,F(\mathbf{r}) = EF(\mathbf{r}) \tag{10.4.70}$$

from which to determine the bound-state energy levels. We have written $F(\mathbf{r})$ to connect with the detailed treatment given below.

What we have in mind is a semi-conductor with valence and conduction bands as shown in Figure 10.7. The shallow states we shall consider, in say

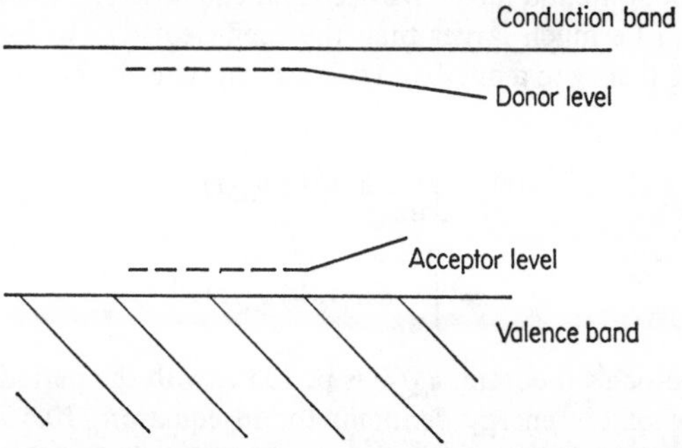

FIGURE 10.7. Shallow donor and acceptor levels in semiconductor.

Ge and Si, have energies measured relative to the band edges of the order of 0.01 to 0.1 eV. We shall see that the corresponding wave functions are quite diffuse, spreading out over many atoms. On the other hand, some impurities in Ge and Si, such as Cu and Au, can bind levels with energies as large as $\frac{1}{2}$ eV. These "deep" impurity states are not well understood and will require a more sophisticated theory than we outline here.

Though the equivalent Hamiltonian method is rather general, we shall derive the theory from the scattering formalism we have set up and used in the Koster–Slater model, for very simple assumptions about the band structure (compare Lukes and Roberts, 1967; also private communication).

(a) *Bound-wave functions for shallow donors.* For bound states, we omit the incident wave in equation (10.4.12) and also, since the state E lies outside the bands, omit the term $i\varepsilon$.

Since the Bloch functions $\phi_{n\mathbf{k}}(\mathbf{r})$ form a complete set, the bound-state wave function $\psi(\mathbf{r})$ may be expanded in the form

$$\psi(\mathbf{r}) = \sum_n \int_{\mathrm{BZ}} d\mathbf{k}\, A_n(\mathbf{k})\, \phi_{n\mathbf{k}}(\mathbf{r}). \tag{10.4.71}$$

Substituting equation (10.4.71) into (10.4.12), we obtain by straightforward calculation the result for the coefficients $A_n(\mathbf{k})$

$$A_n(\mathbf{k}) = \frac{1}{E-E_{n\mathbf{k}}} \int \phi^*_{n\mathbf{k}}(\mathbf{r}')\, V(\mathbf{r}')\, \psi(\mathbf{r}')\, d\mathbf{r}'. \tag{10.4.72}$$

(b) *Conduction band minimum at* $\mathbf{k} = 0$. If the conduction band minimum is at $\mathbf{k} = 0$, then E is nearly equal to $E_c(0)$ for shallow donor levels, where c is the conduction band label. We see from equation (10.4.72) that the term $A_c(\mathbf{k})$ will then be much larger than the coefficients $A_n(\mathbf{k})$ for $n \neq c$ and we can reduce the theory to a one-band calculation. Thus from equation (10.4.71) we have

$$\begin{aligned}\psi(\mathbf{r}) &= \int_{\mathrm{BZ}} d\mathbf{k}\, A_c(\mathbf{k})\, \phi_{c\mathbf{k}}(\mathbf{r}) \\ &= \int_{\mathrm{BZ}} d\mathbf{k}\, A_c(\mathbf{k})\, u_{c\mathbf{k}}(\mathbf{r})\, e^{i\mathbf{k}.\mathbf{r}},\end{aligned} \tag{10.4.73}$$

where, from Bloch's theorem, $u_{c\mathbf{k}}(\mathbf{r})$ is periodic, with the period of the lattice. Also, because of the energy denominator in equation (10.4.72), $A_c(\mathbf{k})$ falls off in magnitude rapidly as we go away from $\mathbf{k} = 0$ and this suggests that we expand $u_{c\mathbf{k}}(\mathbf{r})$ around $\mathbf{k} = 0$, that is,

$$u_{c\mathbf{k}}(\mathbf{r}) = u_{c0}(\mathbf{r}) + \mathbf{k}.(\nabla_{\mathbf{k}} u_{c\mathbf{k}}(\mathbf{r}))_{\mathbf{k}=0} + \ldots, \tag{10.4.74}$$

and then neglect the $\mathbf{k}$-dependent term, since $A_c(\mathbf{k})$ is becoming small when that term becomes important in equation (10.4.74). Then we find from equation (10.4.73) that

$$\psi(\mathbf{r}) = u_{c0}(\mathbf{r})\, F(\mathbf{r}), \tag{10.4.75}$$

where

$$F(\mathbf{r}) = \int_{\mathrm{BZ}} d\mathbf{k}\, A_c(\mathbf{k})\, e^{i\mathbf{k}.\mathbf{r}}. \tag{10.4.76}$$

If, in equation (10.4.72), we now make the approximation that we can expand $u_{c0}(\mathbf{r})$ as

$$u_{c0}(\mathbf{r}) = \sum_{\mathbf{K}_n} v_{\mathbf{K}_n} e^{i\mathbf{K}_n.\mathbf{r}}, \tag{10.4.77}$$

where the $\mathbf{K}_n$'s are reciprocal lattice vectors and that $v_{\mathbf{K}_n} \ll v_0 \simeq 1$ for $\mathbf{K}_n \neq 0$, we find explicitly

$$F(\mathbf{r}) \simeq \int_{\mathrm{BZ}} d\mathbf{k} \frac{1}{E - E_c(\mathbf{k})} \int e^{i\mathbf{k}.(\mathbf{r}-\mathbf{r}')} V(\mathbf{r}') F(\mathbf{r}') \, d\mathbf{r}'. \tag{10.4.78}$$

Since $1/[E - E_c(\mathbf{k})]$ is very large for small $\mathbf{k}$, falling away quite quickly as we go away from $\mathbf{k} = 0$, we can extend the integration with small error over the whole of $\mathbf{k}$-space. We find that the resulting equation is exactly the equivalent integral representation of the effective mass equation (10.4.70).

Obviously, for a parabolic band, with $E_c(k) = \hbar^2 k^2/2m^*$ and neglecting the effect of the Debye screening on the Coulomb potential, we find from equation (10.4.70) hydrogenic bound states for the potential $-e^2/\varepsilon r$ with ε as the static dielectric constant.

It would be interesting to see whether the form (10.4.78) could be checked by including higher terms than v_0 in the expansion (10.4.77), perhaps simply including the first reciprocal lattice vector $\mathbf{K}_1$. Estimates of $v_{\mathbf{K}_1}$ should be possible from band structure calculations on Ge and Si.

With a theory which takes account in the effective mass equation (10.4.70) of the departure from spherical symmetry of the bands, calculations can be carried out which are reasonably realistic for Si and without going into further details we shall record the results below, along with the relevant experimental values (see Kohn, 1957, for a fuller account).

(c) *Some experimental results in* Si. We can pick up not only the ground impurity level for donors in Si, say P, As and Sb, but also some of the excited states. We have collected in Table 10.3 the experimental values of the

TABLE 10.3. Experimental energy differences in eV for excited donor levels in Si for different impurities

Impurity		
P	As	Sb
0.050	0.053	0.047
0.031	0.032	0.034
0.020	0.023	

differences in energy between successive excited states. The fact that the values are the same for P, As and Sb is reassuring. We can, of course, calculate these separations from the effective mass theory developed above and we find values of 0.050 eV and 0.030 eV.

The agreement for the excited states is substantially better than for the ground state. The reasons that seem to explain this are:

(i) The wave functions of the excited states are more diffuse than for the ground state.

(ii) The excited state wave functions involved above have vanishing amplitude at the donor nucleus.

For both these reasons, the orbits of the excited states penetrate rather little into the region near the donor ion, where the potential energy $-e^2/\varepsilon r$ presumably begins to break down, the use of the dielectric constant being dubious.

10.5 Magnetic field-dependent impurity states

In Chapter 4, we saw how the Landau wave functions and energy levels, or, more precisely, the Bloch density matrix, could be calculated for electrons in magnetic fields. Indeed, even for electrons moving in a periodic potential, we were able to calculate the partition function to $O(B^2)$ in the magnetic field, in terms of the zero field Bloch density matrix.

We are here concerned with the question of the effects of impurities on this system: initially say a charged impurity, such as Zn, in Cu metal. Later on, we shall have something to say about the effective mass theory of semiconductors in a magnetic field, because there, in favourable circumstances, we can obtain striking magnetic field-dependent effects in modest fields ~ 10 kilogauss.

We shall be concerned with the self-consistency of the impurity potential in the presence of the magnetic field. This, as we saw earlier, can be very conveniently expressed in terms of a dielectric function $\varepsilon(k)$. However, because the impurity centre is now being screened by the electrons which are moving in a magnetic field B, the dielectric function will become a function $\varepsilon(k, B)$.

Our first task, then, is to calculate $\varepsilon(k, B)$ and at present we can get explicit results, even in zero magnetic field, only when we replace the Bloch waves in the pure crystal by plane waves.

Thus, the method we shall adopt is as follows:

(i) We write down the integral form of the Bloch equation, for free electrons in a uniform magnetic field, but with the perturbation due to the charged defect "switched on".

(ii) Once we have solved for $C(\mathbf{r}\beta)$, the Bloch density to $O(B^2)$ in the field and to first order in the impurity potential $V(\mathbf{r})$, we use the relation (10.2.10) to calculate the electron density $\rho(\mathbf{r}E)$.

(iii) Just as in zero magnetic field, we calculate the self-consistent potential from the Poisson equation, where the displaced charge now depends on the magnetic field.

Having carried out this programme, we shall use the results to examine the change in the Landau orbital diamagnetism for free electrons in a metal due to the presence of scattering centres. Then we shall generalize the theory to all orders in the magnetic field B, in order to make some comments on:

(a) The de Haas–van Alphen effect in dilute alloys.

(b) The Brooks–Herring theory of screening and magnetic field-dependent impurity states in a semi-conductor.

10.5.1 Bloch density matrix to second order in magnetic field

In Chapter 4, we saw how to calculate the Bloch density matrix in a magnetic field. Here we shall briefly develop the low B form to use as the unperturbed problem in the next section.

We simply return to equation (A1.2.4) and insert for C_0 the free-electron form

$$C_0(\mathbf{r}\mathbf{r}_0\,\beta) = (2\pi\beta)^{-\frac{3}{2}} \exp\left(-\frac{|\mathbf{r}-\mathbf{r}_0|^2}{2\beta}\right). \tag{10.5.1}$$

Then we find, to first order in the magnetic field B, the result

$$C_{0B}(\mathbf{r}\mathbf{r}_0\,\beta) = (2\pi\beta)^{-\frac{3}{2}} \exp\left(-\frac{|\mathbf{r}-\mathbf{r}_0|^2}{2\beta}\right) [1 - i\mu_{\mathrm{B}}\, B(x_0\, y - y_0\, x) + O(B^2)], \tag{10.5.2}$$

where μ_{B} is the Bohr magneton.

We now iterate the Bloch equation to get the density matrix C_{0B} correct to second order in B^2, which is, of course, necessary for calculating the field-independent susceptibility.

The result thus obtained is given by

$$\begin{aligned} C_{0B}(\mathbf{r}\mathbf{r}_0\,\beta) = (2\pi\beta)^{-\frac{3}{2}} \exp\left(-\frac{|\mathbf{r}-\mathbf{r}_0|^2}{2\beta}\right) \Big[& 1 - i\mu_{\mathrm{B}}\, B(x_0\, y - y_0\, x) \\ & - \tfrac{1}{2}(\mu_{\mathrm{B}}\, B)^2 (x_0\, y - y_0\, x)^2 - \tfrac{1}{6}(\mu_{\mathrm{B}}\, B\beta)^2 \\ & \times \left\{ 1 + \frac{1}{\beta}[(x-x_0)^2 + (y-y_0)^2] \right\} \Big]. \end{aligned} \tag{10.5.3}$$

This is a special case of the result of Sondheimer and Wilson for free electrons, and can be generalized to all orders in the magnetic field to yield equation (4.15.14). We shall make use of their result below to deal with the de Haas–van Alphen effect in the case of dilute metallic alloys, as well as for the effective mass-theory of donor levels in semi-conductors, already discussed for $B = 0$ in section 10.4.6.

10.5.2 Dielectric function in low magnetic fields for metals

We now use equation (A1.2.4), with C_0 replaced by C_{0B} of equation (10.5.3), and with V as the potential due to the impurity centre in the metal. Then taking the case where $\mathbf{r}_0 = \mathbf{r}$, we obtain for the Bloch density of the defect crystal in a magnetic field the result

$$C(\mathbf{rr}\beta) = C_{0B}(\mathbf{rr}\beta) - \int d\mathbf{r}_1\, V(\mathbf{r}_1)\, [I(B,\beta)]_{\mathbf{r}_0=\mathbf{r}}, \tag{10.5.4}$$

where the quantity $I(B,\beta)$ must reduce to the "kernel" in equation (10.2.11) when we put $B = 0$, and away from $B = 0$ has the form, for low fields,

$$\begin{aligned}[I(B,\beta)]_{\mathbf{r}_0=\mathbf{r}} = {} & \frac{\exp(-2|\mathbf{r}-\mathbf{r}_1|^2/\beta)}{\pi(2\pi\beta)^{3/2}|\mathbf{r}-\mathbf{r}_1|} - \frac{(\mu_B B)^2 \beta^{1/2} \exp(-2|\mathbf{r}-\mathbf{r}_1|^2/\beta)}{3(2\pi)^{5/2}|\mathbf{r}-\mathbf{r}_1|} \\ & - \frac{(\mu_B B)^2 \beta^{-1/2}}{3(2\pi)^{5/2}} \frac{\exp(-2|\mathbf{r}-\mathbf{r}_1|^2/\beta)}{|\mathbf{r}-\mathbf{r}_1|} [(x_1-x)^2+(y_1-y)^2] \\ & + \frac{(\mu_B B)^2}{24\pi^2} \operatorname{erfc}(2^{1/2}\beta^{-1/2}|\mathbf{r}-\mathbf{r}_1|). \end{aligned} \tag{10.5.5}$$

We now use equations (10.5.4) and (10.5.5) in the relation (1.3.3) to derive the electron density $\rho(\mathbf{r}E)$. We then find, after some calculation, with $k = (2E)^{1/2}$,

$$\begin{aligned}\rho(\mathbf{r}E) = \rho_0(\mathbf{r}E) - \int d\mathbf{r}_1\, V(\mathbf{r}_1) \Bigg\{ & \frac{k}{4\pi^3 R^3} \left(\frac{\sin 2kR}{2kR} - \cos 2kR \right) \\ & - \frac{(\mu_B B)^2}{6\pi^3} \left[\frac{\cos 2kR}{kR} + \frac{1}{2} \frac{X^2+Y^2}{R^2} \sin 2kR + \operatorname{si}(2kR) \right] \Bigg\}, \end{aligned} \tag{10.5.6}$$

where $R = |\mathbf{r}-\mathbf{r}_1|$, $X = (x-x_1)$, $Y = (y-y_1)$. As expected, equation (10.5.6) reduces to (2.3.14) in zero field. We see that there is no longer spherical symmetry but rather axial symmetry around the field direction (z-axis).

As before, we must combine equation (10.5.6) with the Poisson equation.

$$\nabla^2 V = 4\pi(\rho_0 - \rho) + 4\pi Z \delta(\mathbf{r}) \tag{10.5.7}$$

for an impurity of excess valence Z to obtain the self-consistent field and density. Introducing the Fourier components $\tilde{\rho}(\mathbf{q})$, etc. through

$$\tilde{\rho}(\mathbf{q}) = \int d\mathbf{r}\, \rho(\mathbf{r})\, \mathrm{e}^{-i\mathbf{q}\cdot\mathbf{r}}, \tag{10.5.8}$$

we find, with β the angle between $\mathbf{q}$ and the magnetic field,

$$\begin{aligned}\tilde{\rho}(\mathbf{q}) = \tilde{\rho}_0(\mathbf{q}) &- \frac{\tilde{V}(\mathbf{q})}{4\pi}\left[\frac{2k}{\pi} + \left(\frac{2k^2}{\pi q} - \frac{q}{2\pi}\right)\ln\left|\frac{2k+q}{2k-q}\right|\right] \\ &+ \frac{2(\mu_B B)^2\tilde{V}(q)}{3\pi^2 k(q^2-4k^2)} + \frac{(\mu_B B)^2\tilde{V}(\mathbf{q})}{12\pi^3}\left\{-\frac{16\pi k\sin^2\beta}{(q^2-4k^2)^2}\right. \\ &\left.-\frac{4\pi}{q^2}(\cos^2\beta - \tfrac{1}{2}\sin^2\beta)\left[\frac{4k}{4k^2-q^2} - \frac{1}{q}\ln\left|\frac{q+2k}{q-2k}\right|\right]\right\} \\ &- \frac{(\mu_B B)^2}{6\pi^3}\tilde{V}(\mathbf{q})\left[\frac{2\pi}{q^3}\ln\left|\frac{2k+q}{2k-q}\right| - \frac{8\pi k}{q^2(4k^2-q^2)}\right],\end{aligned} \tag{10.5.9}$$

while the Poisson equation takes the form

$$-q^2\,\tilde{V}(\mathbf{q}) = 4\pi(\tilde{\rho}_0 - \tilde{\rho}) + 4\pi Z. \tag{10.5.10}$$

Hence we obtain the Fourier components of the screened Coulomb potential in the magnetic field as (see Hebborn and March, 1964)

$$\begin{aligned}\tilde{V}(\mathbf{q})\Bigg\{q^2 &+ \frac{k}{\pi}\left[2 + \left(\frac{q}{2k} - \frac{2k}{q}\right)\ln\left|\frac{2k-q}{2k+q}\right|\right] - \frac{8(\mu_B B)^2}{3\pi k(q^2-4k^2)} \\ &+ \frac{(\mu_B B)^2}{3\pi^2}\left[\frac{16\pi k\sin^2\beta}{(q^2-4k^2)^2} + \frac{4\pi}{q^2}(\cos^2\beta - \tfrac{1}{2}\sin^2\beta)\left(\frac{4k}{4k^2-q^2} - \frac{1}{q}\ln\left|\frac{q+2k}{q-2k}\right|\right)\right] \\ &+ \frac{2(\mu_B B)^2}{3\pi^2}\left[\frac{2\pi}{q^3}\ln\left|\frac{2k+q}{2k-q}\right| - \frac{8\pi k}{q^2(4k^2-q^2)}\right]\Bigg\} = -4\pi Z.\end{aligned} \tag{10.5.11}$$

As before, we define the dielectric function $\varepsilon(\mathbf{q}, B)$ through

$$\tilde{V}(\mathbf{q}) = -\frac{4\pi Z}{q^2\,\varepsilon(\mathbf{q}, B)} \tag{10.5.12}$$

and hence ε is readily determined. The Lindhard dielectric function (2.3.20) is regained from equations (10.5.11) and (10.5.12) with $B = 0$.

We mentioned earlier the calculation of the ion–ion interaction in a metal (see also section 10.12.1). The Friedel oscillations come from pathology in $\varepsilon(\mathbf{q})$ around $q = 2k_f$ and this leads to the Kohn anomaly in phonon spectra discussed in Chapters 3 and 6.

While the expansion in B leading to equation (10.5.11) is only valid for $q < 2k_{\mathrm{f}}$ we see that there are singularities at $q = 2k_{\mathrm{f}}$ occurring in the B-dependent terms. Further analysis leads to the conclusion that whereas, in zero field, the kink in the dispersion relation $\omega(\mathbf{q})$ for phonons is manifested in the group velocity, for $B \neq 0$ it appears in the phase velocity. Defining the surface over which the pathology occurs as the "Kohn sphere" in zero field, the field causes this to deform until, in the extreme high field limit, it becomes two planes perpendicular to the magnetic field.

Unfortunately, unattainably high fields are at present involved to observe such effects in phonon spectra in metals.

However, the field dependence of the dielectric function does affect the calculation of the orbital diamagnetism in a dilute alloy, a calculation to which we shall return in section 10.6.2.

10.5.3 Magnetic field-dependent effective mass theory of shallow donor states in semi-conductors

As a final consequence of our considerations on the dielectric function of electrons in a magnetic field, we shall consider the effect of a magnetic field on the screening of an ionized impurity in a semi-conductor, and subsequently on the shallow donor levels, via the effective mass theory discussed in section 10.4.6.

It turns out that n-type InSb provides a case where magnetic field-dependent effects are observed at quite modest fields. This is shown up clearly in the Hall measurements of Putley (1960), presented in Figure 10.8. We notice, in the range of fields between 4 kilogauss and 8 kilogauss the variation of the shape of the measured curves with field for $T > 2.5$ K, and this can be interpreted as showing an activation energy for Hall conduction which is varying with magnetic field.

It might be thought, at first sight, that this is simply showing us, via effective mass theory of the impurity states, the fact that the ionization energy in a Coulomb potential $e/\varepsilon r$ with ε the dielectric constant is varying quite strongly with field, as has been predicted earlier by Yafet, Keyes and Adams (1956). However, it turns out that the activation energy required to understand the Hall data shown is much less than the ionization energy predicted for a Coulomb field. It is clear, then, that the Brooks–Herring theory of free-carrier screening needs to be generalized to include the effect on the screening of the magnetic field.

(a) *Shielding of ionized impurity in semi-conductor in a magnetic field.* Let us recall that the Bloch density $C(\mathbf{r}\beta)$ is essentially the carrier density in a semi-conductor, apart from a normalization factor. Furthermore, as remarked above, we can obtain $C_{0B}(\mathbf{r}\mathbf{r}_0\beta)$ for arbitrary fields exactly, as shown by Sondheimer and Wilson. The case $\mathbf{r}_0 = \mathbf{r}$ is given already in equation (4.15.17) in the course of our discussion of the de Haas–van Alphen effect, and its

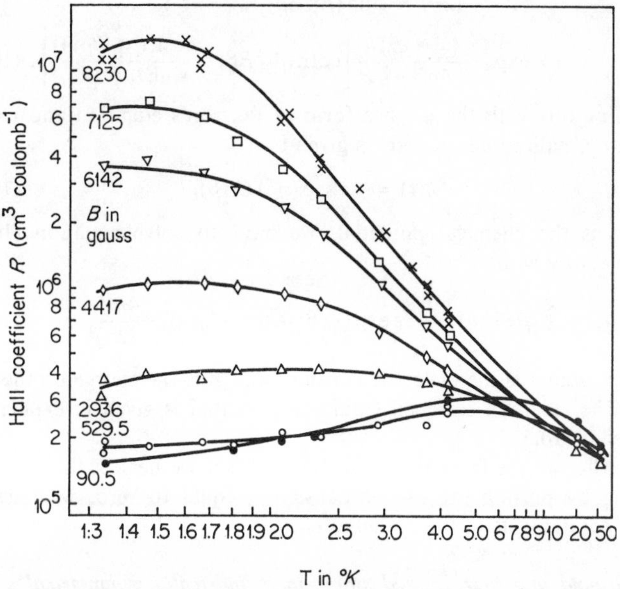

FIGURE 10.8. Hall measurements on n-type InSb for a range of magnetic fields (after Putley, 1960).

generalization may be written (by combining equations (4.15.14) and (4.15.16))

$$C(\mathbf{r}\mathbf{r}_0\beta) = \left(\frac{2\pi m^*}{h^2\beta}\right)^{3/2} \frac{\mu_B^* B\beta}{\sinh(\mu_B^* B\beta)} \exp\left[-\frac{i4\pi^2 m^*}{h^2}\mu_B^* B(xy_0 - x_0 y)\right]$$

$$\times \exp\left[-\frac{2\pi^2 m^*}{h^2\beta}(z-z_0)^2\right] \exp\left\{-\frac{2\pi^2 m^*}{h^2}\mu_B^* B \coth(\mu_B^* B\beta)\right.$$

$$\left.\times [(x-x_0)^2 + (y-y_0)^2]\right\}, \qquad (10.5.13)$$

where $\mu_B^* = e\hbar/2m^*c$.

This expression can be shown to satisfy the Bloch equation and the correct boundary condition. Inserting this in equation (10.5.10), we can calculate $C(\mathbf{r}\mathbf{r}\beta)$ taking into account the ionized impurity potential $V(\mathbf{r})$.

Following the procedure we have used before, we take the Fourier transform of equation (10.5.13) and after some calculation we obtain the desired result

$$\tilde{C}(\mathbf{q}\beta) = \tilde{C}_0(\mathbf{q}\beta) + \frac{e\tilde{V}(q)\,\mu_{\mathrm{B}}^*\,B\beta^{\frac{1}{2}}\,m^{*\frac{3}{2}}}{(2\pi)^{\frac{3}{2}}\,\hbar^3 \sinh(\mu_{\mathrm{B}}^*\,B\beta)} \int_0^1 dy \exp\left[-\frac{q_z^2\hbar^2}{8m^*}\beta(1-y^2)\right]$$
$$\times \exp\left\{-\frac{(q_x^2+q_y^2)\,\hbar^2}{4m^*\,\mu_{\mathrm{B}}^*\,B}\left[\coth(\mu_{\mathrm{B}}^*\,B\beta) - \frac{\cosh(\mu_{\mathrm{B}}^*\,B\beta y)}{\sinh(\mu_{\mathrm{B}}^*\,B\beta)}\right]\right\}. \quad (10.5.14)$$

Combining this with the **q**-space form of Poisson's equation and recalling that the normalized density $n(\mathbf{r})$ is given by

$$n(\mathbf{r}) = -e\exp(\eta\beta)\,C(\mathbf{r}\beta), \quad (10.5.15)$$

where η is the chemical potential, we have to solve equation (10.5.14) simultaneously with

$$-q^2\,\varepsilon\tilde{V}(\mathbf{q}) = 4\pi e\exp(\eta_0\beta)\,[\tilde{C}(\mathbf{q}) - \tilde{C}_0(\mathbf{q})] - \frac{4\pi Ze}{(2\pi)^{\frac{3}{2}}}, \quad (10.5.16)$$

where as usual ε is the dielectric constant and Z is the charge on the donor centre. The resulting screened Coulomb potential is given in explicit form in Appendix 10.5.

Naturally, we see from the result there that if we neglect the free carrier screening by putting the carrier density n_0 equal to zero, we regain the Fourier components of the potential $e/\varepsilon r$.

(b) *Ground and first excited states in cylindrically symmetrical screened potential.* The ground and first excited donor levels in the screened field of Appendix 10.5 have been calculated variationally as a function of n_0 for chosen temperature and magnetic field B, using the variational trial forms recorded in that appendix. The ground-state function Φ_0 has cylindrical symmetry about the field direction while Φ_1 includes the azimuthal angle ϕ. In both Φ_0 and Φ_1, a and b can be used as variational parameters.

The donor levels for $B = 5$ kilogauss and $T = 3$ K are shown in curves 1 and 2 of Figure 10.9, while the curves labelled first Coulomb level and second Coulomb level show the pure hydrogenic results. Screening, of course, reduces the ionization energy as expected. However, detailed analysis suggests that it is the gap between the first and second levels which is more closely connected with the activation energy for Hall conduction in the curves shown in Figure 10.10. This gap is not very sensitive to the carrier density and so it does not provide a critical test of the dielectric function and its magnetic field dependence. In Figure 10.10 the unscreened Coulomb field result is also plotted; this should be regained for very high fields.

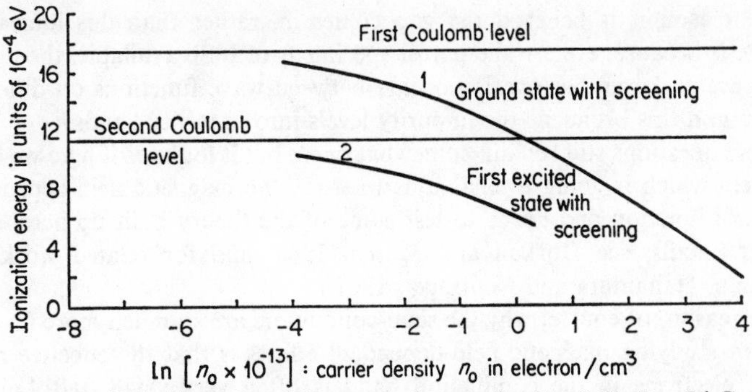

FIGURE 10.9. Ground and first excited states for $T = 3$ K, $B = 5$ kG in hydrogenic potential $e/\varepsilon r$ and screened potential.

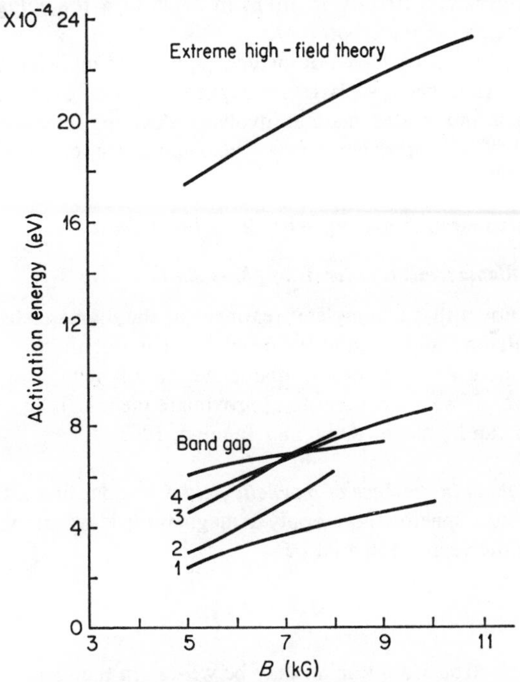

FIGURE 10.10. Activation energy for Hall conduction in n-type InSb in a magnetic field, as determined from Hall measurements of Figure 10.8. Curves 1–4 refer to different samples (see Durkan and March, 1968).

The reason it is believed the gap comes in, rather than this ionization energy, is because, even in the purest specimens of InSb available, the donor orbits are so large that overlap occurs between wave functions on different donors and this broadens the impurity levels into impurity bands.

These questions still remain somewhat open, but it looks as if here we have a system which is going to enable us to study the magnetic field-dependent dielectric function and hence to test some of the theory built up here. (For further details, see Durkan and March, 1968, and for related work see Beckman, Hanamura and Neuringer, 1967.)

The reason, of course, why the semi-conductors are so much more favourable for studying magnetic field-dependent effects is that the effective mass m^* of electrons in the conduction band is often very small (0·014 m for InSb) and the dielectric constant is large ($\varepsilon \sim 15$). Therefore the energy associated with the magnetic field can be made comparable with the effective impurity Rydberg (7×10^{-4} eV in InSb) in fields of a few kilogauss which are readily attained in the laboratory.

In these arguments, we have not included spin, for simplicity, though with the g-factor in InSb being so large we ought to account eventually for spin explicitly. This, and related matters involving shielding, has been discussed by Horing (1969) and other workers (see also von Ortenberg, 1973).

10.6 Conduction electron diamagnetism in metallic alloys

10.6.1 Alloy diamagnetism in terms of phase shifts

It is well known that a complete treatment of the diamagnetism of Bloch electrons is already a very complex problem. It is not therefore surprising that no definitive solution of the diamagnetism of metallic alloys as yet exists. However, if we are content to approximate the matrix by plane waves, then progress can be made (Kohn and Luming, 1963).

(a) *Hamiltonian in presence of magnetic field.* Consider first a free electron gas enclosed in a sphere. If we apply a magnetic field $\mathscr{B}$ along the z-axis, and write for the vector potential $\mathscr{A}$

$$\mathscr{A} = \tfrac{1}{2}\mathscr{B}\times\mathbf{r}, \tag{10.6.1}$$

then the one-electron Hamiltonian may be written in the form

$$H = -\tfrac{1}{2}\nabla^2 + \frac{\mathscr{B}}{2ic}\left(x\frac{\partial}{\partial y} - y\frac{\partial}{\partial x}\right) + \frac{\mathscr{B}^2}{8c^2}(x^2+y^2). \tag{10.6.2}$$

Introducing the angular momentum operator

$$L_z = \frac{1}{i}\left(x\frac{\partial}{\partial y} - y\frac{\partial}{\partial x}\right), \tag{10.6.3}$$

we may write H in the form

$$H = H_0 + H', \tag{10.6.4}$$

where

$$H_0 = -\tfrac{1}{2}\nabla^2 + \frac{B}{2c}L_z \tag{10.6.5}$$

and

$$H' = \frac{B^2}{8c^2}(x^2+y^2). \tag{10.6.6}$$

To obtain the magnetic moment M we may employ Feynman's theorem. If E is the total energy, $M = -\partial E/\partial B$ and to obtain the magnetic moment operator we therefore differentiate the Hamiltonian given by equations (10.6.5) and (10.6.6) with respect to B. We find

$$M = M_0 + M_1, \tag{10.6.7}$$

where

$$M_0 = -\frac{1}{2c}L_z \tag{10.6.8}$$

and

$$M_1 = -\frac{B}{4c^2}(x^2+y^2). \tag{10.6.9}$$

To obtain the expectation value of M to first order in B, it might seem, at first sight, that one could neglect H', which is of second order in B. The Schrödinger equation then separates, and the procedure for calculating the magnetic moment is straightforward.

Actually, such a procedure would not work if we tried to apply it to calculate the Landau diamagnetism of free electrons. The reason for the failure is that while B is small, the perturbation H' is actually $\propto B^2(x^2+y^2)$ and for sufficiently large x or y this term is not negligible.

However, for a localized perturbation $V(r)$ representing the solute atom, it turns out that we can apply the method above, since it is the range of the potential rather than the size of the box which dominates the dilute alloy problem. Hence H' can be neglected in this particular application.

(b) *Pure metal.* Thus, neglecting H', the eigenfunctions of H are

$$\phi_{klm} = j_l(kr)\, Y_{lm}(\theta, \phi), \tag{10.6.10}$$

with corresponding energies

$$E_{klm} = \frac{k^2}{2} + \frac{Bm}{2c}. \tag{10.6.11}$$

Insisting that the wave functions shall vanish on the surface R we have

$$j_l(kR) = 0, \tag{10.6.12}$$

which for values of $l \ll kR$ is equivalent to

$$(kR)^{-1} \sin\left(kR - \frac{l\pi}{2}\right) = 0 \tag{10.6.13}$$

or

$$kR - \frac{l\pi}{2} = n\pi, \tag{10.6.14}$$

where n is an integer.

Now suppose that E_f is the Fermi energy in the presence of the field. Then all one-particle states such that

$$E_{klm} < E_f \tag{10.6.15}$$

are occupied, and all others are empty. As an example, we consider the p-wave, with $l = 1$. Then from equations (10.6.11) and (10.6.15), the maximum values of k associated with $m = 1$, 0 and -1 are given by

$$\left.\begin{aligned} \tfrac{1}{2}k_{1,1}^2 &= E_f - \frac{B}{2c}, \\ \tfrac{1}{2}k_{1,0}^2 &= E_f, \\ \tfrac{1}{2}k_{1,-1}^2 &= E_f + \frac{B}{2c}. \end{aligned}\right\} \tag{10.6.16}$$

But from the result for M_0 in terms of L_z we have

$$\langle M_0 \rangle_{k,l,m} = -\frac{m}{2c} \tag{10.6.17}$$

and we see, therefore, that the total contribution to $\langle M_0 \rangle$ from states with $l = 1$ is

$$\langle M_0 \rangle_1 = \frac{B}{2c} n_1(E_f), \tag{10.6.18}$$

where $n_l(E_f)$ is the density of states with given l, and any compatible value of m, at $k^2/2 = E_f$. Clearly, we can calculate n_l in the absence of the field, since we only need the magnetic moment to first order in the field. Then we find, including both spin directions, that

$$n_l(E_f) = \frac{2R}{\pi k_f}. \tag{10.6.19}$$

Similarly we can treat the term M_1 in the magnetic moment, but this time we can use the unperturbed wave functions. We can obviously then write, from the form of M_1,

$$\langle M_1 \rangle_{l=1} = -\frac{B}{4c^2}\langle x^2+y^2\rangle_1, \tag{10.6.20}$$

where

$$\langle x^2+y^2\rangle_1 = \int (x^2+y^2)\,\rho_1(\mathbf{r})\,d\mathbf{r}. \tag{10.6.21}$$

Here $\rho_1(\mathbf{r})$ is the unperturbed density associated with $l = 1$.

In exact analogy, we find for a general value of l

$$\langle M_0 \rangle_l = \frac{B}{2c^2} n_l(E_f)\frac{(2l+1)(l+1)l}{6} \tag{10.6.22}$$

and

$$\langle M_1 \rangle_l = -\frac{B}{6c^2}\int r^2 \rho_l(r)\,dr. \tag{10.6.23}$$

It must be noted that we can only use the expression for $\langle M_0 \rangle_l$ when $l \ll k_f R$.

(c) *Alloy*. We now switch on the solute potential $V(r)$ which is assumed to have a finite range. As usual, the asymptotic form of the radial eigenfunctions becomes

$$(kr)^{-1}\sin\left[kr - \frac{l\pi}{2} + \eta_l(k)\right] \tag{10.6.24}$$

and the boundary conditions are therefore replaced by

$$kR - \frac{l\pi}{2} + \eta_l(k) = n\pi, \tag{10.6.25}$$

where n is an integer. The density of states is then changed to

$$n_l'(E_f) = 2\left[\frac{R}{\pi k_f} + \frac{1}{\pi}\left(\frac{\partial \eta_l}{\partial E}\right)_{E=E_f}\right]. \tag{10.6.26}$$

We find for the new contributions to the magnetic moment,

$$\langle M_0' \rangle_l = \frac{B}{2c^2} n_l'(E_f) \frac{l(l+1)(2l+1)}{6} \tag{10.6.27}$$

and

$$\langle M_1' \rangle_l = -\frac{B}{6c^2} \int r^2 \rho_l'(r)\, d\mathbf{r}, \tag{10.6.28}$$

where the primes mean that the physical quantities are calculated in the presence of the solute atom. Thus, the additional moment due to the solute atom is found by subtracting the sum over $\langle M_0' \rangle_l$ and $\langle M_1' \rangle_l$ from the unprimed quantities.

The final result may be written

$$\Delta M = \frac{B}{c^2}\left[\frac{1}{\pi}\sum_{l=1}^{\infty} \frac{l(l+1)(2l+1)}{6}\left(\frac{\partial \eta_l}{\partial E}\right)_{E=E_f} - \frac{1}{6}\int \Delta\rho(r) r^2\, d\mathbf{r}\right], \tag{10.6.29}$$

where $\Delta\rho(r)$ is the charge displaced by the solute atom.

In terms of the number of impurity atoms per unit volume N_i, the increase in the magnetic susceptibility per unit volume, $\Delta\chi$ say, is given by

$$\Delta\chi = N_i\left[\left(\frac{e\hbar}{mc}\right)^2 \frac{1}{\pi}\sum_{l=1}^{\infty} \frac{l(l+1)(2l+1)}{6}\left(\frac{\partial \eta_l}{\partial E}\right)_{E=E_f} - \frac{e^2}{6mc^2}\int \Delta\rho(r) r^2\, d\mathbf{r}\right]. \tag{10.6.30}$$

This then is the basic result of the theory (Kohn and Luming, 1963). The second term is simply the Langevin–Pauli result for an "atom", with a density equal to that of the displaced charge. However, the other term is often of comparable magnitude and of opposite sign. Below we discuss the case of a weak solute potential $V(r)$, but for arbitrary field strength.

10.6.2 de Haas–van Alphen effect in dilute metallic alloys

Actually, the quantity $\int V(\mathbf{r})\, d\mathbf{r}$, that is the $\mathbf{q} = 0$ Fourier component of the impurity potential, which evidently enters the calculation of the free energy of a dilute alloy, can be obtained without calculating the full dielectric function, by an argument which essentially counts the electrons in the pure metal and in the alloy, and combines this with the first-order theory of the Bloch density. This, as we saw, is given by

$$C(\mathbf{rr}\beta) = C_{0B}(\mathbf{rr}\beta) - \int d\mathbf{r}_1 \int_0^\beta d\beta_1\, C_{0B}(\mathbf{rr}_1\beta-\beta_1)\, V(\mathbf{r}_1)\, C_{0B}(\mathbf{r}_1\mathbf{r}\beta_1). \tag{10.6.31}$$

Here C_{0B} is the Bloch density in the presence of the magnetic field, while

C is this quantity perturbed by the impurity potential. It should be noted that we have not made any assumptions about small magnetic fields.

If we now integrate the above equation through the crystal, we obtain for the partition function

$$Z(\beta) = Z_{0B}(\beta) - \beta \int C_{0B}(\mathbf{r}\mathbf{r}\beta)\, V(\mathbf{r})\, d\mathbf{r}. \tag{10.6.32}$$

Assuming now that we are dealing with free electrons, C_{0B} is independent of $\mathbf{r}$, and employing equation (10.2.10) relating C and ρ we obtain from equation (10.6.32)

$$\int \rho(\mathbf{r})\, d\mathbf{r} = \int \rho_0(\mathbf{r})\, d\mathbf{r} - \frac{\partial \rho_0}{\partial E} \int d\mathbf{r}\, V(\mathbf{r}). \tag{10.6.33}$$

But for a single impurity centre, with Z as the excess valency, we have

$$\int (\rho - \rho_0)\, d\mathbf{r} = Z \tag{10.6.34}$$

and hence from equations (10.6.33) and (10.6.34) we find

$$\int d\mathbf{r}\, V(\mathbf{r}) = -Z \Big/ \frac{\partial \rho_0}{\partial E}. \tag{10.6.35}$$

For a magnetic field of arbitrary strength B, the quantities C_{0B} and ρ_0 can be obtained from the work of Sondheimer and Wilson (1951) discussed in Chapter 4. $C_{0B}(\mathbf{r}\mathbf{r}\beta)$ is simply the partition function per unit volume, and this is readily obtained as

$$C_{0B}(\mathbf{r}\mathbf{r}\beta) = \frac{1}{(2\pi\beta)^{\frac{3}{2}}} \frac{\beta B}{\sinh \beta B}. \tag{10.6.36}$$

Hence $\rho_0(E)$ may be calculated, with the result

$$\rho_0(E) = \frac{1}{(2\pi)^{\frac{3}{2}}} \left[\frac{4}{3\sqrt{\pi}} E^{\frac{3}{2}} - \frac{1}{6\sqrt{\pi}} B^2 E^{-\frac{1}{2}} - \frac{B^{\frac{3}{2}}}{\pi} \int_0^\infty \left(\frac{1}{y^{\frac{3}{2}}} - \frac{1}{6y^{\frac{1}{2}}} - \frac{1}{y^{\frac{3}{2}} \sinh y} \right) e^{-Ey/B}\, dy + 2B^{\frac{3}{2}} \sum_{r=1}^{\infty} \frac{(-1)^r}{(r\pi)^{\frac{3}{2}}} \sin\left(\frac{r\pi E}{B} - \frac{\pi}{4} \right) \right]. \tag{10.6.37}$$

Thus, for free electrons, $\rho_0(E)$ evidently contains the de Haas–van Alphen oscillations and hence, from equation (10.6.37) we see that oscillations with field enter the $\int V(\mathbf{r})\, d\mathbf{r}$ from equation (10.6.35).

If we neglect the field dependence of $\int V(\mathbf{r})\, d\mathbf{r}$ by replacing $\partial \rho_0/\partial E$ by its value $\partial \rho_0/\partial E|_{B=0}$ then the de Haas–van Alphen oscillations suffer a simple change in period P, say ΔP. This is readily calculated from equation (10.6.35) as

$$\frac{\Delta P}{P} = -\frac{2}{3}\frac{Z}{N} \quad \text{per impurity atom,} \tag{10.6.38}$$

and follows from the rigid band model of section 10.2. However, when we include the field dependence of $\partial \rho_0 / \partial E$ in equation (10.6.38), then it does not appear possible to describe the effect of alloying in terms of a simple change in period as in equation (10.6.38).

It is, of course, clear that it is very important to include Fermi surfaces which are distorted from the Fermi sphere, that is, to rework the problem for the scattering of Bloch waves. This is a formidable task, if V is to be calculated self-consistently in the presence of the magnetic field. Furthermore, of course, for simplicity our considerations here have been restricted to $T = 0$, whereas we must, to make direct contact with experiment, deal with the case of elevated temperatures.

However, there should be considerable information to be gained from de Haas–van Alphen experiments on dilute metallic alloys (see, for example, Springford, 1971).

10.7 Localized magnetic states round impurities in metals

The occurrence of localized magnetic moments on iron-group ions dissolved in non-magnetic metals is frequent. Earlier work of Friedel and his school has now been extended, primarily by Anderson (1961), so that a method has emerged which seems appropriate for discussing electronic states of a metal in which such a moment exists.

We shall be concerned here with Anderson's Hartree–Fock solution for his model: only brief qualitative remarks will be made about the effects of correlation [cf. Schrieffer and Mattis (1965)].

In Anderson's model, the Coulomb correlation integral of electrons in inner shell states is assumed to be a dominant feature and must be included in the Hartree–Fock treatment from the outset. In this case, the Hartree–Fock fields for electrons of different spins differ not only by exchange integrals but by Coulomb terms and it seems as if this circumstance makes localized moments possible in the iron group.

We shall proceed straightaway then to Anderson's Hamiltonian which summarizes his model (see also Rivier and Zitkova, 1971).

10.7.1 Anderson Hamiltonian

We write the Hamiltonian as a sum of four terms:

$$H = H_{\mathrm{f}} + H_{\mathrm{d}} + H_{\mathrm{corr}} + H_{\mathrm{sd}}, \tag{10.7.1}$$

where H_f is the unperturbed energy of the non-interacting electrons, which we can write in the form

$$H_f = \sum_{\mathbf{k}\sigma} \varepsilon_{\mathbf{k}} n_{\mathbf{k}\sigma}. \tag{10.7.2}$$

Here $\varepsilon_{\mathbf{k}}$ is the energy of the Bloch state with momentum **k**. $n_{\mathbf{k}\sigma}$ is the number operator for momentum **k** and spin σ and, in terms of the creation and annihilation operators $a_{\mathbf{k}\sigma}^{\dagger}$ and $a_{\mathbf{k}\sigma}$ it is simply

$$n_{\mathbf{k}\sigma} = a_{\mathbf{k}\sigma}^{\dagger} a_{\mathbf{k}\sigma}. \tag{10.7.3}$$

The band of free electron states is assumed to have a density $g(\varepsilon)$ which is taken to be a relatively slowly varying function of ε.

The second term, H_d, represents the unperturbed energy of the d-states on the impurity atom. To simplify the detail, the calculation will be restricted to a single, non-degenerate level. We then write

$$H_d = E(n_{d\uparrow} + n_{d\downarrow}). \tag{10.7.4}$$

The question may be raised with regard to H_d as to the precise nature of the localized eigenfunction ϕ_d. In most other sections of this chapter we use the picture of the impurity acting as a potential on the band states and deforming them (see also, Wolff, 1961; Stoddart, March and Wiid, 1971).

The difference here is that ϕ_d is assumed to be an inner shell level and the correlation effects are then much larger than for relatively free electrons. (In general, however, ϕ_d ought to be made orthogonal to the Wannier functions belonging to all the free electron-like bands.) This distinction makes it plausible that the omission from the Hamiltonian of the direct perturbation in the energy of the band functions caused by the impurity is not a major defect, when we are focusing attention on the nature of the magnetic states.

The assumption of a d-state distinct from the free electron band is not essential in order to form a virtual state. For, as Wolff has shown, even in the Koster–Slater theory, which is restricted to one band, virtual states can appear.

The third term in the Hamiltonian H_{corr} is the repulsive energy among the d-functions, which can be represented schematically as

$$H_{\text{corr}} = I n_{d\uparrow} n_{d\downarrow}, \tag{10.7.5}$$

where

$$I = \int |\phi_d(\mathbf{r}_1)|^2 |\phi_d(\mathbf{r}_2)|^2 e^2 |\mathbf{r}_1 - \mathbf{r}_2|^{-1} \, d\mathbf{r}_1 \, d\mathbf{r}_2. \tag{10.7.6}$$

The correlation energy arising from the electrons in band states, as well as that arising from the repulsions between d-electrons and band electrons, has been neglected here.

The fourth part of the Hamiltonian is the s–d interaction term

$$H_{sd} = \sum_{k\sigma} V_{dk}(a^\dagger_{k\sigma} a_{d\sigma} + a^\dagger_{d\sigma} a_{k\sigma}). \tag{10.7.7}$$

This is purely one-electron energy, and V_{dk} can, when necessary, be written down in terms of Wannier functions (Anderson, 1961). It will not prove necessary here to examine this.

We shall proceed now to treat the Hamiltonian, as Anderson did in his original paper, entirely in the Hartree–Fock approximation.

10.7.2 Hartree–Fock solution

By Hartree–Fock solution, we mean that the wave function is approximated by an (antisymmetrized) product function

$$\Phi_0 = \prod_{\varepsilon_n < \varepsilon_f} a^\dagger_n \Phi_{vacuum}, \tag{10.7.8}$$

where the one-electron state creation operators $a^\dagger_n$ and energies ε_n are solutions of

$$[H, a^\dagger_{n\sigma}]|_{av} \Phi_0 = \varepsilon_{n\sigma} a^\dagger_{n\sigma} \Phi_0, \tag{10.7.9}$$

where the average denotes the fact that three Fermion terms in the commutator are to be evaluated in terms of average values for the state Φ_0.

For the Hamiltonian of the present model, we can proceed to obtain a solution as follows:

We put

$$a^\dagger_{n\sigma} = \sum_k (n|k)_\sigma a^\dagger_{k\sigma} + (n|d)_\sigma a^\dagger_{d\sigma} \tag{10.7.10}$$

and the equations for the unperturbed operators are

$$[H, a^\dagger_{k\sigma}]|_{av} = \varepsilon_k a^\dagger_{k\sigma} + V_{kd} a^\dagger_{d\sigma} \tag{10.7.11}$$

and

$$[H, a^\dagger_{k\sigma}]|_{av} = [E + I\langle n_{d-\sigma}\rangle] a^\dagger_{d\sigma} + \sum_k V_{dk} a^\dagger_{k\sigma}. \tag{10.7.12}$$

The resulting equations of motion for $(n|k)$ and $(n|d)$ are then readily shown to be

$$\varepsilon_{n\sigma}(n|k)_\sigma = \varepsilon_k(n|\mathbf{k})_\sigma + V_{kd}(n|d)_\sigma \tag{10.7.13}$$

and

$$\varepsilon_{n\sigma}(n|d)_\sigma = [E + I\langle n_{d,-\sigma}\rangle](n|d)_\sigma + \sum_k V_{dk}(n|\mathbf{k})_\sigma. \tag{10.7.14}$$

10.7.3 Solution of equations of motion

We never need to use the individual quantities $(\mathrm{n}|\mathrm{k})_\sigma$ or the energies ε_n, but only certain averages over these, such as the mean density of admixture of the dσ state into the continuum levels of energy ε: that is,

$$g_{\mathrm{d}\sigma}(\varepsilon) = \sum_\mathrm{n} \delta(\varepsilon_\mathrm{n}-\varepsilon)|(\mathrm{n}|\mathrm{d})_\sigma|^2. \tag{10.7.15}$$

This will be the most important quantity, because it determines the d-function occupation number. This as well as the other physical quantities of interest may be obtained most directly by calculating the Green function

$$G(\varepsilon+is) = \frac{1}{\varepsilon+is-H}. \tag{10.7.16}$$

If we use the exact eigenstates then G is diagonal:

$$G^\sigma_{\mathrm{nn}}(\varepsilon+is) = \frac{1}{\varepsilon+is-\varepsilon_{\mathrm{n}\sigma}}, \tag{10.7.17}$$

but in the unperturbed state representation its matrix elements give the required densities. For example,

$$\begin{aligned} g_{\mathrm{d}\sigma}(\varepsilon) &= \frac{1}{\pi}\sum_\mathrm{n}|(\mathrm{d}|\mathrm{n})_\sigma|^2 \lim_{s\to 0}\frac{s}{s^2+(\varepsilon-\varepsilon_{\mathrm{n}\sigma})^2} \\ &= -\frac{1}{\pi}\,\mathrm{Im}\,[G^\sigma_{\mathrm{dd}}(\varepsilon)], \end{aligned} \tag{10.7.18}$$

while the total modified density of states is

$$g_\sigma(\varepsilon) = \frac{1}{\pi}\,\mathrm{Im}\,[\mathrm{Tr}\,G^\sigma(\varepsilon)]. \tag{10.7.19}$$

The equations for the Green function may be derived from the equations of motion for $(\mathrm{n}|\mathrm{k})_\sigma$ and $(\mathrm{n}|\mathrm{d})_\sigma$ using the property

$$\sum_\nu (\varepsilon+is-H)_{\mu\nu}\,G_{\nu\kappa} = \delta_{\mu\kappa}. \tag{10.7.20}$$

They are, writing

$$\left.\begin{aligned} E_\sigma &= E+I\langle n_{\mathrm{d},-\sigma}\rangle \\ e &= \varepsilon+is, \end{aligned}\right\} \tag{10.7.21}$$

and

$$\left.\begin{aligned} (e-E_\sigma)\,G^\sigma_{\mathrm{dd}}-\sum_\mathrm{k} V_{\mathrm{dk}}\,G^\sigma_{\mathrm{kd}} &= 1, \\ (e-\varepsilon_\mathrm{k})\,G^\sigma_{\mathrm{kd}}-V_{\mathrm{kd}}\,G^\sigma_{\mathrm{dd}} &= 0, \\ (e-E_\sigma)\,G^\sigma_{\mathrm{dk}}-\sum_{\mathrm{k}'} V_{\mathrm{dk}'}\,G^\sigma_{\mathrm{k}'\mathrm{k}} &= 0 \end{aligned}\right\} \tag{10.7.22}$$

and

$$(e-\varepsilon_{k'})\,G^{\sigma}_{k'k}-V_{k'd}\,G^{\sigma}_{dk} = \delta_{k'k}. \tag{10.7.23}$$

From equations (10.7.22), we immediately obtain

$$G^{\sigma}_{dd}(e) = \left[e-E_{\sigma}-\sum_{k}|V_{dk}|^2(e-\varepsilon_k)^{-1}\right]^{-1}. \tag{10.7.24}$$

The sum over k can now be evaluated

$$\lim_{s\to 0}\sum_{k}|V_{dk}|^2\frac{(\varepsilon-\varepsilon_k)-is}{(\varepsilon-\varepsilon_k)^2+s^2} = -i\pi\langle V^2_{dk}\rangle_{av}\,g(\varepsilon). \tag{10.7.25}$$

Here we have neglected the effective energy shift of the d-state:

$$\Delta E_d = P\left\{\sum_{k}\frac{|V_{dk}|^2}{\varepsilon-\varepsilon_k}\right\} \tag{10.7.26}$$

because it may be taken into account simply by shifting the assumed unperturbed energy E. If the density of states is reasonably constant, ΔE_d will not change very radically as E_{σ} changes. Thus we see that, apart from the energy shift, G_{dd} behaves as if there were a virtual state at

$$e = E_{\sigma}+i\Delta, \tag{10.7.27}$$

where the width parameter Δ of the virtual state is defined by

$$\Delta = \pi\langle V^2\rangle_{av}\,g(\varepsilon). \tag{10.7.28}$$

We usually assume that Δ is a constant parameter, roughly independent of E_{σ}. The density distribution of the d-state is then, from (10.7.18),

$$g_{d\sigma}(\varepsilon) = \frac{1}{\pi}\frac{\Delta}{(\varepsilon-E_{\sigma})^2+\Delta^2} \tag{10.7.29}$$

This is the basic formula of the self-consistent treatment. It gives us the density of d-state admixture in the continuum states of energy ε.

It is also of interest in the problem of the polarization in the Bloch electron bands to work out G more completely. Thus from equation (10.7.22) we obtain

$$G^{\sigma}_{k'k} = \frac{V_{k'd}\,G^{\sigma}_{dk}}{e-\varepsilon_{k'}} \quad (\mathbf{k}'\neq\mathbf{k}). \tag{10.7.30}$$

This may be substituted into equation (10.7.22) to yield

$$\left[e-E_{\sigma}-\sum_{k'\neq k}|V_{dk'}|^2(e-\varepsilon_{k'})^{-1}\right]G^{\sigma}_{dk} = V_{dk}\,G^{\sigma}_{kk}, \tag{10.7.31}$$

which, using the formula (10.7.25) for summing over **k**, is

$$[e-E_\sigma+i\Delta+|V_{\mathrm{dk}}|^2(e-\varepsilon_{\mathrm{k}})^{-1}]\,G^\sigma_{\mathrm{dk}} = V_{\mathrm{dk}}\,G^\sigma_{\mathrm{kk}}. \tag{10.7.32}$$

Now the diagonal element of equation (10.7.22) gives

$$\begin{aligned} G^\sigma_{\mathrm{kk}} &= \left[e-\varepsilon_{\mathrm{k}}-\frac{|V_{\mathrm{dk}}|^2}{e-E_\sigma+i\Delta+|V_{\mathrm{dk}}|^2(e-\varepsilon_{\mathrm{k}})^{-1}}\right]^{-1} \\ &= (e-\varepsilon_{\mathrm{k}})^{-1}+[|V_{\mathrm{dk}}|^2/(e-\varepsilon_{\mathrm{k}})^2(e-E_\sigma+i\Delta)]. \end{aligned} \tag{10.7.33}$$

The interpretation of this result is that although each G_{kk} has a pole at precisely the same energy, so that its perturbed energy is unshifted, nevertheless a certain amount of its density is to be found in the region of the virtual state. This admixture is given quantitatively by

$$-\frac{1}{\pi}\,\mathrm{Im}\,G_{\mathrm{kk}}(\varepsilon) \simeq \frac{|V_{\mathrm{dk}}|^2}{(\varepsilon-\varepsilon_{\mathrm{k}})^2}\,g_{\mathrm{d}}(\varepsilon) \tag{10.7.34}$$

near to the virtual state. There are also shifts near the pole in the density.

10.7.4 Existence of localized moments

We wish now to discuss the self-consistency conditions for the localized moments.

We start out from our basic result for the density of d-admixture in the continuum states of energy ε. In order to determine the number of d-electrons of a given spin, we integrate this up to the Fermi energy ε_{f}. Then we find

$$\begin{aligned} \langle n_{\mathrm{d}\sigma}\rangle &= \frac{1}{\pi}\int_{-\infty}^{\varepsilon_{\mathrm{f}}}\frac{\Delta\,de}{(e-E_\sigma)^2+\Delta^2} \\ &= \frac{1}{\pi}\cot^{-1}\left(\frac{E_\sigma-\varepsilon_{\mathrm{f}}}{\Delta}\right). \end{aligned} \tag{10.7.35}$$

But now we have that

$$E_\sigma = E+I\langle n_{\mathrm{d},-\sigma}\rangle \tag{10.7.36}$$

and hence we must solve simultaneously the equations

$$\left.\begin{aligned} \langle n_{\mathrm{d}\uparrow}\rangle &= \frac{1}{\pi}\cot^{-1}\left[\frac{E-\varepsilon_{\mathrm{f}}+I\langle n_{\mathrm{d}\downarrow}\rangle}{\Delta}\right], \\ \langle n_{\mathrm{d}\downarrow}\rangle &= \frac{1}{\pi}\cot^{-1}\left[\frac{E-\varepsilon_{\mathrm{f}}+I\langle n_{\mathrm{d}\uparrow}\rangle}{\Delta}\right]. \end{aligned}\right\} \tag{10.7.37}$$

To investigate the solutions, it is convenient to introduce dimensionless variables

$$y = I/\Delta \tag{10.7.38}$$

and

$$x = (\varepsilon_f - E)/I. \tag{10.7.39}$$

y clearly represents the ratio of the Coulomb integral to the width of the virtual state. When the parameter $x = 0$, the empty d-state is at the Fermi level, while $x = 1$ puts $E+I$ at the Fermi level. $x = \frac{1}{2}$ corresponds to the case when E and $E+I$ are symmetrically placed about the Fermi level, and represents the most favourable case for magnetic states.

We then may write in an obvious notation

$$\pi n_{d\pm} = \cot^{-1}[y(n_{d\mp} - x)]. \tag{10.7.40}$$

10.7.5 Limiting cases

(a) *Magnetic moment formation—*$y \gg 1$, x not small or not near 1. In these cases, $\cot^{-1}$ is either close to zero or to π and $n_{d\pm}$ is near to zero or one. Assuming $n_{d+} \sim 1$, $n_{d-} \sim 0$, then

$$\pi n_{d+} \simeq \pi - \frac{1}{y(x - n_{d-})}, \tag{10.7.41}$$

$$\pi n_{d-} \simeq \frac{1}{y(n_{d+} - x)}. \tag{10.7.42}$$

We find then the approximate solution

$$n_{d+} - n_{d-} = 1 - 1/[\pi y x(1-x) - 1] \tag{10.7.43}$$

and obviously there is a resulting magnetic moment.

(b) *Non-magnetic cases.* Here, we have $n_{d+} = n_{d-} = n$, say. Then

$$\cot \pi n = y(n - x). \tag{10.7.44}$$

When n is near to $\frac{1}{2}$ we have

$$\cot \pi n \simeq \pi(\tfrac{1}{2} - n) \tag{10.7.45}$$

and

$$y(n - x) = \pi(\tfrac{1}{2} - n). \tag{10.7.46}$$

Hence

$$n \simeq \frac{1}{2}\frac{1 + 2xy/\pi}{1 + y/\pi}. \tag{10.7.47}$$

In this case, the effective energy of the d-state relative to the Fermi level is

$$E_{\text{eff}} = I(n-x) = \frac{I}{2}\left(\frac{1-2x}{1+y/\pi}\right) = \frac{y\Delta(1-2x)}{2(1+y/\pi)}. \tag{10.7.48}$$

The case when y is large and x is near to zero can be similarly treated.

(c) *Transition from magnetic to non-magnetic behaviour.* Clearly, on the transition curve from magnetic to non-magnetic behaviour, $n_{\text{d}+} = n_{\text{d}-}$, and one condition is

$$\cot \pi n = y(n-x). \tag{10.7.49}$$

The second condition comes from forming $\langle n_{\text{d}+}\rangle - \langle n_{\text{d}-}\rangle$ and reads, if we use the subscript c to denote values on the transition curve

$$\frac{\pi}{\sin^2 \pi n_{\text{c}}} = y_{\text{c}}. \tag{10.7.50}$$

Hence we can obtain immediately

$$y_{\text{c}} = \frac{\pi}{\sin^2 \pi n_{\text{c}}} = \frac{\cot \pi n_{\text{c}}}{n_{\text{c}} - x} \tag{10.7.51}$$

or

$$\sin 2\pi n_{\text{c}} = 2\pi(n_{\text{c}} - x). \tag{10.7.52}$$

For $x \simeq \frac{1}{2}$ and $x \simeq 0$ we find

$$y_{\text{c}} \simeq \pi + \tfrac{1}{4}\pi^3(x-\tfrac{1}{2})^2 + \ldots, \quad x \simeq \tfrac{1}{2}, \tag{10.7.53}$$

and

$$y_{\text{c}}^3 \simeq 4\pi/9x_{\text{c}}^2 + \ldots, \quad x \simeq 0. \tag{10.7.54}$$

Figure 10.11 shows the transition curve as a function of x and π/y, while Figure 10.12 shows n, and, where they are different, $\langle n_{\text{d}+}\rangle$ and $\langle n_{\text{d}-}\rangle$ as functions of π/y for two typical cases, $x = \frac{1}{4}$ and $x = \frac{1}{2}$.

Anderson makes some order of magnitude estimates of I and in the iron group it is expected to be around 10 eV. Also it seems that $\Delta = \pi\langle V^2\rangle_{\text{av}}\, g(\varepsilon) \simeq$ 2–5 eV. This shows that the transition $I/\Delta = y = \pi$ occurs right in the interesting region and that it is quite possible that the transitions from magnetic to non-magnetic localized states observed by Matthias, Peter, Williams, Clogston, Corenzwit and Sherwood (1960) could be due either to changes in the density of states or to changes in x (shifts in the Fermi level).

For rare-earth solutes, in contrast, $I \sim 15$ eV and $V_{\text{av}} \sim 1$ eV at most and magnetic cases are to be expected very frequently.

10.7.6 *Susceptibility and specific heat*

In the non-magnetic region, the localized virtual d-states can have a very considerable effect on the spin susceptibility and on the specific heat. The effect of an external magnetic field is to shift the +spin Fermi level relative to the −spin level by an amount $2\mu H$. This shift will change the occupations of the virtual levels, leading to relative shifts of these levels themselves.

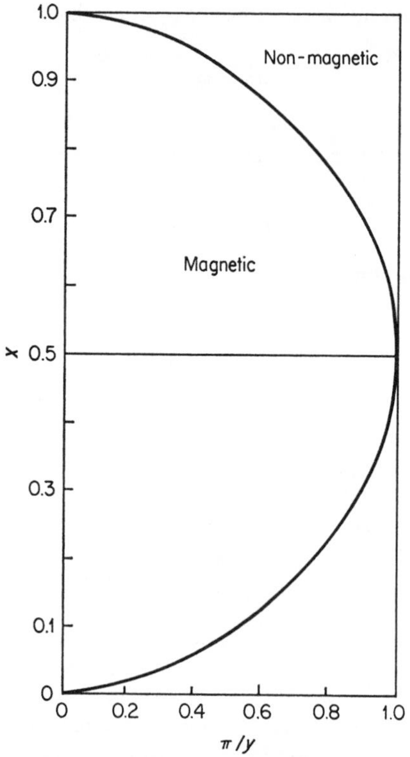

FIGURE 10.11. Regions of magnetic and non-magnetic character in Anderson's Hartree–Fock solution (after Anderson, 1961).

However, if the density of states $g(\varepsilon)$ is not strongly varying in energy, any motion of the virtual levels, it turns out, cannot affect the net polarization of the free electrons. Thus, the free-electron band will contribute essentially its unperturbed susceptibility (see Anderson, 1961).

The changes of the d-electron density are readily found as follows. The shifts of the effective positions of the virtual levels are

$$\delta E_+ = -\mu H + I\langle \delta n_- \rangle \tag{10.7.55}$$

and

$$\delta E_- = \mu H + I\langle \delta n_+ \rangle . \tag{10.7.56}$$

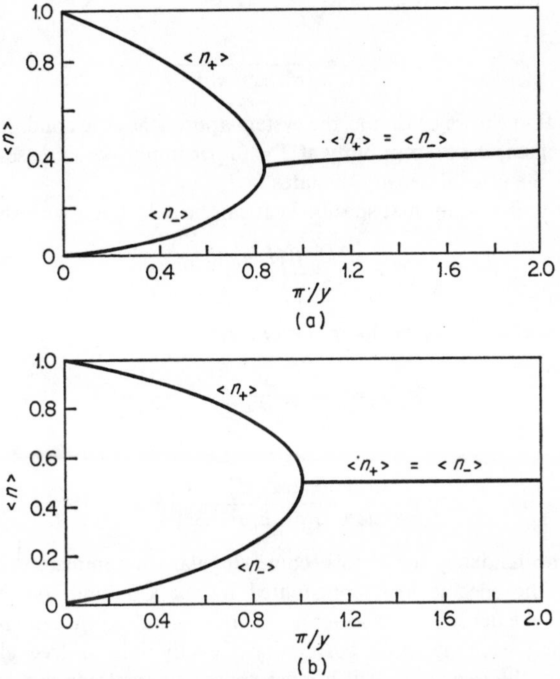

FIGURE 10.12. Occupation numbers of impurity d-state as function of π/y. Region in which local moment can form is that of $\pi/y \lesssim 0{\cdot}8$. Curve (a): $x = \frac{1}{4}$. Curve (b): $x = \frac{1}{2}$ (after Anderson, 1961).

The resulting changes in population may be obtained by differentiation of the equations (10.7.37) for $\langle n_+ \rangle$ and $\langle n_- \rangle$ yielding

$$(\pi/\sin^2 \pi n)\langle \delta n_+ \rangle = -\delta E_+/\Delta \tag{10.7.57}$$

and

$$(\pi/\sin^2 \pi n)\langle \delta n_- \rangle = -\delta E_-/\Delta . \tag{10.7.58}$$

Subtracting these two equations, we find

$$(\pi/\sin^2 \pi n)\langle \delta n_+ - \delta n_- \rangle = y\langle \delta n_+ - \delta n_- \rangle + \frac{2\mu H}{\Delta}. \tag{10.7.59}$$

Thus the susceptibility χ is given by

$$\chi = \frac{\mu\langle \delta n_+ - \delta n_- \rangle}{H}$$

$$= \frac{2\mu^2}{(\pi\Delta/\sin^2 \pi n) - I}. \tag{10.7.60}$$

This equation shows us that as the system approaches the condition (10.7.50) for magnetism, the susceptibility at $T = 0$, per impurity, increases, becoming infinite at the critical density of states.

Similarly, the additional specific heat can be calculated with the result

$$C = \frac{\pi^2 k_B^2 T}{3}\left(\frac{I\, dg_d/d\varepsilon|_{\varepsilon_f}}{1 + Ig_d} + g_d\right), \tag{10.7.61}$$

where g_d is the density of the impurity state

$$g_d(\varepsilon) = \frac{\pi\Delta}{(\varepsilon - E_d)^2 + \Delta^2} = \frac{\pi\Delta}{\sin^2 \pi n} \tag{10.7.62}$$

and

$$\frac{dg_d}{d\varepsilon} = \frac{2\pi\Delta(E_d - \varepsilon)}{[(\varepsilon - E_d)^2 + \Delta^2]^2}. \tag{10.7.63}$$

There is no tendency for C to become singular, in complete contrast to χ. At most, the specific heat contributed by the Coulomb integral term is 20–30% of the density of states term g_d. Thus, as indeed we can also conclude by inserting exchange and correlation directly into a free electron gas calculation, the specific heat is a much more accurate measure of density of states than is the susceptibility.

The situation discussed here ought to be contrasted with that of Zn in Cu, for example. Here, as we have seen, the impurity may be considered simply as an extra positive charge, which is to be screened out by the piling up of the band wave functions around the impurity.

But in the type of situation discussed here, say for Mn in Cu, the charge is expected to be compensated not by a deformation of the Cu conduction band but by emptying levels approximating to the orbitals of the free atom, because these orbitals are of an entirely different symmetry and size from the band wave functions. Then it seems a much better starting-point to use the neutral impurity atom itself added to the matrix. This type of situation, where the atomic properties of the solute are not strongly affected by solution may be rather widespread, especially when the solute and solvent atoms are widely different. In this latter case, one does not expect the free electrons to be strongly perturbed, or to annul effectively the Coulombic effects in the inner shell.

Anderson pointed out in his original paper that the use of the Hartree–Fock approximation in treating the basic Hamiltonian would lead to magnetic states rather too frequently and the true magnetic criterion would be more severe. Some later work of considerable interest has concerned itself with this problem (Hewson, 1965; Schrieffer and Mattis, 1965; Evenson, Schrieffer and Wang, 1970). The powerful functional integral formulation is discussed by Sherrington (1971).

10.8 Kondo effect

A number of alloy systems, for example noble-metal solutes containing a dilute concentration of transition metal atoms, exhibit a minimum in the electrical resistivity at low temperatures (see, for example, Coles, 1958).

In seeking the origin of this effect, Kondo (1964; 1969) found that, when calculated to second order in the exchange interaction J for coupling of the form

$$H' = -J\mathbf{S}.\mathbf{s} \tag{10.8.1}$$

the amplitude for exchange scattering of a conduction electron spin $\mathbf{s}$ at the Fermi surface by a single localized spin $\mathbf{S}$ diverges as $\ln T$ when $T \to 0$. The resistivity minimum then follows from the formula

$$\rho = aT^5 + c\rho_0 + c\rho_1 \ln T, \tag{10.8.2}$$

where c is the concentration.

Here, the various terms arise as follows. The first comes from the phonon-induced scattering by the host lattice, which is assumed to be proportional to T^5 (cf. section 6.5), the term $c\rho_0$ includes the temperature-independent result for non-magnetic impurity scattering and the last term is that found by Kondo. When $J < 0$ (antiferromagnetic coupling), $\rho_1 < 0$ also, and one then finds a resistance minimum at

$$T_{\min} = c^{\frac{1}{5}}\left(\frac{|\rho_1|}{5a}\right)^{\frac{1}{5}}, \tag{10.8.3}$$

in reasonable agreement with experiment.

It is clear that the $\ln T$ term in equation (10.8.2) cannot persist right down to $T=0$. Below a certain critical temperature, the Kondo temperature, given roughly by

$$T_{\mathrm{K}} \simeq T_0 \exp\left(-\frac{1}{N(\varepsilon_f)|J|}\right), \tag{10.8.4}$$

where T_0 is the Fermi temperature of the conduction electrons ($k_{\mathrm{B}} T_0 = \varepsilon_f$) and $N(\varepsilon_f)$ is the density of states at the Fermi level, it is thought that the Kondo resonance in the s–d exchange scattering (or perhaps, even, a bound state is formed), can lead to the spins of the conduction electrons piling up round the impurity, polarized opposite to that of the impurity, to produce a net magnetic moment of zero at $T=0$ (Nagaoka, 1965). This condensation due to the Kondo effect normally occurs at temperatures $T_{\mathrm{K}} \sim 10$ K, but Schrieffer (1967) has conjectured that, for some impurities, T_{K} is beyond the melting point. Thus he suggests the possibility that some of the impurities which, at normal temperatures, are observed to have zero magnetic moment, should be described by the form of the Kondo effect below T_{K}, rather than in terms of the conventional treatments of non-magnetic impurities in metals.

The fact that the ground-state form of the Kondo effect, as we have described it above, is a true many-body effect, has attracted the attention of many theorists. So too has the problem of developing non-perturbative methods to eliminate the divergence Kondo found at third order in the s–d scattering. We have to refer the reader to the article by Suhl (1968), for example, for a more detailed discussion.

We should draw attention to a series of papers by Anderson and his co-workers (Anderson and Yuval, 1969, 1970; Anderson, Yuval and Hamann, 1970a, b; Anderson, 1970). Essentially, by exploiting the equivalence of an anisotropic Kondo problem to a type of one-dimensional statistical problem, which could in turn be solved by deriving scaling laws connecting solutions for different sets of parameters with each other, considerable progress proved possible. Anderson (1970) shows that such scaling laws can be derived directly in the Kondo problem itself, using a technique employed in the theory of the "Coulomb pseudo-potential" in superconductivity, which we referred to in Chapter 8. The relation between the localized moment theory described in section 10.7 and the Kondo effect is discussed very clearly by Hamann (1969).

10.9 Palladium alloys and paramagnon effects

We saw in Chapter 4 that there is considerable interest in the magnetic properties of metallic Pd, because it appears to be a paramagnet in which the Stoner criterion for ferromagnetism is almost fulfilled. It is spoken of as a "nearly ferromagnetic" system and it should therefore occasion no great surprise that, when 3d transition metal impurities such as Fe or Co are introduced into the 4d Pd matrix, the alloy becomes ferromagnetic, for very low impurity concentrations (Crangle, 1960).

Additionally, the measured impurity moment is much greater than would be expected from the spin of the impurity itself, and it seems clear that what is happening is that the Pd matrix must be substantially polarized by the ferromagnetic impurities.

We wish to discuss here a model for the Pd alloy system which has been developed recently by Cole and Turner (1969) to obtain an expression for the low temperature electronic specific heat. The model can also lead to a plausible interpretation of the experimental results on the low-temperature electrical resistivity of the dilute alloys of Pd. The work is not, however, adequate to explain the specific heat, nor the electrical resistivity, in the paramagnetic region of these alloys.

10.9.1 Alloy model Hamiltonian

The Fe or Co impurity produces, as we remarked above, a localized moment when substituted in Pd. The localization of the impurity magnetic moment then polarizes the surrounding host d-holes over a substantial region away from the impurity. This polarization, which can be thought of in terms of a long-range interaction like the Ruderman–Kittel interaction we discussed in Chapter 4, gives rise to long-range ordering of the magnetic moments, and in these materials to ferromagnetism.

Though it is an oversimplified representation, we shall follow Doniach and Wohlfarth (1967) in representing the coupling between the impurity moment and the polarization by the expression

$$-J\sum_i \mathbf{S}_i \cdot \boldsymbol{\sigma}(\mathbf{r}_i) p_i. \qquad (10.9.1)$$

Here $\mathbf{S}_i$ is the spin operator at the impurity site i, which is coupled to the electron polarization cloud with coupling strength J. The factor p_i is unity if the site is occupied by an impurity and $p_i = 0$ otherwise.

Metallic Pd has a broad s-band with a narrow d-band overlapping it. The interaction above will exist in both the s- and d-bands and there is no reason to take the coupling parameters to be the same in both. Because of the extended nature of the s-electron wave functions, we can expect the d-hole impurity spin interaction to be a good deal stronger than that for the interaction of the s-electrons and the impurity.

In addition to these interactions, we must include those present in Pd metal, as well as any effects due to charge differences between the impurity and the host atoms. As we discussed at length for simple metals in section 10.2.1, valence differences can be represented roughly by a scattering potential of screened Coulomb form.

This leads us to summarize the model of the Pd alloys in terms of the following Hamiltonian

$$H_{\text{alloy}} = \sum_{\mathbf{k}\sigma} E^{\text{s}}_{\mathbf{k}} a^{\dagger}_{\mathbf{k}\sigma} a_{\mathbf{k}\sigma} + \sum_{\mathbf{k}\sigma} \varepsilon^{\text{d}}_{\mathbf{k}} c^{\dagger}_{\mathbf{k}\sigma} c_{\mathbf{k}\sigma} + I \sum_{i} n^{\text{d}}_{i\uparrow} n^{\text{d}}_{i\downarrow}$$
$$- V_{\text{sd}} \sum_{i} \boldsymbol{\sigma}_{\text{s}}(\mathbf{r}_i) \cdot \boldsymbol{\sigma}_{\text{d}}(\mathbf{r}_i) + V \sum_{i\sigma} a^{\dagger}_{i\sigma} a_{i\sigma} p_i$$
$$- J' \sum_{i} \mathbf{S}_i \cdot \boldsymbol{\sigma}_{\text{s}}(\mathbf{r}_i) p_i - J \sum_{i} \mathbf{S}_i \cdot \boldsymbol{\sigma}_{\text{d}}(\mathbf{r}_i) p_i. \qquad (10.9.2)$$

We are already familiar with the first four terms from the pure metal case. The a operators are creation and annihilation operators for s-electrons, the c's are those for d-holes. The familiar short-range interaction of Anderson and Hubbard is present of course, though its role is naturally different in the alloy for, with ferromagnetism existing at low temperatures, the number of d-holes with upper spin will be greater than the number with opposed spin. The s–d interaction is evidently still included while the coupling parameter V describes the strength of the impurity scattering potential.

10.9.2 Local spin susceptibility

The excess resistivity of the alloy system we are considering can, it turns out, be written down explicitly in terms of the quantity $A(\mathbf{q}\omega)$ which is the impurity spin spectral density function. For our present purposes, all we need note is that this quantity is related to the local spin susceptibility $\chi^{\text{L}}(\mathbf{q}\omega)$ through (cf. Chapter 4)

$$A(\mathbf{q}\omega) = \frac{1}{\pi} \operatorname{Im} \chi^{\text{L}}(\mathbf{q}\omega). \qquad (10.9.3)$$

Thus, both because of its interest in electrical transport, and because of its intrinsic interest experimentally, we shall consider here the nature of the local spin susceptibility, restricting ourselves however to a general outline of the theory.

The Fourier transform of the spin susceptibility is related to the retarded Green function $\langle [S^{+}_{i}(t), S^{-}_{j}(t')] \rangle$ which describes the propagation through the medium of a local spin at site j at time t' and reintroduced at site i at time t. The elementary excitations of the system, as we saw in Chapter 4, are already built into $\chi^{\text{L}}(\mathbf{q}\omega)$ and, since the s-electron scattering is from such excitations, it is natural that this susceptibility enters the calculation of the electrical transport properties of the alloy.

The work of Cole and Turner (1969) shows that the susceptibility for the alloy d-band is closely related to the susceptibility for the pure Pd d-band which we discussed in Chapter 4. There, we derived the result of Izuyama and co-workers that the susceptibility of pure Pd is

$$\chi_{\mathrm{Pd}}(\mathbf{q}\omega) = \frac{\chi_0(\mathbf{q}\omega)}{1 - I\chi_0(\mathbf{q}\omega)}, \tag{10.9.4}$$

where χ_0 is the usual non-interacting susceptibility given by [cf. equation (4.11.1)]

$$\chi_0(\mathbf{q}\omega) = \frac{1}{N}\sum_{\mathbf{k}} \frac{\langle n_{\mathbf{k}}^{\mathrm{d}}\rangle - \langle n_{\mathbf{k}+\mathbf{q}}^{\mathrm{d}}\rangle}{[\mathscr{E}_{\mathbf{k}+\mathbf{q}}^{\mathrm{d}} - \mathscr{E}_{\mathbf{k}}^{\mathrm{d}} - \omega]}. \tag{10.9.5}$$

In this expression, I is the strength of the short-range interaction between Pd d-holes and $\mathscr{E}_{\mathbf{k}}^{\mathrm{d}}$ are the single-particle energies of the Pd matrix.

Following Cole and Turner (1969), we show in Appendix 10.6 that the susceptibility for the alloy d-band is obtained from this by replacing I by an effective frequency-dependent interaction and the single-particle energies $\mathscr{E}_{\mathrm{d}}^{\mathbf{k}}$ by modified energies $E_{\mathbf{k}\sigma}^{\mathrm{d}}$. The form of the interaction is

$$I(\omega) = I + J^2 c D^0(\omega) \tag{10.9.6}$$

and the modified energies are $E_{k\sigma}^{\mathrm{d}}$, as we show in Appendix 10.6, given by

$$E_{k\sigma}^{\mathrm{d}} = \mathscr{E}_k^{\mathrm{d}} + (I/N)\sum_{\mathbf{k}'} \langle n_{\mathbf{k}',-\sigma}^{\mathrm{d}}\rangle - \sigma J\langle S_z\rangle c. \tag{10.9.7}$$

The function $D^0(\omega)$ in equation (10.9.6) is the non-interacting local spin susceptibility or propagator

$$D^0(\omega) = \frac{2\langle S_z\rangle}{\omega_{\mathrm{L}} - \omega}, \tag{10.9.8}$$

where $\omega_{\mathrm{L}} = J\langle \sigma_{\mathrm{d}}^z\rangle$ is the precessional frequency of the local impurity spins in the field arising from the polarization of the d-band. The expression for the d-band local spin susceptibility is then related to

$$\frac{D^0(\omega)\,[1 - I\chi_0(\mathbf{q}\omega)]}{[1 - I(\omega)\,\chi_0(\mathbf{q}\omega)]}, \tag{10.9.9}$$

where χ_0 is the susceptibility of the non-interacting system in the molecular field of the impurity spins; i.e. we use the non-interacting formula (10.9.5) but replace the energies $\mathscr{E}_k^{\mathrm{d}}$ by $E_{k\sigma}^{\mathrm{d}}$.

What is needed in the present work is the local spin susceptibility when the average is over the system described by the full alloy Hamiltonian (10.9.2). The use of a time-dependent average-field method then leads to the form

$$\frac{D(\mathbf{q}\omega)}{[1-J_{\text{eff}}'^{2}(\mathbf{q}\omega)\,c\chi_{\text{s}}(\mathbf{q}\omega)\,D(\mathbf{q}\omega)]} \tag{10.9.10}$$

in terms of the function $D(\mathbf{q}\omega)$ given by

$$D(\mathbf{q}\omega) = \frac{D^{0}(\mathbf{q}\omega)}{[1-J_{c}^{2}\,c\chi_{\text{d}}(\mathbf{q}\omega)\,D^{0}(\mathbf{q}\omega)]}. \tag{10.9.11}$$

Here χ_{s} and χ_{d} are, of course, the response functions of the s- and d-bands respectively.

Actually, in the approximation used for calculating the conductivity, $\chi^{\text{L}}(\mathbf{q}\omega)$ reduces to that derived by Cole and Turner (1969), and the local spin susceptibility function that is relevant here is that of equation (10.9.9).

In summary then, the local susceptibility we require is given by

$$\chi^{\text{L}}(\mathbf{q}\omega) = \frac{2\langle S_z\rangle}{\omega_{\text{L}}-\omega}\,\frac{[1-I\chi_0(\mathbf{q}\omega)]}{\{1-[I-(2J^2c\langle S_z\rangle/\omega_{\text{L}}-\omega)]\,\chi_0(q\omega)\}}. \tag{10.9.12}$$

The precessional frequency of the local spins is easily reduced to

$$\begin{aligned}\omega_{\text{L}} &= J\langle\sigma_z^{\text{d}}\rangle\\ &= 2J^2\langle S_z\rangle\frac{cN_{\text{d}}(0)}{[1-IN_{\text{d}}(0)]},\end{aligned} \tag{10.9.13}$$

where $N_{\text{d}}(0)$ is the long-wave limit of the d-band susceptibility $\chi_0(\mathbf{q}\omega)$ and (in suitable units) is equal to the density of d-states at the Fermi level. Substituting this value into equation (10.9.12), in the limit as $\omega\to 0$ and $q\to 0$ the denominator is easily seen to vanish. This means that the Stoner criterion for ferromagnetism is obeyed.

We can extract the imaginary part of χ^{L} from equation (10.9.12) as

$$\operatorname{Im}\chi^{\text{L}}(\mathbf{q}\omega) = \frac{2\langle S_z\rangle(\pi/4)(\bar{\omega}/\bar{q})(\omega_{\text{L}}/K_0^2)}{[\omega-(\omega_{\text{L}}\bar{q}^2/12K_0^2)]^2+[(\pi/4)(\bar{\omega}/\bar{q})(\omega_{\text{L}}/K_0^2)]^2}, \tag{10.9.14}$$

where, as in the discussion of Chapter 4, we have introduced the dimensionless quantities $\bar{q}=q/k_{\text{f}}$ and $\bar{\omega}=\omega/E_{\text{f}}$. k_{f} and E_{f} are the Fermi wave vector and energy of the d-band, while K_0^2 is simply $1-IN_{\text{d}}(0)$.

We see that there are resonances at $\omega=(\omega_{\text{L}}/12K_0^2)\bar{q}^2$, with energy-dependent widths given by

$$\Delta\omega = \frac{\pi}{4}\frac{\bar{\omega}}{\bar{q}}\frac{\omega_{\text{L}}}{K_0^2}. \tag{10.9.15}$$

It is of importance to note that the effect of introducing local spins into the system is to modify the form of the spin excitation spectrum from that of the paramagnons in pure Pd, in which case there is a linear dispersion relation, to excitations having the form of damped spin waves, with a quadratic dispersion relation such as we derived for magnons in Chapter 4.

A further quantity we need in the alloy theory, the density of local spin states ρ_L, is readily calculated from $\chi^L(\mathbf{q}\omega)$ using the relation

$$\rho_L(\omega) = \frac{1}{N\pi}\sum_{\mathbf{q}} \operatorname{Im} \chi^L(\mathbf{q}\omega). \tag{10.9.16}$$

In the limit of low concentration, it can be shown that

$$\left.\begin{aligned} \rho_L(\omega) &= 2\langle S_z\rangle \frac{3z}{2}\omega^{\frac{1}{2}}/(k_B T_0)^{\frac{3}{2}} \quad \text{for } 0 \leqslant \omega \leqslant \omega_m \\ &= 0 \quad \text{otherwise,} \end{aligned}\right\} \tag{10.9.17}$$

where z is the number of electrons per atom and where $k_B T_0 = \omega_L/12K_0^2$. The cut-off frequency ω_m is chosen as usual for normalization. In the case of low temperatures there is no problem as we can effectively allow ω_m to tend to infinity eventually.

The precessional frequency ω_L of the local spins is proportional to $\langle S_z\rangle$ which is temperature dependent. For the low-temperature régime we are considering it will be sufficient to take the zero temperature value of ω_L. This is equivalent to taking

$$k_B T_0 = \frac{2J^2 S N_d(0)}{[1-N_d(0)]} c. \tag{10.9.18}$$

By calculating electrical transport properties from this model it is possible to extract values for the parameters V and J_{eff} from the comparison of the theory with experiment, but we shall not go into further detail here.

10.10 Kanzaki method of calculating distortions due to point defects

Earlier in this chapter, we have regarded defects in metals as most usefully described by the charge they displace, or, what is equivalent, a one-body defect potential. The electron theory, based on these displaced charges, is used in section 10.12.1 to calculate the interaction between charged defects. In that theory, no account is included of lattice distortions, and here we wish to present briefly the method of Kanzaki (1957) which replaces, essentially, the effect of the point imperfection by a set of external forces which can then be viewed as characterizing the defect. The calculation of these forces, in the end, must be the job of electron theory. The essential feature of the Kanzaki method is that one can expand the displacements in normal co-ordinates, because we have now a perfect lattice with external Kanzaki forces. The displacements are assumed small in his theory.

We shall present the general theory below, and summarize how he has applied the method to a vacancy in solid argon. Later applications have been made; for example, Hardy and Bullough (1967) have used it to calculate the interaction between two vacancies in copper.

The potential energy Φ of the perfect lattice distorted by applied forces $\mathbf{F}(\mathbf{l})$, components $F_\alpha(\mathbf{l})$, on an atom at site $\mathbf{l}$ in a crystal with one atom per unit cell is first expanded in powers of the components of the displacements $\boldsymbol{\xi}(\mathbf{l})$ up to second order: namely,

$$\Phi-\Phi_0 = -\sum_l \mathbf{F}(\mathbf{l}).\boldsymbol{\xi}(\mathbf{l})$$

$$+\frac{1}{2}\sum_{ll'}\sum_{\alpha\beta}\Phi_{\alpha\beta}(ll')\,\xi_\alpha(\mathbf{l})\,\xi_\beta(\mathbf{l}') \quad (\alpha,\beta = 1,2,3), \qquad (10.10.1)$$

where Φ_0 is the potential energy of the undisplaced perfect lattice and, exactly as in Chapter 3,

$$\Phi_{\alpha\beta}(ll') = \left[\frac{\partial^2\Phi}{\partial x_\alpha^l . \partial x_\beta^l}\right]_0, \qquad (10.10.2)$$

the suffix zero denoting the undisplaced lattice.

The equilibrium value of the displacements can be obtained from the condition

$$\frac{\partial\Phi}{\partial\xi_\alpha(\mathbf{r}l)} = 0, \qquad (10.10.3)$$

which then becomes, using equation (10.10.1),

$$F_\alpha(\mathbf{l}) = \sum_{l'}\sum_{\beta}\Phi_{\alpha\beta}(ll')\,\xi_\beta(\mathbf{l}'). \qquad (10.10.4)$$

These equations are the fundamental ones of the Kanzaki theory, and will now be reduced to a simpler form by introducing normal coordinates. Thus we write the displacements as

$$\boldsymbol{\xi}(\mathbf{l}) = \sum_{\mathbf{q}}\mathbf{Q}(\mathbf{q})\,e^{i\mathbf{q}.\mathbf{l}}, \qquad (10.10.5)$$

where $\mathbf{q}$ ranges over the BZ and $\mathbf{Q}(\mathbf{q})$ are the normal coordinates, just as in Chapter 3.

It is convenient to rewrite the forces in the form

$$\tilde{F}_\alpha(\mathbf{q}) = \sum_l F_\alpha(\mathbf{l})\, e^{i\mathbf{q}.\mathbf{l}} \tag{10.10.6}$$

and introduce the coupling coefficients (the dynamical matrix of Chapter 3)

$$D_{\alpha\beta}(\mathbf{q}) = \sum_{ll'} \Phi_{\alpha\beta}(ll')\, e^{-i\mathbf{q}.(\mathbf{l}'-\mathbf{l})}:\ \Phi_{\alpha\alpha}(ll') = \Phi_{\alpha\beta}(\mathbf{l}-\mathbf{l}'). \tag{10.10.7}$$

Then we can rewrite $\Phi - \Phi_0$ as

$$\Phi - \Phi_0 = \sum_{\mathbf{q}} \left[-\sum_\alpha \tilde{F}_\alpha(\mathbf{q})\, Q_\alpha(\mathbf{q}) + \frac{N}{2} \sum_{\alpha\beta} D_{\alpha\beta}(\mathbf{q})\, Q_\alpha(\mathbf{q})\, Q_\beta(-\mathbf{q}) \right], \tag{10.10.8}$$

where N is the number of atoms in the crystal.

Corresponding to equation (10.10.3), the condition for equilibrium becomes

$$\frac{\partial \Phi}{\partial Q_\alpha(\mathbf{q})} = 0, \tag{10.10.9}$$

which from equation (10.10.8) yields

$$N \sum_\beta D_{\alpha\beta}(-\mathbf{q})\, Q_\beta(\mathbf{q}) = \tilde{F}_\alpha(-\mathbf{q}), \tag{10.10.10}$$

which is essentially the **q**-space form of equation (10.10.4).

For each **q**, equation (10.10.10) gives us three simultaneous equations for the three components $Q_\alpha(\mathbf{q})$. When $Q_\alpha(\mathbf{q})$ is known, the value of the displacement of any lattice point can be found from equation (10.10.5).

Though we shall not, as we said, give a detailed example here, we finally remark that Kanzaki calculated the dynamical matrix to deal with a vacancy in argon from a Lennard–Jones potential $\phi(r)$. He then considered the three configurations illustrated in Figure 10.13, and using the Lennard–Jones potential, we then have, for the three configurations a, b and c, the trivial identity

$$\begin{aligned}\Phi(c) - \Phi(a) &= [\Phi(c) - \Phi(b)] + [\Phi(b) - \Phi(a)] \\ &= -\sum_l \phi[|\mathbf{l} + \boldsymbol{\xi}(\mathbf{l})|] \\ &\quad + \frac{1}{2} \sum_{ll'} \{\phi[|\mathbf{l}' + \boldsymbol{\xi}(\mathbf{l}') - \mathbf{l} + \boldsymbol{\xi}(\mathbf{l})|] - \phi(|\mathbf{l}' - \mathbf{l}|)\}.\end{aligned} \tag{10.10.11}$$

The equilibrium configuration is given by

$$\frac{\partial}{\partial \xi_\alpha(\mathbf{l})} [\Phi(c) - \Phi(a)] = 0. \tag{10.10.12}$$

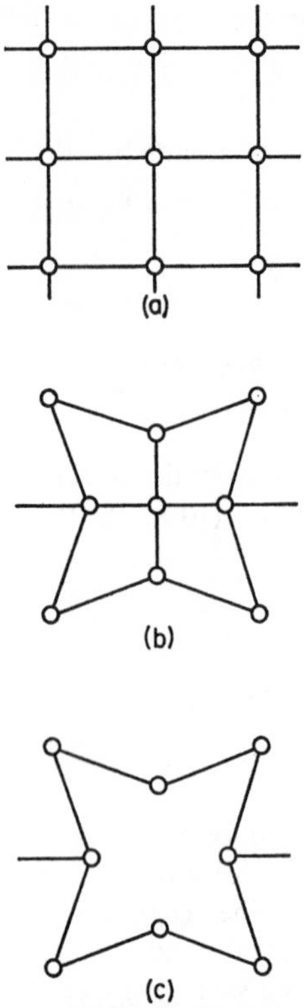

FIGURE 10.13. Vacancy configurations in argon. (a) Perfect lattice. (b) Lattice strained to final equilibrium position, but with no vacancy. (c) Vacancy present in fully relaxed lattice.

Now $\Phi(b)-\Phi(a)$ is the energy when the perfect lattice is displaced in the absence of the Kanzaki forces and hence is given by the second-order term in equation (10.10.10). Thus we must satisfy the condition

$$F_\alpha(\mathbf{l}) = \frac{\partial}{\partial \xi_\alpha(\mathbf{l})} \sum_{l} \phi[\mathbf{l}+\boldsymbol{\xi}(\mathbf{l})] \qquad (10.10.13)$$

yielding the Kanzaki forces from the given pair potential (see also Gahlen, Beeler and Jaffee, 1972).

It is interesting then that there are two ways of characterizing a defect. In the language of electron theory, it is the displaced charge, and it seems clear that from a knowledge of this the Kanzaki forces are calculable. One expects an iterative procedure to be necessary, however, in combining electron theory with the Kanzaki method.

10.11 Electron density distribution near metal surfaces

We could very well view the presence of a surface on a piece of metal as an imperfection, as it obviously causes departures from perfect periodicity. We shall therefore give a brief account of the nature of the electron-density distribution near a metal surface.

The theory will be worked out for an oversimplified model in which the electron density is calculated in a model akin to jellium, with a planar surface. However, to obtain a rough idea of some of the features to be expected, let us calculate the density $\rho(z)$ associated with electrons moving along the z-axis, but confined to the region between 0 and l. Then we may write, for N electrons singly occupying the lowest energy levels

$$\rho(z) = \frac{2}{l}\sum_{1}^{N}\sin^2\frac{n\pi z}{l}. \tag{10.11.1}$$

We wish to examine the effect of the "surface" at $z = 0$ on the density, so that we now allow l to tend to infinity, and replacing the summation by an integration, we find

$$\rho(z) = \rho_0 - \frac{\sin 2k_f z}{2\pi z}, \tag{10.11.2}$$

where ρ_0 is the average density N/l, and k_f is the Fermi momentum. The form of this distribution is such that

(i) The density deep in the interior of the metal oscillates with wavelength π/k_f; these are again the Friedel oscillations characteristic of a localized perturbation in a Fermi gas.

(ii) The density rises from zero to its value ρ_0 when $\sin 2k_f z = 0$, i.e. over a distance $\pi/2k_f$ which is half the de Broglie wavelength for an electron at the Fermi surface.

Though this one-dimensional model is, of course, grossly oversimplified, nevertheless we shall see that these features must be present in general. A simple extension of the above argument can be made by working in three dimensions, but confining the electrons inside a sphere of radius a, say. A very simple calculation then yields the electron density $\rho(r)$, and, for

$a \sim 16$ Bohr radii, the density obtained for 186 electrons is shown in Figure 10.14 (March and Murray, 1960). Though this is a very small piece of solid indeed, we can see the same features, though the oscillations in density in this small system are more nearly the analogue of the shells in atomic densities,

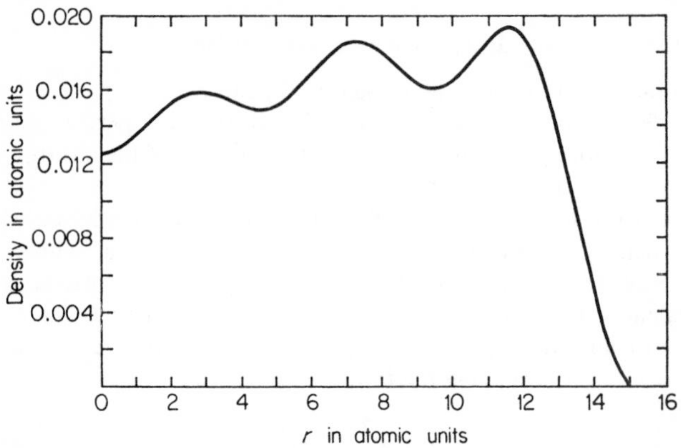

FIGURE 10.14. Electron density for 186 particles confined within a sphere by an infinite potential barrier.

than of the Friedel oscillations. Actually, formula (10.11.2) has its analogue in this problem in that the density of the s-electrons is given by

$$\rho^{N_s}(r) = \sum_1^{N_s} R_n^2(r), \tag{10.11.3}$$

where the normalized radial wave functions are given explicitly by

$$R_n^2(r) = \frac{2\pi a r^2}{1} \sin^2 \frac{n\pi r}{a}, \quad \int_0^a R_n^2 \, 4\pi r^2 \, dr = 1, \tag{10.11.4}$$

and where N_s denotes the number of occupied s-states. It follows that for doubly filled s-states the density is given by

$$\rho^{N_s}(r) = \frac{1}{\pi a r^2} \sum_1^{N_s} \sin^2 \frac{n\pi r}{a}$$

$$= \frac{1}{2\pi a r^2} \left\{ N_s + \tfrac{1}{2} - \frac{\sin [\pi(2N_s+1) r/a]}{2 \sin (\pi r/a)} \right\}. \tag{10.11.5}$$

This formula is useful because it enables us to find the total density at the origin, which is evidently entirely due to s-states. The result is

$$\rho^{N_s}(0) = \frac{N_s(N_s+1)(2N_s+1)\pi}{6a^3}. \qquad (10.11.6)$$

In a large piece of metal, with N electrons, we expect, of course, that at the centre

$$\rho^{N_s}(0) \simeq \frac{3N}{4\pi a^3} \qquad (10.11.7)$$

and for $N \sim 10^6$ particles, $N_s \sim 60$. Even with 60 occupied s-states, we still have a very small number of particles, and this method using a finite spherical metal is tedious to apply in practice.

Therefore we turn to the theory of an inhomogeneous electron gas, in order to get a more practicable approach to the surface problem.

10.11.1 Thomas–Fermi theory

The simplest way to incorporate the effect of inhomogeneity into Fermi gas theory is the Thomas–Fermi model discussed in Chapter 1. Unfortunately, however, this theory is only valid when the electron density changes by but a small fraction of itself over a distance of the order of the de Broglie wavelength for an electron at the Fermi surface. But we have already seen that, near a metal surface, the density varies from zero to its bulk value over this distance and therefore the Thomas–Fermi theory must be refined. Unlike the models used above, which erect an infinite potential barrier at the "surface", and therefore imply an infinite work function, Frenkel (1928), who first considered the problem under discussion, showed that the Thomas–Fermi method led to zero work function.

(a) *Improved theory, including gradient term, plus electron interactions.* Bardeen (1936), within the jellium model with a planar surface, calculated the solutions of the Hartree–Fock equations to obtain a theory of the work function of Na. This is almost the best calculation we have on a surface property, but it is a lengthy matter to apply it to other metals, and we shall describe a rougher approach, based on an extension of the Thomas–Fermi method. In that method, we saw in Chapter 1 that the kinetic energy density is given in terms of the electron density $\rho(\mathbf{r})$ by

$$\text{Kinetic energy density} = c_k[\rho(\mathbf{r})]^{5/3}, \quad c_k = \tfrac{3}{10}(3\pi^2)^{2/3}. \qquad (10.11.8)$$

This is, essentially, kinetic energy due to the exclusion principle. However, when the wave functions and density are varying rapidly, as they are in the vicinity of a metal–vacuum interface, we need to take account of the associated kinetic energy. We saw in Chapter 2 how to do this. Following Kirzhnits (1957), we have that, in Fermi gas theory, the inhomogeneity correction to the kinetic energy (10.11.8) is given by

$$\frac{\hbar^2}{72m}\int\frac{(\nabla\rho)^2}{\rho}\,d\mathbf{r}, \tag{10.11.9}$$

a result derived in Appendix A2.4.

Thus, the total energy of the metal may be written down, in this approximation as

$$E = \int v(\mathbf{r})\,\rho(\mathbf{r})\,d\mathbf{r}+\frac{1}{2}\int\frac{\rho(\mathbf{r})\,\rho(\mathbf{r}')\,d\mathbf{r}\,d\mathbf{r}'}{|\mathbf{r}-\mathbf{r}'|}+c_{\mathrm{k}}\int\rho^{5/3}\,d\mathbf{r}$$
$$+\frac{\hbar^2}{72m}\int\frac{(\nabla\rho)^2}{\rho}\,d\mathbf{r}+E_{\mathrm{exch}}+E_{\mathrm{corr}}. \tag{10.11.10}$$

Here $v(\mathbf{r})$ is the "external potential", in the present case that due to the positive ion background, while E_{exch} and E_{corr} are the energies associated with exchange and correlation interactions between electrons.

As we discussed in Chapter 2 in connection with Dirac–Slater exchange, we shall replace E_{exch} by the local free-electron value, namely

$$E_{\mathrm{exch}} = -c_{\mathrm{e}}\int\rho(\mathbf{r})^{4/3}\,d\mathbf{r}, \quad c_{\mathrm{e}} = \frac{3}{4}\left(\frac{3}{\pi}\right)^{1/3}. \tag{10.11.11}$$

Also, within the same philosophy, we could take over the theory of the correlation energy of a uniform gas into the inhomogeneous gas theory. If we use the convenient interpolation formula of Wigner (1938), which we gave in Chapter 2, we can write

$$E_c = -0.056\int\frac{\rho^{4/3}\,d\mathbf{r}}{0.079+\rho^{1/3}}. \tag{10.11.12}$$

It will be recognized that this is all a special case of the density functional theory set out in Chapter 1, where a quite crude choice has admittedly been made of the energy functionals. We now calculate the density ρ in the inhomogeneous gas from the variational principle

$$\delta[E(\rho)-\mu N] = 0, \tag{10.11.13}$$

where μ is a Lagrange multiplier, which in fact is the chemical potential while

$$N = \int\rho(\mathbf{r})\,d\mathbf{r}, \tag{10.11.14}$$

the total number of electrons. The Euler equation of the variational problem is then found to be (Smith, 1969)

$$\frac{d^2\rho}{dz^2}-\frac{1}{2\rho}\left(\frac{d\rho}{dz}\right)^2 = 36\left[\tfrac{1}{2}(3\pi^2)^{\frac{2}{3}}\rho^{\frac{5}{3}}+(\phi-\mu)\,\rho-\left(\frac{3}{\pi}\right)^{\frac{1}{3}}\rho^{\frac{4}{3}}-\frac{0.056\rho^{\frac{5}{3}}+0.0059\rho^{\frac{4}{3}}}{(0.079+\rho^{\frac{1}{3}})^2}\right], \tag{10.11.15}$$

where $\phi(\mathbf{r})$ is the electrostatic potential given by

$$\phi(\mathbf{r}) = v(\mathbf{r})+\int\frac{\rho(\mathbf{r}')}{|\mathbf{r}-\mathbf{r}'|}\,d\mathbf{r}'. \tag{10.11.16}$$

(b) *Electron work function.* The electron work function W is defined as the difference in energy between the original metal with N electrons and that in which one electron has been removed from the metal and is at rest outside it. Then it follows that

$$W = -\left.\frac{\partial E}{\partial N}\right|_{N=N+} \tag{10.11.17}$$

and, furthermore, from the variational principle

$$\begin{aligned}\delta E &= \mu\delta N\\ &= \frac{\partial E}{\partial N}\delta N.\end{aligned} \tag{10.11.18}$$

Therefore it follows that

$$W = -\frac{\partial E}{\partial N} = -\mu. \tag{10.11.19}$$

Thus we can obtain the work function μ from the Euler equation (10.11.15) by noting that, as $z\to-\infty$, $d^2\rho/dz^2$ and $d\rho/dz\to 0$ and $\rho\to\rho_0$. Thus we find

$$W = -\mu = -\phi(-\infty)-\tfrac{1}{2}(3\pi^2)^{\frac{2}{3}}\rho_0^{\frac{2}{3}}+\left(\frac{3}{\pi}\right)^{\frac{1}{3}}\rho_0^{\frac{1}{3}}+\frac{0.056\rho_0^{\frac{2}{3}}+0.0059\rho_0^{\frac{1}{3}}}{(0.079+\rho_0^{\frac{1}{3}})^2}. \tag{10.11.20}$$

By noting that the one-body potential including exchange and correlation, discussed fully in Chapter 1, is formally defined by

$$V^{(1)} = \frac{\delta}{\delta\rho}[E_{\text{exch}}+E_{\text{corr}}]+\phi, \tag{10.11.21}$$

we may rewrite equation (10.11.20) in the form

$$W+E_{\text{f}} = -V^{(1)}(-\infty). \tag{10.11.22}$$

This is shown schematically in Figure 10.15, and the problem is seen to reduce to estimating $\phi(-\infty)$ and $V_{xc}(-\infty)$.

(c) *Variational density and corresponding potential.* Instead of solving the Euler equation (10.11.15) directly, another approach is open to us, namely

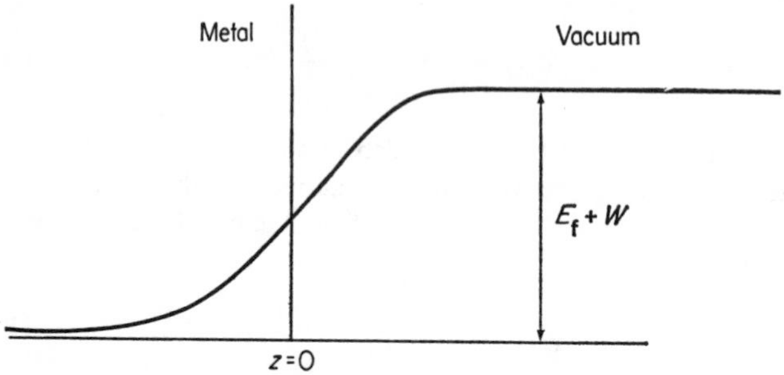

FIGURE 10.15. One-body potential in metal, relating to electronic work function.

a direct use of the variation principle. This is the method adopted by Smith (1969) who makes a choice for the trial form for $\rho(\mathbf{r})$ as

$$\left.\begin{aligned}\rho &= \rho_0 - \tfrac{1}{2}\rho_0 e^{\beta z}, \quad z<0,\\ &= \tfrac{1}{2}\rho_0 e^{-\beta z}, \quad z>0,\end{aligned}\right\} \tag{10.11.23}$$

where β is to be treated as the variational parameter.

It should be noted that the Friedel oscillations are not present in this choice. This is correct within the present framework; they have been lost by making gradient expansions.

Adopting this choice for the density, the corresponding potential is given by

$$\left.\begin{aligned}\phi &= 2\pi\rho_0 e^{\beta z}/\beta^2 - 4\pi\rho_0/\beta^2, \quad z<0,\\ &= -2\pi\rho_0 e^{-\beta z}/\beta^2, \quad z>0.\end{aligned}\right\} \tag{10.11.24}$$

Once β is found by minimizing the energy, the work function can be calculated straightforwardly as outlined above.

The results for the electron densities near the metal surface for Na and Li are shown in Figure 10.16. The corresponding one-body potentials $V^{(1)}$ are shown in Figure 10.17. Finally a comparison between theory and experiment

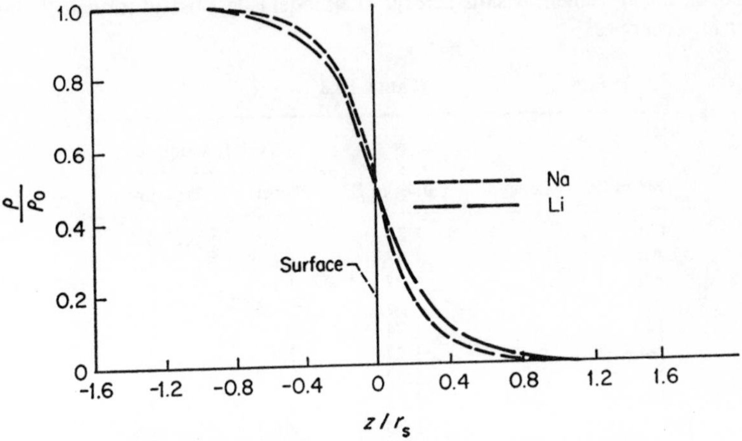

FIGURE 10.16. Electron density distribution near metal surface for jellium model of Na and Li. r_s is interelectronic spacing (after Smith, 1969).

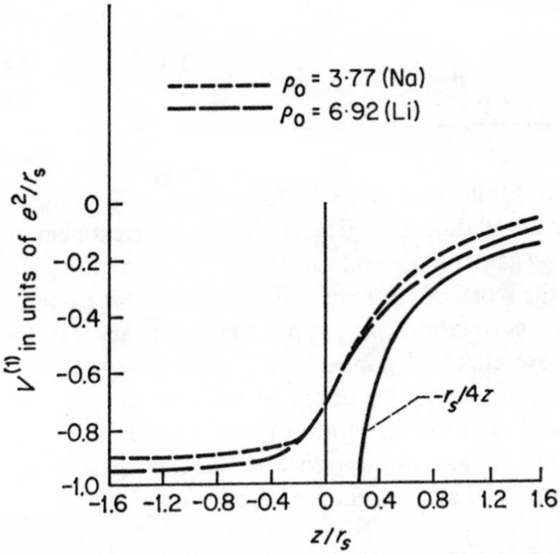

FIGURE 10.17. One-body potentials corresponding to electron densities in Figure 10.16. ρ_0 is recorded in units of 10^{-3} a.u. Form of image potential $(4z)^{-1}$ is also shown (after Smith, 1969).

is shown for the work function in Table 10.3. Considering the crudity of the model, the agreement is satisfactory. The model is less useful for calculating surface energies.

TABLE 10.3

			Work function (eV)	
Metal	Valency	Values of β	Theory	Experimental
Cu		1.23	3.3	4.4
Ag	1	1.23	3.2	4.3
Au		1.23	3.2	4.3
Li		1.24	3.1	2.4
Na		1.27	2.9	2.4
K	1	1.32	2.8	2.2
Rb		1.32	2.7	2.2
Cs		1.33	2.6	1.8
Be		1.26	3.7	3.9
Mg		1.22	3.3	3.6
Zn	2	1.22	3.5	4.2
Cd		1.22	3.4	4.1
Ca		1.24	3.1	2.8
Sn	4	1.22	3.4	4.4
Pb		1.22	3.5	4.0

It will be of some interest to include eventually gradient terms in the exchange and correlation energies in the above treatment of the inhomogeneous electron gas near a metal surface. Methods of doing this are described in Chapter 2, the work of Herman and co-workers for exchange and Ma and Brueckner for correlation providing the necessary starting point. The inclusion of core effects, by means of pseudo-potentials, is also a further step to be taken in refining the above theory (see Lang and Kohn, 1970).

We shall conclude this discussion of surface effects by giving a theory of collective effects in an inhomogeneous electron gas, and then we shall apply this theory to discuss surface collective oscillations.

10.11.2 Green functions in an inhomogeneous system

In equation (2.7.107) we wrote down the relation between one- and two-particle Green functions in a homogeneous electron gas. Our purpose here is to generalize this to deal with an inhomogeneous electron gas such as we encounter in the neighbourhood of a metal surface.

To see the nature of the equation for the one-particle Green function $G(1,2)$, let us first neglect the interactions and suppose the electrons to move in an external potential $V(\mathbf{r}t)$. Then we have the time-dependent Schrödinger equation

$$-\frac{\hbar^2}{2m}\nabla^2\psi(\mathbf{r}t)+V(\mathbf{r}t)\psi(\mathbf{r}t) = i\hbar\frac{\partial}{\partial t}\psi(\mathbf{r}t). \tag{10.11.25}$$

Using this equation for $\psi(\mathbf{r}_1 t_1)\equiv\psi(1)$ and the complex conjugate form for $\psi(2)$, it is straightforward to show that

$$-\frac{\hbar^2}{2m}[\nabla_1^2-\nabla_2^2]G+[V(\mathbf{r}_1 t_1)-V(\mathbf{r}_2 t_2)]G = i\hbar\left[\frac{\partial}{\partial t_1}+\frac{\partial}{\partial t_2}\right]G. \tag{10.11.26}$$

We shall see below that, from the exact relation between G and G_2, which we shall now derive as a generalization of equation (2.7.107), a Hartree field approximation will simply modify equation (10.11.26) by replacing $V(\mathbf{r}t)$ by an effective potential $V_{\text{eff}}(\mathbf{r}t)$. A precise form for this will be found from the argument presented below.

(a) *Relation between one- and two-particle Green functions with interactions present*. We shall work with the thermal Green function

$$G(x_1 x_2) = i\langle T\{\psi(x_1)\psi^\dagger(x_2)\}\rangle, \tag{10.11.27}$$

where $\langle\,\rangle$ denotes both a quantum-mechanical and a thermal average and $x\equiv(\mathbf{r},t)$.

We start out from the Heisenberg equations of motion for the field operators ψ:

$$i\hbar\frac{\partial\psi}{\partial t} = [\psi, H], \tag{10.11.28}$$

where H is the full Hamiltonian, given by

$$H = H_0+v, \tag{10.11.29}$$

v being the two-particle interaction. H_0 satisfies

$$[H_0, \psi(x)] = \frac{\hbar^2}{2m}\nabla^2\psi(x)-V(x)\psi(x) \tag{10.11.30}$$

and

$$[H_0, \psi^\dagger(x)] = -\frac{\hbar^2}{2m}\nabla^2\psi^\dagger(x)+V(x)\psi^\dagger(x). \tag{10.11.31}$$

Also we may write

$$[v, \psi(x)] = -\int v(x-x')\,\psi^\dagger(x')\,\psi(x')\,dx'\,\psi(x) \tag{10.11.32}$$

and

$$[v, \psi^\dagger(x)] = \int \psi^\dagger(x')\,v(x-x')\,\psi(x')\,dx'\,\psi^\dagger(x), \tag{10.11.33}$$

where

$$v(x-x') = v(\mathbf{r}-\mathbf{r}')\,\delta(t-t'). \tag{10.11.34}$$

Therefore, introducing Heaviside functions θ to take account of the time-ordering operator T in the definition (10.11.27) we find, with $\hbar = 1$,

$$\begin{aligned} i\frac{\partial}{\partial t_1}G(x_1 x_2) &= -\frac{\partial}{\partial t_1}[\theta(t_1-t_2)\langle\psi(x_1)\,\psi^\dagger(x_2)\rangle] \\ &\quad +\frac{\partial}{\partial t_1}[\theta(t_2-t_1)\langle\psi^\dagger(x_2)\,\psi(x_1)\rangle] \\ &= -\delta(t_1-t_2)\langle\psi(x_1)\,\psi^\dagger(x_2)+\psi^\dagger(x_2)\,\psi(x_1)\rangle \\ &\quad -\theta(t_1-t_2)\left\langle\frac{\partial\psi(x_1)}{\partial t_1}\psi^\dagger(x_2)\right\rangle+\theta(t_2-t_1)\left\langle\psi^\dagger(x_2)\frac{\partial\psi(x)}{\partial t_1}\right\rangle \\ &= i\langle T\{[\psi(x_1), H], \psi^\dagger(x_2)\}\rangle-\delta(x_2-x_1). \end{aligned} \tag{10.11.35}$$

Substituting for the commutator explicitly and rearranging terms we readily find

$$\begin{aligned} i\frac{\partial}{\partial t_1}G(x_1 x_2)+\frac{1}{2m}\nabla_1^2 G(x_1 x_2) &= -\delta(x_1-x_2)+\int dx\,v(x_1-x)\,G_2(xx_1xx_2) \\ &\quad +V(x_1)\,G(x_1 x_2), \end{aligned} \tag{10.11.36}$$

where

$$G_2(xx_1x'x_2) = i\langle T\{\psi(x)\,\psi(x_1)\,\psi^\dagger(x')\,\psi(x_2)\}\rangle. \tag{10.11.37}$$

We find similarly

$$\begin{aligned} i\frac{\partial}{\partial t_2}G(x_1 x_2)-\frac{1}{2m}\nabla_2^2 G(x_1 x_2) = \delta(x_1-x_2)-\int dx\,v(x_2-x)\,G_2(xx_1 x_3 x_4) \\ -V(\mathbf{r}_2)\,G(x_1 x_2) \end{aligned} \tag{10.11.38}$$

and on adding this equation to (10.11.36) the desired result

$$i\left\{\frac{\partial}{\partial t_1}+\frac{\partial}{\partial t_2}\right\} G(x_1 x_2)-\frac{1}{2m}\{\nabla_1^2-\nabla_2^2\} G(x_1 x_2)$$
$$=\{V(\mathbf{r}_1)-V(\mathbf{r}_2)\} G(x_1 x_2)$$
$$-\int dx\, v(x_1-x)\, G_2(xx_1\, xx_2)+\int dx\, v(x_2-x)\, G_2(xx_1\, xx_2) \qquad (10.11.39)$$

follows.

The single-particle thermal Green function $G_1(1,2)$ depends on position through $\mathbf{r}_1$ and $\mathbf{r}_2$, and not simply on $\mathbf{r}_1-\mathbf{r}_2$, as in the homogeneous electron gas. If we use the Hartree approximation

$$G_2(xx_1\, xx_2) = G(xx)\, G(x_1 x_2) \qquad (10.11.40)$$

for the two-particle Green function, then the equation reduces to:

$$\left\{i\left(\frac{\partial}{\partial t_1}+\frac{\partial}{\partial t_2}\right)+(\nabla_1+\nabla_2)\cdot\left(\frac{\nabla_1-\nabla_2}{2}\right)-[U_{\text{eff}}(x_1)-U_{\text{eff}}(x_2)]\right\} G(x_1 x_2) = 0, \qquad (10.11.41)$$

where the quantity $U_{\text{eff}}(x_1)$ is an effective local potential given in terms of the Coulomb interaction $v(\mathbf{r})$ by $(x_1 \equiv \mathbf{r}_1 t_1)$

$$U_{\text{eff}}(x_1) = U_{\text{ext}}(\mathbf{r}_1 t_1)+i\int d\mathbf{r}'\, v(\mathbf{r}_1-\mathbf{r}')\, G(\mathbf{r}' t_1;\, \mathbf{r}' t_1^+). \qquad (10.11.42)$$

It is now useful to introduce relative and centre-of-mass coordinates

$$\mathbf{r} = \mathbf{r}_1-\mathbf{r}_2, \quad \mathbf{R} = \frac{\mathbf{r}_1+\mathbf{r}_2}{2}, \qquad (10.11.43a)$$

to put

$$t = t_1-t_2, \quad T = \frac{t_1+t_2}{2}, \qquad (10.11.43b)$$

and to write

$$G(\mathbf{r}_1 t_1;\, \mathbf{r}_2 t_2) = G(\mathbf{r}t;\, \mathbf{R}T). \qquad (10.11.44)$$

We shall work with the mixed representation of the density matrix (Wigner distribution function), defined by

$$\rho(\mathbf{p}, \mathbf{R}, T) = \int_{-\infty}^{\infty} \frac{d\omega}{2\pi} G^<(\mathbf{p}\omega;\, \mathbf{R}T)$$
$$= \int d\mathbf{r}\, e^{-i\mathbf{p}\cdot\mathbf{r}}\, G^<(\mathbf{r}t = 0;\, \mathbf{R}T), \qquad (10.11.45)$$

where

$$\left.\begin{aligned} G(\mathbf{r}t;\, \mathbf{R}T) &= iG^{>}(\mathbf{r}t;\, \mathbf{R}T), \quad t>0, \\ &= -iG^{<}(\mathbf{r}t;\, \mathbf{R}T), \quad t<0. \end{aligned}\right\} \tag{10.11.46}$$

We can regard $\rho(\mathbf{p}, \mathbf{R}, T)$ as being the number of particles with momentum $\mathbf{p}$ at position $\mathbf{R}$ and time T. The presentation we use below follows closely that of Harris and Griffin (1970).

Then it is clear that the particle number and current densities are simply

$$\begin{aligned} n(\mathbf{R}T) &= G^{<}(\mathbf{r}=0, t=0;\, \mathbf{R}T) \\ &= \int \frac{d\mathbf{p}}{(2\pi)^3}\, \rho(\mathbf{p}, \mathbf{R}, T) \end{aligned} \tag{10.11.47}$$

and

$$\mathbf{j}(\mathbf{R}T) = \int \frac{d\mathbf{p}}{(2\pi)^3} \left(\frac{\mathbf{p}}{m}\right) \rho(\mathbf{p}, \mathbf{R}, T). \tag{10.11.48}$$

The equation of motion for $\rho(\mathbf{p}, \mathbf{R}, T)$ is simply derived from equation (10.11.41) and has the form

$$\begin{aligned} \left[\frac{\partial}{\partial T} + \frac{\mathbf{p}.\nabla_{\mathbf{R}}}{m}\right] \rho(\mathbf{p}, \mathbf{R}, T) = -i \int d\mathbf{r} \int \frac{d\mathbf{p}'}{(2\pi)^3}\, e^{i(\mathbf{p}'-\mathbf{p}).\mathbf{r}}\, \rho(\mathbf{p}', \mathbf{R}, T) \\ \times \left[U_{\text{eff}}\left(\mathbf{R}+\frac{\mathbf{r}}{2}, T\right) - U_{\text{eff}}\left(\mathbf{R}-\frac{\mathbf{r}}{2}, T\right)\right]. \end{aligned} \tag{10.11.49}$$

This, combined with equation (10.11.42) for U_{eff}, which we can write in the alternative form

$$U_{\text{eff}}(\mathbf{R}T) = U_{\text{ext}}(\mathbf{R}T) + \int d\mathbf{R}'\, v(\mathbf{R}-\mathbf{R}')\, n(\mathbf{R}'T), \tag{10.11.50}$$

give us the basic equations that we must solve for the mixed density matrix $\rho(\mathbf{p}, \mathbf{R}, T)$. These approximations are referred to elsewhere in the book as the self-consistent field method, random phase approximation, or the time-dependent Hartree theory.

If it could be assumed that the one-body effective potential is slowly varying in space, then we obtain the collisionless Boltzmann equation (Vlasov equation; see, for example, Montgomery and Tidman, 1964), namely

$$\left(\frac{\partial}{\partial T} + \frac{\mathbf{p}.\nabla_{\mathbf{R}}}{m}\right) \rho(\mathbf{p}, \mathbf{R}, T) = \nabla_{\mathbf{R}}\, U_{\text{eff}}(\mathbf{R}T) . \nabla_{\mathbf{p}}\, \rho(\mathbf{p}, \mathbf{R}, T). \tag{10.11.51}$$

(b) *Solution for mixed density matrix in time-dependent external field.* We now consider a small disturbance of the equilibrium system, so that we write

$$\rho(\mathbf{p},\mathbf{R},T) = \rho_0(\mathbf{pR})+\delta\rho(\mathbf{p},\mathbf{R},T), \tag{10.11.52}$$

where $\delta\rho$ is proportional to the time-dependent external field $\delta U_{\text{ext}}(\mathbf{R}T)$, given by

$$\delta U_{\text{ext}}(\mathbf{R}T) = U_{\text{ext}}(\mathbf{R}T)-U^0_{\text{ext}}(\mathbf{R}). \tag{10.11.53}$$

The time-independent density matrix is given by

$$\frac{\mathbf{p}.\nabla_{\mathbf{R}}}{m}\rho_0(\mathbf{pR}) = -i\int d\mathbf{r}\int\frac{d\mathbf{p}'}{(2\pi)^3}\rho_0(\mathbf{p}'R)\times\left[U^0_{\text{eff}}\left(\mathbf{R}+\frac{\mathbf{r}}{2}\right)-U^0_{\text{eff}}\left(\mathbf{R}-\frac{\mathbf{r}}{2}\right)\right]e^{i(\mathbf{p}'\mathbf{p}).\mathbf{r}}, \tag{10.11.54}$$

where the static self-consistent field is given in the usual way by

$$U^0_{\text{eff}}(\mathbf{R}) = U^0_{\text{ext}}(\mathbf{R})+\int d\mathbf{R}'\,v(\mathbf{R}-\mathbf{R}')\int\frac{d\mathbf{p}}{(2\pi)^3}\rho_0(\mathbf{pR}'). \tag{10.11.55}$$

That $\rho(\mathbf{p},\mathbf{R},T)$ becomes $\rho_0(\mathbf{pR})$ in the limit $T\to-\infty$ can be regarded as a boundary condition on equation (10.11.49). Then, working to first order in $\delta U_{\text{ext}}(RT)$, the equation of motion for $\delta\rho(\mathbf{p},\mathbf{R},T)$ is given, after Fourier transforming with respect to $\mathbf{R}$ and T, by

$$\left(\omega-\frac{\mathbf{p}.\mathbf{k}}{m}\right)\delta\rho(\mathbf{p},\mathbf{k},\omega) = \int\frac{d\mathbf{p}'}{(2\pi)^3}\left[\rho_0\left(\mathbf{p}-\frac{\mathbf{p}'}{2},\mathbf{k}-\mathbf{p}'\right)-\rho_0\left(\mathbf{p}+\frac{\mathbf{p}'}{2},\mathbf{k}-\mathbf{p}'\right)\right]\delta U_{\text{eff}}(\mathbf{p}'\omega) + \int\frac{d\mathbf{p}'}{(2\pi)^3}\left[\delta\rho\left(\mathbf{p}-\frac{\mathbf{p}'}{2},\mathbf{k}-\mathbf{p}',\omega\right)-\delta\rho\left(\mathbf{p}+\frac{\mathbf{p}'}{2},\mathbf{k}-\mathbf{p}',\omega\right)\right]U^0_{\text{eff}}(\mathbf{p}'), \tag{10.11.56}$$

where

$$\delta U_{\text{eff}}(\mathbf{p}'\omega)\equiv\delta U_{\text{ext}}(\mathbf{p}'\omega)+v(\mathbf{p}')\int\frac{d\mathbf{p}}{(2\pi)^3}\delta\rho(\mathbf{p},\mathbf{p}',\omega). \tag{10.11.57}$$

Integrating equation (10.11.56) over $\mathbf{p}$, and using equations (10.11.57) and (10.11.48), it is easy to show that the density and current fluctuations satisfy the continuity equation

$$\omega\,\delta n(\mathbf{k}\omega) = \mathbf{k}.\delta\mathbf{j}(\mathbf{k}\omega). \tag{10.11.58}$$

We now notice that solutions of equation (10.11.56) in the limit $\delta U_{\text{ext}} \to 0$ correspond to collective modes of the system. Since we expect damping of such modes, we must analytically continue equation (10.11.56) on to the lower half of the complex ω-plane. We then obtain an integral equation for $\delta\rho(\mathbf{p}, \mathbf{k}, \omega)$, namely

$$\delta\rho(\mathbf{p}, \mathbf{k}, \omega) = \int \frac{d\mathbf{p}'}{(2\pi)^3} \frac{\{\rho_0[\mathbf{p}-(\mathbf{p}'/2), \mathbf{k}-\mathbf{p}'] - \rho_0[\mathbf{p}+(\mathbf{p}'/2), \mathbf{k}-\mathbf{p}']\}}{[\omega - (\mathbf{p}\,.\,\mathbf{k}/m)]} \delta U_{\text{eff}}(\mathbf{p}'\omega)$$
$$+ \int \frac{d\mathbf{p}'}{(2\pi)^3} \int \frac{d\mathbf{q}}{(2\pi)^3} U^0_{\text{eff}}(\mathbf{p}')$$
$$\times \frac{\{\delta[\mathbf{q}-\mathbf{p}+(\mathbf{p}'/2)] - \delta[\mathbf{q}-\mathbf{p}-(\mathbf{p}'/2)]\}}{[\omega - (\mathbf{p}\,.\,\mathbf{k}/m)]} \delta\rho(\mathbf{q}, \mathbf{k}-\mathbf{p}', \omega). \tag{10.11.59}$$

We can now effect a solution of this equation by iteration, treating the first term on the right-hand side as the inhomogeneous term of a Fredholm equation for $\delta\rho(\mathbf{p}, \mathbf{k}, \omega)$. Then, carrying out this iteration, and integrating over the variable $\mathbf{p}$ leads to an equation linking the density fluctuation $\delta n(\mathbf{k}\omega)$ and the change in the local field $\delta U_{\text{eff}}(\mathbf{p}\omega)$, namely

$$\delta n(\mathbf{k}\omega) = \int \frac{d\mathbf{p}}{(2\pi)^3} K(\mathbf{k}, \mathbf{p}, \omega)\, \delta U_{\text{eff}}(\mathbf{p}\omega), \tag{10.11.60}$$

where

$$K(\mathbf{k}, \mathbf{p}, \omega) = \int \frac{d\mathbf{q}}{(2\pi)^3} \rho_0(\mathbf{q}, \mathbf{k}-\mathbf{p}) \left\{ \frac{1}{\omega - [\mathbf{q}+(\mathbf{p}/2)]\,.\,(\mathbf{k}/m)} - \frac{1}{\omega - [\mathbf{q}-(\mathbf{p}/2)]\,.\,(\mathbf{k}/m)} \right\}$$
$$+ \int \frac{d\mathbf{q}}{(2\pi)^3} \int \frac{d\mathbf{q}'}{(2\pi)^3} U^0_{\text{eff}}(\mathbf{q})$$
$$\times \left\{ \frac{\rho_0[\mathbf{q}'-(\mathbf{q}/2), \mathbf{k}-\mathbf{q}-\mathbf{p}] - \rho_0[\mathbf{q}'+(\mathbf{q}/2), \mathbf{k}-\mathbf{q}-\mathbf{p}]}{\omega - (\mathbf{q}'/m)\,.\,(\mathbf{k}-\mathbf{q})} \right\}$$
$$\times \left\{ \frac{1}{\omega - [\mathbf{q}'+(\mathbf{q}/2)]\,.\,(\mathbf{k}/m)} - \frac{1}{\omega - [\mathbf{q}'-(\mathbf{q}/2)]\,.\,(\mathbf{k}/m)} \right\}$$
$$+ \text{higher-order iteration contributions.} \tag{10.11.61}$$

For a homogeneous system, where $n_0(\mathbf{r}) = n_0$, only the first term of this expansion remains and we obtain, with $E_p = \varepsilon_p + n_0 v(q=0)$,

$$K(\mathbf{k}, \mathbf{p}, \omega) = (2\pi)^3\, \delta(\mathbf{k}-\mathbf{p}) \int \frac{d\mathbf{q}}{(2\pi)^3} \frac{[f(E_{\mathbf{q}-(\mathbf{k}/2)}) - f(E_{\mathbf{q}+(\mathbf{k}/2)})]}{[\omega - (\mathbf{q}\,.\,\mathbf{k}/m)]}, \tag{10.11.62}$$

where $f(E)$ is the Fermi function. The integral is simply related to the susceptibility $\chi_0(\mathbf{k}\omega)$ discussed in Chapter 4, and we can write

$$K(\mathbf{k}, \mathbf{p}, \omega) = (2\pi)^3\, \delta(\mathbf{k}-\mathbf{p})\, \chi_0(\mathbf{k}\omega). \tag{10.11.63}$$

(c) *High-frequency limit.* For high-energy modes of oscillation, we can assume $\omega \gg \mathbf{q}\,.\,\mathbf{k}/m$ and expand the energy denominators in equation (10.11.61), which will yield an expansion of K in inverse powers of ω. We then find

$$K(\mathbf{k}, \mathbf{p}, \omega) = \frac{\mathbf{k}\,.\,\mathbf{p}}{m\omega^2} n_0(\mathbf{k}-\mathbf{p}) + O\left(\frac{1}{\omega^4}\right). \tag{10.11.64}$$

Remembering that we are setting $\delta U_{\text{ext}} = 0$ to get the collective oscillations, we find the integral equation

$$\delta n(\mathbf{k}\omega) = \frac{1}{\omega^2} \int \frac{d\mathbf{p}}{(2\pi)^3} \frac{\mathbf{p}\,.\,\mathbf{k}}{m} v(\mathbf{p})\, n_0(\mathbf{k}-\mathbf{p})\, \delta n(\mathbf{p}\omega) \tag{10.11.65}$$

for the collective oscillations. This equation (see Sziklas, 1965; Sham, 1968) can be derived by considering the bound plasmon states which arise when a charged impurity is introduced into an electron gas.

One can also obtain from equation (10.11.59) an expression for the current fluctuation $\delta\mathbf{j}(\mathbf{k}\omega)$ associated with the modes described by equation (10.11.65), namely

$$\delta\mathbf{j}(\mathbf{k}\omega) = \frac{1}{\omega} \int \frac{d\mathbf{p}}{(2\pi)^3} \frac{\mathbf{p}}{m} n_0(\mathbf{k}-\mathbf{p})\, v(\mathbf{p})\, \delta n(\mathbf{p}\omega). \tag{10.11.66}$$

The equation to be solved for $\rho_0(\mathbf{pk})$ is

$$-\frac{\mathbf{p}\,.\,\mathbf{k}}{m} \rho_0(\mathbf{pk}) = \int \frac{d\mathbf{p}'}{(2\pi)^3} \left[\rho_0\left(\mathbf{p}-\frac{\mathbf{p}'}{2}, \mathbf{k}-\mathbf{p}'\right) - \rho_0\left(\mathbf{p}+\frac{\mathbf{p}'}{2}, \mathbf{k}-\mathbf{p}'\right)\right] U^0_{\text{eff}}(\mathbf{p}'). \tag{10.11.67}$$

This reduces to the Vlasov equation result when we make a Taylor expansion about $\mathbf{p}' = 0$ to yield

$$-\frac{\mathbf{p}\,.\,\mathbf{k}}{m} \rho_0(\mathbf{p}k) = \int \frac{d\mathbf{p}'}{(2\pi)^3} \mathbf{p}'\nabla_p\, \rho_0(\mathbf{p}, \mathbf{k}-\mathbf{p}')\, U^0_{\text{eff}}(\mathbf{p}'). \tag{10.11.68}$$

Once these equations are solved, then the density distribution is given by

$$n_0(\mathbf{k}) = \int \frac{d\mathbf{p}}{(2\pi)^3} \rho_0(\mathbf{pk}). \tag{10.11.69}$$

We shall discuss these equations a little further in section 10.11.4, when we deal with the surface of a metal, both from the point of view of the electron density near the surface and the related problem of collective surface modes. We want, however, to stress that the theory as yet is still in a fairly primitive state here. Presumably, Ma and Brueckner's work could be derived from such an approach, though it may require transcending the approximation to the two-particle Green function involved in writing down the equation of motion (10.11.87). In essence, the methods of Singwi and co-workers, Ma and Brueckner, and the work of Harris and Griffin referred to above all involve the study of the response of an electron gas to an external potential, and in this sense are very closely related.

10.11.3 Surface excitations

We shall summarize here how the random phase approximation can be used to discuss surface excitations. We follow again the presentation of Harris and Griffin (1970).

The standard method of deriving surface plasmon dispersion relations is to match solutions of Maxwell's equations in two distinct media at a boundary (see Economou, 1969 for a rather full discussion of thin films). But in the presence of a density variation in the surface such as that which we discussed in the previous section, such an approach is not fruitful. To assess the way the density profile affects the surface properties, it seems clear that one needs a microscopic many-body theory.

10.11.4 Surface plasmons

Let us now idealize the problem by assuming that both the electron and ion densities are step functions. We saw earlier that this is not correct, for given a cut-off in the ion density, the electron wave functions leak out of the metal appreciably over a distance ~ 1 Å.

Hence, we can integrate the equation (10.1.65) of Sziklas (1965) and Sham (1968) to obtain, with $v(k) = 4\pi e^2/k^2$,

$$\delta n(\mathbf{k}_\parallel, k_z, \omega) = \frac{\omega_p^2}{\omega^2}\int_{-\infty}^{\infty} \frac{dp_z}{(-2\pi i)} \frac{(k_\parallel^2 + p_z k_z)}{(p_z - k_z + i0^+)(p_z + ik_\parallel)(p_z - ik_\parallel)} \times \delta n(\mathbf{k}_\parallel, p_z, \omega), \tag{10.11.70}$$

where $\mathbf{k}_\parallel = (k_x, k_y)$ and $\omega_p^2 = (4\pi n_0 e^2/m)$. We can then obtain, after a short calculation,

$$\left[1 - \frac{\omega_p^2}{\omega^2}\right] \delta n(\mathbf{k}_\parallel, z, \omega) = -\frac{\omega_p^2}{2\omega^2}\delta n(\mathbf{k}_\parallel, -i\mathbf{k}_\parallel, \omega)\,\delta(z). \tag{10.11.71}$$

The solution corresponding to a surface mode is simply

$$\delta n(\mathbf{k}_\parallel, z, \omega) = \delta n(\mathbf{k}_\parallel, \omega)\,\delta(z), \quad \omega^2 = \omega_p^2/2. \tag{10.11.72}$$

This gives the frequency of the surface plasmon, obtained earlier by Ferrell (1958; see also Stern and Ferrell, 1960).

The fact that the density disturbance associated with the surface plasmon is given by a singular function is clearly unphysical and we interpret the result (10.11.72) as expressing the high degree of localization of the surface plasmon.

For a given density profile $n_0(z)$, such as that we discussed in the previous section, equation (10.11.65) may be written in Fourier transform as

$$\delta n(\mathbf{k}_{\|}, z, \omega)\left[\frac{m\omega^2}{4\pi e^2}-n_0(z)\right]$$
$$=\frac{1}{2}\frac{dn_0(z)}{dz}\int_{-\infty}^{\infty} dz'\, \mathrm{e}^{-k_{\|}|z-z'|}\,\mathrm{sgn}\,(z-z')\,\delta n(\mathbf{k}_{\|}\, z', \omega). \quad (10.11.73)$$

The singular nature of the solution (10.11.70) arises from the assumption of the step function form for the density profile.

It can be shown from the expression for $\delta\mathbf{j}$ that, associated with the delta function density fluctuation at the surface, the current decays exponentially with a skin depth proportional to the wavelength of the excitation.

10.12 Interaction between defects in metals

10.12.1 Electrostatic model

As regards point defects, we have so far assumed that we can neglect any interactions. However, it is an interesting matter in metal crystals to understand how charged impurities interact.

In connection with impurity diffusion, Lazarus (1954) suggested that the interaction could be obtained by an electrostatic model, in which one defect, say of charge Ze, was regarded as sitting in the screened field of a second defect. In the linear theory of section 10.2.1, this model is in fact correct, as we shall now briefly show.

In addition to the interaction of the kind discussed above, we have as a second term to consider the interaction of the displaced charge round one defect with the screened field of the other. Thirdly, we have to deal with the kinetic energy change in the Fermi gas, as we bring the two defects up, in the Fermi gas, from infinity to their finite separation. It is clear how to write down the first two contributions. Let us consider therefore what the kinetic energy change will be. For simplicity we shall deal with the semi-classical theory, but the wave-theory argument can be carried through and again verifies the electrostatic model. The difference, of course, is that the screened

Coulomb potential of semi-classical theory must be replaced by the long-range oscillatory potential, and thus, in the electrostatic model, we can get an oscillatory interaction with distance between charged defects.

In the semi-classical theory, the kinetic energy change can be dealt with as follows.

When the density changes from ρ_0 to $\rho(\mathbf{r})$, the change in kinetic energy is given by

$$\tfrac{3}{10}(3\pi^2)^{\frac{2}{3}} \int (\rho^{\frac{5}{3}} - \rho_0^{\frac{5}{3}})\, d\mathbf{r}. \tag{10.12.1}$$

Assuming the displaced charge $\Delta = \rho - \rho_0$ is small, this may readily be rewritten as

$$E_{\mathrm{f}} \int \Delta\, d\mathbf{r} + \frac{2\pi}{q^2} \int \Delta^2\, d\mathbf{r}. \tag{10.12.2}$$

It is now a relatively easy matter to compute the interaction energy between two impurities in the Fermi gas. Suppose the impurities are represented by point charges Z and Z', the separation being r_0. Then we can calculate the changes in potential and kinetic energies as we bring the impurities together.

The three terms enumerated above can then be written, in an obvious notation, as

(i) $ZZ'e^2 \exp(-qr_0)/r_0,$ (10.12.3)

(ii) $-\dfrac{q^2}{4\pi e^2} \displaystyle\int dr\, V_1 V_2,$ (10.12.4)

(iii) $\left(\dfrac{-q^2}{4\pi e^2}\right)^2 \dfrac{E_{\mathrm{f}}}{3n_0} \left\{ \displaystyle\int d\mathbf{r}\, [(V_1+V_2)^2 - V_1^2 - V_2^2] \right\} = \dfrac{q^2}{4\pi e^2} \displaystyle\int d\mathbf{r}\, V_1 V_2,$ (10.12.5)

where (iii) follows from equation (10.12.2). Thus, the contribution (ii) is just cancelled by the change in kinetic energy, and we are left with the final result

$$\Delta E = ZZ'e^2 \exp(-qr_0)/r_0. \tag{10.12.6}$$

But this is simply the electrostatic energy of an ion of charge Ze, sitting in the electrostatic potential $(Ze/r_0)\exp(-qr_0)$ of the second ion as anticipated above. A general review of this area has been given by March and Rousseau (1970).

10.12.2 Shift in Fermi level as function of concentration

The problem of the perturbation due to an isolated impurity atom in a metal, treated semi-classically in section 1.1, is extended here to the case of a finite concentration of impurity atoms. The system to be dealt with is then that of a metal A containing atoms of a metal B dispersed in the A-matrix as substitutional impurities. At the outset, some assumptions have to be made to simplify the model, as an actual primary solid solution is too complex to be treated. The first assumption concerns the ordering of the impurity atoms in the matrix: the metal is divided into polyhedra, each having at its centre an impurity atom and each polyhedron is replaced by a sphere of the same volume, and it is assumed to a first approximation that all the spheres have the same radius R.

A further assumption is made that the impurity atoms interact only weakly with each other. The interaction between solute atoms is known to be made up of two parts: the elastic and the electronic. The elastic interaction is negligible in the model used here, but the electronic interaction depends on the concentration of solute atoms, the condition of weak interaction therefore sets some restriction on the range of the impurity concentration to be dealt with.

The basic model reduces then to a uniform distribution of weakly interacting impurity atoms, each being at the centre of a spherical cell of solvent metal of radius R. The charge distribution in each cell is made up of a localized charge which may be positive or negative depending on the nature of the impurity, a uniform background of positive charge and a perturbed distribution of electrons, the whole system in the cell being electrically neutral, and it should also be specified that there is the condition of continuity of potential across the boundary of each cell.

The important quantity in the system, as in the case of the isolated impurity, is the perturbing potential, which here determines some important properties of alloys, like the energy of solution, electrical resistivity and thermoelectric power. The determination of the perturbing potential follows similar lines as for the isolated impurity case, the differential equation to be solved being of the form of equation (1.2.5) combined with Poisson's equation.

Friedel (1954) has obtained an approximate solution of the above-mentioned equation in an analytical form by making the assumption that $V \ll E_f$, this assumption corresponds to that of Mott for the isolated impurity case and clearly Friedel's solution should suffer from the same limitations as that of Mott (cf. section 1.1). Friedel has also obtained a second-order approximation from a normalizing condition, but is is hardly more satisfactory. It is possible to solve the equation for the perturbing potential exactly by numerical methods, and it is then found that the deviations between the exact solution and Friedel's first-order and second-order solutions are quite significant.

The differential equation for the perturbing potential energy in the case of a finite concentration of impurities is given by equation (1.2.5), that is,

$$\nabla^2 V = \frac{2^{7/2}}{3\pi}\left[E_f^{3/2} - (E_f + \Delta E_f - V)^{3/2}\right], \tag{10.12.7}$$

where V is the perturbing potential energy and the new Fermi level in the perturbed electron gas has been written as $E_f + \Delta E_f$, where ΔE_f is the shift in the Fermi level and is taken to be small.
The boundary conditions for V are:

$$\left.\begin{aligned} &\text{(a)} \quad V \to -Z/r \quad \text{as } r \to 0, \\ &\text{(b)} \quad V(R) = 0, \\ &\text{(c)} \quad \left(\frac{\partial V}{\partial r}\right)_R = 0. \end{aligned}\right\} \tag{10.12.8}$$

The second condition follows from the charge neutrality of the spheres, and the third one by reason of symmetry at the boundary of two polyhedra (Friedel, 1954).

Friedel (1954) obtained an approximate solution for equation (10.12.7) by making the assumption that $|\Delta E_f - V| \ll E_f$, so that the equation could be linearized to take the form

$$\nabla^2 V = q^2[V - \Delta E_f], \tag{10.12.9}$$

where $q^2 = 2^{5/2} E_f^{1/2}/\pi$, corresponding to Mott's q of equation (10.2.1).

The solution of equation (10.12.9) subject to the boundary conditions (a) and (b) of equation (10.12.8) is

$$V = \Delta E_f - \frac{Z}{r}\,\frac{qR\cosh q(R-r) - \sinh q(R-r)}{qR\cosh qR - \sinh qR} \tag{10.12.10}$$

and the second boundary condition gives for the shift in the Fermi level the value

$$\Delta E_f = Zq/(qR\cosh qR - \sinh qR). \tag{10.12.11}$$

This is plotted as a function of concentration c in Figure 10.18. It is seen that there is no term proportional to c at low concentrations. Optical absorption measurements on Cu–Zn alloys by Biondi and Rayne (1959) confirm the general features of this plot, in spite of the relative crudity of the model.

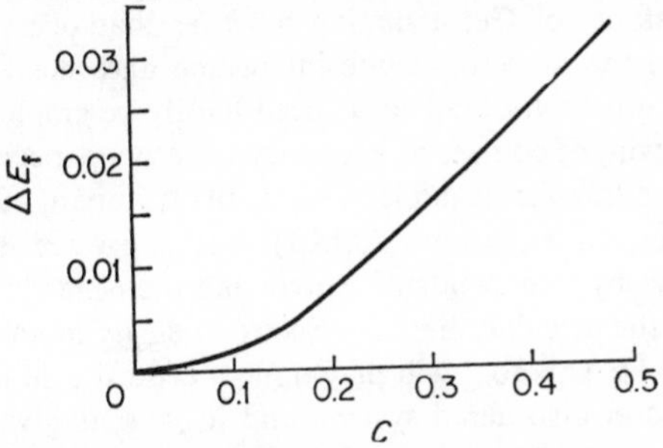

FIGURE 10.18. Change in Fermi level as function of concentration in dilute metallic alloy.

10.13 Electron states in disordered systems

In the previous section, a start has been made on the problem of the interactions between defects. However, in a binary alloy say, we would like to be able to calculate the electron states as a function of concentration, particularly the density of electronic levels and associated physical properties. Furthermore, it is of considerable interest to ask about the nature of the electron states in amorphous solids, for example in amorphous Ge. These are much more difficult problems, but some methods of attack are now available to us.

10.13.1 Virtual crystal approximation

Nordheim suggested in the early days of the electron theory of solids that the electronic structure of a binary A–B alloy might be tackled by considering the atomic potentials A and B as V_A and V_B, and then placing on each lattice site of the crystal the average potential

$$V_{av} = c_A V_A + c_B V_B, \tag{10.13.1}$$

with c_A and c_B the concentration of A and B atoms. With this assumption, the band structure can, of course, be calculated by the usual one-electron methods of Chapter 1. The binary alloy is replaced by a virtual crystal.

We shall see what is the régime of applicability of the virtual crystal method to an A–B alloy below. But in passing it is of interest that Herman and van Dyke (1968) have used such a method to deal with the electron states in amorphous Ge. They have taken account of the lower density of the amorphous phase, increased the average lattice spacing accordingly and recalculated the energy bands. Their approach was then used to calculate the optical constants of Ge, assuming however that one must neglect the selection rule on the **k**-vector. Quite interesting agreement between theory and experiment was then obtained. It need hardly be emphasized that more refined theories will, of course, be necessary for other aspects of the problem of amorphous Ge (see for example Weaire, 1970; Ziman, 1970).

A modification of equation (10.13.1) was suggested subsequently by Korringa and Beeby, independently, involving the t-matrix which we introduced earlier in the calculation of the electron density in an imperfect lattice (see section 10.4.5). This tool will play quite a central role in our discussion of electron states in disordered systems and so we shall give a fairly detailed account of the general properties of the t-matrix below. However, we can remark at this stage that the proposal of Korringa and Beeby was that the t-matrices t_A and t_B be calculated, for scattering off the potentials V_A and V_B respectively, and then an average t-matrix be placed at each site, in a manner analogous to equation (10.13.1). However, another method based on a property of an average t-matrix, the so-called coherent potential method, will play a more prominent part in the ensuing discussion.

10.13.2 T-matrix formalism

(a) *General relations.* Let us define a one-particle *Green function operator* $\mathscr{G}$ such that the Green function is obtained as the matrix-elements of this operator. Thus, following the formalism initiated by Lax (1958),

$$G(\mathbf{k}',\mathbf{k}) = \langle \mathbf{k}'|\mathscr{G}|\mathbf{k}\rangle \tag{10.13.2}$$

is the Green function in momentum space, whilst in configuration space it takes the form

$$G(\mathbf{r}'\mathbf{r}) = \langle \mathbf{r}'|\mathscr{G}|\mathbf{r}\rangle, \tag{10.13.3}$$

where $|\mathbf{r}\rangle$ is an eigenket of the position operator.

This being so, the ket $|\chi_{sc}\rangle$ representing a wave that has suffered scattering is given by

$$|\chi_{sc}\rangle = \mathscr{G}|\chi\rangle \tag{10.13.4}$$

if $|\chi\rangle$ represents the incident wave, for equation (10.13.4) may be written

$$\langle \mathbf{r}|\chi_{sc}\rangle = \int \langle \mathbf{r}|\mathscr{G}|\mathbf{r}'\rangle \langle \mathbf{r}'|\chi\rangle \, d\mathbf{r}', \tag{10.13.5}$$

that is,

$$\chi_{sc}(\mathbf{r}) = \int G(\mathbf{r}'\mathbf{r})\,\chi(\mathbf{r}')\,d\mathbf{r}', \tag{10.13.6}$$

which is the usual relation between the wave functions.

The frequency-dependent Green function operator can be written as

$$\mathcal{G} = \frac{1}{\omega - H_0 - V}, \tag{10.13.7}$$

where V is the potential causing the scattering and H_0 is the Hamiltonian of the unperturbed system. The variable ω takes complex values in order that one may obtain both G^+ and G^- from $\mathcal{G}$.

An equation for $\mathcal{G}$ may be obtained by noting that

$$\frac{1}{\omega - H_0 - V} = \frac{1}{\omega - H_0}\left(1 + V\frac{1}{\omega - H_0 - V}\right) \tag{10.13.8}$$

and, since

$$\mathcal{G}_0 = \frac{1}{\omega - H_0} \tag{10.13.9}$$

is evidently the Green function operator for the unperturbed system, we obtain the Dyson-like equation

$$\mathcal{G} = \mathcal{G}_0 + \mathcal{G}_0 V \mathcal{G}. \tag{10.13.10}$$

We should note that equation (10.13.8) could equally well be written as

$$\mathcal{G} = \left(1 + \frac{1}{\omega - H_0 - V} V\right)\frac{1}{\omega - H_0}, \tag{10.13.11}$$

giving us

$$\mathcal{G} = \mathcal{G}_0 + \mathcal{G} V \mathcal{G}_0. \tag{10.13.12}$$

The first term on the right-hand side of equation (10.13.10) or (10.13.12) corresponds to the unscattered part of the outgoing wave. To be explicit, let us for a moment take time-dependent Green functions. Then

$$|\chi_{sc}(t)\rangle = \mathcal{G}(t)|\chi_0\rangle, \tag{10.13.13}$$

$|\chi_0\rangle$ representing the state at $t = 0$. On using equation (10.13.10), equation (10.13.13) may also be put into the form

$$|\chi_{sc}(t)\rangle = \mathcal{G}_0(t)|\chi_0\rangle + \mathcal{G}_0 V \mathcal{G}|\chi_0\rangle \tag{10.13.14}$$

or

$$|\chi_{sc}(t)\rangle = |\chi(t)\rangle + \mathcal{G}_0 V \mathcal{G}|\chi_0\rangle. \tag{10.13.15}$$

The scattered part of the outgoing wave is evidently given by

$$|\psi\rangle = |\chi_{sc}(t)\rangle - |\chi(t)\rangle = \mathcal{G}_0 V \mathcal{G}|\chi_0\rangle \tag{10.13.16}$$

and if we introduce the *scattering operator* T such that

$$T\mathcal{G}_0 = V\mathcal{G} \tag{10.13.17}$$

we have

$$|\psi\rangle = \mathscr{G}_0 T \mathscr{G}_0 |\chi_0\rangle = \mathscr{G}_0 T |\chi(t)\rangle. \tag{10.13.18}$$

To obtain an equation for T we simply multiply equation (10.13.12) through by V, whereupon we see that

$$T = V + T\mathscr{G}_0 V. \tag{10.13.19}$$

In general V will be the sum of the potentials of many scattering centres:

$$V(\mathbf{r}) = \sum_n V_n(\mathbf{r}), \tag{10.13.20}$$

in which case it is helpful to discuss the total scattering in terms of the scattering from the individual sites. If we write T as the sum

$$T = \sum_n Q_n, \tag{10.13.21}$$

equation (10.13.19) suggests that we define Q_n by

$$Q_n = V_n + Q_n \mathscr{G}_0 V. \tag{10.13.22}$$

On the other hand, the scattering from a single isolated centre is given by

$$T_n = V_n + T_n \mathscr{G}_0 V_n. \tag{10.13.23}$$

To connect Q_n and T_n we observe that the formal solution of equation (10.13.22) is

$$Q_n = V_n \frac{1}{1-\mathscr{G}_0 V}, \tag{10.13.24}$$

whereas equation (10.13.19) gives

$$T = V \frac{1}{1-\mathscr{G}_0 V}. \tag{10.13.25}$$

Thus we see that

$$Q_n \mathscr{G}_0 V = V_n \frac{1}{1-\mathscr{G}_0 V} \mathscr{G}_0 V = V_n \mathscr{G}_0 T, \tag{10.13.26}$$

so that equation (10.13.22) may also be written

$$Q_n = V_n + V_n \mathscr{G}_0 T = V_n + V_n \mathscr{G}_0 \sum_m Q_m. \tag{10.13.27}$$

Taking the Q_n on the right-hand side over to the left we see that

$$(1 - V_n \mathscr{G}_0) Q_n = V_n + V_n \mathscr{G}_0 \sum_{m \neq n} Q_m \tag{10.13.28}$$

or

$$Q_n = \frac{1}{1-V_n \mathscr{G}_0} V_n + \frac{1}{1-V_n \mathscr{G}_0} V_n \mathscr{G}_0 \sum_{m \neq n} Q_m. \tag{10.13.29}$$

But since from equation (10.13.23)

$$T_n = \frac{1}{1 - V_n \mathscr{G}_0} V_n, \tag{10.13.30}$$

equation (10.13.29) is just

$$Q_n = T_n + T_n \sum_{m \neq n} \mathscr{G}_0 Q_m \tag{10.13.31}$$

and Q_n appears as an operator giving the contribution of the nth scatterer to the scattered wave in the presence of the other scatterers. To make this interpretation clear let us write

$$|\psi_n\rangle = \mathscr{G}_0 Q_n |\chi\rangle, \tag{10.13.32}$$

so that the total scattered wave is

$$|\psi\rangle = \sum_n |\psi_n\rangle. \tag{10.13.33}$$

Then by equation (10.13.31),

$$|\psi_n\rangle = \mathscr{G}_0 T_n |\chi\rangle + \mathscr{G}_0 T_n \sum_{m \neq n} |\psi_m\rangle \tag{10.13.34}$$

and, defining $|\chi_n\rangle$ through

$$|\psi_n\rangle = \mathscr{G}_0 T_n |\chi_n\rangle = \mathscr{G}_0 Q_n |\chi\rangle, \tag{10.13.35}$$

we find

$$|\chi_n\rangle = |\chi\rangle + \sum_{m \neq n} \mathscr{G}_0 T_m |\chi_m\rangle. \tag{10.13.36}$$

It is evident that $|\psi_n\rangle$ may be interpreted as that part of the scattered wave contributed by the centre at n, and by equations (10.13.35) and (10.13.36) the wave $|\psi_n\rangle$ incident upon this centre is composed of the original incident wave plus all waves scattered from other centres.

(b) *Averaging in disordered systems.* The Green function operator will evidently depend on the positions $\mathbf{R}_n$ of the scatterers. On the other hand, we know, almost by definition, that the electronic properties of a disordered system do not depend on the particular arrangement of scatterers in a given sample. It therefore suffices to obtain $\langle \mathscr{G} \rangle$, the average over all possible configurations, weighted, of course, with the probabilities of them occurring.

With this in mind, we can write down a sequence of equations for partially averaged operators. While the approximations we shall discuss in detail later are clear in their physical interpretation, this sequence may be referred to if one wishes to see exactly what has been ignored in any approximate solution.

Let $g(\mathbf{R}_1, \mathbf{R}_2)$ be the pair correlation function, that is, the probability that if there is a scatterer at $\mathbf{R}_1$ there is one at $\mathbf{R}_2$; let $g^{(3)}(\mathbf{R}_1, \mathbf{R}_2, \mathbf{R}_3)$ be the probability that if there are scatterers at $\mathbf{R}_1$ and $\mathbf{R}_2$ there is one at $\mathbf{R}_3$; $g^{(4)}$ is similarly defined, and so on. We also define partially averaged operators: $\langle Q_n(\mathbf{R})\rangle$ is the scattering from the nth centre at $\mathbf{R}$ averaged over the positions of all other particles, $\langle Q_n(\mathbf{R}, \mathbf{R}_1)\rangle$ is the averaged scattering operator for the nth centre at $\mathbf{R}$ given that there is a centre at $\mathbf{R}_1$, but information on no other particles, and so on. Then the totally averaged operator is

$$\langle Q_n\rangle = \frac{1}{\mathscr{V}}\int\langle Q_n(\mathbf{R})\rangle\, d\mathbf{R}, \tag{10.13.37}$$

where $\mathscr{V}$ is the volume of the system, while

$$\langle Q_n(\mathbf{R})\rangle = \int\langle Q_n(\mathbf{R};\, \mathbf{R}_1)\rangle\, g(\mathbf{R}, \mathbf{R}_1)\, d\mathbf{R}_1, \tag{10.13.38}$$

$$\langle Q_n(\mathbf{R}, \mathbf{R}_1)\rangle = \int\langle Q_n(\mathbf{R};\, \mathbf{R}_1 \mathbf{R}_2)\rangle\, g^{(3)}(\mathbf{R}, \mathbf{R}_1 \mathbf{R}_2)\, d\mathbf{R}_2 \tag{10.13.39}$$

and so on.

It now follows from equation (10.13.31) that we may write

$$\langle Q_n(\mathbf{R})\rangle = t_n(\mathbf{R}) + t_n(\mathbf{R}) \sum_{m\neq n} \mathscr{G}_0 \int d\mathbf{R}_m \langle Q_m(\mathbf{R}_m, \mathbf{R})\rangle\, g(\mathbf{R}_m, \mathbf{R}), \tag{10.13.40}$$

where

$$\langle Q_n(\mathbf{R}, \mathbf{R}_m)\rangle = t(\mathbf{R}_m) + t(\mathbf{R}_m) \sum_{l\neq n} \mathscr{G}_0 \int d\mathbf{R}_i \langle Q_l(\mathbf{R}_l;\, \mathbf{R}_m, \mathbf{R})\rangle\, g^{(3)}(\mathbf{R}_m, \mathbf{R}, \mathbf{R}_i), \tag{10.13.41}$$

with corresponding equations for the higher-order partially averaged operators. Under the assumpation that the scattering centres are all identical we have replaced T_n by $t(\mathbf{R}_n)$. This is appropriate for a disordered monatomic solid.

(c) *Complete disorder.* (i) *Average t-matrix approximation*: We shall set $V_m = V_n$ for all m, and

$$V_n = V_n(\mathbf{r} - \mathbf{R}_n). \tag{10.13.42}$$

To calculate Q_n from equation (10.13.31) we must evidently approximate to $\Sigma_{m\neq n} Q_m$, when averaging. We shall write

$$\langle \sum_{m\neq n} Q_m\rangle \simeq (N-1)\langle Q\rangle, \tag{10.13.43}$$

an approximation that neglects the short-range order.

Equation (10.13.31) now becomes, neglecting correlations between n and m,

$$\langle Q(\mathbf{R})\rangle = \langle t(\mathbf{R})\rangle + \langle t(\mathbf{R})\rangle \langle \sum_{m\neq n} Q_m\rangle, \tag{10.13.44}$$

which may be combined with (10.13.43) to obtain

$$\langle Q\rangle = \langle t\rangle + \langle t\rangle (N-1)\mathscr{G}_0\langle Q\rangle, \tag{10.13.45}$$

that is,

$$\langle Q\rangle = \frac{\langle t\rangle}{1-N\langle t\rangle \mathscr{G}_0}, \tag{10.13.46}$$

where we have neglected 1 in comparison with N.

From equation (10.13.21) we have

$$\langle T\rangle = \frac{N\langle t\rangle}{1-N\langle t\rangle \mathscr{G}_0} \tag{10.13.47}$$

and, since on combining equations (10.13.10) and (10.13.17) we get

$$\mathscr{G} = \mathscr{G}_0 + \mathscr{G}_0 T \mathscr{G}_0, \tag{10.13.48}$$

$$\langle \mathscr{G}\rangle = \frac{\mathscr{G}_0}{1-N\mathscr{G}_0\langle t\rangle}. \tag{10.13.49}$$

Now equation (10.13.19) yields

$$t(\mathbf{R}) = \frac{V_n(\mathbf{R})}{1-\mathscr{G}_0 V_n(\mathbf{R})}, \tag{10.13.50}$$

so that it is evident from equation (10.13.42) that

$$\langle \mathbf{k}_1 | t(\mathbf{R}) | \mathbf{k}_0\rangle = t(\mathbf{k}_1, \mathbf{k}_0)\, e^{i(\mathbf{k}_0-\mathbf{k}_1).\mathbf{R}}, \tag{10.13.51}$$

where $t(\mathbf{k}_1, \mathbf{k}_0)$ is a matrix element for scattering from the origin, obeying [cf. equation (10.13.23)]

$$t(\mathbf{k}_1\, \mathbf{k}_0) = V_{\mathbf{k}_1,\mathbf{k}_0} + \sum_{\mathbf{k}_2} V_{\mathbf{k}_1,\mathbf{k}_2}\, \mathscr{G}_0(\mathbf{k}_2)\, t(\mathbf{k}_2, \mathbf{k}_0). \tag{10.13.52}$$

Equation (10.13.52) yields

$$\langle \mathbf{k} | \langle t(\mathbf{R})\rangle | \mathbf{k}_0\rangle = \langle \mathbf{k} | \int \frac{t(\mathbf{R})}{\mathscr{V}}\, d\mathbf{R} | \mathbf{k}_0\rangle = t(\mathbf{k}, \mathbf{k}_0)\, \delta_{\mathbf{k},\mathbf{k}_0}, \tag{10.13.53}$$

so that in the present approximation

$$\langle Q(\mathbf{k}\mathbf{k}_0)\rangle = \frac{\langle t(\mathbf{k}, \mathbf{k})\rangle}{1-Nt(\mathbf{k}, \mathbf{k})}\, \delta_{\mathbf{k},\mathbf{k}_0} \tag{10.13.54}$$

and

$$\langle \mathscr{G}(\mathbf{k}_0, \omega)\rangle = \frac{1}{\omega - H_0 - Nt(\mathbf{k}_0, \mathbf{k}_0)}, \qquad (10.13.55)$$

where $t(\mathbf{k}, \mathbf{k})$ is the forward scattering amplitude for a single centre. This is the average t-matrix approximation.

(ii) *Coherent potential approximation*: Equation (10.13.55) is written for the diagonal element only, for the off-diagonal elements of $\mathscr{G}$ are zero—after averaging, only that part of the scattered wave which is coherent with the incident wave survives. Consistent with this is the presence of $t(\mathbf{k}, \mathbf{k})$, representing forward scattering, in the denominator of equation (10.13.55). On the other hand, $t(\mathbf{k}, \mathbf{k})$ relates to the scattering of free electrons from an isolated centre, with no account taken of the fact that the electrons are really travelling through a medium, namely, the entire system of scatterers. This suggests we improve the approximation by replacing equation (10.13.55) by (cf. Faulkner, 1970)

$$\mathscr{G}_{\mathrm{M}} = \frac{1}{\omega - H_0 - (N/\mathscr{V})\, W_{\mathbf{p}}}, \qquad (10.13.56)$$

where $W_{\mathbf{p}}$ represents forward scattering due to a scatterer *in the medium described by* $\mathscr{G}_{\mathrm{M}}$, in other words,

$$W_{\mathbf{k}}/\mathscr{V} = \hat{t}_{\mathbf{k},\mathbf{k}}, \qquad (10.13.57)$$

where [cf. equation (10.13.52)]

$$\hat{t}_{\mathbf{k}\mathbf{k}_1} = V_{\mathbf{k}_1\mathbf{k}_0} + \sum_{\mathbf{k}_2} V_{\mathbf{k}_1\mathbf{k}_2}\mathscr{G}_{\mathrm{M}}(\mathbf{k}_2)\, \hat{t}(\mathbf{k}_2, \mathbf{k}_0). \qquad (10.13.58)$$

We can alternatively fix $W_{\mathbf{p}}$ by a less intuitive criterion which is at first sight quite different. To calculate $\mathscr{G}$ from $\mathscr{G}_0$ we note that equation (10.13.7) can be written as

$$\mathscr{G} = \frac{1}{\omega - H_0 - (N/\mathscr{V})\, W_{\mathbf{p}} - \tilde{V}(\mathbf{r})}, \qquad (10.13.59)$$

where

$$\tilde{V}(\mathbf{r}) = \sum_n \tilde{V}_n(\mathbf{R}_n), \qquad (10.13.60)$$

with

$$\tilde{V}_n(\mathbf{R}_n) = V_n(\mathbf{r} - \mathbf{R}_n) - W_{\mathbf{p}}/\mathscr{V}. \qquad (10.13.61)$$

The Green function operator may now be seen to obey the relation

$$\mathscr{G} = \mathscr{G}_{\mathrm{M}} + \mathscr{G}_{\mathrm{M}}\tilde{V}\mathscr{G} = \mathscr{G}_{\mathrm{M}} + \mathscr{G}_{\mathrm{M}}\tilde{T}\mathscr{G}_{\mathrm{M}}, \qquad (10.13.62)$$

with

$$\tilde{T} = \tilde{V} + \tilde{V}\mathscr{G}_{\mathrm{M}}\tilde{T}. \qquad (10.13.63)$$

Making the same approximations as before we have, in correspondence with equation (10.13.47),

$$\langle \tilde{T} \rangle = \frac{N\langle \tilde{t} \rangle}{1 - N\langle \tilde{t} \rangle \mathscr{G}_M}, \tag{10.13.64}$$

where

$$\tilde{t}(\mathbf{R}) = \tilde{V}_n(\mathbf{R}) + \tilde{V}_n(\mathbf{R})\, \mathscr{G}_M\, \tilde{t}(\mathbf{R}). \tag{10.13.65}$$

Equation (10.13.57) yields

$$\langle \mathscr{G} \rangle = \mathscr{G}_M + \mathscr{G}_M \langle \tilde{T} \rangle \mathscr{G}_M \tag{10.13.66}$$

and if $\mathscr{G}_M$ actually is our approximation to $\langle \mathscr{G} \rangle$, we must evidently set $\langle \tilde{T} \rangle = 0$ which in turn implies, from equation (10.13.64),

$$\langle \tilde{t} \rangle = 0, \tag{10.13.67}$$

that is, the averaged scattering from an effective ion of potential $\tilde{V}_n(\mathbf{r}-\mathbf{R})$ must be zero.

Equation (10.13.67) is the basic equation of the coherent potential method as originally formulated by Soven (1967). We introduce $W_\mathbf{p}$ in equation (10.13.56) as a "coherent potential", constant in space but possibly dependent on the momentum operator $\mathbf{p}$, to replace $V(\mathbf{r})$. We fix $W_\mathbf{p}$ by the condition (10.13.67), which is the requirement that the averaged scattering, from $V_n(\mathbf{r}-\mathbf{R}) - W_\mathbf{p}/\mathscr{V}$, is zero in the medium described by $\mathscr{G}_M$. This also ensures $\langle \mathscr{G} \rangle = \mathscr{G}_M$ in the approximation of complete disorder.

In fact, in the limit as the volume $\mathscr{V} \to \infty$ but the density $N/\mathscr{V}$ of scatterers remains finite, equation (10.13.67) is equivalent to equations (10.13.57) and (10.13.58). To see this we first note that the scattering from a single centre $\tilde{V}_n$ immersed in the medium described by $\mathscr{G}_M$ is given by

$$\tilde{\mathscr{G}}_n = \mathscr{G}_M + \mathscr{G}_M \tilde{V}_n \tilde{\mathscr{G}}_n \tag{10.13.68}$$

or

$$\tilde{\mathscr{G}}_n = \mathscr{G}_M + \mathscr{G}_M \tilde{t} \mathscr{G}_M. \tag{10.13.69}$$

Alternatively, the problem of the scattering by $\tilde{V}_n$ may be solved by including $-W_p/\mathscr{V}$ from the outset: we define

$$\tilde{\mathscr{G}}_M = \frac{1}{\omega - H_0 - [(N-1)/\mathscr{V}]\, W_\mathbf{p}} = \frac{\mathscr{G}_M}{1 + (W_p/\mathscr{V})\, \mathscr{G}_M}, \tag{10.13.70}$$

whereupon we see that

$$\hat{\mathscr{G}}_n = \hat{\mathscr{G}}_M + \hat{\mathscr{G}}_M V_n \tilde{\mathscr{G}}_n. \tag{10.13.71}$$

If we then introduce a scattering $\tilde{t}'$ such that

$$\tilde{t}' \hat{\mathscr{G}}_M = V_n \tilde{\mathscr{G}}_n, \tag{10.13.72}$$

this operator will obey the equation

$$\hat{t}' = V_n + \frac{V_n \mathscr{G}_M \hat{t}'}{1+(W_p/\mathscr{V})\mathscr{G}_M}. \tag{10.13.73}$$

Now

$$\hat{t}\mathscr{G}_M = \left(V_n - \frac{W_p}{\mathscr{V}}\right)\tilde{\mathscr{G}}_n = V_n \tilde{\mathscr{G}}_n - \frac{W_p}{\mathscr{V}}\tilde{\mathscr{G}}_n, \tag{10.13.74}$$

that is,

$$\hat{t}\mathscr{G}_M = \hat{t}'\hat{\mathscr{G}}_M - \frac{W_p}{\mathscr{V}}\tilde{\mathscr{G}}_n. \tag{10.13.75}$$

Since we see from equations (10.13.67) and (10.13.69) that $\langle\tilde{\mathscr{G}}_n\rangle = \mathscr{G}_M$, we find, on averaging equation (10.13.75),

$$0 = \langle\hat{t}\rangle\mathscr{G}_M = \langle\hat{t}'\rangle\hat{\mathscr{G}}_M - \frac{W_p}{\mathscr{V}}\mathscr{G}_M. \tag{10.13.76}$$

As $\mathscr{V} \to \infty$ we may set $\hat{\mathscr{G}}_M = \mathscr{G}_M$ in this equation, to find

$$\langle\hat{t}'\rangle\mathscr{G}_M = \frac{W_p}{\mathscr{V}}\mathscr{G}_M. \tag{10.13.77}$$

Further, in the same limit, equation (10.13.73) for $\hat{t}'$ becomes equivalent to equation (10.13.58) for $\hat{t}$, and so equation (10.13.57) is established.

(d) *Introduction of short-range order.* Suppose we approximate the term $\sum_{m\neq n} Q_m$ in the right-hand side of equation (10.13.31) by

$$\int Q(\mathbf{R}_m)\, g(\mathbf{R}_m - \mathbf{R}_n)\, d\mathbf{R}_m,$$

where we have written the pair correlation function $g(\mathbf{R}_m, \mathbf{R}_n)$ as $g(\mathbf{R}_m - \mathbf{R}_n)$ because the system is assumed to be homogeneous.

This approximation is termed the quasi-crystalline approximation since it is exact for a perfect crystal, when

$$g(\mathbf{R}) = \left\langle \sum_{m\neq n} \delta(\mathbf{R} - \mathbf{R}_n + \mathbf{R}_m) \right\rangle = \sum_{m\neq n} \delta(\mathbf{R} - \mathbf{R}_n + \mathbf{R}_m) \tag{10.13.78}$$

contains all information about the distribution of scatterers. In general it has the defect that we ignore the fact that placing a scatterer at $\mathbf{R}_m$ determines to some extent the distribution of the other scatterers through three-body and higher-order correlation functions. However, the method is the first significant step in taking account of short-range order, and $g(\mathbf{R})$ is obtainable from experiment, as we saw in Chapter 5.

We therefore approximate equation (10.13.31) by

$$Q_n(\mathbf{R}) = t(\mathbf{R}) + t(\mathbf{R})\mathcal{G}_0 \int Q(\mathbf{R}')g(\mathbf{R}-\mathbf{R}')\,d\mathbf{R}'. \qquad (10.13.79)$$

Comparison of this equation with equation (10.13.38) shows us that we are strictly calculating $\langle Q_n(\mathbf{R})\rangle$ under the assumption that

$$\langle Q_m(\mathbf{R}_m;\, \mathbf{R}_n)\rangle = \langle Q_m(\mathbf{R}_m)\rangle. \qquad (10.13.80)$$

For a homogeneous system we would expect that an equation analogous to equation (10.3.51), viz.

$$\langle \mathbf{k}'|\langle Q_n(\mathbf{R})\rangle|\mathbf{k}\rangle = Q(\mathbf{k}',\mathbf{k})\,e^{i(\mathbf{k}'-\mathbf{k})\cdot\mathbf{R}}, \qquad (10.13.81)$$

obtains, and in fact if equation (10.13.79) is solved by iteration it becomes obvious that this is so for the present approximation. Then

$$\langle \mathcal{G}(\mathbf{k},\omega)\rangle = \mathcal{G}_0 + \mathcal{G}_0\langle T\rangle \mathcal{G}_0 = \mathcal{G}_0(\mathbf{k}) + N\mathcal{G}_0(\mathbf{k})\,Q(\mathbf{k},\mathbf{k})\,\mathcal{G}_0(\mathbf{k}). \qquad (10.13.82)$$

We also obtain from equation (10.13.79) the result

$$Q(\mathbf{k}_1,\mathbf{k}_0) = t(\mathbf{k}_1,\mathbf{k}_0) + (N-1)\sum_{\mathbf{k}_2} t(\mathbf{k}_1,\mathbf{k}_2)\,\mathcal{G}_0(\mathbf{k}_2)\,Q(\mathbf{k}_2,\mathbf{k}_0)\,g(\mathbf{k}_0-\mathbf{k}_2), \qquad (10.13.83)$$

where

$$g(\mathbf{k}) = \int e^{i\mathbf{k}\cdot\mathbf{R}}\,g(\mathbf{R})\,d\mathbf{R}. \qquad (10.13.84)$$

Coherent potential approximation: Let us again write $\mathcal{G}$ in the form (10.13.56) and attempt to determine W_{p} within the quasi-crystalline approximation. Now, since we will have

$$\langle \mathcal{G}\rangle = \mathcal{G}_{\mathrm{M}} + N\mathcal{G}_{\mathrm{M}}\langle \tilde{Q}(\mathbf{k}\mathbf{k})\rangle\,\mathcal{G}_{\mathrm{M}}, \qquad (10.13.85)$$

it is evident from the equation for $\tilde{Q}$ corresponding to (10.13.83), namely

$$\tilde{Q}(\mathbf{k},\mathbf{k}_0) = \tilde{t}(\mathbf{k},\mathbf{k}_0) + (N-1)\sum_{\mathbf{k}_1} \tilde{t}(\mathbf{k},\mathbf{k}_1)\,\mathcal{G}_{\mathrm{M}}(\mathbf{k}_1)\,\tilde{Q}(\mathbf{k}_1\,\mathbf{k}_0)\,g(\mathbf{k}_0-\mathbf{k}_1), \qquad (10.13.86)$$

that condition (10.13.67) will no longer ensure that

$$\langle \mathcal{G}\rangle = \mathcal{G}_{\mathrm{M}}. \qquad (10.13.87)$$

It will be seen from equation (10.13.62) that the requirement (10.13.87) is quite generally fulfilled provided

$$\langle \tilde{T}\rangle = 0 \qquad (10.13.88)$$

and on writing [cf. equation (10.13.21)]

$$\tilde{T} = \sum_n \tilde{Q}_n \tag{10.13.89}$$

we must also have

$$\langle \tilde{Q}_n \rangle = 0 \quad (\text{all } n) \tag{10.13.90}$$

if the scatterers are identical. Equation (10.13.90) is the general condition replacing equation (10.13.31). Then, since

$$\tilde{Q}_n = \tilde{V}_n + \sum_m \tilde{V}_n \mathscr{G}_{\mathrm{M}} \tilde{Q}_m \tag{10.13.91}$$

with $\tilde{V}_n$ given by equation (10.13.6), we have

$$0 = \langle V_n \rangle - \frac{W_p}{\mathscr{V}} + \sum_m \langle V_n \mathscr{G}_{\mathrm{M}} \tilde{Q}_m \rangle - \sum_m \frac{W_p}{\mathscr{V}} \mathscr{G}_{\mathrm{M}} \langle \tilde{Q}_m \rangle. \tag{10.13.92}$$

Thus we can write, as a quite general condition,

$$\frac{W_p}{\mathscr{V}} = \langle V_n \rangle + \sum_m \langle V_n \mathscr{G}_{\mathrm{M}} \tilde{Q}_m \rangle. \tag{10.13.93}$$

Although the procedure is clear in principle, that one chooses a W_p to calculate $\tilde{Q}_m$ and self-consists by use of equation (10.13.93), in practice the configurational average must create considerable difficulty, even within the quasi-crystalline approximation. A discussion of the combination of the coherent potential method with the quasi-crystalline approximation is given by Gyorffy (1970) (see also Thornton, 1971; Korringa and Mills, 1972). However, in the random case, detailed progress can be made as we see below.

10.13.3 Model of binary alloys

We shall now go on to discuss the coherent potential method as applied to a model of binary alloys (Velický, Kirkpatrick and Ehrenreich, 1968).

The model, which is closely related to the tight-binding approximation, assumes a single band. A single atomic orbital $|n\rangle$ is associated with each site n, and a single band would result in the case of a pure crystal. Of course, two sub-bands may occur in an alloy under certain conditions, which we discuss below.

The Hamiltonian of the model is taken to be

$$\begin{aligned} H &= \sum_n |n\rangle \varepsilon_n \langle n| + \sum_{n \neq m} |n\rangle t_{mn} \langle m| \\ &\equiv D + W. \end{aligned} \tag{10.13.94}$$

D and W refer to splitting the model Hamiltonian H into a diagonal part D and an off-diagonal part W with respect to a Wannier representation. The matrix elements of the Hamiltonian will, in general, depend on the arrangement of the two types of atom, A and B say, in the crystal.

The assumptions on which the model is based, which are appropriate when the wave functions are highly localized and the atomic potentials on A and B are not too dissimilar, are:

(1) The energies ε_n can be viewed as atomic levels, which have one of the two possible values ε_A and ε_B depending on the type of atom A or B at the site.

(2) t_{mn}, referred to in Chapter 4 as hopping terms, are assumed independent of the composition of the alloy. $\varepsilon_A + W$ can be regarded then as the Hamiltonian of the pure A crystal, and similarly for $\varepsilon_B + W$.

It then follows that the operator W is diagonal in the Bloch representation $|\mathbf{k}\rangle$ and the matrix elements are given by

$$\langle \mathbf{k}|W|\mathbf{k}'\rangle = \delta_{\mathbf{kk}'} \sum_n t_{0n}\, e^{i\mathbf{k}.\mathbf{R}_n} \equiv \delta_{\mathbf{kk}'}\, ws(k), \tag{10.13.95}$$

where w is half the band-width and $s(k)$ is the dispersion relation in dimensionless units. In the same way it is useful to measure ε_A and ε_B in dimensionless units by writing

$$\varepsilon_A = \tfrac{1}{2}w\delta, \quad \varepsilon_B = -\tfrac{1}{2}w\delta \tag{10.13.96}$$

and hence

$$\delta = (\varepsilon_A - \varepsilon_B)/w. \tag{10.13.97}$$

This is a crucial parameter in the theory therefore, which, together with c, the concentration, serves to characterize the problem. Usually below we take $w = 1$ for convenience.

Velický and colleagues have studied the density of states as functions of δ and c in the various régimes of interest. From symmetry, only the quadrant $0 \leqslant c \leqslant \frac{1}{2}$, $\delta \geqslant 0$ need be considered. Then we can usually consider three physical limits:

(i) The line $c = 0$ which corresponds to the pure B crystal, and its neighbourhood to B as host crystal, A being impurities.

(ii) The line $\delta = 0$, corresponding to the pure crystal described by the Hamiltonian W. The vicinity of this line is the virtual crystal region, with a single band which is very like that of the pure crystal.

(iii) The case $\delta = (\varepsilon_A - \varepsilon_B)/w \to \infty$. This can be achieved (a) by letting the band width $w \to 0$, keeping $\varepsilon_A - \varepsilon_B$ finite, (b) by fixing w but letting the level separation tend to infinity.

Limit (a) is elementary because the level spectrum consists simply of the atomic levels ε_A and ε_B. Limit (b) is not simple, however, for, no matter how far the sub-bands are separated, there are always clusters of like atoms which produce molecular or band-like splittings in the energy spectrum which are of the order of w (cf. Harris and Lange, 1967).

(a) *Single-band model: single-site approximation.* We can see from section 10.13.2 that the coherent potential method can be used to define from the average Green function $\langle G\rangle$ a medium Hamiltonian, H_m say, through

$$\mathscr{G}_m = \frac{1}{\omega - H_m}, \tag{10.13.98}$$

where [see equations (10.13.66) and (10.13.67)], $\mathscr{G}_m \equiv \langle G\rangle$. In terms of the original Hamiltonian (10.13.94), we then define a function $u(\omega)$ by the equation

$$H_m = W + \sum_n |n\rangle\, u(\omega)\, \langle n| = W + u(\omega)\,\mathbf{1}, \tag{10.13.99}$$

where this equation also defines the single-site approximation. The self-energy we require is clearly related to u and is $\mathbf{k}$-independent in this model. It will be useful to define solutions for the pure crystal with Hamiltonian W: namely

$$\mathscr{G}^0(\omega) = (\omega - W)^{-1}, \quad \mathscr{G}^0(k, \omega) = [\omega - s(\mathbf{k})]^{-1}, \tag{10.13.100}$$

$$F^0(\omega) = \int_{-\infty}^{\infty} dE(\omega - E)^{-1} n^0(E) = N^{-1} \sum_{\mathbf{k}} [\omega - s(\mathbf{k})]^{-1}, \tag{10.13.101}$$

$n^0(E)$ being the density of states.

The Green function $\mathscr{G}_m$ is given by, from equations (10.13.98), (10.13.99) and (10.13.100),

$$\begin{aligned} \mathscr{G}_m &= [\omega - u(\omega) - W]^{-1} \\ &= \mathscr{G}^0[\omega - u(\omega)]. \end{aligned} \tag{10.13.102}$$

We note also that

$$\langle 0|\mathscr{G}_m(\omega)|0\rangle = F^0[\omega - u(\omega)] \equiv F(\omega), \tag{10.13.103}$$

the last two equations expressing the results for the medium in terms of the perfect crystal with Hamiltonian W. It is this feature that makes the present model tractable.

With H_m as we have defined it above, it follows that

$$H - H_m = \sum_n |n\rangle\, [\varepsilon_n - u(\omega)]\, \langle n|. \tag{10.13.104}$$

But we want to express $H - H_{\mathrm{m}}$ as a sum of single-centre terms:

$$H - H_{\mathrm{m}} = \sum_n \tilde{V}_n. \tag{10.13.105}$$

Evidently we can choose

$$\tilde{V}_n = |n\rangle [\varepsilon_n - u(\omega)] \langle n| = |n\rangle v_n \langle n|. \tag{10.13.106}$$

From equations (10.13.65) and (10.13.103) we can show, by repeated iteration or otherwise, that

$$\tilde{t}_n(\omega) = |n\rangle v_n [1 - v_n F(\omega)]^{-1} \langle n|, \tag{10.13.107}$$

where as before $\tilde{t}_n(\omega)$ represents the scattering off a single-site n, but now in the medium. We can now perform the average over the random configuration of impurity atoms, to obtain

$$\langle \tilde{t}_n(\omega) \rangle = |n\rangle \left[\frac{c(\varepsilon_{\mathrm{A}} - u)}{1 - (\varepsilon_{\mathrm{A}} - u) F} + \frac{(1-c)(\varepsilon_{\mathrm{B}} - u)}{1 - (\varepsilon_{\mathrm{B}} - u) F} \right] \langle n|. \tag{10.13.108}$$

From the defining equation (10.13.67) of the coherent potential method, we must put this quantity $\langle \tilde{t}_n(\omega) \rangle = 0$. It is clear from equations (10.13.98) and (10.13.99) that $u(\omega)$ is the self-energy $\Sigma(\omega)$ and hence from equation (10.13.108) we find

$$\Sigma(\omega) = \frac{(2c-1)\,\delta}{2} - [\varepsilon_{\mathrm{A}} - \Sigma(\omega)]\, F(\omega)\, [\varepsilon_{\mathrm{B}} - \Sigma(\omega)], \tag{10.13.109}$$

where we have used equation (10.13.96) with $w = 1$. We recall that $F(\omega)$ is calculable from the density of states in the perfect crystal with Hamiltonian W.

But to get an explicit solution for $\Sigma(\omega)$, we have to specify this density of states. We shall assume a model for this proposed initially by Hubbard and work out its consequences in some detail.

(b) *Model density of states.* This density of states is defined by

$$\left. \begin{aligned} n^0(E) &= \left(\frac{2}{\pi w^2}\right) (w^2 - E^2)^{\frac{1}{2}} \quad |E| \leqslant w, \\ n^0(E) &= 0 \quad |E| > w, \end{aligned} \right\} \tag{10.13.110}$$

which is such that

$$n^0(E) = n^0(-E).$$

The Koster–Slater function $F^0(\omega)$ defined by equation (10.13.101) is, for this model, readily evaluated as

$$F^0(\omega) = \frac{2}{w^2}[\omega-(\omega^2-w^2)^{\frac{1}{2}}]. \tag{10.13.111}$$

Then it follows that we have to solve a cubic equation for $\Sigma(\omega)$; namely

$$[\omega+\varepsilon]\Sigma^3-[\omega\varepsilon+\tfrac{1}{4}(\delta^2-1)]\Sigma^2 + [\tfrac{1}{4}\delta^2(\omega+\varepsilon)+\tfrac{1}{2}\varepsilon]\Sigma+\tfrac{1}{4}[\delta^2(\omega\varepsilon+\tfrac{1}{4}\delta^2)+\varepsilon^2] = 0, \tag{10.13.112}$$

where

$$\varepsilon = \left(\frac{2c-1}{2}\right)\delta. \tag{10.13.113}$$

This can be solved for $\Sigma(\omega)$ as a function of the two parameters δ and c, the band width w being taken as the unit of energy.

It is clear that the various alloy properties can be calculated and we shall conclude this discussion by referring to the density of states calculated by Velický and co-workers and shown in Figure 10.19.

Figure 10.19(a) shows a case for $\delta = \frac{1}{4}$, and we are seeing behaviour characteristic of the virtual crystal limit. With increasing δ, Figure 10.19(b), corresponding in fact to $\delta = \frac{1}{2}$, shows some distortion of the band shape. In Figure 10.19(c), the levels ε_A and ε_B are sufficiently well separated so that an impurity band splits off the upper edge of the host band. The width and height of the impurity sub-band turn out to be proportional to $c^{\frac{1}{2}}$ for low concentration and this remains approximately true until the bands merge.

It is by no means clear that a model of this kind will work quantitatively in any real binary alloy. It does have the merit, however, that it can be pushed through in detail, and it shows that the coherent potential method nicely reproduces exact limiting forms corresponding to the dilute alloy, the virtual crystal and the atomic limit. The model is not giving the correct results for the case $\delta \geqslant 1$ and w finite however.

Finally we remark that the coherent potential approximation, combined with appropriate potentials for Cu and Ni atoms, has been used by Stocks, Williams and Faulkner (1971) to obtain the density of states of the Cu–Ni alloy system for a range of concentration across the entire alloy diagram. Their calculations show, in a very beautiful way, the main features of the densities of states as inferred from photo-emission and soft X-ray experiments.

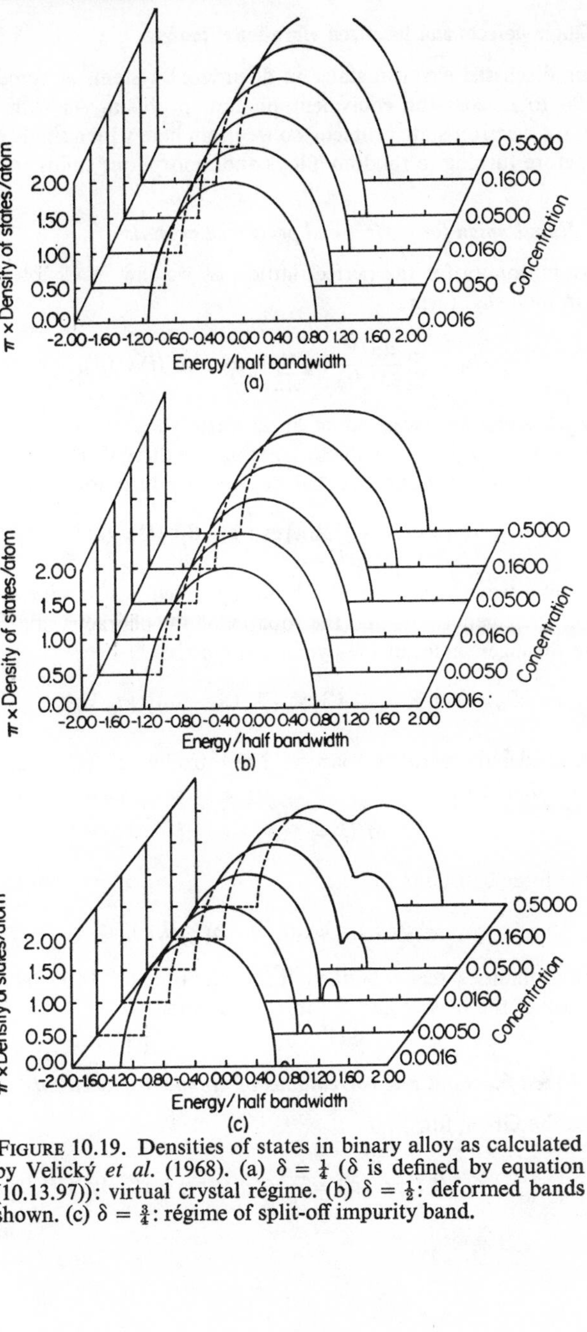

FIGURE 10.19. Densities of states in binary alloy as calculated by Velický *et al.* (1968). (a) $\delta = \frac{1}{4}$ (δ is defined by equation (10.13.97)): virtual crystal régime. (b) $\delta = \frac{1}{2}$: deformed bands shown. (c) $\delta = \frac{3}{4}$: régime of split-off impurity band.

10.14 Single defects and localized vibrational modes

Having discussed electron states in disordered systems at some length, it is natural to discuss the equivalent phonon problem. As with the single defect in the electron-state problem, so we begin here with a single vibrational defect, before turning to random alloys and amorphous solids.

10.14.1 Hamiltonian for perfect and perturbed crystals

The Hamiltonian for the perfect lattice, as we saw in Chapter 3, can be written in the usual form

$$H_0 = \sum_{\alpha l} \frac{p_\alpha(l)^2}{2M_\alpha(l)} + \tfrac{1}{2} \sum_{\alpha\beta ll'} \phi_{\alpha\beta}(l,l')\, u_\alpha(l)\, u_\beta(l'), \tag{10.14.1}$$

where $u_\alpha(l)$ is the displacement of an atom in the lth unit cell around $\mathbf{R}_l$. The subscript α can take $3s$ values, denoting the three Cartesian coordinates of each of s atoms in the cell. As before, we introduce normal coordinates

$$d_j(\mathbf{k}) = N^{-\frac{1}{2}} \sum_{\alpha l} \varepsilon_\alpha^j(\mathbf{k}) \exp(i\mathbf{k}.\mathbf{R}_l)\, M_\alpha^{\frac{1}{2}} u_\alpha(l) \tag{10.14.2}$$

specified by a wave vector $\mathbf{k}$ and a branch j which has $3s$ values. $\varepsilon_\alpha^j(\mathbf{k})$ are, as before, the eigenvectors and the squares of the characteristic frequencies $\omega_j(\mathbf{k})$ are the eigenvalues of the dynamical matrix

$$D_{\alpha\beta}(\mathbf{k}) = \sum_{l'} \phi_{\alpha\beta}(l,l') \exp[i\mathbf{k}.(\mathbf{R}_l - \mathbf{R}_{l'})]\,(M_\alpha M_\beta)^{-\frac{1}{2}}. \tag{10.14.3}$$

Now consider the perturbed lattice. The introduction of the defect causes changes in $M_\alpha(l)$ to

$$M'_\alpha(l) = M_\alpha[1 - \varepsilon_\alpha(l)] \tag{10.14.4}$$

and in the force constants to

$$\phi'_{\alpha\beta}(l,l') = \phi_{\alpha\beta}(l,l') - \Delta\phi_{\alpha\beta}(l,l'). \tag{10.14.5}$$

The Hamiltonian is then obtained from equation (10.14.1) by putting in these values of the masses and the force constants.

10.14.2 Green functions and correlations of atomic displacements

We use the Green function

$$G_{\alpha\beta}(l,l';\, t-t') = \frac{2\pi}{\hbar} \langle\langle u_\alpha(l,t);\, u_\beta(l',t')\rangle\rangle \tag{10.14.6}$$

of the atomic displacements, where the thermal average of the computator on the right is to be taken, with the usual restriction on the time ordering. As before the retarded Green function for a pair of operators $A(t), B(t')$ is given by

$$\langle\langle A(t); B(t')\rangle\rangle = -i\theta(t-t')\langle[A(t), B(t')]\rangle, \tag{10.14.7}$$

while the advanced Green function has $+i\theta(t-t')$ on the right-hand side. The equation of motion for G is then [cf. equation (10.11.35)]

$$\frac{i\hbar^2}{2\pi}\frac{dG}{dt} = \hbar\delta(t-t')\langle[A, B]\rangle + \langle\langle[A, H]; B\rangle\rangle. \tag{10.14.8}$$

For the Green function G, the last term on the right-hand side becomes

$$\frac{i\hbar}{M'_\alpha(l)}\langle\langle p_\alpha(l, t); u_\beta(l', t')\rangle\rangle, \tag{10.14.9}$$

where $p_\alpha(l, t)$ specifies the momentum component in the lth unit cell at time t. Differentiating again, we get an equation to determine the Green function, namely

$$-M'_\alpha(l)\frac{d^2}{dt^2}G_{\alpha\beta}(l, l'; t-t') = 2\pi\delta_{\alpha\beta}\,\delta(l, l')\,\delta(t-t') + \sum_{\gamma l''}\phi'_{\alpha\gamma}(l, l'')\,G_{\gamma\beta}(l'', l'; t-t'). \tag{10.14.10}$$

At this stage we take the Fourier transform using the result

$$G(\omega) = \frac{1}{2\pi}\int G(t)\,e^{i\omega t}\,dt \tag{10.14.11}$$

and writing in explicitly the changes introduced by the defect leads to the equation

$$-M_\alpha(l)\,\omega^2\,G_{\alpha\beta}(l, l'; \omega) + \sum_{\gamma l''}\phi_{\alpha\gamma}(l, l'')\,G_{\gamma\beta}(l'', l'; \omega) + \delta_{\alpha\beta}\,\delta(l, l') = \sum_{\gamma l''}C_{\alpha\gamma}(l, l''; \omega)\,G_{\gamma\beta}(l'', l'; \omega), \tag{10.14.12}$$

where

$$C_{\alpha\beta}(l, l'; \omega) = -M_\alpha(l)\,\varepsilon_\alpha(l)\,\omega^2\,\delta_{\alpha\beta}\,\delta(l, l') + \Delta\phi_{\alpha\beta}(l, l'). \tag{10.14.13}$$

This equation is readily rewritten in the form

$$G_{\alpha\beta}(l, l'; \omega) = G^{\mathrm{p}}_{\alpha\beta}(l, l'; \omega) - \sum_{\gamma\delta l_1 l_2} G^{\mathrm{p}}_{\alpha\gamma}(l, l_1; \omega)\,C_{\gamma\delta}(l_1, l_2; \omega)\,G_{\delta\beta}(l_2, l'; \omega), \tag{10.14.14}$$

where G^p denotes the Green function in the perfect lattice. This is given by

$$G^p_{\alpha\beta}(l,l';\omega) = \sum_{j\mathbf{k}} \frac{\chi^j_\alpha(\mathbf{k},l)\,\chi^{j*}_\beta(\mathbf{k},l')}{\omega^2-\omega_j^2(\mathbf{k})}, \tag{10.14.15}$$

where we have introduced the transformation

$$u_\alpha(l) = \sum_{j\mathbf{k}} \chi^j_\alpha(\mathbf{k}l)\,d_j(\mathbf{k}) \tag{10.14.16}$$

and the $d_j(\mathbf{k})$ are normalized as

$$d_j(\mathbf{k}) = \left[\frac{\hbar}{2\omega_j(\mathbf{k})}\right]^{\frac{1}{2}} [a_j(\mathbf{k})+a_j^\dagger(\mathbf{k})], \tag{10.14.17}$$

where a and $a^\dagger$ are the usual creation and destruction operators for phonons.

For a cubic lattice with s identical atoms per unit cell, of mass M, we have explicitly

$$G^p_{\alpha\beta}(l,l';\omega) = \frac{\delta_{\alpha\beta}}{3sNM} \sum_{j\mathbf{k}} \frac{\exp[-i\mathbf{k}.(\mathbf{R}_l-\mathbf{R}_{l'})]}{\omega^2-\omega_j^2(\mathbf{k})}. \tag{10.14.18}$$

We now consider the case of one defect of a different mass at the origin, in which case we obtain from equation (10.14.14) the result

$$G_{\alpha\alpha}(l,l';\omega) = G^p_{\alpha\alpha}(l,l';\omega) + \frac{M\varepsilon\omega^2 G^p_{\alpha\alpha}(l,0;\omega)\,G^p_{\alpha\alpha}(0,l';\omega)}{1-M\varepsilon\omega^2 G^p_{\alpha\alpha}(0,0;\omega)}. \tag{10.14.19}$$

Putting $l = l' = 0$, we obtain in an obvious notation

$$G(0,\omega) = G^p(0,\omega)/[1-M\varepsilon\omega^2 G^p(0,\omega)]. \tag{10.14.20}$$

As in the Koster–Slater theory of electron states, we see that localized modes can occur if equation (10.4.20) has a pole for real ω_l outside the allowed frequency bands in the perfect crystal. This evidently occurs when

$$M\varepsilon\omega^2 G^p(0,\omega) = 1. \tag{10.14.21}$$

Let us calculate at this stage the mean square displacement of the defect atom. We obtain, summing over three components which are equivalent in a cubic crystal,

$$3\langle u_\alpha(0)\rangle^2 = 3 \lim_{t-t'\to 0} \langle u_\alpha(0,t)\,u_\alpha(0,t')\rangle. \tag{10.14.22}$$

Going from the Green function to a correlation function in the manner of Zubarev (1960) we find

$$\langle u_\alpha(0)\rangle^2 = \lim_{\delta\to 0} \frac{i\hbar}{2\pi} \int_{-\infty}^{\infty} \frac{d\omega}{e^{\beta\omega}-1} [G(0,\omega+i\delta)-G(0,\omega-i\delta)], \tag{10.14.23}$$

which depends on the imaginary part of G. Here β is in fact $\hbar/k_B T$. Using the explicit result (10.14.18) for the perfect lattice Green function, in the form

$$G^p(0,\omega+i\delta) = \frac{1}{M}\left[P\int\frac{\nu(\omega')\,d\omega'}{\omega^2-\omega'^2} - \frac{i\pi[\nu(\omega)+\nu(-\omega)]}{2\omega}\right], \tag{10.14.24}$$

in analogy with Koster–Slater theory, $\nu(\omega)$ being the density of modes per unit frequency range normalized to unity, we find from equations (10.14.22) and (10.14.24) the result

$$3\langle u_\alpha(0)\rangle^2 = \frac{3\hbar}{2M}\int\frac{\coth(\frac{1}{2}\beta\omega)\,\nu(\omega)\,d\omega}{\omega\left[\left\{1-\varepsilon\omega^2 P\int[\nu(\omega')/(\omega^2-\omega'^2)]\,d\omega'\right\}^2+\frac{1}{4}\pi^2\varepsilon^2\omega^2\nu^2(\omega)\right]} \tag{10.14.25}$$

in the absence of localized modes. It is readily shown from equation (10.14.23) that a localized mode adds to equation (10.14.25) a contribution

$$\frac{3\hbar\coth(\frac{1}{2}\beta\omega_l)}{2\omega_l M}\left[\varepsilon^2\omega_l^4\int\frac{\nu(\omega')\,d\omega'}{(\omega_l^2-\omega^2)^2}-\varepsilon\right]^{-1}. \tag{10.14.26}$$

In a similar manner, momentum correlation functions can be derived. If we define (Elliott and Taylor, 1964)

$$F_{\alpha\beta}(l,l';t-t') = \frac{2\pi}{\hbar}(M_\alpha M_\beta)^{-\frac{1}{2}}\langle\langle p_\alpha(l,t);p_\beta(l',t')\rangle\rangle, \tag{10.14.27}$$

the masses being appropriate to the site in the unit cell with the α label, then following a similar procedure to that above it may be shown that

$$\omega^2 F_{\alpha\beta}(l,l';\omega) = \frac{\phi'_{\alpha\beta}(l,l')}{(M_\alpha M_\beta)^{\frac{1}{2}}} + \sum_{\gamma l''}\frac{\phi'_{\alpha\gamma}(l,l'')}{(M_\alpha M_\gamma)^{\frac{1}{2}}}F_{\gamma\beta}(l'',l';\omega)\frac{M_\gamma}{M_\gamma(l'')}. \tag{10.14.28}$$

In the final term, $M_\alpha(l'')$ indicates the actual mass at site γ in cell l'', which may differ from M_γ, that in the perfect crystal.

In the perfect crystal when $\phi'=\phi$, the Green function analogous to equation (10.14.15) is

$$Q_{\alpha\beta}(l,l';\omega) = (M_\alpha M_\beta)^{\frac{1}{2}}\sum_{j\mathbf{k}}\frac{\chi_\alpha^j(\mathbf{k},l)\,\omega_j^2(\mathbf{k})\,\chi_\beta^j(\mathbf{k},l')}{\omega^2-\omega_j^2(\mathbf{k})}, \tag{10.14.29}$$

which is related to equation (10.14.15) by

$$Q_{\alpha\beta}(l,l';\omega) = (M_\alpha M_\beta)^{\frac{1}{2}}\omega^2 G^p_{\alpha\beta}(l,l';\omega) - \delta_{\alpha\beta}\,\delta(l,l'). \tag{10.14.30}$$

Again taking the simple case of one changed mass at the origin of a cubic crystal, the solution of equation (10.14.28) becomes

$$F_{\alpha\alpha}(l,l';\,\omega) = Q_{\alpha\alpha}(l,l';\,\omega)+\frac{\varepsilon Q_{\alpha\alpha}(l,0;\,\omega)\,Q_{\alpha\alpha}(0,l';\,\omega)}{1-\varepsilon-\varepsilon Q_{\alpha\alpha}(0,0;\,\omega)}. \quad (10.14.31)$$

Using the same procedure as before, we find from equation (10.14.31) the result

$$3\langle p_\alpha(0)^2\rangle = \frac{3M}{2}(1-\varepsilon)^2\hbar\int\frac{\coth(\tfrac{1}{2}\beta\omega)\,\omega\nu(\omega)\,d\omega}{\left\{1-\varepsilon\omega^2 P\int[\nu(\omega')/(\omega^2-\omega'^2)]\,d\omega'\right\}^2+\tfrac{1}{4}\pi^2\varepsilon^2\omega^2\nu^2(\omega)} \quad (10.14.32)$$

plus contributions analogous to equation (10.14.26) from the local modes.

The results (10.14.19) and (10.14.31) show directly the nature of the Green functions in the perturbed lattice. The correlation between l and l' consists first of a part giving the same correlation as in the perfect lattice. The second contribution gives the correlation in the perfect lattice between l and 0 and then between 0 and l', together with a factor multiplying these, namely $\varepsilon\omega^2[1-M\varepsilon\omega^2 G^{\mathrm{p}}(0,\omega)]$ which characterizes the influence of the defect at that point. This term is directly related to the scattering of lattice waves by the defect, as we discuss below.

10.14.3 Scattering of lattice waves by defect

We shall now go on to discuss the scattering of lattice waves by defects. We shall see that under certain conditions resonant scattering can occur, and then this can influence the thermal conductivity of insulating crystals. Some brief remarks will then be made on optical absorption by defects.

We wish to calculate the rate of transition of phonons in state $j,\mathbf{k}$ of the perfect lattice into state $j',\mathbf{k}'$ say. This is clearly related to the operator $a_j^\dagger(\mathbf{k}',t')\,a_j(\mathbf{k},t)\,\theta(t'-t)$ which destroys a phonon of type $j,\mathbf{k}$ at time t and creates one of type $j',\mathbf{k}'$ at a subsequent time t'. For dealing with this problem, it is useful to work with the Green function

$$J_{j'j}(\mathbf{k}',\mathbf{k};\,t'-t) = \langle\langle b_{j'}^\dagger(\mathbf{k}',t');\,b_j(\mathbf{k},t)\rangle\rangle, \quad (10.14.33)$$

where b is defined by

$$b_j(\mathbf{k}) = [\omega_j(\mathbf{k})]^{\frac{1}{2}}\,a_j(\mathbf{k}). \quad (10.14.34)$$

Then the Hamiltonian (10.14.1) of the unperturbed crystal takes the form

$$H_0 = \frac{\hbar}{2}\sum_{j\mathbf{k}}[b_j^\dagger(\mathbf{k})\,b_j(\mathbf{k})+b_j(\mathbf{k})\,b_j^\dagger(\mathbf{k})], \quad (10.14.35)$$

while for a single modified mass at the origin the perturbation H_1 is

$$H_1 = p_0^2\left(\frac{1}{M'}-\frac{1}{M}\right) = -\frac{\varepsilon\hbar}{4N(1-\varepsilon)}\sum_{\alpha}\left\{\sum_{j\mathbf{k}}\sigma_\alpha^j(\mathbf{k})\,[b_j(\mathbf{k})-b_j^\dagger(\mathbf{k})]\right\}^2. \tag{10.14.36}$$

Forming the equation of motion again from equation (10.14.8) and taking a Fourier transform like equation (10.14.11), we obtain

$$\begin{aligned}2\pi[\omega+\omega_{j'}(\mathbf{k}')]\,J_{j'j}(\mathbf{k}',\mathbf{k};\,\omega)+\delta_{j'j}\,\delta(\mathbf{k}',\mathbf{k})\,\omega_{j'}(\mathbf{k}')\\ =\frac{\varepsilon\pi}{N(1-\varepsilon)}\sum_{\alpha}\omega_{j'}(\mathbf{k}')\,\sigma_\alpha^{j'}(\mathbf{k}')\sum_{j_1\mathbf{k}_1}\sigma_\alpha^{j_1}(\mathbf{k}_1)\,[K_{j_1j}(\mathbf{k}_1,\mathbf{k};\,\omega)-J_{j_1j}(\mathbf{k}_1,\mathbf{k};\,\omega)],\end{aligned} \tag{10.14.37}$$

where

$$K_{j'j}(\mathbf{k}',\mathbf{k};\,t'-t) = \langle\langle b_{j'}(\mathbf{k}',t);\,b_j(\mathbf{k},t)\rangle\rangle. \tag{10.14.38}$$

The equation for K is, in the same way, found to be

$$\begin{aligned}[\omega-\omega_{j'}(\mathbf{k}')]\,K_{j'j}(\mathbf{k}',\mathbf{k};\,\omega)\\ =\frac{\varepsilon}{2N(1-\varepsilon)}\sum_{\alpha}\omega_{j'}(\mathbf{k}')\,\sigma_\alpha^{j'}(\mathbf{k}')\sum_{j_1\mathbf{k}_1}\sigma_\alpha^{j'}(\mathbf{k}_1)\,[K_{j_1j}(\mathbf{k}_1,\mathbf{k};\,\omega)-J_{j_1j}(\mathbf{k}_1,\mathbf{k};\,\omega)]\end{aligned} \tag{10.14.39}$$

and leads to the relation

$$\begin{aligned}2\pi[\omega+\omega_{j'}(\mathbf{k}')]\,J_{j'j}(\mathbf{k}',\mathbf{k};\,\omega)+\delta_{j'j}\,\delta(\mathbf{k}',\mathbf{k})\,\omega_{j'}(\mathbf{k}')\\ =2\pi[\omega-\omega_{j'}(\mathbf{k}')]\,K_{j'j}(\mathbf{k}',\mathbf{k};\,\omega),\end{aligned} \tag{10.14.40}$$

between J and K. Substituting equation (10.14.40) into equation (10.14.37) gives for J the equation

$$\begin{aligned}2\pi[\omega+\omega_{j'}(\mathbf{k}')]\,J_{j'j}(\mathbf{k}',\mathbf{k};\,\omega)+\delta_{j'j}\,\delta(\mathbf{k}',\mathbf{k})\,\omega_{j'}(\mathbf{k}')\\ =\frac{2\pi\varepsilon}{N(1-\varepsilon)}\sum_{\alpha}\omega_{j'}(\mathbf{k}')\,\sigma_\alpha^{j'}(\mathbf{k}')\\ \left\{\frac{\sigma_\alpha^{j}(\mathbf{k})\,\omega_j(\mathbf{k})}{4\pi[\omega-\omega_j(\mathbf{k})]}+\sum_{j_1\mathbf{k}_1}\frac{\sigma_\alpha^{j'}(\mathbf{k}_1)\,\omega_{j_1}(\mathbf{k}_1)\,J_{j_1j}(\mathbf{k}_1,\mathbf{k};\,\omega)}{\omega-\omega_{j_1}(\mathbf{k}_1)}\right\}.\end{aligned} \tag{10.14.41}$$

The unperturbed solution for J is given by making the right-hand side zero, when we find

$$J^0_{j'j}(\mathbf{k}',\mathbf{k};\,\omega) = -\frac{\delta_{j'j}\,\delta(\mathbf{k}',\mathbf{k})\,\omega_j(\mathbf{k})}{2\pi[\omega+\omega_{j'}(\mathbf{k}')]}. \tag{10.14.42}$$

If J' represents the change in J due to the defect, an equation like (10.14.41) is obtained for J'. Summing these equations with appropriate factors it is then straightforward to show that

$$\sum_{j'\mathbf{k}'} \frac{\sigma_\alpha^{j'}(\mathbf{k}')\,\omega_{j'}(\mathbf{k}')\,J'_{j'j}(\mathbf{k}',\mathbf{k};\,\omega)}{\omega-\omega_{j'}(\mathbf{k}')} = \frac{\varepsilon\sigma_\alpha^{j}(\mathbf{k})\,\omega_j(\mathbf{k})\,Q_{\alpha\alpha}(0,0;\,\omega)}{4\pi[\omega+\omega_j(\mathbf{k})]\,[1-\varepsilon-\varepsilon Q_{\alpha\alpha}(0,0;\,\omega)]}, \tag{10.14.43}$$

where Q is given by equation (10.14.29). Hence we find from equations (10.14.41) and (10.14.30) that

$$J'_{j'j}(\mathbf{k}',\mathbf{k};\,\omega) = \frac{\varepsilon\omega_{j'}(\mathbf{k}')\,\omega_j(\mathbf{k})\sum_\alpha \sigma_\alpha^{j'}(\mathbf{k}')\,\sigma_\alpha^{j}(\mathbf{k})}{4\pi N[\omega+\omega_{j'}(\mathbf{k})]\,[\omega+\omega_j(\mathbf{k})]\,[1-M\varepsilon\omega^2 G^{\mathrm{p}}_{\alpha\alpha}(0,0;\,\omega)]}. \tag{10.14.44}$$

The correlation function is obtained again from a procedure like equation (10.14.29) as

$$\langle a^\dagger_{j'}(\mathbf{k}',t')\,a_j(\mathbf{k},t)\rangle$$
$$= \lim_{\delta\to 0} i\int_{-\infty}^{\infty}\frac{d\omega\,e^{\beta\omega}}{e^{\beta\omega}-1}\left[\frac{J_{j'j}(\mathbf{k}',\mathbf{k};\,\omega+i\delta)-J_{j'j}(\mathbf{k}',\mathbf{k};\,\omega-i\delta)}{[\omega_{j'}(\mathbf{k}')\,\omega_j(\mathbf{k})]^{\frac{1}{2}}}\right]e^{-i\omega(t'-t)} \tag{10.14.45}$$

and the probability of finding the scattered state after time $t'-t$, $P\,(\mathbf{k}\to\mathbf{k}')$ is given by the ratio of the square moduli of the correlations (10.14.45) in the perturbed lattice and the unperturbed crystal.

For the perturbed lattice, equation (10.14.44) is substituted into (10.14.45) and assuming no localized modes the only poles of J are at $-\omega_j(\mathbf{k})$ and $-\omega_{j'}(\mathbf{k}')$ slightly below the real axis in the first term and above it in the second if $\delta>0$. Since we are interested in $t'-t>0$, we must close the contour in the lower half plane to calculate equation (10.14.45), where only the first term has poles. The assumption that $t'-t$ is large then gives for the square modulus of the defect correlation

$$\frac{\pi\varepsilon^2\,\omega_j(\mathbf{k})\,\omega_{j'}(\mathbf{k}')\,n_j(\mathbf{k})\,n_{j'}(\mathbf{k}')\left[\sum_\alpha \sigma_\alpha^{j'}(\mathbf{k}')\,\sigma_\alpha^{j}(\mathbf{k})\right]^2(t'-t)\,\delta[\omega_{j'}(\mathbf{k}')-\omega_j(\mathbf{k})]}{2N^2[1-M\varepsilon^2\,\omega^2_{j'}(\mathbf{k}')\,G^{\mathrm{p}}(0,0;\,\omega_{j'}(\mathbf{k}')+i\delta)]\,[1-M\varepsilon^2\,\omega_j^2(\mathbf{k})\,G^{\mathrm{p}}(0,0,\omega_j(\mathbf{k})+i\delta]^*}, \tag{10.14.46}$$

where

$$n_j(\mathbf{k}) = \{\exp[\beta\omega_j(\mathbf{k})]-1\}^{-1}. \tag{10.14.47}$$

In the unperturbed lattice, the correlation is obtained from equation (10.14.42) as

$$\left[\frac{\omega_j(\mathbf{k})}{\omega_{j'}(\mathbf{k}')}\right]^{\frac{1}{2}} \delta_{j'j}\,\delta(\mathbf{k}'\mathbf{k})\frac{\exp\,[i\omega_j(\mathbf{k})\,(t'-t)]}{\exp\,[\beta\omega_j(\mathbf{k})]-1}, \tag{10.14.48}$$

and hence we obtain the desired result

$$P(\mathbf{k}\rightarrow\mathbf{k}') = \frac{\Pi\varepsilon^2 n_{j'}(\mathbf{k}')\,\omega_{j'}(\mathbf{k}')\,\omega_j(\mathbf{k})\left[\sum_\alpha \sigma_\alpha^{j'}(\mathbf{k}')\,\sigma_\alpha^j(\mathbf{k})\right]^2 (t'-t)\,\delta[\omega_{j'}(\mathbf{k}')-\omega_j(\mathbf{k})]}{2N^2 n_j(\mathbf{k})\,[1-M\varepsilon^2\omega_{j'}^2(\mathbf{k}')\,G^{\mathrm{p}}(0,0;\,\omega_{j'}(\mathbf{k}')+i\delta]\times[1-M\varepsilon^2\omega_j^2(\mathbf{k})\,G^{\mathrm{p}}(0,0;\,\omega_j(\mathbf{k})+i\delta]^*}. \tag{10.14.49}$$

The rate of transition of a particular mode $j,\mathbf{k}$ is obtained by dividing by $t'-t$ and summing over all $j',\mathbf{k}'$ of the same frequency. In a cubic crystal, the sum over σ's averages to $\frac{1}{3}$ to yield the relaxation time $\tau_j(\mathbf{k})$ as

$$\frac{1}{\tau_j(\mathbf{k})} = \frac{\Pi\varepsilon^2\,\omega_j^2(\mathbf{k})\,\nu[\omega_j(\mathbf{k})]}{2N\left[[1-\varepsilon\omega_j^2(\mathbf{k})\,P\displaystyle\int\{\nu(\omega')/[\omega_j^2(\mathbf{k})-\omega'^2]\}\,d\omega'\right]^2+\frac{1}{4}\pi^2\,\varepsilon^2\,\omega^2\,\nu^2(\omega)}. \tag{10.14.50}$$

A formula which is a special case of this, obtained using perturbation theory by Klemens (1958), puts the denominator equal to unity and uses a Debye spectrum (3.8.12) for $\nu(\omega)$. The denominator in equation (10.14.50) can of course lead to a resonance, centred on the frequencies which satisfy the condition

$$M\varepsilon\omega^2\,\mathrm{Re}\,G_{\alpha\alpha}^{\mathrm{p}}(0,0;\,\omega) = 1, \tag{10.14.51}$$

the resonance having a width given by

$$M\varepsilon\omega^2\,\mathrm{Im}\,G_{\alpha\alpha}^{\mathrm{p}}(0,0;\,\omega). \tag{10.14.52}$$

In the neighbourhood of the frequencies given by equation (10.14.52), the lattice modes have enhanced amplitude in the proximity of the defect, and this enhancement is reflected in the increased scattering of these modes.

10.14.4 Influence of resonances on thermal conductivity

Pohl (1962) proposed that such resonances would give observable results in measurement of thermal conductivity, and Walker and Pohl (1963) demonstrated these effects by measurements on doped alkali halide crystals.

If we write the thermal conductivity κ in terms of the relaxation time $\tau_j(\mathbf{k})$ of equation (10.14.50) above, we find (see Carruthers, 1961)

$$\kappa = \frac{1}{3}\sum_{j\mathbf{k}} v_j^2(\mathbf{k})\,\tau_j(\mathbf{k})\,c_j(\mathbf{k}),$$

where $c_j(\mathbf{k})$ is the contribution of phonons of type $j, \mathbf{k}$ to the specific heat, $v_j(\mathbf{k})$ being the group velocity of such phonons and τ and c depend on temperature. For a general branch, the group velocity is not a simple function but at small $\mathbf{k}$ it is essentially constant. The resonance is only important if it occurs at small $\mathbf{k}$ and ω and is dominant at low temperatures. At large ω, the umklapp processes dominate all the other effects.

Following Walker and Pohl (1963), the assumption of a Debye spectrum, which is reasonably accurate at low $\mathbf{k}$ and hence at low temperature, allows one to write

$$\frac{1}{\tau(\omega)} = \frac{1}{\tau_b}+\frac{1}{\tau_i}+\frac{1}{\tau_u} = a+b\omega^4+c\omega^2\exp\left(-\frac{A}{T}\right), \qquad (10.14.53)$$

where contributions from boundary, impurity and umklapp scattering are separated out.

Elliott and Taylor (1964) have carried out calculations using the above

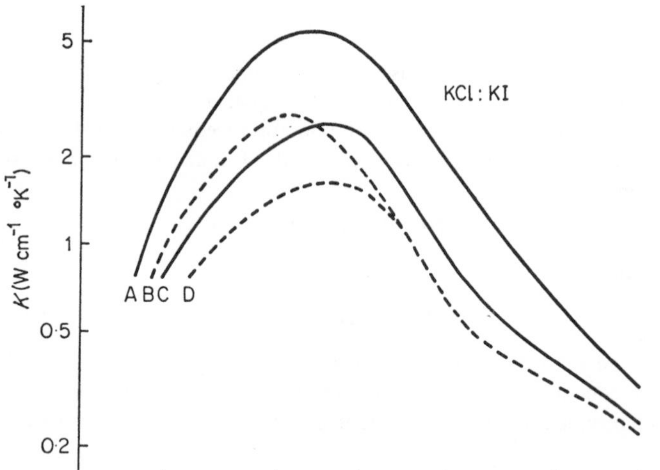

FIGURE 10.20. Thermal conductivity for iodine in KCl (after Elliott and Taylor, 1964). A, Experimental results for pure KCl. B, Experiment for $1{\cdot}25\times10^{19}$ iodine atoms/cm^3. D, Best fit to resonance with choice of $\varepsilon = -3{\cdot}3$ and concentration $1{\cdot}8\times10^{19}$.

theory for iodine in KCl and the results are shown in Figure 10.20. Curve C is obtained with $\varepsilon = -2.58$ and the given concentration of 1.25×10^{19} iodine atoms/cm^3. The pronounced dip from the resonance is in about the correct position at 25 K, but the magnitude of the thermal conductivity is not given quantitatively. Nevertheless, the essential features of the resonance are well described by the theory given above.

10.14.5 Optical absorption arising from vibrations of defects

As we have just seen, there are circumstances in which the vibrational properties of the perfect crystal can be profoundly modified by defects. We shall now consider briefly how these modifications are reflected in the absorption of light by such defect crystals (see Elliott, 1965, *Lattice Dynamics*, p. 459).

Earlier in this volume, we saw that in perfect crystals, the absorption of light due to the creation of single phonons can take place only subject to very stringent selection rules. First, the wave vector $\mathbf{k}$ characterizing the normal modes must be conserved in the physically allowed processes. In practice, when we are concerned with light in the infrared, the photon wave vector is so small that we can regard only phonons with $\mathbf{k} = 0$ as excited. Secondly, the crystals in which absorption takes place must be at least partially ionic, in order that coupling to the light can occur. In ionic crystals, when such absorption occurs, the interaction is very strong and the Reststrahl phenomenon results (Born and Huang, 1954). The information that can be obtained from such one-phonon processes about the vibrational spectra of crystals is therefore rather limited. A good deal more information can be obtained from absorption which is due to the simultaneous creation of two phonons.

On the introduction of a defect, as we saw in this chapter, the wave vectors $\mathbf{k}$ no longer classify the normal modes, these retaining only the local symmetry of the defect. If charged ions are present, then absorption can take place by excitation of single phonons in all modes having appropriate point symmetry over a wide spectral range. The importance of this, of course, is that a measurement of this absorption spectrum will contain detailed information about the response of the lattice defect as a function of frequency.

The systems which can be investigated in practice are unfortunately somewhat limited. Thus experimental work is confined to charged defects in covalent crystals like diamond, or to defects in ionic crystals which produce a spectrum in a régime separate from the Reststrahl region. For metals, there is of course strong electronic absorption in the infrared. The information obtained about the defect from such optical measurements is different from that which is found from a study of its electronic absorption spectrum.

We showed above, for the simplest case of a defect consisting only of one altered mass from M to M', that

$$z\left(\frac{M-M'}{M}\right)\int\frac{\nu(\mu)\,d\mu}{z-\mu} = 1, \quad z = \omega^2 \tag{10.14.54}$$

determines the allowed frequencies of vibration of the perturbed model. The quantity μ is equal to $\omega_i(\mathbf{k})^2$, the square of the frequencies of the perfect lattice modes, while the density of vibrational states in the perfect crystal is $\nu(\mu)$.

Equation (10.14.54), as we have seen, has two types of solution. If z lies outside the allowed range of μ, we get a local mode, which has an amplitude tending to zero far from the defect. If, on the other hand, z lies inside the range of allowed frequencies in the perfect lattice, we get resonant modes, with enhanced amplitude near the defect but which tend to unperturbed lattice modes far from the defect, exactly as for electronic states.

(a) *Types of absorption.* Let us denote by $e_\alpha(l)$ and $u_\alpha(l)$ the charge and the displacement respectively of the α atom in the lth cell. Then the optical absorption is essentially determined by the squared matrix elements of

$$\sum_{\alpha,l} e_\alpha(l)\,u_\alpha(l). \tag{10.14.55}$$

We can then isolate three types of absorption peaks:

(i) Sharp lines at frequencies in forbidden regions of the perfect lattice, due to local modes.

(ii) Resonant peaks of finite width, in an allowed band of the perfect crystal.

(iii) Peaks associated with regions of high density of vibrational states in the perfect lattice.

It is an interesting matter that the total absorption in the local mode peaks is of the same order of magnitude as that coming from the complete sum from all the band modes.

(b) *Simple model of single charged atom.* We shall consider briefly the optical absorption in the case where a single atom is altered in mass and carries charge e. This was first considered by Dawber and Elliott (1963a) (see also Lifshitz, 1956).

In this situation, the local mode gives an absorption

$$K(\omega) = \frac{2\pi^2 De^2\Lambda}{nc}|\chi(0)|^2 s(\omega). \tag{10.14.56}$$

$s(\omega)$ is a normalized shape function, D is the number of defects and $\chi(0)$ is the relative amplitude of the defect atom given by

$$|\chi(0)|^2 = \frac{1}{M}\left[\left(\frac{M-M'}{M}\right)^2 z^2 \int \frac{\nu(\mu)\,d\mu}{(z-\mu)^2} - \left(\frac{M-M'}{M}\right)\right]^{-1}. \quad (10.14.57)$$

Λ is the local field correction, which can be written in terms of the refractive index n as

$$\Lambda = (n^2+2)^2/9. \quad (10.14.58)$$

In this case, the absorption from the band modes is given by

$$K(\omega) = \frac{2\pi^2 De^2 \Lambda}{3nc} |\chi(0)|^2 S(\omega) \quad (10.14.59)$$

where $S(\omega)$ is the density of lattice modes per unit frequency range and

$$|\chi(0)|^2 = \frac{1}{MsN}\left\{\left[1 - \frac{M-M'}{M} zP \int \frac{\nu(\mu)\,d\mu}{z-\mu}\right]^2 + \left[\pi\left(\frac{M-M'}{M}\right) z\nu(z)\right]^2\right\}^{-1}, \quad (10.14.60)$$

which is clearly a function of $z = \omega^2$ (s atoms per unit cell).

Figure 10.21 shows the energy of the local mode plotted as a function of $(M-M')/M$ for a Debye spectrum. The relative amplitude of the defect atom M' $\chi(0)|^2$ is also plotted from equation (10.14.57).

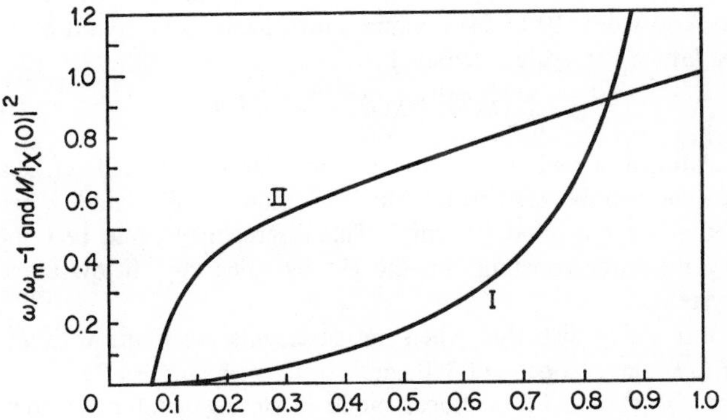

FIGURE 10.21. Isotopic mass defect. Energy of local mode plotted as function of relative mass difference for a Debye spectrum (curve I). Curve II shows relative amplitude of localized mode at defect (after Dawber and Elliott, 1963b).

For small mass differences, that is, for masses near the host mass, the local mode has low energy and is very extended. The relative amplitude (10.14.57) thus tends to zero. For $(M-M')/M \to 1$, that is, for light masses, all the amplitude becomes concentrated on the defect atom.

The main features of these calculations are revealed in the observed spectra from diamond, silicon and in some ionic crystals containing hydrogen. We discuss only the last case below.

(c) *Ionic crystals containing hydrogen.* H^- ions occur substitutionally on the cation sites of common ionic crystals. They are the *U*-centres in alkali halides (see Gourary and Adrian, 1960), with a characteristic ultraviolet electronic absorption. However Schäfer (1960) observed a sharp absorption line in the infrared in the same crystals. This can be associated with the local mode arising from the light impurity. With hydrogen, it seems that it is so light that, to a fair approximation, it oscillates independently in the potential well obtained for a static host lattice.

If the assumption is made that the force constants are unchanged from those of the host, then this potential well may be calculated from a knowledge of the vibrational properties of the perfect lattice. If the additional assumption is made that in the ionic crystal the two types of atom have the same mass (for example, this is approximately true for KCl), then we can solve (10.14.54) for the local mode frequency by expanding in powers of M'/M. To first order, the frequency is given by

$$\omega^2 = \langle\omega^2\rangle\,[(M/M')+L], \quad L = (\langle\omega^4\rangle/\langle\omega^2\rangle^2)-1, \tag{10.14.61}$$

where $\langle\omega^n\rangle$ are the average nth powers of the perfect lattice frequency spectra. Also from (10.14.54) the amount of local mode found on atoms of the host lattice is readily calculated as

$$\sum_{\mathbf{R}\neq 0} |\chi(\mathbf{R})|^2/|\chi(0)|^2 = (M'/M)^2 L. \tag{10.14.62}$$

Calculations carried out on the phonon spectra of KCl (cf. Karo and Hardy, 1966, on NaCl) predict the local mode at 600 cm^{-1}, whereas the observed value is around 500 cm^{-1}. This discrepancy could be removed by reducing the force constants on the H^- by some 20% below those of the host crystal.

It is interesting also that when measurements are made with D^- rather than H^-, D^- gives a peak at 360 cm^{-1}, so that $\omega_H^2/\omega_D^2 = 1.95$. As was to be expected the spread of the local mode is clearly greater for the case of D^-. A value of L about unity results, which seems reasonable. Results for many other alkali halides are also given by Schäfer (1960). Localized modes and resonance states in alkali halides are also reviewed by Klein (1968).

10.15 Vibrational modes in alloys

We have given above in section 10.14 a quantitative discussion of the effect of a single defect on the vibrational spectrum of a crystal lattice.

For a single substitutional impurity, with a different mass, the dominant features are:

(i) For light defects, the introduction of localized modes with frequencies outside the perfect lattice continuum. The band modes are attenuated near the defects in this case.

(ii) For heavy defects, there are no localized modes, and the band modes may have a considerable enhancement by a resonance in a particular part of the spectrum.

With a small finite concentration of impurities, the largest modification of the spectrum arises from these two effects. An impurity band may form out of the localized modes and appreciable enhancement of the density of modes occurs near a resonance (for electrons, cf. Matsubara and Toyozawa, 1961, Edwards, 1962).

The theory given below shows both of these effects and the experimental results confirm the general nature of the theoretical predictions (Taylor, 1967).

10.15.1 Phonons in disordered alloys

As in the case of a single defect, we shall work with the displacement–displacement double-time thermal Green functions $G_{\alpha\beta}(l, l', \omega)$ given by

$$G_{\alpha\beta}(l, l', \omega) = \frac{1}{\hbar}\int_{-\infty}^{\infty} \langle\langle u_\alpha(l, t);\, u_\beta(l', 0)\rangle\rangle\, e^{i\omega t}\, dt, \qquad (10.15.1)$$

where the quantities on the right-hand side (cf. Chapter 4) are explicitly

$$\langle\langle A(t);\, B(0)\rangle\rangle = i\theta(t-t')\langle[A(t), B(0)]\rangle \quad \text{(retarded Green function)} \qquad (10.15.2)$$

$$= -i\theta(t-t')\langle[A(t), B(0)]\rangle \quad \text{(advanced Green function)} \qquad (10.15.3)$$

and θ is the Heaviside function; $\theta(t) = 1$ if $t>0$; $= 0$ if $t<0$. $u_\alpha(l, t)$ with $\alpha = (a, p)$, is the atomic displacement of the pth atom in the unit cell at R_l in the direction $(a = x, y, z)$.

The Hamiltonian for the alloy in the harmonic approximation may be written as [equation (10.14.1)]

$$H = \frac{1}{2}\sum_{\alpha l}\frac{p_\alpha^2(l)}{M_\alpha(l)} + \frac{1}{2}\sum_{\substack{\alpha\beta \\ ll'}} u_\alpha(l, t)\,\Phi_{\alpha\beta}(l, l')\,u_\beta(l't), \qquad (10.15.4)$$

where $\Phi_{\alpha\beta}(l, l')$ is the force constant matrix defined in Chapter 3 or equation (10.10.2), and $M_\alpha(l)$ is the atomic mass.

It will be useful at this stage to separate out from the Hamiltonian (10.15.4), a part representing the perfect crystal. For defect atom of type $A(l_i)$ at l_i, we suppose the mass to change by [cf. equation (10.14.4)]

$$M_\alpha - M_\alpha(l_i) = M_\alpha \, \varepsilon_\alpha^{A(l_i)}. \tag{10.15.5}$$

The changes in the force constants caused by the introduction of the defects are given by

$$\Delta\Phi_{\alpha\beta}^{A(l_i)}(l, l') = \Phi_{\alpha\beta}(l, l') - \Phi_{\alpha\beta}^{0}(l, l'), \tag{10.15.6}$$

the superscript zero on $\Phi_{\alpha\beta}$ referring to the perfect crystal, which has Green function G_p. If we now introduce the defect matrix defined by

$$C_{\alpha\beta}^{A(l_i)}(l, l'; \omega) = \Delta\Phi_{\alpha\beta}^{A(l_i)}(l, l') + M_\alpha \, \varepsilon_\alpha^{A(l_i)} \omega^2 \, \delta_{\alpha\beta} \, \delta(l, l') \, \delta(l, l_i), \tag{10.15.7}$$

then what is essentially Dyson's equation for this problem can be written in the form

$$\mathbf{G}(ll'; \omega) = \mathbf{G}_p(ll'; \omega) + \sum_{l_1 l_2} \mathbf{G}_p(l, l_1; \omega) \, \mathbf{C}(l_1 l_2; \omega) \, \mathbf{G}(l_2, l'; \omega), \tag{10.15.8}$$

where the total defect matrix $\mathbf{C}$ is defined as

$$\mathbf{C}(l, l'; \omega) = \sum_i \mathbf{C}^{A(l_i)}(l, l'; \omega). \tag{10.15.9}$$

In physical problems, such a superposition of independent defect matrices $\mathbf{C}^i$ is only correct at low concentrations, for different atomic species. However, if only isotopic mass changes are assumed, all is well. If we assume that an atom at l_i causes force constant changes only locally, over a set of sites s_i, then we have

$$\mathbf{G}(l, l', \omega) = \mathbf{G}_p(l, l', \omega) + \sum_i \sum_{s_i} \sum_{s_i'} \mathbf{G}_p(l, s_i, \omega) \, \mathbf{C}^{A(l_i)}(s_i; s_i'; \omega) \, \mathbf{G}(s_i', l'; \omega). \tag{10.15.10}$$

At this stage, it is assumed that $\mathbf{G}(l, l'; \omega)$ is dominated by whether or not the sites l or l' are affected by an individual defect matrix. This suggests separating out the Green functions associated with a given defect in the manner indicated below. Thus we write

$$\sum_{s_i'} \left[\mathbf{1}\delta(s_i, s_i') - \sum_{s_i''} G_p(s_i, s_i''; \omega) \, C^{A(l_i)}(s_i'', s_i'; \omega) \right] G(s_i', l'; \omega)$$

$$\equiv \sum_{s_i'} X^{A(l_i)}(s_i, s_i'; \omega) \, G(s_i', l'; \omega) = G^{A(l_i)}(s_i, l'; \omega), \tag{10.15.11}$$

where

$$G^{A(l_i)}(l, l'; \omega) = G_p(l, l'; \omega) + \sum_{j \neq i} \sum_{s_j s_i'} G_p(l, s_j; \omega) \times C^{A(l_j)}(s_j, s_j'; \omega) G(s_j', l'; \omega). \quad (10.15.12)$$

Then we get from equation (10.15.10), on inverting equation (10.15.11), the result

$$G(l, l'; \omega) = G_p(l, l'; \omega) + \sum_i \sum_{s_i s_i'} G_p(l, s_i; \omega) T^{A(l_i)}(s_i, s_i'; \omega) G^{A(l_i)}(s_i', l'; \omega) \quad (10.15.13)$$

and

$$G^{A(l_i)}(l, l'; \omega) = G_p(l, l'; \omega) + \sum_{j \neq i} \sum_{s_j s_j'} G_p(l, s_j; \omega) T^{A(l_j)}(s_j, s_j'; \omega) G^{A(l_j)}(s_j', l'; \omega), \quad (10.15.14)$$

where T is the t-matrix discussed in detail for electron states above, which in the present case describes the scattering of lattice excitations due to the perturbation introduced by the defect atom in an otherwise pure crystal (cf. Klein, 1963).

This is given by

$$T^{A(l_i)}(s_i, s_i'; \omega) = \sum_{s_i''} C^{A(l_i)}(s_i, s_i''; \omega) X^{A(l_1)}(s_i'', s_i'; \omega)^{-1}. \quad (10.15.15)$$

We must now average equations (10.15.14) and (10.15.15) over all configurations of defects. The average of the summation on the right-hand side of equation (10.15.13) can be written in terms of a conditionally averaged G^A

$$\sum_{l_1 l_2} \sum_{l_i} G_p(l, l_1; \omega) \langle T^{A(l_i)}(l_1, l_2; \omega) G^{A(l_i)}(l_2, l'; \omega) \rangle = \sum_{l_1 l_2 l_3} G_p(l, l_1; \omega) \sum_A c^A T^{A(l_3)}(l_1 l_2; \omega) \times \langle G^{A(l_3)}(l_2, l'; \omega) \rangle_{A(l_3)}, \quad (10.15.16)$$

where $\langle \; \rangle$ denotes an average over all configurations and $\langle \; \rangle_{A(l_i)A(l_j)\ldots}$ an average conditional on a defect of type $A(l_i)$ at l_i, etc. The probability that a defect is at l_i is given by the concentration of such defects c^A. The equation for G^A can be dealt with in the same way. The equations which are then obtained for $\langle G \rangle$ and $\langle G^A \rangle$ are given by

$$\langle G(l, l'; \omega) \rangle = G_p(l, l'; \omega) + \sum_{l_1 l_2 l_3} G_p(l, l_1; \omega) \times \sum_A c^A T^{A(l_3)}(l_1, l_2; \omega) \langle G^{A(l_3)}(l_2, l'; \omega \rangle_{A(l_3)} \quad (10.15.17)$$

and

$$\langle G^{A(l_3)}(l,l';\omega)\rangle_{A(l_3)} = G_\mathrm{p}(l,l';\omega)+\sum_{l_1 l_2}\sum_{l_4\neq l_3} G_\mathrm{p}(l,l_1;\omega)$$
$$\times\sum_A c^A T^{A(l_4)}(l_1,l_2;\omega)\langle G^{A(l_4)}(l_2,l';\omega)\rangle_{A(l_3)A(l_4)}. \tag{10.15.18}$$

Except where explicitly shown, the sums on the lattice site labels are now over all sites. A G^A conditionally averaged on r sites is always given in terms of G^A's conditionally averaged on $r+1$ sites, thus leading to an infinite set of equations which can only be terminated by making some physical approximation.

The Green function $G^{A(l_1)}(l,l';\omega)$ can be interpreted as the effective field, as seen by the atoms at the sites involved in the perturbation, caused by a defect at l_1, due to a disturbance stemming from l'. The wave has been allowed to scatter off all other defects in the crystal before it scatters off the perturbation due to the defect at l_1. The t-matrix describes this final scattering explicitly in equation (10.15.17). The effective field Green function will depend, in general, upon the type of defect it refers to as well as the configuration of the other defects in the crystal. We can anticipate that it will be made up of a mean part depending only upon the defect type and a random part that depends mainly upon the local environment. It is necessary to proceed to at least the conditional average on two sites to include this random part.

The simplest way to proceed is to neglect the effect of variations in the local environment and hence the random part of the effective field Green function. Thus, we use the approximation, valid at low concentrations, that

$$\langle G^{A(l_2)}(l,l';\omega)\rangle_{A(l_2)A(l_3)} \approx \langle G^{A(l_2)}(l,l';\omega)\rangle_{A(l\;)}. \tag{10.15.19}$$

Then we find

$$\sum_{l_1}\left[\mathbf{1}\delta(l,l_1)+G_\mathrm{p}(l,l_2;\omega)\sum_A c^A T^{A(l_3)}(l_2,l_1;\omega)\right]$$
$$\times\langle G^{A(l_3)}(l_1,l';\omega)\rangle_{A(l_3)} \equiv \sum_{l_1} Y(l,l_1;\omega)\langle G^{A(l_3)}(l_1,l';\omega)\rangle_{A(l_3)}$$
$$=\langle G(l,l';\omega\rangle, \tag{10.15.20}$$

and, finally,

$$\langle G(l,l';\omega\rangle = G_\mathrm{p}(l,l';\omega)+\sum_{l_1 l_2 l_3 l_4} G_\mathrm{p}(l,l_1;\omega)$$
$$\times\sum_A c^A T^{A(l_3)}(l_1,l_2;\omega)\, Y^{-1}(l_2,l_4;\omega)\langle G(l_4,l';\omega)\rangle. \tag{10.15.21}$$

Using the example considered above of only one type of mass defect in a cubic crystal, this equation takes the simpler form [cf. equation (10.14.19)]

$$\langle G_{\alpha\beta}(ll'; \omega)\rangle = G^{\mathrm{p}}_{\alpha\beta}(l, l'; \omega) + \sum_{\gamma l_1} G^{\mathrm{p}}_{\alpha\gamma}(l, l_1; \omega) \times \frac{M_\gamma c\varepsilon_\gamma \omega^2}{1-(1-c)M_\gamma \varepsilon_\gamma \omega^2 G^{\mathrm{p}}_{\gamma\gamma}(0,0;\omega)} \langle G_{\gamma\beta}(l_1, l'; \omega)\rangle. \tag{10.15.22}$$

This result appears to have been first given by Elliott and Taylor (1967). The presentation adopted above follows Taylor (1967), for it is conveniently generalized, as we see below, to the case of concentrations which are not necessarily small, by a method paralleling the coherent potential theory we discussed above for electron states.

The application of (10.15.22) to obtain the density of states, optical absorption and neutron scattering for an imperfect crystal has been discussed by Elliott and Taylor and we refer the reader to the original paper for the results.

We note, however, that the presence of the factor $(1-c)$ in equation (10.15.22) has the effect for a light defect $(0<\varepsilon<1)$ of centring the impurity band around the local mode frequency due to one mass defect in the crystal, as we anticipated would be the case on physical grounds above.

(a) *Case of large concentrations.* We turn then to what is essentially the coherent potential method for dealing with large concentrations. On iterating equation (10.15.8), it is clear that the general form of the result is

$$\langle G(l, l'; \omega)\rangle = G_{\mathrm{p}}(l, l'; \omega) + \sum_{l_1 l_2} G_{\mathrm{p}}(l, l_1; \omega) \times \Sigma(l_1, l_2; \omega)\langle G(l_2, l'; \omega)\rangle, \tag{10.15.23}$$

the summation being over all lattice sites. What we have done above is essentially to give a low concentration form for the self-energy Σ. Just as in the electron state problem, we now introduce a new self-energy u, which will eventually become the approximation adopted for Σ. Then a new Green function G^0 is defined in terms of u by

$$G^0(l, l'; \omega) = G_{\mathrm{p}}(l, l'; \omega) + \sum_{l_1 l_2} G_{\mathrm{p}}(l, l_1; \omega)\, u(l_1, l_2; \omega)\, G^0(l_2, l'; \omega). \tag{10.15.24}$$

If we now write equation (10.15.8) in terms of G^0 rather than in terms of

G_p, we obtain

$$G(l,l';\omega) = G^0(l,l';\omega)+\sum_{l_1 l_2} G^0(l,l_1;\omega) \times V^{A(l_2)}(l_1,l_2;\omega)\,G(l_2,l';\omega), \qquad (10.15.25)$$

where

$$\begin{aligned} V^{A(l_2)}(l_1,l_2;\omega) &= -u(l_1,l_2;\omega) \quad \text{for a host atom at } l_2 (A=h) \\ &= -u(l_1,l_2;\omega)+M\varepsilon\omega^2\,\delta(l_1,l_2)\,\mathbf{1} \\ &\qquad \text{for a defect atom at } l_2\,(A=d). \end{aligned} \qquad (10.15.26)$$

Equation (10.15.25) has the same form as equation (10.15.8), so we can immediately proceed to the equivalent of equation (10.15.17), namely

$$\langle G(l,l';\omega)\rangle = G^0(l,l';\omega)+\sum_{l_1 l_2} G^0(l,l_1;\omega)\sum_A c^A \times T^{A(l_2)}(l_1,l_2;\omega)\,\langle G^{A(l_2)}(l_2,l';\omega)\rangle_{A(l_2)}. \qquad (10.15.27)$$

It must be noted that T is now to be calculated in terms of G^0 instead of G_p.

If we now identify u with the exact Σ, then G^0 becomes equal to the average exact $\langle G\rangle$ and the scattering term on the right-hand side of equation (10.15.27) is equal to zero. However, as we have seen above, equation (10.15.27) is just the first equation of an infinite set and we must approximate. The method adopted by Taylor (1967), following earlier related work by Lax, is to set the scattering term equal to zero after making the approximation necessary to terminate this infinite set of equations. This will then give the "best" value of u as the desired approximation to the self-energy Σ.

The method will again hinge on neglecting the random part of the effective field Green function, as in equation (10.15.19). But in the present case, as we have seen, the t-matrix is to be calculated using G^0 rather than G_p.

In the present situation, all the functions in equation (10.15.21) depend only on the distance between the two sites indicated, and it is then helpful to Fourier transform. The condition for obtaining u then becomes

$$\sum_A c^A T^A(\mathbf{k},\omega) = 0, \qquad (10.15.28)$$

which should be compared with equation (10.13.67) in the discussion of electron states. The simplest approximation for large concentrations is the virtual crystal method [cf. equation (10.13.1)], where

$$u(\omega) = Mc\varepsilon\omega^2\,\mathbf{1}, \qquad (10.15.29)$$

which is obtained by replacing T^A by V^A in equation (10.15.28). But to obtain a correct description of the local, or resonant, mode, it is necessary, as we remarked above, to proceed beyond the first iteration and obtain the T-matrix.

Using the explicit forms of V in equation (10.15.26), we find with $c^d = c$ and $c^h = 1-c$

$$\mathbf{u}(\mathbf{k}\omega)\left[\mathbf{1}+\frac{1}{N}\sum_{\mathbf{k}}\mathbf{G}^0(\mathbf{k},\omega)\,\mathbf{u}(\mathbf{k},\omega)\right]^{-1} \times\left[\mathbf{1}-(1-c)\,M\varepsilon\omega^2\,\mathbf{G}^0(\omega)+\frac{1}{N}\sum_{\mathbf{k}}\mathbf{G}^0(\mathbf{k},\omega)\,\mathbf{u}(\mathbf{k},\omega)\right] = Mc\varepsilon\omega^2\,\mathbf{1} \tag{10.15.30}$$

where

$$\mathbf{G}^0(\omega) = \mathbf{G}^0(l,l;\,\omega) = \frac{1}{N}\sum_{\mathbf{k}}\mathbf{G}^0(\mathbf{k},\omega). \tag{10.15.31}$$

Thus, just as for the electron states case, $\mathbf{u}(\mathbf{k},\omega)$ is independent of $\mathbf{k}$ and we find

$$\mathbf{u}(\omega)-Mc\varepsilon\omega^2\,\mathbf{1}-\mathbf{u}(\omega)\,[M\varepsilon\omega^2\,\mathbf{1}-\mathbf{u}(\omega)]\,\mathbf{G}^0(\omega) = 0. \tag{10.15.32}$$

G^0 is also related to u by equation (10.15.24) in terms of G_{p} which is given by

$$G^{\mathrm{p}}_{\alpha\beta}(l,l';\,\omega) = \frac{1}{N(M_\alpha M_\beta)^{\frac{1}{2}}} \times\sum_{j\mathbf{k}}\frac{\sigma^{j\dagger}_{\alpha}(\mathbf{k})\,\sigma^{j}_{\beta}(\mathbf{k})\,\mathrm{e}^{-i\mathbf{k}.(\mathbf{R}_l-\mathbf{R}_{l'})}}{\omega^2-\omega_j^2(\mathbf{k})} \tag{10.15.33}$$

$\sigma^j_\alpha(\mathbf{k})$ and $\omega_j(\mathbf{k})$ being as usual the eigenvector and eigenfrequency for the $j,\mathbf{k}$ normal mode of the perfect crystal. In a monatomic cubic crystal we have

$$G^{\mathrm{p}}_{\alpha\beta}(l,l;\,\omega) = \delta_{\alpha\beta}\,G^{\mathrm{p}}(\omega), \tag{10.15.34}$$

and it is then consistent to assume that $\mathbf{u}(\omega)$ is diagonal and to rewrite equation (10.15.32) in the form

$$\left.\begin{aligned} &\tilde{\varepsilon}(\omega)-c\varepsilon = \tilde{\varepsilon}(\omega)\,[\varepsilon-\tilde{\varepsilon}(\omega)]\,\omega^2\,G^0(\omega) \\ &\text{with} \\ &\mathbf{u}(\omega) = M\tilde{\varepsilon}(\omega)\,\omega^2\,\mathbf{1} \quad\text{and}\quad G^0(\omega) = \int\frac{\nu(\omega')\,d\omega'}{\omega^2[1-\tilde{\varepsilon}(\omega)]-\omega'^2}. \end{aligned}\right\} \tag{10.15.35}$$

After solution of the self-consistent equation (10.15.35), the density of states is readily found as

$$\nu'(\omega) = \frac{2}{\pi\omega}\int \nu(\omega')\,d\omega'\,\mathrm{Im}\frac{\omega'^2}{\omega^2[1-\tilde{\varepsilon}(\omega)]-\omega'^2}. \tag{10.15.36}$$

Taylor also gives the one-phonon approximation to the neutron scattering cross-section, but we shall not go into details here.

(b) *Example of gold in copper.* Svensson and colleagues (1965) have studied the case of gold in copper using inelastic neutron scattering, and Taylor has

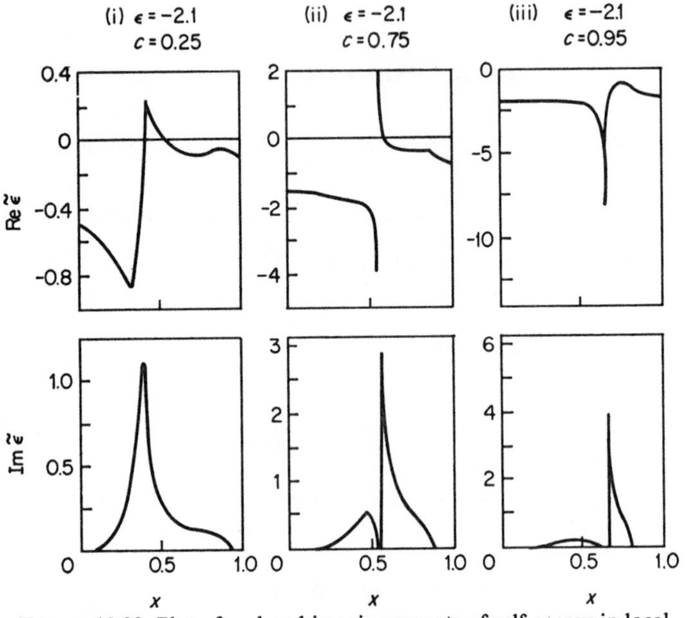

FIGURE 10.22. Plot of real and imaginary parts of self-energy in local phonon mode. Actual quantity plotted is $\tilde{\varepsilon}(\omega)$ [see equation (10.15.35)]. Case is gold in copper: three different concentrations c are shown (after Taylor, 1967).

therefore applied the theory given above to this case, using the density of states for pure copper calculated by Sinha (1966). In this case $\varepsilon = -2.1$, and Taylor has solved equation (10.15.35) by an iterative procedure. The behaviour he finds for $\tilde{\varepsilon}(\omega)$ for three different concentrations is shown in Figure 10.22.

At $c = 0.25$, typical resonant behaviour is clearly shown, and this is familiar already from the small concentration theory discussed previously. Essentially, we are seeing the frequency at which the heavier defects prefer to oscillate. As the concentration increases, this resonant characteristic becomes more exaggerated and moves to higher frequencies. When a gap appears, the real part of $\tilde{\varepsilon}(\omega)$ may diverge as illustrated in Figure 10.22 for $c = 0.75$. At even larger values of c, the real part of $\tilde{\varepsilon}(\omega)$ takes on a different behaviour in the gap with the divergence disappearing, as shown in Figure 10.22 for $c = 0.95$.

Finally, we show in Figure 10.23 the density of states and $\operatorname{Im} G^0$ for three different concentrations. As the concentration c increases, the resonant peak

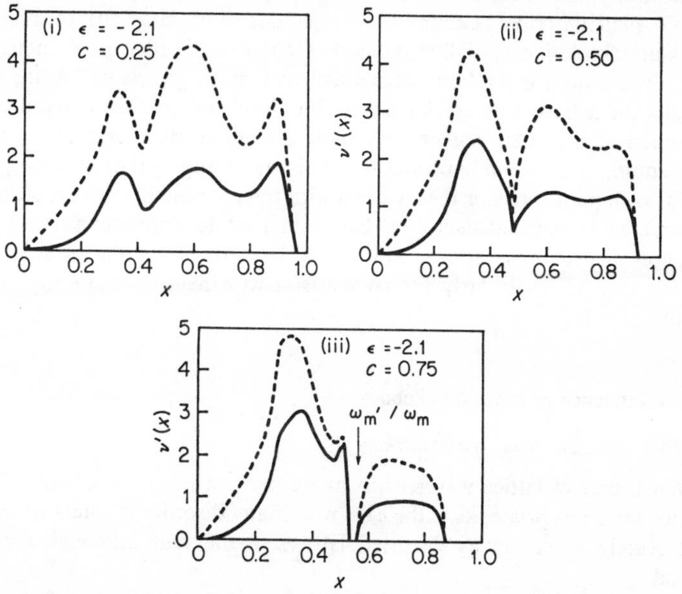

FIGURE 10.23. Plot of phonon density of states (solid lines) for three different concentrations of gold in copper. Resonance peak is associated predominantly with motion of heavy defect atom (after Taylor, 1967).

grows and broadens and eventually begins to show the structure of the pure density of states $\nu(\omega)$. This is not surprising if we view the peak as mainly due to the motions of the heavy defect atoms.

That this coherent potential method is giving good results for the density of states is shown by comparing the predictions of the theory with machine calculations made by Payton and Visscher (1967) for simple cubic crystals. This has been done by Taylor, with very encouraging results, though the spikes the machine calculations revealed for low concentrations of light defects are not given by the coherent potential method. This structure can be shown to arise from small groupings of defects, and such a local environment effect is eliminated by the coherent potential approximation. But this method, stemming from the ideas of Rayleigh, is clearly a very important one in the theory of disordered systems.

Having dealt with disorder on a lattice, as in a random solid alloy, we shall turn finally to discuss how phonons can be treated when there is only short-range order, as in amorphous solids or liquids. Here, the treatments, of necessity, are at the moment much more primitive.

We shall describe essentially two methods. The first, due to Bardasis, Falk and Simkin, essentially looks at the influence of order on phonons, and can give results for random systems such as a dilute gas, as well as the usual results for a harmonic crystal. It has the disadvantage that it requires the assumed atomic pair potential to have a Fourier transform. The second procedure, due to Hubbard and Beeby, is more appropriate to amorphous solids or liquids. Neither theory deals with the phonon lifetime problem in a disordered system satisfactorily, but both lead to approximations to the phonon dispersion relations, which may be valuable at the present stage of the theory, particularly for comparison with neutron-scattering experiments.

10.16 Influence of order on phonons

10.16.1 Bardasis–Falk–Simkin theory

When perfect lattice waves, such as we discussed fully in Chapter 3, are no longer appropriate, as is the case with many disordered solids which are not readily described as distorted lattices, a different approach must be found.

Collective coordinates which have been widely used in liquids are the density fluctuations $\rho_{\mathbf{k}}$ (see, for example, Percus and Yevick, 1958). We shall begin the discussion here by referring especially to the work of Bardasis, Falk and Simkin (1965) and to that of Morgan (1968).

We begin by summarizing what differences exist when we speak of sound waves, or longitudinal phonons, in crystals, liquids and gases, since here we are covering the entire spectrum of disorder. In each case, we can say that we are dealing with any compressional density fluctuation which a system might undergo, in any of these three phases.

One factor which distinguishes the phonons in a crystal from those in a gas is that the total momentum carried by lattice phonons is zero, while for the gas this need not be so. More obviously, differences in the nature of the different phases are clearly reflected in the theoretical treatments. Thus, in the crystalline solid, we have seen that all emphasis is placed on the deviation of the positions of the ions from a set of fixed lattice points, while in a gas it is appropriate to focus attention on the deviation of the density from a uniform value.

We summarize first below a uniform approach to phonon theory, which, nonetheless, still reflects the physical differences of the phonons in crystalline and disordered systems.

(a) *Potential energy*. Let us start out, quite generally, from a system of N particles of mass M in a volume $\mathscr{V}$, and let n be the number density. To be quite specific, we suppose the particles interact via a two-particle central potential $\phi(r)$. The potential energy of the system is

$$\Phi = \frac{1}{2}\sum_{ii'}{}' \phi(\mathbf{R}_{ii'}) = \frac{1}{2}\sum_{ii'} \phi(\mathbf{R}_{ii'}) - \tfrac{1}{2}N\phi(0), \tag{10.16.1}$$

where the position of atom i is $\mathbf{R}_i$ and $\mathbf{R}_{ii'} = \mathbf{R}_i - \mathbf{R}_{i'}$. The prime on the summation means that the term $i = i$ is to be omitted. We now write

$$\mathbf{R}_i = \mathbf{R}_i^0 + \mathbf{r}_i \tag{10.16.2}$$

and expand equation (10.16.1) for small $\mathbf{r}_i$. For the case of the crystalline solid, we have, of course, previously chosen $\mathbf{R}_i^0$ as the lattice sites. It need be of no concern at the moment whether the $\mathbf{R}_i^0$ are on a lattice, or even positions fixed in space.

It will be convenient, though somewhat of a limitation to the realism of the treatment, to suppose that the pair potential $\phi(r)$ has a Fourier transform, defined by

$$\phi(r) = \frac{1}{(2\pi)^3}\int d\mathbf{k}\,\tilde{\phi}(k)\exp(-i\mathbf{k}.\mathbf{r}). \tag{10.16.3}$$

Then we can write equation (10.16.1) in Fourier transform as

$$\Phi = \frac{1}{2}\int \frac{d\mathbf{k}'}{(2\pi)^3}\tilde{\phi}(\mathbf{k}')\sum_{ii'}\exp(-i\mathbf{k}'.\mathbf{R}_{ii'}) - \tfrac{1}{2}N\phi(0). \tag{10.16.4}$$

Equation (10.16.2) may equivalently be expressed as

$$\mathbf{R}_{ii'} = \mathbf{R}_{ii'}^0 + \mathbf{r}_{ii'}, \tag{10.16.5}$$

where $\mathbf{r}_{ii'} = \mathbf{r}_i - \mathbf{r}_{i'}$, and we expand equation (10.16.4), treating $\mathbf{r}_{ii'}$ as small.

Then we obtain

$$\Phi = \frac{1}{2}\int \frac{d\mathbf{k}'}{(2\pi)^3}\tilde{\phi}(k')\sum_{ii'}\exp(-i\mathbf{k}.\mathbf{R}^0_{ii'})$$

$$\times[1-i\mathbf{k}'.\mathbf{r}_{ii'}-\tfrac{1}{2}(\mathbf{k}'.\mathbf{r}_{ii'})^2+\ldots]-\tfrac{1}{2}N\phi(0). \qquad (10.16.6)$$

We are studying longitudinal phonons, and we can write

$$\mathbf{r}_i = \frac{i}{(NM)^{\frac{1}{2}}}\sum_{\mathbf{k}}\frac{\mathbf{k}}{k}q_{\mathbf{k}}\exp(i\mathbf{k}.\mathbf{R}^0_i), \qquad (10.16.7)$$

where $q_{\mathbf{k}}$ satisfies $q^\dagger_{\mathbf{k}} = q_{-\mathbf{k}}$. In terms of the q's, the potential energy becomes

$$\Phi = -\tfrac{1}{2}N\phi(0)+\frac{1}{2}\int\frac{d\mathbf{k}'}{(2\pi)^3}\tilde{\phi}(k')\sum_{ii'}\exp(-i\mathbf{k}'.\mathbf{R}^0_{ii'})$$

$$\times\Bigg[1+\frac{1}{(NM)^{\frac{1}{2}}}\sum_{\mathbf{k}}(\mathbf{k}'.\mathbf{k}/k)q_{\mathbf{k}}\exp(i\mathbf{k}.\mathbf{R}^0_{i'})[\exp(i\mathbf{k}.\mathbf{R}^0_{ii'}-1)]$$

$$+\frac{1}{2NM}\sum_{\mathbf{k}_1\mathbf{k}_2}(\mathbf{k}'.\mathbf{k}_1/k_1)(\mathbf{k}'.\mathbf{k}_2/k_2)q_{\mathbf{k}_1}q_{\mathbf{k}_2}\exp[i\{\mathbf{k}_1+\mathbf{k}_2\}.\mathbf{R}^0_{i'}]$$

$$\times\{\exp[i(\mathbf{k}_1+\mathbf{k}_2).\mathbf{R}^0_{ii'}-\exp(i\mathbf{k}_1.\mathbf{R}^0_{ii'})-\exp(i\mathbf{k}_2.\mathbf{R}^0_{ii'})+1]\}+\ldots\Bigg]. \qquad (10.16.8)$$

We convert sums into integrals by writing

$$\sum_{ab}f(\mathbf{R}^0_{i'},\mathbf{R}^0_{ii'}) = \int d\mathbf{r}\int d\mathbf{R}\, n_{\mathrm{op}}(\mathbf{r})p_{\mathrm{op}}(\mathbf{R})f(\mathbf{rR}) \qquad (10.16.9)$$

where $n_{\mathrm{op}}(\mathbf{r})$ and $p_{\mathrm{op}}(\mathbf{R})$ are defined by

$$n_{\mathrm{op}}(\mathbf{r}) = \sum_{i'}\delta(\mathbf{r}-\mathbf{R}^0_{i'}) \qquad (10.16.10)$$

and

$$p_{\mathrm{op}}(\mathbf{R}) = \sum_{\mathbf{R}_{ii'}{}^0}\delta(\mathbf{R}-\mathbf{R}^0_{ii'}). \qquad (10.16.11)$$

Now an approximation is made to the right-hand side of equation (10.16.9) by replacing $n_{\mathrm{op}}(\mathbf{r})$ and $p_{\mathrm{op}}(\mathbf{R})$ by their expectation values in the ground state, that is, when no excitations are present in the system. Thus $n_{\mathrm{op}}(\mathbf{r})$ is replaced by the number density $n(\mathbf{r})$ and $p_{\mathrm{op}}(\mathbf{R})$ by $g(\mathbf{R})$, the pair correlation function expressing the probability of finding a particle at $\mathbf{R}$, given one at the origin. This approximation is naturally neglecting higher-order correlations in evaluating equation (10.16.9).

(b) *Models.* (i) *Gas*: To see how the method works out in practice, let us consider a case of a dilute gas where we can neglect correlations and write

$$n(\mathbf{r}) = g(\mathbf{R}) = n, \tag{10.16.12}$$

so that

$$\sum_{ii'} f(\mathbf{R}_{i'}^0, \mathbf{R}_{ii'}^0) = n^2 \int d\mathbf{r} \int d\mathbf{R}\, f(\mathbf{r}\mathbf{R}). \tag{10.16.13}$$

Using the result

$$\int d\mathbf{r} \exp(-i\mathbf{k}'.\mathbf{r}) = (2\pi)^3\, \delta(\mathbf{k}') = \mathscr{V}\, \delta_{\mathbf{k}'0}, \tag{10.16.14}$$

the potential energy may be written as

$$\begin{aligned}\Phi = &-\tfrac{1}{2}N\phi(0) + \frac{1}{2}\int \frac{d\mathbf{k}'}{(2\pi)^3}\tilde{\phi}(k')\, n^2 \int d\mathbf{R} \exp(-i\mathbf{k}'.\mathbf{R})\\ &\times \mathscr{V}\Big\{1 + \frac{1}{(NM)^{\frac{1}{2}}}\sum_{\mathbf{k}}\left(\frac{\mathbf{k}'.\mathbf{k}}{k}\right) q_{\mathbf{k}}[\exp(i\mathbf{k}.\mathbf{R}) - 1]\,\delta_{\mathbf{k}0}\\ &+ \frac{1}{2NM}\sum_{\mathbf{k}_1\mathbf{k}_2}\left(\mathbf{k}'.\frac{\mathbf{k}_1}{k_1}\right)\left(\mathbf{k}'.\frac{\mathbf{k}_2}{k_2}\right) q_{\mathbf{k}_1} q_{\mathbf{k}_2}\\ &\times [\exp i(\mathbf{k}_1+\mathbf{k}_2).\mathbf{R} - \exp(-i\mathbf{k}_1.\mathbf{R}) - \exp(i\mathbf{k}_2.\mathbf{R}) + 1]\\ &\times \delta_{\mathbf{k}_1,-\mathbf{k}_2} + \dots\Big\}\end{aligned} \tag{10.16.15}$$

or

$$\Phi = -\tfrac{1}{2}N\phi(0) + \tfrac{1}{2}nN\tilde{\phi}(0) + \frac{1}{2}\sum_{\mathbf{k}} k^2 \frac{n}{M}\tilde{\phi}(k)\, q_{\mathbf{k}}\, q_{-\mathbf{k}} + \dots. \tag{10.16.16}$$

To this order the form (10.16.16) is the same as we would have obtained if we had used the density fluctuations $\rho_{\mathbf{k}}$.

(ii) *Crystal*: For a crystal we have

$$\begin{aligned}\int d\mathbf{r}\, n(\mathbf{r}) \exp(i\mathbf{k}.\mathbf{r}) &= N \sum_{\mathbf{k}} \delta_{\mathbf{k},\mathbf{K}}\\ &= \int d\mathbf{R}\, g(\mathbf{R}) \exp(i\mathbf{K}.\mathbf{R}),\end{aligned} \tag{10.16.17}$$

where the $\mathbf{K}$, as usual, are reciprocal lattice vectors for the lattice defined by $n(\mathbf{r})$.

Using this, and the property that $\exp(i\mathbf{K}.\mathbf{R}^0_{ii'}) = 1$, the potential energy (10.16.8) becomes

$$\Phi = -\tfrac{1}{2}N\phi(0)+\frac{1}{2}\int\frac{d\mathbf{k}'}{(2\pi)^3}\phi(\mathbf{k}')\sum_{\mathbf{R}_{ii'}{}^0}\exp(-i\mathbf{k}'.\mathbf{R}^0_{ii'})$$

$$\times\left\{N+\frac{1}{2M}\sum_{\mathbf{k}_1\mathbf{k}_2}\sum_{\mathbf{K}_1}\left(\mathbf{k}'.\frac{\mathbf{k}_1}{k_1}\right)\left(\mathbf{k}'.\frac{\mathbf{k}_2}{k_2}\right)q_{\mathbf{k}_1}q_{\mathbf{k}_2}\,\delta_{\mathbf{k}_1+\mathbf{k}_2,\mathbf{K}_1}\right.$$

$$\left.\times[2-\exp(-i\mathbf{k}_1.\mathbf{R}^0_{ii'})-\exp(-i\mathbf{k}_1.\mathbf{R}^0_{ii'})]+\dots\right\}$$

$$= -\tfrac{1}{2}N\phi(0)+\tfrac{1}{2}nN\sum_{\mathbf{K}}\tilde{\phi}(\mathbf{K})+\frac{n}{2M}\sum_{\mathbf{k}_1\mathbf{k}_2}\sum_{\mathbf{K}_1\mathbf{K}_2}\left(\mathbf{K}_2.\frac{\mathbf{k}_1}{k_1}\right)\left(\mathbf{K}_2.\frac{\mathbf{k}_2}{k_2}\right)$$

$$\times q_{\mathbf{k}_1}q_{\mathbf{k}_2}\,\delta_{\mathbf{k}_1+\mathbf{k}_2,\mathbf{K}_1}\tilde{\phi}(\mathbf{K}_2)-\frac{n}{2M}\sum_{\mathbf{k}_1\mathbf{k}_2}\sum_{\mathbf{K}_1\mathbf{K}_2}\left[(\mathbf{k}_1+\mathbf{K}_2).\frac{\mathbf{k}_1}{k_1}\right]$$

$$\times\left[(\mathbf{k}_1+\mathbf{K}_2).\frac{\mathbf{k}_2}{k_2}\right]q_{\mathbf{k}_1}q_{\mathbf{k}_2}\,\delta_{\mathbf{k}_1+\mathbf{k}_2,\mathbf{K}_1}\tilde{\phi}(\mathbf{k}_1+\mathbf{K}_2)+\dots. \qquad (10.16.18)$$

For crystals, the summation over $\mathbf{k}$ in the form (10.16.7) of $\mathbf{r}_i$ is restricted to the BZ, so that only $\mathbf{K}_1 = 0$ contributes in equation (10.16.18), since both $\mathbf{k}_1$ and $\mathbf{k}_2$ lie in the BZ. Hence

$$\Phi = -\tfrac{1}{2}N\phi(0)+\tfrac{1}{2}nN\sum_{\mathbf{K}}\tilde{\phi}(\mathbf{K})$$

$$+\frac{n}{2M}\sum_{\mathbf{k}}q_{\mathbf{k}}q_{-\mathbf{k}}\sum_{\mathbf{K}}\left\{\left[\frac{\mathbf{k}}{k}.(\mathbf{k}+\mathbf{K})\right]^2\tilde{\phi}\left(|\mathbf{k}+\mathbf{K}|\right)-\left(\frac{\mathbf{k}}{k}.\mathbf{K}\right)^2\tilde{\phi}(k)\right\}+\dots. \qquad (10.16.19)$$

(c) *Kinetic energy*. We must next turn to deal with the kinetic energy, which reflects the major difference between the crystal and gas. This difference resides in the fact that for a crystal one is interested in the ionic displacement from equilibrium $\mathbf{r}_i$, or equivalently $q_{\mathbf{k}}$, while in the case of a gas one looks instead at the deviation of the density operator

$$\rho(\mathbf{r}) = \sum_i \delta(\mathbf{r}-\mathbf{R}_i) \equiv \int\frac{d\mathbf{k}}{(2\pi)^3}\rho_{\mathbf{k}}\exp(i\mathbf{k}.\mathbf{r}) \qquad (10.16.20)$$

from its average value N. Or, put another way, we are concerned with the $\mathbf{k}\neq 0$ values of

$$\rho_{\mathbf{k}} = \sum_i \exp(-i\mathbf{k}.\mathbf{R}_i) = \rho^{\dagger}_{-\mathbf{k}}. \qquad (10.16.21)$$

Thus the momentum variables that we work with should be canonically conjugate to $q_{\mathbf{k}}$ for the crystal, but to $\rho_{\mathbf{k}}$ for the gas.

(i) *Crystal*: Here one defines

$$\mathbf{P}_i = \frac{1}{i}\left(\frac{M}{N}\right)^{\frac{1}{2}} \sum_{\mathbf{k}} \frac{\mathbf{k}}{k} P_{\mathbf{k}} \exp(-i\mathbf{k}.\mathbf{R}_i^0), \tag{10.16.22}$$

where $\mathbf{k}$ is confined to the BZ and the operators $\mathbf{P}_{\mathbf{k}}$ satisfy

$$\mathbf{P}_{\mathbf{k}}^{\dagger} = \mathbf{P}_{-\mathbf{k}}. \tag{10.16.23}$$

Equations (10.16.7) and (10.16.22) are not strictly correct when applied to a solid, as sums over polarization should be included. We shall give a treatment below, due to Hubbard and Beeby (1969) in which this is done. However, to the extent that only longitudinal phonons are present, the argument here is valid. It is easy to see from the equation for $\mathbf{r}_i$ and the commutation relation

$$[P_{i'\mu}, r_{i\nu}] = \frac{\hbar}{i}\delta_{ii'}\,\delta_{\mu\nu}, \tag{10.16.24}$$

that

$$[P_{\mathbf{k}}, q_{\mathbf{k}'}] = \frac{\hbar}{i}\delta_{\mathbf{k}\mathbf{k}'}. \tag{10.16.25}$$

In terms of the $P_{\mathbf{k}}$ we may write

$$\begin{aligned} T &= -\frac{1}{2N}\sum_{\mathbf{k}_1\mathbf{k}_2}\left(\frac{\mathbf{k}_1}{k_1}.\frac{\mathbf{k}_2}{k_2}\right) P_{\mathbf{k}_1} P_{\mathbf{k}_2} \sum_i \exp(-i\{\mathbf{k}_1+\mathbf{k}_2\}.\mathbf{R}_i^0) \\ &= -\frac{1}{2N}\sum_{\mathbf{k}_1\mathbf{k}_2}\left(\frac{\mathbf{k}_1}{k_1}.\frac{\mathbf{k}_2}{k_2}\right) P_{\mathbf{k}_1} P_{\mathbf{k}_2} N \sum_{K} \delta_{\mathbf{k}_1+\mathbf{k}_2,\mathbf{K}}. \end{aligned} \tag{10.16.26}$$

Using the fact that $\mathbf{k}_1$ and $\mathbf{k}_2$ are both in the BZ,

$$T = \frac{1}{2}\sum_{\mathbf{k}} P_{\mathbf{k}} P_{-\mathbf{k}}, \tag{10.16.27}$$

which should be compared with equation (3.2.21).

(ii) *Gas*: From equation (10.16.22) for $\mathbf{P}_i$ it is clear that the total lattice momentum $\sum_i \mathbf{P}_i$ vanishes for the crystal, for any excitation of the system.

As the total phonon momentum need not vanish for the gas, we cannot use the $\mathbf{P}_k$ defined by equation (10.16.22) and we must, as we remarked above, introduce momenta conjugate to $\rho_{\mathbf{k}}$ rather than $q_{\mathbf{k}}$. Thus we define

$$\mathbf{P}_i = \mathbf{P}_i^\dagger = \frac{1}{i}\left(\frac{M}{N}\right)^{\frac{1}{2}} \sum_{\mathbf{k}} \frac{\mathbf{k}}{k} \exp(-i\mathbf{k}.\mathbf{R}_i)\, P_{\mathbf{k}}$$

$$= \frac{1}{i}\left(\frac{M}{N}\right)^{\frac{1}{2}} \sum_{\mathbf{k}} \frac{\mathbf{k}}{k} P_{\mathbf{k}}^\dagger \exp(i\mathbf{k}.\mathbf{R}_i). \qquad (10.16.28)$$

Summing over $\mathbf{R}_i$ in equation (10.16.28) above, we do not necessarily get zero, in contrast to the crystal.

At this point, we should emphasize that the validity of the above expansion depends crucially on the completeness of the $\mathbf{P}_{\mathbf{k}}$'s. Physically, this amounts to assuming that the only excitations possible are density fluctuations, which are collective in character.

Evaluating the kinetic energy, taking care to deal correctly with the fact that $\mathbf{P}_{\mathbf{k}}$ and $\mathbf{R}_a$ do not commute, we find

$$T = \frac{1}{2M} \sum_i \mathbf{P}_i^\dagger . \mathbf{P}_i$$

$$= \frac{1}{2N} \sum_{\mathbf{k}_1\mathbf{k}_2} \left(\frac{\mathbf{k}_1}{k_1} . \frac{\mathbf{k}_2}{k_2}\right) P_{\mathbf{k}_1}^\dagger \rho_{\mathbf{k}_2-\mathbf{k}_1} P_{\mathbf{k}_2}. \qquad (10.16.29)$$

Utilizing the commutation relation

$$[P_i, f(\mathbf{R}_{i'})] = \delta_{ii'} \frac{\hbar}{i} \nabla_i f(\mathbf{R}_i), \qquad (10.16.30)$$

we find

$$[P_i, \rho_{\mathbf{k}'}] = -\hbar\mathbf{k}' \exp(-i\mathbf{k}'.\mathbf{R}_i). \qquad (10.16.31)$$

This result can now be used to cast the kinetic energy into the desired form as shown in Appendix 10.7. The result is

$$T = \frac{1}{2} \sum_{\mathbf{k}} \left(\bar{P}_{\mathbf{k}} \bar{P}_{-\mathbf{k}} + \frac{\hbar^2 k^2}{2m} q_{\mathbf{k}} q_{-\mathbf{k}}\right)$$

$$-\frac{1}{2} \sum_{\mathbf{k}} \frac{\hbar^2 k^2}{2M} + (\text{terms cubic in the } q\text{'s and } \bar{P}\text{'s}), \qquad (10.16.32)$$

where $\bar{P}_{\mathbf{k}}$ is defined by

$$\bar{P}_{\mathbf{k}} = P_{\mathbf{k}} + \frac{i\hbar k}{2(MN)^{\frac{1}{2}}} \rho_{-\mathbf{k}} \qquad (10.16.33)$$

and satisfies

$$[\bar{P}_{\mathbf{k}}, q_{\mathbf{k}'}] = \frac{\hbar}{i} \delta_{\mathbf{k}\mathbf{k}'} + \text{terms linear in } q. \qquad (10.16.34)$$

(d) *Phonon spectra.* We can now achieve our object of getting the spectrum of elementary excitations in the two cases:

(i) *Crystal*: Adding the potential and kinetic energy given by equations (10.16.19) and (10.16.27) we find

$$H = T+V = \tfrac{1}{2}nN\sum_{\mathbf{K}}\tilde{\phi}(\mathbf{K})-\tfrac{1}{2}N\phi(0)$$
$$+\frac{1}{2}\sum_{\mathbf{k}}\Bigg[P_{\mathbf{k}}P_{-\mathbf{k}}+q_{\mathbf{k}}q_{-\mathbf{k}}\frac{n}{M}$$
$$\times\sum_{\mathbf{K}}\left\{\left[\frac{\mathbf{k}}{k}.(\mathbf{k}+\mathbf{K})\right]^2\tilde{\phi}(|\mathbf{k}+\mathbf{K}|)-\left(\frac{\mathbf{k}}{k}.\mathbf{K}\right)\tilde{\phi}(K)\right\}\Bigg]+\ldots. \tag{10.16.35}$$

But this is just the Hamiltonian for independent harmonic oscillators with frequencies given by

$$\omega^2(\mathbf{k}) = \frac{n}{M}\sum_{K}\left[\!\!\left[\left\{\left[\frac{\mathbf{k}}{k}.(\mathbf{k}+\mathbf{K})\right]\right\}^2\tilde{\phi}(\mathbf{k}+\mathbf{K})-\left(\frac{\mathbf{k}.\mathbf{K}}{k}\right)^2\phi(K)\right]\!\!\right]. \tag{10.16.36}$$

This is the standard result for longitudinal phonons in a crystal (see Pines, 1963).

(ii) *Gas*: Similarly, adding equations (10.16.16) and (10.16.32) we find

$$H = \tfrac{1}{2}nN\tilde{\phi}(0)-\tfrac{1}{2}N\phi(0)-\frac{1}{2}\sum_{k}\frac{\hbar^2k^2}{2M}$$
$$+\frac{1}{2}\sum_{\mathbf{k}}\left\{\bar{P}_{\mathbf{k}}P_{-\mathbf{k}}+\left[\left(\frac{\hbar k^2}{2M}\right)^2+k^2\frac{n}{M}\tilde{\phi}(k)\right]q_{\mathbf{k}}q_{-\mathbf{k}}\right\}+\ldots. \tag{10.16.37}$$

Again we have uncoupled harmonic oscillators, this time with frequencies

$$\omega^2(k) = \left(\frac{\hbar k^2}{2M}\right)^2+k^2\frac{n}{M}\tilde{\phi}(k), \tag{10.16.38}$$

which is to be compared with the Bogliubov and Zubarev (1955) result for the low-density Bose gas.

Setting $\tilde{\phi}(k) = 0$ in (10.16.38) will, of course, give the usual free-particle dispersion relation. This is because, in contrast with our treatment of the solid, we have included the kinetic energy associated with the equilibrium points.

The above argument could be applied to liquids, but would need to be corrected chiefly for the effects of backflow (Feynman and Cohen, 1956). To lowest order, equation (10.16.21) shows us that $\rho_{\mathbf{k}}$ is proportional to $q_{\mathbf{k}}$. To include backflow, higher-order terms involving quadratic terms in the

q's would have to be included. The problem of backflow is less important in a gas, since the atoms are further apart than in a liquid, and essentially negligible in a crystal, where ions remain close to lattice equilibrium positions.

10.16.2 Hubbard–Beeby theory

We shall next discuss the theory of Hubbard and Beeby (1969) designed to describe collective motions in liquids. We shall restrict ourselves here to its application to cold amorphous solids.

The method used is to calculate the density–density response function. Suppose that a weak time-dependent potential field $V(\mathbf{r}t)$ acts on the system. This will give rise to changes in the density and for small V we have

$$\delta\rho(\mathbf{r}t) = \int_{-\infty}^{\infty} dt' \int d\mathbf{r}'\, \chi(\mathbf{r}-\mathbf{r}', t-t')\, V(\mathbf{r}'t'), \qquad (10.16.39)$$

where χ is the density–density response function. If we take the Fourier transform of $\chi(\mathbf{r}t)$ into $\mathbf{k}\omega$ space and write

$$\chi(\mathbf{k}\omega) = \chi'(\mathbf{k}\omega) + i\chi''(\mathbf{k}\omega), \qquad (10.16.40)$$

then using the fluctuation–dissipation theorem discussed in Chapter 7 (see also section 4.6, Volume 1) it can be shown that the van Hove scattering function $S(\mathbf{k}\omega)$ is given by

$$S(\mathbf{k}\omega) = \frac{1}{\pi\rho} \frac{\hbar\omega}{1-e^{\beta\hbar\omega}} \frac{\chi''(\mathbf{k}\omega)}{\omega}. \qquad (10.16.41)$$

One property of $\chi(\mathbf{r}t)$ is that it must vanish when $t<0$ since $\delta\rho(\mathbf{r}t)$ cannot depend on $V(\mathbf{r}t)$ at later times. Hence $\chi(\mathbf{k}\omega)$ satisfies a dispersion relation of the form

$$\chi(\mathbf{k}\omega) = \lim_{\eta\to 0+} \int \frac{d\omega'\chi''(\mathbf{k}\omega')}{\pi(\omega'-\omega-i\eta)}. \qquad (10.16.42)$$

$\chi''(\mathbf{k}\omega)$ is an odd function of ω and its Fourier transform $\chi''(\mathbf{k}t)$ is correspondingly an odd function of t. Equation (10.16.42) implies that the transform $\chi(\mathbf{k}t)$ of $\chi(\mathbf{k}\omega)$ is given by

$$\chi(\mathbf{k}t) = 2i\theta(t)\chi''(\mathbf{k}t): \quad \left.\begin{array}{ll} \theta(t) = 1, & t>0, \\ \theta(t) = 0, & t<0. \end{array}\right\} \qquad (10.16.43)$$

We shall study then the linear response of a disordered stationary array of atoms, as in a cold amorphous solid.

Let us assume that the weak external potential energy has the explicit form

$$V(\mathbf{r}t) = V e^{i(\mathbf{k}.\mathbf{r}-\omega t)}. \qquad (10.16.44)$$

Then the atoms will be displaced from their sites by an amount $\mathbf{u}_i(t)$, $\mathbf{u}_i$ being infinitesimal if V is infinitesimal (as assumed). Clearly, we can calculate the change in the force acting on the ith particle from a knowledge of the pair interaction $\phi(r)$ and the external potential. This enables us to write down the equation of motion for $\mathbf{u}_i$,

$$M\ddot{\mathbf{u}}_i = -\sum_{j\neq i} \nabla\nabla\phi(\mathbf{r}_i-\mathbf{r}_j).(\mathbf{u}_i-\mathbf{u}_j) - iV\mathbf{k}\exp[i(\mathbf{k}.\mathbf{r}_i-\omega t)], \tag{10.16.45}$$

where we have worked to first order in V (and therefore in $\mathbf{u}$).

Secondly, since we want to relate $\delta\rho$ to V, we can calculate the change $\delta\rho_{\mathbf{k}}$ in $\rho_{\mathbf{k}}$ trivially, to the same order, and we find

$$\delta\rho_{\mathbf{k}} = \sum_{i=1}^{N} i\mathbf{k}.\mathbf{u}_i e^{i\mathbf{k}.\mathbf{r}_i}. \tag{10.16.46}$$

In principle, we can now solve the equation of motion for $\mathbf{u}_i$, substitute in the above expression for $\delta\rho_{\mathbf{k}}$ and hence obtain $\chi(\mathbf{k}\omega)$. In practice we cannot do this exactly, but we shall see below that a rather natural approximation brings back our earlier result for crystals.

To tidy up a little, we notice that since (10.16.45) is linear in $\mathbf{u}_i(t)$ its time dependence must be $e^{-i\omega t}$ and we work with a quantity defined by

$$\bar{\mathbf{u}}_i = \frac{\mathbf{u}_i(t)\exp[i(\omega t-\mathbf{k}.\mathbf{r}_i)]}{V}. \tag{10.16.47}$$

Then $\bar{\mathbf{u}}_i$ satisfies the equation

$$M\omega^2\bar{\mathbf{u}}_i = i\mathbf{k} - \sum_j \exp[i\mathbf{k}.(\mathbf{r}_j-\mathbf{r}_i)]\, v_{ij}\bar{\mathbf{u}}_j, \tag{10.16.48}$$

where v_{ij} is simply a shorthand notation for $\nabla\nabla\phi(\mathbf{r}_i-\mathbf{r}_j)$ for $i\neq j$ and $-v_{ii} = \sum_{j(\neq i)} v_{ij}$. Then we also have for the density change

$$\delta\rho_{\mathbf{k}}(t) = -iVe^{-i\omega t}\mathbf{k}.\sum_i \bar{\mathbf{u}}_i. \tag{10.16.49}$$

Now we proceed to iterate the equation of motion and we find

$$\bar{\mathbf{u}}_i = \frac{i\mathbf{k}}{M\omega^2} - \frac{i}{(M\omega^2)^2}\sum_i v_{ij}\, e^{i\mathbf{k}.(\mathbf{r}_j-\mathbf{r}_i)}\mathbf{k}$$

$$+\frac{i}{(M\omega^2)^3}\sum_{j,l} v_{ij}v_{jl}\, e^{-i\mathbf{k}.(\mathbf{r}_j-\mathbf{r}_l)}\mathbf{k}+\ldots. \tag{10.16.50}$$

This equation has now to be averaged over all the configurations. The average of the linear in v_{ij} in equation (10.16.54) is readily achieved, and involves just the pair function $g(r)$.

In the higher terms, we do not know precisely what distribution functions to use [we might use the Kirkwood (1935) approximation but life becomes rapidly unbearable]. The simplest procedure which suggests itself, and leads to $S(k\omega)$ of equation (10.16.61), is to represent the average of any product of the v_{ij} by the corresponding product of the averages of the individual v_{ij}. When we do this, we get for the first-order term

$$\left\langle \sum_j v_{ij}\,e^{i\mathbf{k}.(\mathbf{r}_i-\mathbf{r}_j)} \right\rangle = -\rho\psi(\mathbf{k}), \tag{10.16.51}$$

where explicitly

$$\psi(\mathbf{k}) = \int d\mathbf{r}\, g(\mathbf{r})\,\nabla\nabla\phi(\mathbf{r})\,(1-e^{i\mathbf{k}.\mathbf{r}}). \tag{10.16.52}$$

The second-order term, in the approximation referred to above, is $\rho^2[\psi(\mathbf{k})]^2$, and so on.

The result can be summed, the response function calculated from

$$\chi(\mathbf{k},\omega) = \frac{\langle \delta\rho_{\mathbf{k}} \rangle}{V e^{-i\omega t}} \tag{10.16.53}$$

and we obtain

$$\chi(\mathbf{k},\omega) = \rho\mathbf{k}.[M\omega^2\mathbf{1}-\rho\psi(\mathbf{k})]^{-1}.\mathbf{k}. \tag{10.16.54}$$

The singularities of the response function appear at the frequencies ω given by the roots of the secular equation

$$|M\omega^2\mathbf{1}-\rho\psi(\mathbf{k})| = 0, \tag{10.16.55}$$

which determines the frequencies of the collective phonon modes.

The frequencies are real, and this means that the approximations made do not include phonon lifetime effects, which is, of course, a rather severe limitation.

In the case of a cold crystalline solid, the pair function $g(\mathbf{r})$ can be written as

$$g(\mathbf{r}) = \frac{1}{\rho}\sum_{\mathbf{R}} \delta(\mathbf{r}-\mathbf{R}), \tag{10.16.56}$$

where the sum is over the lattice vectors. Then the result (10.16.55) reduces to

$$\left| M\omega^2\mathbf{1}-\sum_{\mathbf{R}} \phi(\mathbf{R})\,(1-e^{-i\mathbf{k}.\mathbf{R}}) \right| = 0, \tag{10.16.57}$$

which is the usual result giving the phonon dispersion relation, as discussed in Chapter 3. Thus the theory becomes exactly the same as the usual harmonic approximation in the crystalline case.

For isotropic systems, such as an amorphous solid or a liquid, the matrix $\psi(\mathbf{k})$ has a form which, from symmetry considerations, must be

$$\frac{\rho}{M}\psi(\mathbf{k}) = \omega_k^2 \hat{\mathbf{k}}\hat{\mathbf{k}} + \omega_{kt}^2(\mathbf{1} - \hat{\mathbf{k}}\hat{\mathbf{k}}), \qquad (10.16.58)$$

where $\hat{\mathbf{k}}$ is the unit vector $\mathbf{k}/k$. The equation (10.16.55) determining the frequencies of the collective modes then has one eigenvalue ω_k^2 and a pair of degenerate eigenvalues ω_{kt}^2 corresponding to one longitudinal mode and two transverse modes. Only the longitidinal mode contributes in equation (10.16.50) and we find

$$\chi(\mathbf{k}, \omega) = \frac{\rho k^2}{M} \frac{1}{\omega^2 - \omega_k^2}, \qquad (10.16.59)$$

where

$$\omega_k^2 = \frac{\rho}{M}\hat{\mathbf{k}}.\psi(\mathbf{k}).\hat{\mathbf{k}}$$

$$= \frac{\rho}{M}\int \frac{\partial^2 \phi}{\partial z^2} g(r)(1 - \cos kz)\, d\mathbf{r}. \qquad (10.16.60)$$

For small $\mathbf{k}$, the right-hand side is $\propto k^2$, and so $\omega_k = v_s k$ corresponding to a longitudinal sound velocity v_s. The scattering function has the form

$$S(\mathbf{k}\omega) = \frac{\hbar k^2}{2M\omega_k}[(n_k + 1)\,\delta(\omega - \omega_k) + n_k\,\delta(\omega + \omega_k)], \qquad (10.16.61)$$

where

$$n_k = [\exp(\hbar\beta\omega_k) - 1]^{-1}. \qquad (10.16.62)$$

The relation (10.16.55) predicts non-zero frequencies for the transverse modes even for small k. This is reflecting the fact that an amorphous solid has rigidity and so we expect to find transverse modes. The lack of rigidity of a liquid (in contrast) is a consequence of relaxation processes which result from the movement of the particles in a liquid and is clearly not allowed for in the above treatment of stationary atoms.

We expect from equation (10.16.61) that there will be a peak in the neutron scattering at the longitudinal collective mode frequency ω_k.

10.16.3 Thermal conduction in disordered solids

We shall conclude this account of phonons in disordered systems by discussing an application of the theories outlined above to the calculation of the phonon component of the thermal conductivity of structurally disordered solids. The approach follows that of Morgan (1968).

The basic method is to work with the variables $U(\mathbf{k})$ defined by

$$u_\alpha(j) = N^{-\frac{1}{2}} \sum_{\mathbf{k}} U_\alpha(\mathbf{k}) \exp(i\mathbf{k}.\mathbf{r}_j), \tag{10.16.63}$$

where $u_\alpha(j)$ is a Cartesian component of the displacement from equilibrium of the jth atom whose equilibrium position is at $\mathbf{r}_j$.

The plane waves in equation (10.16.63) are chosen to satisfy periodic boundary conditions within a volume containing N atoms and the sum over $\mathbf{k}$ contains just N terms. Thus, for an amorphous solid, the wave vectors are assumed to lie within a Debye sphere.

The basic difficulty in the method lies in transforming from momentum variables $p_\alpha(j)$ to collective momentum $P_\alpha(k)$. We shall not give the details as the calculations are similar to those of Bardasis and colleagues reported above. For the dispersion relation, the result is the same as that which can be obtained by averaging the harmonic equation of motion for the solid (cf. Morgan and Ziman, 1967), the eigenfrequencies being $\omega_\rho(\mathbf{k})$ say.

Morgan shows that from this approach, and following the procedure of Hardy (1966), the standard transport equation comes back, namely

$$-\frac{\partial}{\partial \mathbf{k}}\omega_\rho(\mathbf{k}).\nabla T\frac{\partial}{\partial T}n_0(\mathbf{k}\rho) = -\frac{\partial}{\partial t}n(\mathbf{k}\rho)\bigg|_{\text{coll}}, \tag{10.16.64}$$

where $n(\mathbf{k}\rho)$ is the expected number of phonons in the state $\mathbf{k}\rho$, ∇T is the temperature gradient and $n_0(\mathbf{k}\rho)$ is the Bose function

$$n_0(\mathbf{k}\rho) = [\exp(\hbar\beta\omega_\rho(\mathbf{k}))-1]^{-1}. \tag{10.16.65}$$

The heat current is simply given by

$$Q = \frac{1}{\mathscr{V}}\sum_\rho\sum_k \hbar\omega_\rho(\mathbf{k})\frac{\partial}{\partial \mathbf{k}}\omega_\rho(\mathbf{k})\, n(\mathbf{k}\rho) \tag{10.16.66}$$

and the collision term is to be evaluated from the perturbations to the Hamiltonian representing the independent oscillators, to first order in perturbation theory.

Morgan considers the intrinsic scattering rates between phonon states and shows that the transition probability $Q_{\mathbf{k}\rho}^{\mathbf{k}'\rho'}$ denoting the intrinsic scattering rate from a state $\mathbf{k}\rho$ to a state $\mathbf{k}'\rho'$ involves a term representing scattering due to density fluctuations, while the other terms which come in represent the effect of fluctuations in the elastic properties of the solid.

If we were dealing with Rayleigh scatterers distributed at the equilibrium sites of an amorphous solid, then the long wavelength form of Q is

$$Q_{\mathbf{k}\rho}^{\mathbf{k}'\rho'} \propto \omega^2 S(\mathbf{k}-\mathbf{k}'),$$

where $S(\mathbf{k})$ is the pair correlation function of the sites.

The evaluation of Morgan's result for $Q_{\mathbf{k}\rho}^{\mathbf{k}'\rho'}$ would be of considerable interest for different theoretical models of amorphous solids, but no further results are available at the time of writing.

10.17 Localization in random lattices

We shall conclude this discussion of disordered systems by dealing briefly with the question of whether localization can occur due to disorder.

In the discussion of the T-matrix approach, initially we assumed that electrons were scattered by an assembly of identical potentials, which, however, had to be distributed with the information we had available on the configurational correlation functions. For example, in a liquid metal, the above approach seems reasonable, even if, eventually, one would have to allow the single-centre potentials to depend to some extent on the local environment. We then, of course, dealt with the case of binary alloys where we had two, quite distinct, scattering potentials.

We shall follow Anderson (1958) by discussing the nature of the states which can arise in a disordered system, in which, in terms of the electron states discussed above, the scattering potentials are kept on a lattice, but have a probability distribution of scattering strengths. Anderson's conclusion, which has been the subject of some controversy, but which we believe is essentially correct, is that, under certain circumstances which the theory allows one to specify, states can become localized by the disorder.

Since, however, Anderson's paper is of wider interest than the electron-states problem, for example it could apply to spin diffusion, we shall present it in the manner he did initially, and only at the end recast the conclusions to apply to the problem of electron states. In connection with the above discussion of phonons in disordered systems, similar localization can occur there (see, for example, Dean and Martin, 1960). A general review of the dynamics of disordered lattices has been given by Bell (1972) (see also Hori, 1968).

10.17.1 Anderson's model

In terms of the problem of spin diffusion referred to above, Anderson's model can be stated as follows. We assume that we have sites j distributed, in some way, regularly on a lattice (or for that matter randomly). Let us suppose that spins occupy these sites, and that if a spin occupies site j it has energy ε_j, which has a probability distribution, say $P(\varepsilon)\,d\varepsilon$, which it will be useful later to characterize by a width w. Finally, between the sites we have some interaction, with matrix elements t_{jk}, which transfers the spins from one site to the next.

This model corresponds then to the equation for the time dependence of the probability amplitude a_j that a particle is on site j:

$$i\frac{\partial a_j}{\partial t} = \varepsilon_j a_j + \sum_{k\neq j} t_{jk} a_k, \tag{10.17.1}$$

which can be compared with the Hamiltonian of equation (10.13.94) which we used in the alloy problem.

We want to stress at this point that this time-dependent description works with the "unperturbed" energies ε_j and treats the interaction as causing transitions.

However, another approach is to regard the problem of determining the electron states in the random lattice as a stationary state problem. Thus, suppose the allowed energies are E_α and corresponding wave functions are ψ_α. Then we can regard the wave functions $\psi_\alpha(\mathbf{r})$ as expanded in terms of the original "site" wave functions $\psi_i(\mathbf{r})$ as

$$\psi_\alpha(\mathbf{r}) = \sum_i (\alpha|i)\,\psi_i(\mathbf{r}). \tag{10.17.2}$$

The Green function $G(\mathbf{r}\mathbf{r}'E)$ is obviously given by

$$G(\mathbf{r}\mathbf{r}'E) = \sum_\alpha \frac{\psi_\alpha^*(\mathbf{r})\,\psi_\alpha(\mathbf{r}')}{E-E_\alpha} \tag{10.17.3}$$

and if we now use the representation

$$G(\mathbf{r}\mathbf{r}'E) = \sum_{ij} G_{ij}(E)\,\psi_i^*(\mathbf{r})\,\psi_j(\mathbf{r}') \tag{10.17.4}$$

then we find immediately

$$G_{ij}(E) = \sum_\alpha \frac{(\alpha|i)^*\,(\alpha|j)}{E-E_\alpha}. \tag{10.17.5}$$

If we consider the diagonal element

$$G_{ii}(E) = \sum_\alpha \frac{|(\alpha|i)|^2}{E-E_\alpha} \tag{10.17.6}$$

then this is the quantity with which Thouless (1970) works, in his more rigorous formulation of the Anderson theory.

The question of localization then hinges on whether the quantities $|(\alpha|i)|^2$ are $O(1/N)$ or $O(1)$. We shall see below that, in terms of the probability amplitude a_j in equation (10.17.1), this is tantamount to investigating the form of $a_j(t)$ as $t \to \infty$. This can be conveniently accomplished by working with the Laplace transform $f_j(s)$ defined by

$$f_j(s) = \int_0^\infty e^{-st} a_j(t)\, dt \tag{10.17.7}$$

and then

$$i[sf_j(s) - a_j(0)] = \varepsilon_j f_j + \sum_{k \neq j} t_{jk} f_k. \tag{10.17.8}$$

What we wish to do is to consider what happens to a single spin which we take on site n at $t = 0$, and in particular what the probability amplitude looks like at a later time t.

An alternative form of equation (10.17.8), suitable for iteration, is

$$f_j(s) = \frac{i\delta_{0j}}{is - \varepsilon_j} + \sum_{k \neq j} \frac{1}{i\delta - \varepsilon_j} t_{jk} f_k(s), \tag{10.17.9}$$

where this now includes the boundary condition $a_0(0) = 1$.

As in conventional transport theory (cf. Chapter 6, section 6.10.1), we can proceed to solve iteratively, when we find

$$f_j(s) = \frac{1}{is - \varepsilon_j} t_{j0} f_0(s) + \sum_k \frac{1}{is - \varepsilon_j} t_{jk} \frac{1}{is - \varepsilon_k} t_{k0} f_0(s) + \dots. \tag{10.17.10}$$

In particular, at site $j = 0$ which the spin occupied at $t = 0$ we have

$$f_0(s) = \frac{i}{is - \varepsilon_0} + \sum_k \frac{1}{is - \varepsilon_0} \left(\frac{t_{0k} t_{k0}}{is - \varepsilon_k} + \sum_l t_{0k} \frac{1}{is - \varepsilon_k} t_{kl} \frac{1}{is - \varepsilon_l} t_{l0} + \dots \right). \tag{10.17.11}$$

In terms of conventional Green function language, this procedure is equivalent to working with the self-energy or the reciprocal of the Green function. The advantage then gained is that one encounters no strong poles near the energies of the states localized near the chosen site.

Calling the quantity in the bracket ( ) of equation (10.17.11) $V_c(s)$, and if the first term is a sufficient approximation, then we can write

$$V_c(s) = \sum_k (t_{0k})^2 \left(\frac{-\varepsilon_k}{s^2 + \varepsilon_k^2} - \frac{is}{s^2 + \varepsilon_k^2} \right). \tag{10.17.12}$$

We now observe that if we want the probability amplitude as $t \to \infty$, the connection with $f_j(s)$ is that

$$\lim_{s \to 0+} sf_j(s) = a_j(\infty). \tag{10.17.13}$$

The significance of a_0 at large time is that, if there is localization, $a_0(\infty)$ will remain finite, and so will the a_j's for neighbouring sites.

To study this, we take the limit $s \to 0$ of $V_c(s)$ in equation (10.17.12). The first term yields an energy $\Delta E^{(2)}$, say (it is essentially the second-order perturbation energy). The remaining part of equation (10.17.12) contributes

$$\lim_{s\to 0} V_c(s) = -i\sum_k (t_{0k})^2\,\delta(\varepsilon_k) - \lim_{s\to 0} is \sum_{k,E_k\neq 0} \frac{(t_{0k})^2}{E_k^2}, \tag{10.17.14}$$

which we can write formally as

$$\lim_{s\to 0} V_c(s) = -\frac{i}{\tau} - i\lim_{s\to 0} sK. \tag{10.17.15}$$

If τ^{-1} is finite, K is, of course, indeterminate, but in the form (10.17.15) it will be useful in seeing what can happen as $\tau \to \infty$.

The solution for f_0 is then

$$f_0 = \frac{i}{is(1+K)+(i/\tau)-(\varepsilon_0-\Delta E^{(2)})} \tag{10.17.16}$$

If τ is finite, this clearly represents a state having energy $\varepsilon_0 - \Delta E^{(2)}$, decaying with a relaxation time τ. But if $\tau \to \infty$, then K in (10.17.15) enters the result in an important way and we have

$$f_0(s) = \frac{1}{s(1+K)+i(\varepsilon_0-\Delta E^{(2)})}. \tag{10.17.17}$$

But this yields $a_0(t)$ finite as $t \to \infty$, with simply a reduction in amplitude $1(1+K)$. Clearly then, K is related to the spread of the probability amplitude on to near neighbour sites.

The essence of the procedure is now clear. $V_c(s)$ defined in equation (10.17.11) must be carefully studied, with the probability distribution of ε_j's. The question is then under what conditions the imaginary part of $V_c(s)$ tends to zero as $s \to 0$, as we see from equation (10.17.15).

Treatment of $V_c(s)$ as probability variable. The probability distribution of V_c must now be studied. Two questions come up:

(i) What is the contribution from equation (10.17.12)?

(ii) How do the higher-order terms affect the theory?

To deal with (i), we note that

$$\operatorname{Im} V_c(s) = -s\sum_k \frac{|t_{0k}|^2}{s^2+\varepsilon_k^2} \equiv -sX(s). \tag{10.17.18}$$

Thus for no diffusion, $\lim_{s\to 0} sX(s) \to 0$ assuming that the answer to (ii) shows that the higher-order terms are not affecting the theory in an important way.

The probability distribution $P(X)$ of $X(s)$ can now be calculated, and this is done in Appendix 10.8. The result can be stated as

$$P(X) = \frac{n\langle t\rangle}{wX^{\frac{3}{2}}} \exp\left[-(2n\Gamma(\tfrac{1}{2}))^2 \frac{\langle t\rangle^2}{w^2}\frac{1}{X}\right], \tag{10.17.19}$$

which is valid for large X at least, where n is the density of sites, and $\langle t\rangle = \langle t(\mathbf{r})\rangle$, where $t(r_{ij})$ corresponds to t_{ij}. This result depends, as seen in Appendix 10.8, on the fact that $t(r) \sim 1/r^{3+\epsilon}$ for large r with $\varepsilon > 0$.

This leads to the conclusion that the average of X is divergent and moreover the probability that τ is finite is exactly zero. This conclusion, at first sight, says that there is never transport in this model, in conflict with one's intuition. The answer must lie in the higher terms in the perturbation theory; Anderson argues that these can only affect the above conclusion if, in fact, the conditions are such that the series diverges.

His arguments are detailed and not rigorous and therefore will not be reproduced here. We shall simply summarize his conclusions therefore:

(i) The first few terms of perturbation theory are convergent in the sense that Im (V_c) is finite, with probability unity, at any particular randomly chosen energy.

(ii) The terms of order L or higher in the perturbation theory are, at any particular random value of the energy, smaller than $e^{-\epsilon L}$ with probability $1-e^{-\epsilon L}$, provided that $\langle t\rangle$ is less than a critical value of the order of magnitude of w. Anderson estimates, for sites defined on a simple cubic lattice $w/\langle t\rangle \sim 20$, as a semi-quantitative criterion for such localization to occur.

Thouless and his co-workers have extended our knowledge of the Anderson model in a group of papers to which the reader is referred for further discussion (see, for example, Edwards and Thouless, 1971). There is, by now, rather general agreement that Anderson's criterion is essentially correct, but some uncertainty remains about its quantitative form. The range of localized states is discussed by Abram and Edwards (1972).

Appendixes

APPENDIX A6.1 BOLTZMANN EQUATION FOR ELECTRONS SCATTERED BY PHONONS

To recover the Boltzmann equation for electrons scattered by phonons from equation (6.10.17), we proceed in detail as follows. In that case we have

$$|i\rangle = |q\mathbf{k}\rangle = |q\rangle|\mathbf{k}\rangle, \tag{A6.1.1}$$

where $|q\rangle$ and $|\mathbf{k}\rangle$ are respectively phonon and electron states in the absence of any mutual scattering. To lowest order

$$\langle i'|\frac{\partial\rho_0}{\partial\mathbf{p}}|i\rangle = \langle q'\mathbf{k}'|\frac{\partial\rho_0}{\partial\mathbf{p}}|q\mathbf{k}\rangle \tag{A6.1.2}$$

is also to be calculated in the absence of any scattering, and so we put

$$\rho_0 = \rho_{\mathrm{el}}\,\rho_{\mathrm{ph}} \tag{A6.1.3}$$

and by equation (6.10.32) we may write equation (6.10.17) as

$$0 = \langle q|\rho_{\mathrm{ph}}|q\rangle\frac{\partial}{\partial\mathbf{k}}\langle\mathbf{k}|\rho_{\mathrm{el}}|\mathbf{k}\rangle.\mathscr{E}e$$

$$+\frac{2\pi}{\hbar}\sum_{q'\mathbf{k}'}|V_{\mathbf{k}'\mathbf{k}}^{q'q}|^2(\langle q\mathbf{k}|\rho|q\mathbf{k}\rangle - \langle q'\mathbf{k}'|\rho|q'\mathbf{k}'\rangle)\,\delta(\varepsilon_q - \varepsilon_{q'} + E_{\mathbf{k}} - E_{\mathbf{k}'}), \tag{A6.1.4}$$

where now

$$H = H_{\mathrm{ph}} + H_{\mathrm{el}} + V, \tag{A6.1.5}$$

$$V_{\mathbf{k}'\mathbf{k}}^{q'q} = \langle q'\mathbf{k}'|V|q\mathbf{k}\rangle \tag{A6.1.6}$$

and the $E_{\mathbf{k}}$ and ε_q respectively are eigenvalues of H_{el} and H_{ph}, the Hamiltonian of the electron and phonon systems.

Following the qualitative discussion in the main text, we take the phonon distribution to be independent of the electron distribution, and the probability of finding the phonons in state q to be, with Z_{ph} the phonon partition function,

$$\langle q|\rho_{\mathrm{ph}}|q\rangle = \mathrm{e}^{-\beta\varepsilon_q}/Z_{\mathrm{ph}}. \tag{A6.1.7}$$

Since this distribution is independent of the electron distribution the total density matrix must factorize when we obtain, in terms of the many-electron distribution function F,

$$\langle q\mathbf{k}|\rho|q\mathbf{k}\rangle = \frac{e^{-\beta\varepsilon_q}}{Z_{ph}} F(\mathbf{k}). \tag{A6.1.8}$$

Summing over q, equation (A6.1.4) then reads

$$0 = e\mathscr{E}.\frac{\partial F_0}{\partial \mathbf{k}}(\mathbf{k}) + \frac{2\pi}{\hbar}\sum_{\mathbf{k}'}\sum_{qq'}|V_{\mathbf{k}'\mathbf{k}}^{q'q}|^2\left[\frac{e^{-\beta\varepsilon_q}}{Z_{ph}}F(\mathbf{k}) - \frac{e^{-\beta\varepsilon_{q'}}}{Z_{ph}}F(\mathbf{k}')\right]\delta(\varepsilon_{q'}-\varepsilon_q+E_{\mathbf{k}'}-E_{\mathbf{k}})$$

or

$$0 = e\mathscr{E}.\frac{\partial F_0}{\partial \mathbf{k}}(\mathbf{k}) + \sum_{\mathbf{k}'}[P(\mathbf{k}',\mathbf{k})F(\mathbf{k}) - P(\mathbf{k},\mathbf{k}')F(\mathbf{k}')], \tag{A6.1.9}$$

where

$$P(\mathbf{k}',\mathbf{k}) = \frac{2\pi}{\hbar}\sum_{q'q}|V_{\mathbf{k}'\mathbf{k}}^{q'q}|^2\frac{e^{-\beta\varepsilon_q}}{Z_{ph}}\delta(\varepsilon_{q'}-\varepsilon_q+E_{\mathbf{k}'}-E_{\mathbf{k}}). \tag{A6.1.10}$$

It is to be noted that equation (A6.1.9) is an equation for the many-particle distribution and if the Hartree–Fock method is valid $\mathbf{k}$ stands for $\mathbf{k}_1, \mathbf{k}_2, \ldots, \mathbf{k}_N$. To obtain the single-particle function $f(\mathbf{k}_1)$ we must take the sum

$$f(\mathbf{k}_1) = N\sum_{\mathbf{k}_2,\ldots,\mathbf{k}_N} F(\mathbf{k}_1, \ldots, \mathbf{k}_N). \tag{A6.1.11}$$

Now, in full,

$$\begin{aligned}\sum_{\mathbf{k}'}|V_{\mathbf{k}'\mathbf{k}}^{q'q}|^2 F(\mathbf{k}') &= \sum_{\mathbf{k}'}\langle q'|\langle\mathbf{k}'|V|\mathbf{k}\rangle|q\rangle\langle q|\langle\mathbf{k}|V|\mathbf{k}'\rangle|q'\rangle F(\mathbf{k}')\\ &= \sum_{\mathbf{k}_1',\ldots,\mathbf{k}_N'}\langle q'|\langle\mathbf{k}_1'\ldots\mathbf{k}_N'|V|\mathbf{k}_1\ldots\mathbf{k}_N\rangle|q\rangle\\ &\qquad\times\langle q|\langle\mathbf{k}_1\ldots\mathbf{k}_N|V|\mathbf{k}_1'\ldots\mathbf{k}_N'\rangle|q'\rangle F(\mathbf{k}_1', \ldots, \mathbf{k}_N').\end{aligned} \tag{A6.1.12}$$

We also write

$$V = \sum_{i=1}^{N} v(\mathbf{r}_i), \tag{A6.1.13}$$

a sum of single-particle interactions with the lattice. Let us sample the terms we now obtain: we take $|\mathbf{k}_1\ldots\mathbf{k}_N\rangle$ and $|\mathbf{k}_1'\ldots\mathbf{k}_N'\rangle$ to be Hartree products $\psi_{\mathbf{k}_1}(\mathbf{r}_1)\ldots\psi_{\mathbf{k}_N}(\mathbf{r}_N)$ and $\psi_{\mathbf{k}_1'}(\mathbf{r}_1)\ldots\psi_{\mathbf{k}_N'}(\mathbf{r}_N)$ where no two $\mathbf{k}_i$'s and no two $\mathbf{k}_i'$'s can be the same by the Exclusion Principle. We have then, as a typical term,

$$\begin{aligned}&\int\psi_{\mathbf{k}_1'}^*(\mathbf{r}_1)\ldots\psi_{\mathbf{k}_N'}^*(\mathbf{r}_N)\,v(\mathbf{r}_1)\,\psi_{\mathbf{k}_1}(\mathbf{r}_1)\ldots\psi_{\mathbf{k}_N}(\mathbf{r}_N)\,d\mathbf{r}_1\ldots d\mathbf{r}_N\,F(\mathbf{k}_1, \ldots, \mathbf{k}_N)\\ &\quad= \int\psi_{\mathbf{k}_1'}^*(\mathbf{r}_1)\,v(\mathbf{r}_1)\,\psi_{\mathbf{k}}(\mathbf{r}_1)\,d\mathbf{r}_1\,\delta_{\mathbf{k}_2',\mathbf{k}_2}\ldots\delta_{\mathbf{k}_N',\mathbf{k}_N}F(\mathbf{k}_1, \ldots, \mathbf{k}_N).\end{aligned} \tag{A6.1.14}$$

We see, in short, that by summing equation (A6.1.9) over $\mathbf{k}_2, \ldots, \mathbf{k}_N$ ($\mathbf{k} \equiv \mathbf{k}_1, \mathbf{k}_2, \ldots, \mathbf{k}_N$) we obtain

$$0 = e\mathscr{E} \cdot \frac{\partial f_0}{\partial \mathbf{k}_1}(\mathbf{k}_1) + N \sum_{q'q} \sum_{\mathbf{k}_1'} |V_{\mathbf{k}_1'\mathbf{k}_1}^{q'q}|^2$$

$$\times \left[\frac{e^{-\beta\varepsilon_q}}{Z_{\text{ph}}} \sum_{\mathbf{k}_2,\ldots,\mathbf{k}_N}^{(\neq \mathbf{k}_1')} F(\mathbf{k}_1, \mathbf{k}_2, \ldots, \mathbf{k}_N) - \frac{e^{-\beta\varepsilon_{q'}}}{Z_{\text{ph}}} \sum_{\mathbf{k}_2',\ldots,\mathbf{k}_N'}^{(\neq \mathbf{k}_1)} F(\mathbf{k}_1', \mathbf{k}_2', \ldots, \mathbf{k}_N') \right]$$

$$\times\, \delta(\varepsilon_{q'} - \varepsilon_q + E_{\mathbf{k}_1'} - E_{\mathbf{k}_1}), \tag{A6.1.15}$$

where $E_{\mathbf{k}_1'}$ and $E_{\mathbf{k}_1}$ are single particle eigenvalues and

$$V_{\mathbf{k}_1'\mathbf{k}_1}^{q'q} = \langle q' | \int \psi_{\mathbf{k}_1'}^*(\mathbf{r}_1)\, v(\mathbf{r}_1)\, \psi_{\mathbf{k}_1}(\mathbf{r}_1)\, d\mathbf{r}_1 | q \rangle. \tag{A6.1.16}$$

Now

$$N \sum_{\mathbf{k}_2,\ldots,\mathbf{k}_N}^{(\neq \mathbf{k}_1')} F(\mathbf{k}_1, \ldots, \mathbf{k}_N) = N \sum_{\mathbf{k}_2,\ldots,\mathbf{k}_N} F(\mathbf{k}_1, \ldots, \mathbf{k}_N)$$
$$- N(N-1) \sum_{\mathbf{k}_3,\ldots,\mathbf{k}_N} F(\mathbf{k}, \mathbf{k}_1', \mathbf{k}_3, \ldots, \mathbf{k}_N). \tag{A6.1.17}$$

The first term on the right-hand side is just $f(\mathbf{k}_1)$ and the second term the probability of simultaneously finding particles in states $\mathbf{k}_1$ and $\mathbf{k}_1'$, which is just $f(\mathbf{k}_1) f(\mathbf{k}_1')$. Hence we have

$$N \sum_{\mathbf{k}_2,\ldots,\mathbf{k}_N}^{(\neq \mathbf{k}_1')} F(\mathbf{k}_1, \ldots, \mathbf{k}_N) = f(\mathbf{k}_1) - f(\mathbf{k}_1) f(\mathbf{k}_1') \tag{A6.1.18}$$

and equation (A6.1.15) becomes

$$0 = e\mathscr{E} \cdot \frac{\partial f_0}{\partial \mathbf{k}_1}(\mathbf{k}_1) + \sum_{\mathbf{k}_1'} \{P_s(\mathbf{k}_1', \mathbf{k}_1) f(\mathbf{k}_1)\,[1 - f(\mathbf{k}_1')] - P_s(\mathbf{k}_1, \mathbf{k}_1') f(\mathbf{k}_1')\,[1 - f(\mathbf{k}_1)]\}, \tag{A6.1.19}$$

with

$$P_s(\mathbf{k}_1', \mathbf{k}_1) = \sum_{q'q} |V_{\mathbf{k}_1'\mathbf{k}_1}^{qq'}|^2 \frac{e^{-\beta\varepsilon_q}}{Z_{\text{ph}}}\, \delta(E_{\mathbf{k}_1} - E_{\mathbf{k}_1'} + \varepsilon_q - \varepsilon_{q'}). \tag{A6.1.20}$$

Equation (A6.1.19) is the formula implied in section 6.4.1 [see equations (6.4.2) and (6.4.9)]. Equation (A6.1.20) is a generalization of the form in which we wrote down the transition probability in section 6.5.

APPENDIX A7.1 PHASE–AMPLITUDE DIAGRAMS

The phase–amplitude diagram most commonly met is the Cornu spiral shown in Figure (A7.1), for which

$$\xi_A = \int_0^s \cos\frac{\pi x^2}{2}\,dx,$$

$$\eta_A = \int_0^s \sin\frac{\pi x^2}{2}\,dx \tag{A7.1.1}$$

(the Fresnel integrals), where s is the distance along the spiral from the origin to a point A. We can refer to this diagram in discussing the more general case.

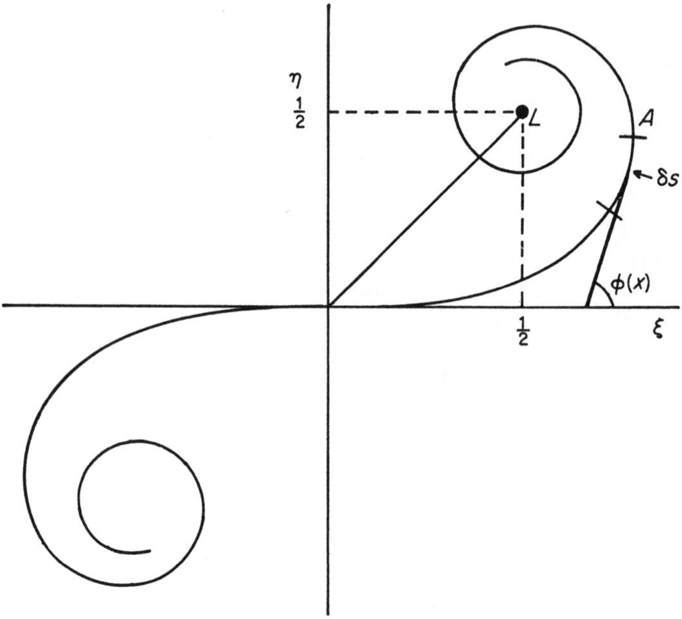

FIGURE A7.1. Cornu spiral.

Any integral $\int f(x)\,dx$ can be written as

$$\int_0^x f(x)\,dx = \int_0^x |f(x)|\,e^{i\phi(x)}\,dx = \int_0^A e^{i\phi}\,ds \tag{A7.1.2}$$

and the phase–amplitude diagram for this integral is a line of length

$$s = \int_0^A ds = \int_0^x |f(x)|\,dx \tag{A7.1.3}$$

and the tangent to which at any point corresponding to the point x makes the angle $\phi(x)$ with the abscissa. One can see that if

$$d\xi = \cos\phi\,ds, \quad d\eta = \sin\phi\,ds, \tag{A7.1.4}$$

$$\left.\begin{aligned} \xi &= \int_0^A \cos\phi\,ds = \int_0^x |f(x)|\cos\phi\,dx, \\ \eta &= \int_0^A \sin\phi\,ds = \int_0^x |f(x)|\sin\phi\,dx. \end{aligned}\right\} \tag{A7.1.5}$$

If these integrals converge as $x \to \infty$, and $\int_0^x |f(x)|\,dx \to \infty$, the diagram will form a spiral with limiting point L given by (A7.1.5) with $x = \infty$.

We also note that

$$\int_0^x f(x)\,dx = \xi + i\eta = (0, A)\,e^{i\phi(A)}, \tag{A7.1.6}$$

where A is the point at a distance *along the spiral* $s = \int_0^A ds = \int_0^x |f(x)|\,dx$ from the origin, $(0, A)$ is the length of the *straight line* from 0 to A, and $\phi(A)$ is the angle this straight line makes with the abscissa.

The phase–amplitude diagram, extensively employed in wave optics, is also useful in the summation of oscillatory components such as contribute in the de Haas–van Alphen effect and cyclotron resonance (see Pippard, 1965), and we shall here discuss the latter.

Suppose we evaluate $\sigma(q, \mathscr{H})/\sigma(q, 0)$. It is then easy to see from equation (7.9.21) that provided $\omega\tau$ is not too large the phase of the integral over k_y involved is given by

$$\frac{2\pi\omega m_c}{e\mathscr{H}} = \frac{2\pi\omega}{e\mathscr{H}}(m_c^0 + \Delta m_c) = \frac{2\pi\omega m_c^0}{e\mathscr{H}} + \phi, \tag{A7.1.7}$$

where m_c^0 is the cyclotron mass for an extremal orbit, by which we define the zero of integration of k_y (see Figure A7.2). The cyclotron mass will be even in k_y about the origin, and so

$$\phi = \frac{2\pi\omega\Delta m_c}{e\mathscr{H}} \propto k_y^2 \tag{A7.1.8}$$

for small values of k_y. Since $|\rho_x(k_y)|$ is stationary at 0, the phase amplitude diagram will start with $\phi \propto s^2$; in other words, the diagram will begin as a

Cornu spiral. Let us suppose the diagram continues as such a spiral. Then if Δm_c varies so much that ϕ_L, the phase at the limiting point L, is greater than about 2π, the line from the limiting spiral to S (the point on the spiral

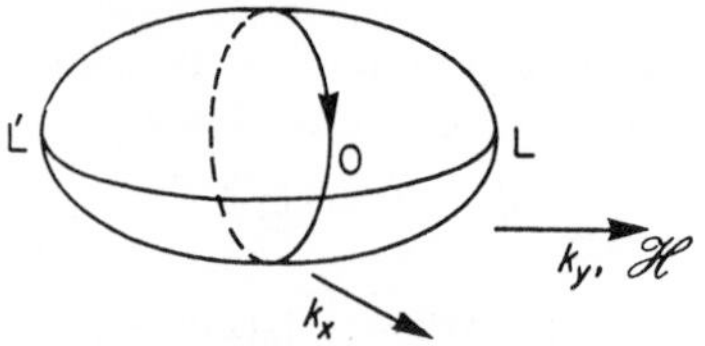

FIGURE A7.2. Fermi surface showing line L′OL along which the integration over k_y is carried out. O represents the origin of k_y, and is the intersection of the line with an extremal orbit.

corresponding to the end of the integration) will be approximately normal to the limiting point (see Figure 7.3) and the integration becomes

$$\exp(i2\pi\omega m_c^0/e\mathscr{H})\,[R_0\exp(i\phi_0)+R_1\exp i(\tfrac{3}{2}\pi+\phi_L)]. \qquad (A7.1.9)$$

Now R_0 and ϕ_0 pertain to the limiting point of the spiral and so are independent of $\mathscr{H}$. We can therefore see that the dominant oscillations are governed by m_c^0, with a small correction, phase $(2\pi\omega m_c^0 c/e\mathscr{H})+\phi_L+\frac{3}{2}\pi$, governed by the value of m_c at the limiting point L on the Fermi surface. A similar analysis obtains when $\mathscr{E}\,\|\,\mathscr{H}$.

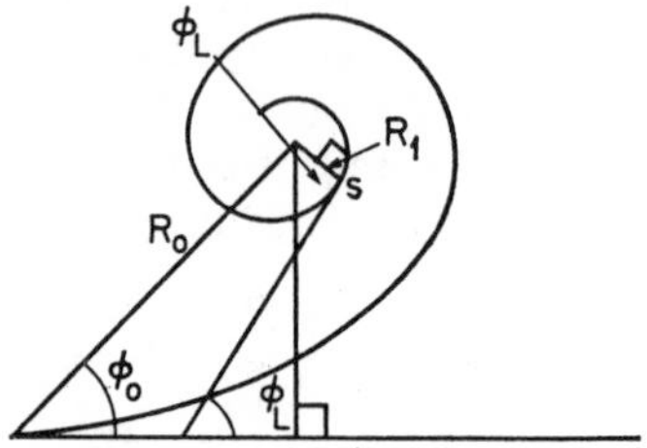

FIGURE A7.3. Phase-amplitude diagram for integration over dk_y in expressions for cyclotron resonance. If it were a Cornu spiral we would have $\phi_0 = \pi/4$.

Actually, the Cornu spiral overestimates the contribution of the limiting point, since the real spiral will close up more rapidly, but the conclusions of the analysis are unchanged.

APPENDIX A8.1 DEFORMATION POTENTIALS

In the main text (section 8.7 and section 9.4) we introduced the concept of a deformation potential in the treatment of the vibrational effects of a lattice. We shall now investigate this in detail and also examine the possibility of obtaining this potential from observation of electron-energy changes on uniform strain of a crystal. We therefore begin by investigating these latter changes and then relate them to the corresponding changes when a long wavelength longitudinal mode is propagating through the crystal.

A8.1.1 Energy changes under uniform strain

Let us first see how a uniform strain, represented by a tensor $Y_{\mu\nu}$, will change the energy of an electron, in state $\mathbf{k}$, to first order.

It must be realized immediately that we cannot use the wave functions $\psi_{\mathbf{k}}(\mathbf{r})$ of the unstrained lattice as the zeroth-order wave functions of the strained lattice as they stand, for they do not satisfy Bloch's theorem in the strained lattice. We accordingly make a change of variable, from $\mathbf{r}$ to $\boldsymbol{\xi}$, which brings the strained configuration into the unstrained one. In terms of this variable, the kinetic energy operator becomes, retaining only derivatives to third order,

$$-\frac{\hbar^2}{2m}\frac{\partial^2}{\partial \mathbf{r}^2} = -\frac{\hbar^2}{2m}\left(\frac{\partial^2}{\partial\xi_\alpha\partial\xi_\alpha} - 2Y_{\alpha\beta}\frac{\partial^2}{\partial\xi_\alpha\partial\xi_\beta} + \frac{\partial Y_{\alpha\beta}}{\partial\xi_\alpha}\frac{\partial}{\partial\xi_\beta} + \ldots\right). \quad \text{(A8.1.1)}$$

Three comments are called for here: (1) we have used the convention that we sum over repeated indices, and will do so throughout this appendix; (2) at present, since we have a uniform strain, the last term shown is zero; (3) for later convenience the strain tensor is unsymmetrized, being defined simply as the derivative of the displacement $\mathbf{u}$ of a point at $\mathbf{r}$ on straining:

$$Y_{\alpha\beta} = \partial u_\alpha/\partial x_\beta. \quad \text{(A8.1.2)}$$

We now turn to the potential energy. If $\mathbf{l}$ represents *all* the lattice vectors, this must take the form such that

$$V(\mathbf{r}, \mathbf{l}) = V(\mathbf{r}+\boldsymbol{\eta};\, \mathbf{l}+\boldsymbol{\eta}), \quad \text{(A8.1.3)}$$

because a rigid displacement $\boldsymbol{\eta}$ of the lattice cannot change the potential. This has the important consequence that

$$V(\mathbf{r}+\boldsymbol{\eta};\, \mathbf{l}) = V(\mathbf{r};\, \mathbf{l}-\boldsymbol{\eta}). \quad \text{(A8.1.4)}$$

Under strain, the new potential is, primes denoting the new lattice vectors,

$$V(\mathbf{r};\, \mathbf{l}') = V(\mathbf{r}, \mathbf{l}) + \sum_{\mathbf{l}} (\mathbf{l}'-\mathbf{l}) \,.\, V_{\mathbf{l}}'(\mathbf{r}) + \ldots. \quad \text{(A8.1.5)}$$

Now it is true that as one moves away from the origin, which can be assumed undisplaced by the strain, $(\mathbf{l}'-\mathbf{l})$ increases without limit in an infinite crystal. However, since

$$V_{\mathbf{l}}'(\mathbf{r}) = \frac{\partial V}{\partial \mathbf{l}}(\mathbf{r}) \tag{A8.1.6}$$

will be localized about $\mathbf{l}$, we can see that if we restrict ourselves to the unit cell about the origin, equation (A8.1.5) will be convergent for physically reasonable potentials. We need only integrate over one unit cell to obtain energies, and so shall not find ourselves running into trouble.

Let us now transform to $\boldsymbol{\xi}$-space. Within the unit cell about the origin, the displacement $(\mathbf{r}-\boldsymbol{\xi})$ is macroscopically small so that, remembering the origin is undisplaced, we write to a good approximation

$$(\mathbf{r}-\boldsymbol{\xi})_\mu = Y_{\mu\nu}\,\xi_\nu. \tag{A8.1.7}$$

Using equations (A8.1.4) and (A8.1.5), the potential is then

$$\begin{aligned} V(\mathbf{r},\mathbf{l}') &= V(\boldsymbol{\xi}+Y\boldsymbol{\xi};\,\mathbf{l}') = V(\boldsymbol{\xi};\,\mathbf{l}'-Y\boldsymbol{\xi}) \\ &= V(\boldsymbol{\xi},\mathbf{l}) + \sum_{\mathbf{l}} (\mathbf{l}'-\mathbf{l}-Y\boldsymbol{\xi}).V_{\mathbf{l}}'(\boldsymbol{\xi}). \end{aligned} \tag{A8.1.8}$$

For the region of $\boldsymbol{\xi}$ with which we are concerned, significant values of $\mathbf{l}'-\mathbf{l}$ will be small, and so we insert into (A8.1.8)

$$(\mathbf{l}'-\mathbf{l})_\alpha = Y_{\alpha\beta}\,l_\beta. \tag{A8.1.9}$$

In full we now have a Hamiltonian

$$H = H_0 + H_1, \tag{A8.1.10}$$

where

$$H_0 = -\frac{\hbar^2}{2m}\frac{\partial^2}{\partial\xi_\alpha\,\partial\xi_\alpha} + V(\boldsymbol{\xi},\mathbf{l}) \tag{A8.1.11}$$

and the perturbation H_1 is

$$H_1 = \frac{\hbar^2}{m} Y_{\alpha\beta}\frac{\partial^2}{\partial\xi_\alpha\,\partial\xi_\beta} + Y_{\alpha\beta}\,X_{\alpha\beta}(\boldsymbol{\xi}), \tag{A8.1.12}$$

where we have introduced a tensor

$$X_{\mu\nu}(\boldsymbol{\xi}) = \sum_{\mathbf{l}} (\mathbf{l}-\boldsymbol{\xi})_\mu\, V_{\mathbf{l}}^{\nu}(\boldsymbol{\xi}). \tag{A8.1.13}$$

Here

$$V_{\mathbf{l}}^{\nu}(\boldsymbol{\xi}) = \frac{\partial V(\boldsymbol{\xi},\mathbf{l})}{\partial l_\nu}. \tag{A8.1.14}$$

We now take matrix elements of (A8.1.10) between $\psi_{\mathbf{k}}^*(\boldsymbol{\xi})$ and $\psi_{\mathbf{k}}(\boldsymbol{\xi})$, which are eigenfunctions of H_0, eigenvalues $E(\mathbf{k})$. We thus obtain the energy

$$E(\mathbf{k})+Y_{\alpha\beta}\langle\psi_{\mathbf{k}}|-\frac{1}{m}p_\alpha p_\beta+X_{\alpha\beta}(\mathbf{r})|\psi_{\mathbf{k}}\rangle. \tag{A8.1.15}$$

We have not yet finished the calculation, because this is *not* the energy for the vector **k** in the strained lattice. Remembering that as a lattice expands its reciprocal lattice *contracts* in the same proportion, this energy corresponds to a **k**-value given by

$$k'_\mu = k_\mu - Y_{\mu\nu}k_\nu. \tag{A8.1.16}$$

To correct (A8.1.15) we use the group-velocity formula

$$\frac{\partial E(\mathbf{k})}{\partial \mathbf{k}} = \frac{\hbar}{m}\langle\psi_{\mathbf{k}}|\mathbf{p}|\psi_{\mathbf{k}}\rangle. \tag{A8.1.17}$$

We can make the required correction to first order using the $E(\mathbf{k})$ relation of the unstrained lattice, and we finally obtain the change in energy on uniform strain as

$$\delta E(\mathbf{k}) = Y_{\alpha\beta}\langle\psi_{\mathbf{k}}|-\frac{1}{m}p_\alpha p_\beta+X_{\alpha\beta}(\mathbf{r})+\frac{\hbar}{m}k_\alpha p_\beta|\psi_{\mathbf{k}}\rangle. \tag{A8.1.18}$$

A8.1.2 Strains by phonons

We wish now to obtain an expression for the electron–phonon matrix element $M_{\mathbf{k}'\mathbf{k}}$ involving the strain tensor Y expressing the deformation by a phonon mode, which will be taken to be of long wavelength so that derivatives of Y can be ignored. The strain tensor will be defined by [cf. equation (A8.1.2)]

$$Y(\mathbf{r}) \equiv \partial\mathbf{u}/\partial\mathbf{r}. \tag{A8.1.19}$$

This definition of Y is not unique, for the only requirement we need to make on $\mathbf{u}(\mathbf{r})$ is that $\mathbf{u}(\mathbf{l})$ is the displacement of the **l**th lattice point when the mode $(\mathbf{q}\sigma)$ is considered. For definiteness we shall take a smooth interpolation between lattice points, viz.

$$\mathbf{u}(\mathbf{r}) = Q_{\mathbf{q}\sigma}\,\boldsymbol{\varepsilon}_{\mathbf{q}\sigma}\,e^{i\mathbf{q}\cdot\mathbf{r}}. \tag{A8.1.20}$$

One could carry out analysis similar to that we performed for a uniform strain, so avoiding a formal defect in the Bloch formula (6.5.5) it can give a non-zero matrix element when all the $\mathbf{u}(\mathbf{l})$'s are equal, which corresponds merely to uniform translation of the lattice. However, we shall not obtain such spurious results if we confine ourselves to actual displacements due to lattice vibrations, and since our purpose is to reach comparison with equation (A8.1.18) by the shortest route, we shall begin with the Bloch formula. For the alternative approach mentioned above one should consult Whitfield (1961), Sham and Ziman (1963) and Herring (1960). This latter author points out that the deformation transformation is also useful in considering two-phonon processes.

As in equation (A8.1.9), we write

$$U_{\mathbf{l}}^{\alpha} = (\mathbf{l}' - \mathbf{l})_{\alpha} = Y_{\alpha\beta}(\mathbf{l})\, l_{\beta}, \tag{A8.1.21}$$

where $\mathbf{l}'$ is now no longer a lattice vector, but the displaced position of the $\mathbf{l}$th lattice point. Thus the Bloch expression (6.5.5) for the electron–phonon matrix element now becomes

$$M_{\mathbf{k}'\mathbf{k}} = \langle \psi_{\mathbf{k}'} | Y_{\alpha\beta}(\mathbf{r}) \sum_{\mathbf{l}} l_{\alpha} V_{\mathbf{l}}^{\beta}(\mathbf{r}) | \psi_{\mathbf{k}} \rangle. \tag{A8.1.22}$$

Here we have written $Y_{\alpha\beta}(\mathbf{r})$ for $Y_{\alpha\beta}(\mathbf{l})$, which will be accurate if Y is slowly varying compared with the localization of $V_{\mathbf{l}}'(\mathbf{r})$ about $\mathbf{l}$. To make contact with equation (A8.1.18) we again introduce the tensor $X_{\alpha\beta}(\mathbf{r})$, in terms of which equation (A8.1.22) is

$$M_{\mathbf{k}'\mathbf{k}} = \langle \psi_{\mathbf{k}'} | Y_{\alpha\beta}(\mathbf{r})\, X_{\alpha\beta}(\mathbf{r}) - u_{\alpha} \frac{\partial V}{\partial x_{\alpha}} | \psi_{\mathbf{k}} \rangle. \tag{A8.1.23}$$

To obtain this equation we have used the property of V expressed in equation (A8.1.4):

$$\begin{aligned} Y_{\alpha\beta}(\mathbf{r}) \sum_{\mathbf{l}} r_{\alpha} V_{\mathbf{l}}^{\beta}(\mathbf{r}; \mathbf{l}) \simeq \sum_{\mathbf{l}} \mathbf{u} . \frac{\partial V}{\partial \mathbf{l}}(\mathbf{r}; \mathbf{l}) &= \mathbf{u} . \frac{\partial V}{\partial \boldsymbol{\eta}}(\mathbf{r}; \mathbf{l} + \boldsymbol{\eta})|_{\eta=0} \\ &= \mathbf{u} . \frac{\partial V}{\partial \boldsymbol{\eta}}(\mathbf{r} - \boldsymbol{\eta}; \mathbf{l})|_{\eta=0} \\ &= -\mathbf{u} . \frac{\partial V}{\partial \mathbf{r}}(\mathbf{r}; \mathbf{l}). \end{aligned} \tag{A8.1.24}$$

To complete the correspondence with (A8.1.18), let us obtain the commutator

$$\mathbf{u} . \frac{\partial}{\partial \mathbf{r}} H_0 - H_0\, \mathbf{u} . \frac{\partial}{\partial \mathbf{r}} = \left[\mathbf{u} . \frac{\partial}{\partial \mathbf{r}}, H_0 \right], \tag{A8.1.25}$$

where

$$H_0 = -\frac{\hbar^2}{2m} \frac{\partial^2}{\partial \mathbf{r}^2} + V(\mathbf{r}; \boldsymbol{\eta}). \tag{A8.1.26}$$

We obtain

$$\left[\mathbf{u} . \frac{\partial}{\partial \mathbf{r}}, H_0 \right] = \mathbf{u} . \frac{\partial V}{\partial \mathbf{r}} + \frac{\hbar^2}{m} \frac{\partial u_{\alpha}}{\partial x_{\beta}} \frac{\partial^2}{\partial x_{\alpha} \partial x_{\beta}} + \frac{\partial^2 u_{\alpha}}{\partial x_{\beta} \partial x_{\beta}} \frac{\partial}{\partial x_{\alpha}}. \tag{A8.1.27}$$

Thus if we neglect second derivatives of u,

$$\langle\psi_{\mathbf{k}'}|\mathbf{u}.\frac{\partial V}{\partial \mathbf{r}}|\psi_{\mathbf{k}}\rangle = \langle\psi_{\mathbf{k}'}|\frac{1}{m}Y_{\alpha\beta}p_\alpha p_\beta|\psi_{\mathbf{k}}\rangle + \langle\psi_{\mathbf{k}'}|\left[\mathbf{u}.\frac{\partial}{\partial \mathbf{r}}, H_0\right]|\psi_{\mathbf{k}}\rangle. \tag{A8.1.28}$$

Evaluating the second term on the right-hand side, we obtain

$$\langle\psi_{\mathbf{k}'}|\mathbf{u}.\frac{\partial}{\partial \mathbf{r}}H_0 - H_0\mathbf{u}.\frac{\partial}{\partial \mathbf{r}}|\psi_{\mathbf{k}}\rangle = -[E(\mathbf{k}') - E(\mathbf{k})]\langle\psi_{\mathbf{k}'}|\mathbf{u}.\frac{\partial}{\partial \mathbf{r}}|\psi_{\mathbf{k}}\rangle. \tag{A8.1.29}$$

Now by the selection rule in electron–phonon interactions [see equation (6.5.6)]

$$[E(\mathbf{k}') - E(\mathbf{k})]\,u_\alpha = \hbar\omega(\mathbf{q}\sigma)\,u_\alpha = \hbar c_\beta q_\beta u_\alpha, \tag{A8.1.30}$$

where c_β is a phase-velocity for the longitudinal mode, represented by (A8.1.20). From this equation,

$$Y_{\alpha\beta} = \partial u_\alpha/\partial x_\beta = iq_\beta u_\alpha. \tag{A8.1.31}$$

Combining (A8.1.23), (A8.1.28), (A8.1.29), (A8.1.30) and (A8.1.31), we finally get

$$M_{\mathbf{k}'\mathbf{k}} = \langle\psi_{\mathbf{k}'}|Y_{\alpha\beta}(\mathbf{r})\,E_{\alpha\beta}(\mathbf{r})|\psi_{\mathbf{k}}\rangle, \tag{A8.1.32}$$

where

$$E_{\alpha\beta}(\mathbf{r}) = -\frac{1}{m}p_\alpha p_\beta + X_{\alpha\beta}(\mathbf{r}) + c_\alpha p_\beta, \tag{A8.1.33}$$

so that the *deformation potential* is defined as

$$\delta V(\mathbf{r}) = Y_{\alpha\beta}(\mathbf{r})\,E_{\alpha\beta}(\mathbf{r}). \tag{A8.1.34}$$

Comparison of (A8.1.33) with (A8.1.18) shows that in fact the operators involved differ in their last terms, which are $(\hbar/m)\,k_\alpha p_\beta$ in (A8.1.18) and $c_\alpha p_\beta$ in (A8.1.33). However, the deformation potential is used mainly for semiconductors [where it was introduced by Bardeen and Shockley (1950)] and the terms vanish at band maximum and minima, so that the usual calculation of mobility is unaffected.

It will be seen that equation (A8.1.18) is most useful when $\mathbf{q} = \mathbf{k}' - \mathbf{k}$ is small, when it will be a good approximation to write equation (A8.1.34) as

$$\delta V(\mathbf{r};\mathbf{k}) = Y_{\alpha\beta}(\mathbf{r})\left[D_{\alpha\beta}(\mathbf{k}) + \left(c_\alpha - \frac{\hbar}{m}k_\alpha\right)p_\beta\right], \tag{A8.1.35}$$

where

$$D_{\alpha\beta}(\mathbf{k}) = \langle\psi_{\mathbf{k}}|-\frac{1}{m}p_\alpha p_\beta + X_{\alpha\beta}(\mathbf{r}) + \frac{\hbar}{m}k_\alpha p_\beta|\psi_{\mathbf{k}}\rangle. \tag{A8.1.36}$$

A8.1.3 Modification of deformation potential

In Chapter 8 we introduced the deformation potential by a discussion of screening effects for a system of ions immersed in a uniform electron gas. There we saw that the deformation potential and energy change were the same in magnitude. However, in this appendix we have taken no account of the possibility that the effect of screening could be different in the deformed and homogeneously strained lattices. Remembering that the deformation potential theorem embodied in equations (A8.1.32)–(A8.1.36) holds only for long wavelengths, we proceed in the spirit of the argument of Chapter 8.

We begin with equation (A8.1.15), the correction of equation (A8.1.33) not being applicable here since the deformation is a local one, and the **k**-space argument applies to the entire crystal. In the absence of a screening correction therefore, the local energies are

$$E_{\mathrm{L}}(\mathbf{k}) = E(\mathbf{k}) + Y_{\alpha\beta}(\mathbf{r})\, D_{\alpha\beta}(\mathbf{k}) - Y_{\alpha\beta}\frac{\hbar}{m}k_\alpha \frac{\partial E}{\partial k_\beta}(\mathbf{k}) \tag{A8.1.37}$$

and so, unless a screening correction were applied, we would obtain a local increase δE_{f} of Fermi level given by

$$n(E_{\mathrm{f}})\,\delta E_{\mathrm{f}} = \frac{1}{4\pi^3}\int_{\mathrm{FS}} \delta k\, dS, \tag{A8.1.38}$$

where

$$\left|\frac{\partial E}{\partial \mathbf{k}}\right| \delta k = Y_{\alpha\beta}(\mathbf{r})\, D_{\alpha\beta}(\mathbf{k}) - Y_{\alpha\beta}\frac{\hbar}{m}.k_\alpha \frac{\partial E}{\partial k_\beta}(\mathbf{k}). \tag{A8.1.39}$$

Now $Y_{\alpha\beta}k_\alpha$ is just the change in k_β on deformation, and so [see equation (8.7.4)], if Δ is the dilatation,

$$\frac{1}{4\pi^3}\int_{\mathrm{FS}} \frac{Y_{\alpha\beta}k_\alpha}{|\partial E/\partial \mathbf{k}|}\frac{\partial E}{\partial k_\beta} dS = ZN\Delta(\mathbf{r}) = Y_{\alpha\alpha}ZN. \tag{A8.1.40}$$

In cancelling off the spurious change δE_{f} in equation (A8.1.38) by a screening potential, we thus find that equation (A8.1.35) is modified to read

$$\delta V_{\mathrm{S}}(\mathbf{r}) = Y_{\alpha\beta}(\mathbf{r})\left[D_{\alpha\beta}(\mathbf{k}) - \frac{1}{4\pi^3 n(E_{\mathrm{f}})}\int_{\mathrm{FS}} D_{\alpha\beta}(\mathbf{k})\, dS + \frac{ZN}{n(E_{\mathrm{f}})}\delta_{\alpha\beta} + \left(c_\alpha - \frac{\hbar}{m}k_\alpha\right)p_\beta\right]. \tag{A8.1.42}$$

One sees that if $D_{\alpha\beta}$ is constant over the Fermi surface, we regain the results of the main text.

APPENDIX A8.2 THERMAL CONDUCTIVITY OF SUPERCONDUCTORS

We give here the argument leading to the formula for the ratio of the thermal conductivity in the superconducting to the normal state (see Rickayzen, 1965).

To obtain this, we must consider the effect of a temperature gradient on the distribution function $f(\mathbf{k})$. We can find this change by solving the transport equation

$$\left.\frac{\partial f(\mathbf{k})}{\partial t}\right|_{\text{temp}} = \left.\frac{\partial f(\mathbf{k})}{\partial t}\right|_{\text{coll}}. \tag{A8.2.1}$$

We saw in Chapter 6 that if we deal with impurity scattering, and V is the impurity potential, then

$$-\left.\frac{\partial f_0(\mathbf{k})}{\partial t}\right|_{\text{coll}} = \frac{2\pi}{\hbar}\sum_{\mathbf{k}'}|\langle\mathbf{k}'|V|\mathbf{k}\rangle|^2\{f_0(\mathbf{k})[1-f_0(\mathbf{k}')]-f_0(\mathbf{k}')[1-f_0(\mathbf{k})]\}\,\delta(E-E'). \tag{A8.2.2}$$

If we now use the BCS wave function to evaluate this, then we find

$$-\left.\frac{\partial f_0(\mathbf{k})}{\partial t}\right|_{\text{coll}} = \frac{2\pi}{\hbar}N(0)\int\frac{d\mu}{2}\int d\varepsilon'\,|V(\mathbf{k}-\mathbf{k}')|^2 \times(uu'-vv')^2[f_0(\mathbf{k})-f_0(\mathbf{k}')]\,\delta(E-E'). \tag{A8.2.3}$$

The solution we are seeking can be expressed in the form

$$f_0(\mathbf{k}) = f(E_{\mathbf{k}})+(\mathbf{k}\,.\,\nabla T)\,c(E)\frac{\partial f(E)}{\partial E}\frac{\varepsilon}{|\varepsilon|}, \tag{A8.2.4}$$

where the second term on the right-hand side is antisymmetric about the Fermi surface. Substituting this in equation (A8.2.3), and then eliminating the function $c(E)$ in favour of f we can write

$$\begin{aligned}-\left.\frac{\partial f_0(\mathbf{k})}{\partial t}\right|_{\text{coll}} &= \frac{2\pi}{\hbar}N(0)\int\frac{d\mu}{2}\int d\varepsilon'\,|V(\mathbf{k}-\mathbf{k}')|^2\frac{\varepsilon^2}{E^2}(\mathbf{k}\,.\,\nabla T)\,c(E)\\ &\quad\times(1-\mu)\frac{\partial f}{\partial E}\delta(E-E')\\ &= \frac{2\pi}{\hbar}N(0)\,[f_0(\mathbf{k})-f(E)]\left|\frac{\varepsilon}{E}\right|\\ &\quad\times\int d\mu\,|V(\mathbf{k}-\mathbf{k}')|^2(1-\mu)\big|_{k'=k=k_f}.\end{aligned} \tag{A8.2.5}$$

In terms of the relaxation time τ_n for electrical conductivity in the normal state we can write

$$-\frac{\partial f_0(\mathbf{k})}{\partial t}\bigg|_{\text{coll}} = \frac{[f_0(\mathbf{k})-f(E)]}{\tau_n}\left|\frac{\varepsilon}{E}\right|. \tag{A8.2.6}$$

Hence, the ratio of the relaxation times in the superconducting and normal states is given by

$$\frac{\tau_s}{\tau_n} = \left|\frac{E}{\varepsilon}\right| \tag{A8.2.7}$$

and the mean free path by

$$l = |v|\tau_s = \left|\frac{1}{\hbar}\frac{\partial\varepsilon}{\partial k}\frac{\varepsilon}{E}\right|\tau_n\left|\frac{E}{\varepsilon}\right| = \frac{1}{\hbar}\left|\frac{\partial\varepsilon}{\partial k}\right|\tau_n. \tag{A8.2.8}$$

This latter quantity is seen to be the same as in the normal state.

We have now to deal with the rate of change of $f(\mathbf{k})$ due to temperature. This is given by

$$v.\nabla f = v.\nabla T\frac{\partial f(\mathbf{k})}{\partial T} = (v.\nabla T)\frac{\partial f}{\partial E}\left(-\frac{E}{T}\right), \tag{A8.2.9}$$

where v is the velocity of a quasi-particle with momentum $\hbar\mathbf{k}$. The velocities of quasi-particles above and below the Fermi surface are given by $v_0 = \nabla_{\mathbf{k}} E/\hbar$ and $-\nabla_{\mathbf{k}} E/\hbar$. Also the total energy carried by a single quasi-particle is, as would be anticipated, $E_{\mathbf{k}}$. It follows that

$$\frac{\partial f_0(\mathbf{k})}{\partial t}\bigg|_{\text{temp}} = -\nabla T.\frac{\nabla_{\mathbf{k}} E}{\hbar}\frac{\partial f}{\partial E}\left(\frac{E}{T}\right) = -\frac{\partial f_1(\mathbf{k})}{\partial t}\bigg|_{\text{temp}}. \tag{A8.2.10}$$

Thus the transport equation is explicitly

$$\frac{1}{\hbar}\nabla_{\mathbf{k}} E.\nabla T\left(\frac{E}{T}\right)\frac{\partial f}{\partial E} = \frac{f_0(\mathbf{k})-f(E)}{\tau_n}\left|\frac{\varepsilon}{E}\right|. \tag{A8.2.11}$$

Hence we have

$$f_0(\mathbf{k}) = f(E)+\frac{\varepsilon}{|\varepsilon|}\frac{\tau_n}{\hbar}\nabla_{\mathbf{k}}\varepsilon.\nabla T\left(\frac{E}{T}\right)\frac{\partial f}{\partial E} \tag{A8.2.12}$$

which is of just the form (A8.2.4). Similarly we have

$$f_1(\mathbf{k}) = f(E)-\frac{\varepsilon}{|\varepsilon|}\frac{\tau_n}{\hbar}\nabla_{\mathbf{k}}\varepsilon.\nabla T\left(\frac{E}{T}\right)\frac{\partial f}{\partial E}. \tag{A8.2.13}$$

The total thermal current (see Chapter 6, section 9) is given by

$$\sum_{\mathbf{k}} [f_0(\mathbf{k}) - f_1(\mathbf{k})]\, Ev_0(\mathbf{k})$$

$$= 2\nabla T \tau_n N(0) \int \frac{d\mu}{2} \int d\varepsilon \frac{\hbar^2 k^2}{m^2} \frac{E}{T} |\varepsilon| \frac{\partial f}{\partial E}. \tag{A8.2.14}$$

Hence the desired result follows, namely

$$\frac{K_s}{K_n} = \int_0^\infty \varepsilon E\left(\frac{\partial f}{\partial E}\right) d\varepsilon \Big/ \int_0^\infty \varepsilon^2 \left(\frac{\partial f}{\partial \varepsilon}\right) d\varepsilon$$

$$= \int_\Delta^\infty E^2 \left(\frac{\partial f}{\partial E}\right) dE \Big/ \int_0^\infty E^2 \left(\frac{\partial f}{\partial E}\right) dE. \tag{A8.2.15}$$

APPENDIX A8.3 NUMERICAL SOLUTION OF ELIASHBERG EQUATIONS AT ELEVATED TEMPERATURES

We want to summarize here the important findings of McMillan (1968) on the strong-coupling theory of superconductors, which we discussed in detail in section 8.10. The equations (8.10.36) and (8.10.37) are readily generalized to elevated temperatures and it is then found (Ambegaokar and Tewordt, 1964) that the integral equations for the normal and pairing self-energies at the transition temperature are

$$\begin{aligned}
\xi(\omega) &= [1 - Z(\omega)]\, \omega \\
&= \int_0^\infty d\omega' \int_0^{\omega_0} d\omega_q\, \alpha^2(\omega_q) F(\omega_q) \\
&\quad \times \{[N(\omega_q) + f(-\omega')][(\omega' + \omega_q + \omega)^{-1} - (\omega' + \omega_q - \omega)^{-1}] \\
&\quad + [N(\omega_q) + f(\omega)][(-\omega' + \omega_q + \omega)^{-1} - (-\omega' + \omega_q - \omega)^{-1}]\}
\end{aligned} \tag{A8.3.1}$$

and

$$\begin{aligned}
\Delta(\omega) = {} & [Z(\omega)]^{-1} \int_0^\infty \frac{d\omega'}{\omega'} \\
& \times \mathrm{Re}\,[\Delta(\omega')] \int_0^{\omega_0} d\omega_q\, \alpha^2(\omega_q) F(\omega_q) \\
& \times \{[N(\omega_q) + f(-\omega')][(\omega' + \omega_q + \omega)^{-1} + (\omega' + \omega_q - \omega)^{-1}] \\
& - [N(\omega_q) + f(\omega')][(-\omega' + \omega_q + \omega)^{-1} + (-\omega' + \omega_q - \omega)^{-1}]\} \\
& - \frac{N(0)\, V_c}{Z(\omega)} \int_0^{E_b} \frac{d\omega'}{\omega'} \times \mathrm{Re}\,[\Delta(\omega')][1 - 2f(\omega')].
\end{aligned} \tag{A8.3.2}$$

Briefly, $F(\omega_q)$ is the phonon density of states, ω_0 is the maximum phonon frequency, $\alpha^2(\omega_q)$ is an average of the electron–phonon interaction and V_c is the matrix element of the screened Coulomb interaction averaged over the Fermi surface. Also, the electronic band width is E_b while $N(\omega)$ and $f(\omega)$ are the Bose and the Fermi functions respectively. The screened Coulomb interaction is represented parametrically through $N(0)V_c$ and E_b, and the electron–phonon interaction by the function $\alpha^2(\omega_q)F(\omega_q)$.

A8.3.1 Approximate solution of equations (A8.3.1) and (A8.3.2)

McMillan has obtained an approximate solution of equations (A8.3.1) and (A8.3.2) by assuming a trial form for $\Delta(\omega)$, inserting this in the right-hand side of equation (A8.3.2) and computing the new $\Delta(\omega)$ by integration. Obviously the trial function should be chosen so that input and output are as nearly consistent as practicable. The trial function assumed was in fact

$$\left.\begin{aligned}\Delta(\omega) &= \Delta_0, \quad 0<\omega<\omega_0,\\ &= \Delta_\infty, \quad \omega_0<\omega,\end{aligned}\right\} \tag{A8.3.3}$$

and $\Delta(0)$ and $\Delta(\infty)$ were then calculated from equation (A8.3.2).

We shall not give the intermediate steps, as McMillan superseded this trial solution by solving equation (A8.3.2) on a computer and we shall summarize his quantitative results below. However, his result for the transition temperature T_c from the strong coupling theory may be written

$$\frac{T_c}{\omega_0} = \exp\left(\frac{-(1+\lambda)}{\lambda-\mu^*-(\langle\omega\rangle/\omega_0)\lambda\mu^*}\right), \tag{A8.3.4}$$

where λ is a dimensionless electron–phonon coupling constant given by

$$\lambda = 2\int_0^{\omega_0}\alpha^2(\omega_q)F(\omega_q)\frac{d\omega_q}{\omega_q} \tag{A8.3.5}$$

and λ corresponds roughly to the $N(0)V$ of the BCS theory. The quantity μ^* in equation (A8.3.4) is, in fact, a Coulomb pseudo-potential (see Morel and Anderson, 1962) given by

$$\mu^* = N(0)V_c/[1+N(0)V_c\ln(E_B/\omega_0)]. \tag{A8.3.6}$$

The connection between equation (A8.3.4) and the corresponding BCS formula is that in weak coupling ($\lambda \ll 1$) they become the same, $\lambda-\mu^*$ taking the place of $N(0)V$. The main features then of the strong coupling theory are:

(i) The interactions are renormalized by the factor $Z = 1+\lambda$.

(ii) The Coulomb interaction changes the gap function in such a way that the phonon contribution is reduced from λ to $\lambda[1-(\langle\omega\rangle)/\omega_0)\mu^*]$, where the average phonon frequency $\langle\omega\rangle$ appearing also in equation (A8.3.4) is simply

$$\langle\omega\rangle = \int_0^{\omega_0} d\omega_q\, \alpha^2(\omega_q)\, F(\omega_q) \Big/ \int_0^{\omega_0} \frac{d\omega_q}{\omega_q}\, \alpha^2(\omega_q)\, F(\omega_q). \tag{A8.3.7}$$

A8.3.2 Computer results

By using the solution described above as a starting point for numerical iteration, McMillan has obtained a computer solution of equation (A8.3.2). Since a specific choice of $F(\omega_q)$ has to be made, McMillan took $F(\omega_q)$ to be the density of phonon states in Nb, taken from the neutron work of Nakagawa and Woods (1963). $\alpha^2(\omega_q)$ was taken to be a constant α^2 over most of the phonon spectrum. However, to eliminate the coupling to the long-wavelength transverse phonons he took $\alpha^2 F(\omega) = 0$ for $\omega < 100$ K.

Using the formula (A8.3.4) to fit his computer data, McMillan then finds the transition temperature to be given by

$$T_c = \frac{\theta}{1.45} \exp\left[-\frac{1.04(1+\lambda)}{\lambda - \mu^*(1+0.62\lambda)}\right], \tag{A8.3.8}$$

where the Debye temperature θ has been used instead of the characteristic phonon frequency.

A8.3.3 Empirical results for electron–phonon coupling constant

McMillan has used the above theory to get information on the electron–phonon coupling constant λ from experiment. Briefly, for metals for which the isotope shift has been measured, the parameter μ^* can be estimated. McMillan uses the value 0.13 for μ^* for all transition metals.

He writes equation (A8.3.8) in a form which enables the electron–phonon coupling constant λ to be obtained from the experimental values of the transition temperature T_c and the Debye temperature, namely

$$\lambda = \frac{1.04 + \mu^* \ln(\theta/1.45T_c)}{(1-0.62\mu^*)\ln(\theta/1.45T_c) - 1.04}. \tag{A8.3.9}$$

The results found by using $\mu^* = 0.13$ for the transition metals and 0.1 for the polyvalent metals are recorded in Table A8.3.1 along with the values of T_c and θ used to obtain λ. McMillan argues that the coupling constants are reliable for weak and intermediate coupling $\lambda < 1$. However, for $\lambda > 1$, the resulting λ is sensitive to the choice of phonon spectrum. McMillan also obtains the band-structure density of states, but we must refer the reader to his paper for the details.

In concluding this discussion, it should be remarked that McMillan's examination of the electron–phonon coupling constant over parts of the periodic table leads to the surprising result that the coupling constant

TABLE A8.3.1

	T_c in °K	Debye temperature in °K	Electron–phonon coupling λ
Be	0.026	1390	0.23
Al	1.16	428	0.38
Zn	0.85	309	0.38
Ga	1.08	325	0.40
Cd	0.52	209	0.38
In	3.40	112	0.69
Sn	3.72	200	0.60
Hg	4.16	72	1.00
Tl	2.38	79	0.71
Pb	7.19	105	1.12
Ti	0.39	425	0.38
V	5.30	399	0.60
Zr	0.55	290	0.41
Nb	9.22	277	0.82
Mo	0.92	460	0.41
Ru	0.49	550	0.38
Hf	0.09	252	0.34
Ta	4.48	258	0.65
W	0.012	390	0.28
Re	1.69	415	0.46
Os	0.65	500	0.39
Ir	0.14	420	0.34

depends mainly on the phonon frequencies and is insensitive to rather large variations in the electronic properties, e.g. the band structure density of states. McMillan has made use of this to predict a maximum transition temperature for a given class of materials.

APPENDIX A8.4 HYDRODYNAMIC DERIVATION OF GINZBURG–LANDAU THEORY

Frohlich (1966) has pointed out that the basic equations of the Ginzburg–Landau theory may be derived from hydrodynamics, in conjunction with the London equations.

The starting point is the equations for the velocity field $\mathbf{v}(\mathbf{r}, t)$ namely

$$\mathbf{v} = \operatorname{grad}\Phi - \frac{e}{mc}\mathscr{A}, \tag{A8.4.1}$$

where $\mathscr{A}$ as usual is the vector potential. For $e/m = 0$, it is obvious that equation (A8.4.1) represents rotation-free flow in which Φ is the velocity potential. If $\rho(\mathbf{r}, t)$ is the mass density of the fluid, then the current density $(e/m)\mathbf{j}$ is given by

$$\frac{e}{m}\mathbf{j} = \frac{e}{m}\rho\mathbf{v}. \tag{A8.4.2}$$

The fields ρ and $\mathbf{j}$ completely describe the dynamic properties of this superconducting fluid.

We now assume they are such as to be described by classical hydrodynamics. Then we have the equations of motion:

$$\rho\frac{\partial \mathbf{v}}{\partial t} + \rho(\mathbf{v}\operatorname{grad})\mathbf{v} = -\operatorname{grad} p + \mathbf{f} \tag{A8.4.3}$$

and

$$\operatorname{div}\mathbf{j} = -\frac{\partial \rho}{\partial t}. \tag{A8.4.4}$$

Here $\mathbf{f}$ is the force density due to electromagnetic fields, p is the pressure, which will be used in the form

$$p = \frac{\rho - \rho_0}{\kappa\rho_0} - \alpha\nabla^2\rho, \tag{A8.4.5}$$

where the mean density ρ_0, the compressibility κ and α are temperature-dependent parameters.

Now we introduce a complex macroscopic "wave function" ψ (compare the main text)

$$\psi = \rho^{\frac{1}{2}} e^{i\phi}, \tag{A8.4.6}$$

where ϕ is the velocity potential in units of a constant γ considered below. γ is chosen such that ϕ is dimensionless.

Then we can readily show that if we take

$$\rho = \psi^*\psi, \quad \mathbf{j} = \frac{\gamma}{2i}[\psi^*\nabla\psi - (\nabla\psi^*)\psi] - \frac{e}{mc}\mathscr{A}\rho, \tag{A8.4.7}$$

the equations (A8.4.1) and (A8.4.2) follow. Furthermore, equations (A8.4.3) and (A8.4.2) follow if ψ satisfies the equation

$$\frac{\gamma}{i}\dot{\psi} = \frac{\gamma^2}{2}\left(\nabla - \frac{ie}{mc\gamma}\mathscr{A}\right)^2\psi - \left[\frac{1}{\kappa\rho_0}(\psi^*\psi - \rho_0) - \frac{\beta}{\rho_0}\nabla^2(\psi^*\psi)\right]\psi. \tag{A8.4.8}$$

To show that (A8.4.3) follows, we form $\partial \mathbf{j}/\partial t$ using equation (A8.4.7) and insert ψ and ψ^* from equation (A8.4.8). Use of equations (A8.4.1), (A8.4.5) (A8.4.6) and (A8.4.7) then leads to equation (A8.4.3) provided we put

$$\beta = \alpha - (\gamma^2/4). \tag{A8.4.9}$$

The Ginzburg–Landau equation follows if we put $\beta = 0$ and give κ and ρ_0 specific temperature-dependent values. The hydrodynamic energy density U can be written in the alternative forms

$$U = \tfrac{1}{2}\rho v^2 + \frac{1}{2\kappa}\left(\frac{\rho-\rho_0}{\rho_0}\right)^2 + \frac{\alpha}{2}\left(\frac{\nabla\rho}{\rho_0}\right)^2 + U_0 \tag{A8.4.10}$$

or

$$U = \frac{\gamma^2}{2}\left[\left(\nabla + \frac{ie}{mc}\mathscr{A}\right)\psi^*\right]\left[\left(\nabla - \frac{ie}{mc\gamma}\mathscr{A}\right)\psi + \frac{1}{2\kappa}\left(\frac{\psi^*\psi-\rho_0}{\rho_0}\right)^2\right.$$

$$\left. + \frac{\beta}{2}\frac{[\nabla(\psi^*\psi)]^2}{\rho_0} + U_0.\right. \tag{A8.4.11}$$

The parameters $\kappa, \rho_0, \alpha, \beta, U_0$ are temperature dependent. The energy $\int U d\mathbf{r}$ is playing the role of a free energy.

Thus, it has been shown that any solution ψ of the wave equation (A8.4.8) can be expressed, using (A8.4.7), as a solution $\rho, \mathbf{v}$ of the hydrodynamic equations (A8.4.1–5). The opposite also holds provided quantization effects on ψ are not involved. When they are, then ϕ and hence Φ/γ become multivalued. Thus the wave equation (A8.4.8) is equivalent to hydrodynamics plus London equations plus flux quantization.

APPENDIX A9.1 POLARON ENERGY IN CONTINUUM POLARIZATION MODEL

We wish here to summarize the argument by which the polaron energy can be calculated from a knowledge of the phonon spectrum in the small polaron régime.

One way of approaching the problem is afforded by the result (8.7.29) in the discussion of superconductivity. If, in this formula, we go to the limit of a narrow band, then $\varepsilon_{\mathbf{k}} = \varepsilon_{\mathbf{k}-\mathbf{q}}$ and we find that the energy obtained by averaging (8.7.28) over electron coordinates contains a term of the form D_q^2/ω_q, where D_q measures the strength of the electron–phonon coupling.

However, we shall obtain essentially this result here by a variational procedure, in which we regard the electron–phonon interaction as described by displaced harmonic oscillator states. By this we mean that we take a wave function for these states of the form

$$\Psi_{n,f} = \frac{1}{\sqrt{n!}}(b^\dagger + f^*)^n \exp(-b^\dagger f)|0\rangle \exp(-\tfrac{1}{2}|f|^2), \qquad (A9.1.1)$$

where $|0\rangle$ is the vacuum state of the phonon system and $b^\dagger$ is the phonon creation operator; f is to be determined.

The wave function of the electron plus its phonon cloud is now to be regarded as describing an electron localized on one site and surrounded by displaced harmonic oscillators. Obviously we must deal with the electron wave function in the spirit of the tight-binding method, and then we can write with ϕ as the localized atomic orbital.

$$|\mathbf{k}n\rangle \equiv \frac{1}{N^{\frac{1}{2}}} \sum_{\mathbf{l}} \exp(i\mathbf{k}.\mathbf{l})\,\phi(\mathbf{r}-\mathbf{l}) \prod \Psi_{n_\mathbf{q} f_\mathbf{q}}. \qquad (A9.1.2)$$

Since states can only change by a phase factor when we add a lattice vector to every coordinate, we must require that

$$f_\mathbf{q}(\mathbf{l}) = d_\mathbf{q} \exp(-i\mathbf{q}.\mathbf{l}). \qquad (A9.1.3)$$

Using the electron–phonon interaction

$$H_{\text{int}} = \sum_{\mathbf{k}} [V_\mathbf{k} b_\mathbf{k} \exp(i\mathbf{k}.\mathbf{r}) + V_\mathbf{k}^\dagger b_\mathbf{k}^\dagger \exp(-i\mathbf{k}.\mathbf{r})] \qquad (A9.1.4)$$

as in equation (9.5.1), we now regard the $d_\mathbf{q}$'s as variational parameters. Minimizing the energy, and neglecting overlaps of the electron wave functions $\phi(\mathbf{r})$ between different sites we find the result

$$d_\mathbf{q} = -\frac{V_\mathbf{q}}{\hbar\omega_\mathbf{k}} \int \phi(\mathbf{r}) \exp(i\mathbf{q}.\mathbf{r})\,\phi(\mathbf{r})\,d\mathbf{r}. \qquad (A9.1.5)$$

We shall assume we can replace the integral by its value unity in the long-wavelength limit, which is clearly a good approximation when the ϕ's are well localized.

We now calculate the expectation value $\langle H_{\text{int}}\rangle$ and we find

$$W_p = \sum_\mathbf{q} \hbar\omega_\mathbf{q} |d_\mathbf{q}|^2 = \sum_\mathbf{q} |V_\mathbf{q}|^2/\hbar\omega_\mathbf{q}, \qquad (A9.1.6$$

which is the desired result.

APPENDIX A9.2 RESPONSE THEORY OF EXCITONIC INSULATORS

In this appendix, we shall follow Jérome, Rice and Kohn (1967) in deriving the results, summarized in the main text, for the response of an excitonic insulator to an electromagnetic field (see also Kohn, 1968).

As in Chapter 9, section 9.9, we introduce Fermion operators such that $a_{\mathbf{k}}^{\dagger}, a_{\mathbf{k}}$ respectively create and destroy electrons of wave-vector $\mathbf{k}$ in band a, while $b_{\mathbf{k}}^{\dagger}, b_{\mathbf{k}}$ respectively create and destroy electrons of wave-vector $\mathbf{k}+\mathbf{q}$ in band b. We next define the field operators

$$\left.\begin{aligned} \psi_a(\mathbf{x}) &= \sum_{\mathbf{k}} a_{\mathbf{k}} e^{i\mathbf{k}.\mathbf{x}}, \\ \psi_b(\mathbf{x}) &= \sum_{\mathbf{k}} b_{\mathbf{k}} e^{i\mathbf{k}.\mathbf{x}}, \end{aligned}\right\} \tag{A9.2.1}$$

in terms of which we can write the Green functions for the two bands as

$$\left.\begin{aligned} \mathscr{G}_a(1',1) &= -i\langle T\tilde{\psi}_a(1')\tilde{\psi}_a^{\dagger}(1)\rangle, \\ \mathscr{G}_b(1',1) &= -i\langle T\tilde{\psi}_b(1')\tilde{\psi}_b^{\dagger}(1)\rangle, \end{aligned}\right\} \tag{A9.2.2}$$

where $\tilde{\psi}(1)$ is a Heisenberg operator with $1 \equiv \mathbf{x}_1, t_1$.

From equation (5.2.8) we can see that we will have a current density

$$\frac{e}{2m_a}[(\nabla_1 - \nabla_{1'})\mathscr{G}_a(1,1')]_{1'\equiv 1} \tag{A9.2.3}$$

associated with the carriers of band a, with a similar expression, but opposite in sign, for band b. In equilibrium the total current will be zero and so, to first order, the current will be linear in the vector potential $\mathscr{A}$ by which we characterize an electromagnetic field perturbing the system. Referring to the definition of a functional derivative‡ (section 2.1.2), it is evident from (A9.2.3) that we now have a current which we can write in the form

$$\mathbf{J}^{\mathrm{P}}(1) = -\frac{e}{2}(\nabla_1 - \nabla_{1'})\left[m_b^{-1}\frac{\delta\mathscr{G}_b(1,1')}{\delta\mathscr{A}(2)} - m_a^{-1}\frac{\delta\mathscr{G}_a(1,1')}{\delta\mathscr{A}(2)}\right]_{1'\equiv 1}\mathscr{A}(2). \tag{A9.2.4}$$

This is the "paramagnetic current" and to this we must add a "diamagnetic" term, which is trivially found as we shall see below. It therefore follows that we can calculate the response function K defined in equation (9.9.14) by evaluating the functional derivatives of the Green functions with respect to $\mathscr{A}$, and this is our first task.

‡ In the notation we use here

$$u(12) = \frac{\delta\rho(1)}{\delta V(2)},$$

where

$$\delta\rho(1) = \int u(12)\,\delta V(2)\,d2.$$

A9.2.1 Functional derivatives of G

In terms of the single-particle energies defined in equations (9.9.1) and (9.9.2) we may write a model Hamiltonian for our system as

$$H = \sum_{\mathbf{k}} \varepsilon_a(\mathbf{k})\, a_{\mathbf{k}} a_{\mathbf{k}}^{\dagger} + \sum_{\mathbf{k}} \varepsilon_b\, b_{\mathbf{k}}^{\dagger} b_{\mathbf{k}} + \frac{1}{2}\sum_{\mathbf{q}} V(\mathbf{q})\,\rho(\mathbf{q})\,\rho(-\mathbf{q}), \qquad \text{(A9.2.5)}$$

where

$$V(\mathbf{q}) = 4\pi e^2/\varepsilon(\mathbf{q})\, q^2, \qquad \text{(A9.2.6)}$$

$\varepsilon(\mathbf{q})$ being an effective dielectric constant and [cf. equation (2.5.22)]

$$\rho(\mathbf{q}) = \sum_{\mathbf{k}} (a_{\mathbf{k}+\mathbf{q}}^{\dagger} a_{\mathbf{k}} + b_{\mathbf{k}+\mathbf{q}}^{\dagger} b_{\mathbf{k}}). \qquad \text{(A9.2.7)}$$

Alternatively, in configuration space, equation (A9.2.5) becomes

$$H = \sum_{i=a,b} \int \psi_i^{\dagger}(\mathbf{x})\,\varepsilon_i(\mathbf{x})\,\psi_i(\mathbf{x})\,d\mathbf{x} + \frac{1}{2}\int \rho(\mathbf{x})\,\rho(\mathbf{x}')\,V(\mathbf{x}-\mathbf{x}')\,d\mathbf{x}\,d\mathbf{x}', \qquad \text{(A9.2.8)}$$

where [cf. equations (9.9.1) and (9.9.2)]

$$\varepsilon_a(\mathbf{x}) = -\tfrac{1}{2}E_g + (2m_a)^{-1}\nabla^2, \quad \varepsilon_b(\mathbf{x}) = -\tfrac{1}{2}E_g - (2m_b)^{-1}\nabla^2 \qquad \text{(A9.2.9)}$$

and

$$\rho(\mathbf{x}) = \sum_{i=a,b} \psi_i^{\dagger}(\mathbf{x})\,\psi_i(\mathbf{x}). \qquad \text{(A9.2.10)}$$

Now defining

$$V(1-1') = V(\mathbf{x}-\mathbf{x}')\,\delta(t_1-t_1'), \qquad \text{(A9.2.11)}$$

the equation of motion for $\tilde{\psi}_b(1)$ is

$$\left[i\frac{\partial}{\partial t_1} - \varepsilon_b(1)\right]\tilde{\psi}_b(1) = V(1-\bar{1})\,\rho(\bar{1})\,\tilde{\psi}_b(1), \qquad \text{(A9.2.12)}$$

where a bar over a variable indicates integration over that variable. From this equation it follows that

$$\left[i\frac{\partial}{\partial t_1} - \varepsilon_b(1)\right]\mathscr{G}_b(1,1') = \delta(1,1') - iV(1-\bar{1})\,\langle T\rho(\bar{1})\,\tilde{\psi}_b(1)\,\tilde{\psi}_b^{\dagger}(1')\rangle \qquad \text{(A9.2.13)}$$

and this equation of motion for $\mathscr{G}_b$ may be rewritten as

$$\left[i\frac{\partial}{\partial t_1} - \varepsilon_b(1)\right]\mathscr{G}_b(1,1') - iV(1-\bar{1})\,F^{\dagger}(\bar{1},1')\,F(1,\bar{1}) = \delta(1,1'), \qquad \text{(A9.2.14)}$$

where F is an "anomalous Green function" similar to that entering the theory of superconductivity (see section 8.10):

$$\left.\begin{aligned} F(1,1') &= -i\langle T\tilde{\psi}_b(1)\,\tilde{\psi}_a^\dagger(1')\rangle, \\ F^\dagger(1,1') &= -i\langle T\tilde{\psi}_a(1)\,\psi_b^\dagger(1')\rangle. \end{aligned}\right\} \tag{A9.2.15}$$

We can find an equation of motion for F just as we found that for $\mathscr{G}_b$; it is

$$\left[i\frac{\partial}{\partial t_1}-\varepsilon_a(1)\right]F^\dagger(1,1')-iV(1-\bar{1})\,F^\dagger(1,\bar{1})\,\mathscr{G}_b(\bar{1},1')=0. \tag{A9.2.16}$$

If we now introduce a perturbation U, equations (A9.2.14) and (A9.2.16) remain valid provided ε_a and ε_b are replaced by operators containing the perturbation:

$$\left[i\left(\frac{\partial}{\partial t_1}\right)\delta(1,\bar{1})-h_b(1,\bar{1};\,U)\right]\mathscr{G}_b(\bar{1},1';\,U)-iV(1-\bar{1})\,F^\dagger(\bar{1},1;\,U)\,F(1,\bar{1};\,U) = \delta(1,\bar{1}) \tag{A9.2.17}$$

and

$$\left[i\left(\frac{\partial}{\partial t_1}\right)\delta(1,\bar{1})-h_a(1,\bar{1};\,U)\right]F^\dagger(\bar{1},1';\,U)-iV(1-\bar{1})\,F^\dagger(1,\bar{1};\,U)\,\mathscr{G}_b(\bar{1},1';\,U) = 0, \tag{A9.2.18}$$

where

$$h_i(1,1';\,U)=\varepsilon_i(1)\,\delta(1,1')-U(1,1')-v(1-\bar{1})\,\langle\rho(\bar{1};\,U)\rangle\,\delta(1,1'). \tag{A9.2.19}$$

The last term of h_i is the Hartree field due to the perturbation, with $v(1-\bar{1})$ representing the unscreened interaction.

To find the functional derivatives, equations (A9.2.17) and (A9.2.18) are expanded to first order in U. The details of this procedure are given by Jérome *et al.*, the final results being

$$\begin{aligned} \frac{\delta\mathscr{G}_b(1,1')}{\delta U(2,2')} &= -\mathscr{G}_b(1,\bar{1})\,\mathscr{G}_b(\bar{2},1')\frac{\delta h_b(\bar{1},\bar{2})}{\delta U(2,2')}-F(1,\bar{1})\,F^\dagger(\bar{2},1')\frac{\delta h_a(\bar{1},\bar{2})}{\delta U(2,2')} \\ &\quad +iV(\bar{1}-\bar{2})\frac{\delta F(\bar{1},\bar{2})}{\delta U(2,2')}\mathscr{G}_b(1,\bar{1})\,F^\dagger(\bar{2},1') \\ &\quad -iV(\bar{1}-\bar{2})\frac{\delta F^\dagger(\bar{1},\bar{2})}{\delta U(2,2')}\mathscr{G}_b(\bar{2},1')\,F(1,\bar{1}) \end{aligned} \tag{A9.2.20}$$

and

$$\frac{\delta F^{\dagger}(1,1')}{\delta U(2,2')} = -\mathscr{G}_a(1,\bar{1})\,F^{\dagger}(\bar{2},1')\frac{\delta h_a(\bar{1},\bar{2})}{\delta U(2,2')} - \mathscr{G}_b(\bar{2},1')\,F(1,\bar{1})\frac{\delta h_b(\bar{1},\bar{2})}{\delta U(2,2')}$$

$$+ iV(\bar{1}-\bar{2})\frac{\delta F^{\dagger}(\bar{1},\bar{2})}{\delta U(2,2')}\mathscr{G}_b(\bar{2},1')\,\mathscr{G}_a(1,\bar{1})$$

$$+ iV(\bar{1}-\bar{2})\frac{\delta F(\bar{1},\bar{2})}{\delta U(2,\bar{2}')}F^{\dagger}(1,\bar{1})\,F^{\dagger}(\bar{2},1'). \qquad \text{(A9.2.21)}$$

The functional derivatives of $\mathscr{G}_a$ and F may be obtained from the above two equations using the symmetry with respect to a and b bands. All the quantities in these equations are to be evaluated at $U = 0$. $\mathscr{G}$ and F are readily found from equations (A9.2.14) and (A9.2.16) by taking a Fourier transform. Introducing a "gap function" $\Delta(\mathbf{p})$ defined by

$$\Delta(\mathbf{p}) \equiv i\sum_{\mathbf{k}}\int\frac{d\omega}{2\pi}V(\mathbf{p}-\mathbf{k})\,F(\mathbf{k},\omega), \qquad \text{(A9.2.22)}$$

$$\Delta^{\dagger}(\mathbf{p}) \equiv i\sum_{\mathbf{k}}\int\frac{d\omega}{2\pi}V(\mathbf{p}-\mathbf{k})\,F^{\dagger}(\mathbf{k},\omega), \qquad \text{(A9.2.23)}$$

where

$$F(1,1') = \sum_{\mathbf{k}}\int\frac{d\omega}{2\pi}F(\mathbf{k},\omega)\exp\left[i\mathbf{p}(\mathbf{x}_1-\mathbf{x}_{1'})\right]\exp\left[-i\varepsilon(t_1-t_{1'})\right], \qquad \text{(A9.2.24)}$$

we write equations (A9.2.14) and (A9.2.16) in Fourier transform as

$$\left.\begin{aligned}&[\varepsilon-\varepsilon_b(\mathbf{p})]\,\mathscr{G}_b(\mathbf{p},\varepsilon)-\Delta(\mathbf{p})\,F^{\dagger}(\mathbf{p},\varepsilon)=1\\&[\varepsilon-\varepsilon_a(\mathbf{p})]\,F(\mathbf{p},\varepsilon)-\Delta^{\dagger}(\mathbf{p})\,\mathscr{G}_b(\mathbf{p},\varepsilon)=0\end{aligned}\right\}. \qquad \text{(A9.2.25)}$$

From these equations follows that at $T = 0$ the Green functions have the form

$$\mathscr{G}_b(\mathbf{p},\varepsilon) = \frac{\varepsilon-\varepsilon_a(\mathbf{p})}{[\varepsilon-\varepsilon_a(\mathbf{p})]\,[\varepsilon-\varepsilon_b(\mathbf{p})]-|\Delta(\mathbf{p})|^2}, \qquad \text{(A9.2.26)}$$

$$F(\mathbf{p},\varepsilon) = \frac{\Delta^{\dagger}(\mathbf{p})}{[\varepsilon-\varepsilon_a(\mathbf{p})]\,[\varepsilon-\varepsilon_b(\mathbf{p})]-|\Delta(\mathbf{p})|^2}, \qquad \text{(A9.2.27)}$$

with similar equations for $\mathscr{G}_a$ and F. Inserting equation (A9.2.27) into (A9.2.23), we obtain equation (9.9.7) for the gap function.

Equations (A9.2.26) and (A9.2.27) may be rewritten in terms of the u_k and v_k, introduced in equation (9.9.5), and the energy $E_{\mathbf{k}}$, instead of Δ. We then find

$$\mathscr{G}_b(\mathbf{p},\varepsilon)=\frac{u_p^2}{\varepsilon-\zeta_{\mathbf{p}}-E_{\mathbf{p}}+i\delta}+\frac{v_p^2}{\varepsilon-\zeta_{\mathbf{p}}+E_{\mathbf{p}}-i\delta}, \tag{A9.2.28}$$

$$F(\mathbf{p},\varepsilon)=F^{\dagger}(\mathbf{p},\varepsilon)=\frac{u_p v_p}{\varepsilon-\zeta_{\mathbf{p}}-E_{\mathbf{p}}+i\delta}-\frac{u_p v_p}{\varepsilon-\zeta_{\mathbf{p}}+E_{\mathbf{p}}-i\delta} \tag{A9.2.29}$$

and

$$\mathscr{G}_a(\mathbf{p},\varepsilon)=\frac{v_p^2}{\varepsilon-\zeta_{\mathbf{p}}-E_{\mathbf{p}}+i\delta}+\frac{u_p^2}{\varepsilon-\zeta_{\mathbf{p}}+E_{\mathbf{p}}-i\delta}. \tag{A9.2.30}$$

Here, as in the main text,

$$\left.\begin{aligned} E_{\mathbf{k}} &= \xi_{\mathbf{k}}^2+|\Delta_{\mathbf{k}}|^2,\\ \xi &= \tfrac{1}{2}(\varepsilon_a-\varepsilon_b), \quad \zeta=\tfrac{1}{2}(\varepsilon_a+\varepsilon_b), \end{aligned}\right\} \tag{A9.2.31}$$

while

$$\left.\begin{aligned} u_{\mathbf{k}} &= \left[\frac{1}{2}\left(1-\frac{\xi_{\mathbf{k}}}{E_{\mathbf{k}}}\right)\right]^{\frac{1}{2}},\\ v_{\mathbf{k}} &= \left[\frac{1}{2}\left(1-\frac{\xi_{\mathbf{k}}}{E_{\mathbf{k}}}\right)\right]^{\frac{1}{2}}\frac{\Delta(\mathbf{k})}{|\Delta(\mathbf{k})|}. \end{aligned}\right\} \tag{A9.2.32}$$

These values also result from minimization of the expectation value of the total energy with respect to a BCS type wave function.

A9.2.2 Response at zero temperature

We now introduce the vector potential $\mathscr{A}(\mathbf{x},t)$ into the Hamiltonian, whereupon

$$h_i(1,1')=\varepsilon_i[\mathbf{p}_1-e\mathscr{A}(1)]\,\delta(1,1')-v(1-\bar{1})\langle\rho(\bar{1};\mathscr{A})\rangle\,\delta(1,1') \tag{A9.2.33}$$

and

$$\frac{\delta h_i(1,1')}{\delta\mathscr{A}(2)}=-\frac{e}{2m_i i}\{\nabla_2-\nabla_{2'}\}\,\delta(1-2)\,\delta(1'-2')|_{2'=2}-v(1-\bar{1})\left[\frac{\delta\langle\rho(\bar{1})\rangle}{\delta\mathscr{A}(2)}\delta(1,1')\right]. \tag{A9.2.34}$$

On inserting this result into equations (A9.2.20) and (A9.2.21) we find

$$\begin{aligned}\frac{\delta\mathscr{G}_i(1,1')}{\delta\mathscr{A}(2)}=\sum_{\mathbf{p},\mathbf{q}}\int\frac{d\varepsilon\,d\omega}{4\pi^2}\Lambda^i_{\mathbf{p}}(\mathbf{q})\\ \times\exp[i\mathbf{q}.(\mathbf{x}_1-\mathbf{x}_2)]\exp[i\omega(t_2-t_1)]\\ \times\exp[i\mathbf{p}.(\mathbf{x}_1-\mathbf{x}_{1'})]\exp[i\varepsilon(t_{1'}-t_1)]\end{aligned} \tag{A9.2.35}$$

and

$$\frac{\delta F^{\dagger}(1,1')}{\delta \mathscr{A}(2)} = \sum_{\mathbf{p},\mathbf{q}} \int \frac{d\varepsilon\, d\omega}{4\pi^2} \boldsymbol{\lambda}_{\mathbf{p}}^{i}(\mathbf{q}) \times \exp[i\mathbf{q}.(\mathbf{x}_1-\mathbf{x}_2)] \exp[i\omega(t_2-t_1)] \times \exp[i\mathbf{p}.(\mathbf{x}_1-\mathbf{x}_{1'})] \exp[i\varepsilon(t_{1'}-t_1)], \quad \text{(A9.2.36)}$$

where

$$\boldsymbol{\Lambda}_{\mathbf{p}}^{b}(\mathbf{q}) = -\frac{e}{2m_b}(2\mathbf{p}+\mathbf{q})\mathscr{G}_b(\mathbf{p}+\mathbf{q})\mathscr{G}_b(\mathbf{p}) + \frac{e}{2m_a}(2\mathbf{p}+\mathbf{q})F(\mathbf{p}+\mathbf{q})F^{\dagger}(\mathbf{p})$$
$$+i\sum_{\mathbf{k}} \int \frac{dv}{2\pi} V(\mathbf{k})[\boldsymbol{\lambda}_{\mathbf{p}-\mathbf{k}}(\mathbf{q})F^{\dagger}(\mathbf{p})\mathscr{G}_b(\mathbf{p}+\mathbf{q}) + \boldsymbol{\lambda}_{\mathbf{p}-\mathbf{k}}^{*}F(\mathbf{p}+\mathbf{q})\mathscr{G}_b(\mathbf{p})]$$
$$-i\sum_{\mathbf{k}} \int \frac{dv}{2\pi} v(\mathbf{q})[\mathscr{G}_b(\mathbf{p}+\mathbf{q})\mathscr{G}_b(\mathbf{p}) + F(\mathbf{p}+\mathbf{q})F^{\dagger}(\mathbf{p})]\{\boldsymbol{\Lambda}_{\mathbf{k}}^{a}(\mathbf{q}) + \Lambda_{\mathbf{k}}^{b}(\mathbf{q})\} \quad \text{(A9.2.37)}$$

and

$$\boldsymbol{\lambda}_{\mathbf{p}}^{a}(\mathbf{q}) = \frac{e}{2m_a}(2\mathbf{p}+\mathbf{q})\mathscr{G}_a(\mathbf{p}+\mathbf{q})F^{\dagger}(\mathbf{p}) - \frac{e}{2m_b}(2\mathbf{p}+\mathbf{q})\mathscr{G}_b(\mathbf{p})F(e+\mathbf{q})$$
$$+i\sum_{\mathbf{k}} \int \frac{dv}{2\pi} V(\mathbf{k})[\boldsymbol{\lambda}_{\mathbf{p}-\mathbf{q}}(\mathbf{q})F^{\dagger}(\mathbf{p}+\mathbf{q})F^{\dagger}(\mathbf{p}) + \boldsymbol{\lambda}_{\mathbf{p}-\mathbf{k}}^{*}(\mathbf{q})\mathscr{G}_a(\mathbf{p}+\mathbf{q})\mathscr{G}_b(\mathbf{p})]$$
$$-i\sum_{\mathbf{k}} \int \frac{dv}{2\pi} v(\mathbf{q})[F^{\dagger}(\mathbf{p}+\mathbf{q})\mathscr{G}_b(\mathbf{p}) + F^{\dagger}(\mathbf{p})\mathscr{G}_a(\mathbf{p}+\mathbf{q})][\boldsymbol{\Lambda}_{\mathbf{k}}^{a}(\mathbf{q}) + \boldsymbol{\Lambda}_{\mathbf{k}}^{b}(\mathbf{q})]. \quad \text{(A9.2.38)}$$

Equation (A9.2.4) can be rewritten, in Fourier transform, in terms of the $\boldsymbol{\Lambda}$'s, the result being

$$\mathbf{J}^{\mathrm{P}}(\mathbf{q}) = -ie\sum_{\mathbf{p}} \int \frac{de}{4\pi}(2\mathbf{p}+\mathbf{q})\left\{\frac{\boldsymbol{\Lambda}_{\mathbf{p}}^{b}(\mathbf{q})}{m_b} - \frac{\boldsymbol{\Lambda}_{\mathbf{p}}^{a}(\mathbf{q})}{m_a}\right\}\mathscr{A}(\mathbf{q}). \quad \text{(A9.2.39)}$$

We can now investigate whether there is a Meissner effect in the system and also if it is, in fact, an insulator at $T = 0$. From equation (8.11.28) there is no Meissner effect if

$$\lim_{\mathbf{q}\to 0} \lim_{\omega\to 0} K(\mathbf{q},\omega) = 0. \quad \text{(A9.2.40)}$$

On the other hand, if no scalar potential is applied, the electric field is given by $\mathscr{E} = -\partial\mathscr{A}/\partial t$ and we must take the limit $\mathbf{q}\to 0$ before $\omega\to 0$ to find the d.c. conductivity. Thus the system is an insulator if

$$\lim_{\omega\to 0}\lim_{\mathbf{q}\to 0} K(\mathbf{q},\omega) = 0. \tag{A9.2.41}$$

In fact an examination of the integrals of equations (A9.2.37) and (A9.2.38) shows them to be regular in the limit $q,\omega\to 0$ at $T=0$ (but not, it turns out, at other temperatures), the order in which we take the limits being immaterial. We therefore take a static external field; then for simplicity taking $m_a = m_b$ we find

$$\mathbf{J}^{\mathrm{P}}(\mathbf{q}) = \frac{e}{2m}\sum_{\mathbf{p}}(2\mathbf{p}+\mathbf{q})\left[\frac{\mu_{\mathbf{p}}(\mathbf{q})}{E_{\mathbf{p}}+E_{\mathbf{p}+\mathbf{q}}}\right]\left[\nu_{\mathbf{p}}(\mathbf{q})\,\beta_{\mathbf{p}}(\mathbf{q})+\frac{e}{m}(2\mathbf{p}+\mathbf{q})\,\mathscr{A}(\mathbf{q})\,\mu_{\mathbf{p}}(\mathbf{q})\right], \tag{A9.2.42}$$

where

$$\left.\begin{aligned}\mu_{\mathbf{p}}(\mathbf{q}) &= u_{\mathbf{p}}\,v_{\mathbf{p}+\mathbf{q}}+v_{\mathbf{p}}\,u_{\mathbf{p}+\mathbf{q}},\\ \nu_{\mathbf{p}}(\mathbf{q}) &= u_{\mathbf{p}}\,u_{\mathbf{p}+\mathbf{q}}+v_{\mathbf{p}}\,v_{\mathbf{p}+\mathbf{q}}\end{aligned}\right\} \tag{A9.2.43}$$

and

$$\beta_{\mathbf{p}}(\mathbf{q}) = i\sum_{\mathbf{k}}\int\frac{d\nu}{2\pi}\,V(\mathbf{p}-\mathbf{q})\,[\boldsymbol{\lambda}_{\mathbf{k}}^{*}(\mathbf{q})+\boldsymbol{\lambda}_{\mathbf{k}}(\mathbf{q})].\mathscr{A}(\mathbf{q})], \tag{A9.2.44}$$

obeying the inhomogeneous integral equation

$$\beta_{\mathbf{p}'}(\mathbf{q}) = \sum_{\mathbf{p}} V(\mathbf{p}-\mathbf{p}')\left[\frac{\nu_{\mathbf{p}}(\mathbf{q})}{E_{\mathbf{p}}+E_{\mathbf{p}+\mathbf{q}}}\right]\left[\nu_{\mathbf{p}}(\mathbf{q})\,\beta_{\mathbf{p}}(\mathbf{q})+\frac{e}{m}(2\mathbf{p}+\mathbf{q}).\mathscr{A}(\mathbf{q})\,\mu_{\mathbf{p}}(\mathbf{q})\right]. \tag{A9.2.45}$$

In the limit $q\to 0$, this is perfectly regular and so $\beta_{\mathbf{p}}(\mathbf{q})$ is independent of the polarization, or direction of $\mathbf{q}$ relative to that of $\mathscr{A}$, as the limit is taken. But if we choose a time-dependent vector potential of the form

$$\mathscr{A}(\mathbf{q}) = \mathbf{q}\mathscr{A}_0 \tag{A9.2.46}$$

the response of the system must be zero if the theory obeys the gauge invariance requirement, the right-hand side of equation (A9.2.46) being the gradient of a vector. That the above equations are in fact gauge invariant is readily checked. With $\mathscr{A}$ given by equation (A9.2.46) one finds

$$\beta_{\mathbf{p}'}(\mathbf{q}) = 2e\mathscr{A}_0(\Delta_{\mathbf{p}'}-\Delta_{\mathbf{p}'+\mathbf{q}}) \tag{A9.2.47}$$

and then

$$\mathbf{J}^{\mathrm{P}}(\mathbf{q}) = \frac{e^2\mathscr{A}_0}{m}\sum_{\mathbf{p}}(2\mathbf{p}+\mathbf{q})\frac{\mu_{\mathbf{p}}(\mathbf{q})}{E_{\mathbf{p}}+E_{\mathbf{p}+\mathbf{q}}}\{\nu_{\mathbf{p}}(\mathbf{q})(\Delta_{\mathbf{p}}-\Delta_{\mathbf{p}+\mathbf{q}})+[\xi(\mathbf{p}+\mathbf{q})-\xi(\mathbf{p})]\mu_{\mathbf{p}}(\mathbf{q})$$

$$= -\frac{2e^2}{m}\mathscr{A}_0\sum_{\mathbf{p}}(\mathbf{p}+\tfrac{1}{2}\mathbf{q})(u_{\mathbf{p}}^2 v_{\mathbf{p}+\mathbf{q}}^2 - v_{\mathbf{p}}^2 u_{\mathbf{p}+\mathbf{q}}^2)$$

$$= -\frac{2e^2}{m}\mathscr{A}_0\sum_{\mathbf{p}}(\mathbf{p}+\tfrac{1}{2}\mathbf{q})(v_{\mathbf{p}+\mathbf{q}}^2 - v_{\mathbf{p}})$$

$$= \frac{2e^2}{m}\mathscr{A}_0\mathbf{q}\sum_{\mathbf{p}}v_{\mathbf{p}}^2. \tag{A9.2.48}$$

This we can evidently rewrite as

$$\mathbf{J}^{\mathrm{P}}(\mathbf{q}) = \frac{2e^2}{m}\mathscr{A}n_{\mathrm{c}}, \tag{A9.2.49}$$

where

$$n_{\mathrm{c}} = \sum_{\mathbf{p}} v_{\mathbf{p}}^2 = \sum_{\mathbf{p}}\langle a_{\mathbf{p}}a_{\mathbf{p}}^{\dagger}\rangle = \sum_{\mathbf{p}}\langle b_{\mathbf{p}}^{\dagger}b_{\mathbf{p}}\rangle \tag{A9.2.50}$$

is the number of carriers in either band.

Now since

$$m\mathbf{v} = \mathbf{p} - e\mathscr{A}/c \tag{A9.2.51}$$

$\mathbf{J}^{\mathrm{P}}(\mathbf{q})$ arising from the $\mathbf{p}$ term, must be corrected by the "diamagnetic" current $\mathbf{j}^{\mathrm{D}}$ arising from the $e\mathscr{A}/c$ term and it is evident that

$$\mathbf{j}^{\mathrm{D}} = -\frac{2e^2}{m}n_{\mathrm{c}}\mathscr{A}. \tag{A9.2.52}$$

Thus, the total current $\mathbf{J}^{\mathrm{P}}+\mathbf{j}^{\mathrm{D}}$ given by the sum of equations (A9.2.49) and (A9.2.52) is zero. This establishes both the gauge invariance and the result

$$K(\mathbf{q}=0;\ \omega=0;\ T=0) = 0. \tag{A9.2.53}$$

A9.2.3 Conductivity at finite temperature

When $T \neq 0$, the paramagnetic current is again given by equation (A9.2.39), $\mathscr{G}_a$, $\mathscr{G}_b$ and F in equations (A9.2.37) and (A9.2.38) now being temperature-dependent Green functions [the angle brackets of equations (A9.2.2) and (A9.2.15) representing thermal averages]. Further, the forms we found for the equations of motion of the Green function are also unchanged, but since they only yield the real parts, must be supplemented by the relation [cf. equation (2.8.74), the following simply being the $T \neq 0$ generalization. See also equation (3.14.15)],

$$\operatorname{Re}\mathscr{G}(\varepsilon) = -\pi^{-1}\int_{-\infty}^{\infty} \coth\frac{\beta x}{2}\frac{\operatorname{Im}\mathscr{G}(\mathbf{x})}{\varepsilon - x}\,dx. \tag{A9.2.54}$$

One then finds (if $m_a = m_b$) that equations (A9.2.28) to (A9.2.30) become

$$\mathscr{G}_b(\mathbf{p},\varepsilon) = \frac{u_{\mathbf{p}}^2(1-n_{\mathbf{p}})}{\varepsilon - E_{\mathbf{p}} + i\delta} + \frac{v_{\mathbf{p}}^2 n_{\mathbf{p}}}{\varepsilon + E_{\mathbf{p}} + i\delta} + \frac{v_p^2(1-n_p)}{\varepsilon + E_{\mathbf{p}} - i\delta} + \frac{u_p^2 n_p}{\varepsilon - E_{\mathbf{p}} - i\delta}, \tag{A9.2.55}$$

$$F(\mathbf{p},\varepsilon) = F^{\dagger}(\mathbf{p},\varepsilon) = u_{\mathbf{p}} v_{\mathbf{p}}\left(\frac{1-n_{\mathbf{p}}}{\varepsilon - E_{\mathbf{p}} + i\delta} - \frac{n_{\mathbf{p}}}{\varepsilon + E_{\mathbf{p}} + i\delta} - \frac{1-n_{\mathbf{p}}}{\varepsilon + E_{\mathbf{p}} - i\delta} + \frac{n_{\mathbf{p}}}{\varepsilon_{\mathbf{p}} - E_{\mathbf{p}} - i\delta}\right), \tag{A9.2.56}$$

$$\mathscr{G}_a(\mathbf{p},\varepsilon) = \frac{v_{\mathbf{p}}^2(1-n_{\mathbf{p}})}{\varepsilon - E_{\mathbf{p}} + i\delta} + \frac{u_{\mathbf{p}}^2 n_{\mathbf{p}}}{\varepsilon + E_{\mathbf{p}} + i\delta} + \frac{u_{\mathbf{p}}^2(1-n_{\mathbf{p}})}{\varepsilon + E_{\mathbf{p}} - i\delta} + \frac{v_{\mathbf{p}}^2 n_{\mathbf{p}}}{\varepsilon - E_{\mathbf{p}} - i\delta}, \tag{A9.2.57}$$

where

$$n_{\mathbf{p}} = \frac{1}{e^{\beta E_{\mathbf{p}}} + 1} \tag{A9.2.58}$$

and the gap function is given by

$$\Delta_{\mathbf{p}} = \sum_{\mathbf{k}} V(\mathbf{p}-\mathbf{k})\frac{\Delta_{\mathbf{k}}}{2E_{\mathbf{k}}}\tanh\left(\frac{\beta E_{\mathbf{k}}}{2}\right). \tag{A9.2.59}$$

From these equations we obtain the paramagnetic current in similar fashion to the case when $T = 0$ and thereby obtain equations (9.9.21) and (9.9.22) of the main text for the paramagnetic part K^{P} of the response function, while the diamagnetic contribution is readily found to be simply

$$\begin{aligned} K_{ij}^{\mathrm{D}}(0,0) &= \frac{e^2}{m}\sum_{\mathbf{p}} \langle a_{\mathbf{p}} a_{\mathbf{p}}^{\dagger} + b_{\mathbf{p}}^{\dagger} b_{\mathbf{p}}\rangle\,\delta_{ij} \\ &= \frac{e^2}{m}\sum_{\mathbf{p}} (2v_{\mathbf{p}}^2 + [u_{\mathbf{p}}^2 - v_{\mathbf{p}}^2]\,n_{\mathbf{p}})\,\delta_{ij}, \end{aligned} \tag{A9.2.60}$$

which is the desired equation (9.2.25).

APPENDIX A9.3 SMALL POLARON MOTION IN MOLECULAR CRYSTAL MODEL

We reproduce here some of the detail omitted from the discussion of the small polaron in the main text (section 9.7.1). We also take the opportunity to refer to later work‡ on the lowering of the transition temperature below

‡ We are grateful to Dr. I. G. Austin for drawing our attention to these developments.

that given in the main text (Lang and Firsov, 1964; de Wit, 1968; Bosman and van Daal, 1970) and to further investigations of transport properties (Reik and Heese, 1967; Holstein and Friedman, 1968; Emin and Holstein, 1969).

We first evaluate the expression (9.7.17) for the matrix element $\langle p\{n_q\}|V|p'\{n_q'\}\rangle$ using the result that, to second order in $\delta = \Delta\xi_q^{(m)} - \Delta\xi_q^{(m+1)}$,

$$\int \chi_{n_q}^*(\xi^{(m)})\,\chi_{n_q}(\xi_q^{(m+1)})\,d\xi_q = 1+\frac{\delta^2}{4}(2n_q+1). \qquad \text{(A9.3.1)}$$

The result may be written in the form

$$\langle p\{n_q\}|V|p'\{n_q'\}\rangle = -J\sum_{r=\pm1}\delta_{p,p'+r}\prod_q\left[\left\{1-\left(\frac{4}{N}\right)(n_q'+\tfrac{1}{2})\gamma_q'\right.\right.$$
$$\left.\times\cos^2\left[q\left(p'+\frac{r}{2}\right)+\frac{\pi}{4}\right]\right\}\delta_{n_q,n_q'}$$
$$\left.-\left\{\left(\frac{8}{N}\right)^{\frac{1}{2}} r\mu_q\gamma_q^{\frac{1}{2}}\left(\frac{n_q'+\frac{1}{2}\pm\frac{1}{2}}{2}\right)\cos\left[q\left(p'+\frac{r}{2}\right)+\frac{\pi}{4}\right]\right\}\delta_{n_q,n_q'}\right], \qquad \text{(A9.3.2)}$$

where the factor μ_q is plus one for positive q and minus one for negative q. Here, two-quantum jumps, corresponding to $n_q \to n_q \pm 2$ have been ignored, since they contribute to order $1/N$ to the matrix element, compared to $1/N^{\frac{1}{2}}$ for one phonon processes $n_q \to n_q \pm 1$.

A9.3.1 Low temperatures

(a) $T=0$. If $T=0$, then $n_{-q}=0$ for all q and (A9.3.2) simplifies to

$$\langle p,\{n_q\}\equiv 0|V|p',\{n_q'\}\equiv 0\rangle = -J\sum_{r=\pm1}\delta_{p,p'+r}\exp\left\{-\sum_q \gamma_q/N\right\}. \qquad \text{(A9.3.3)}$$

Inserting this into (9.7.16) for $c(p) = c(p,\{n_q\}\equiv 0)$ and assuming that the time dependence of $c(p)$ is of the form $e^{-i\mathscr{E}t/\hbar}$ we find

$$c_k(p) = e^{ikp} \qquad \text{(A9.3.4)}$$

and

$$\mathscr{E}(p) = -J[c(p+1)+c(p-1)]e^{-S_0} = -2J\cos k\, e^{-S_0}, \qquad \text{(A9.3.5)}$$

where

$$S_0 = \sum_q \frac{\gamma_q}{N} = \frac{1}{\pi}\int_0^\pi \gamma_q\,dq. \qquad \text{(A9.3.6)}$$

Then from (9.7.15)

$$a_n(k,\{n_q\}\equiv 0) = e^{ikn}\prod_q \chi_0\left[\left(\frac{M\omega_0}{\hbar}\right)^{\frac{1}{2}}(\xi_q - \Delta\xi_q^{(n)})\right] \qquad \text{(A9.3.7)}$$

and, correspondingly, the ground-state energies of the polaron band are

$$E(k,\{n_q\}\equiv 0) = \mathscr{E}(\{n_q\}\equiv 0) - 2J\cos k\,\mathrm{e}^{-S_0}, \qquad \text{(A9.3.8)}$$

the half-width of the band being

$$\Delta E(k,\{n_q\}\equiv 0) = 2J\mathrm{e}^{-S_0}. \qquad \text{(A9.3.9)}$$

(b) $T>0$. We ignore non-diagonal transitions while allowing some of the n_q's to differ from zero. The diagonal transitions are then of the form

$$\langle p\{n_q\}|V|p'\{n_q\}\rangle = -J\sum_{r=\pm 1}\delta_{p,p+}$$

$$\times\exp\left\{-\sum_q\left[\frac{2(1+2n_q)}{N}\right]\gamma_q\cos^2\left[q\left(p+\frac{r}{2}\right)+\frac{\pi}{4}\right]\right\}.$$

(A9.3.10)

Inserting this into (9.7.16) for $c(p,\{n_q\})$ we find

$$c_k(p,\{n_q\}) = \exp(ikp)\exp\left[\frac{it}{\hbar}\{2J\cos k\exp[-2S(\{n_q\})]\}\right] \qquad \text{(A9.3.11)}$$

and hence, for $a_n(k,\{n_q\})$, equation (9.7.18) of the main text.

The transition probability for $k\to k'$ due to non-diagonal transitions is given by equation (9.7.23) of the main text. At sufficiently low temperatures, the lowest order transition dominates. Since energy is conserved in (9.7.23), one-phonon processes are not allowed, the lowest-order processes being of the form $k, n_{q_1}, n_{q_2}, \ldots, \to k',\ n_{q_1}-1, n_{q_2}+1, \ldots$. The corresponding transition probability turns out to be proportional to $n_{q_1}(n_{q_2}+1)$ and therefore vanishes as $T\to 0$.

A9.3.2 Elevated temperatures

Let us consider both diagonal and non-diagonal transitions in the calculation of transition probability of a localized polaron. In practice we require the temperature average of this:

$$W(p\to p';T) = Z^{-1}\sum_{\{n_q\}}W(p\to p',\{n_q\})\exp\left[\sum_q\beta\hbar\omega_q(n_q+\tfrac{1}{2})\right],$$

(A9.3.12)

where Z is the phonon partition function

$$Z = \sum_{\{n_q\}}\exp\left[-\sum_q\beta\hbar\omega_q(n_q+\tfrac{1}{2})\right] \qquad \text{(A9.3.13)}$$

and where

$$W(p \to p'; \{n_q\}) = \sum_{\{n_q'\}} W(p, \{n_q\} \to p', \{n_q'\}) \tag{A9.3.14}$$

is the transition probability with arbitrary final set $\{n_q'\}$ of phonon occupation numbers. The transition probability for a given final set $\{n_q'\}$ is given by the standard formula

$$W(p; \{n_q\} \to p', \{n_q'\}) = \frac{2}{\hbar} |\langle p, \{n_q\}| V |p'\{n_q'\}\rangle|^2 \frac{\partial}{\partial t} \Omega \left[\sum_q \hbar\omega_q (n_q - n_q') \right]. \tag{A9.3.15}$$

Here Ω is the quantity

$$\Omega(x) = \frac{1 - \cos(xt/\hbar)}{x^2/\hbar^2}. \tag{A9.3.16}$$

Following Holstein, the total thermally averaged transition probability is found from (A9.3.12) by inserting (A9.3.2) into (A9.3.15) and using the result

$$\frac{\partial}{\partial t} \Omega(x) = \frac{1}{2} \int_{-t}^{t} e^{ixt'/\hbar}\, dt'. \tag{A9.3.17}$$

One obtains

$$W(p \to p', T) = \frac{2tJ^2}{\hbar^2} e^{-2S(T)} + \frac{J^2}{\hbar^2} e^{-2S(T)} \times \int_{-t}^{t} \left\{ \exp\left[\left(\frac{2\gamma_q}{N} \right) \operatorname{cosech} \left(\frac{\beta\hbar\omega_q}{2} \right) \cos\omega_q t' \right] - 1 \right\} dt', \tag{A9.3.18}$$

with $S(T)$ given by equation (9.7.33) of the main text. The first term on the right-hand side is the contribution of diagonal transitions, and is proportional to t. This result, unexpected at first sight, is due to the fact that in diagonal transitions the energy is *exactly* conserved. Equation (A9.3.18) gives the transitions per unit time, so after time $\tau(p, T)$, the lifetime of the localized state, the number of diagonal transitions will be

$$2\tau^2(p, T) J^2/\hbar^2 e^{-2S(T)} \ll 1,$$

provided

$$\hbar/\tau(p, T) \gg J e^{-S(T)}, \tag{A9.3.19}$$

the condition that we are above the transition temperature. If we accordingly neglect diagonal transitions, and use the high temperature condition (9.7.28) of the main text, we can extend the t'-integral of (A9.2.18) to infinity, and evaluate it by the method of steepest descents. The result is

$$W(p\to p\pm 1;\, T) = \frac{J^2}{\hbar^2}\left[\frac{1}{2\pi^2}\int_0^\pi 2\gamma_q\,\omega_q^2 \operatorname{cosech}(\tfrac{1}{2}\beta\hbar\omega_q)\,dq\right]^{-\frac{1}{2}}$$

$$\times\exp\left\{\pi^{-1}\int_0^\pi 2\gamma_q \tanh(\tfrac{1}{4}\beta\hbar\omega_q)\,dq\right\}. \quad \text{(A9.3.20)}$$

In the classical limit

$$\beta\hbar\omega_q = \hbar\omega_q/k_B T \ll 1 \quad \text{(A9.3.21)}$$

equation (A9.3.20) simplifies to (9.7.20) of the main text.

APPENDIX 10.1 DEBYE SHIELDING AND IMPURITY SCATTERING IN SEMI-CONDUCTORS

We start from the first-order equation [equation (10.2.13)]

$$\nabla^2 V = \frac{4me^2 n_0}{\hbar^2\varepsilon}\int \frac{d\mathbf{r}'\, V(\mathbf{r}')\exp(-2mk_B T|\mathbf{r}'-\mathbf{r}|^2/\hbar^2)}{|\mathbf{r}-\mathbf{r}'|} \quad \text{(A10.1.1)}$$

and take the Fourier transform. Then we find, almost immediately,

$$\tilde{V}(\mathbf{k}) = -\frac{Ze^2}{\varepsilon}\left[\pi k^2 + \frac{mn_0 e^2}{\pi\hbar^2\varepsilon}G(k)\right]^{-1}, \quad \text{(A10.1.2)}$$

where $G(k)$ is explicitly given by

$$G(\mathbf{k}) = \int d\mathbf{r}\exp(2\pi i\mathbf{k}.\mathbf{r})\frac{\exp[-(2mk_B T/\hbar^2)r^2]}{r}$$

$$= \frac{2}{k}\int_0^\infty dr \sin 2\pi kr \exp\left(\frac{-2mk_B T}{\hbar^2}r^2\right)$$

$$= \frac{\pi h^2}{mk_B T}\int_0^1 dr \exp\left[\frac{\pi^2 k^2\hbar^2}{2mk_B T}(r^2-1)\right]. \quad \text{(A10.1.3)}$$

At this stage, we expand $G(k)$ in powers of T^{-1}, when we obtain from equations (A10.1.2) and (A10.1.3) the result

$$\tilde{V}(\mathbf{k}) = -\frac{Ze^2}{K}\left[\pi k^2 + \frac{q_D^2}{4\pi}\left\{1-\frac{\gamma}{3}\left(\frac{2\pi k}{q_D}\right)^2 + \frac{\gamma^2}{15}\left(\frac{2\pi k}{q_D}\right)^4 + \dots\right\}\right]^{-1}, \quad \text{(A10.1.4)}$$

where

$$\gamma = \frac{\pi h^2 n_0 e^2}{\varepsilon m(k_B T)^2}. \quad \text{(A10.1.5)}$$

If $\gamma \ll 1$, the Brooks–Herring solution [cf. equation (10.2.14)] is a good approximation to the solution of equation (A10.1.1).

A10.1.1 Impurity scattering in semi-conductors

We shall summarize at this point the way in which the Brooks–Herring screened potential can be used to discuss impurity scattering in semi-conductors.

For the general case, when we include the Fermi function rather than Maxwell–Boltzmann statistics, the conductivity can be expressed in the form

$$\sigma = \frac{16\sqrt{(2)}e^2\pi m^{\frac{1}{2}}(k_B T)^{\frac{3}{2}}}{3h^3}\int_0^\infty \frac{\tau\varepsilon^{\frac{3}{2}}\exp(\varepsilon-\eta)\,d\varepsilon}{[\exp(\varepsilon-\eta)+1]^2}. \tag{A10.1.6}$$

The relaxation time τ is given [cf. equation (6.4.24)] by

$$\tau^{-1} = Nv\int_0^\pi (1-\cos\theta)\,I(\theta)\,2\pi\sin\theta\,d\theta, \tag{A10.1.7}$$

where N is the concentration of impurities and v is the velocity of the incident carrier.

In Born approximation, it is quite straightforward to show that, in terms of the Debye length q_D^{-1} [determined through equation (10.2.14)] we have

$$I(\theta) = \left[\frac{e^2/\varepsilon}{mv^2(1-\cos\theta)+q_D^2\,h^2/8\pi^2 m}\right]^2. \tag{A10.1.8}$$

Hence, from equations (A10.1.8), (A10.1.7) and (A10.1.6), the conductivity can be obtained.

APPENDIX A10.2 GREEN FUNCTION FOR KOSTER–SLATER IMPURITY MODEL

We shall obtain here an expression for the density of states in the Koster–Slater impurity model. Essentially we shall restrict ourselves to terms linear in the concentration c. Then we can carry out the calculation for a single impurity, and simply multiply by the number of impurities.

The method we shall use is to analyse the Green function of the defect lattice by writing

$$G(\mathbf{r}\mathbf{r}'E) = -i\sum_{\mathbf{q}_1\mathbf{q}_2\gamma}\psi^*_{\mathbf{q}_1\gamma}(\mathbf{r})\,\psi_{\mathbf{q}_2\gamma}(\mathbf{r}')\,\tilde{G}_\gamma(\mathbf{q}_1\,\mathbf{q}_2\,E), \tag{A10.2.1}$$

where the ψ's are the Bloch functions of the unperturbed lattice.

The density of states can be calculated from

$$n(E) = -\frac{1}{2\pi i}\int d\mathbf{r}\,[G(\mathbf{rr}E) - G^*(\mathbf{rr}E)] \tag{A10.2.2}$$

and we see that we only need $\tilde{G}_\gamma(\mathbf{q}_1\,\mathbf{q}_2\,E)$ on the diagonal, say $\tilde{G}_\gamma(\mathbf{k}E)$.

In the Koster–Slater one-band model, we can deal simply with $\tilde{G}_\gamma(\mathbf{k}E)$ by writing it as

$$\tilde{G}_\gamma(\mathbf{q}E) = \frac{1}{E - E_\gamma(\mathbf{q}) - \Sigma(\mathbf{q}E)}. \tag{A10.2.3}$$

Using the methods of Chapter 2, with a modification given below, we can in fact now write down a diagrammatic series for $\Sigma(\mathbf{q}E)$. In the case when the Koster–Slater assumption is made for the matrix elements

$$V_{\mathbf{kk}'} = V/N \tag{A10.2.4}$$

this series can be summed as we show below.

We want to stress, that, to $O(c)$, the same answer could be got for the density of states by averaging over random impurities, though in that case the average Green function is diagonal in $\mathbf{q}$, which is not true for the single impurity (see Dawber and Turner, 1960).

To obtain the desired results, we note that the perturbation series for the diagonal element of the perturbed Green function may be represented in diagrammatic fashion as in Figure A10.2.1(a), as explained in Chapter 2, section 2.11.4, with the external interaction lines representing matrix elements $V_{\mathbf{kk}}$ of the impurity potential, the Fermion lines being associated, in this case, with Bloch propagators. The resulting series can be summed to give an integral equation as seen in Chapter 2, section 2.11.4 but here we wish to sum the series in a way such that $G(\mathbf{kk})$ has the original Bloch form, except for a modification of the Bloch energies by a self-energy Σ; that is, we wish to obtain G in the form

$$G(\mathbf{kk}E) = \frac{G_0(\mathbf{k}E)}{1 - \Sigma(\mathbf{k}E)\,G_0(\mathbf{k}E)} = \frac{1}{E - E(\mathbf{k}) - \Sigma(\mathbf{k}E)}. \tag{A10.2.5}$$

To do this we pick out the terms of the sums over $\mathbf{k}_1, \mathbf{k}_2$ for which $\mathbf{k} = \mathbf{k}_1, \mathbf{k}_2$ etc., and redraw the series as in Figure A10.2.1(b), the primes on $\mathbf{k}_1', \mathbf{k}_2'$ indicating the omission of $\mathbf{k}$ in the sums. It is now easy to see that the series can now be put in the form shown in Figure A10.2.2(a), with Σ, the sum of all diagrams, none of which has an internal line labelled by $\mathbf{k}$, represented by Figure A10.2.2(b). Equation (5) follows immediately from Figure A10.2.2(a), and, further, the series for Σ can also be summed when V is given by equation

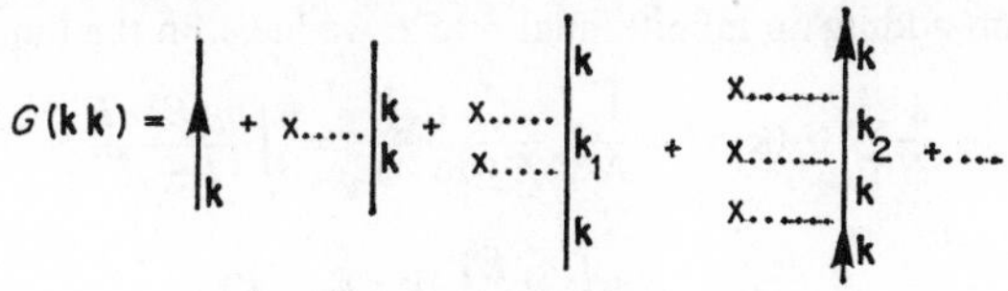

FIGURE A10.2.1(a).

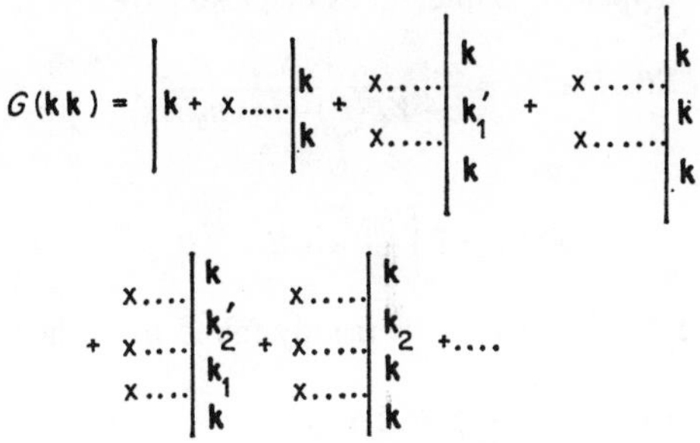

FIGURE A.10.2.1(b).

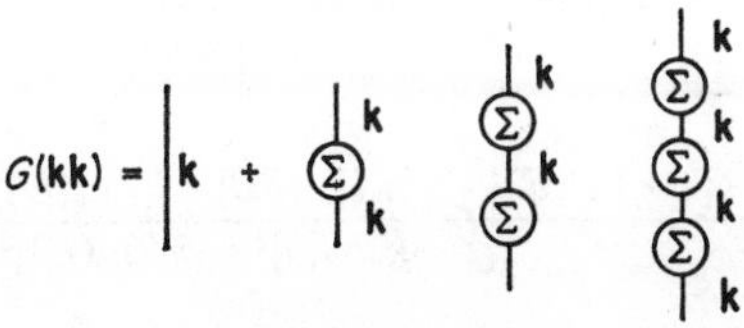

FIGURE A10.2.2(a).

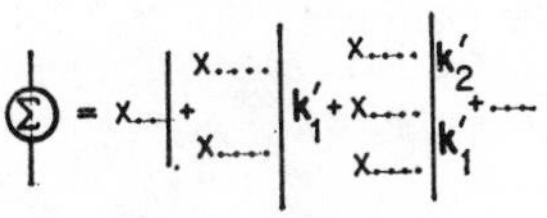

FIGURE A10.2.2(b).

(A10.2.4); we have

$$\Sigma(\mathbf{k}E) = \frac{V}{N} + \frac{V}{N}\sum_{\mathbf{k}'}{}' G_0(\mathbf{k}'E)\frac{V}{N} + \frac{V}{N}\left[\frac{V}{N}\sum_{\mathbf{k}'}{}' G_0(\mathbf{k}'E)\right]^2 + \dots$$

$$= \frac{V/N}{1-(V/N)\sum_{\mathbf{k}_1'}{}' G_0(\mathbf{k}'E)}. \tag{A10.2.6}$$

We can rewrite this in terms of the unperturbed density of states $n_0(E)$ by noting that on adding an infinitesimal δ to E we have, in the limit as $N \to \infty$,

$$\frac{1}{N}\sum_{\mathbf{k}'}{}' G_0(\mathbf{k}'E) = \frac{1}{N}\sum_{\mathbf{k}} \frac{1}{E \pm i\delta - E_{\mathbf{k}}} = \int \frac{n_0(E)\,dE}{E \pm i\delta}$$

$$= P\int \frac{n_0(E)}{E-\varepsilon}\,dE \pm i\pi n_0(E). \tag{A10.2.7}$$

Then writing the principal-value integral as $P(E)$ we have

$$\Sigma(\mathbf{k}, E - i\delta) = \frac{V/N}{1 - VP(E) - i\pi V n_0(E)}, \tag{A10.2.8}$$

where

$$P(E) = P\int \frac{n_0(E)}{E-\varepsilon}\,dE. \tag{A10.2.9}$$

The change in the density of states is readily found from the formula

$$n(E) = \operatorname{Im} \frac{1}{\pi N}\sum_{\mathbf{k}} G(\mathbf{k}, E_+)$$

$$= \operatorname{Im} \frac{1}{\pi N}\sum_{\mathbf{k}} \frac{1}{E - E_{\mathbf{k}} - \Sigma(\mathbf{k}, E - i\delta)} \tag{A10.2.10}$$

and yields

$$n(E) = n_0(E) - \frac{\{[1 - VP(E)]\,V[dn_0(E)/dE] + V^2 n_0(E)\,[dP(E)/dE]\}}{[1 - VP(E)]^2 + [\pi V n_0(E)]^2}. \tag{A10.2.11}$$

Hence the electronic specific heat can be obtained from the Koster–Slater model as referred to in the main text.

APPENDIX A10.3 DISPLACED CHARGE ROUND VACANCY IN A METAL

To obtain equation (10.4.60) of the main text from equation (10.4.47), we must perform the sum of the $\mathbf{R}_i$'s for the $(N+1)$th term in equation (10.4.48). In terms of A_L and b_L defined in equations (10.4.59) and (10.4.57), the $(N+1)$th term is readily rewritten in the form

$$\sum_{\mathbf{R}_1\mathbf{R}_2\ldots\mathbf{R}_N} (\mathbf{A}+\mathbf{b}).\mathbf{G}^0(\mathbf{R}_1).\mathbf{t}\mathbf{G}^0(\mathbf{R}_2).\mathbf{t}\ldots\mathbf{G}_0(\mathbf{R}_N).(\mathbf{A}+\mathbf{b})\,\delta_{\mathbf{R}_1+\ldots+\mathbf{R}_N,0}, \tag{A10.3.1}$$

where

$$(\mathbf{A}+\mathbf{b})_L = A_L(\mathbf{r})+b_L(\mathbf{r}). \tag{A10.3.1}$$

We now write the δ function in (A10.3.1) as

$$\delta_{\Sigma\mathbf{R}_i,0} \equiv \frac{1}{\Omega_{\mathrm{B}}}\int \exp(i\mathbf{m}\,.\,\Sigma\,\mathbf{R}_i)\,d\mathbf{m} \tag{A10.3.2}$$

and introduce

$$\mathbf{G}(\mathbf{m}) = \frac{1}{\Omega_{\mathrm{B}}}\sum_{\mathbf{R}_i\neq 0} \exp(-i\mathbf{m}\,.\,\mathbf{R}_i)\,\mathbf{G}(\mathbf{R}_i), \tag{A10.3.3}$$

whereupon (A10.3.1) may be written as

$$\sum_{\mathbf{R}_1\ldots\mathbf{R}_N}\int d\mathbf{m}\int_{\Omega_{\mathrm{B}}} d\mathbf{m}_1\ldots d\mathbf{m}_N(\mathbf{A}+\mathbf{b})\,.\,\mathbf{G}^0(\mathbf{m}_1)\,\mathbf{t}\,.\,\mathbf{G}^0(\mathbf{m}_2)\,t\ldots\mathbf{G}^0(\mathbf{m}_N)$$

$$\times(\mathbf{A}+\mathbf{b})\prod_{i=1}^{N}\exp[i(\mathbf{m}_i-\mathbf{m})\,.\,\mathbf{R}_i]. \tag{A10.3.4}$$

Since the $\mathbf{m}_i$ are restricted to the BZ we may put

$$\sum_{\mathbf{R}_1\ldots\mathbf{R}_N}\prod_{i=1}^{N}\exp[i(\mathbf{m}_i-\mathbf{m})\,.\,\mathbf{R}_i] = \prod_{i=1}^{N}\delta(\mathbf{m}_i-\mathbf{m})\,\Omega_{\mathrm{B}}^N \tag{A10.3.5}$$

to obtain

$$\Omega_{\mathrm{B}}^N\int d\mathbf{m}(\mathbf{A}+\mathbf{b})\,.\,\mathbf{G}^0(\mathbf{m})\,\mathbf{t}\,.\,\mathbf{G}^0(\mathbf{m})\,\mathbf{t}\ldots t\mathbf{G}^0(\mathbf{m})\,(\mathbf{A}+\mathbf{b}), \tag{A10.3.6}$$

the dots indicating that $\mathbf{G}^0.\mathbf{t}$ appears N times.

The series involving $\Omega_{\mathrm{B}}\,\mathbf{G}^0(\mathbf{m})\,.\,\mathbf{t}$ can now be summed into the form $[1-\Omega_{\mathrm{B}}\,\mathbf{G}^0(\mathbf{m})\,\mathbf{t}]^{-1}$ and equation (10.4.60) follows almost immediately.

We need next to establish the result (10.4.63). To do this, we make use of the identity

$$\Omega_{\mathrm{B}}\,\mathbf{G}(\mathbf{m})\,.\,\mathbf{t}[1-\Omega_{\mathrm{B}}\,\mathbf{G}(\mathbf{m})\,.\,\mathbf{t}]^{-1}\mathbf{G}(\mathbf{m}) = \frac{t^{-1}}{\Omega_{\mathrm{B}}}[t^{-1}-\Omega_{\mathrm{B}}\,\mathbf{G}(\mathbf{m})]^{-1}\mathbf{t}^{-1}-\frac{t^{-1}}{\Omega_{\mathrm{B}}}-\mathbf{G}(\mathbf{m}), \tag{A10.3.7}$$

which is readily proved by expanding both sides.

The last term on the right-hand side of equation (A10.3.7) vanishes on integration over the BZ. The second can be shown (Beeby, 1967) to cancel the first two terms on the right-hand side of equation (10.4.60) in calculating the density $\rho(\mathbf{r})$.

To obtain $A_L(\mathbf{r})+b_L(\mathbf{r})$, we use the expression

$$\frac{t(\mathbf{r},\mathbf{r}')}{v(\mathbf{r})} = \delta(\mathbf{r}-\mathbf{r}')+\int \mathscr{G}_0(\mathbf{r}-\mathbf{r}'')\,t(\mathbf{r}'',\mathbf{r}')\,d\mathbf{r}'' \tag{A10.3.8}$$

to obtain

$$b_L(\mathbf{r}) = \int \left[\frac{t(\mathbf{r},\mathbf{r}_1')}{v(r)} - \delta(\mathbf{r}_1-\mathbf{r}_1')\right] Y_L(\mathbf{r}')j_l[\sqrt{(E)}\,r_1']\,dr_1'$$

$$= \frac{1}{v(r)}\int t(\mathbf{r},\mathbf{r}_1')\,Y_L(\mathbf{r}_1')j_l(\sqrt{(E)}\,r_1')\,d\mathbf{r}_1' - A_l(\mathbf{r}). \tag{A10.3.9}$$

Using equation (A10.3.7), it is then straightforward to establish equation (10.4.63) which is the desired result. This equation can be put into a quite simple form. Let us take the imaginary part of the expression

$$t_l(r,r') = \frac{v(r)}{r^2}\,\delta(r-r') + \frac{2}{\pi}\int v(r)\frac{j_l(kr)j_l(kr'')}{E-k^2+i\varepsilon}k^2\,dk\,t_l(r''r)\,r''^2\,dr'', \tag{A10.3.10}$$

the integral equation for the t-matrix. This may be written in the form

$$\operatorname{Im} t_l(r'r) = S_l(r)\int \operatorname{Re} t_l(r',r)j_l[\sqrt{(E)}\,r]\,r^2\,dr, \tag{A10.3.11}$$

where S_l obeys the integral equation

$$S_l(r) = -\sqrt{(E)}\,v(r)j_l[\sqrt{(E)}\,r] + \frac{2}{\pi}\int v(r)j_l(kr)(E-k^2)^{-1}j_l(kr')$$
$$\times k^2\,dk\,S_l(r')\,r''^2\,dr', \tag{A10.3.12}$$

the right-hand side of which is just $v(r)$ multiplied by the integral equation for the wave function. Furthermore, one can also show by taking the real part of equation (A10.3.10) and writing it as an equation for

$$\int \operatorname{Re} t_l(rr')j_l[\sqrt{(E)}\,r']\,r'^2\,dr'$$

that

$$\int t_l(r,r')j_l[\sqrt{(E)}\,r']\,r'^2\,dr' \propto S_l(r) \tag{A10.3.13}$$

and hence

$$\frac{A_l(\mathbf{r})+b_l(\mathbf{r})}{t_l(E)} = \frac{Y_L(\mathbf{r})}{v(r)}\,\frac{S_l(r)}{\int S_l(r)j_l[\sqrt{(E)}\,r]\,r^2\,dr}. \tag{A10.3.14}$$

It immediately follows that we can choose the normalization of the radial part of the wave functions such that

$$\frac{A_L(\mathbf{r})+b_L(\mathbf{r})}{t_l(E)} = R_L(\mathbf{r}) \tag{A10.3.15}$$

and equation (10.4.64) follows.

To obtain equation (10.4.65) we note that if there is a vacancy at the origin, $t = 0$ there. In the sums over lattice vector in equation (10.4.48), and so equation (A10.3.4), we must therefore omit $\mathbf{R}_i = 0$. Otherwise the method is similar to that used before, the final result being

$$\sigma_{\text{vac}}(\mathbf{r}, E) = -\frac{\text{Im}}{\pi}\left[\sum_{LL'} A_L(\mathbf{r})\left\{\frac{1}{\Omega_B}\int_{\Omega_B}[t^{-1}-\Omega_B\,\mathbf{G}(\mathbf{m})]^{-1}\right\}^{-1}_{LL'} d\mathbf{m}\right] A_{L'}(\mathbf{r}). \tag{A10.3.16}$$

For the detailed derivation of this result, reference may be made to Beeby's original paper (1967).

If the potential is such that only $t_0(E)$ is non-zero, we find

$$\int[t_{00}^{-1}(E)-\Omega_B\,\mathscr{G}_{00}(\mathbf{m}, E)]^{-1}\,d\mathbf{m} = \int[t_0^{-1}(E)-f]^{-1}\,\delta[f-\Omega_B\,\mathscr{G}_0(\mathbf{m}, E)]\,d\mathbf{m}\,df$$

$$= \int[t_0^{-1}(E)-f]^{-1}\,D(f, E)\,df, \tag{A10.3.17}$$

where $D(f, E)$ is defined by this equation. On taking real and imaginary parts, we see that

$$\int D(f, E)\left\{\frac{P}{t_0^{-1}(E)-f} - i\pi\delta[t_0^{-1}(E)-f]\right\}df = F(E)-i\pi\eta(E), \tag{A10.3.18}$$

where $\eta(E) = D[t_0^{-1}(E), E]$ is the density of states of the host lattice and $F(E)$ its Hilbert transform. Equation (A10.3.6) then yields equation (10.4.67).

APPENDIX A10.4 EQUIVALENT HAMILTONIANS AND EFFECTIVE-MASS THEORY

Let us suppose we know the $E(\mathbf{k})$ relationship for the electrons in a crystal, and we now wish to treat the effect of a small external perturbation. It would evidently be very advantageous if, instead of solving the entire problem anew, one could utilize the $E(\mathbf{k})$ relationship to write down an effective Hamiltonian, for the problem, in which the periodic potential does not explicitly appear. At its simplest, this effective Hamiltonian might be

$$H_{\text{eff}} = -\frac{\hbar^2\nabla^2}{2m^*} + H_1(\mathbf{r}), \tag{A10.4.1}$$

where H_1 represents the external perturbation (in particular, an electric field) and m^* is an *effective* mass taking account of the effect of the crystal potential on the motions of the electrons. We shall see below that under certain circumstances such an effective Hamiltonian can be obtained; indeed this was briefly discussed in Chapter 1 using the crystal momentum representation. Here we shall work in the Wannier representation.

Let the original Hamiltonian be

$$H = H_0 + H_1(\mathbf{r}), \tag{A10.4.2}$$

where

$$H_0 = -\frac{\hbar^2 \nabla^2}{2m} + V(\mathbf{r}), \tag{A10.4.3}$$

$V(\mathbf{r})$ being the periodic potential. We have now to solve for the eigenvalues E of the equation

$$H\psi = (H_0 + H_1)\psi = E\psi. \tag{A10.4.4}$$

Employing a Wannier representation and taking matrix elements with respect to the Wannier function

$$w_{\mathbf{R}} = w(\mathbf{r}-\mathbf{R}), \tag{A10.4.5}$$

we have

$$\langle w_{\mathbf{R}}^* H_0 \psi\rangle + \langle w_{\mathbf{R}}^* H_1 \psi\rangle = E\langle w_{\mathbf{R}}^* \psi\rangle. \tag{A10.4.6}$$

We shall suppose the variation of $H_1(\mathbf{r})$ to be negligible over the region in which $w(\mathbf{r}-\mathbf{R})$ is appreciable. We can then write

$$\langle w_{\mathbf{R}}^* H_1 \psi\rangle = H_1(\mathbf{R})\chi(\mathbf{R}), \tag{A10.4.7}$$

where

$$\chi(\mathbf{R}) = \langle w_{\mathbf{R}}^* \psi\rangle \tag{A10.4.8}$$

and in this approximation equation (A10.4.6) becomes

$$\langle w_{\mathbf{R}}^* H_0 \psi\rangle + H_1(\mathbf{R})\chi(\mathbf{R}) = E\chi(\mathbf{R}). \tag{A10.4.9}$$

Suppose, further, that H_1 is such that no interband transitions take place. Then

$$\psi = \sum_{\mathbf{k}} c_{\mathbf{k}} \psi_{\mathbf{k}} \tag{A10.4.10}$$

where the functions $\psi_{\mathbf{k}}$ all pertain to the same band of the unperturbed problem. We then find

$$\chi(\mathbf{R}) = \sum_{\mathbf{k}} c_{\mathbf{k}} \langle w_{\mathbf{R}}^* \psi_{\mathbf{k}}\rangle = \sum_{\mathbf{k}} c_{\mathbf{k}} E_{\mathbf{k}} e^{i\mathbf{k}.\mathbf{R}}, \tag{A10.4.11}$$

while

$$\langle w_{\mathbf{R}}^* H_0 \psi\rangle = \sum_{\mathbf{k}} c_{\mathbf{k}} E_{\mathbf{k}} \langle w_{\mathbf{R}}^* \psi_{\mathbf{R}}\rangle = \sum_{\mathbf{k}} c_{\mathbf{k}} E_{\mathbf{k}} e^{i\mathbf{k}.\mathbf{R}}. \tag{A10.4.12}$$

It is now evident that if R can be regarded as a continuous variable we can put

$$\langle w_{\mathbf{R}}^* H_0 \psi \rangle = E\left(\frac{1}{i}\frac{\partial}{\partial \mathbf{R}}\right)\chi(\mathbf{R}). \tag{A10.4.13}$$

Equation (A10.4.9) may now be written as

$$E\left(\frac{1}{i}\frac{\partial}{\partial \mathbf{R}}\right)\chi(\mathbf{R}) + H_1(\mathbf{R})\chi(\mathbf{R}) = E\chi(\mathbf{R}), \tag{A10.4.14}$$

so that we see that our original aim is accomplished—we have obtained an effective Hamiltonian of the form

$$H_{\text{eff}} = E\left(\frac{1}{i}\frac{\partial}{\partial \mathbf{R}}\right) + H_1(\mathbf{R}). \tag{A10.4.15}$$

Two points concerning the above may be noted. First, it is readily seen that time-dependent problems are treated in this framework simply by using the equation

$$H_{\text{eff}}\chi = i\hbar\frac{\partial}{\partial t}\chi. \tag{A10.4.16}$$

Secondly, $\mathbf{R}$ and $\hbar\mathbf{k}$ are canonically conjugate variables. This can be seen by the fact that $(1/i)(\partial/\partial\mathbf{R})$ replaces $\mathbf{k}$ in the effective Hamiltonian and that the mean value of $\mathbf{k}$ is

$$\langle \mathbf{k} \rangle = \sum_{\mathbf{k}} \mathbf{k}|c_{\mathbf{k}}|^2 = \frac{1}{i}\sum_{\mathbf{R}} \chi^*(\mathbf{R})\frac{\partial}{\partial \mathbf{R}}\chi(\mathbf{R}) = \left\langle \chi^*(\mathbf{R})\frac{1}{i}\frac{\partial}{\partial \mathbf{R}}\chi(\mathbf{R})\right\rangle. \tag{A10.4.17}$$

If we treat the introduction of a magnetic field by the customary procedure of replacing $(\hbar/i)(\partial/\partial\mathbf{r})$ by $(\hbar/i)(\partial/\partial\mathbf{r}) - e\mathscr{A}/c$, we obtain the Onsager effective Hamiltonian $E[(1/i)(\partial/\partial\mathbf{R}) - e\mathscr{A}(\mathbf{R})/\hbar c]$, which conforms with our earlier discussion of this problem in Chapter 4. In view of our earlier discussion of the magnetic field problem we shall say no more about it here, except that the Onsager Hamiltonian is derivable using the Wannier representation if the Wannier functions are suitably modified to take account of gauge invariance (see, for example, Wannier, 1962).

Equation (A10.4.14) can be strictly solved as a differential equation only if $\chi(\mathbf{R})$ is slowly varying, because the values of $\mathbf{R}$ strictly form the discrete set of lattice vectors. Since

$$\chi(\mathbf{r}) = \sum_{\mathbf{k}} c_{\mathbf{k}} e^{i\mathbf{k}.\mathbf{r}} \tag{A10.4.18}$$

will be slowly varying only if the significant values of $\mathbf{k}$ in this sum are close to zero. About $\mathbf{k} = 0$, we expect the $E(\mathbf{k})$ relationship to be quadratic—most simply, we have

$$E = E_0 + \frac{h^2 k^2}{2m^*} + \ldots. \qquad \text{(A10.4.19)}$$

We can then write equation (A10.4.15) as

$$-\frac{\hbar^2}{2m^*}\left(\frac{\partial^2}{\partial X^2}+\frac{\partial^2}{\partial Y^2}+\frac{\partial^2}{\partial Z^2}\right)\chi + H_1(\mathbf{R})\chi = (E-E_0)\chi \qquad \text{(A10.4.20)}$$

and the proportionality constant m^* takes the role of effective mass for free electrons acted upon by the external perturbation.

Stationary values of the $E(\mathbf{k})$ relation are often important at values $\mathbf{k}_0$ of $\mathbf{k}$ other than $\mathbf{k}_0 = 0$. To obtain χ slowly varying, we now define

$$\chi(\mathbf{R}) = \langle w_{\mathbf{R}}^* \, \mathrm{e}^{-i\mathbf{k}_0.\mathbf{R}} \psi \rangle = \sum_{\mathbf{k}} c_{\mathbf{k}} \, \mathrm{e}^{i(\mathbf{k}-\mathbf{k}_0).\mathbf{R}} \qquad \text{(A10.4.21)}$$

and again obtain equation (A10.4.14), except that we have an energy operator

$$E\left(\mathrm{e}^{-i\mathbf{k}_0.\mathbf{R}} \frac{1}{i}\frac{\partial}{\partial \mathbf{R}} \mathrm{e}^{i\mathbf{k}_0.\mathbf{R}}\right). \qquad \text{(A10.4.22)}$$

Alternatively we may write E as a function of $\mathbf{k}-\mathbf{k}_0$ and replace $\mathbf{k}-\mathbf{k}_0$ by $(1/i)(\partial/\partial\mathbf{k})$. The simplest case occurs if

$$E = E_0 + \frac{\hbar^2}{2m^*}(\mathbf{k}-\mathbf{k}_0)^2, \qquad \text{(A10.4.23)}$$

when we may again obtain equation (A10.4.20).

Equation (A10.4.19) obtains for cubic crystals, for example, but more generally we must write

$$E = E_0 + \frac{\hbar^2}{2}\sum_{\alpha\beta}\left(\frac{1}{m^*}\right)_{\alpha\beta}(\mathbf{k}-\mathbf{k}_0)_\alpha(\mathbf{k}-\mathbf{k}_0)_\beta + \ldots \qquad \text{(A10.4.24)}$$

and we have a reciprocal effective mass tensor with components $(1/m^*)_{\alpha\beta}$. This can always be diagonalized by choosing principal axes, so that the effective Hamiltonian may be written, again ignoring magnetic fields,

$$H_{\text{eff}} = -\frac{\hbar^2}{2}\left(\frac{1}{m_{11}^*}\frac{\partial^2}{\partial X^2}+\frac{1}{m_{22}^*}\frac{\partial^2}{\partial Y^2}+\frac{1}{m_{33}^*}\frac{\partial^2}{\partial Z^2}\right). \qquad \text{(A10.4.25)}$$

To choose principal axes may not bring any advantage, however, for there may be pockets of electrons, at two or more local minima of the $E(\mathbf{k})$ relation, which need to be treated simultaneously, and the principal axes of one local minimum may not be those of another. This occurs, for example, in germanium, a situation in which we may utilize effective mass theory if we investigate the Faraday effect, for example.

Actually, it is not usually necessary to use the quantum-mechanical form of the effective Hamiltonian, which remains only as a formal basis for treating the Bloch electrons as if they were free. One simply uses the Lorentz force-like equation ($\Delta\mathbf{k} = \mathbf{k} - \mathbf{k}_0$)

$$\hbar\Delta\dot{\mathbf{k}} = e\mathscr{E} + (e/c)(\mathbf{v}\times\mathscr{H}). \tag{A10.4.26}$$

This equation is probably most simply solved by introducing the velocity on the left-hand side. (Additionally, we then solve directly for the quantity required in a calculation of the conductivity.) Now one may write $\hbar\mathbf{k}$ exactly as the classical momentum, except that one uses an effective mass tensor:

$$\hbar\Delta k_\alpha = \sum_\beta m^*_{\alpha\beta} v_\beta. \tag{A10.4.27}$$

However, it must be noted that, when principal axes are not chosen, $m^*_{\alpha\beta}$ is not simply $(1/m^*)_{\alpha\beta}$, but the (α, β) component of the inverse matrix of the matrix with components $(1/m^*)_{\mu\nu}$, that is,

$$\delta_{\alpha\beta} = \sum_\gamma m^*_{\alpha\gamma}\left(\frac{1}{m^*}\right)_{\gamma\beta}. \tag{A10.4.28}$$

To see this we note from equation (A10.4.24) that, on using

$$\mathbf{v} = \frac{1}{\hbar}\frac{\partial E}{\partial \mathbf{k}}, \tag{A10.4.29}$$

we find

$$v_\gamma = \hbar \sum_\beta \left(\frac{1}{m^*}\right)_{\gamma\beta} \Delta k_\beta, \tag{A10.4.30}$$

so that on using equation (A10.4.28)

$$\sum_\gamma m^*_{\alpha\gamma} v_\gamma = \hbar \sum_\beta \sum_\gamma m^*_{\alpha\gamma}\left(\frac{1}{m^*}\right)_{\gamma\beta} \Delta k_\beta = \hbar \sum_\beta \delta_{\alpha\beta}\Delta k_\beta = \Delta k_\alpha. \tag{A10.4.31}$$

Finally, it is to be noted that equation (A10.4.26) can be rewritten with a single component of the velocity on the left-hand side. Combination of equations (A10.4.26) and (A10.4.27) gives

$$\sum_\beta m^*_{\alpha\beta} v_\beta = F_\alpha, \tag{A10.4.32}$$

where F_α is the force component. From this we see that

$$\sum_\alpha \left(\frac{1}{m^*}\right)_{\gamma\alpha} F_\alpha = \sum_\beta \sum_\alpha \left(\frac{1}{m^*}\right)_{\gamma\alpha} m^*_{\alpha\beta} v_\beta = \sum_\beta \delta_{\alpha\beta} v_\beta, \tag{A10.4.33}$$

that is,

$$v_\alpha = \sum_\beta \left(\frac{1}{m^*}\right)_{\alpha\beta} F_\beta. \tag{A10.4.34}$$

APPENDIX A10.5 SCREENING OF CHARGED DEFECT IN STRONG MAGNETIC FIELD

We saw in the main text that the electron density $n(\mathbf{r})$ was related to the diagonal element $C(\mathbf{r}\beta)$ of the Bloch density matrix by

$$n(\mathbf{r}) = -e\exp\ (\eta\beta)\, C(\mathbf{r}\beta), \tag{A10.5.1}$$

where η is the chemical potential.

Using the first-order theory in the potential V for the density matrix, we obtain from Poisson's equation, written in q-space

$$-q^2\varepsilon\tilde{V}(q) = 4\pi e\exp(\eta_0\beta)\,[\tilde{C}(q) - \tilde{C}_0(q)] - \frac{4\pi Ze}{(2\pi)^{\frac{3}{2}}}. \tag{A10.5.2}$$

We then find explicitly

$$\tilde{V}(q) = \frac{4\pi Ze}{2\pi^{\frac{3}{2}}\left[\varepsilon q^2 + 4\pi e^2 n_0\beta \int_0^1 dy\exp[(-q_z^2\hbar^2/8m^*)\beta(1-y^2)] \times \exp[-(q_x^2+q_y^2)\hbar^2/4m^*\mu_{\mathrm{B}}^* B]\{\coth(\mu_{\mathrm{B}}^* B\beta) - [\cosh(\mu_{\mathrm{B}}^* B\beta y)/\sinh(\mu_{\mathrm{B}}^* B\beta)]\}\right]}. \tag{A10.5.3}$$

The trial wave functions used in the variational calculation of the ground and first excited states have the forms

$$\phi_0 = A\exp\left(-\frac{x^2+y^2}{a^2}\right)\exp\left(-\frac{z^2}{b^2}\right) \tag{A10.5.4}$$

and

$$\phi_1 = \frac{\exp(-i\phi)}{(2\pi)^{\frac{1}{2}}}\frac{\exp(-z^2/b^2)}{b^{\frac{1}{2}}}\left(\frac{2}{\pi}\right)^{\frac{1}{4}}\frac{4r}{2^{\frac{1}{2}}a^2}\exp\left(-\frac{r^2}{a^2}\right). \tag{A10.5.5}$$

The curves shown in Figure 10.5 were obtained from equations (A10.5.3)–(A10.5.5).

APPENDIX A10.6 LOW-TEMPERATURE RESISTIVITY OF Pd ALLOYS

If the Curie temperature T_c is such that $T \ll T_c$, then we shall now show that for the model of Pd alloys discussed in Chapter 10 the excess resistivity may be written in the form

$$\Delta\rho(T) = \Delta\rho(0) + dc(T/T_c)^{\frac{3}{2}}, \tag{A10.6.1}$$

where $\Delta\rho(0)$ is the zero temperature value of the excess resistivity and the coefficient d is determined by the strength of the effective interaction J'_{eff}, the screened Coulomb potential scattering and the magnitude of the spin on the impurity.

Cole and Turner (1968) have shown that $\Delta\rho$ is given by $(T \ll T_c)$

$$\Delta\rho = \frac{3\pi m^* c}{2\hbar e^2 E_f}\left(\frac{\Omega_0}{N}\right)\left\{\left[V^2 + J'^2_{\text{eff}}\langle S_z^2\rangle + J'^2_{\text{eff}}\int d\omega \sum_{\mathbf{q}} A(\mathbf{q}\omega)\operatorname{cosech}\beta\omega\right] - \frac{4J'^2_{\text{eff}}V^2\langle S_z\rangle^2}{\left[V^2 + J'^2_{\text{eff}}\langle S_z^2\rangle + J'^2_{\text{eff}}\int d\omega \sum_{\mathbf{q}} (\mathbf{q}\omega)\operatorname{cosech}\beta\omega\right]}\right\} \tag{A10.6.2}$$

in the low-temperature region under discussion, where the quantity $A(\mathbf{q}\omega)$ is the impurity spin spectral density function. Thus the problem of calculating the temperature dependence of the resistivity $\Delta\rho$ is that of determining the energy and wave-number-dependent spectral density. This quantity is related to the spin susceptibility $\chi^{\mathrm{L}}(\mathbf{q}\omega)$ through

$$A(\mathbf{q}\omega) = \frac{1}{\pi}\operatorname{Im}\chi^{\mathrm{L}}(\mathbf{q}\omega). \tag{A10.6.3}$$

The Fourier transform of the spin susceptibility is related to the retarded Green function $\langle[S_i^+(t), S_j^-(t')]\rangle$ which describes the propagation through the medium of a local spin at site j at time t and reintroduced at site i at time t. The elementary excitations of the system, as we saw in Chapter 4, are clearly built into $\chi^{\mathrm{L}}(\mathbf{q}\omega)$ and, since the s-electron scattering is from such excitations, it is natural that this susceptibility enters the calculation of the transport properties of the alloy.

We follow the methods introduced in Chapters 2 and 4 for calculating response functions. Thus we employ a time-dependent average field method. It is clear that, provided the s-electron polarization is small, the major contributions to the susceptibility are from the d-band spin system. Then we get $A(\mathbf{q}\omega)$ in the usual manner.

The theory is usefully expressed in terms of the density of local spin states $\rho_{\mathrm{L}}(\omega)$ defined as

$$\rho_{\mathrm{L}}(\omega) = \frac{1}{N}\sum_{\mathbf{q}} A(\mathbf{q}\omega). \tag{A10.6.4}$$

In the limit of low concentration, this represents the density of spin-wave states. It turns out that this is because the calculation of the spectral density function $A(\mathbf{q}\omega)$ reveals spin-wave excitations with damping, the width being proportional to the concentration. The excitations are in marked contrast to those in pure Pd, the paramagnons of Chapter 4 becoming spin waves.

Since the formula (A10.6.2) for $\Delta\rho$ involves the mean values of S_z and S_z^2, we shall first deal with these quantities. In fact, it is not difficult to show that we can write these mean values in terms of the spin correlation function $\langle S^-S^+\rangle$. The results are

$$\langle S_z\rangle = S - \frac{\langle S^-S^+\rangle}{2S} \tag{A10.6.5}$$

and

$$\langle S_z^2\rangle = S^2 - \frac{(2S-1)}{2S}\langle S^-S^+\rangle. \tag{A10.6.6}$$

Now in terms of the density of spin states, we have

$$\langle S^-S^+\rangle = \frac{1}{N}\int \sum_{\mathbf{q}} A(\mathbf{q}\omega)\, n(\omega)\, d\omega = \int \rho_L(\omega)\, n(\omega)\, d\omega, \tag{A10.6.7}$$

where $n(\omega)$ is written for the Bose distribution $n(\omega) = [\exp(\beta\omega - 1)]^{-1}$. It will turn out that $\rho_L(\omega)$ is proportional to $\langle S_z\rangle$ and it is therefore useful to define a function Φ which will be central to the transport theory by

$$\int \rho_L(\omega)\, n(\omega)\, d\omega = 2\langle S_z\rangle\,\Phi(T). \tag{A10.6.8}$$

Then the mean values we require may be expressed as

$$\langle S_z\rangle = S - \frac{\langle S_z\rangle}{S}\Phi(T) \quad \text{or} \quad \langle S_z\rangle = S\Big/\left[1+\frac{\Phi(T)}{S}\right] \tag{A10.6.9}$$

and

$$\langle S_z^2\rangle = S^2 + (2S-1)\,\Phi(T)\Big/\left[1+\frac{\Phi(T)}{S}\right]. \tag{A10.6.10}$$

A second temperature-dependent function related to the spin spectral density function will also be important in our analysis. This is defined by

$$\frac{1}{N}\int \sum_{\mathbf{q}} A(q\omega)\,\operatorname{cosech}\beta\omega\, d\omega = \int \rho_L(\omega)\,\operatorname{cosech}\beta\omega\, d\omega = 2\langle S_z\rangle\,\Gamma(T)$$

$$= 2S\Gamma(T)\Big/\left[1+\frac{\Phi(T)}{S}\right]. \tag{A10.6.11}$$

The resistivity can now be expressed in terms of these two functions.

A10.6.1 Local spin susceptibility for alloy

In Chapter 4 we derived the result of Izuyama *et al.* that the susceptibility for the Pd d-band could be expressed in the form

$$\chi_{\mathrm{Pd}}(\mathbf{q}\omega) = \int dt \int d\mathbf{r}\, e^{i\mathbf{q}\cdot\mathbf{r}}\, e^{-i\omega t}\{-i\theta(t)\langle[\sigma_d^-(\mathbf{r}t), \sigma_d^+(0,0)]\rangle\}$$

$$= \frac{\chi_0(\mathbf{q}\omega)}{1 - I\chi_0(\mathbf{q}\omega)}, \tag{A10.6.12}$$

where χ_0 is the usual non-interacting susceptibility given by [see equation (4.11.1)]

$$\chi_0(\mathbf{q}\omega) = \frac{1}{N}\sum_{\mathbf{k}} \frac{\langle n_{\mathbf{k}}^{\mathrm{d}}\rangle - \langle n_{\mathbf{k}+\mathbf{q}}^{\mathrm{d}}\rangle}{[\mathscr{E}_{\mathbf{k}+\mathbf{q}}^{\mathrm{d}} - \mathscr{E}_{\mathbf{k}}^{\mathrm{d}} - \omega]}. \tag{A10.6.13}$$

I is the strength of the short-range interaction between Pd d-holes and $\mathscr{E}_{\mathbf{k}}^{\mathrm{d}}$ are the single-particle energies of the Pd matrix.

A10.6.2 Low-temperature resistivity

Using the density of spin states given in Chapter 4, we can now calculate the low-temperature forms of the temperature-dependent functions $\Phi(T)\, 4\Gamma(T)$ which we need to calculate the resistivity. We find then

$$\Phi(T) = \frac{3z}{2}\frac{1}{(k_{\mathrm{B}}T_0)^{\frac{3}{2}}}\int_0^\infty \omega^{\frac{1}{2}}\, n(\omega)\, d\omega = \frac{3z}{2}\frac{\pi^{\frac{1}{2}}}{2}\zeta\left(\frac{3}{2}\right)\left(\frac{T}{T_0}\right)^{\frac{3}{2}} = \alpha_1\left(\frac{T}{T_0}\right)^{\frac{3}{2}} \tag{A10.6.14}$$

and

$$\Gamma(T) = \frac{3z}{2}\frac{1}{(k_{\mathrm{B}}T_0)^{\frac{3}{2}}}\int_0^\infty \omega^{\frac{1}{2}} \operatorname{cosech}\beta\omega\, d\omega = \frac{3z}{2}\pi^{\frac{1}{2}}\left[1 - \frac{1}{2^{\frac{3}{2}}}\right]\left(\frac{T}{T_0}\right)^{\frac{3}{2}}. \tag{A10.6.15}$$

Similarly we find for $\langle S_z^2\rangle$ and $\langle S_z\rangle$ the explicit low-temperature forms

$$\langle S_z^2\rangle = S^2 - (2S-1)\,\alpha_1(T/T_0)^{\frac{3}{2}} \tag{A10.6.16}$$

and

$$\langle S_z\rangle = S - \alpha_1(T/T_0)^{\frac{3}{2}}. \tag{A10.6.17}$$

Also we have

$$\int \rho_{\mathrm{L}}(\omega) \operatorname{cosech}\beta\omega\, d\omega = 2S\alpha_2(T/T_0)^{\frac{3}{2}} \tag{A10.6.18}$$

and hence from equation (A10.6.2) we find for $\Delta\rho$ the result

$$\Delta\rho(T) = \frac{3\pi m^* c}{2\hbar e^2 E_{\mathrm{f}}}\left(\frac{\Omega_0}{N}\right)\left[V^2 + J'^2_{\mathrm{eff}}\left[S^2-(2S-1)\,\alpha_1\left(\frac{T}{T_0}\right)^{3/2}+2S\alpha_2\left(\frac{T}{T_0}\right)^{3/2}\right]\right.$$
$$\left. - \frac{4V^2J'^2_{\mathrm{eff}}[S-\alpha_1(T/T_0)^{3/2}]^2}{\{V^2+J'^2_{\mathrm{eff}}[S^2-(2S-1)\,\alpha_1(T/T_0)^{3/2}+2S\alpha_2(T/T_0)^{3/2}]\}}\right]. \tag{A10.6.19}$$

Expanding this to first order in $(T/T_0)^{3/2}$ we can write $\Delta\rho(T)$ as in equation (A10.6.1) where $\Delta\rho(0)$, the zero-temperature value of the resistivity, is given by

$$\Delta\rho(0) = \frac{3\pi m^* c}{2\hbar e^2 E_{\mathrm{f}}}\left(\frac{\Omega_0}{N}\right)\frac{(V^2-J'^2_{\mathrm{eff}}S^2)^2}{V^2+J'^2_{\mathrm{eff}}S^2}. \tag{A10.6.20}$$

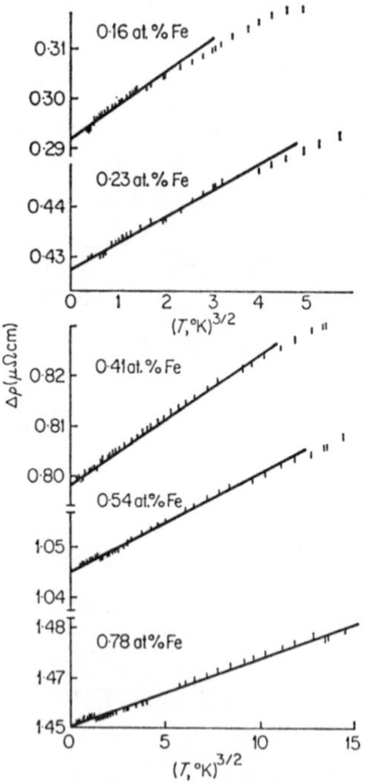

FIGURE A10.6.1. Resistivity of Pd with small concentrations of Fe, as a function of temperature (after Williams, 1970).

The coefficient d in equation (A10.6.1) is then obtained explicitly as

$$d = \frac{3\pi m^*}{2\hbar e^2 E_f}\left(\frac{\Omega_0}{N}\right)\frac{2J'^2_{\mathrm{eff}}S}{V^2+J'^2_{\mathrm{eff}}S^2}$$
$$\times\left\{4V^2(V^2+J'^2_{\mathrm{eff}}S^2)\,\alpha_1-\left[\frac{(2S-1)}{2S}\alpha_1-\alpha_2\right][4V^2J'^2_{\mathrm{eff}}S^2+(V^2+J'^2_{\mathrm{eff}}S^2)]\right\}. \tag{A10.6.21}$$

Thus for temperatures very much less than the transition temperature, this theory predicts a $\frac{3}{2}$ power law for the excess resistivity. The concentration dependence of the $T^{\frac{3}{2}}$ term is $c^{-\frac{1}{2}}$.

A10.6.3 Comparison with experiment

We finally consider briefly the available low-temperature measurements to test this theory.

The results of Williams and Loram (1970) for the dilute Pd–Fe system are plotted against $T^{\frac{3}{2}}$ in Figure A10.6.1. The fit with the theory is quite good in

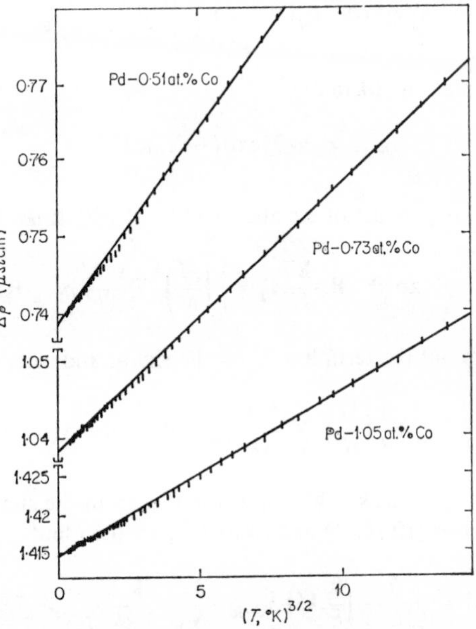

FIGURE A10.6.2. Resistivity of Pd with small concentrations of Co, as a function of temperature (after Williams, 1970).

the sense that the $T^{\frac{3}{2}}$ dependence is confirmed. The same is true for the Pd–Co system studied by Williams (1970), results being shown in Figure A10.6.2.

Confirmation that theory and experiment are in reasonable agreement in this low-temperature region is also got by combining the later treatment near to T_c with these results to extract the parameters V and J_{eff}. The results show that V is fairly constant but that there is a spread in J'_{eff}.

APPENDIX A10.7 PHONONS IN DISORDERED SYSTEMS

We wish to derive equation (10.16.32) for the kinetic energy in terms of collective coordinates.

To do this, we return to equation (10.16.28), multiply on the left by $\exp(i\mathbf{k}'.\mathbf{R}_i)$ and then sum over i when we find

$$\sum_i \exp(i\mathbf{k}'.\mathbf{R}_i)\,P_i = \frac{1}{i}\left(\frac{M}{N}\right)^{\frac{1}{2}} \sum_{\mathbf{k}} \frac{\mathbf{k}}{k}\rho_{\mathbf{k}-\mathbf{k}'}P_{\mathbf{k}}, \tag{A10.7.1}$$

where, as in equation (10.16.21),

$$\rho_{\mathbf{k}} = \sum_i \exp(-i\mathbf{k}.\mathbf{R}_i). \tag{A10.7.2}$$

Taking the scalar product of equation (A10.7.1) with $\mathbf{k}'/k'$, we find

$$\sum_i \exp(i\mathbf{k}'.\mathbf{R}_i)\frac{\mathbf{k}'}{k'}.P_i = \frac{1}{i}\left(\frac{M}{N}\right)^{\frac{1}{2}} \sum_{\mathbf{k}} \frac{\mathbf{k}.\mathbf{k}'}{kk'}\rho_{\mathbf{k}-\mathbf{k}'}P_{\mathbf{k}}. \tag{A10.7.3}$$

We now take out the term $\mathbf{k} = \mathbf{k}'$ on the right-hand side, when we find

$$\frac{1}{i}(MN)^{\frac{1}{2}}P_{\mathbf{k}'} + \frac{1}{i}\left(\frac{M}{N}\right)^{\frac{1}{2}} \sum_{\mathbf{k}}{}' \frac{\mathbf{k}.\mathbf{k}'}{kk'}\rho_{\mathbf{k}-\mathbf{k}'}P_{\mathbf{k}} = \sum_i \exp(i\mathbf{k}'.\mathbf{R}_i)\frac{\mathbf{k}'}{k'}.P_i,$$

the prime denoting that $\mathbf{k} = \mathbf{k}'$ is now omitted from the sum.

Using equations (10.16.31) and (A10.7.4), we then find

$$[P_{\mathbf{k}}, \rho_{\mathbf{k}'}] = \frac{\hbar}{i(MN)^{\frac{1}{2}}}\left(\frac{\mathbf{k}}{k}.\mathbf{k}'\right)\rho_{\mathbf{k}'-\mathbf{k}} - \frac{1}{N}\sum_{\mathbf{k}''}{}' \left(\frac{\mathbf{k}.\mathbf{k}''}{kk''}\right)\rho_{\mathbf{k}''-\mathbf{k}}[P_{\mathbf{k}''}, \rho_{\mathbf{k}'}]. \tag{A10.7.5}$$

This is an integral equation for the commutator, which can be solved to yield

$$[P_{\mathbf{k}}, \rho_{\mathbf{k}'}] = \frac{\hbar}{i}\left(\frac{N}{M}\right)^{\frac{1}{2}} k\delta_{\mathbf{k},\mathbf{k}'}. \tag{A10.7.6}$$

Thus $P_{\mathbf{k}}$ is conjugate to $\rho_{\mathbf{k}}$, though $P_{\mathbf{k}}^{\dagger} \neq P_{-\mathbf{k}}$. In order to preserve this type of relation, we can redefine the $P_{\mathbf{k}}$'s through

$$\bar{P}_{\mathbf{k}} = P_{\mathbf{k}} + \frac{i\hbar k}{2(MN)^{\frac{1}{2}}}\rho_{-\mathbf{k}}. \tag{A10.7.7}$$

Then we have

$$\bar{P} = \bar{P}_{-\mathbf{k}} \tag{A10.7.8}$$

and

$$[\bar{P}_{\mathbf{k}}, \rho_{\mathbf{k}'}] = \frac{\hbar}{i}\left(\frac{N}{M}\right)^{\frac{1}{2}} k\delta_{\mathbf{k},\mathbf{k}'}. \tag{A10.7.9}$$

Thus, the $\bar{P}_{\mathbf{k}}$'s have the desired properties, and we can now write the kinetic energy (10.16.29) in the form

$$T = \frac{1}{2N}\sum_{\mathbf{k}_1\mathbf{k}_2}\left(\frac{\mathbf{k}_1}{k_1}\cdot\frac{\mathbf{k}_2}{k_2}\right)\left\{\bar{P}_{-\mathbf{k}_1}\rho_{\mathbf{k}_1-\mathbf{k}_2}\bar{P}_{\mathbf{k}_2} + \frac{\hbar^2}{4MN}k_1 k_2 \rho_{\mathbf{k}_1}\rho_{\mathbf{k}_2-\mathbf{k}_1}\rho_{-\mathbf{k}_2}\right\}$$

$$+\frac{i\hbar}{2(MN)^{\frac{1}{2}}} \times [k_1\rho_{\mathbf{k}_1}\rho_{\mathbf{k}_2-\mathbf{k}_1}\bar{P}_{\mathbf{k}_2} - k_2\bar{P}_{-\mathbf{k}_1}\rho_{\mathbf{k}_2-\mathbf{k}_1}\rho_{-\mathbf{k}_2}]. \tag{A10.7.10}$$

If we now make use of the commutation relations (A10.7.9), equation (A10.7.10) becomes

$$T = \frac{1}{2N}\sum_{\mathbf{k}_1\mathbf{k}_2}\left(\frac{\mathbf{k}_1}{k_1}\cdot\frac{\mathbf{k}_2}{k_2}\right) \times \left[\bar{P}_{-\mathbf{k}_1}\rho_{\mathbf{k}_2-\mathbf{k}_1}\bar{P}_{\mathbf{k}_2} + \frac{\hbar^2 k_1 k_2}{4MN}\rho_{\mathbf{k}_1}\rho_{\mathbf{k}_2-\mathbf{k}_1}\rho_{-\mathbf{k}_2}\right]. \tag{A10.7.11}$$

At this stage we expand the kinetic energy just as we did for the potential energy in the main text. Then, from equation (A10.6.2) we have, with $\mathbf{R}_i = \mathbf{R}_i^0 + \mathbf{r}_i$,

$$\rho_{\mathbf{k}} = \sum_i \exp(-i\mathbf{k}.\mathbf{R}_i^0)[1 - i\mathbf{k}.\mathbf{r}_i + \ldots]$$

$$= N\delta_{\mathbf{k},0} + \left(\frac{N}{M}\right)^{\frac{1}{2}} kq_{\mathbf{k}} + \text{terms quadratic in the } q\text{'s}. \tag{A10.7.12}$$

Inserting equation (A10.7.12) into equation (A10.7.11) we obtain the desired result (10.16.32).

APPENDIX A10.8 ANDERSON MODEL OF LOCALIZATION

We wish to calculate the probability distribution of $X(s)$ and following Anderson we use the method of Markoff (see, for example, Chandrasekhar, 1943). This method involves finding the Fourier transform of the probability distribution $P(X)$ given by

$$P(X) = \int_{-\infty}^{\infty} \exp(ixX)\,\phi(x)\,dx, \tag{A10.8.1}$$

where, with n the density of sites,

$$\phi(x) = \exp\left[-n\left\langle\int\left\{1-\exp\left[\frac{ixt^2(r)}{E^2+s^2}\right]d\mathbf{r}\right\}\right\rangle\right], \tag{A10.8.2}$$

the average being taken over the distribution $P(E)$ of E. Thus, we focus on the integral

$$I = \int_{-\infty}^{\infty} dE\,P(E)\int_0^{\infty} 4\pi r^2\,dr\left\{1-\exp\left[\frac{ixt^2(r)}{E^2+s^2}\right]\right\}. \tag{A10.8.3}$$

The behaviour of $P(X)$ for large X depends on the form of I for small x. If we consider the case $s = 0$, for sufficiently small x and a finite E of the order of w say, one can expand $\exp(ixt^2/E^2)$ in a power series in x, and complete the integration over r,‡ to obtain a leading term x for small x. Then, only the behaviour for small E is important, and the dependence of $P(E)$ on E can be neglected, $P(E)$ being replaced by a constant $1/w$. The result for I can then be written

$$I = 2\left(\frac{x}{i}\right)^{\frac{1}{2}}\Gamma(\tfrac{1}{2})\frac{\langle t(r)\rangle}{w} \tag{A10.8.4}$$

and line-broadening theory (see, for example, Margenau, 1935) allows the result (10.17.19) to be obtained. For a full discussion of localization due to disorder, reference should be made to Mott and Davis (1971).

‡ Provided $t(r)$ is finite and falls off more rapidly then $r^{-3/2}$.

GENERAL APPENDIX‡

Crystal Symmetry

APPENDIX G1 INTRODUCTION

In this appendix we shall discuss the application of group theory to the classification of states of propagation of waves in periodic structures. To facilitate the discussion we shall have the particular problem of electronic states in mind in Parts A, B and C, while in Part D we particularize the argument to lattice vibrations.

The space groups were defined in Appendix A1.3 of Vol. 1 and we first summarize some of the conclusions we reached there.

Every operator bringing a lattice into coincidence with itself may be written $(\alpha|\mathbf{T})$, where α is a unitary transformation, to be followed by translation through $\mathbf{T}$. The rule for multiplying two such operators is

$$(\alpha|\mathbf{T}_1)(\beta|\mathbf{T}_2) = (\alpha\beta|\mathbf{T}_1+\alpha\mathbf{T}_2). \tag{G1.1}$$

To find the essential degeneracy at a particular point $\mathbf{k}$ of the BZ we need only consider the relevant subgroup of the space group of all such operators; namely of those operators containing transformations such that

$$\alpha\mathbf{k} = \mathbf{k}+\mathbf{K}, \tag{G1.2}$$

$\mathbf{K}$ being a reciprocal lattice vector.

The translations $\mathbf{T}$ are such that

$$\mathbf{T} = \boldsymbol{\tau}(\alpha)+\mathbf{R}, \tag{G1.3}$$

where $\mathbf{R}$ is a lattice vector and $\boldsymbol{\tau}(\alpha)$ is determined by α. If $\boldsymbol{\tau}(\alpha) = 0$ for all α, the lattice is termed symmorphic, and we may restrict ourselves to the point group, i.e. the group of operators $\alpha, \beta, \ldots$. It is to the symmorphic lattices that we first direct our attention.

‡ We are indebted to Dr. Peter Grout for reading this Appendix and for his valuable criticism.

A point worthy of mention before we begin, however, is that we ought to show that every irreducible matrix representation, of the symmetry group of the Hamiltonian H, is actually obtainable from a set of degenerate eigenfunctions of H. This can in fact be shown, but it will be convenient to defer the proof to the end of this appendix.

PART A. POINT GROUPS

APPENDIX G2 CLASSES

Any crystallographic point group consists of proper rotations, which can only be multiples of rotations through 60° or 90°, and improper rotations, which can only be proper rotations through multiples of 60° or 90° followed by the operation of inversion. We use the notation:

E	identity,
C_6	rotation through 60°,
C_3 or C_6^2	rotation through 120°,
C_2 or C_6^3 or C_4^2	rotation through 180°,
C_6^4 or C_3^2	rotation through 240°,
C_6^5	rotation through 300°,
C_4	rotation through 90°,
C_4^3	rotation through 270°,
J	inversion through the origin.

The significance of the various notations for rotations through 120°, 180° and 240° will become apparent a little later.

When two rotations A and B, either proper or improper, are performed about physically equivalent axes (i.e. axes such that A becomes B if the coordinate system is suitably changed), A and B are said to belong to the same *class*. Suppose X is an operator denoting the change of coordinate system. Then if $\phi(\mathbf{r}) = B\psi(\mathbf{r})$ we must have

$$X\phi(\mathbf{r}) = AX\psi(\mathbf{r}) \tag{G2.1}$$

or

$$\phi(\mathbf{r}) = X^{-1}AX\psi(\mathbf{r}), \tag{G2.2}$$

that is,

$$B = X^{-1}AX. \tag{G2.3}$$

Equation (G2.3) constitutes a general definition: a set of members of a group which are all obtainable from one another by (similarity) transformations of the form (G2.3), the operator X also being a member of the group, constitute a *class*. Each member belongs to one, and only one, class.

Classes of cubic group

The classes of a point group of a cube may be denoted by E, C_4, C_4^2, C_2, C_3, J, JC_4, JC_4^2, JC_2 and JC_3. Referring to Figure G1 they are:

E	the identity E 1 element,
C_4	rotations of $\pm 90°$ about an axis such as Δ 6 elements,
C_4^2	rotations of 180° about an axis such as Δ 3 elements,
C_2	rotation of 180° about an axis such as Σ 6 elements,
C_3	rotation of $\pm 120°$ about an axis such as Λ 8 elements,
J	inversion with respect to the origin 1 element,
JC_4	(6 elements), JC_4^2 (3 elements),
JC_2	(6 elements), JC_3 (8 elements).

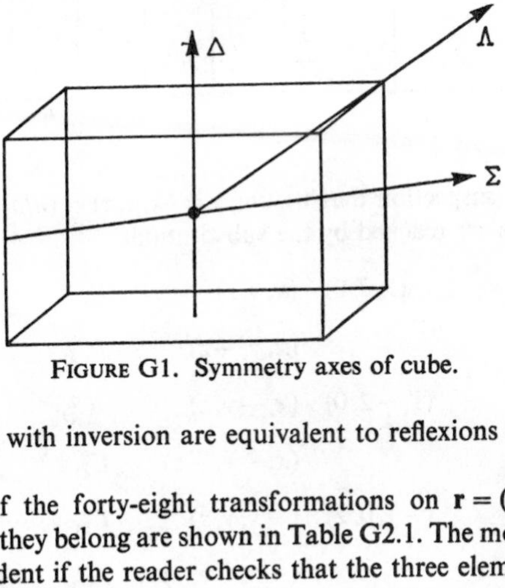

FIGURE G1. Symmetry axes of cube.

The rotations with inversion are equivalent to reflexions about symmetry planes.

The effects of the forty-eight transformations on $\mathbf{r} = (x, y, z)$ with the classes to which they belong are shown in Table G2.1. The mode of tabulation will become evident if the reader checks that the three elements of the class C_4^2 transform the vector $\mathbf{r} = (x, y, z)$ into vectors $(-x, y, -z)$, $(-x, -y, z)$ and $(x, -y, -z)$.

We can use this table to find subgroups, appropriate to high symmetry points. We take the point W as example (see Figure G2) which has coordinates $(\pi/a)(1, 2, 0)$ where a is the edge of the basic cube. The three other equivalent points, which differ from this by reciprocal lattice vectors, are $(\pi/a)(1, -2, 0)$, $(\pi/a)(-1, 0, 2)$ and $(\pi/a)(-1, 0, -2)$.

TABLE G2.1. Table of coordinate transformations and classes of full cubic rotation group

Coordinate in y-position	Sign of third coordinate	Coordinate in x-position					
		$+x$	$-x$	$+y$	$-y$	$+z$	$-z$
$+x$	$+$	—	—	JC_2	C_4	C_3	JC_3
	$-$	—	—	C_2	JC_4	JC_3	C_3
$-x$	$+$	—	—	C_4	JC_2	JC_3	C_3
	$-$	—	—	JC_4	C_2	C_3	JC_3
$+y$	$+$	E	JC_4^2	—	—	JC_2	C_4
	$-$	JC_4^2	C_4^2	—	—	C_4	JC_2
$-y$	$+$	JC_4^2	C_4^2	—	—	C_2	JC_4
	$-$	C_4^2	J	—	—	JC_4	C_2
$+z$	$+$	JC_2	C_2	C_3	JC_3	—	—
	$-$	C_4	JC_4	JC_3	C_3	—	—
$-z$	$+$	C_4	JC_4	JC_3	C_3	—	—
	$-$	JC_2	C_2	C_3	JC_3	—	—

We can see by inspection that writing $\mathbf{r} = (x, y, z) = (\pi/a)(1, 2, 0)$ the four equivalent points are reached by the substitutions:

$$\begin{array}{lll}
(1,2,0) & (x,y,z) & E, \\
 & (x,y,-z) & JC_4^2, \\
(1,-2,0) & (x,-y,z) & JC_4^2, \\
 & (x,-y,-z) & C_4^2, \\
(-1,0,2) & (-x,y,z) & C_2, \\
 & (-x,-z,y) & JC_4, \\
(-1,0,-2) & (-x,z,-y) & JC_4, \\
 & (-x,-z,-y) & C_2.
\end{array}$$

The class to which each transformation belongs is read off from Table G2.1. A slight complication occurs where a class splits into two or more classes. Examples will be given later.

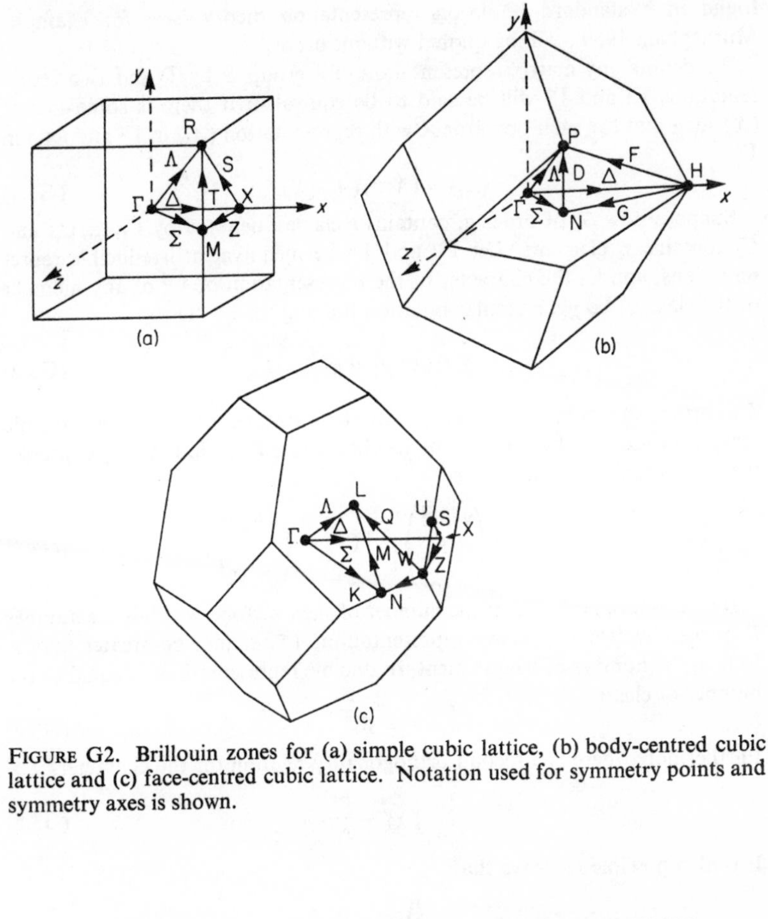

FIGURE G2. Brillouin zones for (a) simple cubic lattice, (b) body-centred cubic lattice and (c) face-centred cubic lattice. Notation used for symmetry points and symmetry axes is shown.

APPENDIX G3 CHARACTERS OF REPRESENTATION

In representation theory we have seen that the trace χ of a matrix is termed the character of the matrix. Its importance derives from its invariance under a similarity transformation (so that, for example, in any given matrix representation the character of every member of a given class is the same) and that the set of characters for a given irreducible representation specifies the representation uniquely to within a similarity transformation. This is a consequence of one of the theorems of representation theory; theorems relevant to our purposes are summarized below. Most of the results, to be found in a standard book on representation theory (see, for example, Murnaghan, 1950), will be quoted without proof.

We denote any matrix representation of a group G by Γ and two representations Γ^1 and Γ^2 will be said to be equivalent if there is some matrix (X) such that for each operator A with representation (A_1) in Γ^1 and (A_2) in Γ^2

$$(A_1) = (X^{-1})(A_2)(X). \tag{G3.1}$$

Suppose now G, of order g, contains r classes, denoted by $C_1, \ldots, C_r$, and C_i contains s_i elements. Let Γ^p and Γ^q be inequivalent irreducible representations, and let the character of the representation of Γ^p of any member of the class C_i be χ_i^p, a similar notation holding for Γ^q. Then

$$\sum_{i=1}^{r} s_i \chi_i^{p*} \chi_i^q = g\delta_{p,q}. \tag{G3.2}$$

This orthogonality relationship substantiates our statement that an irreducible representation may be specified by its characters. Equation (G3.2) expresses the orthogonality of the vectors

$$|\Gamma^p\rangle = \begin{pmatrix} \sqrt{s_1}\,\chi_1^p \\ \vdots \\ \sqrt{s_r}\,\chi_r^p \end{pmatrix}, \quad |\Gamma^q\rangle = \begin{pmatrix} \sqrt{s_1}\,\chi_1^q \\ \vdots \\ \sqrt{s_r}\,\chi_r^q \end{pmatrix}. \tag{G3.3}$$

It is therefore evident that the number of such vectors, which is the number R of inequivalent irreducible representations Γ^p, cannot be greater than r. In fact, the number of inequivalent irreducible representations is equal to the number of classes:

$$r = R. \tag{G3.4}$$

Furthermore, there is a second orthogonality relation for the characters:

$$\sum_{p=1}^{r} \chi_i^p \chi_j^p = \frac{g}{s_i}\delta_{ij}. \tag{G3.5}$$

It is also possible to prove that

$$\sum_{p=1}^{R} d_p^2 = g \tag{G3.6}$$

and

$$\sum_{p=1}^{R} d_p \chi_j^p = 0, \quad j = 2, 3, \ldots, R, \tag{G3.7}$$

where d_p denotes the dimensions of the matrices in the irreducible representation Γ^p, and χ_1^p has been reserved for the character of the representation for the identity E. Since this must be just a unit matrix it is evident that

$$\chi_1^p = d_p. \tag{G3.8}$$

Criteria for irreducibility

The equations we have already given can provide criteria for irreducibility. For example, the group of the point W in the BZ of the f.c.c. lattice has eight elements and five classes. Thus

$$\sum_{p=1}^{5} d_p^2 = 8, \tag{G3.9}$$

and it is evident that a 3×3 matrix representation cannot satisfy this equation. In fact, equation (G3.9) gives the dimensions of the irreducible representations uniquely: it can only be satisfied by $d_1 = d_2 = d_3 = d_4 = 1$, $d_5 = 2$ (the labelling subscript is here arbitrary, of course).

Equation (G3.2) provides us with a necessary and sufficient condition for irreducibility: a matrix group is irreducible if and only if

$$\sum_{i=1}^{R} s_i |\chi_i|^2 = g. \tag{G3.10}$$

The necessity follows immediately from equation (G3.2). If a matrix is reducible, giving each inequivalent representation, Γ^p, c_p times on its diagonal, we may write

$$\chi_i = \sum_p c_p \chi_i^p. \tag{G3.11}$$

Then equation (G3.5) gives

$$\sum_i s_i |\chi_i|^2 = \sum_p |c_p|^2 \sum_i s_i |\chi_i^p|^2 = g \sum_p |c_p|^2. \tag{G3.12}$$

Thus equation (G3.10) is obviously sufficient, only holding if $c_p = 1$ for one particular Γ^p and no others.

A second criterion, which can be proved without any great difficulty, is that a matrix group is irreducible if and only if the only matrices commuting with all its elements are scalar.

APPENDIX G4 CHARACTER TABLES

A character table is a square array,‡ the rows pertaining to the irreducible representations, and the columns to the classes. Equation (G3.2) shows that the rows are orthogonal to one another, and equation (G3.5) shows that the columns also form orthogonal vectors.

‡ When certain characters are trivially obtained from others, as is the case of space groups, we tabulate only what is essential and the table is not square (see below).

As examples of character tables we give in Tables G4.1 to G4.10 those for all symmetry points and lines of s.c., b.c.c. and f.c.c. lattices, shown in Figure G2. Symmetry lines are shown there by arrows, and symmetry points by dots. (For an exhaustive account of crystal symmetry, see Bradley and Cracknell, 1972).

TABLE G4.1. Character table for point group of a cube—symmetry points Γ, H, R (the number in front of the class symbol denoting the number of elements in the class)

Γ, H, R	E	$8C_3$	$3C_4^2$	$6C_4$	$6C_2$	J	$8JC_3$	$3JC_4^2$	$6JC_4$	$6JC_2$
Γ_s	1	1	1	1	1	1	1	1	1	1
Γ_p	3	0	−1	1	−1	−3	0	1	−1	1
Γ_d	2	−1	2	0	0	2	−1	2	0	0
Γ'_d	3	0	−1	−1	1	3	0	−1	−1	1
Γ_f	1	1	1	−1	−1	−1	−1	−1	1	1
Γ'_f	3	0	−1	−1	1	−3	0	1	1	−1
Γ_g	3	0	−1	1	−1	3	0	−1	1	−1
Γ_h	2	−1	2	0	0	−2	1	−2	0	0
Γ_i	1	1	1	−1	−1	1	1	1	−1	−1
Γ_j	1	1	1	1	1	−1	−1	−1	−1	−1

TABLE G4.2. Character table for groups of symmetry lines Δ and T

Δ, T	E	$2C_4$	C_4^2	$2JC_4^2$	$2JC_2$
Δ_s	1	1	1	1	1
Δ_p	2	0	−2	0	0
Δ_d^1	1	−1	1	1	−1
Δ_d^2	1	−1	1	−1	1
Δ_g	1	1	1	−1	−1

TABLE G4.3. Character table for groups of symmetry lines Λ and F

Λ, F	E	$2C_3$	$3JC_2$
Λ_s	1	1	1
Λ_p	2	−1	0
Λ_f	1	1	−1

TABLE G4.4. Character table for groups of symmetry lines Σ, D, G, S, Z and points U and K

D	E	C_4^2	JC_2	JC_2'
Z	E	C_4^2	JC_4^2	$JC_4^{2\prime}$
Σ, G, S	E	C_2	JC_4^2	JC_2
Σ_s	1	1	1	1
Σ_p^1	1	−1	1	−1
Σ_p^2	1	−1	−1	1
Σ_d	1	1	−1	−1

Notes: (1) Point K is on Σ and transforms as such; U is on S and transforms as such; the points have greater significance however, being equivalent, although no symmetry operator brings one into the other. (2) The splitting of class JC_2 for D, and JC_4^2 for Z, is discussed in section G4(ii).

TABLE G4.5. Character table of group for line Q (and reflexion planes)

Q	E	C_2
Q_s	1	1
Q_p	1	−1

Note: If the group is for a reflexion, plane C_2 is, of course, replaced by the operator representing the reflexion.

TABLE G4.6. Character table of group for point L

	E	$2C_3$	$3C_2$	J	$2JC_3$	$3JC_2$
L_s	1	1	1	1	1	1
L_p^1	1	1	1	−1	−1	−1
L_p^2	2	−1	0	−2	1	0
L_d	2	−1	0	2	−1	0
L_f	1	1	−1	−1	−1	1
L_g	1	1	−1	1	1	−1

TABLE G4.7. Character table of group for point N

	E	C_4^2	C_2	C_2^1	J	JC_4^2	JC_2	JC_2^1
N_s	1	1	1	1	1	1	1	1
N_p^1	1	1	1	1	−1	−1	−1	−1
N_p^2	1	−1	1	−1	−1	1	−1	1
N_p^3	1	−1	−1	1	−1	1	1	−1
N_p^4	1	1	−1	−1	−1	−1	1	1
N_d^1	1	−1	1	−1	1	−1	1	−1
N_d^2	1	−1	−1	1	1	−1	−1	1
N_d^3	1	1	−1	−1	1	1	−1	−1

Note: Here the class C_2 of the full cubic group has been split into two. This is discussed in section G4.2(ii).

TABLE G4.8. Character table of group for point P

P	E	$8C_3$	$3C_4^2$	$6JC_4$	$6JC_2$
P_s	1	1	1	1	1
P_p	3	0	−1	−1	1
P_d	2	−1	2	0	0
P_f	3	0	−1	1	−1
P_g	1	1	1	−1	−1

TABLE G4.9. Character table of group for point W

	E	C_4^2	$2C_2$	$2JC_4^2$	$2JC_4$
W_s	1	1	1	1	1
W_p^1	1	1	−1	1	−1
W_p^2	2	−2	0	0	0
W_d	1	1	1	−1	−1
W_f	1	1	−1	−1	1

TABLE G4.10. Character table of group for points X and M

X, M	E	C_4^2	$2C_4^{2\prime}$	$2C_4$	$2C_2$	J	JC_4^2	$2JC_4^{2\prime}$	$2JC_4$	$2JC_2$
X_s	1	1	1	1	1	1	1	1	1	1
X_p^1	1	1	−1	1	−1	−1	−1	1	−1	1
X_p^2	2	−2	0	0	0	−2	2	0	0	0
X_d^1	1	1	1	−1	−1	1	1	1	−1	−1
X_d^2	1	1	−1	−1	1	1	1	−1	−1	1
X_d^3	2	−2	0	0	0	2	−2	0	0	0
X_f^1	1	1	1	−1	−1	−1	−1	−1	1	1
X_f^2	1	1	−1	−1	1	−1	−1	1	1	−1
X_g	1	1	−1	1	−1	1	1	−1	1	−1
X_h	1	1	1	1	1	−1	−1	−1	−1	−1

Note: In this table the classes C_4^2 and JC_4^2 of the full group are split into two. This is discussed in section G4.2.

The 32 possible crystallographic point groups are tabulated in various places, but we shall discuss the derivation of our examples in some detail to provide insight into their nature as well as the theory used. For this reason we present derivations which depend on the particular nature of the groups. For general methods of derivation, one may consult Lomont (1959) or Zak (1969). We might remark immediately that it is important to obtain a clear picture of the symmetry operations involved, the correspondence between operators of subgroups of different symmetry points being of great relevance, as we shall see in section G7.

As an example, it is worth noting that the group of the point P of the BZ of the b.c.c. lattice (Figure G2), is the set of invariant operations of a regular tetrahedron. This can be seen from Figure G3 where the tetrahedron is

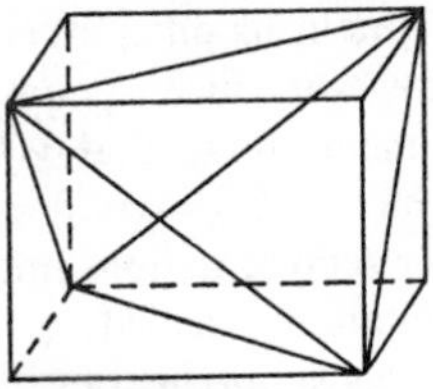

FIGURE G3. Regular tetrahedron formed by diagonals of faces of cube.

inscribed in a cube. The point groups of some **k**-vectors are most easily seen by studying a diagram and others by consideration of the coordinates of **k**. (For a treatment using stereograms extensively, see Jones, 1960.)

We might also remark that, since our treatment is illustrative and not exhaustive, we have chosen notations, both for the operators and the labelling of representations, which are immediately descriptive rather than fitting naturally in a general notational scheme.

(i) Character table for full cubic group

The full cubic group contains the inversion. It is therefore sufficient to discuss the group $G_p(P)$ consisting of proper rotations only (the representations of improper rotations only differ from those of the proper ones in sign, if at all).

Now, there being five classes in the group $G_p(P)$, there are five irreducible representations and a little trial and error on the relation for the dimensions d_p of these, namely

$$\sum_{p=1}^{5} d_p^2 = g = 24, \tag{G4.1}$$

shows that there must be two one-dimensional representations, one two-dimensional representation and two three-dimensional representations. One of the one-dimensional representations is immediate: this is the unit representation in which every element is represented by $+1$. We now seek the other, forming a group homomorphic to G_p.

We first note that an element of C_2 is its own inverse, so that $\chi^2(C_2) = 1$. Similarly, $\chi^2(C_4) = 1$, and since an element of C_4^2 is a product of elements of C_4, $\chi(C_4^2) = 1$. Now $\chi^3(C_3) = 1$, and since we have established that the other operators are represented by ± 1, it readily follows that $\chi(C_3)$ must also be real: hence $\chi(C_3) = 1$. The representation we seek must therefore have $\chi(C_4) = -1$ or $\chi(C_2) = -1$ or $\chi(C_4) = \chi(C_2) = -1$. Suppose $\chi(C_4) = +1$ and $\chi(C_2) = -1$. It must follow, since $\chi(C_3) = \chi(C_4^2) = 1$, that $a_4 b_2 = d_2$, where a_4 is any element of C_4 and b_2 and d_2 are elements of C_2. Now by letting a_4 take all six possibilities, we obtain six different elements of C_2, none of which is equal to b_2, also in C_2. However, since C_2 contains just six elements, this is impossible. A similar argument shows that $\chi(C_4) = -1$, $\chi(C_2) = 1$ cannot obtain. We must therefore have $\chi(C_4) = \chi(C_2) = -1$.

We now turn to the other representations and make use of the character table for invariant representations yielded by spherical harmonics (Table G5.1). We shall show how this is obtained in section G5. We take the rows in order and testing for reducibility by the equation

$$\sum s_i |\chi_i|^2 = g \sum_p |c_p|^2 \tag{G4.2}$$

we find that the row p in fact provides an irreducible representation. The third row d, when tested in the same way, gives $\sum_p c_p^2 = 2$, showing that there must be two representations, one of which is three-dimensional and the other two-dimensional. Using the orthogonality relation

$$\sum_i s_i \chi_i^p \chi_i^q = g\delta_{pq}, \tag{G4.3}$$

we obtain

$$(5\times 3)+3(-1\times 1)+6(1\times -1)+6(-1\times 1) = 0, \tag{G4.4}$$

showing that row d is orthogonal to row p. Hence row d contains a three-dimensional representation, the characters of which we do not yet know. We therefore turn to the row f. Testing by means of the relation

$$\sum_i s_i \chi_p^i \chi_i = gc_p, \tag{G4.5}$$

where c_p is the number of times a representation with characters χ yields Γ^p, we find that the three-dimensional representation we discussed above appears. Subtracting corresponding characters, we see that we have a four-dimensional representation with characters

E	$8C_3$	$3C_4^2$	$6C_4$	$6C_2$
4	1	0	-2	0

Using (G4.2) we find $\sum_p c_p^2 = 2$ and so we have two inequivalent representations, one of which must be three-dimensional and the other one-dimensional. Using the orthogonality relation it is easily shown that the one-dimensional representation is not the unit representation. Subtracting the other one-dimensional representation, we find the characters of the other three-dimensional irreducible representation to be

E	$8C_3$	$3C_4^2$	$6C_4$	$6C_2$
3	0	-1	-1	1

Subtracting these characters from those for the row d we obtain those for the two-dimensional irreducible representation as

E	$8C_3$	$3C_4^2$	$6C_4$	$6C_2$
2	-1	$+2$	0	0

We include the elements obtained from G_p by introduction of the inversion by noting that J must be a scalar matrix with diagonal elements ± 1. The improper rotations, therefore, are represented by the same matrices as the corresponding proper rotations apart from a change of sign if the diagonal elements of J are -1. We thus obtain Table G4.1.

We have therefore been able to obtain the character table without going through the whole laborious procedure of finding explicit irreducible representations. The method we have employed is, however, by no means the only one available. Others may be found in sources already cited.

(ii) Character tables of subgroups

The nomenclature we have used for the representations is further explained in section G5; it gives the lowest spherical harmonic appearing in an expansion of the wave function. Such expansions are used in the cellular method which we discuss in section G5. In applying this method to a band structure calculation we saw in the main text that it was natural to treat the $\mathbf{k} = 0$ state first, when it is desirable to know the character table of the full point group G. For such groups the character table may be worked by methods such as that already described.

It is unnecessary to repeat the whole procedure for every symmetry point or line, for the character table of a subgroup of G can be obtained from the full character table for G if the decomposition into classes is known. Let us take as example the group for the point W in the BZ of the f.c.c. lattice. This, as we have seen, has classes E (one element), C_4^2 (one element), C_2 (two elements), JC_4^2 (two elements) and JC_4 (two elements). Now all inequivalent irreducible manifolds of the full group G are of necessity submanifolds of a subgroup. It therefore remains to test, where necessary, if these latter manifolds are reducible when one considers the subgroup only. This test is of course unnecessary for one-dimensional manifolds of G, so we begin by writing down the relevant characters of these by reference to Table G4.1.

We obtain the following array:

E	C_4^2	$2C_2$	$2JC_4^2$	$2JC_4$
1	1	1	1	1
1	1	−1	−1	1
1	1	−1	1	−1
1	1	1	−1	−1

These are all different, and by reference to the equation

$$\sum_{i=1}^{N} d_p^2 = g = 8, \tag{G4.6}$$

it is seen that we have obtained all one-dimensional representations. It remains to find the characters of the two-dimensional representation. It is perhaps best to work systematically down the table for G; we must not assume that the two-dimensional representation we seek can be obtained from one of the two-dimensional representations classified in Table G4.1. We take, therefore, the first n-dimensional representation ($n > 1$) we encounter on looking down Table G4.1 giving characters

E	JC_4^2	$2C_2$	$2JC_4^2$	$2JC_4$
3	−1	−1	1	−1

We now use equation (G3.12), when we obtain

$$9+1+2+2+2 = 16 = 2g.$$

Hence $\sum |c_p|^2 = 2$, and the representation must be reducible to a two-dimensional representation plus a one-dimensional one. We must next subtract the characters for the one-dimensional representations in turn until we obtain a set of characters satisfying (G3.10). Since we have found that all characters of one-dimensional representations of E and C_4^2 are $+1$, the characters of E and C_4^2 must be $+2$ and -2 respectively. But $2^2+2^2 = 8 = g$, and we immediately see that other characters for this representation must be zero. Collecting all our information together, we obtain Table G4.11.

This table is the same as Table G4.9 except for the labelling of the rows. The way in which the rows are relabelled to obtain Table G4.9 is shown in section G5.

TABLE G4.11. Character table of group for point W, with rows labelled according to correspondence with table for the full point group

W	E	C_4^2	$2C_2$	$2JC_4^2$	$2JC_4$
Γ_s	1	1	1	1	1
Γ_p	2	−2	0	0	0
Γ_f	1	1	−1	−1	1
Γ_i	1	1	−1	1	−1
Γ_j	1	1	1	−1	−1

(iii) Classes of subgroups

In subgroups for most of the symmetry points and lines we have considered, elements belong to the same classes if they belong to the same classes of the full point group. Two exceptions are those for points N and X, as indicated in Tables G4.7 and G4.10. Let us see how these exceptions arise.

In short, the reason for possible exceptions is that if A and B belong to the same class of G, $A = X^{-1}BX$, where X is a member of G; if A and B are also members of a subgroup G_s of G, but X is *not* a member of G_s, A and B are not members of the same class of G_s. This is worth examining in detail for the two examples mentioned.

Point N

In Figure G4 we show the BZ of the b.c.c. lattice with N on a face perpendicular to the plane of the drawing. The axes of the proper rotations are

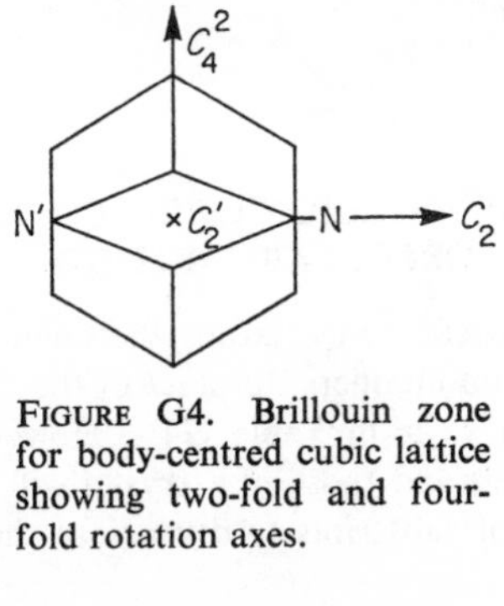

FIGURE G4. Brillouin zone for body-centred cubic lattice showing two-fold and four-fold rotation axes.

marked. Rotations through about C_4^2 and C_2' bring N into N', where N' is equivalent to N, differing from it by a lattice vector. The axis C_2' is into the plane of the drawing and is marked by the cross; the other twofold axis C_2 passes through N and N', and rotation about it leaves N unchanged. The inversion coupled with these rotations gives the remainder of the elements of the group. It is fairly obvious that no operation of the group can bring about interchange of axes of C_2' and C_2. The inequivalence of C_2 and C_2' is even more evident if we consider elongation of the diagram along the axis C_2. Such an elongation does not affect the result of any operation of the subgroup but changes the length of the C_2 axis inside the figure.

Point X

The axes for the proper rotations bringing X into an equivalent point, itself or X', are shown in Figure G5. The inequivalence of the rotation C_4^2

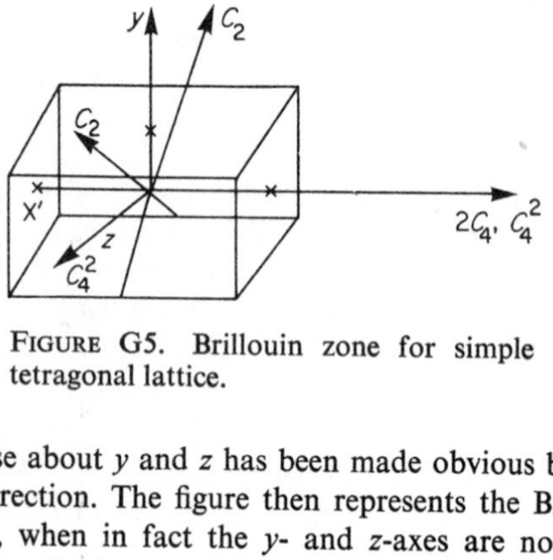

FIGURE G5. Brillouin zone for simple tetragonal lattice.

about X and those about y and z has been made obvious by elongating the figure in the x-direction. The figure then represents the BZ for the simple tetragonal lattice, when in fact the y- and z-axes are no longer fourfold rotation axes but twofold.

APPENDIX G5 ATOMIC CORRESPONDENCE AND CELLULAR METHOD

When atoms come together in a lattice the quantum numbers l and m cease to be good quantum numbers. In place of the $s, p, d, \ldots$ levels we have levels labelled by the Γ's, as in Table G4.1. However, a correspondence between the atomic levels and those in a crystal still exists, which we have signified by the choice of subscripts to distinguish the Γ's. Of the methods for calculating Bloch functions $\psi_{\mathbf{k}}$, the one in which this atomic correspondence is most evident is the cellular method and we shall discuss the atomic correspondence further to show how the cellular method can utilize the information of a character table.

Under any proper or improper rotation the spherical harmonic $Y_{lm}(\theta, \phi) = P_l^m(\cos\theta)\, e^{im\phi}$ becomes a combination of $Y_{lm'}(\theta, \phi)$'s. Thus the Y_{lm} of a given l form a manifold invariant under a point group. To obtain the character of any operator, we take it as a rotation about the polar axis through the angle ϕ_0. Then $P_l^m(\cos\theta)\, e^{im\phi} \to P_l^m(\cos\theta)\, e^{im(\phi+\phi_0)}$. Taking all values of m, we therefore obtain a diagonal representation

$$\begin{pmatrix} e^{il\phi_0} & 0 & 0 & \cdots \\ 0 & e^{i(l-1)\phi_0} & 0 & \cdots \\ \cdot & \cdot & \cdot & \cdot \\ \cdots & \cdots & 0 & e^{-il\phi_0} \end{pmatrix}, \tag{G5.1}$$

with character

$$\chi(\phi_0) = \sum_{m=-l}^{l} e^{im\phi_0} = e^{-il\phi_0}(1 + e^{i\phi_0} + \ldots + e^{2il\phi_0}) \tag{G5.2}$$

or

$$\chi_l(\phi_0) = \pm \frac{\sin(l+\frac{1}{2})\phi_0}{\sin \phi_0/2}. \tag{G5.3}$$

The minus sign is taken for improper rotations when l is odd. For the classes of the cubic group, we find from (G5.3) the characters in Table G5.1 which are taken up to the level j ($l = 7$).

TABLE G5.1

	E	$8C_3$	$3C_4^2$	$6C_4$	$6C_2$	J	$8JC_3$	$3JC_4^2$	$6JC_4$	$6JC_2$
s	1	1	1	1	1	1	1	1	1	1
p	3	0	−1	1	−1	−3	0	1	−1	1
d	5	−1	1	−1	1	5	−1	1	−1	1
f	7	1	−1	−1	−1	−7	−1	1	1	1
g	9	0	1	1	1	9	0	−1	−1	−1
h	11	−1	−1	−1	−1	−11	1	1	1	1
i	13	1	1	1	1	13	1	1	1	1
j	15	0	−1	−1	−1	−15	0	1	1	1

We can find what irreducible representations are contained in the representations tabulated here by writing, for example

$$\chi_i^f = \sum_p c_p^f \chi_i^p \tag{G5.4}$$

and using the orthogonality relationship, for rows of Table G5.1, embodied in equation (G3.5). One finds that the atomic levels change in the following manner:

an s-level becomes Γ_s,

a p-level becomes Γ_p,

a d-level splits into $\Gamma_d + \Gamma'_d$,

an f-level splits into $\Gamma_p + \Gamma_f + \Gamma'_f$, etc.

From this one can see the way in which we have labelled the irreducible representations.

We have not described how this calculation of the splitting is obtained in detail because there is no difference in principle from the shorter example described below. Let us consider the point W. From Table G5.1 we obtain Table G5.2 when we pick out the columns corresponding to the operations of the point group of W (cf. Table G4.11).

TABLE G5.2. Characters of atomic levels for group of point W

W	E	C_4^2	$2C_2$	$2JC_4^2$	$2JC_4$
s	1	1	1	1	1
p	3	−1	−1	1	−1
d	5	1	1	1	−1
f	7	−1	−1	1	1
g	9	1	1	−1	−1
h	11	−1	−1	1	1
i	13	1	1	1	1
j	15	−1	−1	1	1

(i) Non-degenerate level W_p^1

We now examine the level labelled as Γ_i in Table G4.11. Now if any one of the various spherical harmonics s, p, d appear in the expansion of $\psi_{\mathbf{k}}$ ($\mathbf{k}$ at W) for this representation, the corresponding row in Table G5.2 cannot be orthogonal to the row Γ_i of Table G4.11. In examining the orthogonality we must include as weights the number of elements in the class in accordance with equation (G3.2), of course. Let us take the row Γ_i of Table G4.11 with the row s of Table G5.2. We have

$$(1\times 1)+(1\times 1)+2(-1\times 1)+2(1\times 1)+2(-1\times 1)$$
$$= 1+1-2+2-2 = 0.$$

Hence for this irreducible representation, $\psi_{\mathbf{k}}$ has no s-term. Taking the row Γ_i of Table G4.11 with the row p of Table G5.2 we have

$$(3\times 1)+(+1\times -1)+2(-1\times -1)+2(1\times 1)+2(-1\times -1)$$
$$=3-1+2+2+2=8.$$

$\Psi_{\mathbf{k}}$ therefore must have a p-term. In fact this representation is usually labelled W_p^1, the p denoting the spherical harmonic of lowest order appearing in the expansion of $\Psi_{\mathbf{k}}$. Correspondingly, the usual labellings of the other rows of Table G4.11 are $W_s \equiv \Gamma_s$, $W_d \equiv \Gamma_j$, $W_f \equiv \Gamma_f$, $W_p^2 \equiv \Gamma_p$.

Use of equation (G3.2) tells us nothing further about the W_p^1 representation beyond the fact that the wave function contains spherical harmonics for all l-values other than the s-term. Many further spherical harmonics can be eliminated, however. To illustrate the kind of procedure involved, let us take polar coordinates, with x as polar axis, as in Figure G6 which shows

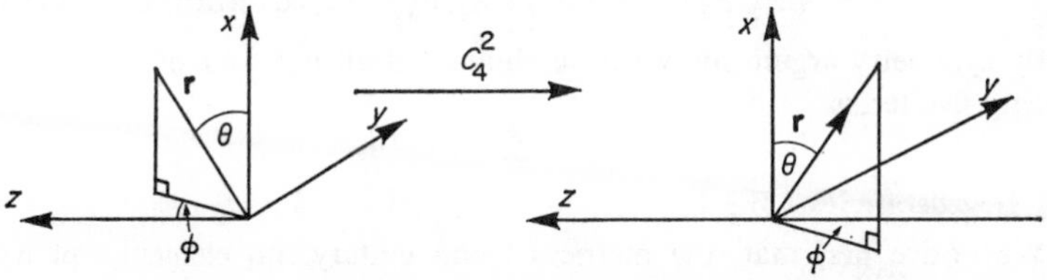

FIGURE G6. Transformation of vector under operation C_4^2.

as an example how the angular coordinates $\begin{pmatrix}\theta\\ \phi\end{pmatrix}$ transform under the operation C_4^2 of the point W as listed in section G2. Figure G6 then shows that

$$C_4^2\begin{pmatrix}\theta\\ \phi\end{pmatrix}=\begin{pmatrix}\theta\\ \phi+\pi\end{pmatrix}. \tag{G5.5}$$

We similarly find

$$JC_4^2\begin{pmatrix}\theta\\ \phi\end{pmatrix}=\begin{pmatrix}\theta\\ \pi-\phi\end{pmatrix} \quad\text{and}\quad \begin{pmatrix}\theta\\ -\phi\end{pmatrix}, \tag{G5.6}$$

$$JC_4\begin{pmatrix}\theta\\ \phi\end{pmatrix}=\begin{pmatrix}\pi-\theta\\ \phi\pm\frac{1}{2}\pi\end{pmatrix}, \tag{G5.7}$$

$$C_2\begin{pmatrix}\theta\\ \phi\end{pmatrix}=\begin{pmatrix}\pi-\theta\\ \frac{1}{2}\pi-\phi\end{pmatrix} \quad\text{and}\quad \begin{pmatrix}\pi-\theta\\ \phi-\frac{1}{2}\pi\end{pmatrix}. \tag{G5.8}$$

Let us now take the wave function for the non-degenerate level W_p^1 in the form

$$\psi = \sum_{lm} P_l^m(\cos\theta)\,[A_{lm}\cos m\phi + B_{lm}\sin m\phi]\,R_l(r). \tag{G5.9}$$

Now equation (G5.6) shows that one of the elements of JC_4^2 changes ϕ to $-\phi$ without change of θ; however, Table G4.9 shows ψ to be unchanged. We thus see immediately that the sines must vanish. Considering the other operations, we readily find that m can only be even, and since $P_l^m(x)$ is then even in x if l is even and odd in x when l is odd, the terms vanish unless

$$m = 0, 4, 8, \ldots \quad \text{when } l \text{ is odd,}$$

$$m = 2, 6, 10, \ldots \quad \text{when } l \text{ is even.}$$

Thus, including all possible terms up to $l = 4$,

$$\psi = A_{10}P_1^0(\cos\theta)\,R_1(r) + A_{22}P_2^2(\cos\theta)\cos 2\phi R_2(r)$$

$$+ A_{30}P_3^0(\cos\theta)\,R_3(r) + A_{42}P_4^2(\cos\theta)\cos 2\phi R_4(r). \tag{G5.10}$$

By symmetry arguments we have eliminated all but four of the possible twenty-five terms.

(ii) Degenerate level W_p^2

We notice first that, the matrices being unitary, no element can have modulus greater than one. The character Table G4.9 then immediately tells us that the matrix for the element of C_4^2 is just $\begin{pmatrix}-1 & 0\\ 0 & -1\end{pmatrix}$. Hence C_4^2 merely changes the sign of the function, and so equation (G5.15) shows that in the expansion (G5.9) all m's must be odd. Next, taking the operation $\begin{pmatrix}\theta\\ \phi\end{pmatrix} \to \begin{pmatrix}\theta\\ -\phi\end{pmatrix}$ of JC_4^2, equation (G5.9) becomes

$$\psi = \sum_{lm} P_l^m(\cos\theta)\,[A_{lm}\cos m\phi - B_{lm}\sin m\phi]\,R_l(r). \tag{G5.11}$$

Addition of this to the original equation shows that we can take as the wave function

$$\psi_1(\mathbf{r}) = \sum_{lm} P_l^m(\cos\theta)\,A_{lm}\cos m\phi R_l(r) \quad \text{(all } m \text{ odd).} \tag{G5.12}$$

The other operations do not further reduce the number of terms, but serve to show that the second eigenfunction orthogonal to this is

$$\psi_2 = \sum_{lm} \pm P_l^m(\cos\theta)\,A_{lm}\sin m\phi R_l(r) \quad \text{(all } m \text{ odd).} \tag{G5.13}$$

The plus sign is taken for l even, $m = 3, 7, 11$, and l odd, $m = 1, 5, 9, \ldots$; the minus sign is taken for l odd, $m = 3, 7, 11, \ldots$, and l even, $m = 1, 5, 9, \ldots$. Finally we note that using the results

$$e^{i\phi} = (y+iz)/(1-x^2)^{\frac{1}{2}}, \quad x = \cos\theta, \tag{G5.14}$$

we can show that all terms appearing in expansions such as (G5.13) are of the form

$$(d^m/dx^m)\, P_l(x)\, [(y+iz)^m + (y-iz)^m] \tag{G5.15}$$

and can thereby construct the coefficients of $R_l(r)$ in terms of the direction cosines x, y and z. An example of this is given in section 1.11.1(c) of the main text.

APPENDIX G6 FREE ELECTRON CORRESPONDENCE AND EXPANSION IN PLANE WAVES

If we take a general vector **k**, we can construct forty-eight plane waves $e^{i\alpha\mathbf{k}\cdot\mathbf{r}}$, α running over all operations of the cubic group. Combinations of these give all possible irreducible representations as is readily seen. Taking matrix elements of members of the group between plane waves, we obtain forty-eight dimensional matrices. Each row of each 48×48 matrix has just one non-zero element, and only the matrix representing the identity element has an element on the diagonal. (Such a representation for a group is called the *regular representation.*) Hence the character of every class except E is zero, and use of equation (G3.12) shows that every irreducible representation is contained in the decomposition of the regular representation. To set up a representation of the full group in this way may be useful if the character table is unknown, but we now wish to turn to special **k**-values. Let us take the point Γ in the f.c.c. lattice. For free electrons we have for the lowest band a non-degenerate level (reciprocal lattice vector $\mathbf{K}_0 = 0$). The next level at Γ is eightfold degenerate, with plane waves $e^{i\mathbf{K}_1\cdot\mathbf{r}}$, $\mathbf{K}_1$ being of the form $(\pm 1, \pm 1, \pm 1)$. From these plane waves we readily obtain an 8×8 matrix representation of the group. These irreducible representations into which the representation can be decomposed may be found using the character table G4.1 in a way which is by now familiar. We find the representation to be of the form $\Gamma_s + \Gamma_d + \Gamma'_d + \Gamma_f$.

The above brief summary is all that should be required after the detailed examples of section G5, but such a procedure can be useful in a crystal with nearly free electron band structure. We should also remark that spherical harmonics expressed in direction cosines $x/r, y/r, z/r$ can be obtained from small-argument expansions of combinations of plane-waves realizing irreducible representations.

Another use of such combinations is to test the symmetry of a wave function about different sites of a diatomic crystal, as shown in section 1.11.1(c) of the main text.

We can also determine features of the momentum eigenfunction expansion of $\psi_{\mathbf{k}}(\mathbf{r})$ at high symmetry points from the use of the character table. This is, of course, an unnecessary procedure if we have a simple band and we have already determined the symmetry of the Wannier function or momentum eigenfunction as described in the main text. However, such general symmetry properties may not be known, and we may be forced to investigate the special case. Suppose $\mathbf{k}$ and $\mathbf{k}+\mathbf{K}$ are equivalent points such that $\alpha^{\dagger}\mathbf{k} = \mathbf{k}+\mathbf{K}_1$, α being a member of the point group. We have

$$\psi_{\mathbf{k}}(\mathbf{r}) = \sum_{\mathbf{K}} v_{\mathbf{K}}(\mathbf{k})\, e^{i(\mathbf{k}+\mathbf{K}).\mathbf{r}} \tag{G6.1}$$

and

$$\psi_{\mathbf{k}}(a^{\dagger}\mathbf{r}) = \sum_{\mathbf{K}} v_{\mathbf{K}}(\mathbf{k})\, e^{ia(\mathbf{k}+\mathbf{K}).\mathbf{r}} \tag{G6.2}$$

$$= \sum_{\mathbf{K}} v_{\alpha^{\dagger}\mathbf{K}}(\mathbf{k})\, e^{i\alpha\mathbf{k}.\mathbf{r}}\, e^{i\mathbf{K}.\mathbf{r}}$$

$$= \sum_{\mathbf{K}} v_{\alpha^{\dagger}\mathbf{K}}(\mathbf{k})\, e^{i\mathbf{k}.\mathbf{r}}\, e^{i(\mathbf{K}+\mathbf{K}_1).\mathbf{r}}. \tag{G6.3}$$

Also let us suppose that the character table gives us the information

$$\psi_{\mathbf{k}}(\alpha^{\dagger}\mathbf{r}) = \chi^{i}(\alpha)\,\psi_{\mathbf{k}}(\mathbf{r}). \tag{G6.4}$$

Then by comparison of (G6.1) and (G6.4) we obtain

$$v_{\alpha^{\dagger}\mathbf{K}}(\mathbf{k}) = \chi^{i}(\alpha)\, v_{\mathbf{k}+\mathbf{K}_1}(\mathbf{k}). \tag{G6.5}$$

Since $v_{\mathbf{K}}(\mathbf{k})$ is a function of $\mathbf{k}+\mathbf{K}$, this equation can be rewritten as

$$v(\mathbf{k}+\alpha^{\dagger}\mathbf{K}) = \chi^{i}(\alpha)\, v(\mathbf{k}+\mathbf{K}+\mathbf{K}_1) = \chi^{i}(\alpha)\, v(\alpha\mathbf{k}+\mathbf{K}) \tag{G6.6}$$

or

$$v(\mathbf{k}+\mathbf{K}) = \chi^{i}(\alpha)\, v(\alpha\mathbf{k}+\alpha\mathbf{K}). \tag{G6.7}$$

From the discussion in section 1.8.1 of Volume 1, it can be seen that this *may* be a property of the momentum eigenfunction for all $\mathbf{k}+\mathbf{K}$.

APPENDIX G7 COMPATIBILITY RELATIONS

The point Γ is on the symmetry line Λ and so the representation of the group of the point Γ, $G(\Gamma)$, must yield the representations of the group of the line Λ, $G(\Lambda)$. Further, because of the continuity in $\mathbf{k}$ of the wave functions of a given band, the wave functions for the band, labelled by any $\mathbf{k}$ along Λ, must transform according to the representation according to which the wave functions at $\mathbf{k} = 0$ transform. The same holds, for example, for the line Λ and the point P. In this manner we can obtain information on the relationship between representations at Γ and P. As examples of such *compatibility relations* we summarize those for s.c., b.c.c. and f.c.c. structures in Tables G7.1 to G7.7.

That the compatibility relations implicit in the definition of a "simple band" (for which we discussed, in Chapter 1, the symmetry of the Wannier function) exist is almost immediately obvious. However, a systematic tabulation of the compatibility relations is by no means superfluous, for we can learn from it whether all non-degenerate bands must be simple or not, and gain general information on the character of a band in which essential degeneracy exists. We shall discuss this when we have explained how the tables are constructed.

(i) Compatibility tables

We shall illustrate how the tables are obtained by considering the line Λ and point P for the b.c.c. lattice. The group of Λ has elements $E, 2C_3, 3JC_2$. Its character table is Table G4.3. We also form a table for the representations of P by extracting from Table G4.8 the characters given there for E, $2C_3$ and $3JC_2$ and in this way we obtain Table G7.1. We proceed in

TABLE G7.1. Table of characters, for point P, of elements of group of line Λ

	E	$2C_3$	$3JC_2$
P_s	1	1	1
P_p	3	0	1
P_d	2	-1	0
P_f	3	0	-1
P_g	1	1	-1

exactly the same manner as that by which we determined the atomic correspondence in section G5. By comparison with Table G4.3 we see immediately that P_s corresponds to the representation Λ_f. For the level P_p, we find that the row of Table G7.1 is orthogonal to the row Λ_s of Table G4.3; as we move move from P along Λ the triply degenerate level P_p splits into the non-degenerate level Λ_s and doubly degenerate level Λ_p. The meaning of these tables is that only the representations shown for the various points are compatible with the listed symmetry corresponding to the line on which they lie.

We should mention that while a picture of the symmetry operations is always desirable, it becomes necessary when classes of the group are subdivided in the subgroup as, for example, when we require the correspondence of the operators for the line Σ and point M. These are shown in Figures G7 and G8.

TABLE G7.2. Compatibility relations for line Λ and F

Points	Line
$\Gamma_s, \Gamma_f; L_s, L_f; P_s; R_s, R_f$	Λ_s
$\Gamma_d, \Gamma_h; L_p^2, L_d; P_d; R_d, R_h$	Λ_p
$\Gamma_i, \Gamma_j; L_p, L_g; P_g; R_i, R_j$	Λ_f
$\Gamma_p, \Gamma_d; P_p$	$\Lambda_s+\Lambda_p$
$\Gamma_f, \Gamma_g; P_f$	$\Lambda_p+\Lambda_f$

Note: The compatibility relations between F and H are the same as those between Λ and Γ, and those between F and P are the same as those between Λ and P.

TABLE G7.3. Compatibility relations for lines Δ and T

Points	Line
$\Gamma_s; X_s, {}^*X_p$	Δ_s
${}^*X_p^2, {}^*X_d^3$	Δ_p
$\Gamma_i; X_d, {}^*X_f^2$	Δ_d
$\Gamma_f; {}^*X_d^2, X_f'$	Δ_d^2
$\Gamma_j; {}^*X_g, X_h$	Δ_g
Γ_p	$\Delta_s+\Delta_p$
Γ_d	$\Delta_s+\Delta_d^1$
Γ_d'	$\Delta_p+\Delta_d^2$
Γ_f'	$\Delta_p+\Delta_d^1$
Γ_g	$\Delta_p+\Delta_g$
Γ_h	$\Delta_d^2+\Delta_g$

Note: (1) An asterisk indicates that the representation is one in which characters of C_4^2 and $C_4^{2\prime}$ are different. Asterisks in subsequent tables have a similar significance. (2) The compatibility relations between T and R are the same as those between Δ and Γ, and those between T and M are the same as those between Δ and X.

TABLE G7.4. Compatibility relations for lines Σ, G and S

Points	Line
Γ_s; M_s, $*M_d^2$; N_s, $*N_p^2$	Σ_s
Γ_i; M_d, $*M_g$; $*N_p^3$, N_d^3	Σ_p
Γ_f; $*M_p^1$, M_f^1; N_p^4, $*N_d^2$	Σ_p^2
Γ_j; $*M_f^2$, M_n; N_p^1, $*N_d^1$	Σ_d
Γ_p	$\Sigma_s+\Sigma_p^1+\Sigma_p^2$
Γ_d, $*M_p^2$	$\Sigma_s+\Sigma_p^1$
Γ_d'	$\Sigma_s+\Sigma_p^2+\Sigma_d$
Γ_f'	$\Sigma_s+\Sigma_p^1+\Sigma_d$
Γ_g	$\Sigma_p+\Sigma_p^2+\Sigma_d$
Γ_h; $*M_d^3$	$\Sigma_p^2+\Sigma_d$

Notes: The compatibility relations between G and H are the same as those between Σ and Γ; those between G and N are the same as those between Σ and N. The compatibility relations between S and X are the same as those between Σ and M. Note that because K is on Σ and U is on S, and K and U are equivalent points, the representations of Σ and S must be the same for the f.c.c. lattice.

TABLE G7.5. Compatibility relations for line D

Points	Line
N_s, N_p^4; P_s	D_s
$*N_p^3$, $*N_d$	$*D_p^1$
$*N_p^2$, $*N_d^2$	$*D_p^2$
N_p; N_d^3; P_g	D_d
P_p	$D_s+D_p^1+D_p^2$
P_d	D_s+D_d
P_f	$D_p^1+D_p^2+D_d$

TABLE G7.6. Compatibility relations for line Q

Points	Line
L_s, L_p^1; W_s, W_d	Q_s
L_f, L_g; W_p^1, W_f	Q_p
L_p^2, L_d; W_p^2	Q_s+Q_p

TABLE G7.7. Compatibility relations for line Z

Points	Line
$M_s, M_d^1; W_s, W_p^1; X_s, X_d^1$	Z_s
$*M_d^2, *M_d; *X_d^2, *X_d$	$*Z_p^1$
$*M_p^1, *M_f^2; *X_p^1, *X_f^2$	$*Z_p^2$
$M_f^1, M_h; W_d, W_f; X_f^1, X_h$	Z_d
$*M_p^2; *X_p^2$	Z_s+Z_p
$*M_d^3; *X_d^3$	$Z_p^2+Z_d$
W_p^2	$Z_p+Z_p^2$

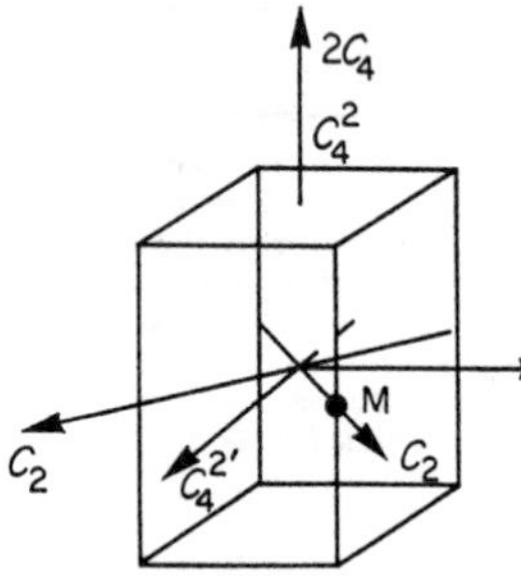

FIGURE G7. Proper rotations of group for point M for cubic lattice.

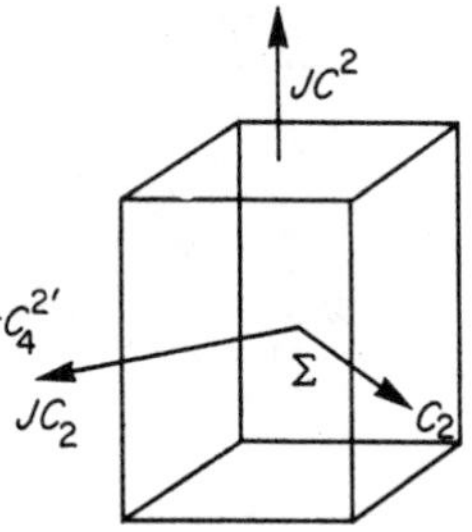

FIGURE G8. Axes corresponding to operators for groups of line Σ.

(ii) Use of compatibility relations

Let us take a b.c.c crystal and examine the possible nature of bands for which there is no accidental degeneracy within the BZ and no essential degeneracy at $\mathbf{k} = 0$. We first take the representation Γ_s. From Table G7.2 it can be seen that the representation along Λ must be labelled by Λ_s, compatible with the representation P_s at P only. From the same table, we see that P_s is compatible only with the representation F_s for the line F; however, F_s is compatible not only with the one-dimensional representations H_s and H_f, but also with the degenerate representations H_p and H_d. (Thus as we move away from H along the line F, the degenerate levels split, and one of these levels is F_s.) Considering all lines in a similar manner, we can draw a diagram as in Figure G9. We see that $*N_p^2$, N_p^4, H_d and H_f are not possible—$*N_p^2$, compatible with Σ_s, is not compatible with D_s, etc., and in this way we obtain Figure G10.

It will be seen that there is one possible diagram for a completely non-degenerate band if the representation at $\mathbf{k} = 0$ is Γ_s, and further, if Γ_s is the

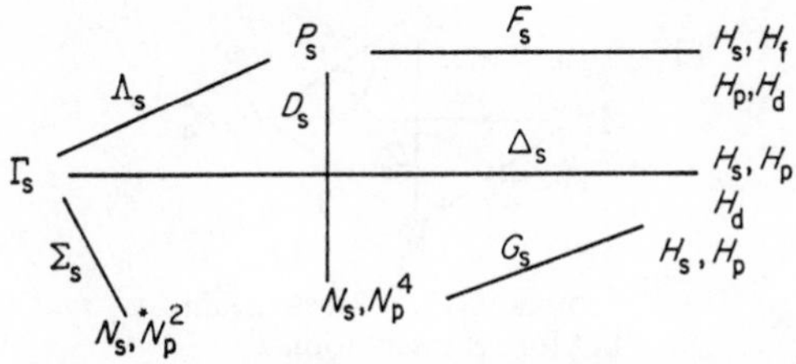

FIGURE G9. Diagram showing feasible symmetry relations for body-centred cubic crystals when representation Γ_s is considered. All possible representations at point H which can be reached along various symmetry axes are shown on extreme right of figure.

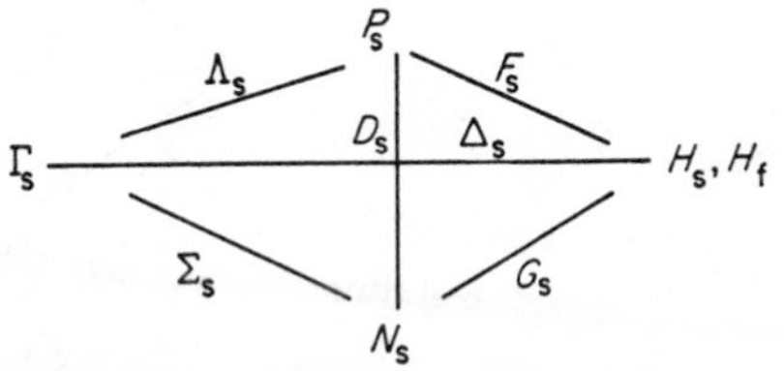

FIGURE G10. Shows compatible representations at H and along symmetry axes when representation Γ_s is considered.

representation at $\mathbf{k} = 0$, the only degeneracy possible is triple degeneracy at the point H, wave functions transforming as H_p. The other three possibilities with single degeneracy at Γ are shown in Figures G11–G13.

In similar fashion one can show that, in the f.c.c. lattice, non-degeneracy at $\mathbf{k} = 0$ implies single degeneracy *throughout the band*, but there are two possibilities for the representation structure for a given one-dimensional

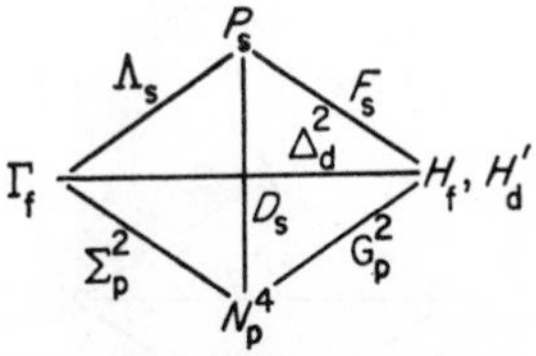

FIGURE G11. Same as Figure G10 except that representation Γ_f is considered.

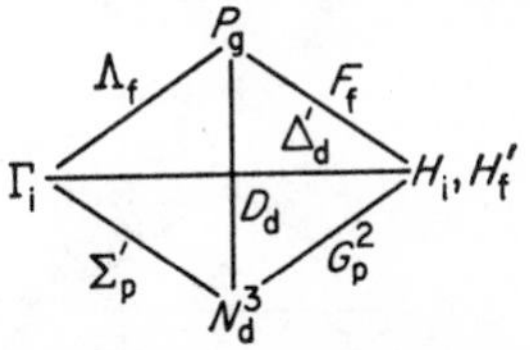

FIGURE G12. Same as Figures G10 and G11 but for representation Γ_i.

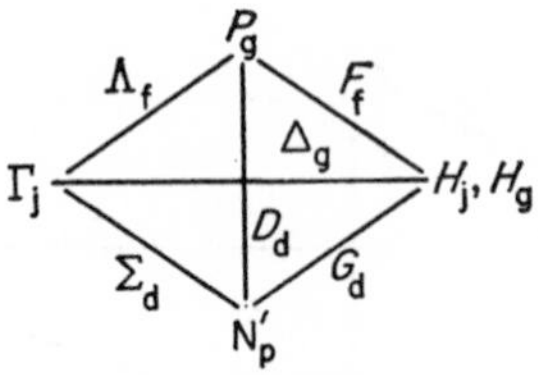

FIGURE G13. Same again but for representation Γ_j.

representation at Γ. These structures are shown in Figures G11–G17. It should be noted that representations of Σ and *S* are always the same. These diagrams show clearly that the equivalence of *K* and *U* implies that Σ and *S* may be regarded as the same line.

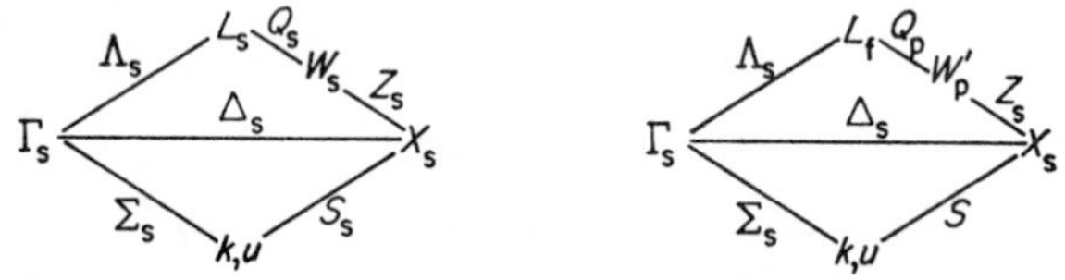

FIGURE G14. Shows compatibility relations for face-centred cubic crystals when representation Γ_s is considered.

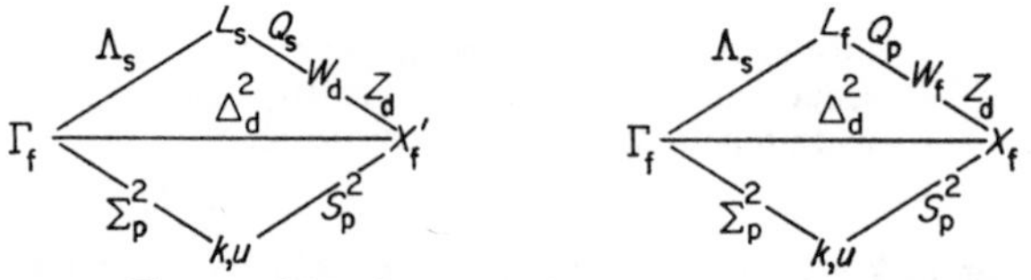

FIGURE G15. Same as Figure G14 but for Γ_f.

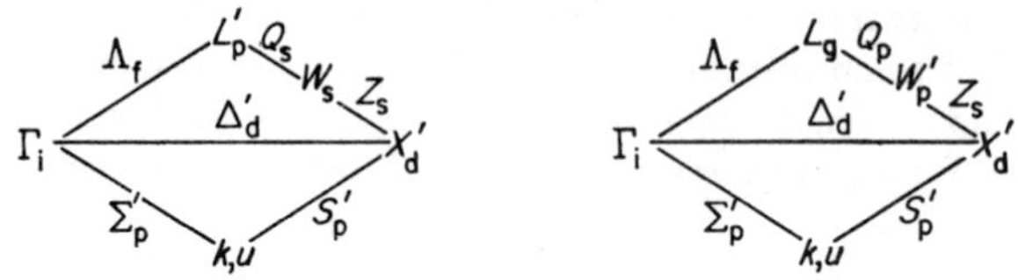

FIGURE G16. Same again but for Γ_i.

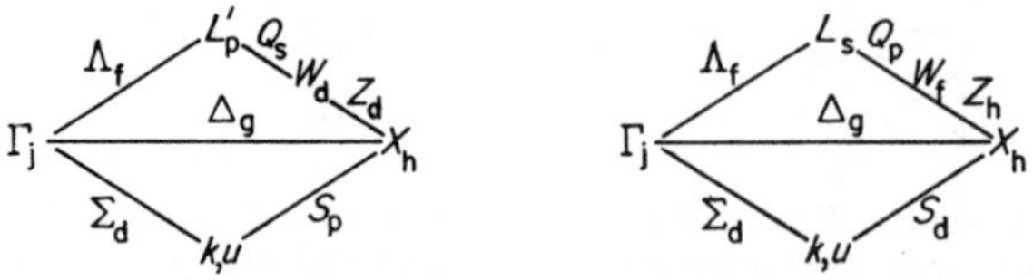

FIGURE G17. Same again, but for Γ_j.

(iii) Simple bands

We have given the name *simple band* to one where it is possible to write, for each transformation α of the point group,

$$\psi_{\alpha\mathbf{k}}(\mathbf{r}) = e^{i\theta(\alpha)}\,\psi_{\mathbf{k}}(\alpha^{\dagger}\,\mathbf{r}), \tag{G7.1}$$

where $e^{i\theta(\alpha)}$ is a phase-factor *independent* of $\mathbf{k}$. The phase-factor is determined as follows: let the character for a member b of the group of Γ be $\chi(b)$; then

$$\psi_0(\mathbf{r}) = e^{i\theta(\alpha)}\,\psi_0(\alpha^{\dagger}\,\mathbf{r}) = e^{i\theta(\alpha)}\,\chi(\alpha^{\dagger})\,\psi_0(\mathbf{r}), \tag{G7.2}$$

so that

$$e^{i\theta(\alpha)} = \chi^{-1}(\alpha^{\dagger}) = \chi(\alpha). \tag{G7.3}$$

Such bands are always possible and the representations at high symmetry points must be such that the characters of the classes appearing are the same as those in the representation of the group of Γ. We thus see that, whereas the only non-degenerate bands of the b.c.c. lattice are simple bands, other possibilities for the f.c.c. lattice cannot be excluded. However, the representation structure of the simple bands can easily be picked out in the diagrams of Figures G11–G13 by looking at the characters of C_2 for the group of Q. We find $+1$ for Q_s and -1 for Q_p. The former is the same as for Γ_s and Γ_j and the latter the same as for Γ_f and Γ_i.

(iv) Influence of accidental degeneracy

Let us suppose we have accidental degeneracy at a certain point P on the line Δ (considered as the line k_x in the (k_x, k_y)-plane) between inequivalent representations. We suppose Γ_1, Δ_1 and X_1 are compatible, and the set Γ_2, Δ_2, X_2 are compatible. Along Δ, the energy of a given representation must be analytic, and we have curves as shown in Figure G18. Suppose the degeneracy occurs at $P \equiv (k'_x, 0)$. Now application of degenerate perturbation theory for the energy of the point (k_x, k_y) shows the degeneracy must disappear

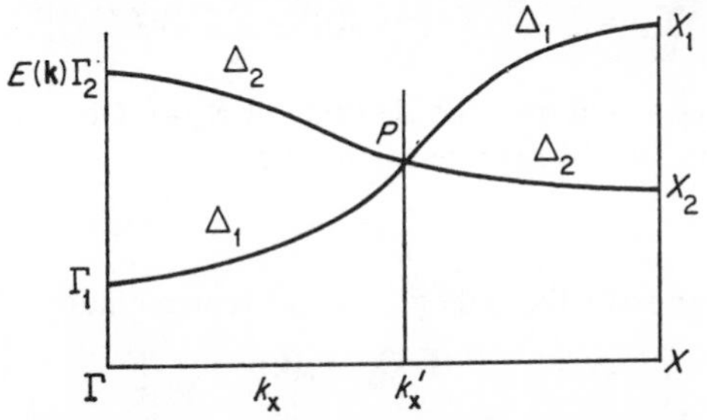

FIGURE G18. Illustrating influence of accidental degeneracy on compatibility relations.

there; since energies must be assigned to bands in a way such that the energy of a band is never discontinuous in **k**, it must follow in the case under consideration that Γ_1 and X_2, although incompatible, must belong to the same band. Such a circumstance, presumably rather rare, will destroy our description of bands in terms of compatible representations.

PART B. NON-SYMMORPHIC LATTICES

APPENDIX G8 SPACE GROUP OF k AND CRITERIA FOR IRREDUCIBILITY

In considering non-symmorphic lattices, we must again obtain the form of the wave functions labelled by **k** in accordance with Bloch's theorem. To do so, we consider the subgroup $G_s(\mathbf{k})$ of the full space group, $G_s(\mathbf{k})$ containing all elements $(A|\boldsymbol{\tau}+\mathbf{R})$ of the full space group which are such that $A\mathbf{k}+\mathbf{K}=\mathbf{k}$. We also require the matrix representations to obey the condition

$$(A|\boldsymbol{\tau}+\mathbf{R}) = e^{i\mathbf{k}.\mathbf{R}}(A|\boldsymbol{\tau}), \tag{G8.1}$$

which may be regarded as a statement of Bloch's theorem.

Let us now define a set $S_p(\mathbf{k})$ of the same order as $G_p(\mathbf{k})$ in the following way. Take each element A say of the point group G_p. Find an element in the space group which contains A as its rotational part, say $(A|\mathbf{t})$. This collection of elements constitutes the desired (but non-unique) set. By equation (G8.1) it is sufficient to find representations of the elements of $S_p(\mathbf{k})$; from these the representations of the whole of $G_s(\mathbf{k})$ are easily generated. Let g be the order of $S_p(\mathbf{k})$ and $G_p(\mathbf{k})$. Then $G_s(\mathbf{k})$ is of order $\sum_{\mathbf{R}} g$. The criterion for irreducibility now becomes, since

$$\chi(A|\boldsymbol{\tau}+\mathbf{R}) = e^{i\mathbf{k}.\mathbf{R}}\chi(A|\boldsymbol{\tau}),$$

$$\sum_{\mathbf{R}} g = \sum_{\mathbf{R}}\sum_{S}|\chi_i(A|\boldsymbol{\tau}+\mathbf{R})|^2 = \sum_{\mathbf{R}}\sum_{S}|\chi_i(A|\boldsymbol{\tau})|^2, \tag{G8.2}$$

where S denotes a sum over the elements of $S_p(\mathbf{k})$. Obviously the criterion for irreducibility may be rewritten simply as

$$g = \sum_{S}|\chi_i(A|\boldsymbol{\tau})|^2. \tag{G8.3}$$

Similarly, we can write the orthogonality of representations

$$\Gamma_p(\mathbf{k}), \Gamma_q(\mathbf{k})$$

as

$$\delta_{pq}\sum_{\mathbf{R}} g = \sum_{\mathbf{R}}\sum_{S}\chi_p^*(A|\boldsymbol{\tau})\,e^{-i\mathbf{k}.\mathbf{R}}\chi_q(A|\boldsymbol{\tau})\,e^{i\mathbf{k}.\mathbf{R}} \tag{G8.4}$$

or

$$\delta_{pq} g = \sum_{S} \chi_p^*(A|\boldsymbol{\tau}) \chi_q(A|\boldsymbol{\tau}). \tag{G8.5}$$

One may also prove that if $d_p(\mathbf{k})$ is the dimension of a representation belonging to point $\mathbf{k}$ (all representations of elements $(E|\mathbf{R})$ are scalar matrices which have diagonal elements $e^{i\mathbf{k}.\mathbf{R}}$), then

$$\sum_{p} d_p^2(\mathbf{k}) = g. \tag{G8.6}$$

We repeat that g is the order of $S_p(\mathbf{k})$ and $G_p(\mathbf{k})$. One may prove equation (G8.6) in the following way. If $|\Gamma_p(\mathbf{k})\rangle$ is a character vector of the form (G3.3) we can expand every c-dimensional vector in terms of such vectors. Here c is the number of classes in the space group. Let us therefore write

$$\sum_{\mathbf{k}p} c_p(\mathbf{k}) |\Gamma_p(\mathbf{k})\rangle = |e^{i\mathbf{k}'.\mathbf{R}}\rangle, \tag{G8.7}$$

where $|e^{i\mathbf{k}.\mathbf{R}}\rangle$ is a vector whose elements are zero except for those corresponding to the position of characters of $(E|\mathbf{R})$, which are $e^{i\mathbf{k}.\mathbf{R}}$. The corresponding elements in $|\Gamma_p(\mathbf{k})\rangle$ are $e^{i\mathbf{k}.\mathbf{R}} d_p(\mathbf{k})$. Hence, using the equations

$$\sum_{\mathbf{R}} e^{i(\mathbf{k}_1-\mathbf{k}_2).\mathbf{R}} = \sum_{\mathbf{R}} \delta(\mathbf{k}_1, \mathbf{k}_2) \tag{G8.8}$$

and

$$\langle \Gamma_p(\mathbf{k}_1) | \Gamma_q(\mathbf{k}_2) \rangle = \sum_{\mathbf{R}} g \delta_{pq} \delta(\mathbf{k}_1, \mathbf{k}_2), \tag{G8.9}$$

we have

$$c_p(\mathbf{k}) \langle \Gamma_p(\mathbf{k}) | \Gamma_p(\mathbf{k}) \rangle = \langle \Gamma_p(\mathbf{k}) | e^{i\mathbf{k}'.\mathbf{R}} \rangle = d_p(\mathbf{k}) \sum_{\mathbf{R}} e^{i(\mathbf{k}'-\mathbf{k}).\mathbf{R}} \tag{G8.10}$$

or

$$c_p(\mathbf{k}) = d_p(\mathbf{k}) \delta(\mathbf{k}', \mathbf{k})/g.$$

Hence

$$\sum_{\mathbf{R}} 1 = \langle e^{i\mathbf{k}'.\mathbf{R}} | e^{i\mathbf{k}'.\mathbf{R}} \rangle = \sum_{p} |c_p(\mathbf{k})|^2 \langle \Gamma_p(\mathbf{k}) | \Gamma_p(\mathbf{k}) \rangle = \sum_{p} \frac{|d_p(\mathbf{k}')|^2}{g} \sum_{\mathbf{R}} 1 \tag{G8.11}$$

and equation (G8.6) is proved.

APPENDIX G9 REPRESENTATIONS OF SPACE GROUP OBTAINABLE FROM THOSE OF POINT GROUPS

In this section we present the rest of the general theory we shall require to obtain the character tables for the specific example of diamond.

A procedure which covers situations where the methods we use for diamond are inadequate will be illustrated later (section G11).

(i) Representations for interior points k

Suppose in a space group there exist translations $\boldsymbol{\tau}_1$ and $\boldsymbol{\tau}_2$ associated with the same point-group operator α in the composite operators $(\alpha|\boldsymbol{\tau}_1)$ and $(\alpha|\boldsymbol{\tau}_2)$. We know that $\boldsymbol{\tau}_2 = \boldsymbol{\tau}_1+\mathbf{R}$, and so can obtain all elements of the group by multiplication between the sets $S(T)$, the group of pure translations $(E|\mathbf{R})$ and the set $S_p(\mathbf{k})$ already defined. Now suppose we wish to obtain $G_s(\mathbf{k})$. The representations of $(E|\mathbf{R})$ are scalar, with diagonal elements $e^{i\mathbf{k}.\mathbf{R}}$, and if $\alpha\beta = \gamma$ and $(\alpha|\boldsymbol{\tau}_1)$, $(\beta|\boldsymbol{\tau}_2)$ and $(\gamma|\boldsymbol{\tau}_3)$ are three elements of $S_p(\mathbf{k})$, the representations are such that

$$(\alpha|\boldsymbol{\tau}_1)(\beta|\boldsymbol{\tau}_2) = (\gamma|\boldsymbol{\tau}_1+\alpha\boldsymbol{\tau}_2) = (\gamma|\boldsymbol{\tau}_3)\,e^{i\mathbf{k}.\mathbf{R}}, \tag{G9.1}$$

where

$$\mathbf{R} = \boldsymbol{\tau}_1+\alpha\boldsymbol{\tau}_2-\boldsymbol{\tau}_3. \tag{G9.2}$$

From G9.1 we see that the matrices $(\alpha|\boldsymbol{\tau}_1)$ form a representation of the point group only if $e^{i\mathbf{k}.\mathbf{R}} = 1$. However, equation (G9.1) does suggest to us that the matrices $(\alpha|\boldsymbol{\tau}_1)\,e^{-i\mathbf{k}.\boldsymbol{\tau}_1}$ might form a representation of $G_p(\mathbf{k})$.

If so, then corresponding to the equation

$$\alpha\beta = \gamma \tag{G9.3}$$

we must have the equation

$$(\alpha|\boldsymbol{\tau}_1)(\beta|\boldsymbol{\tau}_2)\,e^{-i\mathbf{k}.(\boldsymbol{\tau}_1+\boldsymbol{\tau}_2)} = (\gamma|\boldsymbol{\tau}_3)\,e^{-i\mathbf{k}.\boldsymbol{\tau}_3}. \tag{G9.4}$$

However, from equations (G9.1) and (G9.2) we obtain

$$\begin{aligned}(\alpha|\boldsymbol{\tau}_1)(\beta|\boldsymbol{\tau}_2)\,e^{-i\mathbf{k}.(\boldsymbol{\tau}_1+\boldsymbol{\tau}_2)} &= (\gamma|\boldsymbol{\tau}_1+\alpha\boldsymbol{\tau}_2)\,e^{-i\mathbf{k}.(\boldsymbol{\tau}_1+\boldsymbol{\tau}_2)}\\ &= (\gamma|\boldsymbol{\tau}_3)\,e^{-i\mathbf{k}.\boldsymbol{\tau}_3}e^{i\mathbf{k}.(\alpha\boldsymbol{\tau}_2-\boldsymbol{\tau}_2)}.\end{aligned} \tag{G9.5}$$

Hence, we have a representation of the point group if

$$e^{i\alpha^{\dagger}\mathbf{k}.\boldsymbol{\tau}_2} = e^{i\mathbf{k}.\boldsymbol{\tau}_2}. \tag{G9.6}$$

Since the members α of $G_p(\mathbf{k})$ are such that

$$\alpha^{\dagger}\mathbf{k} = \mathbf{k} \tag{G9.7}$$

provided $\mathbf{k}$ is in the interior of the BZ, equation (G9.6) is satisfied for these points, whether or not the space group is symmorphic. However, if we consider a point on the surface of the BZ for which

$$\alpha^{\dagger}\mathbf{k} = \mathbf{k}+\mathbf{K} \quad (\mathbf{K}=0), \tag{G9.8}$$

the above considerations may not be valid. From equation (G9.5) we see that if $\mathbf{k}$ is in the interior of the BZ, every representation of the space group of $\mathbf{k}$ gives a representation of the point group. It follows trivially that, in such circumstances, every irreducible representation of $G_s(\mathbf{k})$ is given by

$$(\alpha|\boldsymbol{\tau}) = e^{i\mathbf{k}.\boldsymbol{\tau}}\,\Gamma_i(\alpha), \tag{G9.9}$$

where $\Gamma_i(\alpha)$ is a member of an irreducible representation of the point group $G_p(\mathbf{k})$. That the representation of $(\alpha|\boldsymbol{\tau})$ is irreducible follows from the criterion that the only matrices commuting with all $\Gamma_i(\alpha)$ are scalar and therefore so are the matrices commuting with $e^{i\mathbf{k}.\boldsymbol{\tau}}\Gamma_i(\alpha)$.

(ii) Representation for surface points

The set $S_p(\mathbf{k})$ may form a group isomorphic with the point group $G_p(\mathbf{k})$, in which case no further considerations are necessary. This will turn out to be true of the point L of the BZ of diamond. We shall also find in diamond that representations for the surface line S may be formed in a very similar way to those for interior points, although equation (G9.7) does not hold. This line realizes a possibility where, for all α associated with primitive translations only,

$$\alpha\mathbf{k} = \mathbf{k} \tag{G9.10}$$

and for pairs A and B each associated with $\boldsymbol{\tau}$ in the form $(A|\boldsymbol{\tau}), (B|\boldsymbol{\tau})$,

$$A\boldsymbol{\tau} = B\boldsymbol{\tau}. \tag{G9.11}$$

(Here one is considering a crystal with only one non-equivalent non-primitive translation $\boldsymbol{\tau}$.)

We then find, on writing

$$\mathbf{R} = A\boldsymbol{\tau}+\boldsymbol{\tau}, \tag{G9.12}$$

that

$$(\alpha|0) = \Gamma_i(\alpha) \tag{G9.13}$$

and

$$(A|\boldsymbol{\tau}) = \Gamma_i(A)\,e^{i\mathbf{k}.\mathbf{R}/2} \tag{G9.14}$$

forms a representation of $S_p(\mathbf{k})$ if Γ_i is a representation of $G_p(\mathbf{k})$. The verification of this is easily performed, and we need not labour it. The proof that all irreducible representations of $S_p(\mathbf{k})$ may be obtained in the form (G9.13) and (G9.14) proceeds exactly as illustrated for interior points in the last section.

We are now in a position to obtain character tables for the space groups $G_s(\mathbf{k})$ of diamond.

APPENDIX G10 DIAMOND LATTICE

The diamond lattice is composed of two interpenetrating f.c.c. lattices, cube side a. Setting up Cartesian axes along cube edges, the lattices have the same orientation but one is displaced by $\frac{1}{4}a(1, 1, 1)$ relative to the other. The unit cube is shown in Figure G19(a), the Bravais lattice evidently being f.c.c. The atoms of the two interpenetrating f.c.c. lattices are differentiated by colour. The "white" atoms are drawn in the diagram at centres of sub-cubes, and the sub-cube containing atom A is shown in Figure G19(b); this figure

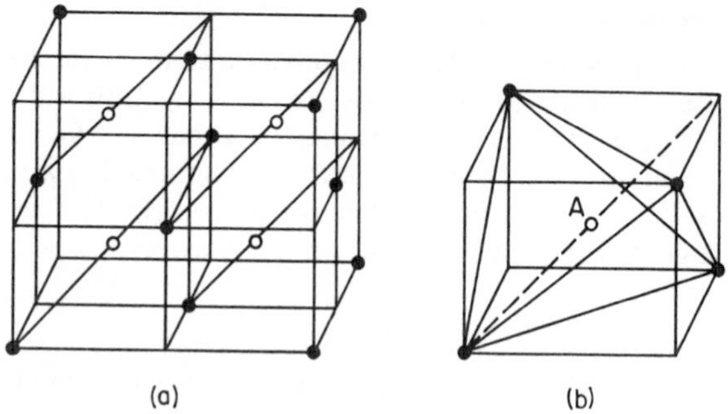

FIGURE G19. Showing for diamond lattice: (a) Unit cube. (b) Tetrahedral environment for an atom.

was previously used in our discussion of point groups to illustrate the subgroup, half the order of the point full cubic group, of the point P of the b.c.c. lattice. The environment of a diamond atom evidently has tetrahedral symmetry, and the operations $6JC_4, 3C_4^2, 6J_dC_2, 8C_3$ are all associated with pure translations. To see what other symmetry elements are present, we note that, taking atom A as origin, inversion J about this atom leaves unchanged the "white" f.c.c. sub-lattice, but transfers black atoms to previously unoccupied corners of the sub-cubes. Now translation of the whole lattice through $\boldsymbol{\tau} = \frac{1}{4}a(1, 1, 1)$ replaces the "white" lattice by the black lattice but brings the "white" lattice to the previously unoccupied sub-cube corners. Thus $(J|\boldsymbol{\tau})$, i.e. translation through $\boldsymbol{\tau}$ followed by inversion about the origin, interchanges "white" and "black" atoms. Since these atoms are not truly physically distinguishable, $(J|\boldsymbol{\tau})$ is an element of the diamond space group. Premultiplying by the elements of the tetrahedral group, we find all the elements of the cubic group which are associated with $\boldsymbol{\tau} = \frac{1}{4}a(1, 1, 1)$. We may thus choose the set $S_p(\Gamma)$ $(S_p(\mathbf{k})$ for $\mathbf{k} = 0)$ as

$$(E|0),\quad 6(C_4|\boldsymbol{\tau}),\quad 3(C_4^2|0),\quad 6(C_2|\boldsymbol{\tau}),\quad 8(C_3|0)$$
$$(J|\boldsymbol{\tau}),\quad 6(JC_4|0),\quad 3(JC_4^2|\boldsymbol{\tau}),\quad 6(JC_2|0),\quad 8(JC_3|\boldsymbol{\tau}).$$

Before presenting the character tables we should first remark that there is a second description of the symmetry system common in the literature. This is where the origin for the point group operation is taken midway between two atoms, say at $\frac{1}{8}a(1, 1, 1)$. In this description the non-primitive translations are combined with the members of the tetrahedral point group.

(i) Character tables

In this section we give the character tables, showing how they may be derived in the following section. We tabulate those of the set $S_p(\mathbf{k})$, noting that $\chi(\alpha|\boldsymbol{\tau}+\mathbf{R}) = \chi(\alpha|\boldsymbol{\tau})e^{i\mathbf{k}.\mathbf{R}}$. We shall no longer use the notation for representations based on atomic correspondence since the symmetry of the wave functions can now not be displayed by spherical harmonics in a simple way.

Point Γ

This table is identical with that of Table G4.1, after substituting $(A|\mathrm{T})$ for α ($\mathrm{T} = 0$ or $\boldsymbol{\tau}$)

TABLE G10.1

Δ	$(E\|0)$	$2(C_4\|\boldsymbol{\tau})$	$(C_4^{2\prime}\|0)$	$2(JC_4^{2\prime}\|\boldsymbol{\tau})$	$2(JC_2\|0)$
Δ_1	1	$e^{i\mathbf{k}.\boldsymbol{\tau}}$	1	$e^{i\mathbf{k}.\boldsymbol{\tau}}$	1
Δ_2	2	0	-2	0	0
Δ_3	1	$-e^{i\mathbf{k}.\boldsymbol{\tau}}$	1	$e^{i\mathbf{k}.\boldsymbol{\tau}}$	-1
Δ_4	1	$-e^{i\mathbf{k}.\boldsymbol{\tau}}$	1	$-e^{i\mathbf{k}.\boldsymbol{\tau}}$	1
Δ_5	1	$e^{i\mathbf{k}.\boldsymbol{\tau}}$	1	$-e^{i\mathbf{k}.\boldsymbol{\tau}}$	-1

TABLE G10.2

Λ	$(E\|0)$	$2(C_3\|0)$	$3(JC_2\|0)$
Λ_1	1	1	1
Λ_2	1	1	-1
Λ_3	2	-1	0

TABLE G10.3

Σ, K, U	$(E\|0)$	$(C_2\|\boldsymbol{\tau})$	$(JC_4^2\|\boldsymbol{\tau})$	$(JC_2\|0)$
Σ_1	1	$e^{i\mathbf{k}.\boldsymbol{\tau}}$	$e^{i\mathbf{k}.\boldsymbol{\tau}}$	1
Σ_2	1	$-e^{i\mathbf{k}.\boldsymbol{\tau}}$	$e^{i\mathbf{k}.\boldsymbol{\tau}}$	-1
Σ_3	1	$-e^{i\mathbf{k}.\boldsymbol{\tau}}$	$-e^{i\mathbf{k}.\boldsymbol{\tau}}$	1
Σ_4	1	$e^{i\mathbf{k}.\boldsymbol{\tau}}$	$-e^{i\mathbf{k}.\boldsymbol{\tau}}$	-1

Note: K, although a surface point, has the same point group as Σ and so no further comment is needed. U is also included since it is equivalent to K.

Surface lines Z and surface points W and X

The character tables of $S_p(\mathbf{k})$ are quite different from those of the corresponding point groups, double degeneracy always occurring. We list them here and show how they can be derived in the next section.

TABLE G10.4

Z	$(E\|0)$	$(C_4^2\|0)$	$(JC_4^2\|\tau)$	$(JC_4^{2'}\|\tau)$
Z_1	2	0	0	0

TABLE G10.5

W	$(E\|0)$	$(C_4^2\|0)$	$2(C_2\|\tau)$	$2(JC_4^2\|\tau)$	$(JC_4\|0)$	$(JC_4^3\|0)$
W_1	2	0	0	0	$1+i$	$1-i$
W_2	2	0	0	0	$-1-i$	$-1+i$

TABLE G10.6

X	$(E\|0)$	$(C_4^2\|0)$	$2(C_4^{2'}\|0)$	$2(C_4\|\tau)$	$2(C_2\|\tau)$
X_1	2	2	0	0	0
X_2	2	2	0	0	0
X_3	2	-2	0	0	2
X_4	2	-2	0	0	-2

X	$(J\|\tau)$	$(JC_4^2\|\tau)$	$2(JC_4^{2'}\|\tau)$	$2(JC_4\|0)$	$2(JC_2\|0)$
X_1	0	0	0	0	2
X_2	0	0	0	0	-2
X_3	0	0	0	0	0
X_4	0	0	0	0	0

Surface lines S and Q and surface points U and L

The other symmetry points and lines on the surface are not always associated with double degeneracy. Again we list the character tables and defer their derivation until the next section.

TABLE G10.7

S, U	$(E\|0)$	$(C_2\|\tau)$	$(JC_4^2\|\tau)$	$(JC_2\|0)$
S_1	1	$e^{i\mathbf{k}.\mathbf{R}/2}$	$e^{i\mathbf{k}.\mathbf{R}/2}$	1
S_2	1	$-e^{i\mathbf{k}.\mathbf{R}/2}$	$e^{i\mathbf{k}.\mathbf{R}/2}$	-1
S_3	1	$-e^{i\mathbf{k}.\mathbf{R}/2}$	$-e^{i\mathbf{k}.\mathbf{R}/2}$	1
S_4	1	$e^{i\mathbf{k}.\mathbf{R}/2}$	$-e^{i\mathbf{k}.\mathbf{R}/2}$	-1

Notes: If we choose $\boldsymbol{\tau} = a(\frac{1}{4}, \frac{1}{4}, \frac{1}{4})$, and $\mathbf{k} = (2\pi/a, k_y, k_y)$ we take $\mathbf{R} = a(0, \frac{1}{2}, \frac{1}{2})$. Thus at X, when $\mathbf{k} = (2\pi/a, 0, 0)$, this table becomes identical with Table G10.6.

Points L and line Q

These tables are exactly those of point group Tables G4.5 and G4.6 after the substitution of rotations α for $(\alpha|0)$ and rotations A for $(A|\boldsymbol{\tau})$.

(ii) Derivation of character tables for diamond

We shall derive these character tables using the general methods already presented together with considerations of compatibility of representations. We remind the reader that this compatibility is necessary when a particular point on a symmetry line is of higher symmetry than the line itself. The representations for the point must yield those for the line, and, in particular, if no representation for the line is one-dimensional, neither can any be for the point.

The points and lines needing special attention are S and U, L and Q, W and Z. The construction of character tables for interior points has been fully dealt with already in section G9(i).

Character table for $S_p(S)$

The general vector of S is $\mathbf{k} = (2\pi/a, k_y, k_y)$, and the operations C_2, JC_4^2, JC_2 of the point group all transform this to $\mathbf{k}+\mathbf{K}$, where $\mathbf{K} = -(4\pi/a, 0, 0)$. The procedure of section G9(ii) is valid here and we must therefore have

$$e^{i(\mathbf{k}+\mathbf{K}).\boldsymbol{\tau}}e^{i\mathbf{k}.\boldsymbol{\tau}} \equiv e^{2i\mathbf{k}.\boldsymbol{\tau}}. \tag{G10.1}$$

Now

$$\boldsymbol{\tau} = \tfrac{1}{4}a(1,1,1) \quad \text{so that} \quad 2\boldsymbol{\tau} = a(\tfrac{1}{2},\tfrac{1}{2},\tfrac{1}{2}) = \mathbf{R}+a(\tfrac{1}{2},0,0).$$

Then, with the value of $\mathbf{k}$ given above,

$$e^{2i\mathbf{k}.\boldsymbol{\tau}} = e^{i\mathbf{k}.\mathbf{R}}e^{i\pi/a}. \tag{G10.2}$$

Point L

The set $S_p(\mathbf{k})$ is obtained by combining the operations $(E|0)$, $2(C_3|0)$ and $3(JC_2|0)$ (the group of Λ) with $(J|\boldsymbol{\tau})$ to produce $(J|\boldsymbol{\tau})$, $2(JC_3|\boldsymbol{\tau})$ and $3(JC_2|\boldsymbol{\tau})$. Now $(J|\boldsymbol{\tau})$ commutes with all members of $G_s(\Lambda)$: let α be any member of this group; then $\alpha\boldsymbol{\tau} = \boldsymbol{\tau}$ since $\boldsymbol{\tau} = \frac{1}{4}a(1,1,1)$ *is in the direction of* Λ. Thus

$$(J|\boldsymbol{\tau})(\alpha|0) = (J\alpha|\boldsymbol{\tau}) = (\alpha|0)(J|\boldsymbol{\tau}). \tag{G10.3}$$

It follows that the set $S_p(L)$ forms a group isomorphic to $G_p(L)$. For suppose α, β belong to $G_p(\Lambda)$, and

$$\alpha J\beta = A, \tag{G10.4}$$

then

$$(\alpha|0)(J\beta|\boldsymbol{\tau}) = (\alpha J\beta|\boldsymbol{\tau}) = (A|\boldsymbol{\tau}). \tag{G10.5}$$

Similarly, if

$$(J\alpha)(J\beta) = \gamma, \tag{G10.6}$$

then

$$\begin{aligned}(J\alpha|\boldsymbol{\tau})(J\beta|\boldsymbol{\tau}) &= (J|\boldsymbol{\tau})(\alpha|0)(J\beta|\boldsymbol{\tau}) = (J|\boldsymbol{\tau})(\alpha J\beta|\boldsymbol{\tau}) \\ &= (J\alpha J\beta|J\boldsymbol{\tau}+\boldsymbol{\tau}) = (\gamma|0).\end{aligned} \tag{G10.7}$$

It follows that the character table of $S_p(L)$ is the character table of $G_p(L)$. Similarly, $S_p(Q)$ is isomorphic to $G_p(Q)$: this follows immediately since $S_p(Q)$ is a subgroup of $S_p(L)$.

Line Z

That there cannot be one-dimensional representations at this point can be shown quite simply. Let us suppose there are; let such a representation be $\chi(E|0) = 1$, $\chi(C_4^2|0) = \pm 1$, $\chi(JC_4^2|\boldsymbol{\tau})$ and $\chi(JC_4^{2'}|\boldsymbol{\tau})$. (Note that the representation of $(C_4^2|0)$ must be ± 1 since $(E|0)$ and $(C_4^2|0)$ form a group of order two.) Let us take C_4^2 such that $C_4^2(x, y, z) = C_4^2(-x, y, -z)$. Now we must have

$$(C_4^2|0)(JC_4^2|\boldsymbol{\tau}) = (JC_4^2|C_4^2\boldsymbol{\tau}) = (JC_4^{2'}|\boldsymbol{\tau})\,e^{i\mathbf{k}.\mathbf{R}}, \tag{G10.8}$$

where

$$\mathbf{R} = C_4^2\boldsymbol{\tau} - \boldsymbol{\tau} = (-\tfrac{1}{4}a, \tfrac{1}{4}a, -\tfrac{1}{4}a) - (\tfrac{1}{4}a, \tfrac{1}{4}a, \tfrac{1}{4}a) = -(\tfrac{1}{2}a, 0, \tfrac{1}{2}a). \tag{G10.9}$$

Hence, putting $\mathbf{k} = (2\pi/a, k_y, 0)$, equation (G10.8) reads, in terms of characters,

$$\chi(C_4^2|0)\,\chi(JC_4^2|\boldsymbol{\tau}') = -\chi(JC_4^{2'}|\boldsymbol{\tau}). \tag{G10.10}$$

However,

$$(JC_4^2|\boldsymbol{\tau})(C_4^2|0) = (JC_4^{2'}|\boldsymbol{\tau}) \tag{G10.11}$$

implies

$$\chi(C_4^2|0)\,\chi(JC_4^2|\boldsymbol{\tau}) = \chi(JC_4^{2'}|\boldsymbol{\tau}). \tag{G10.12}$$

Equations (G10.10) and (G10.12) are incompatible. One-dimensional representations being thus excluded, the equation $\sum_p d_p^2 = g = 4$ shows us we must have just one two-dimensional representation; since of necessity $\chi(E|0) = 2$, equation (G8.3) gives us Table G10.4.

Point W

This point is on the line Z and so each of its representations must give the representation of Z, a two-dimensional one. $S_p(W)$ is of order 8, and the equation $\sum_p d_p^2 = 8$ then shows that there must be just two inequivalent representations, both of order 2. By trial, one may readily verify that the two members of the class $(JC_4|0)$ of the point group are no longer members of the same class in $G_s(W)$, and it follows that the sub-group $(E|0)$, $(C_4^2|0)$, $(JC_4|0)$ and $(JC_4^3|0)$ is *cyclic*—every member may be expressed as C^n (some power of a single element C of the group). The character table for this cyclic group of order 4 is given in Table G10.8.

TABLE G10.8. Table for cyclic group of order 4 for point W

C^4	$(E\|0)$	$(C_4^2\|0)$	$(JC_4\|0)$	$(JC_4^3\|0)$
W_1	1	1	1	1
W_2	1	-1	$-i$	i
W_3	1	1	-1	-1
W_4	1	-1	i	$-i$

The character table for $G_s(W)$ must realize these representations in a form such that $\chi(C_4^2|0) = 0$, to make the representations compatible with those for Z. As a consequence of the structure of Table G10.8, it follows that the characters for one representation are $1+i$, $1-i$, and for the other $-1+i$, $-1-i$. To assign these characters to the two elements in question, we must find the value of $e^{i\mathbf{k}.\mathbf{R}}$ in

$$\chi(\{JC_4|0\}) = e^{i\mathbf{k}.\mathbf{R}}\chi(\{JC_4^3|0\}). \tag{G10.13}$$

Now $\boldsymbol{\tau} = \frac{1}{4}a(1,1,1)$ and $W = (2/a)(2,1,0)$; we readily find by manipulation of the elements of $S_p(W)$ that

$$\mathbf{R} = \tfrac{1}{2}a(0,1,1),$$

and so

$$e^{i\mathbf{k}.\mathbf{R}} = e^{i\pi/2} = i. \tag{G10.14}$$

Table G10.8 is thereby established.

Point X

The group $G_s(X)$ cannot have one-dimensional representations or it will not yield that of Z. Equation (G3.6), which is now $\sum_p |d_p|^2 = 16$, then tells us there are four inequivalent representations, all of order 2.

In finding the character tables for the point W, we utilized the knowledge we already had of the representations for the line Z. We shall similarly investigate the information available for X from the representations for the lines Δ, S and Z, these lines containing the point X. We further note that the set $S_p(X)$ contains the sub-group H, with classes $\{E|0\}$, $\{C_4^2|0\}$, $2\{C_4^{2\prime}|0\}$, $2\{JC_4'|0\}$, $2\{JC_2|0\}$. One easily shows that this group is isomorphic with the point group of the line Δ, and so has characters as displayed in Table G10.9. Let us now set out the elements of $S_p(X)$ displaying to which of the sets $S_p(\Delta)$, $S_p(S)$, $S_p(Z)$ and group H they belong.

TABLE G10.9. Character table of sub-group H of set $S_p(X)$

	$(E\|0)$	$2(C_4^{2'}\|0)$	$(C_4^2\|0)$	$2(JC_4\|0)$	$2(JC_2\|0)$
H_X^1	1	1	1	1	1
H_X^2	2	0	−2	0	0
H_X^3	1	−1	1	1	−1
H_X^4	1	−1	1	−1	1
H_X^5	1	1	1	−1	−1

TABLE G10.10

X	$(E\|0)$	$(C_4^2\|0)$	$(C_4^{2'}\|0)$	$(C_4\|\tau)$	$(C_2\|\tau)$
	Δ	Δ	—	Δ	—
	S	—	—	—	S
	Z	—	Z	—	—
	H	H	H	—	—

X	$(J\|\tau)$	$(JC_4^2\|0)$	$(JC_4^{2'}\|\tau)$	$(JC_4\|0)$	$(JC_2\|0)$
	—	—	Δ	—	Δ
	—	—	S	—	S
	—	Z	Z	—	—
	—	—	—	H	H

We first look at the representations of $S_p(X)$ yielding the one-dimensional representations of $S_p(\Delta)$. From Tables G10.1 and G10.9, we see that these must also yield the two-dimensional representations of H.

From the table for Z, the character of $2(C_4^2|0)$ must be zero, and the Table G10.9 shows that the representations of $S_p(X)$ must yield representations in combinations $H_X^1+H_X^3$ or $H_X^1+H_X^4$ or $H_X^5+H_X^3$ or $H_X^5+H_X^4$. Similarly, the table for Z tells us that the character of $H2(JC_4'|\tau)$ must be zero, so that Table G10.1 tells us $S_p(X)$ must yield representations in combinations $\Delta_1+\Delta_4$ or $\Delta_1+\Delta_5$ or $\Delta_3+\Delta_4$ or $\Delta_3+\Delta_5$.

Some of the combinations of Δ are incompatible with the combinations of H_X since they yield different characters for $2(JC_2|0)$.

In Table G10.11 we give these combinations which are not so incompatible, and the characters of the elements of $S_p(X)$ they yield.

TABLE G10.11

X	$(E\|0)$	$(C_4^2\|0)$	$2(C_4^2\|0)$	$2(C_4\|\tau)$	$2(JC_4^{2'}\|\tau)$	$2(JC_4\|0)$	$2(JC_2\|0)$
$H_X^1+H_X^3, \Delta_1+\Delta_3$	2	2	0	$2e^{i\mathbf{k}.\mathbf{R}}$	0	2	0
$H_X^1+H_X^3, \Delta_3+\Delta_4$	2	2	0	$-2e^{i\mathbf{k}.\mathbf{R}}$	0	2	0
$H_X^1+H_X^4, \Delta_1+\Delta_4$	2	2	0	0	0	0	2
$H_X^5+H_X^3, \Delta_3+\Delta_5$	2	2	0	0	0	0	-2
$H_X^5+H_X^4, \Delta_1+\Delta_5$	2	2	0	$2e^{i\mathbf{k}.\mathbf{R}}$	0	-2	0
$H_X^5+H_X^4, \Delta_3+\Delta_4$	2	2	0	$-2e^{i\mathbf{k}.\mathbf{R}}$	0	-2	0

Now if the representation is irreducible, we must satisfy the equation (summing over the elements of $S_p(X)$)

$$\sum_A |\chi(A)|^2 = 16. \qquad \text{(G10.15)}$$

However, if we calculate the sum from the rows of Table G10.11 we find that the sum exceeds 16 except for the third and fourth rows, where the sum is 16. This means that these two rows give us all non-zero characters for two irreducible representations of $S_p(X)$, and, moreover, these two irreducible representations are the only two yielding one-dimensional irreducible representations of H and $S_p(\Delta)$.

To obtain the other two representations, we turn to the table for S. When $\mathbf{k}$ is at X, $e^{i\mathbf{k}.\mathbf{R}/2} = 1$ in this table; choosing combinations of representations such that the character of $(JC_4^2|\tau)$ is zero, we find that when the character of $2(JC_2|0)$ is zero the character of $2(C_2|\tau)$ is ± 2; to satisfy equation (G10.4) the character of $(J|\tau)$ must be zero, and we have found all possible characters of the set $S_p(X)$.

APPENDIX G11 ALTERNATIVE METHODS OF DERIVATIONS OF CHARACTER TABLES

From the requirement that certain representations be compatible, we have been able to set up all the character tables for the high symmetry points of diamond without detailed examination of the nature of the matrix representations of the full space groups and, indeed, the above method of derivation enables us to write down the compatibility relations for the diamond structure very easily. However, one may be forced to use other modes of attack on some structures, and we shall now illustrate the method used by Herring (1942) in his pioneering work on non-symmorphic crystals. We shall choose for illustration the line Z, for, as the reader will realize, the above arguments concerning the points X and W hinge on the assertion that every state on the line Z is doubly degenerate.

It is simpler to illustrate the symmetry operations for this line on a cube than on the BZ of the f.c.c. lattice, and this is done in Figure G20. Instead of the notation JC_4^2 and $JC_4^{2\prime}$ for mirror planes, we shall use m_1 and m_2.

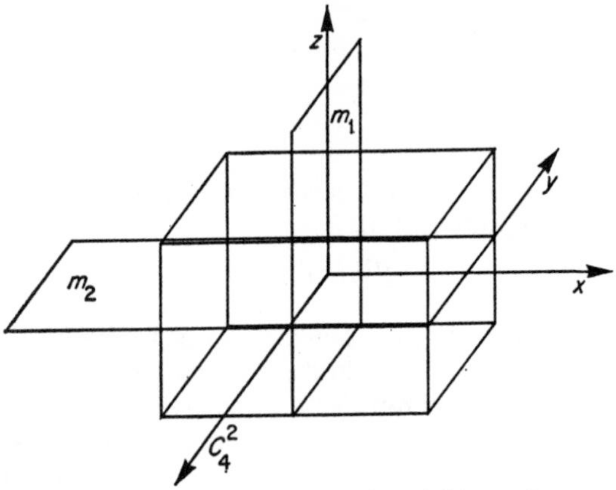

FIGURE G20. Mirror planes and four-fold rotation axes for cubic lattice.

The multiplication table for the point group, of order 4, is shown in Table G11.1.

TABLE G11.1

E	C_4^2	m_1	m_2
C_4^2	E	m_2	m_1
m_1	m_2	E	C_4^2
m_2	m_1	C_4^2	E

From Figure G20 one easily verifies that

$$\left.\begin{aligned} C_4^2(x,y,z) &= (-x,y,-z),\\ m_1(x,y,z) &= (-x,y,z),\\ m_2(x,y,z) &= (x,y,-z). \end{aligned}\right\} \tag{G11.1}$$

Thus, taking $\boldsymbol{\tau} = \frac{1}{4}a(1,1,1)$, we have

$$\left.\begin{aligned} C_4^2\boldsymbol{\tau} &= \boldsymbol{\tau} - \tfrac{1}{2}a(1,0,1),\\ m_1\boldsymbol{\tau}+\boldsymbol{\tau} &= \tfrac{1}{2}a(0,1,1),\\ m_2\boldsymbol{\tau}+\boldsymbol{\tau} &= \tfrac{1}{2}a(1,1,0). \end{aligned}\right\} \tag{G11.2}$$

Hence from Table G11.1, we find Table G11.2 as the multiplication table for corresponding elements of the space group G_p. Here the lattice vectors $\mathbf{R}_1, \mathbf{R}_2, \mathbf{R}_3$ are defined as

$$\mathbf{R}_1 = -\tfrac{1}{2}a(1,0,1), \quad \mathbf{R}_2 = \tfrac{1}{2}a(0,1,1), \quad \mathbf{R}_3 = \tfrac{1}{2}a(1,1,0). \qquad \text{(G11.3)}$$

The class structure of the group is easy to verify, and the character table is shown in Table G11.4. Now Table G11.3 is, as labelled, the multiplication

TABLE G11.2

$(E\|0)$	$(C_4^2\|0)$	$(m_1\|\boldsymbol{\tau})$	$(m_2\|\boldsymbol{\tau})$
$(C_4^2\|0)$	$(E\|0)$	$(m_2\|\boldsymbol{\tau}+\mathbf{R}_1)$	$(m_1\|\boldsymbol{\tau}+\mathbf{R}_1)$
$(m_1\|\boldsymbol{\tau})$	$(m_2\|\boldsymbol{\tau})$	$(E\|\mathbf{R}_2)$	$(C_4^2\|\mathbf{R}_2)$
$(m_2\|\boldsymbol{\tau})$	$(m_1\|\boldsymbol{\tau})$	$(C_4^2\|\mathbf{R}_3)$	$(E\|\mathbf{R}_3)$

TABLE G11.3. Multiplication table of matrices representing elements of $G_s(Z)$

(E)	$(\bar{E})$	(C_4^2)	$(\overline{C_4^2})$	(m_1)	$(\overline{m_1})$	(m_2)	$(\overline{m_2})$
$(\bar{E})$	(E)	$(\overline{C_4^2})$	(C_4^2)	$(\overline{m_1})$	(m_1)	$(\overline{m_2})$	(m_2)
(C_4^2)	$(\overline{C_4^2})$	(E)	$(\bar{E})$	$(\overline{m_2})$	(m_2)	$(\overline{m_1})$	(m_1)
$(\overline{C_4^2})$	(C_4^2)	$(\bar{E})$	(E)	(m_2)	$(\overline{m_2})$	(m_1)	$(\overline{m_1})$
(m_1)	$(\overline{m_1})$	(m_2)	$(\overline{m_2})$	(E)	$(\bar{E})$	(C_4^2)	$(\overline{C_4^2})$
$(\overline{m_1})$	(m_1)	$(\overline{m_2})$	(m_2)	$(\bar{E})$	(E)	$(\overline{C_4^2})$	(C_4^2)
(m_2)	$(\overline{m_2})$	(m_1)	$(\overline{m_1})$	$(\overline{C_4^2})$	(C_4^2)	$(\bar{E})$	(E)
$(\overline{m_2})$	(m_2)	$(\overline{m_1})$	(m_1)	(C_4^2)	$(\overline{C_4^2})$	(E)	$(\bar{E})$

TABLE G11.4. Character table of group isomorphic to matrices of $G_s(Z)$

	(E)	$(\bar{E})$	$(C_4^2), (\overline{C_4^2})$	$(m_1), (\overline{m_1})$	$(m_2), (\overline{m_2})$
	1	1	1	1	1
	1	1	1	-1	-1
	1	1	-1	1	-1
	1	1	-1	-1	1
	2	-2	0	0	0

table of a set of *matrices* where $(\bar{E}) = -(E)$, etc. They must form a matrix representation of the group to which the set is isomorphic. However, we see from Table G11.4, that $\bar{E} = -(E)$ only if the irreducible representation is 2×2.

Let us recapitulate the argument before proceeding further. The set of elements $S_p(Z)$, of order 4, is not a group; however, the *matrices* $(\alpha|\boldsymbol{\tau})$

representing them such that

$$(\alpha|\boldsymbol{\tau}+\mathbf{R}) = e^{i\mathbf{k}.\mathbf{R}}(\alpha|\boldsymbol{\tau}) \tag{G11.4}$$

do form part of a group of order 8.

Looking at the character table for this group, one can see that irreducible matrices obeying (G11.4) must be of order 2.

We have in fact dropped a factor $e^{i\mathbf{k}.\mathbf{a}/2}$ in our argument; from equation (G11.4) one can see that this means that we have particularized to the point X on the line Z. These representations being $\Gamma(\alpha|0)$ and $\Gamma(A|\boldsymbol{\tau})$, we assert that for a general point on Z the representation is $\Gamma(A|0)\,e^{i\mathbf{k}.\mathbf{a}/2}$. This is easy to prove, the considerations being similar to those of section G9(i) for interior points. The general result is the following:

Representation for line obtainable from those of a point

If $\Gamma_i(\alpha|\boldsymbol{\tau})$ is a representation of a space group for a point $\mathbf{k}$, that for $\mathbf{k}+\mathbf{k}'$ is $\Gamma_i(\alpha|\boldsymbol{\tau})\,e^{i\mathbf{k}'.\boldsymbol{\tau}}$ provided $\alpha\mathbf{k}' = \mathbf{k}$ for all $\mathbf{k}$. We then have the requirement

$$\Gamma_i(\alpha|\boldsymbol{\tau}+\mathbf{R})\,e^{i\mathbf{k}'.(\mathbf{R}+\boldsymbol{\tau})} = \Gamma_i(\alpha|\boldsymbol{\tau})\,e^{i\mathbf{k}'.\boldsymbol{\tau}}e^{i(\mathbf{k}+\mathbf{k}').\mathbf{R}}, \tag{G11.5}$$

since

$$\Gamma_i(\alpha|\boldsymbol{\tau}+\mathbf{R}) = \Gamma_i(\alpha|\boldsymbol{\tau})\,e^{i\mathbf{k}.\mathbf{R}}. \tag{G11.6}$$

Also, corresponding to

$$(\alpha|\boldsymbol{\tau})(\beta|\boldsymbol{\tau}) = (\alpha\beta|\alpha\boldsymbol{\tau}+\boldsymbol{\tau}) \tag{G11.7}$$

we have

$$\Gamma_i(\alpha|\boldsymbol{\tau})\,\Gamma_i(\beta|\boldsymbol{\tau})\,e^{i\mathbf{k}'.\boldsymbol{\tau}}e^{i\mathbf{k}'.\boldsymbol{\tau}} = \Gamma_i(\alpha\beta|\,\boldsymbol{\tau}+\alpha\boldsymbol{\tau})\,e^{i\mathbf{k}'.\alpha\boldsymbol{\tau}}e^{i\mathbf{k}'.\boldsymbol{\tau}}, \tag{G11.8}$$

since

$$\alpha\mathbf{k}' = \mathbf{k}. \tag{G11.9}$$

Our result is thus established.

Now since $e^{ika/4} = e^{i\mathbf{k}'.\boldsymbol{\tau}}$, where $\mathbf{k}' = (0, k_y, 0)$, and (G11.9) holds for every operator of the point group of Z we have shown that the representation on the line Z is

$$\Gamma(\alpha|\boldsymbol{\tau}) = \Gamma_X(\alpha|\boldsymbol{\tau})\,e^{\frac{1}{4}ik_ya}, \tag{G11.10}$$

where $\Gamma_X(\alpha|\boldsymbol{\tau})$ is the representation of $G_s(Z)$ particularized to X.

APPENDIX G12 HEXAGONAL LATTICE

We begin by listing the symmetry properties of the simple lattice. We shall omit a table of the influence of operations for coordinates, since their effects on high symmetry points may always be easily seen from a diagram or model. The operations of the full point group are E, $2C_6$, $2C_6^2$, C_6^3, $3C_2$, $3C_2'$, J, $2JC_6$, $2JC_6^2$, JC_6^3, $3JC_2$ and $3JC_2'$. Figure G21(a) shows the rotations and mirror plane perpendicular to the z-axis, and Figure G21(b) shows the

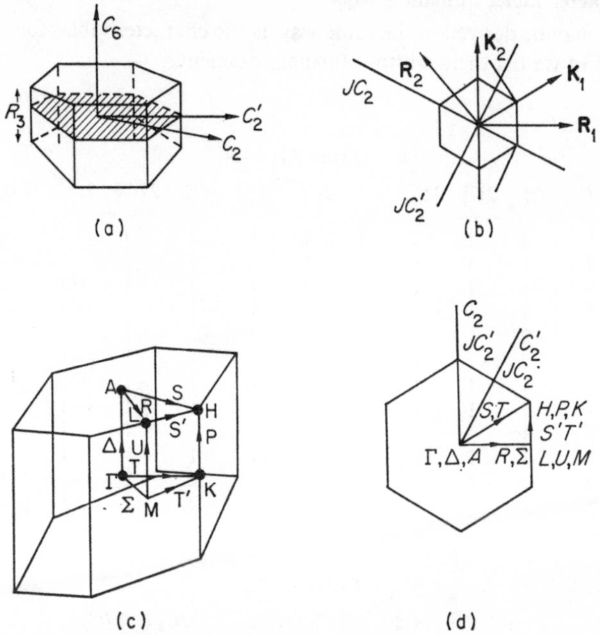

FIGURE G21. Hexagonal lattice. (a) Rotations and mirror plane perpendicular to z-axis. The shaded plane is JC_6^3. (b) Showing other mirror planes. (c) Brillouin zone (basal plane). (d) Brillouin zone—c-axis.

other mirror planes. The correspondence between the directions of rotations and mirror planes associated with them by the inversion should be noted, as should be the orientation of the lattice to its reciprocal lattice, also hexagonal. This orientation is indicated by the vectors $\mathbf{R}_1$, $\mathbf{R}_2$, $\mathbf{K}_1$, $\mathbf{K}_2$ drawn, obeying the relation

$$\mathbf{K}_i . \mathbf{R}_j = \delta_{ij} \quad (i = 1, 2, 3). \tag{G12.1}$$

These vectors may be written

$$\mathbf{R}_1 = s\hat{\mathbf{x}}, \quad \mathbf{R}_2 = \tfrac{1}{2}\sqrt{(3)}\, s\hat{\mathbf{y}} - \tfrac{1}{2}s\hat{\mathbf{x}}, \tag{G12.2}$$

$$\mathbf{K}_1 = (4\pi/s\sqrt{3})\,(\tfrac{1}{2}\sqrt{(3)}\,\hat{\mathbf{x}} + \tfrac{1}{2}\hat{\mathbf{y}}), \quad \mathbf{K}_2 = (4\pi/s\sqrt{3})\,\hat{\mathbf{y}}. \tag{G12.3}$$

In Figures G21(c) and G21(d) the BZ is shown; the elements of the point groups of the **k**-vectors are given in the character tables that follow.

(i) Character tables of point groups

These may be derived in the same way as the character tables for the cubic lattices and we list them without further comment.

TABLE G12.1

Γ, A	E	C_6^3	$2C_6^2$	$2C_6$	$3C_2$	$3C_2'$	J	JC_6^3	$2JC_6^2$	$2JC_6$	$2JC_2$	$2JC_2'$
Γ_1	1	1	1	1	1	1	1	1	1	1	1	1
Γ_2	1	1	1	1	−1	−1	1	1	1	1	−1	−1
Γ_3	1	−1	1	−1	1	−1	1	−1	1	−1	1	−1
Γ_4	1	−1	1	−1	−1	1	1	−1	1	−1	−1	1
Γ_5	2	−2	−1	1	0	0	2	−2	−1	1	0	0
Γ_6	2	2	−1	−1	0	0	2	2	−1	−1	0	0
Γ_7	1	1	1	1	1	1	−1	−1	−1	−1	−1	−1
Γ_8	1	1	1	1	−1	−1	−1	−1	−1	−1	1	1
Γ_9	1	−1	1	−1	1	−1	−1	1	−1	1	−1	1
Γ_{10}	1	−1	1	−1	−1	1	−1	1	−1	1	1	−1
Γ_{11}	2	−2	−1	1	0	0	−2	2	1	−1	0	0
Γ_{12}	2	2	−1	−1	0	0	−2	−2	1	1	0	0

TABLE G12.2

Δ	E	C_6^3	$2C_6^2$	$2C_6$	$3JC_2$	$3JC_2'$
Δ_1	1	1	1	1	1	1
Δ_2	1	1	1	1	−1	−1
Δ_3	1	−1	1	−1	−1	1
Δ_4	1	−1	1	−1	1	−1
Δ_5	2	−2	−1	1	0	0
Δ_6	2	2	−1	−1	0	0

TABLE G12.3

Σ, R	E	C_2'	JC_2	JC_6^3
S, S', T, T'	E	C_2	JC_2'	JC_6^3
U	E	C_6^3	JC_2'	JC_2
Σ_1	1	1	1	1
Σ_2	1	−1	1	−1
Σ_3	1	1	−1	−1
Σ_4	1	−1	−1	1

TABLE G12.4

H, K	E	$2C_6^2$	$3C_2$	JC_6^3	$2JC_6$	$3JC_2'$
H_1	1	1	1	1	1	1
H_2	1	1	−1	1	1	−1
H_3	1	−1	−1	−1	−1	−1
H_4	1	−1	−1	−1	−1	−1
H_5	2	−1	0	2	−1	0
H_6	2	−1	0	−2	1	0

TABLE G12.5

L, M	E	C_6^3	C_2	C_2'	J	JC_6^3	JC_2	JC_2'
L_1	1	1	1	1	1	1	1	1
L_2	1	1	−1	−1	1	1	−1	−1
L_3	1	−1	1	−1	1	−1	1	−1
L_4	1	−1	−1	1	1	−1	−1	1
L_5	1	1	1	1	−1	−1	−1	−1
L_6	1	1	−1	−1	−1	−1	1	1
L_7	1	−1	1	−1	−1	1	−1	1
L_8	1	−1	−1	1	−1	1	1	−1

TABLE G12.6

P	E	$2C_6^2$	$3JC_2'$
P_1	1	1	1
P_2	1	1	−1
P_3	2	−1	0

(ii) Hexagonal close-packed structure

In Figure G22(a) the "white" atoms are atoms at positions $\pm\frac{1}{2}\mathbf{R}_3 = \frac{1}{2}\boldsymbol{\tau}$. Let us first consider this as a two-dimensional lattice, retaining the same rotations for reflexions as before. One immediately sees that the rotations $2C_6^2$ and reflexions $3JC_2$ are associated with pure translations, whereas the operations $2C_6$, C_6^3 and $3JC_2'$ do not bring the lattice into an equivalent position. However, in association with the vector $\boldsymbol{\tau} = \frac{2}{3}\mathbf{R}_1 + \frac{1}{3}\mathbf{R}_2$ they do interchange white and black "atoms". The white circles representing atoms at $\pm\frac{1}{2}\mathbf{R}_3$ above or below the plane of black atoms, we obtain the symmetry operations $2(C_6|\boldsymbol{\tau})$, $(C_6^3|\boldsymbol{\tau})$ and $3(JC_2'|\boldsymbol{\tau})$ for the third lattice, where

$$\boldsymbol{\tau} = \tfrac{2}{3}\mathbf{R}_1 + \tfrac{1}{3}\mathbf{R}_2 + \tfrac{1}{2}\mathbf{R}_3. \qquad \text{(G12.4)}$$

From Figure G22(b) it is evident that $(JC_6^3|0)$ is also a symmetry operation of the lattice, and consequently so is

$$(C_6^3|\tau)(JC_6^3|0) = (J|\tau). \tag{G12.5}$$

By operating on $2(C_6^2|0)$ and $3(JC_2|0)$ with $(J|\tau)$, we generate the operations $2(JC_6^2|\tau)$ and $3(C_2|\tau)$, and by the same means from $2(C_6|\tau)$ and $3(JC_2'|\tau)$ we generate $2(JC_6|0)$ and $3(C_2'|0)$. Hence we have all twenty-four operators

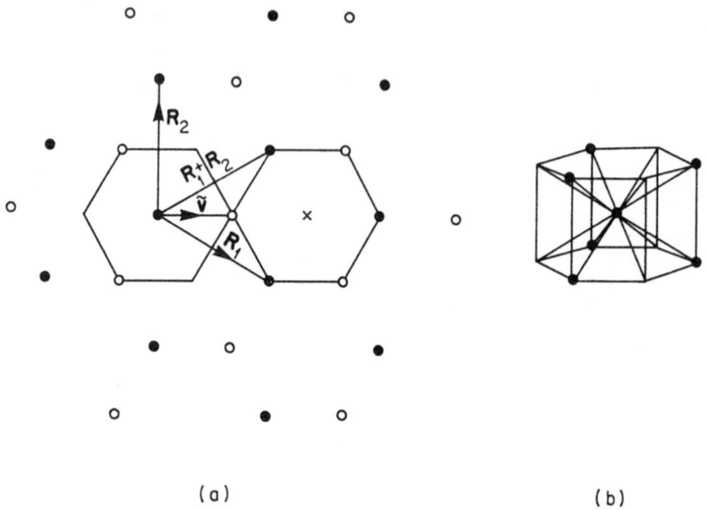

FIGURE G22. Hexagonal close-packed structure. (a) Looking perpendicular to basal plane. (b) Hexagonal close-packed unit cell.

of the point group appearing: associated with pure translations—E, $2C_6^2$, $3C_2'$, $2JC_6^2$, $3JC_2$, JC_6^3; associated with τJ, $2C_6$, C_6^3, $3C_2$, $2JC_6^2$, $3JC_2'$.

(iii) Character tables for hexagonal close-packed lattice

Point* Γ, *Lines* Δ, Σ, *T

The character tables are obtained from those of the corresponding point group of the simple lattice: every character of an operation α is unchanged if α appears as $(\alpha|0)$, every character of an operation α is multiplied by $e^{i\mathbf{k}\cdot\tau}$, with

$$\tau = \tfrac{2}{3}\mathbf{R}_1 + \tfrac{1}{3}\mathbf{R}_2 + \tfrac{1}{2}\mathbf{R}_3 \quad \text{if } \alpha \text{ appears as } (\alpha|\tau).$$

TABLE G12.7. Character table of point A of h.c.p. lattice

A	$(E\|0)$	$(C_6^3\|\tau)$	$2(C_6^2\|0)$	$2(C_6\|\tau)$	$3(C_2\|\tau)$	$3(C_2'\|\tau)$
A_1	2	0	2	0	0	0
A_2	2	0	2	0	0	0
A_3	4	0	−2	0	0	0

A	$(J\|\tau)$	$(JC_6^3\|0)$	$2(JC_6^2\|\tau)$	$2(JC_6\|0)$	$3(JC_2\|0)$	$3(JC_2'\|\tau)$
A_1	0	0	0	0	2	0
A_2	0	0	0	0	−2	0
A_3	0	0	0	0	0	0

TABLE G12.8. Character table of point H of h.c.p. lattice

H	$(E\|0)$	$(C_6^2\|2\mathbf{R}_1)$	$(C_6^4\|\mathbf{R}_1)$	$3(C_2\|\tau)$
H_1	2	2	2	0
H_2	2	−1	−1	0
H_3	2	−1	−1	0

H	$(JC_6'\|\mathbf{R}_1)$	$(JC_6^{-1}\|2\mathbf{R}_1)$	$(JC_6^3\|0)$	$3(JC_2'\|\tau)$
H_1	0	0	0	0
H_2	$i\sqrt{3}$	$-i\sqrt{3}$	0	0
H_3	$-i\sqrt{3}$	$i\sqrt{3}$	0	0

Note: The three elements $(C_2|\tau)$ do not belong to the same class; neither do the three $(JC_2'|\tau)$. Further details of the structure of this group will be found in section (iv) below.

Point K

The character table is obtained from that of the point group of the simple lattice by changing the labelling of classes in the following fashion: E is replaced by $(E|0)$, $2C_6^2$ is replaced by $(C_6^2|2\mathbf{R}_1)$, C_6^4 by $(C_6^4|2\mathbf{R}_1+2\mathbf{R}_2)$, $3(JC_2'|0)$ is replaced by $(m_1|\tau')$, $(m_2|\tau'')$, $(m_3|\tau)$ defined in the note to the character table of P below, $3C_2$ is replaced by $(C_2'|\tau')$, $(C_2^2|\tau'')$, $(C_2^3|\tau)$, where C_i is in the plane m_i, $(JC_6^3|0)$ replaces JC_6^3, $2JC_6$ is replaced by $(JC_6'|2\mathbf{R}_1+2\mathbf{R}_2)$ and $(JC_6^{-1}|2\mathbf{R}_1)$. The simpler example of point P will make this perfectly clear. For point K, $e^{i\mathbf{k}\cdot\mathbf{R}_3/2}=1$.

TABLE G12.9. Character table for line P of h.c.p. lattice

P	$(E\|0)$	$(C_6^2\|2\mathbf{R}_1)$	$(C_6^4\|2\mathbf{R}_1+2\mathbf{R}_2)$	$(m_1\|\tau')$	$(m_2\|\tau'')$	$(m_3\|\tau)$
P_1	1	1	1	$e^{i\mathbf{k}\cdot\mathbf{R}_3/2}$	$e^{i\mathbf{k}\cdot\mathbf{R}_3/2}$	$e^{i\mathbf{k}\cdot\mathbf{R}_3/2}$
P_2	1	1	1	$-e^{i\mathbf{k}\cdot\mathbf{R}_3/2}$	$-e^{i\mathbf{k}\cdot\mathbf{R}_3/2}$	$-e^{i\mathbf{k}\cdot\mathbf{R}_3/2}$
P_3	2	−1	−1	0	0	0

Note: Here $\tau'=\tau-\mathbf{R}_1$, $\tau''=\tau+\mathbf{R}_2$; the mirror planes m_1, m_2 and m_3 are shown in Figure G23.

TABLE G12.10. Character table for point L of h.c.p. lattice

L	$(E\|0)$	$(C_2\|\tau)$	$(C_2'\|0)$	$(C_6^3\|\tau)$	$(J\|\tau)$	$(JC_2\|0)$	$(JC_2'\|\tau)$	$(JC_6^3\|0)$
L_1	2	0	0	0	0	2	0	0
L_2	2	0	0	0	0	−2	0	0

Line R

This line has symmetry elements $(E|0)$, $(C_2'|0)$, $(JC_2|0)$ and $(JC_6^3|0)$ and has a character table identical to that of Table G12.3 of the simple lattice.

TABLE G12.11. Character table for lines S and S'

S, S'	$(E\|0)$	$(JC_6^3\|0)$	$(C_2\|\tau)$	$(JC_2'\|\tau)$
S_1	2	0	0	0

Lines T′ and U

The character tables may be obtained from those of the simple lattice by leaving characters associated with $(\alpha|0)$ unchanged and multiplying those associated with $(\alpha|\boldsymbol{\tau})$ by $e^{i\mathbf{k}.\boldsymbol{\tau}}e^{i\pi/3}$, where $\mathbf{k}$ is taken on the face perpendicular to $\mathbf{K}_1$.

(iv) Derivation of character tables for hexagonal lattice

Character tables for interior points are all trivially found from those of the simple lattice by the means described in section G9.1 and, again, we need only pay detailed attention to surface points and lines. As with diamond we shall, where possible, use compatibility conditions to derive the characters for points where degeneracy always exists.

Lines R, T′ and U

Then line R involves only symmetry elements associated with pure translations, and so its character table is identical with that of the simple lattice. For the line T, we have four symmetry elements, $(E|0)$, $(JC^3|0)$, $(JC_2'|\boldsymbol{\tau})$ and $(C_2|\boldsymbol{\tau})$, where

$$\left.\begin{aligned} JC_6^3\mathbf{k} &= \mathbf{k}, \\ JC_2'\mathbf{k} = C_2\mathbf{k} &= \mathbf{k}-\mathbf{K}_1. \end{aligned}\right\} \tag{G12.6}$$

Thus we can follow the prescription of section G10(ii) for finding the character table from those of the point group of the simple lattice, multiplying characters of elements A associated with $\boldsymbol{\tau}$ by a factor

$$e^{i\mathbf{k}(A\boldsymbol{\tau}+\boldsymbol{\tau})/2} = e^{i\mathbf{k}.\boldsymbol{\tau}}e^{-i\mathbf{K}_1.\boldsymbol{\tau}/2}. \tag{G12.7}$$

The same situation holds for the line U.

Lines S and S′

Exactly the same considerations apply for the lines S and S' as those for the line Z of diamond in sections G10 and G11 and no more need be said.

Point A

We could follow here a procedure similar to that of section G11; we would obtain a group of order 48 with 15 classes and, by construction of the character table, could obtain the relevant representations. For the full results of this procedure the reader may consult the works cited in the Introduction to this appendix; here we shall follow the algebraically simpler method utilizing our knowledge of the representations on lines Δ, R and S. We shall also need to examine the sub-group H of $G_s(A)$ composed of operators of $G_s(A)$ associated with zero translation.

The elements of $G_s(A)$ include all elements of the point group of the lattice in their composition, and the sub-group H in question is that composed of classes $(E|0)$, $3(JC_2|0)$, $3(C_2'|0)$, $(JC_6^3|0)$, $2(C_6^2|0)$ and $2(JC_6|0)$. In structure the character table is similar to the point group of the symmetry point H, and when Table G12.4 is modified to account for the necessary changes in class, Table G12.12 results.

TABLE G12.12. Subgroup of A of elements associated with zero translation

A	$(E\|0)$	$2(C_6^2\|0)$	$3(C_2'\|0)$	$(JC_6^3\|0)$	$2(JC_6\|0)$	$3(JC_2\|0)$
H_1	1	1	1	1	1	1
H_2	1	1	−1	1	1	−1
H_3	1	1	1	−1	−1	−1
H_4	1	1	−1	−1	−1	1
H_5	2	−1	0	2	−1	0
H_6	2	−1	0	−2	1	0

Let us consider representations of $G_s(A)$ yielding the two-dimensional representations of the above sub-group. Now we know that since all representations of $G_s(A)$ yield those of S; no two-dimensional representation of $G_s(A)$ can have $\chi(JC_6^3|0)\neq 0$. It follows that the two-dimensional representations of H must result from irreducible representations of $G_s(A)$ of dimensions greater than two. Now, we remember that the sums of the squares of the dimensions of the irreducible representations must add up to the order of the point group: $\sum_p d_p^2 = 24$. Since we know that every representation of $G_s(A)$ must be at *least* two-dimensional (from our knowledge of the irreducible representations for S), it now follows uniquely that $G_s(A)$ must have one four-dimensional representation and two two-dimensional ones.

The four-dimensional representation must yield H_5 and H_6 of Table G12.12; by addition of the characters of these we obtain characters

$(E\|0)$	$2(C_6^2\|0)$	$3(C_2'\|0)$	$(JC_6^3\|0)$	$2(JC_6\|0)$	$3(JC_2\|0)$
4	−2	0	0	0	0

Now we must satisfy $\sum_i |\chi_i|^2 = g = 24$; $4^2+2(2^2) = 24$; so that no element $(\alpha|\boldsymbol{\tau})$ can have non-zero character, and we have obtained line A_3 of Table G12.7.

We must now find the characters for both two-dimensional representations. From the character table of S we have $\chi(JC_4^3|0) = 0$. Then if we add two lines of the character table of R to make this so and $\chi(JC_2|0) = 0$ also, we find $\chi(C_2')\neq 0$. Adding two lines of the character table of Δ to make $\chi(JC_6^3|0) = \chi(JC_2|0) = 0$, we find $|\chi(C_6^3|\boldsymbol{\tau})| = |\chi(C_6|\boldsymbol{\tau})| = 2$. The sum $\sum_i|\chi_i|^2$ now exceeds 24, and so our supposition that $\chi(JC_2|0) = 0$ is incorrect. If we assume the contrary, maintain the condition $\chi(JC_6^3|0) = 0$ and ensure that $\sum_i|\chi_i|^2 = 24$, we are led to lines A_1 and A_2 without further possibilities.

Point L

The point group is of order 8; since L is on S', so that representations must be at least two-dimensional, it uniquely follows that there are just two inequivalent irreducible representations, both two-dimensional. Thus $\chi(E) = 2$, L also lies on both R and U and from their character tables (see Table G12.3 for simple lattice) we must have $\chi(C_6^3|\boldsymbol{\tau}) = \chi(C_2|\boldsymbol{\tau})$. From the same character tables, if these were non-zero they would each be equal to two, and the sum $\sum_i|\chi_i|^2$ would then exceed $g = 8$. Hence, since we know from the character table of S that $\chi(JC_6^3|0) = 0$, it follows from the tables for R and U that $\chi(JC_2|0) = \pm 2$. To ensure $\sum_i|\chi_i|^2 = 8$, every other character must be zero, and character Table G12.10 follows.

Line P

This and the point K are of interest for two reasons: we now have to distinguish between elements $(\alpha|0)$ and $(\beta|0)$, where α and β are members of the same class of the point group; it is useful to make a change of origin in considering the symmetry elements.

The group of the point P is shown in Figure G23. The reflexion planes $3JC_2'$ have been labelled m_1, m_2, m_3.

Referring to Figure G23 we can see that

$$\left.\begin{array}{c} C_6^2\boldsymbol{\tau} = \boldsymbol{\tau}-\mathbf{R}_1, \quad C_6^4\boldsymbol{\tau} = \boldsymbol{\tau}-\mathbf{R}_1-\mathbf{R}_2, \\ m_1\boldsymbol{\tau}+\boldsymbol{\tau} = \mathbf{R}_1+\mathbf{R}_2+\mathbf{R}_3, \quad m_2\boldsymbol{\tau}+\boldsymbol{\tau} = \mathbf{R}_1+\mathbf{R}_3, \quad m_3\boldsymbol{\tau}+\boldsymbol{\tau} = \mathbf{R}_3. \end{array}\right\} \tag{G12.8}$$

Looking at these equations, to obtain the character table of $S_p(P)$ from the corresponding table of the point group does not appear very simple. However, if we look at Figure G22(a) the point shown by a cross seems a natural origin to take in the discussion of symmetry properties, since relative to it the non-primitive translation becomes $\boldsymbol{\tau}' = \frac{1}{2}\mathbf{R}_3$, perpendicular to the basal

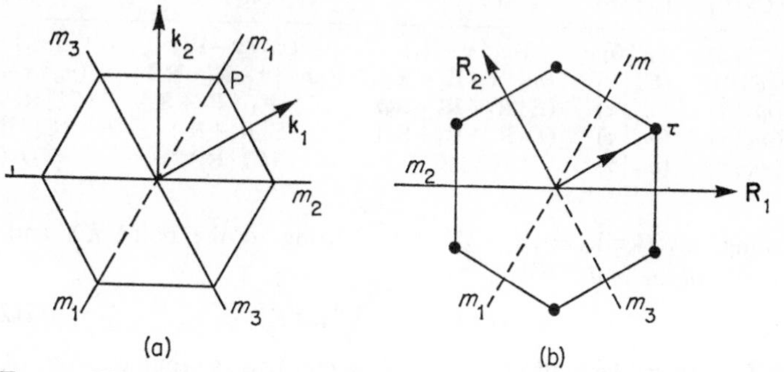

FIGURE G23. Symmetry operations for group of point P of hexagonal lattice. (a) Brillouin zone. (b) Unit cell.

plane. Indeed, the discussion of symmetry for the point P now becomes very easy, for $e^{i\alpha\mathbf{k}.\boldsymbol{\tau}'} = e^{i\mathbf{k}.\boldsymbol{\tau}}$ for any operator α of the point group, and any $\mathbf{k}$ on P. The discussion becomes exactly that of section G9.1 for interior points, and to obtain the character table we merely multiply those elements corresponding to $(\alpha|\boldsymbol{\tau}')$ by $e^{i\mathbf{k}.\boldsymbol{\tau}'} = e^{i\mathbf{k}.\frac{1}{2}\mathbf{R}_3}$.

The natural centre about which to expand a wave function is, however, an atom, and in Table G12.7 a transformation has been made to the original atom. For this procedure, we refer back to section G8(i). We remember that when changing to a new origin having position $\mathbf{a}$ relative to the old, the translation $\mathbf{T}$ in $(A|\mathbf{T})$ changes from $\mathbf{T}'$ to

$$\mathbf{T} = A\mathbf{a} - \mathbf{a} + \mathbf{T}'. \tag{G12.9}$$

From Figure G23(b) it is evident that

$$\mathbf{a} = -\tfrac{4}{3}\mathbf{R}_1 - \tfrac{2}{3}\mathbf{R}_2 = \mathbf{R}_3 - 2\boldsymbol{\tau}. \tag{G12.10}$$

Relative to X, the symmetry elements appear as $2(C_6^2|0)$ and $3(JC_2'|\boldsymbol{\tau}')$ where $\boldsymbol{\tau}' = \frac{1}{2}\mathbf{R}_3$. Relative to the origin at an atom, therefore, the same symmetry elements appear as $(C_6^3|2\mathbf{R}_1)$, $(C_6^4|2\mathbf{R}_1+2\mathbf{R}_2)$, $(m_1|\boldsymbol{\tau}-\mathbf{R}_1)$, $(m_2|\boldsymbol{\tau}+\mathbf{R}_2)$ and $(m_3|\boldsymbol{\tau})$. This is how the labelling of Table G12.8 has arisen. Let us also briefly consider the procedure of section G11 for this case (making no change of origin), as it exemplifies a situation where the matrix group obtained, containing the original symmetry elements, is more than double the order of the point group. Using equation (G12.8) we can set up the multiplication table for the set $(E|0)$, $2(C_6^2|0)$, $3(JC_2'|0)$, when we obtain Table G12.13.

TABLE G12.13. Multiplication table of group of point P

$(E\|0)$	$(C_6^2\|0)$	$(C_6^4\|0)$	$(m_1\|\boldsymbol{\tau})$	$(m_2\|\boldsymbol{\tau})$	$(m_3\|\boldsymbol{\tau})$
$(C_6^2\|0)$	$(C_6^4\|0)$	$(E\|0)$	$(m_3\|\boldsymbol{\tau}-\mathbf{R}_1)$	$(m_1\|\boldsymbol{\tau}-\mathbf{R}_1)$	$(m_2\|\boldsymbol{\tau}-\mathbf{R}_1)$
$(C_6^4\|0)$	$(E\|0)$	$(C_6^2\|0)$	$(m_2\|\boldsymbol{\tau}-\mathbf{R}_1-\mathbf{R}_2)$	$(m_3\|\boldsymbol{\tau}-\mathbf{R}_1-\mathbf{R}_2)$	$(m_1\|\boldsymbol{\tau}-\mathbf{R}_1-\mathbf{R}_2)$
$(m_1\|\boldsymbol{\tau})$	$(m_2\|\boldsymbol{\tau})$	$(m_3\|\boldsymbol{\tau})$	$(E\|\mathbf{R}_1+\mathbf{R}_2+\mathbf{R}_3)$	$(C_6^2\|\mathbf{R}_1+\mathbf{R}_2+\mathbf{R}_3)$	$(C_6^4\|\mathbf{R}_1+\mathbf{R}_2+\mathbf{R}_3)$
$(m_2\|\boldsymbol{\tau})$	$(m_3\|\boldsymbol{\tau})$	$(m_1\|\boldsymbol{\tau})$	$(E\|\mathbf{R}_1+\mathbf{R}_2+\mathbf{R}_3)$	$(E\|\mathbf{R}_1+\mathbf{R}_3)$	$(C_6^2\|\mathbf{R}_1+\mathbf{R}_3)$
$(m_3\|\boldsymbol{\tau})$	$(m_1\|\boldsymbol{\tau})$	$(m_2\|\boldsymbol{\tau})$	$(C_6^2\|\mathbf{R}_3)$	$(C_6^4\|\mathbf{R}_3)$	$(E\|\mathbf{R}_3)$

Taking now $\mathbf{k}\equiv\mathbf{k}_0=\frac{1}{3}\mathbf{K}_1+\frac{1}{3}\mathbf{K}_2$ (specializing to the point K), and the condition on the *matrices*

$$(\alpha|\boldsymbol{\tau}+\mathbf{R})=e^{i\mathbf{k}.\mathbf{R}}(\alpha|\boldsymbol{\tau}), \tag{G12.11}$$

Table G12.14 can be built up. We can now construct a group of order 18, for each matrix (A) representing an element of $S_p(P)$ creating three elements, corresponding to (A), $(A)\omega$ and $(A)\omega^2$, where ω is the cube root of unity. The result is, as one would expect in this particular case, rather simple. It is not difficult to show that this group is isomorphic to a group obtained by direct multiplication of the point group of P with the cyclic group of order 3, whose representation table is given below. (Table G12.15.)

TABLE G12.14. Multiplication table for representing group of P

E	C_6^2	C_6^4	m_1	m_2	m_3
C_6^2	C_6^4	E	$m_3\,\omega^2$	$m_1\,\omega^2$	$m_2\,\omega^2$
C_6^4	E	C_6^2	$m_2\,\omega$	$m_3\,\omega$	$m_1\,\omega$
m_1	m_2	m_3	$E\omega$	$C_6^2\,\omega^2$	$C_6^4\,\omega^2$
m_2	m_3	m_1	$C_6^4\,\omega$	$E\omega$	$C_6^2\,\omega$
m_3	m_1	m_2	C_6^2	C_6^4	E

Note: $\omega=e^{i2\pi/3}$.

TABLE G12.15. Character table of group of order 3

	E	a^1	a^2
C_1	1	1	1
C_2	1	ω	ω^2
C_3	1	ω^2	ω

The character table of the group of order 18 is obtained by multiplication of corresponding elements of this table with that for the point group. The resulting representations are irreducible because only scalar matrices commute with them. To obtain representations for $G_s(P)$ we would pick out those giving Table G12.14.

Point K

The symmetry elements are $(E|0)$, $2(C_6^2|0)$, $(m_1|\boldsymbol{\tau})$, $(m_2|\boldsymbol{\tau})$ and $(m_3|\boldsymbol{\tau})$ (as shown in Figure G22), $(JC_6^3|0)$, $3(C_2|\boldsymbol{\tau})$ and $2(JC_6|0)$. The character table is easily derived after a change of origin; as described for the line P. Noting that we can write $C_2^i = JC_6^3 m_i$, where the axis of C_2^i is in the plane of m_i, $JC_6' = (JC_6^3)(C_6^4)$ and $JC_6^{-1} = (JC_6^3)(C_6^2)$, and JC_6^3 changes no vector such as $\mathbf{a}$ parallel to the basal plane, we obtain the prescription for the character table of K which we gave in section G12(iii).

Point H

Since H is on S, it follows that every representation is at least of order two. Further, the point group is of order 12, so that to satisfy $\sum_p d_p^2 = 12$, there must be just three inequivalent irreducible representations of $G_s(H)$, all of order 2.

H is on P, and one representation is immediately obtained by adding row P_1 to row P_2 of Table G12.9. The characters of all other elements must be zero to satisfy $\sum|\chi_i|^2 = g = 12$. The other two representations must yield representation P_3 of Table G12.9. H is also on S, which tells us that $\chi(JC_3|0) = 0$, $\chi(C_2|\boldsymbol{\tau}) = 0$ and $\chi(JC_2'|\boldsymbol{\tau}) = 0$. (Remember that, for example, although $(m_1|\boldsymbol{\tau})$ and $(m_2|\boldsymbol{\tau})$ do not belong to the same class of $G_s(H)$, since m_1 and m_2 belong to the same class of $G_p(H)$, $\chi(m_1|\boldsymbol{\tau})$ and $\chi(m_2|\boldsymbol{\tau})$ only differ by a phase factor $e^{i\mathbf{k}.\mathbf{R}}$.) We are left only with the characters of $\chi(JC_6'|0)$ and $(JC_6^{-1}|0)$, the squares of which must total 6. Now

$$(JC_6')^{-1} = (JC_6^{-1}), \tag{G12.12}$$

which means, since the matrix representations are unitary,

$$\chi^*(JC_6'|0) = \chi(JC_6^{-1}|0).$$

Since we know that

$$|\chi(JC_6'|0)|^2 + |\chi(JC_6'|0)|^2 = 6, \tag{G12.13}$$

we can also write (θ being as yet unknown)

$$\chi(JC_6'|0) = e^{i\theta}\sqrt{3}. \tag{G12.14}$$

Now we make use of the result that JC_6' and JC_6^{-1} are members of the same class of the point group, and, in particular, $m_1 JC'_6 m_1 = JC_6^{-1}$, so that, using the results

$$m_1 \boldsymbol{\tau}' + \boldsymbol{\tau}' = \mathbf{R}_3 \tag{G12.15}$$

and

$$JC'_6 \boldsymbol{\tau}' = \boldsymbol{\tau}' - \mathbf{R}_3, \tag{G12.16}$$

we have

$$\begin{aligned}(m_1 \ \boldsymbol{\tau}')^{-1}(JC'_6|0)(m_1 \ \boldsymbol{\tau}') &= (m_1|\boldsymbol{\tau}' - \mathbf{R}_3)(JC'_6|0)(m_1|\boldsymbol{\tau}') \\ &= (JC_6^{-1}|-\mathbf{R}_3)\end{aligned} \tag{G12.17}$$

and so

$$\chi(JC'_6|0) = \chi(JC_6^{-1}|-\mathbf{R}_3) = -\chi(JC_6^{-1}|0). \tag{G12.18}$$

Hence

$$e^{i\theta}\sqrt{3} = \chi(JC'_6|0) = -\chi(JC_6^{-1}|0) = -\chi^*(JC'_6|0) = e^{-i\theta}\sqrt{3}$$

or

$$(e^{i\theta})^2 = -1.$$

Hence since we have used r, *relative to X as origin*,

$$\chi(JC'_6|0) = \pm i\sqrt{3}.$$

Moving back to the origin at an atom, as for point K, we obtain Table G12.8. It should be noted that in this table we have used the result that for $\mathbf{k}$ at H

$$\chi(\alpha|\mathbf{t}+\mathbf{R}_i) = e^{i\pi/3}\chi(\alpha|\mathbf{t}) \quad (i = 1 \text{ or } 2). \tag{G12.19}$$

So far this discussion has ignored the presence of a spin–orbit coupling term in the Hamiltonian. Also, we have not considered formally the consequences of time-reversal symmetry on the electron states. These two points will be dealt with below.

PART C. SPIN–ORBIT COUPLING AND TIME-REVERSAL SYMMETRY

APPENDIX G13 TIME-REVERSAL DEGENERACY

The full four-component Dirac equation can be reduced in the non-relativistic limit to a two-component equation which is, ignoring any term without a classical analogue,

$$\left\{\frac{p^2}{2m} + V(\mathbf{r}) + \frac{h^2}{4m^2c^2}[\nabla V \times \mathbf{p}\,.\,\boldsymbol{\sigma}]\right\}\Psi = E\Psi. \tag{G13.1}$$

Hence $\boldsymbol{\sigma}$ is represented by the Pauli spin matrices (1, 2, 3 denoting Cartesian components) given in equation (1.11.140). Bloch's theorem in terms of two-component wave functions is

$$\Psi = \begin{pmatrix} u_{\mathbf{k}}^1(\mathbf{r}) \\ u_{\mathbf{k}}^2(\mathbf{r}) \end{pmatrix} e^{i\mathbf{k}.\mathbf{r}}. \tag{G13.2}$$

In equation (G13.1), the term representing spin-orbit coupling is

$$\frac{\hbar^2}{4m^2c^2}[\nabla V \times \mathbf{p}.\boldsymbol{\sigma}] = -i\mathbf{N}.\boldsymbol{\sigma}, \tag{G13.3}$$

with

$$N_1 = \left(\frac{\hbar}{2mc}\right)^2 \left(\frac{\partial V}{\partial y}\frac{\partial}{\partial z} - \frac{\partial V}{\partial z}\frac{\partial}{\partial y}\right), \quad \text{etc.} \tag{G13.4}$$

If we take the complex conjugate of the equation

$$H\Psi = \frac{\hbar}{i}\frac{\partial \Psi}{\partial t} \tag{G13.5}$$

to obtain

$$H^*\Psi^* = -\frac{\hbar}{i}\frac{\partial \Psi^*}{\partial t}, \tag{G13.6}$$

we note that, making the transformation $t \to -t$, we regain the original equation. Thus we have degenerate solutions $\Psi(t)$ and $\Psi^*(t)$, at least if H is real. This is not true, however, for the Hamiltonian of (G13.1). Nevertheless, such degeneracy still exists, as shown in the main text, and Ψ and $\sigma_2 \Psi^*$ are the degenerate eigenfunctions, where

$$\sigma_2 \Psi^* = i\begin{pmatrix} -u_{\mathbf{k}}^{2*}(\mathbf{r}) \\ u_{\mathbf{k}}^{1*}(\mathbf{r}) \end{pmatrix} e^{-i\mathbf{k}.\mathbf{r}} \tag{G13.7}$$

and leads to the result we derived previously,

$$E(\mathbf{k}) = E(-\mathbf{k}). \tag{G13.8}$$

APPENDIX G14 DEGENERACY FOR A GENERAL k-VALUE

In the absence of spin–orbit coupling, there is double degeneracy for any general **k**-value, the eigenfunctions being

$$\begin{pmatrix} u_{\mathbf{k}} \\ 0 \end{pmatrix} e^{i\mathbf{k}.\mathbf{r}} \quad \text{and} \quad \begin{pmatrix} 0 \\ u_{\mathbf{k}} \end{pmatrix} e^{i\mathbf{k}.\mathbf{r}}. \tag{G14.1}$$

We shall now show that this double degeneracy still exists when spin–orbit coupling is present *provided the crystal has an inversion centre*:

$$V(\mathbf{r}) = V(-\mathbf{r}).$$

Writing

$$U_{\mathbf{k}}(\mathbf{r}) = \begin{pmatrix} u_{\mathbf{k}}^1(\mathbf{r}) \\ u_{\mathbf{k}}^2(\mathbf{r}) \end{pmatrix}, \tag{G14.2}$$

and the Hamiltonian of equation (G13.1) in the form

$$H = H_0 - i\mathbf{N}.\boldsymbol{\sigma}, \tag{G14.3}$$

one finds that

$$\left\{-\frac{\hbar^2}{2m}\nabla^2 - \frac{\hbar^2}{m}i\mathbf{k}.\nabla + V(\mathbf{r})\right\} U_{\mathbf{k}}(\mathbf{r}) - i\left\{i\left(\frac{\hbar}{2mc}\right)^2 \left(k_z\frac{\partial V}{\partial y} - k_y\frac{\partial V}{\partial z} + N_x\right)\right.$$

$$\left.\times \begin{pmatrix} 0 & 1 \\ 1 & 0 \end{pmatrix} + \ldots\right\} U_{\mathbf{k}}(\mathbf{r}) = \left(E - \frac{\hbar^2 k^2}{2m}\right) U_{\mathbf{k}}(\mathbf{r}). \tag{G14.4}$$

Hence, provided $V(\mathbf{r}) = V(-\mathbf{r})$,

$$U_{\mathbf{k}}(\mathbf{r}) = U_{-\mathbf{k}}(-\mathbf{r}), \tag{G14.5}$$

we have a solution of the form

$$\Psi'_{-\mathbf{k}}(\mathbf{r}) = \begin{pmatrix} u_{\mathbf{k}}^1(-\mathbf{r}) \\ u_{\mathbf{k}}^2(-\mathbf{r}) \end{pmatrix} e^{-i\mathbf{k}.\mathbf{r}}. \tag{G14.6}$$

For the singly degenerate case, this would be the same equation as (G13.7), apart from a possible phase factor $e^{i\theta}$. This is impossible, implying

$$u_{\mathbf{k}}^1(-\mathbf{r}) = -u_{\mathbf{k}}^{2*}(\mathbf{r})\, e^{i\theta} \tag{G14.7}$$

and

$$u_{\mathbf{k}}^2(-\mathbf{r}) = u_{\mathbf{k}}^{1*}(\mathbf{r})\, e^{i\theta}, \tag{G14.8}$$

that is,

$$u_{\mathbf{k}}^1(-\mathbf{r}) = u_{\mathbf{k}}^{2*}(\mathbf{r})\, e^{i\theta}, \tag{G14.9}$$

which is inconsistent with (G14.7) unless $\Psi'_{-\mathbf{k}} = 0$. This proof of the double degeneracy for each $\mathbf{k}$ holds only if there is an inversion centre, however.

APPENDIX G15 DEGENERACIES AT HIGH SYMMETRY POINTS

The transformation of a *spinor* (a two-component wave function such as (A14.2)) under rotation a is not effected by simply changing $\mathbf{r}$ to $a\mathbf{r}$; it is well known that a half-integral spin is formally a consequence of the change of sign of the spinor when the coordinate axes are rotated through 2π.

To find the operator $\mathscr{A}$ corresponding to the coordinate transformation a, we proceed as in the one-component case, seeking unitary transformations under which the Hamiltonian remains invariant.

(i) Transformation of spin–orbit coupling Hamiltonian

Let us make a unitary transformation a with associated functional operator A. We first see how **N**, defined by (G13.4), transforms. If $\mathbf{r}' = a\mathbf{r}$ and $\psi(\mathbf{r})$ is any function,

$$A\frac{\partial\psi}{\partial x} = \frac{\partial\psi}{\partial x'}(\mathbf{r}')\bigg|_{\mathbf{r}'=a\mathbf{r}} = \frac{\partial}{\partial(ax')}\psi(a\mathbf{r})\bigg|_{\mathbf{r}'=\mathbf{r}},$$

so that

$$A\frac{\partial}{\partial x}A^{\dagger} = \frac{\partial}{\partial(ax)}, \quad \text{etc.,} \tag{G15.1}$$

where

$$x_i' = \sum_j a_{ij}x_j. \tag{G15.2}$$

Thus

$$\frac{\partial}{\partial x_i'}V(\mathbf{r}')\big|_{a\mathbf{r}} = \frac{\partial}{\partial x_i}V(a\mathbf{r}) = \sum_j a_{ij}\frac{\partial}{\partial x_j}V(a\mathbf{r})$$

or

$$\frac{\partial}{\partial x_i'}V(\mathbf{r}')\big|_{a\mathbf{r}} = \sum_j a_{ij}\frac{\partial}{\partial x_j}V(\mathbf{r}) \tag{G15.3}$$

for the transformations a, A with which we are concerned, because $V(\mathbf{r})$ must remain invariant under them. Hence

$$\mathbf{N}' = A\mathbf{N}A^{\dagger} = A(\nabla V\times\mathbf{p})A^{\dagger}$$

$$= \sum_j\left(a_{1j}\frac{\partial V}{\partial x_j}\hat{x}+a_{2j}\frac{\partial V}{\partial x_j}\hat{y}+a_{3j}\frac{\partial V}{\partial x_j}\hat{z}\right)\times\sum_j\left(a_{1j}\frac{\partial}{\partial x_j}\hat{x}+a_{2j}\frac{\partial}{\partial x_j}\hat{y}+a_{3j}\frac{\partial}{\partial x_j}\hat{z}\right). \tag{G15.4}$$

Then, for example,

$$N_x' = N_1' = \sum_{jk} a_{2j}a_{3k}\left(\frac{\partial V}{\partial x_j}\frac{\partial}{\partial x_k}-\frac{\partial V}{\partial x_k}\frac{\partial}{\partial x_j}\right). \tag{G15.5}$$

Since the term for which $j = k$ is zero, this may be rewritten as

$$N_1' = \frac{1}{2}\sum_{j\neq k}(a_{2j}a_{3k}-a_{3j}a_{2k})\left(\frac{\partial V}{\partial x_j}\frac{\partial}{\partial x_k}-\frac{\partial V}{\partial x_k}\frac{\partial}{\partial x_j}\right) = \sum_j b^{1j}N_j, \tag{G15.6}$$

where b^{1i} is the co-factor of a_{1i}. We can generalize this result to

$$N'_i = \sum_j b^{ij} N_j. \qquad \text{(G15.7)}$$

We must now examine the transformation of the spin vector $\boldsymbol{\sigma}$. For *proper* rotations

$$\sigma'_i = \sum_j a_{ij}\sigma_j \quad (\det|a_{ij}| = +1). \qquad \text{(G15.8)}$$

In considering improper rotations, however, it must be remembered that $\boldsymbol{\sigma}$ is a polar vector representing angular momentum, and thus it remains unchanged on reflexion. Since an improper rotation is a proper rotation plus a reflexion, we therefore have

$$\sigma'_i = -\sum_j a_{ij}\sigma_j \quad (\det|a_{ij}| = -1). \qquad \text{(G15.9)}$$

Combining equations (G15.6) with equations (G15.7) and (G15.8) we therefore have

$$\begin{aligned}\mathbf{N}'.\boldsymbol{\sigma}' &= \sum_i N'_i \sigma'_i = \pm \sum_{ijk} b^{ij} a_{ik} N_j \sigma_k \\ &= \pm \sum_{ij} b^{ij} a_{ij} N_j \sigma_j \pm \sum_i \sum_{j\neq k}{}' b^{ij} a_{ik} N_j \sigma_k, \end{aligned} \qquad \text{(G15.10)}$$

the lower sign referring to improper rotations.

Now the unitary condition $a^\dagger a = 1$ implies that the columns of (a) form orthonormal vectors and that $a^{ij} = \pm a_{ji}$; hence

$$\sum_i b^{ij} a_{ik} = \pm \sum_i a_{ji} a_{ik} = 0 \quad (j=k). \qquad \text{(G15.11)}$$

Thus equation (G15.10) becomes

$$\mathbf{N}'.\boldsymbol{\sigma}' = \pm \sum_{ij} b^{ij} a_{ij} N_j \sigma_j = \sum_{ij} \pm \det|a_{ij}| N_j \sigma_j = \mathbf{N}.\boldsymbol{\sigma}. \qquad \text{(G15.12)}$$

It follows that the total Hamiltonian $H = H_0 - i\mathbf{N}.\boldsymbol{\sigma}$ is invariant for any unitary transformation in which $V(\mathbf{r})$ is unchanged. It remains now to find the operator $\mathscr{A}$ such that

$$\pm \mathscr{A}^\dagger \mathbf{N}.\boldsymbol{\sigma}\mathscr{A} = 1. \qquad \text{(G15.13)}$$

The part of $\mathscr{A}$ acting on functions of $\mathbf{r}$ is of course A, but we must also find the operator acting on $\boldsymbol{\sigma}$. From the above discussion it can be seen that we need only consider proper rotations. We therefore seek a unitary 2×2 matrix P such that

$$\sum_j a_{ij}\sigma_j = P^\dagger \sigma_i P. \qquad \text{(G15.14)}$$

We first note that it is possible to write any unitary 2×2 matrix in the form

$$P = \begin{pmatrix} a & b \\ -b^* e^{i\theta} & a^* e^{i\theta} \end{pmatrix}, \quad \begin{matrix} |a|^2 + |b|^2 = 1, \\ \det|P| = e^{i\theta}. \end{matrix} \qquad \text{(G15.15)}$$

The solutions of (G15.14) are then given by

$$\left.\begin{aligned} a^2 &= e^{i\theta}\left[\tfrac{1}{2}(a_{11}+a_{12})+i\tfrac{1}{2}(a_{12}-a_{21})\right],\\ b^2 &= e^{i\theta}\left[\tfrac{1}{2}(a_{22}-a_{11})+i\tfrac{1}{2}(a_{12}+a_{21})\right],\\ ab &= e^{i\theta}\,\tfrac{1}{2}[a_{13}-ia_{23}], \quad |a|^2-|b|^2 = a_{33}. \end{aligned}\right\} \tag{G15.16}$$

(a_{jk}) must represent a clockwise rotation through an angle ϕ about some axis. Writing the direction cosines of this axis as (l, m, n), equation (G15.16) results in the equations

$$\left.\begin{aligned} a &= \pm(\cos\tfrac{1}{2}\phi - in\sin\tfrac{1}{2}\phi)\,e^{i\theta},\\ b &= \pm(m+il)\sin\tfrac{1}{2}\phi\, e^{i\theta}. \end{aligned}\right\} \tag{G15.17}$$

It is clear that for any given θ, P has two solutions: $P = \pm Q$, and for either choice we can take

$$\mathscr{A} = \mathbf{A}\times\mathbf{P}. \tag{G15.18}$$

It is also clear that, θ being the same for both choices, its value is immaterial. We conveniently fix it at 2π for all transformations so that the totality of transformations $\mathscr{A}$ form a group. This is called the *double group* since the two possible choices of P make $\{\mathscr{A}\}$ twice the order of (A).

(ii) Matrix representations

Let us write the spin functions as

$$|1\rangle = \begin{pmatrix}1\\0\end{pmatrix}, \quad |2\rangle = \begin{pmatrix}0\\1\end{pmatrix}.$$

Then any solution of H can be written in the form $\psi_1(\mathbf{r})|1\rangle+\psi_2(\mathbf{r})|2\rangle$; in other words, every solution is a linear combination of the set S formed by all possible products of a complete set of functions $\{\psi_i(\mathbf{r})\}$ with $|1\rangle$ and $|2\rangle$. Hence the result of Appendix (1.3) means that S must realize all invariant matrix representations of $\{\mathscr{A}\}$. We can therefore obtain all inequivalent matrix representations of $\{\mathscr{A}\}$ in the following way. We take a set $\psi_1(\mathbf{r}_1), \ldots, \psi_n(\mathbf{r}_N)$ of functions realizing the n-dimensional irreducible representation (A) of A. We multiply these by $|1\rangle$ and $|2\rangle$, leading to the representation

$$P = \begin{pmatrix} a & b\\ c & d\end{pmatrix} = \begin{pmatrix}\langle 1|P|1\rangle & \langle 1|P|2\rangle\\ \langle 2|P|1\rangle & \langle 2|P|2\rangle\end{pmatrix}. \tag{G15.19}$$

Then the set $\psi_1(\mathbf{r})|1\rangle, \ldots, \psi_n(\mathbf{r})|1\rangle, \psi_1(\mathbf{r})|2\rangle, \ldots, \psi_n(\mathbf{r})|2\rangle$ will give the matrix representation,

$$\begin{pmatrix} a(A) & b(A)\\ c(A) & d(A)\end{pmatrix}. \tag{G15.20}$$

It is possible that this matrix is reducible. Thus, for example, suppose (A) is three-dimensional, and (G15.20) is reducible to a 4×4 matrix and a 2×2 matrix. Then the sixfold degenerate state existing in the absence of spin–orbit coupling will be split by its introduction into a fourfold degenerate state and a twofold degenerate one. Further details of the matrix representations of double groups will be given later in this appendix.

APPENDIX G16 TIME-REVERSAL SYMMETRY IN ABSENCE OF SPIN EFFECTS

If spin-interaction effects can be ignored, the time-reversal operator T gives us just the complex conjugate of a wave function: $T\psi^* = \psi$, and in Chapter 1 we saw that we could take

$$\psi_{-\mathbf{k}}(\mathbf{r}) = \psi_{\mathbf{k}}^*(\mathbf{r}), \tag{G16.1}$$

which we may regard as a consequence of time-reversal symmetry. In this section we shall discuss this effect further, assuming spin–orbit coupling to be negligible.

(i) Prediction of time-reversal degeneracy from character tables

For every representation $\Gamma\equiv\{\Gamma(\alpha)\}$ of a group G there exists a second representation simply formed by taking the complex conjugates $\Gamma^*(\alpha)$ of all the matrices. It is evident that if the set of wave functions ψ transform according to Γ, then the complex conjugates ψ^* transform according to Γ^*, and so information on the effect of time reversal is obtainable by examination of the relationship of Γ^* to Γ. (To avoid possible confusion, we should perhaps remark here that in general we cannot identify the group G with $G_s(\mathbf{k})$, since $\psi_{\mathbf{k}}^* = \psi_{-\mathbf{k}}$. The group we are considering can be thought of as the full space group, if the reader wishes to make a particular identification with an element of the content of previous sections.)

Time-reversal degeneracy exists if the ψ^*'s are linearly independent of the ψ's. However, if Γ can be transformed to real form, the ψ's spanning the irreducible manifold realizing Γ can be taken as real, so that time-reversal introduces no additional degeneracy. If Γ can be transformed to real form, Γ^* is, of course, an equivalent representation. If, on the other hand Γ^* is inequivalent, the degeneracy is doubled. A third possibility exists where Γ and Γ^* are equivalent but cannot be transformed to real form. The ψ^*'s cannot be then linearly dependent on the ψ's for if they were we could combine the two sets into real wave functions giving a real representation equivalent to Γ, as one can readily show.

In summary, we have three possibilities:

(a) Γ can be transformed to real form: time-reversal introduces no additional degeneracy.

(b) Γ and Γ* are inequivalent: time-reversal doubles the degeneracy predicted from spatial symmetry.

(c) Γ and Γ* are equivalent but cannot be transformed to real form: time-reversal again doubles the degeneracy.

Wigner (1932) has shown that the characters of Γ suffice to tell us which of these three possibilities it realizes. His test takes the form

$$\sum_A \chi(A^2) = \begin{cases} g, & \text{case } (a), \\ 0, & \text{case } (b), \\ -g, & \text{case } (c), \end{cases} \tag{G16.2}$$

the summation being over all elements A of G.

We shall not prove this result as it is a consequence of general representation theory. We wish instead to consider how the test can be adapted to our particular problem.

(ii) Test for time-reversal degeneracy at particular k-values

Herring's test for time-reversal degeneracy at a particular **k**-value takes the form

$$\sum_A \chi(A^2) = \begin{cases} h, & \text{case (a)}, \\ 0, & \text{case (b)}, \\ -h, & \text{case (c)}. \end{cases} \tag{G16.3}$$

The characters χ are those of an irreducible representation of $G_s(\mathbf{k})$, *not* the full space group. When $-\mathbf{k}$ is equivalent to $\mathbf{k}$, the summation is over all h elements of $G_s(\mathbf{k})$. In general, however, the summation is over all h elements A which transform $\psi_{\mathbf{k}}$ to $\psi_{-\mathbf{k}}$; A^2 is an element of $G_s(\mathbf{k})$ and so there is no inconsistency in saying the χ's are those for a representation of $G_s(\mathbf{k})$.

We shall now show how the test of Herring (1940) follows from Wigner's. Consider an irreducible representation $\Gamma(G_s)$ of the full group which yields the irreducible representation Γ of $G_s(\mathbf{k})$. It is not difficult to show that if there are n inequivalent vectors $\mathbf{k}_1, \ldots, \mathbf{k}_n$ in the star of $\mathbf{k}$, the order of the point group of $\mathbf{k}$, $G_p(\mathbf{k})$, is g_p/n, where g_p is the order of the full point group of the crystal, and, further, that the dimensionality of $\Gamma(G_s)$ is $nd(\mathbf{k})$ where $d(\mathbf{k})$ is the dimension of Γ. The irreducible manifold is spanned by Bloch functions, with $\mathbf{k}$-vectors $\mathbf{k}_1, \ldots, \mathbf{k}_n$, and so the representation of $(E|\mathbf{R})$

must be of the form

$$\begin{pmatrix} (e^{i\mathbf{k}_1\cdot\mathbf{R}}) & \cdots & 0 \\ 0 & (e^{i\mathbf{k}_2\cdot\mathbf{R}}) & 0 \\ \cdot & \cdot\ \cdot\ \cdot\ \cdot & \cdot \\ 0 & \cdots & (e^{i\mathbf{k}_n\cdot\mathbf{R}}) \end{pmatrix}. \tag{G16.4}$$

Here $(e^{i\mathbf{k}_i\cdot\mathbf{R}})$ is a diagonal sub-matrix of order $d(\mathbf{k})$. Now the point group of $-\mathbf{k}$ is the same as that of $\mathbf{k}$, since if $\mathbf{k}=\mathbf{k}+\mathbf{K}$, $-\mathbf{k}=-\mathbf{k}-\mathbf{K}$, and both $\mathbf{K}$ and $-\mathbf{K}$ are reciprocal lattice vectors. Thus the matrix corresponding to (A13.4) in an irreducible representation of the full group yielding an irreducible representation of $G_s(-\mathbf{k})$ is

$$\begin{pmatrix} (e^{-i\mathbf{k}_1\cdot\mathbf{R}}) & 0 \\ \cdot\ \cdot\ \cdot & \cdot\ \cdot\ \cdot \\ 0 & (e^{-i\mathbf{k}_n\cdot\mathbf{R}}) \end{pmatrix}. \tag{G16.5}$$

Now suppose there is *no* operator of the point group turning $\mathbf{k}$ into $-\mathbf{k}$, so that $-\mathbf{k}$ does not belong to the star of $\mathbf{k}$. Then no $\mathbf{k}_i=-\mathbf{k}_j$ for any i and j and (G16.4) cannot be equal to (G16.5). However, the set $(\psi_{-\mathbf{k}_i})$ transforms as $\Gamma^*(G_s)$, so that $\Gamma^*(G_s)$ and $\Gamma(G_s)$ are inequivalent. Case (b) obtains, and Herring's test is obviously true here: there are *no* elements A, and so the sum in (G16.3) is zero.

We next examine the situation where there are elements of the group transforming $\mathbf{k}$ to $-\mathbf{k}$, and so elements of the space group transforming $\psi_{\mathbf{k}}$ to $\psi_{-\mathbf{k}}$. We shall not attempt to set up representations for the full space group, because since $\psi_{\mathbf{k}}^*=\psi_{-\mathbf{k}}$ it suffices to examine representations of the group of operators which turn $\mathbf{k}$ into an equivalent vector or vector equivalent to $-\mathbf{k}$. This group we denote by $G_s^{\pm *}(\mathbf{k})$. The representation is spanned by the degenerate set $\psi_{\mathbf{k}}^1, \ldots, \psi_{\mathbf{k}}^n$ and the equally degenerate set $\psi_{-\mathbf{k}}^1, \ldots, \psi_{-\mathbf{k}}^n$. Let us take any operator J, of the space group, which turns $\mathbf{k}$ into $-\mathbf{k}$ and define

$$\psi_{-\mathbf{k}}^i = J\psi_{\mathbf{k}}^i. \tag{G16.6}$$

Then for any element α of $G_s(\mathbf{k})$,

$$\langle \psi_{-\mathbf{k}}^{i*}\,\alpha\psi_{-\mathbf{k}}\rangle = \langle \psi_{\mathbf{k}}^{i*}(J^\dagger\,\alpha J)\,\psi_{\mathbf{k}}\rangle. \tag{G16.7}$$

Now $J^{-1}\alpha J$ is an element of $G_s(\mathbf{k})$, and so if the matrix representation of α is $\Gamma(\alpha)$ for $G_s(\mathbf{k})$, the corresponding representation of α for $G_s(-\mathbf{k})$ is $\Gamma'(\alpha)=\Gamma(J^{-1}\alpha J)$. The set of matrices Γ' is obviously irreducible because it is the same set as before, and the multiplication table is correctly given because

$$\begin{aligned}\Gamma'(\alpha)\,\Gamma'(\beta) &= \Gamma(J^{-1}\alpha J)\,\Gamma(J^{-1}\beta J) \\ &= \Gamma(J^{-1}\alpha\beta J) = \Gamma'(\alpha\beta).\end{aligned} \tag{G16.8}$$

Thus we reach the conclusion that the irreducible representation of $G_s^{\pm}(\mathbf{k})$ has as its matrices representing the elements of $G_s(\mathbf{k})$,

$$\Gamma_2(\alpha) = \begin{pmatrix} \Gamma(\alpha) & 0 \\ 0 & \Gamma(J^{-1}\alpha J) \end{pmatrix}. \tag{G16.9}$$

As for the other elements of $G_s^{\pm}(\mathbf{k})$, which we shall denote by capital letters, one may verify that these are represented by

$$\Gamma_2(A) = \begin{pmatrix} 0 & \Gamma(AJ) \\ \Gamma(J^{-1}A) & 0 \end{pmatrix}. \tag{G16.10}$$

The set of matrices defined in (G16.9) and (G16.10) is irreducible, for

$$\sum_\alpha |\chi_2(\alpha)|^2 + \sum_A |\chi_2(A)|^2 = \sum_\alpha |\chi_2(\alpha)|^2 = 2\sum_\alpha |\chi(\alpha)|^2 = 2g. \tag{G16.11}$$

g is the order of $G_s(\mathbf{k})$ and so $G_s^{\pm}(\mathbf{k})$ is of order $2g$; $\chi(\alpha)$ represents the trace of $\Gamma(\alpha)$. We have also used the result

$$\sum_\alpha \chi^*(\alpha)\chi(J^{-1}\alpha J) = 0, \tag{G16.12}$$

which follows from the fact that Γ and Γ' are inequivalent representations, the former corresponding to $+\mathbf{k}$ and the latter to $-\mathbf{k}$. The matrices representing A [equation (G16.10)] are unique given that those representing α are given by (G16.9) because the characters of a representation specify it uniquely to within an equivalence transformation. We now apply Wigner's test to our group $G_s^{\pm}(\mathbf{k})$. We obtain for the left-hand side of equation (G16.6)

$$\sum_\alpha \chi(\alpha^2) + \sum_\alpha \chi(J^{-1}\alpha J J^{-1}\alpha J) + \sum_A \chi(A^2) + \sum_A \chi(J^{-1}A^2 J)$$

$$= 2\sum_\alpha \chi(\alpha^2) + 2\sum_A \chi(A^2). \tag{G16.13}$$

We know that since $\psi_{-\mathbf{k}} = \psi_{\mathbf{k}}^*$ transforms as Γ^*, Γ and Γ^* are inequivalent, and case (b) obtains for $G_s(\mathbf{k})$:

$$\sum_\alpha \chi(\alpha^2) = 0. \tag{G16.14}$$

We know also that the order of the group $G_s^{\pm}(\mathbf{k})$ is $2g$, where g is the order of $G_s(\mathbf{k})$, and so also the number of elements A, which turn $\psi_{\mathbf{k}}$ into $\psi_{-\mathbf{k}}$.

(iii) Degeneracy over entire BZ face

Let us take as an example of the foregoing any point on the hexagonal face of the BZ of the h.c.p. structure. There are only two elements in $S_p(k)$, which we can take as $(E|0)$ and $(JC_6^3|0)$. There is no degeneracy predicted from spatial symmetry, the character table being Table G16.1.

Now the h.c.p. structure has as a symmetry element

$$(J|\boldsymbol{\tau}) \quad (\boldsymbol{\tau} = \tfrac{2}{3}\mathbf{R}_1+\tfrac{1}{3}\mathbf{R}_2+\tfrac{1}{2}\mathbf{R}_3),$$

where J is the inversion. In such structures the choice of this element to

TABLE G16.1

	$(E\|0)$	$(JC_6^3\|0)$
	1	1
	1	−1

generate the A's of equation (G16.10) from those of $G_s(\mathbf{k})$ is an obvious one. These elements are

$$\left.\begin{aligned}(J|\boldsymbol{\tau})(E|-\mathbf{R}) &= (J|\boldsymbol{\tau}+\mathbf{R}) \quad (\text{all } \mathbf{R}),\\ (J|\boldsymbol{\tau})(JC_6^3|-\mathbf{R}) &= (C_6^3|\boldsymbol{\tau}+\mathbf{R}) \quad (\text{all } \mathbf{R}).\end{aligned}\right\} \tag{G16.15}$$

Now

$$\begin{aligned}(J|\boldsymbol{\tau}+\mathbf{R})(J|\boldsymbol{\tau}+\mathbf{R}) &= (E|0),\\ (C_6^3|\boldsymbol{\tau}+\mathbf{R})(C_6^3|\boldsymbol{\tau}+\mathbf{R}) &= (E|(2n+1)\mathbf{R}_3) \quad (\mathbf{R} = l\mathbf{R}_1+m\mathbf{R}_1+n\mathbf{R}_3).\end{aligned} \tag{G16.16}$$

Thus we obtain, applying Herring's test:

$$\sum_A \chi(A^2) = \sum_{\mathbf{R}} (E|0)+\sum_{\mathbf{R}} (E|0)\, e^{i\mathbf{k}(2n+1)\mathbf{R}_3}. \tag{G16.17}$$

Now $e^{i\mathbf{k}.\mathbf{R}_3} = -1$, so that $e^{i\mathbf{k}(2n+1).\mathbf{R}_3} = -1$. Hence

$$\sum_A \chi(A^2) = 0. \tag{G16.18}$$

Case (b) obtains, and there is essential degneracy because of time-reversal symmetry.

This result is obtainable in a quite different way. We shall now prove that time-reversal doubles the degeneracy over the entire face of a BZ, provided this face has a twofold screw axis perpendicular to it. In Figure G24 F

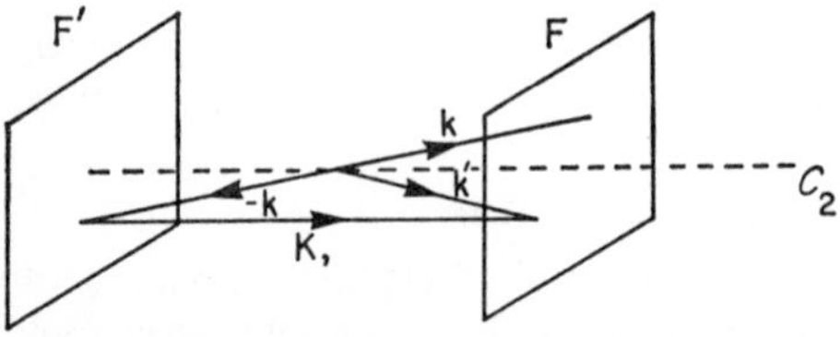

FIGURE G24. Brillouin zone for hexagonal close-packed lattice showing face F perpendicular to two-fold axis C_2.

represents a face perpendicular to the twofold screw axis C_2 and $\mathbf{k}$ is any vector to this face. The vector $-\mathbf{k}$ reaches a face F' normally connected to F by a reciprocal lattice vector $\mathbf{K}$. It is evident from the figure that

$$\mathbf{k}' = C_2\mathbf{k} = \mathbf{K}-\mathbf{k}. \tag{G16.19}$$

Using equation (G16.1) and the periodicity of the wave function in $\mathbf{K}$, we may also write

$$\psi_{\mathbf{k}'}(\mathbf{r}) = \psi_{\mathbf{k}'-\mathbf{K}}(\mathbf{r}) = \psi_{-\mathbf{k}}(\mathbf{r}) = \psi_{\mathbf{k}}^*(\mathbf{r}) \tag{G16.20}$$

and so

$$(C_2|\boldsymbol{\tau})\psi_{\mathbf{k}}(\mathbf{r}) = \psi'_{\mathbf{k}}(\mathbf{r}) = \psi_{\mathbf{k}}(C_2\mathbf{r}+\boldsymbol{\tau}). \tag{G16.21}$$

If we perform exactly the same operations again, we get

$$(C_2|\boldsymbol{\tau})\{(C_2|\boldsymbol{\tau})\psi_{\mathbf{k}}^*(\mathbf{r})\}^* = \psi_{\mathbf{k}}(\mathbf{r}+2\boldsymbol{\tau}) = \psi(\mathbf{r}+\mathbf{R}_3) = e^{i\mathbf{k}.\mathbf{R}_3}\psi_{\mathbf{k}}(\mathbf{r}). \tag{G16.22}$$

Using this result we can show that $\psi'_{\mathbf{k}}(\mathbf{r})$ and $\psi_{\mathbf{k}}(\mathbf{r})$, although of the same energy, are linearly independent. For suppose such were not the case. We could then write $\psi'_{\mathbf{k}}(\mathbf{r}) = \chi\psi_{\mathbf{k}}(\mathbf{r})$, with χ some constant, so that

$$\begin{aligned}(C_2|\boldsymbol{\tau})\{(C_2|\boldsymbol{\tau})\psi_{\mathbf{k}}^*(\mathbf{r})\}^* &= (C_2|\boldsymbol{\tau})\psi_{\mathbf{k}}^*(\mathbf{r}) \\ &= \chi^*(C_2|0)\psi_{\mathbf{k}}^*(\mathbf{r}) = |\chi|^2\psi_{\mathbf{k}}(\mathbf{r}).\end{aligned} \tag{G16.23}$$

From (G16.23) we would have

$$1 = |\chi|^2 = e^{i\mathbf{k}.\mathbf{R}_3} \quad (\mathbf{R}_3 = 2\boldsymbol{\tau}), \tag{G16.24}$$

which is impossible, because $\mathbf{R}_3$ must be one of the three shortest lattice vectors.

APPENDIX G17 INCLUSION OF SPIN–ORBIT COUPLING

We now use the two-component formalism

$$\Psi_{\mathbf{k}} = \begin{pmatrix}\psi_{\mathbf{k}}^1 \\ \psi_{\mathbf{k}}^2\end{pmatrix}, \tag{G17.1}$$

which we introduced earlier in this appendix. We saw that, when using this formalism, the action of the time-reversal operator T could be represented as

$$T\Psi = \sigma_2\Psi^*, \quad \sigma_2 = \begin{pmatrix}0 & -i \\ i & 0\end{pmatrix}. \tag{G17.2}$$

We must discuss the three possibilities of section G16 anew, since Ψ transforms according to the *double group*, also introduced in section G15. The possibilities are

(a) The representations Γ of the double group can be made real.

(b) The representations Γ and Γ^* are inequivalent.

(c) The representations Γ and Γ^* are equivalent, but cannot be made real.

Let us first note that

$$T(T\Psi) = T(\sigma_2 \Psi^*) = \sigma_2 \sigma_2^* \Psi = -\Psi,$$

that is,

$$T^2 \Psi = -\Psi \tag{G17.3}$$

and secondly, if for a member A of the double group

$$A\Psi_i(\mathbf{r}) = \sum_j a_{ij} \Psi_j(\mathbf{r}), \tag{G17.4}$$

$$AT\Psi_i = TA\Psi_i = T\sum_j a_{ij} \Psi_j = \sum_j a_{ij}^* T\Psi_j, \tag{G17.5}$$

so that if the set $\{\Psi_i\}$ transforms as Γ, the set $\{T\Psi_i\}$ transforms as Γ^*. (That A and T commute may be assumed from the fact that both are unitary operators commuting with the Hamiltonian, which may also be verified using the explicit forms of A and T appearing in section G15.)

We can now consider the possibilities (a), (b) and (c) above.

(a) If the $T\Psi_i$ were linearly dependent on the Ψ_i, $T\Psi_i = \sum_j c_{ij} \Psi_j$, the matrix (c_{ij}) would transform Γ to Γ^*: $\Gamma^* = (c)\,\Gamma(c^\dagger)$, i.e. $\Gamma^*(c) = (c)\,\Gamma$. Suppose now we have already chosen the Ψ_i to make Γ real. Then $\Gamma(c) = (c)\,\Gamma$. Since the only matrices commuting with all matrices of an irreducible representation are scalar, we would obtain $T\Psi_i = c\Psi_i$. However, applying equation (G17.2) we see that this leads to

$$-\Psi_i = T^2 \Psi_i = c^* T\Psi_i = |c|^2 \Psi_i, \quad \text{a contradiction.}$$

In case (a), therefore, time-reversal doubles the degeneracy.

(b) It is immediately evident that there must be double degeneracy for this case also.

(c) Here we make use of the theorem of matrix algebra that if $\Gamma^* = (c)\,\Gamma(c^\dagger)$, but Γ cannot be transformed to real form, (c) is *skew symmetric*:‡ $c_{ij} = -c_{ji}$. Let us now take the functions $\Phi_i = \sum_j c_{ij} \Psi_j$ which must transform as Γ^* if the Ψ_j transform as Γ.

‡ We prove this result in section G17(i).

We have

$$\Phi_i = \sum_j c_{ij} \Psi_j, \tag{G17.6}$$

so that

$$T\Phi_i = \sum_j c_{ij}^* T\Psi_j \tag{G17.7}$$

and

$$\Psi_i = \sum_j c_{ij}^* \Phi_j, \tag{G17.8}$$

where

$$c_{ij} = -c_{ji}. \tag{G17.9}$$

Let us now take the set of functions $\{T\Psi_i + \Phi_i\}$. These, of course, transform as Γ^*. Now, in addition,

$$T(T\Psi_i + \Phi_i) = T\Phi_i - \Psi_i, \tag{G17.10}$$

by equation (G17.3). Using equations (G17.7) and (G17.8) we therefore have

$$T(T\Psi_i + \Phi_i) = \sum_j (c_{ij}^* T\Psi_j - c_{ji}^* \Phi_j) = \sum_j c_{ij}^* (T\Psi_j + \Phi_j). \tag{G17.11}$$

The set of functions $T\Psi_i + \Phi_i$, therefore, not only transform among themselves as Γ^*, but transform among themselves under the action of T. The result is that, given Γ and Γ^* equivalent but inequivalent to any real form, one can always choose a set of functions, to realize the representation, which transform among themselves under the action of time-reversal, and time-reversal degeneracy is not predicted.

(i) Test for time-reversal degeneracy

The proof of Herring's test, it will be observed, does not depend on any assumptions as to the nature of the inclusion of spin: it is primarily a test of the nature of the representations of the group, independent of the degeneracy conditions.

Thus we may repeat it unchanged:

$$\sum_A \chi(A^2) = \begin{cases} +h, & \text{case (a)}, \\ 0, & \text{case (b)}, \\ -h, & \text{case (c)}. \end{cases} \tag{G17.12}$$

However, when applied to double groups the deductions concerning degeneracy differ from those of G16. We now have (a) and (b): Time-reversal doubles the degeneracy. (c): There is no additional degeneracy due to time-reversal.

We shall give character tables for the double groups of our explicit examples of cubic and hexagonal structures in section G19. Before we go on to look at specific double groups, however, we should like to amplify the outline of the reasoning leading to the conclusion that time-reversal degeneracy is absent in case (c). The exact reversal of the conclusions of section G16 for cases (a) and (c) when double groups are considered is so remarkable that it seems a pity to base our reasoning on a statement, without proof, of the result that, if Γ cannot be transformed to real form, and

$$\Gamma^* = (c)\,\Gamma(c^\dagger), \tag{G17.13}$$

c is skew symmetric. We shall therefore now prove this result, keeping close to quantum-mechanical formalism.

In the following, all functions will be considered as general ones to which the time-reversal operator T is applicable. In addition, we shall denote the simple operation of complex conjugation by T':

$$T'\psi = \psi^*. \tag{G17.14}$$

Let us now take case (c), where we know that the sets $\{\psi_i\}$ and $\{T'\psi_i\} = \{\psi_i^*\}$ (transforming as Γ and Γ^* respectively) are orthogonal, from section G17. Let us also make the transformation

$$\phi_i = \sum_j c_{ij}\,\psi_j, \tag{G17.15}$$

where ϕ_i transforms as Γ^*. Then the set $\{\chi_i\}=\{\psi_i^*+\phi_i\}$ transforms as Γ^* also. The set $\{T'\chi_i\}$ must also be orthogonal to the set $\{\chi_i\}$, for exactly the same reason as $\{T'\psi_i\}$ is orthogonal to $\{\psi_i\}$. Now from (G17.14)

$$\psi_i = \sum_j c_{ji}^*\,\phi_j, \tag{G17.16}$$

whence

$$T'\chi_i = \phi_i^* + \psi_i = \sum_j (c_{ij}^*\,\psi_j^* + c_{ji}^*\,\phi_j). \tag{G17.17}$$

Further

$$0 = \langle \chi_j^*\, T'\chi_i\rangle = \langle (\psi_j + \phi_j^*)\, T'\chi_i\rangle$$
$$= c_{ij}^* + c_{ji}^*,$$

that is,

$$c_{ij} + c_{ji} = 0. \tag{G17.18}$$

The essential difference between the operators T' and T is that

$$T'^2\psi = \psi \quad \text{but} \quad T^2\psi = -\psi. \tag{G17.19}$$

APPENDIX G18 CONSTRUCTION OF DOUBLE GROUPS

We wish now to discuss rules for setting up the irreducible representations of the double groups introduced in section G17.

We will consider a group of proper rotations. Every other point group is either isomorphic to such a group or a direct product of the group with the inversion.

First we note that because of the duality of sign in equation (G15.17) every member A of the point group splits into two, $\mathscr{A}$ and $\bar{\mathscr{A}}$, and a class c splits into two classes $\mathscr{C}$ and $\bar{\mathscr{C}}$. It is clear that all the representations of the point group form the representations of the double group, where $\mathscr{C}$ and $\bar{\mathscr{C}}$ are given the same representations. However, we are not here interested in these, but require the additional representations which come from the reduction, if possible, of matrices of equation (G15.20). From this equation we see that

$$\chi(\mathscr{A}) = -\chi(\bar{\mathscr{A}}), \tag{G18.1}$$

so that $\mathscr{A}$ and $\bar{\mathscr{A}}$ can belong to the same class only if

$$\chi(\mathscr{A}) = 0. \tag{G18.2}$$

This is possible if

$$a = -d \tag{G18.3}$$

in equation (G15.20). From equation (G15.15) and (G15.17) we see that, taking the rotation about the two-axis for convenience, we must have $\phi = 180°$, when the matrix P of equation (G15.15) becomes [putting $\theta = 0$ in equation (G15.17)]

$$P = \begin{pmatrix} -i & 0 \\ 0 & i \end{pmatrix}. \tag{G18.4}$$

Now $-P$, the matrix taken to form $\bar{A}$, also corresponds to the matrix P for a rotation of 180° about $-Z$. Hence $\bar{A}$ and A *will* belong to the same class if there exists a rotation taking Z to $-Z$, but only then. To be more explicit; we can see from equation (G15.18) that we must have a transformation matrix $\mathscr{B} = P_1 \times B$ such that

$$\bar{\mathscr{A}} = \mathscr{B}^\dagger A \mathscr{B}, \tag{G18.5}$$

where

$$A = B^\dagger AB \tag{G18.6}$$

and

$$-P = P_1^\dagger P P_1. \tag{G18.7}$$

From equation (G15.15) the matrix P must then pertain to a rotation of 180° about y [so that in equation (G15.17) $l = n = 0$, $m = \pm 1$]. Such a rotation on the coordinates (x,y,z) gives $B(x,y,z)=(-x,y,-z)$, whereas $A(x,y,z)=(-x,-y,z)$. Hence

$$B^{\dagger}AB(x,y,z)=B^{\dagger}A(-x,y,-z)=B^{\dagger}(x,-y,-z)$$

$$=(-x,-y,z)=A(x,y,z) \tag{G18.8}$$

and so equation (G18.6) does indeed hold.

To complete our examination of the structure of the double group we note from Appendix G15 that the inversion J is represented in the double group as a scalar matrix, diagonal elements ± 1, and $\bar{J}=-J$.

Rules for construction of representations of double groups

From the foregoing we may summarize the following rules.

Every point group consists of proper rotations only, or consists of the product of proper rotations (excluding the identity) with the group of the inversion and identity, or is isomorphic to a group of proper rotations, becoming the latter group if we replace every improper rotation by the corresponding proper one. We therefore examine the group of proper rotations. We note that representations of the double group are obtainable from that of the space group by assigning the matrix representing A to both $\mathscr{A}$ and $\bar{\mathscr{A}}$. The characters for the extra representations, which are the ones of significance here, may be obtained by methods similar to those presented in earlier sections aided by the following rules:

(i) $\mathscr{A}$ and $\bar{\mathscr{A}}$ have for proper rotations in any of the extra representations, characters

$$\chi(\mathscr{A})=-\chi(\mathscr{A}) \tag{G18.9}$$

and, in particular,

(ii) if the rotation is through 180°,

$$\chi(\mathscr{A})=0, \tag{G18.10}$$

when it is possible that $\mathscr{A}$ and $\bar{\mathscr{A}}$ belong to the same class, the condition being

(iii) $\mathscr{A}$ and $\bar{\mathscr{A}}$ belong to the same class if and only if the corresponding proper rotation is through 180° and there exists another proper rotation, in the group, through 180° about an axis perpendicular to the original one.

(iv) The inversion is represented by a scalar matrix with diagonal elements ± 1 and $\bar{J}=-J$.

(v) It follows that, for reflexions, $\mathscr{A}$ and $\bar{\mathscr{A}}$ are in the same class only if there is a reflexion plane perpendicular to the given plane or a rotation through 180° about an axis in the plane.

APPENDIX G19 CHARACTER TABLES OF DOUBLE GROUPS FOR CUBIC AND HEXAGONAL LATTICES

We complete our examination of the s.c., b.c.c., f.c.c., diamond and h.c.p. structures by presenting the relevant tables of additional representations of the double groups (Koster, 1957).

(i) Double groups of symmorphic cubic structures

The following are tables of the extra representations of double groups associated with the point groups of high symmetry points in the s.c., b.c.c. and f.c.c. structures.

TABLE G19.1

Γ, H, R	E	$\bar{E}$	$8C_3$	$8\bar{C}_3$	$(3C_4^2, 3\bar{C}_4^2)$	$6C_4$	$6\bar{C}_4$	$(6C_2, 6\bar{C}_2)$
Γ_{11}^+	2	−2	1	−1	0	$\sqrt{2}$	$-\sqrt{2}$	0
Γ_{12}^+	2	−2	1	−1	0	$-\sqrt{2}$	$\sqrt{2}$	0
Γ_{13}^+	4	−4	−1	1	0	0	0	0
Γ_{11}^-	2	−2	1	−1	0	$\sqrt{2}$	$-\sqrt{2}$	0
Γ_{12}^-	2	−2	1	−1	0	$-\sqrt{2}$	$\sqrt{2}$	0
Γ_{13}^-	4	−4	−1	1	0	0	0	0

Γ, H, R	J	$\bar{J}$	$8JC_3$	$8\overline{JC}_3$	$(3JC_4^2, 3\overline{JC}_4^2)$	$6JC_4$	$6\overline{JC}_4$	$(6JC_2, 6\overline{JC}_2)$
Γ_{11}^+	2	−2	1	−1	0	2	−2	0
Γ_{12}^+	2	−2	1	−1	0	−2	2	0
Γ_{13}^+	4	−4	−1	1	0	0	0	0
Γ_{11}^-	−2	2	−1	1	0	−2	2	0
Γ_{12}^-	−2	2	−1	1	0	2	−2	0
Γ_{13}^-	−4	4	1	−1	0	0	0	0

TABLE G19.2

Δ, T	E	$\bar{E}$	$2C_4$	$2\bar{C}_4$	$(C_2, \bar{C}_2)$	$(2JC_4^2, 2\overline{JC}_4^2)$	$(2JC_2, 2\overline{JC}_2)$
Δ_6	2	−2	$\sqrt{2}$	$-\sqrt{2}$	0	0	0
Δ	2	−2	$-\sqrt{2}$	$\sqrt{2}$	0	0	0

TABLE G19.3

Λ, F	E	$\bar{E}$	$2C_3$	$2\bar{C}_3$	$3JC_2$	$3\overline{JC}_2$
Λ_4	1	-1	-1	1	i	$-i$
Λ_5	1	-1	-1	1	$-i$	i
Λ_6	2	-2	1	-1	0	0

TABLE G19.4

D	E	$\bar{E}$	C_4^2	$\bar{C}_4^2$	$(JC_2, \overline{JC}_2)$	$(JC_2', \overline{JC}_2')$
Z	E	$\bar{E}$	C_4^2	$\bar{C}_4^2$	$(JC_4^2, \overline{JC}_4^2)$	$(JC_4^{2\prime}, \overline{JC}_4^{2\prime})$
Σ, G, S	E	$\bar{E}$	C_2	$\bar{C}_2$	$(JC_4^2, \overline{JC}_4^2)$	$(JC_2, \overline{JC}_2)$
Σ_5	2	-2	0	0	0	0

TABLE G19.5

Q	E	$\bar{E}$	C_2	$\bar{C}_2$
Q_3	1	-1	i	$-i$
Q_4	1	-1	$-i$	i

TABLE G19.6

W	E	$\bar{E}$	$(C_4^2, \bar{C}_4^2)$	$(2C_2, 2\bar{C}_2)$	$2JC_4$	$2\overline{JC}_4$	$(2JC_4^2, 2\overline{JC}_4^2)$
W_6	2	-2	0	0	$\sqrt{2}$	$-\sqrt{2}$	0
W_7	2	-2	0	0	$-\sqrt{2}$	$\sqrt{2}$	0

TABLE G19.7

MX	E	$\bar{E}$	$2C_4$	$2\bar{C}_4$	$(C_4^2, \bar{C}_4^2)$	$(2C_4^{2\prime}, 2\bar{C}_4^{2\prime})$	$(2C_2, 2\bar{C}_2)$
X_{11}^+	2	-2	$\sqrt{2}$	$-\sqrt{2}$	0	0	0
X_{12}^+	2	-2	$-\sqrt{2}$	$\sqrt{2}$	0	0	0
X_{11}^-	2	-2	$\sqrt{2}$	$-\sqrt{2}$	0	0	0
X_{12}^-	2	-2	$-\sqrt{2}$	$\sqrt{2}$	0	0	0

MX	J	$\bar{J}$	$2JC_4$	$2\overline{JC}_4$	$(JC_4^2, \overline{JC}_4^2)$	$(2JC_4^{2\prime}, 2\overline{JC}_4^{2\prime})$	$(2JC_2, 2\overline{JC}_2)$
X_{11}^+	2	-2	$\sqrt{2}$	$-\sqrt{2}$	0	0	0
X_{12}^+	2	-2	$-\sqrt{2}$	$\sqrt{2}$	0	0	0
X_{11}^-	-2	2	$-\sqrt{2}$	$\sqrt{2}$	0	0	0
X_{12}^-	-2	2	$\sqrt{2}$	$-\sqrt{2}$	0	0	0

TABLE G19.8

L	E	$\bar{E}$	$2C_3$	$2\bar{C}_3$	$3C_2$	$3\bar{C}_2$	J	$\bar{J}$	$2JC_3$	$2\overline{JC}_3$	$3JC_2$	$2\overline{JC}_2$
L_7^+	1	−1	−1	−1	i	$-i$	1	−1	−1	1	i	$-i$
L_8^+	1	−1	−1	1	$-i$	i	1	−1	−1	1	$-i$	i
L_9^+	2	−2	1	−1	0	0	2	−2	1	−1	0	0
L_7^-	1	−1	−1	1	i	$-i$	−1	1	1	−1	$-i$	i
L_8^-	1	−1	−1	1	$-i$	i	−1	1	1	−1	i	$-i$
L_9^-	2	−2	1	−1	0	0	−2	2	−1	1	0	0

TABLE G19.9

N	E	$\bar{E}$	$(C_4^2, \bar{C}_4^2)$	$(C_2, \bar{C}_2)$	$(C_2', \bar{C}_2')$
N^+	2	−2	0	0	0
N^-	2	−2	0	0	0

N	J	$\bar{J}$	$(JC_4^2, \overline{JC_4^2})$	$(JC_2, \overline{JC}_2)$	$(JC_2', \overline{JC_2'})$
N^+	2	−2	0	0	0
N^-	−2	2	0	0	0

TABLE G19.10

P	E	$\bar{E}$	$8C_3$	$8\bar{C}_3$	$(3C_2, 3\bar{C}_2)$	$(6JC_2, 6\overline{JC}_2)$	$6JC_4$	$6\overline{JC}_4$
P_6	2	−2	1	−1	0	0	$\sqrt{2}$	$-\sqrt{2}$
P_7	2	−2	1	−1	0	0	$-\sqrt{2}$	$\sqrt{2}$
P_8	4	−4	−1	1	0	0	0	0

(ii) Double groups of diamond

When the character table may be directly constructed from that of the point group it is unnecessary to go into further detail, since the essentials have already been given. We list characters only for those groups where this is not so, and only as many characters as are essential.

TABLE G19.11

Z	$(E\|0)$	$(\overline{E\|0})$	$(C_4^2\|0)$	$(JC_4^2\|\tau)$	$(JC_4^{2\prime}\|\tau)$
Z_2	1	−1	i	i	−1
Z_3	1	−1	i	$-i$	1
Z_4	1	−1	$-i$	i	1
Z_5	1	−1	$-i$	$-i$	−1

TABLE G19.12

W	$(E\|0)$	$(\overline{E\|0})$	$(C_4^2\|0)$	$(C_2\|\tau)$	$(C_2'\|\tau)$
W_3	1	-1	$-i$	-1	i
W_4	1	-1	$-i$	-1	i
W_5	1	-1	i	-1	$-i$
W_6	1	-1	$-i$	-1	$-i$
W_7	2	-2	$2i$	0	0

W	$(JC_4^2\|\tau)$	$(JC_4^{2\prime}\|\tau)$	$(JC_4\|0)$	$(JC_4^3\|0)$
W_3	$\sqrt{\frac{1}{2}}(1+i)$	$\sqrt{\frac{1}{2}}(1-i)$	$\sqrt{\frac{1}{2}}(1-i)$	$\sqrt{\frac{1}{2}}(1+i)$
W_4	$-\sqrt{\frac{1}{2}}(1+i)$	$\sqrt{\frac{1}{2}}(i-1)$	$\sqrt{\frac{1}{2}}(i-1)$	$-\sqrt{\frac{1}{2}}(1+i)$
W_5	$-\sqrt{\frac{1}{2}}(1+i)$	$\sqrt{\frac{1}{2}}(i-1)$	$\sqrt{\frac{1}{2}}(1-i)$	$\sqrt{\frac{1}{2}}(1+i)$
W_6	$\sqrt{\frac{1}{2}}(1+i)$	$\sqrt{\frac{1}{2}}(i-1)$	$\sqrt{\frac{1}{2}}(i-1)$	$-\sqrt{\frac{1}{2}}(1+i)$
W_7	0	0	0	0

TABLE G19.13

X	$(E\|0)$	$(\overline{E\|0})$	$(E\|\mathbf{R})$	All others
X_5	4	-4	$4e^{i\mathbf{k}\cdot\mathbf{R}}$	0

(iii) Double groups of h.c.p. structure

We give only the surface representations, which cannot readily be obtained from the representations of the corresponding point groups.

TABLE G19.14

A	$(E\|0)$	$(\overline{E\|0})$	$(C_6, C_6^{-1}\|\tau)$	$(C_6^2, C_6^{-2}\|0)$	$(C_6^3\|\tau)$	$(3C_2\|\tau)$	All others of the set $S_p(A)$
A_3	2	-2	0	-2	0	$2i$	0
A_4	2	-2	0	-2	0	$-2i$	0
A_5	4	-4	0	2	0	0	0

TABLE G19.15

H	$(E\|0)$	$(\overline{E\|0})$	$(C_6^2\|\mathbf{R}_1)$	$(C_6^4\|2\mathbf{R}_1)$	$(C_2^2\|\tau')$	$(C_2^2\|\tau'')$	$(C_2^3\|\tau)$
H_4	1	-1	-1	-1	i	i	i
H_5	1	-1	-1	-1	i	i	i
H_6	1	-1	-1	-1	$-i$	$-i$	$-i$
H_7	1	-1	-1	-1	$-i$	$-i$	$-i$
H_8	2	-2	1	1	0	0	0
H_9	2	-2	1	1	0	0	0

TABLE G19.15 (*contd.*)

H	$(JC_6\mid 2\mathbf{R}_1)$	$(JC_6^{-1}\mid \mathbf{R}_1)$	$(JC_6^3\mid 0)$	$(m_1\mid \boldsymbol{\tau}')$	$(m_2\mid \boldsymbol{\tau}'')$	$(m_3\mid \boldsymbol{\tau})$
H_4	i	$-i$	$-i$	1	1	1
H_5	$-i$	i	i	-1	-1	-1
H_6	i	$-i$	$-i$	-1	-1	-1
H_7	$-i$	i	i	1	1	1
H_8	i	$-i$	$2i$	0	0	0
H_9	$-i$	i	$-2i$	0	0	0

Note: An explanation of the notation will be found in section G19(ii).

TABLE G19.16

L	$(E\mid 0)$	$(\overline{E\mid 0})$	$(C_2\mid \boldsymbol{\tau})$	All others of the set $S_p(L)$
L_3	2	-2	$2i$	0
L_4	2	-2	$-2i$	0

TABLE G19.17

S	$(E\mid 0)$	$(\overline{E\mid 0})$	$(C_2\mid \boldsymbol{\tau})$	$(JC_6^3\mid 0)$	$(JC_2'\mid \boldsymbol{\tau})$
S_2	1	-1	i	$-i$	1
S_3	1	-1	i	i	-1
S_4	1	-1	$-i$	i	1
S_5	1	-1	$-i$	$-i$	-1

PART D. LATTICE VIBRATIONS

APPENDIX G20 NON-SYMMORPHIC CRYSTALS

Since the application of group theory to the classification of the vibrational modes of a symmorphic lattice was discussed in Chapter 3 of Vol. 1, we shall illustrate here the application for non-symmorphic crystals only, taking the specific example of the direction Σ in the BZ of the diamond structure.

In Appendix A3.2 of Vol. 1 we found that, corresponding to the operation $(a\mid\mathbf{t})$, the matrix transforming the polarization vectors is of the form

$$A_{\alpha\beta}^{ss'} = e^{i\mathbf{k}(\mathbf{s}'-\mathbf{s})}\, a_{\alpha\beta}\, \delta_{s's''} \quad [s' = (a\mid\mathbf{t})^{\dagger}\,\mathbf{s}]. \tag{G20.1}$$

Here α, β are Cartesian components, and s, s' label atoms with positions $\mathbf{s}, \mathbf{s}'$. It is sufficient to take $\mathbf{s}$ and $\mathbf{s}'$ within the unit cell.

(i) Application to line Σ of diamond

We shall illustrate matters by finding what the implications of symmetry are for modes labelled by **k** lying on the line Σ in the BZ of diamond. We shall proceed largely by trial and error. For a more systematic, but not necessarily simpler, method, the reader may consult, for example, the article by Maradudin and Vosko (1968).

Diamond has two atoms per unit cell, separated by the non-primitive translation $\boldsymbol{\tau} = \frac{1}{4}a(1,1,1)$. We shall take one atom at the origin (so that for $s = 1$, $\mathbf{s} = 0$), and the other (for which $s = 2$) at $\mathbf{s} = \boldsymbol{\tau}$. In writing down the matrices $A^{ss'}_{\alpha\beta}$ we shall label the first three rows with $\alpha = 1,2,3$ and $s = 1$, the second three rows with $\alpha = 1,2,3$ and $s = 2$, and similarly label the columns. Thus the first three components of $|\,\varepsilon_{\mathbf{k}\sigma}\rangle$ will refer to Cartesian displacements of atom $s = 1$ and the fourth, fifth and sixth to displacements of atom $s = 2$: we therefore write

$$|\,\varepsilon_{\mathbf{k}\sigma}\rangle = \begin{pmatrix} \boldsymbol{\varepsilon}^1_{\mathbf{k}\sigma} \\ \hline \boldsymbol{\varepsilon}^2_{\mathbf{k}\sigma} \end{pmatrix}, \tag{G20.2}$$

where $\boldsymbol{\varepsilon}^1_{\mathbf{k}\sigma}$ is the polarization vector for the atom at the origin and $\boldsymbol{\varepsilon}^2_{\mathbf{k}\sigma}$ the polarization vector at $\boldsymbol{\tau}(s = 2)$.

For **k**-vectors on the line Σ the rotational parts a of the operators $(a|\,\mathbf{t})$ are the elements of the point group of the line, the matrix representations being, other than the identity element (see Chapter 3),

$$a = \begin{pmatrix} 0 & 1 & 0 \\ 1 & 0 & 0 \\ 0 & 0 & 1 \end{pmatrix}, \quad b = \begin{pmatrix} 0 & 1 & 0 \\ 1 & 0 & 0 \\ 0 & 0 & -1 \end{pmatrix}, \quad c = \begin{pmatrix} 1 & 0 & 0 \\ 0 & 1 & 0 \\ 0 & 0 & -1 \end{pmatrix}. \tag{G20.3}$$

In terms of these, we find from equation (G20.1),

$$E = \left(\begin{array}{c|c} 1 & 0 \\ \hline 0 & 1 \end{array}\right), \quad A_1 = \left(\begin{array}{c|c} a & 0 \\ \hline 0 & a \end{array}\right),$$

$$A_2 = \left(\begin{array}{c|c} 0 & \rho^* b \\ \hline \rho b & 0 \end{array}\right), \quad A_3 = \left(\begin{array}{c|c} 0 & \rho^* c \\ \hline \rho c & 0 \end{array}\right), \tag{G20.4}$$

where

$$\rho = e^{i\mathbf{k}.\boldsymbol{\tau}}. \tag{G20.5}$$

Now in Chapter 3 we found that there are three orthogonal vectors $\boldsymbol{\varepsilon}_1 = (1,1,0)$, $\boldsymbol{\varepsilon}_2 = (0,0,1)$ and $\boldsymbol{\varepsilon}_3 = (1,-1,0)$ separately invariant (apart from phase factors) under the transformations a, b, c. For example, one easily verifies that

$$a\boldsymbol{\varepsilon}_1 = \boldsymbol{\varepsilon}_1, \quad b\boldsymbol{\varepsilon}_1 = \boldsymbol{\varepsilon}_1, \quad c\boldsymbol{\varepsilon}_1 = \boldsymbol{\varepsilon}_1 \tag{G20.6}$$

and so it follows that

$$A_1\begin{pmatrix}\boldsymbol{\varepsilon}_1\\ \rho\boldsymbol{\varepsilon}_1\end{pmatrix} = \begin{pmatrix}\boldsymbol{\varepsilon}_1\\ \rho\boldsymbol{\varepsilon}_1\end{pmatrix}, \quad B\begin{pmatrix}\boldsymbol{\varepsilon}_1\\ \rho\boldsymbol{\varepsilon}_1\end{pmatrix} = \begin{pmatrix}\boldsymbol{\varepsilon}_1\\ \rho\boldsymbol{\varepsilon}_1\end{pmatrix},$$

$$C\begin{pmatrix}\boldsymbol{\varepsilon}_1\\ \rho\boldsymbol{\varepsilon}_1\end{pmatrix} = \begin{pmatrix}\boldsymbol{\varepsilon}_1\\ \rho\boldsymbol{\varepsilon}_1\end{pmatrix} \tag{G20.7}$$

and also

$$A_1\begin{pmatrix}-\boldsymbol{\varepsilon}_1\\ \rho\boldsymbol{\varepsilon}_1\end{pmatrix} = \begin{pmatrix}-\boldsymbol{\varepsilon}_1\\ \rho\boldsymbol{\varepsilon}_1\end{pmatrix}, \quad A_2\begin{pmatrix}-\boldsymbol{\varepsilon}_1\\ \rho\boldsymbol{\varepsilon}_1\end{pmatrix} = -\begin{pmatrix}-\boldsymbol{\varepsilon}_1\\ \rho\boldsymbol{\varepsilon}_1\end{pmatrix},$$

$$A_3\begin{pmatrix}-\boldsymbol{\varepsilon}_1\\ \rho\boldsymbol{\varepsilon}_1\end{pmatrix} = -\begin{pmatrix}-\boldsymbol{\varepsilon}_1\\ \rho\boldsymbol{\varepsilon}_1\end{pmatrix}. \tag{G20.8}$$

Hence

$$|\varepsilon_{\mathbf{k}}(L1)\rangle = \begin{pmatrix}\boldsymbol{\varepsilon}_1\\ \rho\boldsymbol{\varepsilon}_1\end{pmatrix}, \quad |\varepsilon_{\mathbf{k}}(L2)\rangle = \begin{pmatrix}-\boldsymbol{\varepsilon}_1\\ \rho\boldsymbol{\varepsilon}_1\end{pmatrix} \tag{G20.9}$$

are longitudinal vectors separately invariant under the group of Σ. This does not imply we have exactly longitudinal eigenvectors, however. Consider $\begin{pmatrix}\boldsymbol{\varepsilon}_2\\ \rho\boldsymbol{\varepsilon}_2\end{pmatrix}$. Now

$$a\boldsymbol{\varepsilon}_2 = \boldsymbol{\varepsilon}_2, \quad b\boldsymbol{\varepsilon}_2 = -\boldsymbol{\varepsilon}_2, \quad c\boldsymbol{\varepsilon}_2 = -\boldsymbol{\varepsilon}_2 \tag{G20.10}$$

and so

$$A_1\begin{pmatrix}\boldsymbol{\varepsilon}_2\\ \rho\boldsymbol{\varepsilon}_2\end{pmatrix} = \begin{pmatrix}\boldsymbol{\varepsilon}_2\\ \rho\boldsymbol{\varepsilon}_2\end{pmatrix}, \quad A_2\begin{pmatrix}\boldsymbol{\varepsilon}_2\\ \rho\boldsymbol{\varepsilon}_2\end{pmatrix} = -\begin{pmatrix}\boldsymbol{\varepsilon}_2\\ \rho\boldsymbol{\varepsilon}_2\end{pmatrix},$$

$$A_3\begin{pmatrix}\boldsymbol{\varepsilon}_2\\ \rho\boldsymbol{\varepsilon}_2\end{pmatrix} = -\begin{pmatrix}\boldsymbol{\varepsilon}_2\\ \rho\boldsymbol{\varepsilon}_2\end{pmatrix}. \tag{G20.11}$$

Comparing this with (G20.8), we see that $\alpha\begin{pmatrix}\boldsymbol{\varepsilon}_3\\ -\rho\boldsymbol{\varepsilon}_1\end{pmatrix} + \beta\begin{pmatrix}\boldsymbol{\varepsilon}_2\\ -\rho\boldsymbol{\varepsilon}_2\end{pmatrix}$ is an invariant vector, where α and β are any constants. We similarly find $\gamma\begin{pmatrix}\boldsymbol{\varepsilon}_1\\ \rho\boldsymbol{\varepsilon}_1\end{pmatrix} + \delta\begin{pmatrix}\boldsymbol{\varepsilon}_2\\ -\rho\boldsymbol{\varepsilon}_2\end{pmatrix}$ is also an invariant vector, and $\alpha, \beta, \gamma, \delta$ are to be determined by explicit use of the dynamical matrix.

On the other hand, two non-degenerate transverse modes are readily constructed. We can verify that

$$a\boldsymbol{\varepsilon}_3 = -\boldsymbol{\varepsilon}_3, \quad b\boldsymbol{\varepsilon}_3 = -\boldsymbol{\varepsilon}_3, \quad c\boldsymbol{\varepsilon}_3 = \boldsymbol{\varepsilon}_3. \tag{G20.12}$$

Hence

$$A_1\begin{pmatrix}\boldsymbol{\varepsilon}_3\\ \rho\boldsymbol{\varepsilon}_3\end{pmatrix} = -\begin{pmatrix}\boldsymbol{\varepsilon}_3\\ \rho\boldsymbol{\varepsilon}_3\end{pmatrix}, \quad A_2\begin{pmatrix}\boldsymbol{\varepsilon}_3\\ \rho\boldsymbol{\varepsilon}_3\end{pmatrix} = -\begin{pmatrix}\boldsymbol{\varepsilon}_3\\ \rho\boldsymbol{\varepsilon}_3\end{pmatrix},$$

$$A_3\begin{pmatrix}\boldsymbol{\varepsilon}_3\\ \rho\boldsymbol{\varepsilon}_3\end{pmatrix} = \begin{pmatrix}\boldsymbol{\varepsilon}_3\\ \rho\boldsymbol{\varepsilon}_3\end{pmatrix} \tag{G20.13}$$

and also

$$A_1\begin{pmatrix}\boldsymbol{\varepsilon}_3\\ -\rho\boldsymbol{\varepsilon}_3\end{pmatrix} = -\begin{pmatrix}\boldsymbol{\varepsilon}_3\\ -\rho\boldsymbol{\varepsilon}_3\end{pmatrix}, \quad A_2\begin{pmatrix}\boldsymbol{\varepsilon}_3\\ -\rho\boldsymbol{\varepsilon}_3\end{pmatrix} = \begin{pmatrix}\boldsymbol{\varepsilon}_3\\ -\rho\boldsymbol{\varepsilon}_3\end{pmatrix},$$

$$A_3\begin{pmatrix}\boldsymbol{\varepsilon}_3\\ -\rho\boldsymbol{\varepsilon}_3\end{pmatrix} = -\begin{pmatrix}\boldsymbol{\varepsilon}_3\\ -\rho\boldsymbol{\varepsilon}_3\end{pmatrix}. \tag{G20.14}$$

We see that $\begin{pmatrix}\boldsymbol{\varepsilon}_3\\ \rho\boldsymbol{\varepsilon}_3\end{pmatrix}$ and $\begin{pmatrix}\boldsymbol{\varepsilon}_3\\ -\rho\boldsymbol{\varepsilon}_3\end{pmatrix}$ transform differently from the vectors previously investigated and we can uniquely identify them as non-degenerate transverse acoustic and transverse optical modes, repectively. The former is the acoustic mode since as $\mathbf{k}\to 0$ it gives a rigid translation of the lattice.

(ii) Time-reversal degeneracy

If

$$D(\mathbf{k})|\,\varepsilon_{\mathbf{k}\sigma}\rangle = \omega^2(\mathbf{k}\sigma)|\,\varepsilon_{\mathbf{k}\sigma}\rangle, \tag{G20.15}$$

we also have

$$D^*(\mathbf{k})|\,\varepsilon_{\mathbf{k}\sigma}\rangle^* = \omega^2(\mathbf{k}\sigma)|\,\varepsilon_{\mathbf{k}\sigma}\rangle^* \tag{G20.16}$$

and since (Chapter 3, section 3.2)

$$D^*(\mathbf{k}) = D(-\mathbf{k}), \tag{G20.17}$$

$$\omega^2(-\mathbf{k}\sigma) = \omega^2(\mathbf{k}\sigma), \tag{G20.18}$$

whether the crystal has an inversion centre or not.

Having determined the degenerate polarization vectors implied by spatial symmetry, one can readily determine if there is additional degeneracy implied by time-reversal if $-\mathbf{k}$ is in the star of $\mathbf{k}$. We take any of the polarization vectors, say $|\,\varepsilon_{\mathbf{k}\sigma}\rangle$, determine $|\,\varepsilon_{-\mathbf{k}\sigma}\rangle$ by operating on $|\,\varepsilon_{\mathbf{k}\sigma}\rangle$ with the appropriate transformation and then take the complex conjugate $|\,\varepsilon_{-\mathbf{k}\sigma}\rangle^*$. This is an eigenvector of $D(\mathbf{k})$ and so if no extra degeneracy exists must be a linear combination of $|\,\varepsilon_{\mathbf{k}\sigma}\rangle$ and those of the polarization vectors for $\mathbf{k}$ of the same eigenvalue. For example, one finds no extra degeneracy on the line Σ.

Alternatively, we may apply Wigner's time-reversal test. For a discussion of this applied to lattice vibrations, see Maradudin and Vosko (1968).

PART E.

APPENDIX G21 REALIZATION OF ALL IRREDUCIBLE MANIFOLDS

So far it has been a matter of assumption that the dimension of every particular irreducible representation of the symmetry group G of the Hamiltonian H is the same as some actual essential degeneracy of the eigenstates of H. This is because we have not yet shown that the eigenvectors of H realize all irreducible representations, and we now address ourselves to this matter.

(i) Projection operators

Let us define the operators

$$\varepsilon^j_{rs} = \frac{n_j}{N}\sum A a^{j*}_{rs}, \tag{G21.1}$$

where the summation is over all N elements of the group G, (a^j_{rs}) is the matrix corresponding to A in the jth irreducible representation of the group, and n_j is the order of the representation.

One readily sees that

$$(\varepsilon^i_{rs})^\dagger = \varepsilon^i_{sr} \tag{G21.2}$$

and

$$\varepsilon^i_{rs} H = H\varepsilon^i_{rs}. \tag{G21.3}$$

It may also be shown, using the general theory of irreducible representations (see, for example, Johnston, 1960), that

$$\varepsilon^i_{rs}\,\varepsilon^j_{uv} = \delta_{ij}\,\delta_{su}\,\varepsilon^j_{rv}. \tag{G21.4}$$

For any vector ψ one also readily shows from the definition (G21.1) of ε^j_{rt} that

$$A\varepsilon^j_{rt}\psi = \sum_s a^j_{sr}\,\varepsilon^j_{st}\psi$$

for all A in G. Provided $\varepsilon^j_{rt}\psi$ is non-zero, this equation, together with (G21.3), shows that the set $\{\varepsilon^j_{st}\psi\}$, $s = 1,\ldots, N_j$, spans the sub-space in which the jth representation exists.

(ii) Independence of elements of a group

From the above we see that the eigenstates of H must realize all representations provided that for each ε^j_{st} there is some vector ψ for which $\varepsilon^j_{st}\psi$ is non-zero. We shall now in fact show that the elements A of a group are independent in that

$$\sum_A \alpha(A)\, A\psi = 0 \tag{G21.5}$$

for all vectors ψ only if the coefficients $\alpha(A)$ are zero for all A.

Let $\{\psi_n\}$ be any set of orthonormal vectors, and write

$$A\psi_n = \sum_m d_{mn}(A)\, \psi_m, \tag{G21.6}$$

where, because A is unitary,

$$\sum_n d_{nm'}(A)\, d_{mn}(A) = \delta_{m'm}. \tag{G21.7}$$

We also suppose $C = AB$, whereupon we find

$$\begin{aligned} C\psi_n &= \sum_m d_{mn}(C)\, \psi_m = \sum_{m'} d_{m'n}(B)\, A\psi_{m'} \\ &= \sum_m \sum_{m'} d_{m'n}(B)\, d_{mm'}(A)\, \psi_m, \end{aligned} \tag{G21.8}$$

that is,

$$d_{mn}(C) = \sum_{m'} d_{m'n}(B)\, d_{mm'}(A). \tag{G21.9}$$

We now suppose (G21.5) is true for all vectors ψ. From (G21.6) we find, for all n,

$$\sum_A \alpha(A) \sum_m d_{mn}(A)\, \psi_m = 0 \tag{G21.10}$$

or

$$\sum_A \alpha(A)\, d_{mn}(A) = 0 \quad \text{(all } m, n). \tag{G21.11}$$

It must immediately follow that

$$\begin{aligned} 0 &= \left| \sum_A \alpha(A)\, d_{mn}(A) \right|^2 \\ &= \sum_A |\alpha(A)|^2 |d_{mn}(A)|^2 + \sum_A \alpha^*(A)\, d_{mn}(A) \sum_{B \neq A}{}' \alpha(B)\, d_{mn}(B). \end{aligned} \tag{G21.12}$$

Summing over m, this may also be written, on using (G21.7) and (G21.9)

$$0 = \sum_A |\alpha(A)|^2 + \sum_A \alpha^*(A) \sum_{B \neq A}{}' \alpha(B)\, d_{mn}(AB). \tag{G21.13}$$

But since we are supposing, for all ψ_n, that

$$\sum_B \alpha(B)\, B\psi_m = 0, \tag{G21.14}$$

it immediately follows that, for all A,

$$\sum_B \alpha(B)\, AB\psi_m = 0 \tag{G21.15}$$

and hence, we also have, corresponding to (G21.12),

$$\sum_B \alpha(B)\, d_{mn}(AB) = 0 \quad (\text{all } m, n). \tag{G21.16}$$

Equation (G21.13) may now be rewritten as

$$0 = \sum_A |\alpha(A)|^2 \{1 - d_{mn}(A^2)\}, \tag{G21.17}$$

and taking the real part we find

$$0 = \sum_A |\alpha(A)|^2 \{1 - \operatorname{Re} d_{mn}(A^2)\} \geqslant \sum_A |\alpha(A)|^2 \{1 - |d_{mn}(A)|^2\}. \tag{G21.18}$$

However, $|d_{mm}(A^2)| \leqslant 1$ and so either $\alpha(A) = 0$ for all A, or $d_{mm}(A^2) = 1$ for all m. The latter alternative can occur only if A^2 is the unit operator, and so our assertion is proved.

Problems

Chapter 6

1. The deviation of the Hall constant from its free-electron value decreases for Li, Na, K and Rb under pressure but increases for Cs beyond 11,000 or 12,000 kg/cm².
What can be deduced about the energy surfaces of Cs?

2. Discuss under what conditions the magnetoresistance of divalent metals can saturate and also metals of odd valency if the primitive unit cell contains an even number of electrons.

3. Give the full analytic expression for $\sigma(\omega)$, the frequency-dependent conductivity in the presence of a magnetic field, when the energy surfaces are ellipsoidal.

4. Show that the heat lost per unit time when a current flows in a wire is $\sigma\mathscr{E}^2$.

5. Find the energy operator for thermal transport within the harmonic approximation.

6. From equation (6.11.47), find explicit forms of $Q(E)$ used in the discussion of the range of validity of the Boltzmann equation.

7. Show that the quasi-particle energies $\varepsilon_0(\mathbf{k})$ are altered by the electron–phonon interaction to become

$$\varepsilon(\mathbf{k}) = \varepsilon_0(\mathbf{k}) + \frac{1}{2\pi^3}\int d\mathbf{q}\,|M_{\mathbf{k},\mathbf{k}-\mathbf{q}}|^2 \times \left[\frac{1-n_{\mathbf{q}}}{\varepsilon_0(\mathbf{k})-\varepsilon_0(\mathbf{k}-\mathbf{q})+\hbar\omega_{\mathbf{q}}} + \frac{n_{\mathbf{k}-\mathbf{q}}}{\varepsilon_0(\mathbf{k})-\varepsilon_0(\mathbf{k}-\mathbf{q})-\hbar\omega_{\mathbf{q}}}\right],$$

where $M_{\mathbf{k},\mathbf{k}-\mathbf{q}}$ is the matrix element for scattering, by a phonon of energy $\hbar\omega_{\mathbf{q}}$, of an electron from state $\mathbf{k}$ to $\mathbf{k}-\mathbf{q}$.
[*Hint*: One satisfactory approach to this problem comes from the diagrammatic treatment of electron–phonon interaction discussed in Chapter 8, section 8.10.1.]

8. Treat the problem of scattering of phonons by a point imperfection by analogy with the scattering of sound waves in a gas by a sphere. Hence show that the scattering cross-section, in the long wavelength limit, is inversely proportional to the fourth power of the wavelength.
[*N.B.* Since phonon wavelengths are usually many times the lattice spacing, such Rayleigh scattering is almost always to be expected.]

9. The basic formula given by Ziman (1961) for the resistivity ρ_l of a liquid metal, in Born approximation, is

$$\rho_l = \frac{3\pi}{\hbar e^2 v_f^2 \rho_0}\int_0^1 S(K)\,|U(K)|^2\, 4\left(\frac{K}{2k_f}\right)^3 d\left(\frac{K}{2k_f}\right),$$

where $v_f = \hbar k_f/m$ is the Fermi velocity, $S(K)$ is the structure factor, $U(K)$ is the scattering potential of a single ion and ρ_0 is the number of ions per unit volume. In a solid, for temperatures high compared with the Debye temperature, a semi-quantitative approximation is obtained by replacing $S(K)$ by its long wavelength limit $S(0)$. This in turn is given by the thermodynamic result

$$S(0) = \rho_0 k_B T K_T,$$

where K_T is the isothermal compressibility. Introducing the velocity of sound v_s, and putting the ratio of specific heats equal to unity, this becomes

$$S(0) \doteqdot k_B T/M v_s^2,$$

where M is the ionic mass. Using the Bohm–Staver formula for v_s, and recalling that elementary Debye theory gives for the Debye temperature

$$\theta = \frac{h}{k_B}\left(\frac{3}{4\pi}\right)^{\frac{1}{3}} \frac{v_s}{v^{\frac{1}{3}}},$$

where v is the atomic volume, show that

(i) the resistivity of the solid ρ_s may be written

$$\rho_s = K_i k_B T/M\theta^2,$$

where K_i depends on the scattering of a single ion, and is proportional to

(ii)

$$\int_0^1 |U(K)|^2 \left(\frac{K}{2k_f}\right) d\left(\frac{K}{2k_f}\right).$$

Using a screened Coulomb potential for $U(K)$, with the Thomas–Fermi screening radius, show that the results may be expressed in universal form, by plotting $C\rho_s k_f^3/T$ against $k_f a_0$, where $C = 4\hbar e^2/27\pi^3 m k_B$. Sketch the result.

Why is this theory not applicable to real metals, at least under normal pressures which can be achieved in the laboratory?

10. Examine the consistency of equations (6.2.42) and (6.2.59) for the absolute thermopower S.

Chapter 7

1. Show that the Kramers–Krönig relations between the real and imaginary parts of the frequency-dependent conductivity $\sigma(\omega)$ follow, provided we can interchange the order of integration in the Fourier representation

$$\sigma(\omega) = \int_0^\infty dt \exp(-i\omega t) \int_{-\infty}^{\infty} \sigma(\omega') \exp(i\omega' t)\, d\omega'.$$

2. Derive the analogue of Fowler's law of thermionic emission for a semi-conductor.

3. Obtain the Alfvén wave solution for a general mass tensor.

4. To approximate the frequency dependence of the conductivity in a non-degenerate semi-conductor, average the Drude–Zener result over the Maxwellian velocity distribution.

From this result, and assuming the mean free path l is independent of velocity v, show that

$$\sigma(\omega) = \sigma_0(\alpha + i\beta),$$

where

$$\sigma_0 = 4ne^2\, l/3(2\pi m k_B T)^{\frac{1}{2}},$$

n being the carrier density, and α and β have the forms

$$\alpha(u) = 1 - u^2 + u^4 \exp[u^2]\, E_1(u^2),$$

$$\beta(u) = -\pi^{\frac{1}{2}}\{\tfrac{1}{2}u - u^3 + \pi^{\frac{1}{2}} u^4 \exp[u^2]\, F(u)\}.$$

Here

$$u = \omega l\left(\frac{m}{2k_B T}\right)^{\frac{1}{2}},$$

$$E_1(x) = \int_x^\infty \frac{\exp(-t)}{t}\, dt$$

and $1 - F(x)$ is the Gauss error function defined by

$$\mathrm{Erf}(x) = \frac{2}{\sqrt{\pi}} \int_0^\infty \exp(-t^2)\, dt.$$

[*N.B.* These results are appropriate to a single band model with lattice scattering.]

5. Under what conditions in the anomalous skin effect does the surface resistance of a polycrystalline specimen give the total surface area of the Fermi surface?

6. In the observation of Azbel'–Kaner cyclotron resonance on a thin parallel-sided plate of thickness d, the resonance signals vanish when the steady applied magnetic field is below a certain value H_c. Obtain H_c in terms of d and the caliper diameter of the Fermi surface in the direction perpendicular to the magnetic field and surface of the plate.

Chapter 8

1. In coordinate space, the relation between current and vector potential in a superconductor can be written as

$$\mathbf{j}(\mathbf{r}) = \int K(\mathbf{r} - \mathbf{r}')\, \mathbf{A}(\mathbf{r}')\, d\mathbf{r}'.$$

Show that

(i) in the London theory, the Fourier transform of $K(\mathbf{r})$ is a constant;

(ii) in a non-local theory, the range of K is determined by the coherence length.

2. An alternative approach to flux quantization to that adopted in the main text is to construct the free energies F_n and F_s of the normal and superconducting states as functions of the flux Φ. Then to distinguish the states, we must have, because of the Meissner effect

$$\partial F_s / \partial \Phi = 0.$$

To carry this through, start from the energy levels of independent electrons in a hollow cylinder (Byers and Yang, 1961) given by

$$E_n = \frac{1}{2m}\left[p_r^2+p_z^2+\frac{\hbar^2}{r^2}\left(n+\frac{e}{ch}\Phi\right)^2\right],$$

where p_r and p_z are the momenta in the radial and z-directions and n is an integer.

[*Hint*: Be careful to think in terms of the level spectrum of pairs of electrons.]

3. Experiment shows that the transition temperature of a superconductor is not dramatically sensitive to impurity concentration.

(i) Construct, from the complete set of Bloch functions $\phi_{\mathbf{k}}$ in the pure superconductor, formally exact eigenfunctions when the impurities are present.

(ii) Use time-reversal symmetry arguments to construct another set of eigenfunctions, degenerate with the set (i).

(iii) Add the impurity term to the Hamiltonian (8.7.17) and rewrite this in terms of new operators $d_{n\sigma}$, related to $a_{\mathbf{k}\sigma}$ by

$$a_{\mathbf{k}\sigma} = \sum_n d_{n\sigma}\,\phi(n\mathbf{k}),$$

where $\phi(n\mathbf{k})$ is the Fourier transform of $\phi_{\mathbf{k}}(\mathbf{r})$.

(iv) Show that, in terms of these new operators, the Hamiltonian has the same form as if there were no impurities, assuming $\omega_{\mathbf{q}}$ and $D_{\mathbf{q}}$ in equation (8.7.17) have negligible $\mathbf{q}$ dependence.

(v) Contrast this situation with that in which the impurities are introduced into the BCS reduced Hamiltonian. This latter procedure is erroneous, and would lead to too strong a dependence of T_c on impurity concentration. The correct treatment outlined above is due to Anderson.

4. Prove that, to go from a wave function $|\chi\rangle$ for a superconductor to that in which the phase χ is changed to $\chi+\delta\chi$, we must multiply $|\chi\rangle$ by $\exp(iN\delta\chi)$, where N is the number operator for Cooper pairs.

[*Hint*: From equation (8.13.6) the order parameter $\Delta(\mathbf{r})$ is directly related to the wave function $\phi(\mathbf{r})$ of a Cooper pair. This, in turn, can be related to $\chi(\mathbf{r})$ through equation (8.14.1). Write the Cooper pair wave function in terms of field operators $\psi_\downarrow\,\psi_\uparrow$ analogous to the creation and annihilation operators in equations (8.10.8) and (8.10.9). Hence show that

$$\langle\chi|\exp(-iN\delta\chi)\,\psi_\downarrow\,\psi_\uparrow\exp(iN\delta\chi)|\chi\rangle \propto \exp[i(\chi+\delta\chi)].$$

5. In connection with the Josephson junction, show from the discussion of the Ferrell–Prange equation, or otherwise, that

$$\operatorname{grad}\phi = \frac{2ed}{\hbar c}(\mathscr{H}\times\mathbf{n}),$$

where ϕ is the phase, d is the effective thickness of the sheet of flux, $\mathscr{H}$ is the field in the barrier and $\mathbf{n}$ is a unit vector normal to the barrier.

[*N.B.* From the discussion in the main text, $d\simeq\lambda_1+\lambda_2$, where λ_1 and λ_2 are the penetration depths on the two sides of the barrier. A more precise result is $d = \lambda_1+\lambda_2+t$ where t is the thickness of the barrier.]

Chapter 9

1. Extend the treatment of the Wannier exciton in section 9.8.1 to the case of ellipsoidal band edges, making use of the components of the reciprocal mass tensors.

2. Explain why, when excitons are formed with emission and absorption of phonons, we have an exciton band spectrum with a well-defined lower limit, rather than a line spectrum.

3. Using perturbation theory, make an estimate of the number of acoustic phonons accompanying a slow electron moving through a polar crystal.

4. At very low temperatures, the only states accessible to the polaron are those of small momentum in which any internal degrees of freedom are not excited.

Take the polaron energy spectrum to have the form

$$E = E_0 + (p^2/2m^*) + cp^4 + \ldots.$$

Show that the partition function Z of the decoupled polaron–phonon system is given by

$$\ln Z = \ln Z_0 - \beta(E_0 - E_g) + \frac{3}{2}\ln\frac{m^*}{m} - \frac{15m^{*2}c}{\beta}, \quad \beta = (k_B T)^{-1}.$$

Here Z_0 is the partition function for the entire non-interacting electron–phonon system and E_g is the non-interacting ground state energy.

How could this expression be used to give another definition of the polaron mass?

Chapter 10

1. In a dilute metallic alloy, use the free electron model to show that, to first order in the concentration, the change in the density of states at the Fermi level is the same in the localized screening model as in the rigid-band model.

2. Use the free-electron model of the vacancy formation energy E_v in a metal, to first-order in the valence Z, to relate E_v to the Debye temperature.

[*Hint*: Use the Bohm–Staver formula for the velocity of sound in a metal.]

3. In the Koster–Slater model, if we scatter Bloch electrons from a random array of impurities, the average Green function can be written (see Appendix 10.2)

$$\langle G \rangle = \frac{1}{E - E(\mathbf{k}) - \Sigma(\mathbf{k}E)},$$

where $\Sigma(\mathbf{k}E)$ is given in equation (A10.2.6).

Use this result to obtain an expression for the probability of scattering of a Bloch electron from state $\mathbf{k}$ to $\mathbf{k}'$.

4. Obtain the change in the density of states at the Fermi level for an impurity potential $V(\mathbf{r})$ in a periodic lattice in terms of the local density of electrons $\rho(\mathbf{r}E)$ and the scattering potential.

5. Show that the "electrostatic model" of section 10.12.1 for the interaction between point charges in a Fermi gas is still valid when the density-potential relation of equation (10.2.3) is used instead of the linearized Thomas–Fermi theory.

Show that the result is unchanged to first order when a weak lattice potential is switched on.

6. Calculate the binding energy of the hydrogen atom in a magnetic field H as a function of H using the variational trial function (A10.5.4). Comment on the validity of the variationally calculated binding energy at low field strengths.

7. In Problem 4.2 we saw that the Knight shift is proportional to the density of the valence electrons $\rho(0)$ at the nucleus for electrons on the Fermi surface. If, on alloying, this density changes by $c\Delta\rho(0)$ to first order in the concentration c, the proportional change in the Knight shift is

$$\frac{\Delta K}{K} = c\frac{\Delta\rho(0)}{\rho(0)}.$$

Under the assumptions that the impurities are at random, that the Pauli susceptibility is unchanged, that multiple scattering can be neglected and that in the absence of the impurities the valence-electron wave functions can be treated as plane waves, show that

$$\frac{\Delta K}{K} = c\sum_l (\alpha_l \sin^2 \eta_l + \beta_l \sin^2 \eta_l),$$

where the η_l are the phase-shifts on scattering by impurities, and the coefficients α_l and β_l depend on the lattice structure through

$$\alpha_l = (2l+1)\sum_{\mathbf{R}} [n_l^2(kR) - j_l^2(kR)],$$

$$\beta_l = -(2l+1)\sum_{\mathbf{R}} j_l(kR)\, n_l(kR),$$

the $\mathbf{R}$'s being lattice vectors, and n and j are spherical Neumann and Bessel functions. The above form the basis of the treatment of Blandin and Daniel (1959) of the Knight shift in dilute alloys.

References

The numbers in *italics* after the entries refer to the page in the book where the reference occurs, enabling the list of references to be used as an author index as well.

Abram, R. A. and Edwards, S. F. (1972). *J. Phys. C*, **5**, 1183, 1196. (*1133*)

Abrikosov, A. A. (1957). *Soviet Phys.–JETP*, **5**, 1174; *J. Phys. Chem. Solids*, **2**, 199. (*925*)

Allotey, F. K. (1967). *Phys. Rev.*, **157**, 467. (*831*)

Ambegaokar, V. and Tewordt, L. (1964). *Phys. Rev.*, **134**, A805. (*1151*)

Anderson, P. W. (1958). *Phys. Rev.*, **109**, 1492. (*1129*)

Anderson, P. W. (1960). See Baym (1969). (*916*)

Anderson, P. W. (1961). *Phys. Rev.*, **124**, 41. (*1036, 1044, 1045*)

Anderson, P. W. (1970). *Comments on Solid State Physics*, **2**, 193; see also an exact numerical solution of a special case of the Kondo problem by K. G. Wilson (1973). *Proceedings of Nobel Symposium*, **24**, distributed by Academic Press, New York. (*1048*)

Anderson, P. W. and Yuval, G. (1969). *Phys. Rev. Letters*, **23**, 89. (*1048*)

Anderson, P. W. and Yuval, G. (1970). *Phys. Rev.*, **B1**, 1522. (*1048*)

Anderson, P. W., Yuval, G. and Hamann, D. R. (1970a). *Phys. Rev.*, **B1**, 4464. (*1048*)

Anderson, P. W., Yuval, G. and Hamann, D. R. (1970b). *Solid State Commun.*, **8**, 1033. (*1048*)

Appel, J. (1968). *Solid State Phys.*, **21**, 193 (Eds. F. Seitz, D. Turnbull and H. Ehrenreich), Academic Press, New York. (*943, 961, 962*)

Ashcroft, N. W. and Lekner, J. (1966). *Phys. Rev.*, **145**, 83. (*735*)

Ashcroft, N. W. and Wilkins, J. W. (1965). *Physics Letters*, **14**, 285. (*868*)

Austin, I. G. and Mott, N. F. (1969). *Advan. Phys.*, **18**, 41. (*940*)

Azbel', M. I. and Kaner, E. A. (1956). *Soviet Phys.–JETP*, **3**, 772. (*842*)

Azbel', M. I. and Kaner, E. A. (1963). *Soviet Phys.–JETP*, **39**, 80. (*847*)

Bardasis, A., Falk, D. S. and Simkin, D. A. (1965). *J. Phys. Chem. Solids*, **26**, 1269. (*1116*)

Bardeen, J. (1936). *Phys. Rev.*, **49**, 653. (*1059*)

Bardeen, J., Cooper, L. N. and Schrieffer, J. R. (1957). *Phys. Rev.*, **108**, 1175. (*874, 887*)

Bardeen, J., Rickayzen, G. and Tewordt, L. (1959). *Phys. Rev.*, **113**, 982. (*895*)

Bardeen, J. and Shockley, W. (1950). *Phys. Rev.*, **80**, 72. (*1147*)

Baym, G. (1964). *Phys. Rev.*, **135**, A1691. (*698*)

Baym, G. (1969). *Mathematical Methods in Solid State and Superfluids* (Conference Proceedings, Eds. R. L. Clark and G. H. Derrick), Oliver and Boyd, Edinburgh, p. 121. (*913*)

Baynham, A. C. and Boardman, A. D. (1970). *Advan. Phys.*, **19**, 575. (*913*)

Baynham, A. C. and Boardman, A. D. (1971). *Plasma Effects in Semiconductors—Helicon and Alfven Waves*, Taylor and Francis, London. (*913*)

Beaglehole, D. (1965). *Optical Properties and Electronic Structure of Metals and Alloys* (Ed. F. Abeles), North-Holland, Amsterdam, p. 154. (*811*)
Beeby, J. L. (1967). *Proc. Roy. Soc.*, **A302**, 113. (*1012, 1018, 1175*)
Bell, R. J. (1972). *Rep. Prog. Phys.*, **35**, 1315. (*1129*)
Bell, R. L. and Spicer, W. E. (1970). *Proc. IEEE (USA)*, **58**, 1788. (*821, 823*)
Berglund, C. N. (1965). *Optical Properties and Electronic Structure of Metals and Alloys* (Conference Proceedings, Ed. F. Abeles), North-Holland, Amsterdam, p. 386. (*822, 824, 826*)
Berglund, C. N. and Spicer, W. E. (1964). *Phys. Rev.*, **136**, A1044. (*822, 824, 826*)
Biondi, M. A. and Rayne, J. A. (1959). *Phys. Rev.*, **115**, 1522. (*1076*)
Bogliubov, N. N. and Zubarev, D. N. (1955). *Soviet Phys.–JETP*, **1**, 83. (*1123*)
Born, M. and Huang, K. (1954). *Dynamical Theory of Crystal Lattices*, University Press, Oxford. (*868, 1103*)
Bosman, A. J. and van Daal, H. J. (1970). *Advan. Phys.*, **19**, 1. (*1166*)
Bradley, C. J. and Cracknell, A. P. (1972). *The Mathematical Theory of Symmetry in Solids*, Clarendon Press, Oxford. (*1198*)
Burdick, G. A. (1963). *Phys. Rev.*, **129**, 138. (*828*)

Callaway, J. and Hughes, A. J. (1967). *Phys. Rev.*, **156**, 860. (*985*)
Carruthers, P. (1961). *Rev. Mod. Phys.*, **33**, 92. (*1102*)
Chandrasekhar, S. (1943). *Rev. Mod. Phys.*, **15**, 1. (*1190*)
Chambers, R. G. (1952). *Proc. Phys. Soc.*, **A65**, 458. (*702*)
Chester, G. V. and Thellung, A. (1961). *Proc. Phys. Soc.*, **77**, 1005. (*762*)
Cohen, M. H. and Keffer, F. (1955). *Phys. Rev.*, **99**, 1128. (*977*)
Cole, H. S. D. and Turner, R. E. (1969). *J. Phys. C*, **2**, 124. (*1049, 1051, 1183*)
Coles, B. R. (1958). *Advan. Phys.*, **7**, 40. See also Volume 5 of Rado–Suhl work. (*1047*)
Corbato, F. J. (1959). *Proceedings of the Third Conference on Carbon*, Pergamon Press Ltd., London, p. 173. (*797*)
Coulson, C. A. and Taylor, R. (1952). *Proc. Phys. Soc.*, **A65**, 815. (*797*)
Crangle, J. (1960). *Phil. Mag.*, **5**, 335. (*1049*)

Darby, J. K. and March, N. H. (1964). *Proc. Phys. Soc.*, **84**, 591. (*715, 717*)
Davis, L. (1939). *Phys. Rev.*, **56**, 93. (*722*)
Dawber, P. G. and Elliott, R. J. (1963a). *Proc. Phys. Soc.*, **81**, 453. (*1104*)
Dawber, P. G. and Elliott, R. J. (1963b). *Proc. Roy. Soc.*, **A273**, 222. (*1105*)
Dawber, P. G. and Turner, R. E. (1966). *Proc. Phys. Soc.*, **88**, 217. (*1005, 1172*)
Dean, P. and Martin, J. L. (1960). *Proc. Roy. Soc.*, **A259**, 409. (*1129*)
Deutsch, T. W., Paul, W. and Brooks, H. (1961). *Phys. Rev.*, **124**, 753. (*723*)
de Wit, H. J. (1968). *Philips Res. Rept*, **23**, 449. (*1166*)
Dingle, R. B. (1955). *Phil. Mag.*, **46**, 831. (*989*)
Doniach, S. (1970). *Phys. Rev.*, **B2**, 3898. (*828*)
Doniach, S. and Sunjic, M. (1970). *J. Phys. C*, **3**, 285. (*828*)
Doniach, S. and Wohlfarth, E. P. (1967). *Proc. Roy. Soc.*, **A296**, 442. (*1049*)
Donovan, B. (1967). *Elementary Theory of Metals*, Pergamon Press, Oxford. (*786, 835, 853*)
Douglass, D. H. and Falicov, L. M. (1964). *Progress in Low Temperature Physics*, **4**, North-Holland, Amsterdam. (*919*)
Drain, L. E. (1968). *J. Phys. C*, **1**, 1690. (*1000*)
Dunifer, G., Schultz, S. and Schmidt, P. H. (1968). *J. Appl. Phys.*, **39**, 397. (*867*)

Durkan, J. and March, N. H. (1968). *J. Phys. C*, **1**, 1118. (*1029, 1030*)
Dyson, F. J. (1955). *Phys. Rev.*, **98**, 349. (*850, 864*)

Economou, E. N. (1969). *Phys. Rev.*, **182**, 539. (*1072*)
Edwards, D. M. (1969). *J. Phys. C*, **2**, 84. (*859, 860, 866*)
Edwards, J. T. (1972). *J. Phys. C*, **6**, 49. (*1133*)
Edwards, J. T. and Thouless, D. J. (1971). *J. Phys. C*, **4**, 453. (*1133*)
Edwards, S. F. (1962). *Proc. Roy. Soc.*, **A267**, 518. (*1107*)
Edwards, S. F. and Beeby, J. L. (1962). *Proc. Roy. Soc.*, **A274**, 395. (*1107*)
Ehrenreich, H. (1966). *Optical Properties and Electronic Structure of Metals and Alloys* (Ed. F. Abeles), North-Holland, Amsterdam. (*809*)
Eliashberg, G. M. (1960). *Soviet Phys.–JETP*, **11**, 696. (*896, 900*)
Elliott, R. J. (1963). *Polarons and Excitons* (Eds. C. G. Kuper and G. D. Whitfield), Oliver and Boyd, Edinburgh, p. 269; see also *Polarons* (Ed. J. Deureese), North-Holland, Amsterdam, 1972. (*969*)
Elliott, R. J. (1965). *Lattice Dynamics* (Ed. R. F. Wallis), Pergamon Press, Oxford, p. 459. (*1103*)
Elliott, R. J. and Taylor, D. W. (1964). *Proc. Phys. Soc.*, **83**, 189. (*1097, 1102*)
Elliott, R. J. and Taylor, D. W. (1967). *Proc. Roy. Soc.*, **A296**, 161. (*1111*)
Emin, D. and Holstein, T. (1969). *Ann. Phys.*, **53**, 439. (*1167*)
Englert, F. (1959). *J. Phys. Chem. Solids*, **11**, 78. (*974*)
Evenson, W. E., Schrieffer, J. R. and Wang, S. Q. (1970). *J. Appl. Phys.*, **41**, 1199. (*1047*)

Fabian, D. J. (1969). (Ed.). *Soft X-Ray Emission Spectra* (Strathclyde Conference). (*831*)
Fabian, D. J., Watson, L. M. and Marshall, C. A. W. (1971). *Rep. Prog. Phys.*, **34**, 601. (*831*)
Faulkner, J. S. (1970). *Phys. Rev.*, **B1**, 934. (*1084*)
Ferrell, R. A. (1950). *Phys. Rev.*, **111**, 1214. (*1073*)
Ferrell, R. A. and Prange, R. E. (1963). *Phys. Rev. Letters*, **10**, 479. (*935*)
Feynman, R. P. and Cohen, M. (1956). *Phys. Rev.*, **102**, 1189. (*1123*)
Flynn, C. P. (1972). *Point Defects and Diffusion*, Clarendon Press, Oxford. (*997*)
Fowler, R. H. (1931). *Phys. Rev.*, **38**, 45. (*815*)
Frenkel, J. (1928). *Z. Physik.*, **51**, 232. (*1059*)
Frenkel, J. (1931). *Phys. Rev.*, **37**, 17, 1276. (*969*)
Frenkel, J. (1936). *Physik. Z. Soviet Union*, **9**, 158. (*969*)
Friedel, J. (1954). *Advan. Phys.*, **3**, 446. (*1075, 1076*)
Fröhlich, H. (1937). *Proc. Roy. Soc.*, **160**, 230. (*939*)
Fröhlich, H. (1950). *Phys. Rev.*, **79**, 845. (*874*)
Fröhlich, H. (1954). *Advan. Phys.*, **3**, 325. (*943*)
Fröhlich, H. (1966). *Proc. Phys. Soc.*, **87**, 330. (*1154*)
Fröhlich, H. (1968). *Helv. Phys. Acta*, **41**, 838. (*918*)
Fumi, F. G. (1955). *Phil. Mag.*, **46**, 1007. (*997*)

Gartenhaus, S. (1964). *Elements of Plasma Physics*, Holt, Rinehart and Winston, Inc., New York. (*855*)
Gaumer, R. E. and Heer, C. V. (1960). *Phys. Rev.*, **118**, 955. (*715*)
Gehlen, P. C., Beeler, J. R. and Jaffee, R. I. (1972). *Interatomic Potentials and Simulation of Lattice Defects*, Plenum Press, New York. (*1056*)
Geilikman, B. T. (1958). *Soviet Phys.–JETP*, **7**, 721. (*895*)

Giovannini, B., Peter, M. and Koide, S. (1966). *Phys. Rev.*, **149**, 251. (*1036*)
Gobeli, G. W., Kane, E. O. and Allen, F. G. (1966). *Optical Properties and Electronic Structure of Metals and Alloys* (Ed. F. Abeles), North-Holland, Amsterdam, p. 347. (*822*)
Goodman, B. B. (1962). *IBM J. Res. Develop.*, **6**, 63. (*919*)
Gor'kov, L. P. (1959). *Soviet Phys.–JETP*, **9**, 1364. (*919*)
Gourary, B. S. and Adrian, F. J. (1960). *Solid State Phys.*, **10**, 127 (Eds. F. Seitz and D. Turnbull), Academic Press, New York; see also Lidiard (1973). *Orbital Theories of Molecules and Solids* (Ed. N. H. March), Clarendon Press, Oxford. (*1106*)
Greene, M. P. and Kohn, W. (1965). *Phys. Rev.*, **137**, A513. (*715, 717*)
Greenwood, D. (1960). *Proc. Phys. Soc.*, **71**, 585. (*741*)
Grimes, C. C. (1967). *Phys. Rev. Letters*, **19**, 1529. (*867*)
Grimes, C. C. and Kip, A. F. (1963). *Phys. Rev.*, **132**, 1991. (*867*)
Grimvall, G. (1969). *Solid State Commun.*, **7**, 1629; (1970). *Phys. Kondens. Mater.*, **11**, 279. (*717, 868*)
Griswold, T. W., Kip, A. F. and Kittel, C. (1952). *Phys. Rev.*, **88**, 95. (*858*)
Grosjean, M. (1959). See Appel (1968). (*951*)
Gross, E. F. (1962). *Soviet Phys. Usp.*, **5**, 195. (*974*)
Gurari, M. (1953). *Phil. Mag.*, **44**, 329. (*946*)
Gyorffy, B. L. (1970). *Phys. Rev.*, **B1**, 3290. (*1088*)

Haken, H. and Schottky, W. (1958). *Z. Phys. Chem. N.F.*, **16**, 218. (*974*)
Halperin, B. I. and Rice, T. M. (1968). *Solid State Phys.*, **21**, 116 (Eds. F. Seitz, D. Turnbull and H. Ehrenreich), Academic Press, New York. (*984*)
Hamann, D. R. (1969). *Phys. Rev. Letters*, **23**, 95. (*1048*)
Hanamura, E., Beckman, O. and Neuringer, L. (1967). *Phys. Rev. Letters*, **18**, 773. (*1030*)
Hardy, R. J. (1966). *J. Math. Phys.*, **7**, 9435. (*1128*)
Hardy, J. R. and Bullough, R. (1967). *Phil. Mag.*, **15**, 237. (*1054*)
Harris, A. B. and Lange, R. V. (1967). *Phys. Rev.*, **157**, 295. (*1090*)
Harris, J. and Griffin, A. (1970). *Can. J. Phys.*, **48**, 2592. (*1068, 1072*)
Harris, R. (1970). *J. Phys. C*, **3**, 172. (*1018*)
Harrison, W. A. (1966). *Pseudopotentials in the Theory of Metals*, Benjamin, New York; (1970). *Solid State Theory*, McGraw-Hill, New York. (*985*)
Hebborn, J. E. and March, N. H. (1964). *Proc. Roy. Soc.*, **A280**, 85. (*1025*)
Heller, W. R. and Marcus, A. (1951). *Phys. Rev.*, **84**, 809. (*978*)
Herman, F., Kortum, R. L., Kuglin, C. D., van Dyke, J. P. and Skillman, J. (1968). *Methods of Computational Physics* (Eds. B. Adler, S. Fernbach and M. Rotenberg), Academic Press, New York. (*821, 822*)
Herman, F. and Spicer, W. E. (1968). *Phys. Rev.*, **174**, 906. (*823*)
Herman, F. and van Dyke, J. P. (1968). *Phys. Rev. Letters*, **21**, 1575. (*1077*)
Herring, C. (1937). *Phys. Rev.*, **52**, 361. (*1253*)
Herring, C. (1942). *J. Franklin Inst.*, **233**, 525. (*1231*)
Herring, C. (1960). *Proc. Intern. Conf. Semicond. Phys., Prague*, p. 60. (*1146*)
Hewson, A. C. (1965). *Phys. Letters*, **19**, 5. (*1047*)
Hodges, L., Ehrenreich, H. and Lang, N. D. (1966). *Phys. Rev.*, **139**, A489. (*828*)
Holstein, T. (1954). *Phys. Rev.*, **96**, 535. (*840*)

Holstein, T. (1959). *Ann. Phys.*, **8**, 325, 343. (*962, 963, 969*)
Holstein, T. and Friedman, L. (1968). *Phys. Rev.*, **165**, 1019. (*1167*)
Hopfield, J. J. (1965). *Phys. Rev.*, **139**, A419. (*813*)
Hopfield, J. J. (1969). *Comments on Solid State Physics*, **1**, 111. (*832*)
Hori, J. (1968). *Spectral Properties of Disordered Chains and Lattices*, Pergamon Press, London. (*1129*)
Horing, N. J. (1969). *Ann. Phys.*, **54**, 405. (*1030*)
Howard, R. E. and Lidiard, A. B. (1964). *Rep. Prog. Phys.*, **27**, 161. (*997*)
Hubbard, J. and Beeby, J. L. (1969). *J. Phys. C*, **2**, 556. (*1121, 1124*)

Isaacs, L. L. and Massalski, T. B. (1965). *Phys. Rev.*, **138**, A134. (*1006*)
Izuyama, T., Kim, D. J. and Kubo, R. (1963). *J. Phys. Soc., Japan*, **18**, 1025. (*1051, 1173*)

Jérome, D., Rice, T. M. and Kohn, W. (1967). *Phys. Rev.*, **158**, 462. (*979, 981, 1157*)
Johnston, D. F. (1960). *Rep. Prog. Phys.*, **23**, 66. (*1271*)
Jones, H. (1955). *Proc. Phys. Soc.*, **68**, 1191. (*732*)
Jones, H. (1956). *Handbuch der Physik*, **19**, 297. (*681*)
Jones, H. (1960). *The Theory of the Brillouin Zones and Electronic States in Crystals*, North-Holland, Amsterdam. (*1201*)
Josephson, B. D. (1962). *Phys. Letters*, **1**, 251. (*933*)
Josephson, B. D. (1964). *Rev. Mod. Phys.*, **36**, 216. (*936*)

Kanzaki, H. (1957). *J. Phys. Chem. Solids*, **2**, 24. (*1053*)
Karo, A. M. and Hardy, J. R. (1966). *Phys. Rev.*, **141**, 696. (*1106*)
Kirkwood, J. G. (1935). *J. Chem. Phys.*, **3**, 300. (*1126*)
Kirznits, D. A. (1957). *Soviet Phys.–JETP*, **5**, 64. (*1060*)
Kittel, C. (1963). *Quantum Theory of Solids*, John Wiley and Sons, New York and London. (*993*)
Klein, M. V. (1963). *Phys. Rev.*, **131**, 1500. (*1114*)
Klein, M. V. (1968). *Physics of Color Centers* (Ed. W. B. Fowler), New York, Academic Press. (*1106*)
Kleiner, W. H., Roth, L. M. and Autler, S. H. (1964). *Phys. Rev.*, **133**, A1226. (*925*)
Klemens, P. G. (1951). *Proc. Roy. Soc.*, **A208**, 108. (*772, 776*)
Klemens, P. G. (1958). *Solid State Phys.*, **7**, 1 (Eds. F. Seitz and D. Turnbull), Academic Press, New York. (*1101*)
Knox, R. S. (1963). *Theory of Excitons* (*Solid State Phys.*, *Suppl.* 5), Academic Press, New York. (*974, 976, 978*)
Kohler, M. (1941). *Ann. Phys. Lpz.*, **40**, 601. (*762*)
Kohn, W. (1957). *Solid State Phys.*, **5**, 257 (Eds. F. Seitz and D. Turnbull), Academic Press, New York. (*1021*)
Kohn, W. (1968). *Many-body Physics* (Eds. B. I. Halperin and T. M. Rice), C. Dewitt and R. Balian. (*1157*)
Kohn, W. and Luming, M. (1963). *J. Phys. Chem. Solids*, **24**, 851. (*1030, 1034*)
Kohn, W. and Luttinger, J. M. (1957). *Phys. Rev.*, **108**, 590. (*741*)
Kohn, W. and Vosko, S. H. (1960). *Phys. Rev.*, **119**, 912. (*1009*)
Kondo, J. (1964). *Prog. Theor. Phys.*, **41**, 1199. (*1047*)

Kondo, J. (1969). *Solid State Phys.*, **23**, 183 (Eds. F. Seitz, D. Turnbull and H. Ehrenreich), Academic Press, New York. (*1047*)
Korringa, J. and Mills, R. L. (1972). *Phys. Rev.*, **B5**, 1654. (*1088*)
Koster, G. F. (1957). *Solid State Phys.*, **5**, 173 (Eds. F. Seitz and D. Turnbull), Academic Press, New York. (*1263*)
Koster, G. F. and Slater, J. C. (1954). *Phys. Rev.*, **96**, 1208. (*1002*)
Kubo, R. (1958). *Lectures in Theoretical Physics*, Boulder, I, p. 120. (*741*)

Landau, L. D. (1933). *Phys. Z. Sowjetunion*, **3**, 644. (*939, 940*)
Landau, L. D. and Lifshitz, E. M. (1958). *Statistical Physics*, Pergamon Press, London. (*788*)
Lang, I. G. and Firsov, Yu. A. (1964). *Soviet Phys. Solid State*, **5**, 2049. (*1166*)
Lang, N. D. and Kohn, W. (1970). *Phys. Rev.*, **B1**, 4555. (*1064*)
Langreth, D. C. and Kadanoff, L. P. (1964). *Phys. Rev.*, **133**, A1070. (*951, 953, 958*)
Lax, M. (1952). *Rev. Mod. Phys.*, **85**, 621. (*1078*)
Lax, M. (1958). *Phys. Rev.*, **109**, 1921. (*752*)
Lazarus, D. (1954). *Phys. Rev.*, **93**, 973. (*1073*)
Leder, L. B. and Suddeth, J. A. (1960). *J. Appl. Phys.*, **31**, 1422. (*798*)
Lee, T. D., Low, F. E. and Pines, D. (1953). *Phys. Rev.*, **90**, 297. (*946, 951*)
Lee, T. D. and Pines, D. (1952). *Phys. Rev.*, **88**, 960. (*946*)
Lien, W. H. and Phillips, N. E. (1964). *Phys. Rev.*, **133**, A1370. (*867*)
Lifshitz, I. M. (1960). *Nuovo Cimento*, 3 Suppl., 716. (*1104*)
Lomont, J. S. (1959). *Applications of Finite Groups*, Academic Press, New York. (*1201*)
London, F. (1950). *Superfluids, Vol. I*, John Wiley and Sons, New York. (*919*)
Long, P. D. and Turner, R. E. (1970). *J. Phys. C*, **3**, S127. (*1183*)
Low, F. E. and Pines, D. (1955). *Phys. Rev.*, **98**, 414. (*961*)
Lukes, T. and Roberts, M. (1967). *Proc. Phys. Soc.*, **92**, 758. (*1019*)
Luttinger, J. M. (1964). *Phys. Rev.*, **135**, A1505. (*759*)

MacDonald, D. K. C. (1962). *Thermoelectricity*, John Wiley and Sons, New York. (*773*)
McMillan, W. L. (1968). *Phys. Rev.*, **167**, 331. (*900, 1150–1154*)
Mansfield, R. (1956). *Proc. Phys. Soc.*, **B69**, 76. (*989*)
Maradudin, A. A. and Vosko, S. H. (1968). *Rev. Mod. Phys.*, **40**, 1. (*1268, 1270*)
March, N. H. and Murray, A. M. (1960). *Proc. Roy. Soc.*, **A256**, 400. (*1058*)
March, N. H. and Rousseau, J. S. (1971). *Crystal Lattice Defects*, **2**, 1, Gordon and Breach, Belfast. (*1074*)
March, N. H., Young, W. H. and Sampanthar, S. (1967). *The Many Body Problem in Quantum Mechanics*, The University Press, Cambridge. (*881, 916*)
Margenau, H. (1935). *Phys. Rev.*, **48**, 755. (*1190*)
Martin, D. L. (1961). *Proc. Roy. Soc.*, **A263**, 378; *Phys. Rev.*, **124**, 438. (*867*)
Matsubara, T. and Toyozawa, Y. (1961). *Prog. Theor. Phys.*, **26**, 739. (*1107*)
Matthias, B. T., Peter, M., Williams, H. J., Clogston, A. M., Corenzwit, E. and Sherwood, R. C. (1960). *Phys. Rev. Letters*, **5**, 542. (*1043*)
Mavroides, J. G. (1972). *Optical Properties of Solids* (Ed. F. Abeles), North-Holland, Amsterdam. (*851*)
Mermin, N. D. and Canel, E. (1964). *Ann. Phys.*, **26**, 247. (*859*)
Migdal, A. B. (1958). *Soviet Phys.–JETP*, **7**, 996. (*715*)

Montgomery, D. C. and Tidman, D. A. (1964). *Plasma Kinetic Theory*, McGraw-Hill, New York. (*1068*)
Morel, P. and Anderson, P. W. (1962). *Phys. Rev.*, **108**, 1094. (*900, 906, 1152*)
Morgan, G. J. (1968). *J. Phys. C*, **1**, 347. (*1116, 1127*)
Morgan, G. J. and Ziman, J. M. (1967). *Proc. Phys. Soc.*, **91**, 689. (*1128*)
Mott, N. F. (1961). *Phil. Mag.*, **6**, 287. (*978*)
Mott, N. F. and Davis, E. A. (1971). *Electronic Processes in Noncrystalline Materials*, Oxford University Press. (*1190*)
Murnaghan, F. D. (1938). *The Theory of Group Representations*, Johns Hopkins Press, Baltimore. (*1196*)

Nagoaka, Y. (1965). *Phys. Rev.*, **138**, A1112. (*1048*)
Nakagawa, Y. and Woods, A. D. B. (1963). *Phys. Rev. Letters*, **11**, 271. (*1153*)
Nakajima, S. and Watabe, M. (1963). *Prog. Theor. Phys.*, **29**, 341. (*715*)
Nambu, Y. (1960). *Phys. Rev.*, **117**, 648. (*896, 898*)
Nettel, S. J. (1961). *Phys. Rev.*, **171**, 425. (*967*)
Nilsson, P. O., Norris, C. and Walldén, L. (1970). *Phys. Kondens. Mater.*, **11**, 220; see also Christensen, N. E. and Seraphim, B. O. (1971). *Phys. Rev.*, **B4**, 3321. (*828*)
Nozieres, P. (1969). *Phys. Rev.*, **178**, 1097. (*832*)
Nyquist, H. (1928). *Phys. Rev.*, **32**, 110. (*754–756*)

Pauli, W. (1928). *Festschrift, Zum 60, Geburtstage, A., Sommerfelds*, Hurzel, Leipzig, p. 30. (*736*)
Payton, D. N. and Visscher, W. M. (1967). *Phys. Rev.*, **154**, 802. (*1116*)
Peierls, R. E. (1932). *Ann. Physik*, **10**, 97. (*969*)
Peierls, R. E. (1955). *Quantum Theory of Solids*, Clarendon Press, Oxford. (*747, 766, 770*)
Percus, J. K. and Yevick, G. K. (1958). *Phys. Rev.*, **110**, 1. (*1116*)
Philipp, H. R. (1965). *Optical Properties and Electronic Structure of Metals and Alloys* (Conference Proceedings, Ed. F. Abeles), North-Holland, Amsterdam. (*797, 799, 800, 809, 810*)
Phillips, J. C. (1966). *Solid State Phys.*, **18**, 56 (Eds. F. Seitz and D. Turnbull), Academic Press, New York. (*806, 828*)
Pines, D. (1963). *Polarons and Excitons* (Eds. C. G. Kuper and G. D. Whitfield), Oliver and Boyd, Edinburgh, p. 33. (*944*)
Pines, D. (1963). *Elementary Excitations in Solids*, Benjamin Inc., New York. (*1123*)
Pines, D. and Nozieres, P. (1966). *The Theory of Quantum Liquids*, Benjamin Inc., New York. (*859*)
Pippard, A. B. (1953). *Proc. Roy. Soc.*, **A216**, 547. (*876*)
Pippard, A. B. (1957). *Phil. Trans. Roy. Soc. London*, **250**, 325. (*834, 837*)
Pippard, A. B. (1960). *Rep. Prog. Phys.*, **23**, 176. (*835*)
Pippard, A. B. (1965). *The Dynamics of Conduction Electrons*, Blackie and Sons, London and Glasgow. (*841, 847, 1141*)
Platzman, P. M. and Wolff, P. A. (1967). *Phys. Rev. Letters*, **18**, 280. (*862*)
Platzman, P. M. and Wolff, P. A. (1972). *Waves in Solids*. (*857, 865*)
Pohl, R. O. (1962). *Phys. Rev. Letters*, **8**, 481. (*1101*)
Putley, E. H. (1960). *Proc. Phys. Soc.*, **76**, 802. (*1026, 1027*)

Raimes, S. and Cooper, J. R. A. (1957). *Phil. Mag.*, **4**, 145. (*723*)
Reik, H. G. and Heese, D. (1967). *J. Phys. Chem. Solids*, **28**, 581. (*1167*)
Reuter, G. E. H. and Sondheimer, E. H. (1948). *Proc. Roy. Soc.*, **A195**, 336. (*837*)
Rickayzen, G. (1965). *Theory of Superconductivity*, John Wiley and Sons, New York. (*877, 916, 1149*)
Rivier, N. and Zitkova, J. (1971). *Advan. Phys.*, **20**, 143. (*1036*)
Rooke, G. A. (1968). *J. Phys. C*, **1**, 767. (*831, 832*)

Saint-James, D. and de-Gennes, P. G. (1963). *Phys. Letters*, **7**, 306. (*921*)
Scalapino, D. J., Schrieffer, J. R. and Wilkins, J. W. (1966). *Phys. Rev.*, **148**, 263. (*898, 900, 905*)
Satterthwaite, C. B. (1962). *Phys. Rev.*, **125**, 873. (*894*)
Schaefer, G. (1960). *J. Phys. Chem. Solids*, **12**, 233. (*1106*)
Schafroth, M. R. (1951). *Helv. Phys. Acta*, **24**, 645. (*909*)
Schrieffer, J. R. (1964). *Theory of Superconductivity*, W. A. Benjamin, New York. (*877*)
Schrieffer, J. R. (1967). *J. Appl. Phys.*, **5**, 631. (*1048*)
Schrieffer, J. R. and Mattis, D. C. (1965). *Phys. Rev.*, **140**, A1412. (*1046, 1047*)
Schuey, R. T. (1968). *Phys. Kondens. Mater.*, **5**, 198. (*831*)
Schultz, S., Dunifer, G. and Lathan, A. C. (1966). *Phys. Letters*, **23**, 192. (*858*)
Schultz, S. and Dunifer, G. (1967). *Phys. Rev. Letters*, **18**, 283. (*865, 867*)
Schultz, T. D. (1959). *Phys. Rev.*, **116**, 526. (*961*)
Schultz, T. D. (1963). *Polarons and Excitons* (Eds. C. G. Kuper and G. D. Whitfield), Oliver and Boyd, London, p. 71. (*961*)
Schumacher, R. T. and Slichter, C. P. (1956). *Phys. Rev.*, **101**, 58. (*867*)
Schumacher, R. T. and Vehse, W. E. (1963). *J. Chem. Phys. Solids*, **24**, 297. (*867*)
Segall, B. (1961). *Phys. Rev.*, **124**, 1797. (*831, 832*)
Segall, B. (1962). *Phys. Rev.*, **125**, 109. (*812*)
Sewell, G. L. (1958). *Phil. Mag.*, **3**, 1361. (*965*)
Sham, L. J. (1968). *Localized Excitations in Solids* (Ed. R. F. Wallis), Plenum Press, New York. (*1071, 1072*)
Sham, L. J. (1969). *Phys. Rev.*, **188**, 1431. (*868, 870*)
Sham, L. J. and Ziman, J. M. (1963). *Solid State Phys.*, **15**, 221 (Eds. F. Seitz and D. Turnbull), Academic Press, New York. (*712, 1146*)
Sherrington, D. (1971). *J. Phys. C*, **4**, 2771. (*868*)
Sherrington, D. (1971). *J. Phys. C*, **4**, 401. (*1047*)
Sinha, S. K. (1966). *Phys. Rev.*, **143**, 422. (*1114*)
Slater, J. C. and Shockley, W. (1936). *Phys. Rev.*, **50**, 705. (*804, 805*)
Smith, J. R. (1969). *Phys. Rev.*, **181**, 522. (*1061, 1063*)
Smith, N. V. (1969). *Phys. Rev. Letters*, **23**, 1452. (*822, 825, 827*)
Sokolov, A. V. (1967). *Optical Properties of Metals*, Blackie and Sons, London and Glasgow. (*816*)
Sondheimer, E. H. and Wilson, A. H. (1951). *Proc. Roy. Soc.*, **A210**, 173. (*1023, 1035*)
Soven, P. (1967). *Phys. Rev.*, **156**, 809. (*1085*)
Spicer, W. E. (1966). *Optical Properties and Electronic Structure of Metals and Alloys* (Conference Proceedings, Ed. F. Abeles), North Holland, Amsterdam, p. 296. (*823, 825*)
Springford, M. (1971). *Advan. Phys.*, **20**, 493. (*1036*)

Stern, E. A. and Ferrell, R. A. (1960). *Phys. Rev.*, **120**, 130. (*1073*)
Stocks, G. M., Williams, R. W. and Faulkner, J. S. (1971). *Phys. Rev.*, **B4**, 4390. (*1092*)
Stoddart, J. C., March, N. H. and Stott, M. J. (1969). *Phys. Rev.*, **186**, 683. (*1001, 1008*)
Stoddart, J. C., March, N. H. and Wiid, D. (1972). *Int. J. Quantum Chem.*, **2**, 101. (*1037*)
Stott, M. J., Baranovsky, S. and March, N. H. (1970). *Proc. Roy. Soc.*, **A316**, 201. (*998*)
Stott, M. J. and March, N. H. (1966). *Phys. Letters*, **23**, 408. (*831*)
Stott, M. J. and March, N. H. (1968). *Proc. Conf. on Soft X-ray Emission, Strathclyde.* (*831*)
Suhl, H. (1968). *Phys. Rev. Letters*, **20**, 656. (*1048*)
Svensson, E. L., Brockhouse, B. N. and Rowe, J. M. (1965). *Solid State Commun.*, **3**, 245. (*1114*)
Sziklas, E. A. (1965). *Phys. Rev.*, **138**, A1070. (*1071, 1072*)

Taylor, D. W. (1967). *Phys. Rev.*, **156**, 1017. (*1107, 1111, 1112, 1114, 1115*)
Tewordt, L. (1963). *Phys. Rev.*, **129**, 657. (*895*)
Thomas, H. (1957). *Z. Physik*, **147**, 395. (*816, 817*)
Thornton, D. E. (1971). *Phys. Rev.*, **B4**, 3371. (*1088*)
Thouless, D. J. (1970). *J. Phys. C*, **3**, 1559. (*1130*)
Tyablikov, S. V. (1952). *Zh Eksperim. i Teor. Fiz.*, **23**, 381; (1954). *Zh Eksperim. i Teor. Fiz.*, **26**, 545. (*946*)

Van Hove, L. (1955). *Physica*, **21**, 517. (*753*)
Van Hove, L. (1957). *Physica*, **23**, 441. (*744, 753*)
Velicky, B., Kirkpatrick, S. and E. Ehrenreich, H. (1968). *Phys. Rev.*, **175**, 747. (*1088–1093*)
von Ortenberg, M. (1973). *J. Phys. Chem. Solids*, **34**, 397. (*1030*)

Walker, C. T. and Pohl, R. O. (1963). *Phys. Rev.*, **131**, 1433. (*1101, 1102*)
Wannier, G. H. (1937). *Phys. Rev.*, **52**, 191. (*969*)
Wannier, G. H. (1962). *Rev. Mod. Phys.*, **34**, 645. (*1179*)
Weaire, D. (1970). *Phys. Rev. Letters*, **26**, 1541; see also Weaire, D. and Thorpe, M. F. (1971). *Phys. Rev.*, **B4**, 2508.
Whitfield, G. D. (1961). *Phys. Rev.*, **121**, 720. (*1146*)
Wigner, E. P. (1932). *Nachr. Ges. Wiss. Gottingen*, 546. (*1253*)
Wigner, E. P. (1938). *Trans. Faraday Soc.*, **34**, 678. (*1060*)
Williams, G. (1970). *J. Phys. Chem. Solids*, **31**, 529. (*1186, 1187*)
Williams, G. and Loram, J. W. (1969). *J. Phys. Chem. Solids*, **30**, 1827. (*1186*)
Wilson, A. H. (1953). *The Theory of Metals*, Cambridge University Press. (*681, 775*)
Wolff, P. A. (1961). *Phys. Rev.*, **124**, 1030. (*1037*)
Wooten, F., Heun, T. and Stuart, R. N. (1966). *Optical Properties and Electronic Structures of Metals and Alloys* (Conference Proceedings, Ed. F. Abeles), North-Holland, Amsterdam, p. 333. (*819*)

Yafet, Y., Keyes, R. W. and Adams, E. N. (1956). *J. Phys. Chem. Solids*, **1**, 137. (*1026*)

Yang, C. N. (1962). *Rev. Mod. Phys.*, **34**, 694. (*928*)
Young, C. Y. and Sham, L. J. (1969). *Phys. Rev.*, **188**, 1108. (*717, 867*)
Young, W. H., Meyer, A. and Kilby, G. E. (1967). *Phys. Rev.*, **160**, 482. (*733*)

Zak, J. (1969). *The Irreducible Representations of Space Groups*,, Benjamin Inc., New York. (*1201*)
Ziman, J. M. (1960). *Electrons and Phonons*, Clarendon Press, Oxford. (*681*)
Ziman, J. M. (1965). *Proc. Phys. Soc.*, **86**, 337. (*991*)
Ziman, J. M. (1970). *J. Phys. C*, **4**, 3129. (*1078*)
Ziman, J. M. (1972). *Principles of Solids*, 2nd ed., Cambridge University Press. (*868, 994*)
Zubarev, D. N. (1960). *Soviet Phys. Usp.*, **3**, 320. (*1096*)

Index